VIP, PACAP, AND RELATED PEPTIDES
THIRD INTERNATIONAL SYMPOSIUM

ANNALS OF THE NEW YORK ACADEMY OF SCIENCES

Volume 865

VIP, PACAP, AND RELATED PEPTIDES
THIRD INTERNATIONAL SYMPOSIUM

Edited by Wolf-Georg Forssmann and Sami I. Said

The New York Academy of Sciences
New York, New York
1998

Library of Congress Cataloging-in-Publication Data

VIP, PACAP, and related peptides : third international symposium / edited by Wolf-Georg Forssmann, Sami I. Said.
  p. cm. — (Annals of the New York Academy of Sciences, ; v. 865)
  "The papers in this volume resulted from a conference entitled 3rd International Symposium on VIP, PACAP, and Related Peptides, held September 17–10, 1997, in Freiburg, Germany"—Contents p.
  Includes bibliographical references and index.
  ISBN 1-57331-153-7 (cloth : alk. paper). — ISBN 1-57331-154-5 (pbk. : alk. paper)
    1. Peptide hormones—Physiological effect—Congresses. 2. Vasoactive intestinal peptides—Physiological effect—Congresses. 3. Adenylate cyclase—Agonists—Congresses. 4. Pituitary hormones—Physiological effect—Congresses.
I. Forssmann, W.-G. (Wolf-Georg), 1939– II. Said, Sami. III. International Symposium on VIP, PACAP, and Related Peptides: (3rd : 1997 : Freiburg, Germany) IV. Series.
Q11.N5 vol. 865
[QP572.V28]
500 s—dc21
[572'.65]   98-31190
                                                                                CIP

CDP/PCP

Printed in the United States of America
ISBN 1-57331-153-7 (cloth)
ISBN 1-57331-154-5 (paper)
ISSN 0077-8923

ANNALS OF THE NEW YORK ACADEMY OF SCIENCES

Volume 865
December 11, 1998

VIP, PACAP and Related Peptides
Third International Symposium[a]

Editors
WOLF-GEORG FORSSMANN AND SAMI I. SAID

Conference Organizers
WOLF-GEORG FORSSMANN, JAN FAHRENKRUG, STEPHEN RAWLINGS, WOLFGANG E.
SCHMIDT, JENS JUUL HOLST, FRANK SUNDLER, AND MARC LABURTHE

International Advisory Board
AKIRA ARIMURA, DOMINIQUE BATAILLE, JEAN CHRISTOPHE, JAN FAHRENKRUG,
WOLF-GEORG FORSSMANN, VIKTOR MUTT, GABRIEL ROSSELIN, SAMI I. SAID, AND
NOBORU YANAIHARA

CONTENTS

Preface. *By* WOLF-GEORG FORSSMANN and SAMI I. SAID xv

Part I. Regulation of Gene Expression and Processing of VIP-PACAP Peptides

Special Lecture: Signaling by the Cytokine Receptor Superfamily. *By* J. N. IHLE,
  W. THIERFELDER, S. TEGLUND, D. STRAVAPODIS, D. WANG, J. FENG, and
  E. PARGANAS 1

Cis-Regulatory Elements Controlling Basal and Inducible
  VIP Gene Transcription. *By* S. H. HAHM and L. E. EIDEN 10

p22/PACAP Response Gene 1 (PRG1): A Putative Target Gene for the Tumor
  Suppressor p53. *By* H. SCHÄFER, A. TRAUZOLD, T. SEBENS, W. DEPPERT,
  U. R. FÖLSCH, and W. E. SCHMIDT 27

Solution Structure Comparison of the VIP/PACAP Family of Peptides by NMR
  Spectroscopy. *By* V. WRAY, K. NOKIHARA, and S. NARUSE 37

[a]The papers in this volume resulted from a conference entitled **3rd International Symposium on VIP, PACAP, and Related Peptides**, held September 17-20, 1997 in Freiburg, Germany.

Cloning and Functional Characterization of PACAP-Specific Receptors in
Zebrafish. *By* Y. WEI, S.C. MARTIN, G. HEINRICH and S. MOJSOV 45

Part II. Biochemistry of VIP-PACAP-Related Peptide Receptors

Induction of Type I PACAP Receptor Expression by the New Zinc Finger
Protein Zac1 and p53. *By* A. HOFFMANN, E. CIANI, S. HOUSSAMI,
P. BRABET, L. JOURNOT, and D. SPENGLER 49

Cloning and Functional Characterization of the Human VIP1/PACAP
Receptor Promoter. *By* A. COUVINEAU, J. J. MAORET, C.
ROUYER-FESSARD, I. CARRERO, and M. LABURTHE .59

Desensitization of the Human Vasoactive Intestinal Peptide Receptor
(hVIP$_2$/ PACAP R): Evidence for Agonist-Induced Receptor
Phosphorylation and Internalization. *By* T. P. MCDONALD,
D. M. DINNIS, C. F. MORRISON, and A. J. HARMAR .64

Rat Aortic Smooth-Muscle Cell Proliferation Is Bidirectionally Regulated in
a Cell Cycle–Dependent Manner via PACAP/VIP Type 2 Receptor.
By A. MIYATA, K. SATO, J. HINO, H. TAMAKAWA, H. MATSUO, and
K. KANGAWA 73

Identification of Binding Domains of Pituitary Adenylate Cyclase Activating
Polypeptide (PACAP) for its Type 1 Receptor by Photoaffinity Labeling.
By Y. J. CAO, E. KOJRO, M. JASIONOWSKI, L. LANKIEWICZ, Z. GRZONKA,
and F. FAHRENHOLZ 82

**Part III. Signal Transduction and Intracellular Metabolism Induced by
VIP and PACAP**

The Neurotrophic Activity of PACAP on Rat Cerebellar Granule Cells Is Associated
with Activation of the Protein Kinase A Pathway and *c-fos* Gene Expression.
By D. VAUDRY, M. BASILLE, Y. ANOUAR, A. FOURNIER, H. VAUDRY, and
B. J. GONZALEZ 92

PACAP-38 Protects Cerebellar Granule Cells from Apoptosis. *By* L. JOURNOT,
M. VILLALBA, and J. BOCKAERT 100

PACAP Protects Hippocampal Neurons against Apoptosis: Involvement of
JNK/SAPK Signaling Pathway. *By* S. SHIODA, H. OZAWA, K. DOHI,
H. MIZUSHIMA, K. MATSUMOTO, S. NAKAJO, A. TAKAKI, C. J. ZHOU,
Y. NAKAI, and A. ARIMURA 111

Stimulatory Transducing Systems in Pancreatic Islet Cells. *By* S. EMAMI,
K. REGNAULD, N. FERRAND, A. ASTESANO, M. PESSAH, H. PHAN,
C. BOISSARD, J.-M. GAREL, and G. ROSSELIN 118

Part IV. Physiology of Effector Mechanisms

Miniglucagon: A Local Regulator of Islet Physiology. *By* S. DALLE, P. BLACHE,
D. LE-NGUYEN, L. LE BRIGAND, and D. BATAILLE .132

Pituitary Adenylate Cyclase Activating Polypeptide Induces Degranulation of Rat
    Peritoneal Mast Cells via High-Affinity PACAP Receptor-Independent
    Activation of G Proteins. *By* J. SEEBECK, M. L. KRUSE,
    A. SCHMIDT-CHOUDHURY, and W. E. SCHMIDT 141

The Pituitary Adenylate Cyclase Activating Polypeptide Type 1 Receptor
    (PAC$_1$-R) Is Expressed on Gastric ECL Cells: Evidence by
    Immunocytochemistry and RT-PCR. *By* N. ZENG, T. KANG,
    R. M. LYU, H. WONG, Y. WEN, J. H. WALSH, G. SACHS, and
    J. R. PISEGNA 147

A Critical View of the Methods for Characterization of the VIP/PACAP
    Receptor Subclasses. *By* P. ROBBERECHT and M. WAELBROECK 157

Part V. Neurobiology of VIP-PACAP-Related Peptides

Mechanisms of Pituitary Adenylate Cyclase Activating Polypeptide (PACAP)-
    Induced Depolarization of Sympathetic Superior Cervical Ganglion (SCG)
    Neurons. *By* V. MAY, M. M. BEAUDET, R. L. PARSONS, J. C. HARDWICK,
    E. A. GAUTHIER, P. J. DURDA, and K. M. BRAAS 164

Sympathetic Neurons of the Chick Embryo Are Rescued by PACAP from
    Apoptotic Death. *By* A. R. WAKADE and D. LEONTIV 176

Differential Display PCR Reveals Induction of Immediate Early Genes
    by Vasoactive Intestinal Peptide in PC12 Cells. *By* A. ESCHELBACH,
    A. HUNZIKER, and L. KLIMASCHEWSKI 181

Developmental Regulation of Pituitary Adenylate Cyclase Activating
    Polypeptide (PACAP) and Its Receptor 1 in Rat Brain: Function
    of PACAP as a Neurotrophic Factor. *By* D. LINDHOLM,
    Y. SKOGLÖSA, and N. TAKEI 189

Pituitary Adenylate Cyclase Activating Peptide (PACAP) in the Retinohypothalamic
    Tract: A Daytime Regulator of the Biological Clock. *By* J. HANNIBAL,
    J. M. DING, D. CHEN, J. FAHRENKRUG, P. J. LARSEN, M. U. GILLETTE,
    and J. D. MIKKELSEN 197

VIP Neurotrophism in the Central Nervous System: Multiple Effectors and
    Identification of a Femtomolar-Acting Neuroprotective Peptide.
    By D. E. BRENNEMAN, G. GLAZNER, J. M. HILL, J. HAUSER,
    A. DAVIDSON, and I. GOZES 207

VIP and PACAP in the CNS: Regulators of Glial Energy Metabolism and
    Modulators of Glutamatergic Signaling. *By* P. J. MAGISTRETTI,
    J.-R. CARDINAUX, and J.-L. MARTIN 213

Glutamate Toxicity in the Lung and Neuronal Cells: Prevention or Attenuation
    by VIP and PACAP. *By* S. I. SAID, K. DICKMAN, R. D. DEY,
    A. BANDYOPADHYAY, P. DE STEFANIS, S. RAZA, H. PAKBAZ,
    and H. I. BERISHA 226

Part VI. Pharmacology of VIP-PACAP Agonists and Antagonists

Autoantibody Catalysis: No Longer Hostage to Occam's Razor.
By S. PAUL 238

Analogues of VIP, Helodermin, and PACAP Discriminate between Rat and Human
VIP$_1$ and VIP$_2$ Receptors. By P. GOURLET, A. VANDERMEERS,
J. VAN RAMPELBERGH, P. DE NEEF, J. CNUDDE, M. WAELBROECK,
and P. ROBBERECHT 247

Maxadilan Is a Specific Agonist and Its Deleted Peptide (M65) Is a Specific
Antagonist for PACAP Type 1 Receptor. By D. UCHIDA, I. TATSUNO,
T. TANAKA, A. HIRAI, Y. SAITO, O. MORO, and M. TAJIMA 253

Importance of Conserved Cysteines in the Extracellular Loops of Human PACAP/
VIP$_1$ Receptor for Ligand Binding and Stimulation of cAMP Production.
By S. M. KNUDSEN, J. W. TAMS, B. S. WULFF, and
J. FAHRENKRUG 259

Multiple Actions of a Hybrid PACAP Antagonist: Neuronal Cell Killing and
Inhibition of Sperm Motility. By I. GOZES, O. PERL, R. ZAMOSTIANO,
S. RUBINRAUT, M. FRIDKIN, L. SHOCHAT, and L. M. LEWIN 266

Special Lecture: The PACAP Ligand/Receptor System Regulates Cerebral Cortical
Neurogenesis. By E. DICICCO-BLOOM, N. LU, J. E. PINTAR,
and J. ZHANG 274

Part VII. Progress in Clinical Research of VIP-PACAP Peptides

VIP and Breast Cancer. By T. W. MOODY, J. LEYTON, I. GOZES, L. LANG, and
W. C. ECKELMAN 290

Coordinated Role of Vasoactive Intestinal Peptide and Nitric Oxide in
Cardioprotection. by D. K. DAS, R. KALFIN, N. MAULIK, and
R. M. ENGELMAN 297

The Protective Effect of Vasoactive Intestinal Peptide (VIP) on Stress-Induced Gastric
Ulceration in Rats. By N. TUNÇEL, N. ERKASAP, V. ŞAHINTÜRK,
D. DOĞRUKOL AK, and M. TUNÇEL 309

Is There Appetite after GLP-1 and PACAP? By J. CHRISTOPHE 323

On the Treatment of Diabetes Mellitus with Glucagon-like Peptide-1.
By J. J. HOLST, C. DEACON, M. B. TOFT-NIELSEN,
and L. BJERRE-KNUDSEN 336

Binding Sites for VIP in the Reorganizing Mucosa of the Irradiated Bowel.
By U. HÖCKERFELT, M. HANSSON, S. GULBENKIAN, L. FRANZÉN,
R. HENRIKSSON, and S. FORSGREN 344

Main Sensory Neuropeptides, but not VIP and NPY, Are Involved in Bone
Remodeling during Orthodontic Tooth Movement in the Rat.
By L. I. NOREVALL, L. MATSSON, and S. FORSGREN 353

Special Lecture: PACAP$_{27}$ and Other Neuropeptides in the Inferior Mesenteric
   Ganglion. *By* L. G. ERMILOV and J. H. SZURSZEWSKI 360

Part VIII. Poster Papers

Regulation of Gene Expression and Processing of VIP-PACAP Peptides

Induction of Multiple Pituitary Adenylate Cyclase Activating Polypeptide (PACAP)
   Transcripts through Alternative Cleavage and Polyadenylation of proPACAP
   Precursor mRNA. *By* S. A. HARAKALL, C. A. BRANDENBURG,
   G. A. GILMARTIN, V. MAY, and K. M. BRAAS 367

Biochemistry of VIP-PACAP-Related Peptide Receptors

A Model of the Receptors in the VIP Receptor Family. *By* J. W. TAMS,
   S. M. KNUDSEN, and J. FAHRENKRUG 375

Site-Directed Mutagenesis of Human VIP1 versus VIP2 Receptors.
   By P. NICOLE, K. DU, A. COUVINEAU, and M. LABURTHE 378

Constitutive Activation of the Human VIP1 Receptor. *By* P. GAUDIN,
   C. ROUYER-FESSARD, A. COUVINEAU, J. J. MAORET, and
   M. LABURTHE 382

Construction of Chimeras between Human VIP1 and Secretin Receptors:
   Identification of Receptor Domains Involved in Selectivity
   towards VIP, Secretin, and PACAP. *By* K. DU, P. NICOLE,
   A. COUVINEAU, and M. LABURTHE 386

Involvement of a Pit-1 Binding Site in the Regulation of the Rat Somatostatin
   Receptor 1 Gene Expression. *By* H. BAUMEISTER and W. MEYERHOF 390

Relaxant Responses of VIP and PACAP in Rat Ileum: Receptors and Adaptive
   Supersensitivity. *By* E. EKBLAD, M. EKELUND, and F. SUNDLER 393

VIP$_1$ and VIP$_2$ Receptors but not PVR1 Mediate the Effect of VIP/PACAP on
   Cytokine Production in T Lymphocytes. *By* X. JIANG, H. Y. WANG,
   J. YU, and D. GANEA 397

Distribution and Ultrastructural Localization of Pituitary Adenylate Cyclase-
   Activating Polypeptide (PACAP) and Its Receptor in the Rat Retina.
   By T. SEKI, S. SHIODA, Y. NAKAI, A. ARIMURA, and R. KOIDE 408

Autoradiographic Visualization of the Receptor Subclasses for Vasoactive Intestinal
   Polypeptide (VIP) in Rat Brain. *By* P. VERTONGEN, S. N. SCHIFFMANN,
   P. GOURLET, and P. ROBBERECHT 412

Pituitary Adenylate Cyclase-Activating Polypeptide Receptors in the Fetal Human
   Adrenal Gland. *By* L. YON, L. BREAULT, V. CONTESSE, G. BELLANCOURT,
   C. DELARUE, A. FOURNIER, J. G. LECHOUX, N. GALLO-PAYET, and
   H. VAUDRY. 416

Neuropeptides in Developmental Tumors of the Central and Peripheral Nervous
System. *By* M. C. FRÜHWALD, M. S. O'DORISIO, S. L. COTTINGHAM,
S. J. QUALMAN, and T. M. O'DORISIO 420

Signal Transduction and Intracellular Metabolism Induced by VIP and PACAP

PACAP Increases Cytosolic Calcium in Vasopressin Neurons: Synergism with
Noradrenaline. *By* S. SHIODA, T. YADA, S. NAKAJO, Y. NAKAI, and
A. ARIMURA 427

Effect of PACAP-27 on Adenylate Cyclase in Ductal and Acinar Cells of Rat
Submandibular Gland. *By* E. KABRÉ, N. CHAIB, H. AMSALLEM, A. MORAN,
M. C. VANDERMEERS, and J. P. DEHAYE 431

Distribution and Ultrastructural Localization of PACAP Receptors in the Rat
Pancreatic Islets. *By* M. MUROI, S. SHIODA, T. YADA, C. J. ZHOU,
Y. NAKAI, S. NAKAJA, and A. ARIMURA 438

Protein Kinase A Inhibition and PACAP-Induced Insulin Secretion in HIT-T15
Cells. *By* K. FILIPSSON and B. AHRÉN 441

Physiology of Effector Mechanisms

PACAP and GLP-1 Protect Islet β-Cells against Ca^{2+} Toxicity Induced by
High K^+. *By* K. YAEKURA, K. YANAGIDA, and T. YADA 445

Autocrine Action of PACAP in Islets Augments Glucose-Induced Insulin Secretion.
By T. YADA, M. SAKURADA, M. NAKATA, S. SHIODA, K. YAEKURA, and
M. KIKUCHI 451

A Target Cell to Oxyntomodulin and Glicentin: The Antral Smooth Muscle Cell.
By G. RODIER, R. MAGOUS, T. MOCHIZUKI, J. P. BALI, D. BATAILLE, and
C. JARROUSSE 458

Effects of PACAP/VIP/Secretin on Pancreatic and Gastrointestinal Blood Flow in
Conscious Dogs. *By* S. NARUSE, O. ITO, M. KITAGAWA, H. ISHIGURO,
M. NAKAJIMA, and T. HAYAKAWA 463

Pituitary Adenylate Cyclase Activating Polypeptide Stimulates Insulin Secretion
in a Glucose-Dependent Manner *In Vivo*. *By* B. AHRÉN and
K. FILIPSSON 466

Effects of Pituitary Adenylate Cyclase-Activating Polypeptide (PACAP) on cAMP
Formation and Growth Hormone Release from Chicken Anterior Pituitary
Cells. *By* K. PEETERS, L. LANGOUCHE, F. VANDESANDE, V. M. DARRAS, and
L. R. BERGHMAN 471

Localization of Pituitary Adenylate Cyclase-Activating Polypeptide in the Central
Nervous System of the European Eel *Anguilla anguilla*: Stimulatory Effect
of PACAP on GH Secretion. *By* M. MONTERO, L. YON, K. ROUSSEAU,
A. ARIMURA, A. FOURNIER, S. DUFOUR, and H. VAUDRY 475

Effect of Pituitary Adenylate Cyclase-Activating Polypeptide (PACAP) on Tyrosine
Hydroxylase Gene Expression in the Rat Adrenal Medulla. *By* M. HONG,
L. YON, A. FOURNIER, H. VAUDRY, and G. PELLETIER 478

The Stimulatory Effect of VIP and PACAP on Adrenal Aldosterone Release.
By M. RADZIKOWSKA, E. WASILEWSKA-DZIUBINSKA, and
B. BARANOWSKA 482

Evidence for PACAP To Be an Autocrine Factor on Gonadotrope Cells.
By A. RADLEFF-SCHLIMME, S. LEONHARDT, W. WUTTKE, and
H. JARRY 486

Peptidergic Component of Non-Adrenergic Non-Cholinergic Relaxation of the Rat
Gastric Fundus. *By* D. CURRÒ and P. PREZIOSI 492

Effect of VIP and PACAP on Vascular and Luminal Release of Serotonin from
Isolated Perfused Rat Duodenum. *By* M. FUJIMIYA, H. YAMAMOTO, and
A. KUWAHARA 495

Sites of Actions of Contractile and Relaxant Effects of Pituitary Adenylate Cyclase
Activating Peptide (PACAP) in the Internal Anal Sphincter Smooth Muscle.
By S. RATTAN and S. CHAKDER 503

PACAP Inhibits Spontaneous Contractions in the Intestine of the Atlantic Cod,
Gadus morhua. By C. OLSSON and S. HOLMGREN 512

Effect of Sodium Depletion by Frusemide on Tissue Concentrations and
Metabolism of VIP. *By* K. A. DUGGAN and V. Z. C. YE 515

Neurobiology of VIP-PACAP-Related Peptides

Neuromodulator Role of VIP in Recovery of Rat Behavior and Brain
Neurotransmitter Levels after Frontal Lobectomy. *By* M. KARGANOV,
G. ROMANOVA, W. BRASLAWSKY, D. TARSHITZ, and G. TELEGDY 519

Immunohistochemical Evidence for PACAP and VIP Interaction with Met-
Enkephalin and CRF Containing Neurons in the Bed Nucleus of the
Stria Terminalis. *By* T. KOZICZ, S. VIGH, and A. ARIMURA 523

Distribution and Somatotopical Localization of Pituitary Adenylate Cyclase
Activating Polypeptide (PACAP) in the Trigeminal Ganglion of Cats
and Rats. *By* M. KAUSZ, A. ARIMURA, and K. KÖVES 529

Prenatal Expression of Pituitary Adenylate Cyclase Activating Polypeptide
(PACAP) in Autonomic and Sensory Ganglia and Spinal Cord of Rat
Embryos. *By* H. S. NIELSEN, J. HANNIBAL, and J. FAHRENKRUG 533

VIP and NPY Expression during Differentiation of Cholinergic and Noradrenergic
Sympathetic Neurons. *By* B. SCHÜTZ, M. K. H. SCHÄFER, L. E. EIDEN, and
E. WEIHE 537

PACAP in Visceral Afferent Nerves Supplying the Rat Digestive and Urinary
Tracts. *By* J. FAHRENKRUG and J. HANNIBAL 542

Pharmacology of VIP-PACAP Agonists and Antagonists

Galanin Controls Excitability of the Brain. *By* S. A. CHEPURNOV,
N. E. CHEPURNOVA, and R. K. BERDIEV 547

Vasoactive Intestinal Peptide Supports Spontaneous and Induced Migration of
Human Keratinocytes and the Colonization of an Artificial Polyurethane
Matrix. *By* U. WOLLINA 551

The Stimulatory Effect of VIP on Progesterone Release in Rats after Adrenalectomy,
Ovariectomy, and Hysterectomy: Influence of VIP and PACAP38 on
Progesterone Release in Rats after Ovariectomy and Hysterectomy.
By E. WASILEWSKA-DZIUBINSKA, M. RADZIKOWSKA, and
B. BARANOWSKA 556

Characterization of a PACAP-like Immunoreactive Component in Red Ginseng Root.
By N. TAKASHIMA, Y. ARAKAWA, K. KATAOKA, N. KUROKAWA, C. YANAIHARA,
and N. YANAIHARA 561

The Effect of Vasoactive Intestinal Polypeptide and Pituitary Adenylate Cyclase
Activating Polypeptide on Tolerance to Morphine and Alcohol in Mice.
By G. SZABÓ, M. MÁCSAI, E. SCHEK, and G. TELEGDY 566

The Effect of Vasoactive Intestinal Peptide (VIP) and Inhibition of Nitric Oxide on
Renal Tissue Injury of Rats Exposed to Hemorrhagic Ischemia and
Retransfusion: A Possible Interaction Mechanism among Mast Cells
and Tissue Histamine. *By* S. H. ERDEN, N. TUNÇEL, Y. AYDIN,
V. ŞAHINTÜRK, M. KOŞAR, and M. TUNÇEL 570

Enhancement of Systemic and Pulmonary Vasoconstriction by β-Amyloid Peptides
and Its Suppression by Vasoactive Intestinal Peptide. *By* S. I. SAID,
S. RAZA, and H. I. BERISHA 582

Progress in Clinical Research of VIP-PACAP Peptides

The Effect of Vasoactive Intestinal Peptide (VIP) and Inhibition of Nitric Oxide
Synthase on Survival Rate in Rats Exposed to Endotoxin Shock.
By N. TUNÇEL and F. C. TÖRE 586

Role of PACAP in the Regulation of Gonadotroph Hormone Secretion during
Ontogenesis: A Single Neonatal Injection of PACAP Delays Puberty and
Its Intracerebroventricular Administration before the Critical Period of
Proestrous Stage Blocks Ovulation in Adulthood. *By* K. KÖVES, J. MOLNÁR,
O. KÁNTOR, A. LAKATOS, K. FOGEL, M. KAUSZ, M.C. VANDERMEERS-PIRET,
A. SOMOGYVÁRI-VIGH, and A. ARIMURA 590

Is an Intravenous Bolus Injection Required prior to Initiating Slow Intravenous Infusion
of PACAP38 for Prevention of Neuronal Death Induced by Global Ischemia?:
The Possible Presence of a Binding Protein for PACAP38 in Blood.
By A. SOMOGYVÁRI-VIGH, J. SVOBODA-TEET, S. VIGH, and
A. ARIMURA 595

Effect of Rat Glicentin on Intestinal Adaptation in Small Intestine–Resected Rats.
By Y. HIROTANI, M. TAKI, K. KATAOKA, N. KUROKAWA, T. SATOH,
K. SASAKI, C. YANAIHARA, W. Q. LUO, and N. YANAIHARA 601

Index of Contributors 607

Financial assistance was received from:
- AMYLIN PHARMACEUTICALS, SAN DIEGO, CALIFORNIA, USA
- BAYER AG, WUPPERTAL, GERMANY
- BOEHRINGER MANNHEIM GMBH, MANNHEIM, GERMANY
- DR. FALK PHARMA GMBH, FREIBURG, GERMANY
- HAEMOPEP PHARMA GMBH, HANNOVER, GERMANY
- HOECHST MARION ROUSSEL, FRANKFURT, GERMANY
- HOECHST MARION ROUSSEL INC., CINCINNATI, OHIO, USA
- IMMUNODIAGNOSTIK GMBH, BENSHEIM, GERMANY
- ITOHAM FOODS INC., IBARAKI, JAPAN
- MADAUS AG, KÖLN, GERMANY
- NEOSYSTEM, STRASBOURG, FRANCE
- PENINSULA LABORATORIES INC., BELMONT, CALIFORNIA, USA
- PERKIN ELMER/APPLIED BIOSYSTEMS GMBH, WEITERSTADT, GERMANY
- SCHERING AG, BERLIN, GERMANY
- SOLVAY PHARMACEUTICALS GMBH, HANNOVER, GERMANY
- TAP HOLDINGS INC., DEERFIELD, ILLINOIS, USA
- WHERL GMBH, WOLFENBÜTTEL, GERMANY

Preface

WOLF-GEORG FORSSMANN

Lower Saxony Institute for Peptide Research, Hannover, Germany

SAMI I. SAID

State University of New York at Stony Brook, and Veterans Affairs Medical Center, Northport, New York, USA

Research on VIP, PACAP, and related peptides continues to generate much interest among investigators in diverse fields of science, biology and medicine. One expression of this interest has been the holding of international conferences on a bi-yearly basis, beginning shortly after the discovery of VIP, to present new findings and plan future work. This volume of the *Annals* contains papers given at the most recent conference, held in the attractive university town of Freiburg, close to the Black Forest in Southwestern Germany, September 17-20, 1997.

The meeting, organized by Wolf-Georg Forssmann and his staff in Hannover, Germany, notably Anja Freesemann, aided by an international panel of advisors, was attended by 121 participants from Europe, Asia, the Middle East, Australia, and the United States. Topics addressed included these aspects of peptide biology: regulation of gene expression and processing, biochemistry and signal transduction of receptors, physiological roles, neurobiology, pharmacology of agonists and antagonists, and relationship to disease and its management.

This is the third in this series of conferences to be published by the New York Academy of Sciences. We thank the Academy for providing the unique medium of the *Annals* for disseminating this information. We are particularly grateful to Bill Boland, Executive Editor, and Joyce Hitchcock and Mary K. Brennan, who handled the editing of this volume. We are also pleased to acknowledge the invaluable support and assistance of Anja Freesemann and her colleagues in Hannover, Germany, and Rosalind Antoniazzi, in Stony Brook, New York, for their tireless help in gathering and assembling the manuscripts, preparing the Table of Contents, and communicating with the authors and the publisher. Finally, our thanks to all those who participated in the conference and contributed to this volume.

Signaling by the Cytokine Receptor Superfamily[a]

JAMES N. IHLE,[b] WILLIAM THIERFELDER, STEPHAN TEGLUND, DIMITRIOS STRAVAPODIS, DEMIN WANG, JIAN FENG, AND EVAN PARGANAS

Howard Hughes Medical Institute, Department of Biochemistry, Jude Children's Research Hospital, Memphis, Tennessee 38105 USA

ABSTRACT: A variety of cytokines that regulate functions of multiple lineages share the utilization of receptors that are structurally and functionally related and are referred to as the cytokine receptor superfamily. These receptors associate with one or more of the four mammalian *Janus* kinases (Jaks) and ligand-induced receptor aggregation results in their activation. Critical roles for Jak3 and Jak2 are demonstrated by the phenotypes of mice that lack each gene. Among the substrates of the Jaks are one or more of the seven members of the signal transducers and activators of transcription (Stats). Each Stat family member plays a critical role in the biological functions of specific cytokines as demonstrated by the phenotype of mice lacking one or more of these genes.

The growth and differentiation of the hematopoietic lineages requires one or more of a variety of cytokines. Most of the cytokines that influence hematopoiesis are members of a structurally and functionally related family. Functionally this group of cytokines is related through their utilization of receptors that are similar in structure and functionally related. This group is referred to as the cytokine receptor superfamily. In this regard these cytokines are unique from other growth factors or cytokines that utilize quite distinct families of receptors, such as the G-protein coupled, serpentine receptors, the serine-threonine kinase receptors or the receptor tyrosine kinases, among others. In the past several years, considerable progress has been made with regard to understanding the structure of members of the cytokine receptor superfamily and their mode of action. In this review, the role of the *Janus* protein tyrosine kinases (Jaks) will be considered. Downstream from the Jaks, a variety of signaling pathways have been implicated in cytokine responses. The role of several of these pathways is as yet unclear although considerable information is available for the function of members of the signal transducers and activators of transcription (Stats). The role of the Stats in various cytokine responses will be detailed.

[a]This work was supported by the National Cancer Institute Cancer Center Support (CORE) grant P30 CA21765 (J.N.I.), by support from AMGEN (J.N.I.), and by the American Lebanese Syrian Associated Charities (ALSAC).

[b]Corresponding author: James N. Ihle, Ph.D., Howard Hughes Medical Institute, Department of Biochemistry, St. Jude Children's Research Hospital, 332 N. Lauderdale, Memphis, TN 38105; Tel.: 901-495-3422; Fax: 901-525-8025; E-mail: james.ihle@stjude.org

ESSENTIAL ROLE FOR *JANUS* KINASES IN CYTOKINE
RECEPTOR SIGNALING

Studies directed to determining the critical signal transducing properties of members of the cytokine receptor superfamily have demonstrated that the ability to couple ligand binding to the induction of tyrosine phosphorylation is critical for all the biological responses. Over the past several years, a variety of cytoplasmic protein tyrosine kinases have been implicated in cytokine receptor signaling and it has become evident that the critical kinases are members of the *Janus* family of protein kinases (Jaks). The Jaks were initially identified through approaches to identify novel tyrosine kinases and therefore their role in cytokine signaling was not immediately appreciated. To date four mammalian Jaks have been identified, consisting of the ubiquitously expressed Jak1, Jak2, and Tyk2, and Jak3, which is primarily expressed in hematopoietic cells. Studies with a variety of receptors have demonstrated that each receptor associates with a specific Jak or Jaks and that this association occurs through a critical membrane proximal region of one or more of the receptor chains.[1] Conversely, the rather large non-catalytic amino-terminal domain of the Jaks is required for the interaction with the receptor proximal domain although, for reasons that are not known, it has not been possible to define a more precise domain on the Jaks.

Our studies have primarily focused on Jak2 and its interaction with the receptor for erythropoietin. However, Jak2 is utilized by a variety of highly related receptors including those for growth hormone, prolactin, and thrombopoietin, as well as interacting with the IL-6 receptor family signal transducing chain or one of the chains of the IFNγ receptor chains. Following Epo binding, Jak2 is rapidly tyrosine phosphorylated and its kinase activity activated.[2] A variety of studies support the hypothesis that ligand binding induces the aggregation of the Epo receptor and the associated Jaks and thereby allows a transphosphorylation that is critical for amplification of kinase activity. Support for this model has been derived from studies[3] that have identified the sites of Jak2 autophosphorylation and the biological significance of these phosphorylation events. In particular, there are 10 major sites of autophosphorylation. Among these are Y^{1007} and Y^{1008}, which lie within the postulated activation loop of the kinase domain. Mutation of these sites has demonstrated that Y^{1007} is absolutely essential for Jak2 catalytic activity, while mutation of Y^{1008} to F has little, if any, effect on activity. The amino acid sequence within this region is very similar to that of the receptor tyrosine kinases for FGF and insulin. Therefore it is postulated that the mechanisms of inhibition and activation may be similar to the model that has been derived from their crystal structures.[4,5]

The role of the Jaks in cytokine receptor signaling has been primarily derived from the analysis of receptor mutants and a correlation between the ability to activate the Jaks and to elicit a functional response. These studies have demonstrated a consistent correlation in which mutations that disrupt Jak association with the receptors also inactivate the ability of the receptor to function in all biological assays. More recently, however, mutant mice have been made that are deficient in specific Jaks. Our studies initially focused on Jak3, since Jak3, unlike the other Jaks, is primarily expressed in hematopoietic cells. Among the cytokine receptors, Jak3 is only activated by the group of lymphocyte cytokines that share in common the utilization of the common γ chain (γ_c), which was initially identified in the receptor for IL-2. Studies have demonstrated that Jak3 specifically associates with the γ_c chain through the relatively short cytoplasmic domain of γ_c.[6,7] In addition, these receptor complexes contain a ligand-specific receptor chain that associates with Jak1. Both Jak1 and Jak3 are required for the function of all the receptors in this group of cytokines, which includes IL-2, IL-4, IL-7, IL-9, IL-13, and IL-15.

To assess the role of Jak3 in this group of receptors, mutant mice were derived by homologous targeting of the gene embryonal stem cells (ES cells). Mice that were homozygously deleted for Jak3 were born viable and without gross morphological abnormalities. However, the mice had a classic phenotype of severe combined immunodeficiency (SCID) associated with a 50–100-fold reduction in the numbers of T and B cells.[8–10] This phenotype is similar to the phenotype seen in mice lacking the γ_c receptor chain, the IL-7 receptor, or IL-7. Together, the results demonstrate that among the cytokines that utilize γ_c and Jak3, IL-7 plays a critical role in the expansion of early lymphoid progenitors. Although greatly reduced, T and B cells are present, indicating that some level of normal differentiation can occur. However, the functions associated with IL-2, IL-4, and the other cytokines that activate Jak3 are deficient. The studies support an essential role, *in vivo*, for Jak3 in the function of this family of cytokine receptors.

The engagement of receptors, such as the IL-7 receptor, results in the tyrosine phosphorylation of multiple proteins and the activation of numerous signaling pathways. For example, one of the major substrates of Jak1/Jak3 in the response to IL-7 is Stat5. It could be envisioned that all these pathways are critical, in combination, to the functions of IL-7. Therefore it was quite surprising when it was shown that the function of IL-7 in early T cell expansion was rescued by a transgene expressing Bcl-2.[11,12] It is important to point out however that the B cell lineage was not rescued in these mice, indicating that other pathways are needed either uniquely or in combination with Bcl-2.

More recently we have derived mice that are deficient for Jak2 (unpublished data). These mice are not viable, dying *in utero* at approximately 10–12 days of gestation. Importantly, the phenotype is very similar to that observed in mice that are deficient for Epo or for the Epo receptor.[13] The primary defect is the inability to initiate definitive erythropoiesis, although primitive erythropoiesis is intact. Using stem cells from these mice, colony assays demonstrate their inability to respond to a number of cytokines that had been shown to activate Jak2. Therefore the studies are consistent with the hypothesis that Jak2 is critical for the function of a number of cytokines, including Epo. It will clearly be of interest to examine mice that lack Jak1 or Tyk2 as well.

MULTIPLE, NON-ESSENTIAL SIGNALING PATHWAYS IN CYTOKINE SIGNALING

Cytokines activate a number of identical signaling pathways. Although it is relatively easy to show that another cytokine might activate a particular pathway, it has been much more difficult to identify a clear biological function for a particular response. Our efforts have focused on the role of the activation of various pathways within the context of the Epo receptor and have used various receptor mutants to assess the biological role of the activation of various pathways.

The Epo receptor was initially shown to consist of two functional domains.[14] The membrane distal half of the receptor was found to be dispensable and, indeed, in certain cell lines, negatively influenced receptor function. In contrast, the membrane proximal region of the cytoplasmic domain of the receptor was essential for all receptor functions examined. The membrane proximal region was subsequently shown to be essential for Jak2 binding.[2] Point mutations within this region that inactivate the receptor also disrupt Jak2 binding and activation. As noted above, the activation of Jak2 is tightly regulated through transphosphorylation of Y^{1007} within the activation loop of the kinase domain. Thus the current model envisions ligand-driven aggregation of the receptor and its associated Jak2, which, in turn, allows transphosphorylations to occur resulting in the activation of Jak2. The activated kinase then goes on to phosphorylate itself at multiple sites, the receptor, and substrates that are recruited into the activated receptor complex.

Epo, like many cytokines, induces the activation of a variety of signaling events. A critical problem has been to define the significance of these various pathways in Epo function. Somewhat surprisingly, the activation of most signaling pathways requires the distal half of the cytoplasmic domain of the receptor. For example, the activation of the ras pathway is correlated with the recruitment and phosphorylation of the adaptor protein termed SHC. Carboxyl-truncation of the Epo receptor results in the loss of SHC phosphorylation and the lack of activation of components within the pathway, including the activation of the mitogen-activated kinases ERK1 and ERK2.[15] The significance of the activation of this pathway is not known however since no physiological function has been ascribed to the distal region of the cytoplasmic domain of the receptor. In a similar manner, the distal region is required for activation of phospholipase C γ_1[16] as well as for the recruitment and phosphorylation of the p85 regulatory subunit of phosphatidylinositol 3-kinase.[17]

The distal region of the cytoplasmic domain is also required for the recruitment of the tyrosine-specific phosphatase SHP-1, initially termed HCP, to the receptor complex.[18] Based on the observation that mice that genetically lack SHP-1 have a dramatic, and lethal, amplification of the hematopoietic lineage,[19] it is reasonable to propose that SHP-1 recruitment is essential for downregulation of the receptor complex. This may be mediated through the dephosphorylation of the regulatory tyrosine in the activation loop of Jak2.[20]

In addition to the above signaling pathways, Epo induces the tyrosine phosphorylation and activation of Stat5a and Stat5b. These proteins are the products of two highly related genes that co-localize genetically and are thought to have arisen as a recent gene duplication event.[21,22] In common with the above, Stat5a and Stat5b are initially recruited to the receptor complex through the interaction of the Stat SH2 domain with specific sites of tyrosine phosphorylation within the receptor.[23-26] Several functions have been proposed to be dependent upon the Stat5 proteins based on studies with receptor mutants or dominant negatives. For example, there is considerable data to support the non-redundant role of the Stat5 proteins in inducing the expression of the oncostatin M (OSM) and cytokine-inducible SH2-containing (CIS) genes. It is also clear that the Stat5 proteins can induce c-fos expression although it is also mediated through the activation of the ras pathway.

In contrast to the above, there is considerable debate concerning the significance of the activation of the Stat5 proteins for Epo-induced mitogenesis or differentiation. Our studies have failed to identify a critical non-redundant role for Stat5a/Stat5b,[25] while other studies with receptor mutants[23,26] or with dominant negatives have suggested a role.[27] The differing results may reflect differences in the cell lines that were used in the studies. Irrespective, as noted below, we have recently obtained mice in which both Stat5a and Stat5b are disrupted. While there are clear phenotypes, none involve hematopoiesis.

From the extensive studies with Epo receptor mutants, two conclusion can be made: (1) only the membrane proximal domain of the cytoplasmic domain is required for mitogenesis, and (2) mutations that disrupt Jak2 activation uniformly eliminate a mitogenic response and that the mitogenic response (as well as the induction of several immediate early genes such as c-myc) cannot be ascribed to any known signaling pathway.

STAT PROTEINS: MEDIATORS OF SPECIFIC RECEPTOR FUNCTION

Among the variety of substrates of tyrosine phosphorylation or signaling pathways that have been identified, the Stats have been of considerable interest for a number of reasons. The Stat family of transcription factors was initially identified by the Darnell group in their efforts to identify the transcription factors that mediated the activation of genes following interferon IFN binding.[28,29] From these studies, Stat1 and Stat2 were identified in the transcription complex induced by IFNα/β in association with p48, a member of the IRF family of DNA binding proteins. The Darnell group also demonstrated that the nuclear translocation and activation of the DNA binding activity of the Stat proteins required tyrosine phosphorylation and the concomitant dimerization of the proteins. The current model for Stat activation therefore consists of the initial recruitment to the receptor complex, through interaction of the SH2 domain of the Stat with sites of tyrosine phosphorylation on the receptor. This interaction provides the specificity with which specific cytokines activate specific Stats. The next step involves the tyrosine phosphorylation of the Stat by the associated Jaks as detailed below. Once phosphorylated, the Stats homodimerize, or heterodimerize, translocate to the nucleus and activate, or repress, gene expression. The direct, linear pathway from the cell surface to gene expression is one of the most appealing aspects of the Jak-Stat pathway.

Following the identification of Stat1 and Stat2, it became obvious that additional family members must exist that were activated in the responses to a variety of cytokines.[30] This conclusion led quickly to the cloning of five additional family members by a variety of approaches. Stat3 was cloned both biochemically and by homology. Stat4 was cloned by homology. The two highly related Stat5a and Stat5b genes were cloned biochemically. Finally, Stat6 was cloned biochemically and by searching databases of expressed genes for Stat family members. Following this rather rapid wave of cloning, no additional family members have been identified for the last three years, suggesting that perhaps all the family members that exist have been identified.

The concept that the Stats are substrates for the Jaks is supported by a variety of types of experiments including receptor mutants and the ability of purified Jaks to correctly phosphorylate purified Stat proteins. We have examined the requirements of the Jaks and Stats for this interaction (unpublished data). Somewhat remarkably, the ability of a wild-type Jak to phosphorylate any of the Stat proteins is dependent upon a functional SH2 domain since mutation of the critical phosphotyrosine binding Arg results in a protein that cannot be phosphorylated. However, the Jak kinase domain alone is able to phosphorylate such mutants and the wild-type phenotype is restored by adding back a portion of the pseudokinase domain. These results suggested that the pseudokinase domain can negatively affect the catalytic function of the kinase domain, possibly through steric interactions, and that this can be overcome by interaction of the SH2 domain of the Stat5a site within the kinase domain. It can be envisioned that this interaction tethers the substrate to the kinase domain and overcomes the inhibition. Recent studies suggest that a single site near the catalytic pocket of the kinase domain is involved in this interaction. In any case, the results illustrate another level of regulation of kinase activity that insures that Stat activation only occurs within the context of a receptor complex and activated Jaks.

The function of many of the Stat proteins has been assessed through the creation of mutant mice in which the individual genes have been disrupted. Mice have been derived that are deficient in Stat1.[31,32] Such mice are viable and have no gross abnormalities. However, the mice are extremely sensitive to viral infections, consistent with a critical role of Stat1 in the function of IFNs, which are the primary cytokines capable of activating Stat1. In contrast, mice that are deficient in Stat3 display a very early embryonic lethality, the basis of which was not identified.[33] It is possible that the lethality is related to a function associated with leukemia inhibitory factor (LIF). In particular, female LIF-deficient mice are infertile due to a failure in embryo implantation and this defect can be overcome by giving the females LIF.[34] Since LIF induces the activation of Stat3, it is conceivable that the two phenotypes are related such that the maternal LIF is unable to initiate an embryonic signal dependent upon Stat3.

Our efforts have focused on assessing the function of Stat4, the two related Stat5 proteins, and Stat6 through the derivation of mice deficient in each of the genes or combinations. We initially cloned Stat4 by homology and for a considerable time it was an orphan Stat for which we could not identify a ligand that would induce its tyrosine phosphorylation and activation.[35] However, it was ultimately shown that among the approximately 50 cytokines that utilize receptors of the cytokine receptor superfamily, IL-12 induced the activation of Stat4.[36,37] In addition to its activation by IL-12, Stat4 had a number of interesting properties, including the loss of expression in differentiating hematopoietic cells and high levels of expression during the terminal stages of spermatogenesis. Yet the Stat4-deficient mice were not deficient in either hematopoiesis or in spermatogenesis. Indeed, the primary defects in these mice were specifically related to the biological functions of IL-12.[38,39] The defects included a loss of upregulation of NK cytolytic activity of spleen cells in response to IL-12. In addition, IL-12–induced Th$_1$ differentiation was disrupted. Indeed, the phenotype of the mice was virtually identical to that of mice that lack IL-12.[40]

Stat6 was initially cloned as an IL-4–induced Stat-like activity[41] and by searching databases of expressed sequences.[42] The induction of tyrosine phosphorylation of Stat4 is primarily seen in the response of lymphocytes to IL-4 or to IL-13. The specificity of Stat4 activation by IL-4 is related to the presence of a docking site for the IL-4 SH2 domain within the IL-4 receptor α chain. To assess the biological function of Stat6, we[43] and others[44,45] have generated mice that lack Stat4. The Stat4-deficient mice are viable and lack any gross abnormalities. However, virtually all the biological functions associated with IL-4 or IL-13 are deficient in these mice. For example, splenic T cells are unable to respond to IL-4 by generating Th$_2$ cells. More strikingly, a variety of stimuli are unable to induce IgE production in these mice. This observation supports a previous hypothesis that proposed that Stat6 regulated the transcription of the non-rearranged IgE isotype region of the heavy chain locus and that such transcription was required for efficient rearrangement during the process of class switching. In addition, a number of cell surface antigens that are normally upregulated during immune responses as a consequence of IL-4 production are not upregulated in the deficient mice. Again, somewhat remarkably, the phenotype was very similar to the phenotype of mice in which the IL-4 gene had been deleted.

Unlike the above Stat family members, Stat5 is activated in the response to a wide variety of cytokines. Based on this observation, it has been proposed that Stat5 may function in a more general biological response associated with cytokines, such as the control of proliferation or differentiation. Stat5 activity is due to the products of two very highly related genes, termed Stat5a and Stat5b. These two genes co-localize genetically and are very closely linked to the Stat3 gene on mouse chromosome 11. It is hypothesized that the two genes are the result of a recent gene duplication. Our interest in Stat5a and Stat5b are derived from the observation that Epo predominantly induces the activation of Stat5a and Stat5b, as described above.

In order to study the functions of the Stat5 genes, we have derived mice that are deficient in Stat5a, Stat5b, or in both Stat5a and Stat5b (Stat5a/b). The Stat5a/b mice were derived by double targeting in ES cells, since their genetic co-localization would not allow a purely genetic approach. Perhaps the most striking observation from these studies was the ability to obtain viable mice for all the mutations. In particular, Stat5a/b mice show none of the phenotype seen in Epo- or Epo receptor–deficient mice or in the Jak2-deficient mice. Moreover, analysis of the erythroid lineage in each of the mutant strains failed to show any significant effects of the mutations on erythropoiesis. Thus the data support our receptor mutant studies, which indicated that Stat5 activation was not required for the functions of the Epo receptor that could be assessed with cell lines.

Although erythropoiesis was not affected, the mutant mice had a variety of phenotypes that are currently being studied in detail. However, the effects did include a lactation defect in the Stat5a-deficient mice comparable to a recent report.[46] This phenotype is consistent with a critical role for Stat5a signaling through the prolactin receptor during mammary gland development during pregnancy and lactation. The Stat5b mice have phenotypes that are consistent with a role in the functions of growth hormone. For example, the male mice are smaller and the sexually dimorphic pattern of expression of a number of genes that are regulated by growth hormone in the liver are altered. Again, our results have been consistent with a recent report of the properties of a Stat5b-deficient strain of mice.[47] It is interesting to note that the phenotypes we observed in the Stat5a-deficient mice are distinct from the phenotypes we observed in the Stat5b-deficient mice. Whether this is due to a functional difference between the two Stat5 gene products is not clear. In particular, the liver primarily expresses Stat5b, while the mammary gland preferentially expresses Stat5a. Therefore, the distinct phenotypes might relate more to the levels of expression than to functional differences.

However, the most important mutant is one in which the two Stat5 genes are both disrupted and therefore any redundancy does not complicate the phenotype. These mice show additional phenotypes not seen in either single mutant. First, it is important to note that hematopoiesis is not detectably disrupted in the Stat5a/b-deficient mice. Again this observation is extremely important since a variety of indirect studies suggested that the Stat5a proteins were required for both proliferation and differentiation. The distinct phenotypes however include a dwarfism in both male and female mice. More strikingly, the female Stat5a/b-deficient mice are infertile and it is hypothesized that this may be due to the interruption of the prolactin signaling in the corpus luteum of the ovary.

Perhaps the most striking phenotype is in lymphoid cells. In particular, lymphopoiesis occurs normally such that normal numbers and phenotypes of thymocytes and splenocytes are derived. However, the peripheral T cells are defective and fail to respond to IL-2. The mice ultimately develop a lymphoproliferative disease and in this regard resemble quite strikingly mice that are deficient in the IL-2 receptor β chain[48] or IL-2.[49] Studies are ongoing to further characterize the basis of the defects.

SUMMARY AND CONCLUSIONS

The last several years have provided unique insights into the evolution and function of a large group of cytokines that share the utilization of receptors of the cytokine receptor superfamily. It is somewhat remarkable that there are no known members of this large family of structurally and functionally related receptors or cytokines found in *Drosophila*. From this one might suggest the hypothesis that this family of receptors, and the cognate ligands, have evolved over relatively recent times to mediate newly acquired physiological functions. This can be easily envisioned to be the case for the cytokines that affect lymphoid functions, but is less obvious for the cytokines that affect other lineages of cells.

This family of receptors has in common the utilization of members of the Jak family of protein tyrosine kinases to mediate their functions. Again, it is somewhat remarkable that this is one of the smallest families of tyrosine kinases and, in this case, only a single member has been identified in *Drosophila* and that the mechanisms by which its activity is regulated are unknown.[50] In this regard, it is quite striking that Jak3 is dedicated solely to the function of a subfamily of cytokines that primarily affect lymphoid functions as evidenced by the knockout mice. Clearly the phenotype of mice lacking other Jak family members will be of considerable interest and will provide additional insights into the evolution and function of this family of cytokines.

The various studies emphasize two broad spectrums of functions for the cytokine receptor superfamily. One function is to couple ligand binding to induction of proliferation or to maintenance of cell viability. This function is best illustrated by the requirement of the IL-7 system for the amplification of the early lymphoid compartment, but is also likely to be a major function of cytokines such as Epo. The mechanisms by which any of the cytokine receptor superfamily members mediate this effect are unknown other than to say that Jak activation is essential. Indeed, the identification of the undoubtably novel signaling pathways involved in this response will be a major challenge to those studying this receptor family.

The second spectrum of functions mediated by cytokine receptors involves the physiologically important and specific functions. Many of these functions are mediated by the Stats, as clearly illustrated in the knockouts that have been done. Thus, the Stat family has been a fertile source of diversification of cytokine function. As above, it is important to note that there exists only a single Stat protein in *Drosophila* that is in the same pathway as the single Jak.

REFERENCES

1. IHLE, J. N. 1995. The Janus protein tyrosine kinase family and its role in cytokine signaling. Adv. Immunol. **60**: 1–35.
2. WITTHUHN, B. *et al.* 1993. JAK2 associates with the erythropoietin receptor and is tyrosine phosphorylated and activated following EPO stimulation. Cell **74**: 227–236.
3. FENG, J. *et al.* 1997. Activation of Jak2 catalytic activity requires phosphorylation of Y^{1007} in the kinase activation loop. Mol. Cell. Biol. **17**: 2497–2501.
4. HUBBARD, S. R. *et al.* 1994. Crystal structure of the tyrosine kinase domain of the human insulin receptor. Nature **372**: 746–754.
5. MOHAMMADI, M. *et al.* 1996. Structure of the FGF receptor tyrosine kinase domain reveals a novel autoinhibitory mechanism. Cell **86**: 577–587.
6. MIYAZAKI, T. *et al.* 1994. Functional activation of Jal1 and Jak3 by selective association with IL-2 receptor subunits. Science **266**: 1045–1047.
7. RUSSELL, S. M. *et al.* 1994. Interaction of IL-2 receptor beta and gamma$_c$ chains with JAK1 and JAK3, respectively: Defective gamma$_c$-JAK3 association in XSCID. Science **266**: 1042–1045.
8. NOSAKA, T. *et al.* 1995. Defective lymphoid development in mice lacking Jak3. Science **270**: 800–802.
9. PARK, S. Y. *et al.* 1995. Developmental defects of lymphoid cells in Jak3 kinase-deficient mice. Immunity **3**: 771–782.
10. THOMIS, D. C. *et al.* 1995. Mice lacking Jak3 have defects in B lymphocyte maturation and T lymphocyte activation. Science **270**: 794–797.
11. LAGASSE, E. *et al.* 1997. Enforced expression of Bcl-2 in monocytes rescues macrophages and partially reverses osteopetrosis in *op/op* mice. Cell **89**: 1021–1031.
12. MARASKOVSKY, E. *et al.* 1997. Bcl-2 can rescue T lymphocyte development in interleukin-7 receptor-deficient mice but not in mutant *rag-1$^{-/-}$* mice. Cell **89**: 1011–1019.
13. WU, H. *et al.* 1995. Generation of committed erythroid BFU-E and CFU-E progenitors does not require erythropoietin or the erythropoietin receptor. Cell **83**: 59–67.
14. D'ANDREA, A. D. *et al.* 1991. The cytoplasmic region of the erythropoietin receptor contains nonoverlapping positive and negative growth-regulatory domains. Mol. Cell Biol. **11**: 1980–1987.
15. MIURA, Y. *et al.* 1994. Activation of the mitogen-activated protein kinase pathway by the erythropoietin receptor. J. Biol. Chem. **269**: 29962–29969.
16. REN, H. Y. *et al.* 1994. Erythropoietin induces tyrosine phosphorylation and activation of phospholipase C-gamma 1 in a human erythropoietin-dependent cell line. J. Biol. Chem. **269**: 19633–19638.
17. MIURA, O. *et al.* 1994. Erythropoietin-dependent association of phosphatidylinositol 3-kinase with tyrosine-phosphorylated erythropoietin receptor. J. Biol. Chem. **269**: 614–620.
18. YI, T. *et al.* 1995. Hematopoietic cell phosphatase (HCP) associates with the erythropoietin receptor following Epo induced receptor tyrosine phosphorylation: Identification of potential binding sites. Blood **85**: 87–95.
19. SHULTZ, L. D. *et al.* 1993. Mutations at the murine motheaten locus are within the hematopoietic cell protein tyrosine phosphatase (Hcph) gene. Cell **73**: 1445–1454.
20. KLINGMULLER, U. *et al.* 1995. Specific recruitment of the hematopoietic protein tyrosine phosphatase SH-PTP1 to the erythropoietin receptor causes inactivation of JAK2 and termination of proliferative signals. Cell **80**: 729–738.
21. AZAM, M. *et al.* 1995. Purification of interleukin-3 stimulated DNA binding factors demonstrates the involvement of multiple isoforms of Stat5 in signaling. EMBO J. **14**: 1402–1411.
22. COPELAND, N. G. *et al.* 1995. Distribution of the mammalian Stat gene family in mouse chromosomes. Genomics **29**: 225–228.
23. GOBERT, S. *et al.* 1996. Identification of tyrosine residues within the intracellular domain of the erythropoietin receptor crucial for STAT5 activation. EMBO J. **15**: 2434–2441.

24. KLINGMULLER, U. *et al.* 1996. Multiple tyrosine residues in the cytosolic domain of the erythropoietin receptor promote activation of STAT5. Proc.Natl. Acad. Sci. USA **93**: 8324–8328.

25. QUELLE, F. W. *et al.* 1996. Erythropoietin induces activation of Stat5 through association with specific tyrosines on the receptor that are not required for a mitogenic response. Mol. Cell. Biol. **16**: 1622–1631.

26. DAMEN, J. E. *et al.* 1995. Tyrosine 343 in the erythropoietin receptor positively regulates erythropoietin-induced cell proliferation and Stat5 activation. EMBO J. **14**: 5557–5568.

27. MUI, A.-F. *et al.* 1996. Suppression of interleukin-3-induced gene expression by a C-terminal truncated Stat5: role of Stat5 in proliferation. EMBO J. **15**: 2425–2433.

28. Darnell, J. E., Jr. *et al.* 1994. Jak-STAT pathways and transcriptional activation in response to IFNs and other extracellular signaling proteins. Science **264**: 1415–1421.

29. SCHINDLER, C. *et al.* 1995. Transcriptional responses to polypeptide ligands: the JAK-STAT pathway. Annu. Rev. Biochem. **64**: 621–651.

30. IHLE, J. N. 1996. STATs: signal tranducers and activators of transcription. Cell **84**: 331–334.

31. DURBIN, J. E. *et al.* 1996. Targeted disruption of the mouse Stat1 gene results in compromised innate immunity to viral disease. Cell **84**: 443–450.

32. MERAZ, M. A. *et al.* 1996. Targeted disruption of the Stat1 gene in mice reveals unexpected physiologic specificity in the JAK-STAT signaling pathway. Cell **84**: 431–442.

33. TAKEDA, K. *et al.* 1997. Targeted disruption of the mouse Stat3 gene leads to early embryonic lethality. Proc. Natl. Acad. Sci. USA **94**: 3801–3804.

34. ESCARY, J. L. *et al.* 1993. Leukemia inhibitory factor is necessary for maintenance of haematopoietic stem cells and thymocyte stimulation. Nature **363**: 361–364.

35. YAMAMOTO, K. *et al.* 1994. Stat4: A novel GAS binding protein expressed in early myeloid differentiation. Mol. Cell. Biol. **14**: 4342–4349.

36. JACOBSON, N. G. *et al.* 1995. Interleukin 12 activates Stat3 and Stat4 by tyrosine phosphorylation in T cells. J. Exp. Med. **181**: 1755–1762.

37. BACON, C. M. *et al.* 1995. Interleukin-12 induces tyrosine phosphorylation of JAK2 and TYK2: differential use of Janus tyrosine kinases by interleukin-2 and interleukin-12. J. Exp. Med. **181**: 399–404.

38. KAPLAN, M. H. *et al.* 1996. Impaired IL-12 responses and enhanced development of Th2 cells in Stat4-deficient mice. Nature **382**: 174–177.

39. THIERFELDER, W. E. *et al.* 1996. Stat4 is required for IL-12 mediated responses of NK and T-cells. Nature **382**: 171–174.

40. WOLF, S. F. *et al.* 1994. Interleukin-12: a key modulator of immune function. Stem Cells **12**: 154–168.

41. HOU, J. *et al.* 1995. Identification and purification of human Stat proteins activated in response to interleukin-2. Immunity **2**: 321–329.

42. QUELLE, F. W. *et al.* 1995. Cloning of murine Stat6 and human Stat6, stat proteins that are tyrosine phosphorylated in responses to IL-4 and IL-3 but are not required for mitogenesis. Mol. Cell. Biol. **15**: 3336–3343.

43. SHIMODA, K. *et al.* 1996. Lack of IL-4-induced Th2 response and IgE class switching in mice with disrupted Stat6 gene. Nature **380**: 630–633.

44. KAPLAN, M. H. *et al.* 1996. Stat6 is required for mediating responses to IL-4 and for the development of Th2 cells. Immunity **4**: 313–319.

45. KOPF, M. *et al.* 1993. Disruption of the murine IL-4 gene blocks Th2 cytokine responses. Nature **362**: 245–248.

46. LIU, X. *et al.* 1-15-1997. Stat5a is mandatory for adult mammary gland development and lactogenesis. Genes Dev. **11**: 179–186.

47. UDY, G. B. *et al.* 1997. Requirement of STAT5b for sexual dimorphism of body growth rates and liver gene expression. Proc. Natl. Acad. Sci. USA **94**: 7239–7244.

48. SUZUKI, H. *et al.* 1995. Deregulated T cell activation and autoimmunity in mice lacking interleukin-2 receptor β. Science **268**: 1472–1476.

49. SCHORLE, H. *et al.* 1991. Development and function of T cells in mice rendered interleukin-2 deficient by gene targeting. Nature **352**: 621–624.

50. BINARI, R. *et al.* 1994. Stripe-specific regulation of pair-rule genes by *hopscotch*, a putative Jak family tyrosine kinase in *Drosophila*. Genes & Dev. **8**: 300–312.

Cis-Regulatory Elements Controlling Basal and Inducible VIP Gene Transcription

SUNG HO HAHM AND LEE E. EIDEN[a]

Section on Molecular Neuroscience, Laboratory of Cellular and Molecular Regulation, National Institute of Mental Health, National Institutes of Health, Bethesda, Maryland 20892 USA

ABSTRACT: The *cis*-acting elements of the VIP gene important for basal and stimulated transcription have been studied by transfection of VIP-reporter gene constructs into distinct human neuroblastoma cell lines in which VIP transcription is constitutively high, or can be induced to high levels by protein kinase stimulation. The 5.2 kb flanking sequence of the VIP gene conferring correct basal and inducible VIP gene expression onto a reporter gene in these cell lines was systematically deleted to define its minimal components. A 425-bp fragment (-4656 to -4231) fused to the proximal 1.55 kb of the VIP promoter-enhancer was absolutely required for cell-specific basal and inducible transcription. Four additional components of the VIP gene were required for full cell-specific expression driven by the 425 bp TSE (region A). Sequences from -1.55 to -1.37 (region B), -1.37 to -1.28 (region C), -1.28 to -.094 (region D), and the CRE-containing proximal 94 bp (region E) were deleted in various combinations to demonstrate the specific contributions of each region to correct basal and inducible VIP gene expression. Deletion of region B, or mutational inactivation of the CRE in region E, resulted in constructs with low transcriptional activity in VIP-expressing cell lines. Deletion of regions B and C together resulted in a gain of transcriptional activity, but without cell specificity. All five domains of the VIP gene were also required for cell-specific induction of VIP gene expression with phorbol ester. Gelshift analysis of putative regulatory sequences in regions A–D suggests that both ubiquitous and neuron-specific *trans*-acting proteins participate in VIP gene regulation.

The VIP gene responds to intra- and extracellular stimuli in three distinct modes. First, the gene is active in VIPergic cells and silenced in non-VIPergic cells in a neuroanatomically precise fashion. Second, the VIP gene, like all neuropeptide genes, is episodically activated upon secretion of peptide from the cell, a process referred to as "stimulus-secretion-synthesis coupling." Third, the VIP gene is upregulated by various second messenger pathways stimulated by cytokines, neurotrophins, and other peptides during development, inflammation, neuronal injury, and maintenance of endocrine homeostasis—i.e., physiological gene regulation.

The VIP gene therefore possesses intrinsic response elements for multiple intra- and extracellular stimuli, acting in various combinations to (1) establish and (2) maintain the VIP phenotype, and (3) permit adaptation to physiological and endocrine stressors. These three modes of regulation of the VIP gene provide appropriate levels of VIP and VIP-associated

[a]Corresponding author: Lee E. Eiden, Ph.D., Chief, Section on Molecular Neuroscience, Laboratory of Cellular and Molecular Regulation, Building 36, Rm. 3A-10, National Institute of Mental Health, National Institutes of Health, Bethesda, MD 20892; Tel.: 301-496-4110; Fax: 301-402-1748; E-mail: eiden@codon.nih.gov

peptide (PHM in the human, PHI in the rodent VIP gene) in the VIPergic neurons and endocrine cells that regulate autonomic, endocrine, and brain function.

We have analyzed VIP biosynthesis and gene transcription in chromaffin and neuroblastoma cells to determine how discrete regulatory cassettes within the VIP gene interact in developmental, secretion-coupled, and adaptive transcriptional modes. Here, we review and present new data on VIP gene regulation during development, in stimulus-secretion-synthesis coupling, and during adaptation to signaling pathway activation in primary bovine chromaffin cells, PC12 cells, developing spinal cord cells in culture, and human neuroblastoma cells. We review the evidence that five discrete regions of the VIP gene regulate its tissue-specific expression and its response to a variety of first-messenger–initiated signal transduction pathways. The unique responses of the VIP gene to calcium influx, cyclic AMP elevation, protein kinase C activation, and cytokine signaling, alone and in combination, may underlie the differential expression of VIP and other neuropeptide genes in the central and peripheral nervous systems. The same *cis*-active elements mediate both constitutive and inducible VIP gene transcription. This observation suggests that common signal tranduction pathways may participate first in cell-specific establishment of the VIP phenotype *in vivo* during development, and again in modulation of the rate of VIP gene transcription in response to homeostatic demands in the adult animal. Complex interactions among proteins binding at the several discrete clusters of *cis*-active elements of the VIP gene may be required to integrate the many signals that converge on VIPergic cells regulating VIP biosynthesis and secretion in response to physiological and developmental demands.

VIP GENE REGULATION IN CHROMAFFIN AND PC12 CELLS

Chromaffin cells of the adrenal medulla are of neural crest origin and express a variety of neuropeptides including VIP.[1] An advantage of using a single cell background to study the regulation of multiple neuropeptides is that differential biosynthetic responses of neuropeptide genes co-expressed in chromaffin cells indicate unique features of the genes themselves, rather than differing abundance of transcription factors, receptors, kinases, or phosphatases.

What all neuropeptides have in common is that they are secreted in response to cell depolarization, arising either from spontaneous neuronal activity or that evoked by cell-specific secretagogues.[2] Secretory loss of neuropeptide from the neuroendocrine cell must be compensated for by increased gene transcription. Since neuropeptides represent proteins synthesized specifically to be exported from the cell in a regulated fashion, the genes encoding their prohormones appear to have developed mechanisms for coupling enhanced biosynthesis to evoked secretion.[3]

As expected, VIP and enkephalin biosynthesis are similarly regulated by secretagogues that act by increasing calcium influx via voltage-dependent channels.[4] Both show a two- to threefold elevation in total peptide synthesis (cellular plus secreted peptide levels) after cell depolarization with elevated extracellular potassium, for example, which causes substantial secretion of both enkephalin and VIP from chromaffin cells. The multi-factor pathway leading to upregulation of VIP and other neuropeptides by calcium following cell depolarization is only partly elucidated at this time, although early bifurcation of calcium-dependent pathways triggering secretion and biosynthesis can be identified by substitution of barium for calcium in the secretory, but not the biosynthetic, pathway in chromaffin cells.[4]

A unique pattern of VIP biosynthesis regulation is seen in response to stimulation by PACAP in chromaffin cells. Both VIP and enkephalin peptides are released from chromaffin cells in response to PACAP, and release is in both cases abolished by blockade of voltage-dependent calcium channels (Hahm, Hsu, and Eiden, in preparation). Upregulation of VIP biosynthesis by PACAP in chromaffin cells is also antagonized by blockade of voltage-dependent calcium channels with D-600, consistent with an action of PACAP to increase intracellular calcium through voltage-dependent calcium channel activation as reported in other cell types.[5,6] Enkephalin biosynthesis activation by PACAP, on the other hand, is not dependent on calcium influx, and may involve primarily PACAP's action on the adenylate cyclase system directly (Hahm, Hsu, and Eiden, in preparation).[7]

These data suggest that the VIP and enkephalin genes may be targeted by different combinations of signal transduction pathways activated by PACAP, with VIP gene induction dependent on multiple converging signals including calcium and enkephalin gene induction perhaps dependent only on activation of protein kinase A pathways, which do not require calcium for full activation. The upregulation of both enkephalin and VIP by PACAP is blocked by ascomycin and cyclosporin A, inhibitors of calcium-dependent phosphatase I (Eiden, Hsu, and Hahm, unpublished observations). It is not yet known if blockade of the effect of PACAP is at a transcriptional or translational level. In fact, the action of ascomycin and cyclosporin A could even be at the level of post-translational processing of VIP or enkephalin, since the radioimmunoassay used to detect both VIP and met-enkephalin in these experiments is highly specific for the processed form of each neuropeptide. Neuropeptide elevation by enhanced processing alone, with no increase in either transcription or translation, has been previously reported for reserpine-induced increases in the levels of met-enkephalin in chromaffin cells,[8] although not for VIP.[4]

The cytokines IL-1β and TNF-alpha also differentially affect VIP biosynthesis relative to other chromaffin cell neuropeptides. Thus, IL-1β and TNF at nanomolar concentrations increase VIP biosynthesis and synergistically amplify PMA- and forskolin-stimulated elevation of VIP levels, without stimulating the biosynthesis of enkephalin.[9] The response of the VIP gene to IL-1β and TNF is consistent with the presence within the VIP gene of a cytokine response element containing a STAT binding site,[10] shown by Symes and coworkers to function as a CNTF-, LIF-, oncostatin M-, and IL-6-responsive element in neuroblastoma cell lines expressing the appropriate cytokine/neurotrophin receptors.[11-14] These data underscore the unique combinatorial interactions among signaling pathways that lead to upregulation of VIP biosynthesis, and their potential for amplifying neuropeptide-unique expression patterns in cells stimulated simultaneously via multiple signaling pathways.

Similar combinatorial effects of first messenger stimulation are observed in the rat pheochromocytoma cell line PC12. Treatment of PC12 cells cultured at low density under standard culture conditions with forskolin or NGF at concentrations giving maximal stimulation of the protein kinase A signaling pathway and full neuronal differentiation, respectively, causes no detectable increase in the levels of immunoreactive VIP. However, co-treatment with the same concentrations of these agents elicited a greater than tenfold increase in VIP levels in PC12 cells (FIG. 2), indicating that simultaneous stimulation of separate intracellular signaling pathways by NGF and protein kinase A exerts synergistic effects on VIP expression in these cells. Convergence at the level of CREB as a general mechanism for this synergistic interaction could be envisaged if forskolin and NGF activated separate pathways leading to CREB phosphorylation, for example RSK2/CREB kinase and CAM kinase,[15,16] to increased fractional phosphorylation, or increased duration of activation,[17] of CREB in VIPergic cells.

In summary, multiple neuropeptides in chromaffin cells respond to cell depolarization, e.g., with elevated K^+, or acetylcholine, with a stereotyped severalfold increase in neuropeptide biosynthesis, concomitant with regulated neuropeptide secretion from the cell. Other first messengers, such as histamine, PACAP, and cytokines, differentially affect neuropeptide expression in chromaffin cells (TABLE 1). These patterns may in turn reflect differential responses of neuropeptide genes to the protein kinase A and C signaling pathways as well as calcium-activated signaling pathways, since direct stimulation of protein kinases

A and C with forksolin or phorbol ester, respectively, results in demonstrably more dramatic effects, including synergistic effects of activation of multiple protein kinases (FIG. 1). Finally, growth factors and cytokines may exert both direct effects on the VIP gene and mutually permissive effects in combination with protein kinase A signaling pathway activation to further fine-tune the level of expression of VIP relative to other neuropeptides in neuroendocrine cells (FIGS. 1 and 2).

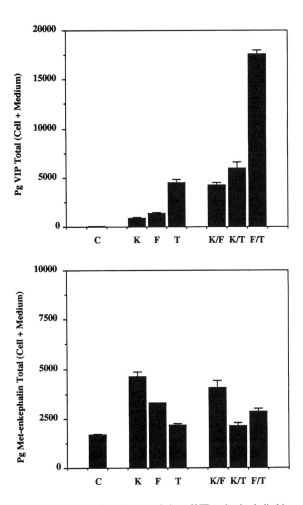

FIGURE 1. Second-messenger interactions in upregulation of VIP and enkephalin biosynthesis in chromaffin cells. Total (cell plus medium) VIP and enkephalin immunoreactivity was measured 72 hours after treatment of chromaffin cells with vehicle (C), 40 mM K$^+$ (K), 25 µM forskolin (F), or 0.1 µM tetradecanoyl phorbol acetate (T) alone or in combination (K/F, K/T, and F/T) at the same concentrations.

TABLE 1. VIP and met-enkephalin: differential upregulation by first-messenger stimulation of chromaffin cells

	VIP	Enkephalin
Experiment I		
Untreated	30 ± 5	8,842 ± 243
10 µM histamine	912 ± 18*	10,827 ± 770*
Experiment II		
Untreated	92 ± 3	5,961 ± 408
10 nM PACAP	4,891 ± 526***	23,993 ± 1748***
Experiment III		
Untreated	86 ± 7	3,090 ± 137
1 nM IL-1β	113 ± 5*	2,535 ± 197*

VIP and met-enkephalin total (cell plus medium) peptide content of chromaffin cell cultures, expressed as pg/well, measured by radioimmunoassay 72 hours after addition of histamine, PACAP, or IL-1β. Culture conditions and radioimmunoassay as described previously.[47] Values expressed as the mean ± s.e.m. of triplicate wells. *$p < 0.05$; **$p < .01$; ***$p < .001$; Student's t-test.

VIP GENE REGULATION IN NEURONS OF THE CENTRAL NERVOUS SYSTEM

Depolarization of spinal cord cells in primary culture elevates enkephalin and VIP biosynthesis, as it does in chromaffin cells.[18,19] Since spinal cord neurons in primary culture exhibit spontaneous electrical activity, regulation of VIP and other neuropeptides by cell depolarization is manifested by decreased expression following blockade of spontaneous electrical activity with tetrodotoxin (TTX), rather than increased gene expression following treatment with elevated extracellular potassium, as in chromaffin cells.[19] Prolonged treatment with TTX during the "critical period" in which spinal cord neurons

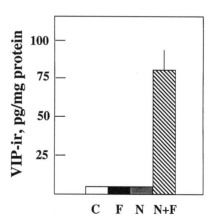

FIGURE 2. Synergistic upregulation of VIP biosynthesis by NGF and forskolin in PC12 cells. PC12 cells were cultured as described previously,[48] and VIP content assayed by radioimmunoassay 5 days after addition of 50 ng/ml nerve growth factor (NGF), 25 µM forskolin, or both. Values represent the mean ± s.e.m. of triplicate determinations from a single experiment. Values less than ~5 pg/mg protein are at the limit of detection of the radioimmunoassay and thus represent overestimates of the amount of immunoreactive VIP present.

first begin to elaborate neurotransmitters and neuropeptides results in a profound reduction in neuropeptide gene expression[18,19], suggesting that the intracellular mechanisms that govern stimulus-secretion-synthesis coupling following chromaffin cell depolarization may also operate to establish chemically distinct neuronal phenotypes in spontaneously depolarizing neurons during development. Thus, despite the very different milieu of endocrine and neuronal cells with respect to first-messenger stimulation, as well as differences in levels of intrinsic electrical activity, responses of the VIP gene to secretory activity appear to transcend cell type.

VIP GENE REGULATION IN HUMAN NEUROBLASTOMA CELL LINES

Human neuroblastomas are peripheral neuroendocrine tumors of neural crest origin. Some of these have been used to establish cell lines that express neuron-specific genes, such as neuropeptides, and can be employed in the study of cell-specific mechanisms of gene regulation *in vitro*. The SH-EP neuroblastoma cell subline, for example, constitutively expresses VIP at high levels. The SK-N-SH, SH-SY-5Y, and SH-IN neuroblastoma cell lines express low to moderate levels of VIP, but the VIP gene can be strongly upregulated by stimulation of the protein kinase A or C signaling pathways.[20–22]

We have previously demonstrated that 5.2 kb of the 5′ flanking sequence of the VIP gene confers complete cell-specific regulation of this gene with respect to constitutive expression, and induction by stimulation of protein kinases A and C, when transfected with a reporter gene into the human neuroblastoma cell lines SH-IN and SH-EP.[22,23] Deletional analysis of this construct revealed that removal of 2.7 kb of upstream sequence resulted in loss of both constitutive and inducible regulation of the VIP gene in these cell lines. The human and mouse VIP gene flanks, including up to about 2.5 kb of each 5′ flank, have been sequenced.[24,25] Recently we have completed the primary sequencing of the human VIP gene 5′ flank in its entirety,[23] i.e., up to the 5′ boundary of the smallest DNA fragment shown to confer cell-specific regulation of the gene (FIG. 3). Domains of the gene containing sequences found as core elements in various *cis*-active DNA domains in other genes are indicated in FIGURE 3, as are domains directly shown to be functional in transcriptional regulation in the VIP gene itself. The Oct-1-like sites of domain A have only recently been directly implicated by site-directed mutagenesis in the tissue-specific regulation of the VIP gene.[25a] These sequences comprise potential binding sites for POU-domain proteins in the Oct-1/Oct-2 subfamily.[26,27] The Oct-1-like motifs found in the VIP gene are quite homologous to similar regions in the GnRH gene 5′ flank that confer cell-specific transcriptional activity in GnRH-containing neurons and cell lines.[28] In domain B, partially overlapping with domain C, E-boxes that bind both cell-specific and non-specifically expressed bHLH proteins flank sequences similar to those for the binding of MEF-2, a family of proteins that function as neuronal/myocyte enhancer factors in collaboration with E-box-binding bHLH myogenic (e.g., myoD), and neurogenic (e.g., MASH-1) *trans*-acting factors.[29–32] The cytokine response element (CyRE) is that region that confers CNTF-, LIF-, and IL-6-inducible VIP gene transcription in neuroblastoma cells.[11–14] The CyRE is contained in domain C and the 5′ portion of domain D in FIGURE 3. A consensus binding element for STAT-1 and STAT-3 transcription factors, which are regulated by the Janus kinase signal transduction pathway(s) activated by cytokine receptor occupany[10,33,34] is also present in domain C as part of the CyRE, as is an overlapping sequence for binding of ETS proto-oncogene transcription factors, which are reported to collaborate with AP-1 transcription factors to confer transcriptional activation by phorbol esters.[25,35] Domain C also contains a dyad symmetry element conserved in both human and mouse VIP genes, whose function however remains unclear.[25] A non-canonical AP-1 consensus element in domain D has been shown to bind FOS and JUN-immunoreactive proteins in

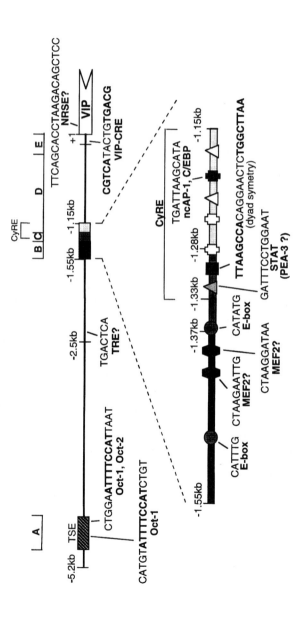

FIGURE 3. Regulatory anatomy of the human VIP gene. Potential *cis*-active regions of the VIP gene (A–E) that, fused to a luciferase reporter gene downstream of the VIP core promoter, first exon and first intron, have allowed functional analysis of the contribution of each region to correct expression of the VIP gene in SH-EP, SK-N-SH, and HeLa cells. In addition to the potential *cis*-active regulatory elements depicted here in the 5′ flank of the VIP gene, a potential neuronally restrictive silencer element (NRSE/RE-1) is present within the first exon of the human VIP gene (positions +99 to +119). This element is present in all VIP constructs tested in SH-EP, HeLa, and SK-H-SH cells (FIG. 4).

gelshift assays.[12] The proximal promoter region of the VIP gene, here designated domain E, contains an element, the VIP-CRE, that has been demonstrated to impart cyclic AMP responsiveness to the VIP gene in PC12 cells, and responsiveness to PMA in HeLa cells when placed upstream, in triplicate, of a heterologous promoter.[21]

We have been attempting to assess systematically the contributions of the various domains of the VIP gene to full cell-specific expression, prompted in large part by our observations that the CRE is required not only for cyclic AMP responsivity in SK-N-SH cells, but also for full expression of the gene even in SH-EP cells, in which stimulation of cyclic AMP provides little additional upregulation of gene transcription. Thus, block mutations of the CRE sufficient to completely abrogate forskolin-stimulated VIP gene transcription in PC12 cells also considerably downregulated TSE-dependent constitutive activity of the gene in human neuroblastoma cells.[23] Cell-specific regulation mediated by the VIP-CRE independent of its function in cAMP-dependent transcriptional activation is similar to the apparently Ser-133-phosphorylation-independent role of CREB in mediating cell-specific somatostatin gene transcription via binding at the CRE and interaction with additional transcription factors of the LIM family, recruited to the somatostatin promoter/enhancer.[36] Simultaneously, deletion of the 2.7 kb upstream region at both its 5′ and 3′ boundaries defined a 425 bp fragment of the VIP gene, which we now refer to as the VIP gene TSE, for tissue specifier element, that imparts tissue-specific expression to reporter constructs, but only, contrary to previous reports,[37] in the context of at least 1.55 kb of the promoter-proximal 5′ flank of the gene. One reason for this may be that the Taq fragment employed by Agoston and colleagues is larger than the 425 bp TSE defined by Hahm and Eiden,[23] and contains additional sequences allowing cell-specific expression. A second possibility is that while the TSE by itself supports transcription from the VIP core promoter in SH-EP, SH-IN, and other neuroblastoma cells, it also does so in non-neuronal cells in the absence of additional downstream sequences, and is thus not cell specific. This, in fact, appears to be the case (vide infra).

Both induction of transcription by PMA in SK-N-SH cells and cell-specific constitutive expression in SH-EP cells depend on all five domains A through E of the VIP gene (FIG. 4). Upregulation of the VIP gene by PMA, contrary to expectation, is not conferred by domain E, i.e., by the VIP-CRE, but rather appears to depend on the presence of domain D, which harbors the non-canonical AP-1 site characterized by Symes and coworkers.[12] Constructs containing A, B, D, and E, but not domain C remain to be tested to determine the contribution of the PEA-3 consensus sequence to PMA induction of the VIP gene. In the absence of the TSE (domain A), regardless of the presence of any combination of domains B–E, uniformly low transcriptional activity, comparable to the activity of the intact 5.2 kb construct in HeLa cells, is observed in all cell lines tested (data not shown). Fusion of the TSE alone (domain A) to the CRE-containing promoter/enhancer of the VIP gene (domain E) does indeed result in significant transcriptional activity in SH-EP cells (between 30 and 60% of full expression seen with the 5.2 kb construct). However, it does not impart significant transcriptional activation in SK-N-SH cells, either with or without PMA treatment. More significantly, there is activation of this construct in HeLa cells as well, demonstrating that cell-specific repressor as well as inducible elements are required.

It is noteworthy that removal of sequences from -2.5 to -1.55 kb does not affect either PMA induction or constitutive expression of the VIP gene, even though this fragment contains a consensus TRE (TPA- or PMA-responsive element) (FIG. 3). This "element" may simply represent a fortuitous sequence incorrectly positioned in the gene for contributing to transcriptional regulation, even upon binding AP-1. Alternatively, the activity of neighboring PMA-responsive elements may mask the contribution of this element in neuroblastoma cells, whereas in other cell lineages combinatorial interaction of this with other TREs in the gene may be required for full VIP expression. These questions will undoubtedly be

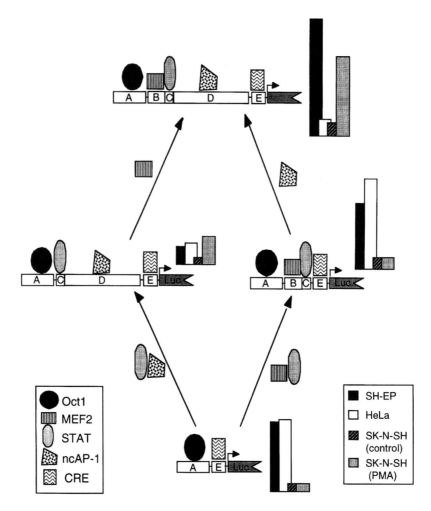

FIGURE 4. Interactions among VIP *cis*-active elements. Shown semi-quantitatively are the levels of transcription in SH-EP, HeLa, and SK-N-SH cells, and SK-N-SH cells treated with 0.1 µM PMA, of constructs containing domains A and E; A, C, D, and E; A, B, C, and E; and A, B, C, D, and E of the VIP gene. Domains contain respectively Oct-1 (A), MEF-2 (B), STAT (C), AP-1 (D), and CRE (E) consensus sequences, and are therefore potentially capable of binding Oct-1, MEF-2, STAT, AP-1, and CREB trans-acting factors, among others.

answered when full batteries of VIP transgenic mice harboring mutations in this domain can be created and analyzed.

Deletion of the sequence from -1.55 to -1.37 results in a VIP gene construct which, even when fused to the TSE, has little transcriptional activity in SH-EP cells.

Paradoxically, further removal of the -1.37 to -1.28 sequence and fusion of the TSE to the -1.28 sequence result in constructs that are now active in SH-EP cells, but are also transcriptionally active in HeLa cells, suggesting that a cell-specific repressor activity is contained in the -1.37 to -1.28 domain (domain C) of the VIP gene.

The data described above are summarized in semi-quantitative fashion in FIGURE 4. Several general features of the structural basis for VIP gene regulation emerge from this analysis. First, the TSE is absolutely required for cell-specific VIP gene expression and induction. The TSE requires, in turn, contributions from domains B–E for full activity. Second, the MEF2-containing domain B of the VIP gene (-1.55 to -1.37) is needed both for full cell-specific constitutive expression in EP cells and for PMA induction in SK-N-SH cells. In SK-N-SH cells, PMA induction is already conferred in the absence of domain B, yet is amplified significantly in its presence. Third, domain C, containing the STAT binding site, appears to possess cell-specific silencer activity insofar as transcription increases in HeLa and SH-EP cells simultaneously in TSE-containing constructs when this element is removed. It is important to note, on the other hand, that the construct A-B-D-E, in which the contribution of domain D to repressor activity in HeLa cells independent of domain C could be assessed, has not yet been assayed for transcriptional activity in vitro. Fourth, the ncAP1-containing domain D is absolutely required both for PMA induction in SK-N-SH cells and for repression of transcriptional activity in HeLa cells, although significant basal activity in SH-EP cells is maintained in the absence of this element. The collaboration of domains C and D in transcriptional repression in HeLa cells is suggested by the data presented here, but is not yet directly proven.

Data obtained with VIP gene constructs fused to reporter genes and transfected into VIP-expressing and non-expressing cells in vitro can be compared to the limited data obtained thus far with the structural requirements for correct tissue-specific expression of the VIP gene in vivo in transgenic animals. A VIP gene consisting in about 2 kb of 5' flank, and absent the TSE, appears to confer neuron-specific regulation on the VIP gene in both the central and peripheral nervous systems of transgenic mice.[38,39] Since the ratio of endogenous VIP to exogenous VIP gene expression varied throughout the nervous system, and there was ectopic expression of the transgene in both Schwann and smooth muscle cells, these authors concluded that additional restrictive and cell-specifier elements were missing from the VIP transgene employed. However, a VIP construct containing the 5.2 kb of 5' flanking sequence shown to impart cell-specific expression in neuroblastoma cells in vitro provided low levels of expression of reporter gene expression in the intestine and had no detectable transcriptional activity in the brain.[37] This apparent discrepancy may well be resolved with VIP gene constructs driving reporter genes encoding proteins whose presence, abundance, and cellular localization can be more easily assessed than either chloramphenicol acetyltransferase or β-galactosidase, the reporter genes employed in the studies carried out with the VIP gene to date. It is also worth noting that the VIP-expressing cells employed by us and others are predominantly neuroblastoma and pheochromocytoma cells representing peripheral autonomic lineages. The sequence requirements for correct expression of the VIP gene in VIP-expressing cells of the central nervous system and the anterior pituitary have not been systematically examined.

As summarized in FIGURE 4, differential expression in HeLa and EP cells and full inducibility with PMA in SK-N-SH cells are observed only with VIP gene constructs in which all five domains A–E are present. Three mechanistic possibilities exist to explain these data. First, different factors may be recruited to domains A–E in EP cells depending on whether additional elements and the proteins that bind to them are also present. This would be the case if, for example, binding of a tissue-specific trans-acting factor to the MEF2 element was a sufficiently low-affinity interaction to require stabilization provided by interaction with another protein bound to a nearby site. MEF2-bHLH interactions leading to transcription activation in an *in vitro* model system suggest that this is a viable possibility.[24] Second, non–cell-specific factors in both EP and HeLa cells that lack full intrinsic activation potential and are normally excluded from binding to the holo-VIP gene may be recruited to domains presented "out of context" in transfected transgene constructs, explaining moderate but not full transcriptional activity of A-C-D-E in the absence of domain B (FIG. 4). Third, cell-specific factors that do not directly contact DNA may be recruited to a binding surface made up of some combination of cell-specific and non-specific proteins, and comprising proteins bound to *cis*-active elements in all five domains A–E. Finally, the same binding proteins may be post-translationally modified in a cell-specific manner by stimulation of specific signal transduction systems in HeLa versus EP cells, after the manner in which the Oct1/Oct2 transcriptional co-activator BOB.1 is activated only after its phosphorylation by BOB's kinase, which itself is regulated in a cell-specific fashion.[40]

Gelshift analyses of elements within each of the five regulatory domains of the VIP gene give some evidence in support of the existence of cell-specific factors, as well as ubiquitous factors, capable of binding to potential *cis*-active motifs in the VIP gene (FIG. 5). A 50-nucleotide (nt) probe containing both of the MEF2-like motifs within domain B exhibits a gelshift band specific to SH-EP, compared to HeLa cell nuclear protein extracts (FIG. 5). The octamer-like motif of domain A also exhibits SH-EP nuclear extract–specific gelshift (FIG. 5). Domain A contains two octamer-like AT-rich sequences (FIG. 3). Detailed gelshift analysis of these octamer-like sequences revealed that both motifs bind ubiquitously

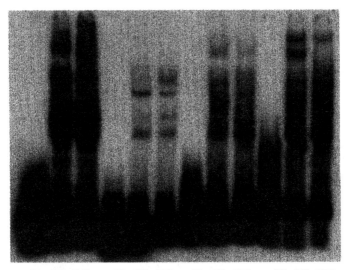

FIGURE 5. Electrophoretic mobility shift assay (EMSA) for cell-specific and non-specific factors binding domains of the VIP gene in SH-EP and HeLa cells. Twenty to fifty nucleotide probes labeled with ^{32}P with ^{32}P-gamma-ATP and T4 polynucleotide kinase were incubated with nuclear extracts from SH-EP (E) and HeLa (H) cells and electrophoresed under non-denaturing conditions along with probe not incubated with any nuclear extract (P).

expressed Oct-1 POU homeodomain protein based on gel supershift with Oct-1 antibodies.[25a] The downstream octamer-like motif is also capable of binding Oct-2. Sequences immediately 5′ to the downstream octamer-like motif are protected specifically by SH-EP nuclear proteins in an *in vitro* footprinting assay. This suggests that cell-specific factors may interact with ubiquitously expressed Oct-1 and Oct-2 proteins within domain A to allow cell-specific expression of the VIP gene. The E-box located upstream of the MEF2 motifs of domain B binds proteins in both HeLa and SH-EP cells (FIG. 5). Whether these are the same or different nuclear proteins could not be determined based on apparent mobility alone. Two of the four bands present in HeLa cell extract/E-box gelshifts appeared to be unique to HeLa cells, and may be candidates for proteins mediating repression of VIP gene transcription in HeLa cells, perhaps in association with proteins binding to elements within the two domains (C and D) likely on functional grounds to be critical for mediating repression of the VIP gene in non–VIPergic cells. It has already been demonstrated that domains D and E contain elements that bind AP-1 and CREB, respectively. These transcription factors are thought not to be expressed in a cell-specific manner, and the mechanisms whereby they impart cell-specific regulation are not yet clear.

FIGURE 6 summarizes the first, second, and third messenger systems, and the *trans*-acting factors, that could act on the VIP gene to orchestrate physiological expression and upregulation. Appropriate VIP gene response to a given signal is likely to be cell-dependent and state-dependent, because so many different proteins may be bound to the VIP gene in a given cell. Thus, a regulatory complex is created that can recruit phosphorylated and dephosphorylated proteins in different combinations. This provides a mechanism to allow VIP gene regulation to recapitulate the multiple inputs to the cell, so gene output represents a faithful response to the inputs that define the VIPergic neuron's synaptic connections, neuroanatomical location, and stage of development.

The complexity of convergent regulation of the production of VIP during critical times of development and endocrine homeostasis, and in normal nervous system functioning, must reflect in microcosm the complexity of the environmental inputs to the VIP-expressing cell itself. Further study should therefore reveal important signal transduction pathways, from extracellular ligand to *cis*-active element, involved in VIP physiology at the level of VIP gene regulation. The first messenger PACAP and the cellular phosphatase calcineurin have already revealed themselves as heretofore undiscovered actors in VIP regulation *in vitro*. It remains to be determined what their roles *in vivo* will be.

FUTURE PERSPECTIVES

A full accounting of the regulatory control of the VIP gene would include identification of the first messengers, signaling pathways, *trans*-acting factors, and *cis*-active elements on the VIP gene responsible for (1) constitutive expression in VIPergic neuroendocrine cells, and not non-neural tissues or non–VIPergic neurons, (2) upregulation of VIP gene expression under physiological conditions including injury, altered endocrine state, and circadian rhythm, and (3) regulation of VIP gene expression during stimulus-secretion-synthesis coupling. Clearly, much additional work is required to fill in the complete picture of VIP gene regulation. Here, we have reviewed the evidence that individual domains of the VIP gene responsible for upregulation by protein kinase A, protein kinase C, and Janus kinase signaling pathways interact with each other, and with an upstream tissue-specifier element, to allow cell-specific constitutive and regulated expression of the VIP gene *in vitro*. We have reviewed the evidence that a variety of first messengers, including growth factors, secretagogues, cytokines, and the novel neurotrophin PACAP upregulate the VIP gene alone and in combination with other extracellular signals. It now remains to be determined which of these first messengers actually mediate VIP gene regulation *in vivo* under a

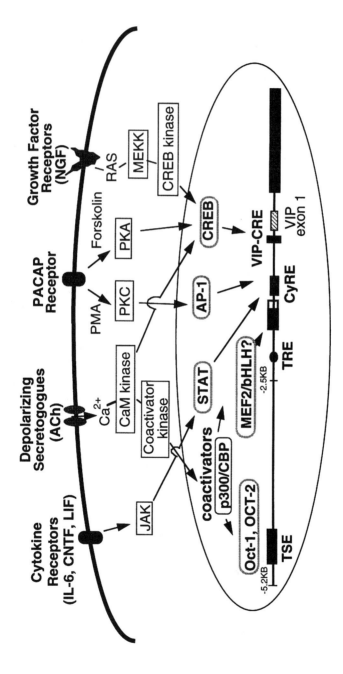

FIGURE 6. Cell type–specific and inducible expression of the human VIP gene. Multiple first messengers are capable of stimulating, individually or in combination, several signal transduction pathways that are in turn known to activate, by phosphorylation or dephosphorylation, *trans*-acting factors including STAT-1 and -3, AP-1, CREB, and Oct-1 that bind to consensus elements located on the VIP gene. Note that at least one of these elements, the AP-1-binding site, or TRE, located at -2.3 kb on the VIP gene, has been shown in the experiments described here to be non-functional in the context of transcription from the human VIP gene, and a second, the CREB binding site or CRE located at position -.094 kb on the VIP gene, has been shown not to subserve PMA upregulation of transcription despite the fact that it does mediate phorbol ester–dependent transcriptional upregulation of the VIP core-promoter in HeLa cells.[21]

given physiological condition in a particular area of the nervous system, and which signaling pathway or pathways each first messenger activates to accomplish upregulation of the VIP gene.

Additional fundamental questions about the mechanisms of regulation of the VIP gene remain unanswered. Do the same or different factors regulate cell-specific expression of the cholinergic gene locus and VIP, co-expressed in the autonomic and enteric nervous systems and in many areas of the brain? The cholinergic gene locus contains a neuronally restrictive silencer element[41–43] that is conserved in human and rodent[44,45] and is functionally active *in vitro*.[45,46] We report here that a similar consensus NRSE sequence exists in the first exon of the human VIP gene (FIG. 3). Does this element participate, along with domains A–E, in co-expression of VIP and acetylcholine in the peripheral and in some regions of the central nervous system? If so, how is this element overridden in non-neuronal tissues such as the anterior pituitary gland, where VIP is also expressed? Does occupancy of *cis*-active sites involve a regulatory cascade such that distinct heterodimeric *trans*-acting proteins successively occupy the E-box/MEF2 and/or other domains of the gene in a temporal sequence during development and maturation of the nervous system? How specific subdomains of regions A–E function in PACAP-regulated VIP gene transcription and in activity-dependent regulation during development is under investigation, as are the elements that subserve depolarization-induced upregulation of the VIP gene during calcium-dependent stimulus-secretion-synthesis coupling. The overall transcriptional activity of the VIP gene, including both constitutive expression and upregulation in response to neuronal depolarization, first-messenger stimulation, neurotrophin signaling, and cell-specific expression of lineage-related transcription factor cascades, depends on complex interactions between multiple *cis*-acting regulatory regions, here designated A–E in the VIP gene promoter/enhancer. All of these interactions will need to be fully delineated to understand the complex, highly integrative regulatory network contained within the 5 kb of the VIP promoter-enhancer that enables it to receive signals from the external and internal environments of the VIPergic cell and transduce these into appropriate transcriptional output from the VIP gene.

ACKNOWLEDGMENTS

We thank Chang-Mei Hsu for carrying out PC12 cell culture and VIP and enkephalin radioimmunoassays and Colin Dias for gelshift analysis.

REFERENCES

1. EIDEN, L. E., R. L. ESKAY, J. SCOTT, H. POLLARD & A. J. HOTCHKISS. 1983. Primary cultures of bovine chromaffin cells synthesize and secrete vasoactive intestinal polypeptide (VIP). Life Sci. **33**: 687–693.

2. MACARTHUR, L. & L. E. EIDEN. 1996. Neuropeptide genes: targets of activity-dependent signal transduction. Peptides **17**: 721–728.

3. EIDEN, L. E., P. GIRAUD, J. DAVE, J. A. HOTCHKISS & H.-U. AFFOLTER. 1984. Nicotinic receptor stimulation activates both enkephalin release and biosynthesis in adrenal chromaffin cells. Nature **312**: 661–663.

4. WASCHEK, J. A., R. M. PRUSS, R. E. SIEGEL, L. E. EIDEN, M.-F. BADER & D. AUNIS. 1987. Regulation of enkephalin, VIP and chromogranin A biosynthesis in actively secreting chromaffin cells: multiple strategies for multiple peptides. Ann. N.Y. Acad. Sci. **493**: 308–323.

5. TATSUNO, I., T. YADA, S. VIGH, H. HIDAKA & A. ARIMURA. 1992. Pituitary adenylate cyclase-activating polypeptide (PACAP) and vasoactive intestinal peptide (VIP) increase cytosolic free calcium concentration in cultured rat hippocampal neurons. Endocrinology **133**: 73–81.

6. CHIK, C. L., B. LI, T. OGIWARA, A. K. HO & E. KARPINSKI. 1996. PACAP modulates L-type $Ca2^+$ channel currents in vascular smooth muscle cells: involvement of PKC and PKA. FASEB J. **10**: 1310–1317.

7. MIYATA, A., A. ARIMURA, R. R. DAHL, N. MINAMINO, A. UEHARA, L. JIANG, M. D. CULLER & D. H. COY. 1989. Isolation of a novel 38 residue-hypothalamic polypeptide which stimulates adenylate cyclase in pituitary cells. Biochem. Biophys. Res. Commun. **164**: 567–574.

8. EIDEN, L. E., P. GIRAUD, H.-U. AFFOLTER, E. HERBERT & A. J. HOTCHKISS. 1984. Alternate modes of enkephalin biosynthesis regulation by reserpine and cyclic AMP in cultured chromaffin cells. Proc. Natl. Acad. Sci. USA **81**: 3949–3953.

9. ESKAY, R. L. & L. E. EIDEN. 1992. Interleukin-1 alpha and tumor necrosis factor alpha differentially regulate enkephalin, vasoactive intestinal polypeptide, neurotensin, and substance P biosynthesis in chromaffin cells. Endocrinology **130**: 2252–2258.

10. DARNELL JR., J. E. 1997. STATs and gene regulation. Science **277**: 1630–1634.

11. SYMES, A. J., M. S. RAO, S. E. LEWIS, S. C. LANDIS, S. E. HYMAN & J. S. FINK. 1993. Ciliary neurotrophic factor coordinately activates transcription of neuropeptide genes in a neuroblastoma cell line. Proc. Natl. Acad. Sci. USA **90**: 572–576.

12. SYMES, A., T. GEARAN, J. EBY & J. S. FINK. 1997. Integration of Jak-Stat and AP-1 signaling pathways at the vasoactive intestinal peptide cytokine response element regulates ciliary neurotrophic factor-dependent transcription. J. Biol. Chem. **272**: 9648–9654.

13. SYMES, A., S. LEWIS, L. CORPUS, P. RAJAN, S. E. HYMAN & J. S. FINK. 1994. STAT proteins participate in the regulation of the vasoactive intestinal peptide gene by the ciliary neurotrophic factor family of cytokines. Mol. Endocrinol. **8**: 1750–1763.

14. SYMES, A. J., P. RAJAN, L. CORPUS & J. S. FINK. 1995. C/EBP-related sites in addition to a Stat site are necessary for ciliary neurotrophic factor-leukemia inhibitory factor-dependent transcriptional activation by the vasoactive intestinal peptide cytokine response element. J. Biol. Chem. **270**: 8068–8075.

15. SEGAL, R. A. & M. E. GREENBERG. 1996. Intracellular signaling pathways activated by neurotrophic factors. Ann. Rev. Neurosci. **19**: 463–489.

16. XING, J., D. D. GINTY & M. E. GREENBERG. 1996. Coupling of the RAS-MAPK pathway to gene activation by RSK2, a growth-factor-regulated CREB kinase. Science **273**: 959–963.

17. BITO, H., K. DEISSEROTH & R. W. TSIEN. 1996. CREB phosphorylation and dephosphorylation: A Ca^{2+}- and stimulus duration-dependent switch for hippocampal gene expression. Cell **87**: 1203–1214.

18. FOSTER, G. A., L. E. EIDEN & D. E. BRENNEMAN. 1989. Regulation of discrete sub-populations of transmitter-identified neurons after inhibition of electrical activity in cultures of mouse spinal cord. Cell Tissue Res. **256**: 543–552.

19. AGOSTON, D. V., L. E. EIDEN, D. E. BRENNEMAN & I. GOZES. 1991. Spontaneous electrical activity regulates vasoactive intestinal peptide expression in dissociated spinal cord cell cultures. Mol. Brain Res. **10**: 235–240.

20. TSUKADA, T., S. J. HOROVITCH, M. R. MONTMINY, G. MANDEL & R. H. GOODMAN. 1985. Structure of the human vasoactive intestinal polypeptide gene. DNA **4**: 293–300.

21. TSUKADA, T., J. S. FINK, G. MANDEL & R. H. GOODMAN. 1987. Identification of a region in the human vasoactive intestinal polypeptide gene responsible for regulation by cyclic AMP. J. Biol. Chem. **262**: 8743–8747.

22. WASCHEK, J. A., C.-M. HSU & L. E. EIDEN. 1988. Lineage-specific regulation of the vasoactive intestinal peptide gene in neuroblastoma cells is conferred by 5.2 kilobases of 5'-flanking sequence. Proc. Natl. Acad. Sci. USA **85**: 9547–9551.

23. HAHM, S. H. & L. E. EIDEN. 1996. Tissue-specific expression of the vasoactive intestinal peptide gene requires both an upstream tissue specifier element and the 5' proximal cAMP-responsive element. J. Neurochem. **67**: 1872–1881.

24. YAMAGAMI, T., K. OHSAWA, M. NISHIZAWA, C. INQUE, E. GOTOH, N. YANAIHARA, H. YAMAMOTO & H. OKAMOTO. 1988. Complete nucleotide sequence of human vasoactive intestinal peptide/PHM-27 gene and its inducible promoter. Ann. N. Y. Acad. Sci. **527**: 87–102.

25. SENA, M., D. T. BRAVO, D. V. AGOSTON & J. A. WASCHEK. 1994. High conservation of upstream regulatory sequences on the human and mouse vasoactive intestinal peptide (VIP) genes. DNA Seq. **5**: 25–29.

25a. HAHM, S. H. & L. E. EIDEN. 1998. Five discrete cis-active domains direct cell type-specific transcription of the vasoactive intestinal peptide (VIP) gene. J. Biol. Chem. **273**: 17083–17094.

26. HERR, W., R. A. STURM, R. G. CLERC, L. M. CORCORAN, D. BALTIORE, P. A. SHARP, H. A. INGRAHAM, M. G. ROSENFELD, M. FINNEY, G. RUVKUN & H. R. HORVITZ. 1988. The POU domain: A large conserved region in the mammalian Pit-1, Oct-1,Oct-2 and Caenorrhabditis elegans Unc-86 gene products. Genes & Dev. **2**: 1513–1516.
27. HERR, W. & M. A. CLEARLY. 1995. The POU domain: versatility in transcriptional regulation by a flexible two-in-one DNA-binding domain. Genes & Dev. **9**: 1679–1693.
28. CLARK, M. E. & P. L. MELLON. 1995. The POU homeodomain transcription factor Oct-1 is essential for activity of the gonadotropin-releasing hormone neuron-specific enhancer. Mol. Cell Biol. **15**: 6169–6177.
29. GUILLEMOT, F., L.-C. LO, J. E. HOHNSON, A. AUERBACH, D. J. ANDERSON & A. L. JOYNER. 1993. Mammalian achaete-scute homolog 1 is required for the early development of olfactory and autonomic neurons. Cell **75**: 463–476.
30. GUILLEMOT, F. 1995. Analysis of the role of basic-helix-loop-helix transcription factors in the development of neural lineages in the mouse. Biol. Cell **84**: 3–6.
31. BLACK, B. L., K. L. LIGON, Y. ZHANG & E. N. OLSON. 1997. Cooperative transcriptional activation by the neurogenic basic helix-loop-helix protein MASH1 and members of the myocyte enhancer factor-2 (MEF2) family. J. Biol. Chem. **271**: 26659–26663.
32. LEIFER, D., Y.-L. LI & W. WEHR. 1997. Myocyte-specific enhancer binding factor 2C expression in fetal mouse brain development. J. Mol. Neurosci. **8**: 131–143.
33. KISHIMOTO, T., S. AKIRA & T. TAGA. 1992. Interleukin-6 and its receptor: A paradigm for cytokines. Science **258**: 593–597.
34. IHLE, J. N. 1996. STATs: Signal transducers and activators of transcription. Cell **84**: 331–334.
35. WASYLYK, B., C. WASYLYK, P. FLORES, A. BEGUE, D. LEPRINCE & D. STEHELIN. 1990. The c-ets proto-oncogenes encode transcription factors that cooperate with c-Fos and c-Jun for transcriptional activation. Nature **346**: 191–193.
36. LEONARD, J., P. SERUP, G. GONZALEZ, T. EDLUND & M. MONTMINY. 1992. The LIM family transcription factor Isl-1 requires cAMP response element binding protein to promote somatostatin expression in pancreatic islet cells. Proc. Natl. Acad. Sci. USA **89**: 6247–6251.
37. AGOSTON, D. V., D. T. BRAVO & J. A. WASCHEK. 1990. Expression of a chimeric VIP gene is targeted to the intestine in transgenic mice. J. Neurosci. Res. **27**: 479–486.
38. TOLENTINO, P. J., P. DIKKES, L. TSURUDA, K. EBERT, J. S. FINK, L. VILLA-KOMAROFF & E. D. LAMPERTI. 1995. Quantitative analysis of the expression of a VIP transgene. Mol. Brain Res. **33**: 47–60.
39. TSURUDA, L. M., E. D. LAMPERTI, S. E. LEWIS, P. J. TOLENTINO, P. DIKKES, L. VILLA-KOMAROFF, K. M. EBERT & J. S. FINK. 1996. Region-specific central nervous system expression and axotomy-induced regulation in sympathetic neurons of a VIP-β-galactosidase fusion gene in transgenic mice. Mol. Brain Res. **42**: 181–192.
40. ZWILLING, S., A. DIECKMANN, P. PFISTERER, P. ANGEL & T. WIRTH. 1997. Inducible expression and phosphorylation of coactivator BOB.1/OBF.1 in T cells. Science **277**: 221–226.
41. CHONG, J. A., J. TAPIA-RAMIREZ, S. KIM, J. J. TOLEDO-ARAI, Y. ZHENG, M. C. BOUTROS, Y. M. ALTSHULLER, M. A. FROHMAN, S. D. KRANER & G. MANDEL. 1995. REST: a mammalian silencer protein that restricts sodium channel gene expression to neurons. Cell **80**: 949–957.
42. SCHOENHERR, C. J. & D. J. ANDERSON. 1995. The neuron-restrictive silencer factor (NRSF): a coordinate repressor of multiple neuron-specific genes. Science **267**: 1360–1363.
43. SCHOENHERR, C. J., A. J. PAQUETTE & D. J. ANDERSON. 1997. Identification of potential target genes for the neuron-restrictive silencer element. Proc. Natl. Acad. Sci. USA **93**: 9881–9886.
44. ERICKSON, J. D., E. WEIHE, M.K.-H. SCHÄFER, E. NEALE, L. WILLIAMSON, T. I. BONNER, J.-H. TAO-CHENG & L. E. EIDEN. The VAChT/ChAT "cholinergic gene locus": new aspects of genetic and vesicular regulation of cholinergic function. *In* Cholinergic Mechanisms: From Molecular Biology to Clinical Significance. J. Klein & K. Löffelholz, Eds. 69–82. Elsevier. Amsterdam.
45. LÖNNERBERG, P., C. J. SCHOENHERR, D. J. ANDERSON & C. F. IBANEZ. 1996. Cell type-specific regulation of choline acetyltransferase gene expression. Role of the neuron-restrictive silencer element and cholinergic-specific enhancer sequences. J. Biol. Chem. **271**: 33358–33365.

46. HAHM, S. H., L. CHEN, C. PATEL, J. ERICKSON, T. I. BONNER, E. WEIHE, M.K.-H. SCHÄFER & L. E. EIDEN. 1997. The human cholinergic gene locus: Upstream sequencing, in vivo transcription patterns, and role of the NRSE/RE-1 and other control elements in VAChT expression. J. Mol. Neurosci. **9**: 223–236.

47. EIDEN, L. E. & A. J. HOTCHKISS. 1983. Cyclic adenosine monophosphate regulates vasoactive intestinal polypeptide and enkephalin biosynthesis in cultured bovine chromaffin cells. Neuropeptides **4**: 1–9.

48. RAUSCH, D. M., A. L. IACANGELO & L. E. EIDEN. 1988. Glucocorticoid- and nerve growth factor-induced changes in chromogranin A expression define two different neuronal phenotypes in PC12 cells. Mol. Endocrinol. **2**: 921–927.

p22/PACAP Response Gene 1 (PRG1): A Putative Target Gene for the Tumor Suppressor p53

HEINER SCHÄFER,[b] ANNA TRAUZOLD, THORSTEN SEBENS,
WOLFGANG DEPPERT,[a] ULRICH R. FÖLSCH, AND WOLFGANG
E. SCHMIDT[b]

*Laboratory of Molecular Gastroenterology, First Department of
Medicine, Christian-Albrechts-University of Kiel, D-24105 Kiel,
Germany*

*[a]Heinrich Pette Institute for Experimental Virology & Immunology,
University of Hamburg, Hamburg, Germany*

ABSTRACT: In this study we describe a novel putative p53-responsive gene, designated p22/PACAP response gene 1 (PRG1), recently identified as a proliferation-associated early-response gene in rats. By means of electrophoretic mobility shift assay and CAT-reporter gene assay, we could demonstrate that the p53 binding site residing in the promoter of p22/PRG1 is functional *in vitro*. Furthermore, in clone 6 cells expression of p22/PRG1 is induced in parallel to p21/Waf1 under conditions permitting mutant p53 to adopt wild-type configuration. An increase of p22/PRG1 transcription was also observed in γ-irradiated rat splenocytes, which undergo p53-dependent apoptosis. Our findings demonstrate that p22/PRG1 fulfills all essential criteria as a p53 target gene and might be implicated in p53-dependent apoptosis.

Over the past decade, the tumor suppressor gene p53 has been recognized as an essential mediator of cell cycle control and apoptosis in mammals. Loss of p53 functions is regarded as a major cause of cancer.[1,2] It has been shown that p53 primarily functions as a transcription factor, initiating the action of p21/WAF1,[3] an important inhibitor of the cell cycle (G1–S transition), or of the *bax* gene,[4] which is implicated in apoptosis by antagonizing the *bcl-2* gene. p53 itself is induced and activated by cellular stress, mainly DNA-damaging insults leading to single-strand and double-strand DNA breaks.[5,6]

Upon induction of p53, the mandate to the cell to undergo growth arrest mainly depends on p21/Waf1 as an immediate p53 target gene, but several other p53-responsive genes, i.e., PCNA[7] and GADD45,[8] are also involved. The capacity of p53 to elicit apoptosis mainly depends on transcriptional induction of the *bax* gene,[4] but additional cellular factors and conditions, which still need to be defined, contribute to p53-dependent apoptosis. Therefore, the mechanisms underlying p53-dependent apoptosis are subject to intensive investigation.

In this study, we describe a novel growth-associated early-response gene, designated p22/PACAP response gene 1 (PRG1),[9] which surprisingly possesses a functional p53

[b]Corresponding authors: Heiner Schäfer and Wolfgang E. Schmidt, First Department of Medicine, Gastrointestinal Unit, Schittenhelmstrasse 12, D-24105 Kiel, Germany; Tel.: 49 431 597-1443/-1395; Fax: 49 431 597-1427; E-mail: weschmidt@1med.uni-kiel.de and hschaef@1med.uni-kiel.de

binding site in its promoter region, and provide experimental data qualifying p22/PRG1 as a novel p53-responsive gene. We also discuss an involvement of p22/PRG1 in p53-dependent apoptosis.

EXPERIMENTAL CONDITIONS

Cell Culture

HeLa and Clone 6 cells were cultured in DMEM supplemented with 5% fetal calf serum at 37°C (HeLa cells) or 39°C (Clone 6 cells). Fresh rat splenocytes were cultured in DMEM plus concanavalin A at 37°C.

RT-PCR

Total RNA was isolated from γ-irradiated splenocytes using the QIAshredder and RNeasy kits (Qiagen, Hilden, Germany). A 2 μg sample of total RNA was heat-denaturated (75°C, 3 min) and submitted to oligo-T primed (1 μM) reverse transcription using 200 units M-MLV reverse transcriptase / 1 μg total RNA. Reverse transcription (final volume 30 μl) was carried out at 37°C for 60 min in the presence of 60 units RNase inhibitor (RNAsin, Promega, Mannheim, Germany), 200 μM dNTPs (GIBCO/BRL, Eggenstein, Germany), and 1 mM DTT. cDNA synthesis was terminated by heating to 95°C for 5 min. Routinely, 3 μl of the first strand cDNA mixture were submitted to PCR (30 μl) using 0.75 units Taq-Polymerase at 1.5 mM $MgCl_2$ and the following primers (0.5 μM): p22/PRG1,[9] ATGTGCCATTCGCGTAACCAC (sense), GAATGCCGCCGGGTGTTGCT (antisense); p21/Waf1, GGATGTCCGTCAGAACCCATGC (sense), TCCTCTTGGAGAAGATCAGCCG (antisense). Cycling conditions (Thermal cycler 2400, Perkin Elmer) were: 95°C, 2 min; 95°C, 1 min / 62°C, 30 sec / 72°C, 30 sec, 3 cycles; 95°C, 1 min / 60°C, 30 sec / 72°C, 30 sec, 3 cycles; 95°C, 1 min / 58°C, 30 sec / 72°C, 30 sec, 12 or 15 cycles and 72°C, 10 min. Amplimer set primers (Clontech, Heidelberg, Germany) were used for the analysis of β-actin mRNA as control. PCR products were analyzed by PAGE and ethidium bromide staining.

Western Blot Analysis

Cells were lysed in 2 × SDS-PAGE sample buffer, heated (95°C) for 5 min, and submitted to 12.5% SDS-PAGE. Then, separated proteins were semi-dry electroblotted (125 mA, 30 min) onto PVDF membranes. Upon blocking overnight with 5% non-fat milk powder, 0.1% Tween 20 in TBS (blocking solution) at 4°C, blots were exposed to primary antibodies diluted in blocking solution as follows: p22/PRG1, affinity-purified rabbit antiserum against a synthetic p22/PRG1-peptide (67-85) at a 500-fold dilution for 30 min; p53, moAb Ab3 (Calbiochem, Bad-Soden, Germany) at 100-fold dilution for 60 min; p21/Waf1, antiserum Ab5 (Calbiochem) at 200-fold dilution for 60 min. After extensive washing with blocking solution, blots were exposed to the appropriate second antibody and developed using the CDP star immuno detection kit (NE Biolabs, Schwalbach, Germany).

Electrophoretic Mobility Shift Assay for p53

Nuclear extracts from SV-80 cells cultured in RPMI 40 medium supplemented with 10% FCS were prepared as described.[10] Equal amounts of protein (15 μg) were incubated with a ^{32}P-labeled oligonucleotide (20 bp) containing the p53 binding site derived from the p22/PRG1 promoter using a buffer containing 50 mM HEPES/HCl, pH 7.8, 250 mM KCl, 5 mM EDTA, 25 mM spermidine, 50% glycerol, 2 µl BSA (10 mg/ml), 2 µl polydI-dC (1 mg/ml), 2 µl DTT (50 mM), and 4 µl H$_2$O. For control, an excess of unlabeled p53 sequence was added. After a 20-min incubation at room temperature, samples were mixed with 1/10 volume of PAGE-loading buffer (50% glycerol, 0.1% bromophenolblue) and submitted to native PAGE (5% acrylamide/bisacrylamide). Gels were dried and exposed to X-ray film (Hyperfilm, Amersham, Braunschweig, Germany) for 16–24 hours. For super-shift assay, a monoclonal p53 antibody (Pab421, Calbiochem) was preincubated (30 min, RT) with the nuclear extract prior to incubation with the probe.

Transfection with the Chloramphenicol Acetyl Transferase (CAT) Reporter Gene and Assay for CAT Activity

HeLa cells were cultured in 6 cm culture dishes until 80% confluency. Then, cells were rinsed once with culture medium without serum for 30 min. During this time, 2 µg of CAT-reporter gene-plasmids containing PRG1 promoter fragments were incubated at room temperature in 600 µl serum-free medium supplemented with 20 µl of Lipofectamine®(GIBCO-BRL). For cotransfections of HeLa cells, various amounts (0.2–2 µg) of the following plasmids were added to the CAT plasmids: pCMV-p53 (wt), pCMV-C5 (mutant p53), and pCMV-lacZ (mock). Afterwards, plasmid mixtures were brought to 3 ml with serum-free medium, added to the cells, and incubated for 6 h under standard culture conditions. Upon replacement with medium plus serum, cells were cultured for 24–36 hours. CAT expression was determined by means of colorimetric evaluation of cell lysates using a β-galactosidase assay (Promega) and a CAT-ELISA (Boehringer, Mannheim, Germany), respectively. Data were expressed as ng CAT protein / mU β-galactosidase activity.

RESULTS AND DISCUSSION

The p22/PRG1 promoter contains a 20 bp DNA motif strongly matching the consensus sequence for p53 recognition[11] that consists of a tandem repeat of the 10 bp palindromic sequence RRRCA/TA/TGYYY. This p53 binding site exhibits a remarkable structural identity (16/18) with a p53 binding site located in the p21 promoter[12] and shares all crucial nucleotides present in p53 binding sites of other p53 target genes including bax, mdm2, PCNA, and GADD45 (Fig. 1).

The p53 Binding Site of the p22/PRG1 Promoter Specifically Binds to wt p53 in Vitro

In order to elucidate whether the p53 binding residing in the p22/PRG1 promoter binds to wild-type p53 *in vitro*, electrophoretic mobility shift assays (EMSA) were performed. Coincubation of a labeled oligonucleotide containing this p53 binding site with a nuclear extract from human SV-80 fibroblasts hyperexpressing wild-type p53 produced a strongly labeled protein/DNA complex (Fig. 2A). Labeling of this complex was dose-dependently diminished in the presence of the unlabeled homologous oligonucleotide at a 10- and 50-fold molar excess. A similar reduction in labeling occurred in the presence of an unlabeled oligonucleotide containing the p53 binding site of the p21/WAF-1 promoter. By contrast, no label displacement was observed with a p53-unrelated sequence like AP1. The specificity of the labeled protein-DNA complex was confirmed by the addition of a p53-specific mono-clonal antibody producing a ternary protein-DNA complex visualized as a supershifted band

```
MDM2:      G G T C A A G T T G - G G A C A C G T C C

Bax:       T C A C A A G T T A G A G A C A A G C C T

GADD45:    C A G C A T G C T T - A G A C A T G G T T

PCNA:      G A A C A A G T C C - G G G C A T A T G T

p21:       G A A C A T G T C C C - A A C A T G T T G

p22/PRG1: C C A C A T G T C C C - G A C A T G T G C

p53 cons.  R R R C W W G Y Y Y - R R R C W W G Y Y Y
```

FIGURE 1. Comparison of the p53 binding site residing in the p22/PRG1 promoter with the p53 binding sites of p21/Waf1, PCNA, mdm2, *Bax*, and GADD45. The consensus sequence for p53 binding is shown below.

(FIG. 2B). Typically, the antibody interacting with the regulatory C-terminal domain of wild-type p53 enhances p53 binding to the binding site of the p22/PRG1 promoter.[13] In contrast, no supershift was observed when adding an unrelated monoclonal antibody (data not shown).

These findings indicate that the p53 binding site derived from the PRG1 promoter specifically binds to wild-type p53 in a similar fashion as the corresponding binding site of the p21/Waf1 promoter.

wt p53-Dependent Transactivation of the p22/PRG1 Promoter

To demonstrate that this p53 binding site also confers p53-inducible transcriptional activity to the p22/PRG1 promoter, HeLa cells were cotransfected with a CMV early-promoter driven expression plasmid for wild-type p53 (pCMV-p53) and plasmids encoding the reporter gene CAT linked to various shortened parts of the p22/PRG1 promoter (pos. -589, -365, and -186) as well as a promoter fragment (-589) carrying a deletion between position -284 and -262. As shown in FIGURE 3A, HeLa cells exhibit a three- to fourfold higher

FIGURE 2. Wild-type p53 binds to the p53 binding site of the p22/PRG1 promoter. Using nuclear extracts from human SV80 fibroblasts, electrophoretic mobility shift assays (EMSA) were performed with a ^{32}P-labeled oligonucleotide derived from positions -288 to -260 of the p22/PRG1 promoter. (**A**) EMSAs were conducted with tracer alone (lane 2), in the presence of a 100-fold excess of the unlabeled homologous oligonucleotide (lane 3), a 50-fold excess of the unlabeled homologous oligonucleotide (lane 4), a 100-fold excess of an unlabeled oligo containing the p53 binding site of the p21/Waf1 promoter (lane 5), a 100-fold excess of a control oligo, AP1 (lane 6), or with the ^{32}P-labeled p53 binding site of the p21/Waf1 promoter as probe (lane 7). (**B**) EMSAs were conducted with tracer alone (lane 4), in the presence of a 50-fold excess of the unlabeled homologous oligonucleotide (lane 1), the p53 binding site of the p21/Waf1 promoter (lane 2), or in the presence of a monoclonal antibody against the C-terminal domain of wild-type p53 (lane 3).

CAT expression when cotransfected with p22/PRG1 promoter fragments containing the p53 binding site and the pCMV-p53 vector compared to HeLa cells cotransfected with pCMV-lacZ (mock). In contrast, no difference in CAT expression was observed in HeLa

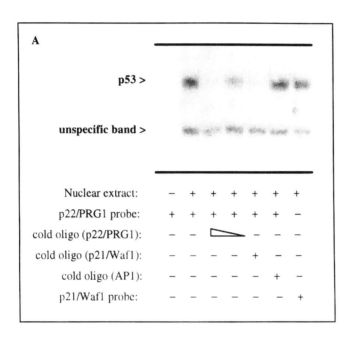

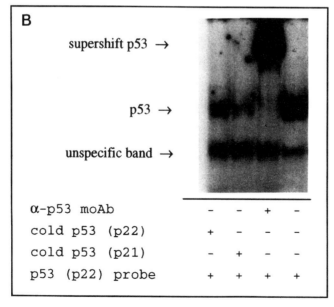

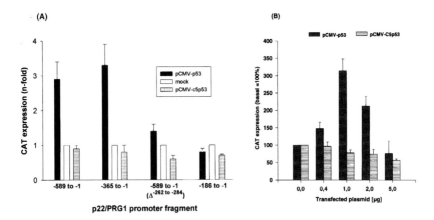

FIGURE 3. Effect of p53 on transcriptional activity of the p22/PRG1 promoter. (**A**) Cotransfection of pCMV-wtp53 or pCMV-C5p53 and CAT-reporter gene plasmids containing p22/PRG1 promoter fragments. HeLa cells were grown in 6-cm dishes and transiently transfected with 1 μg of pCMV-p53 or pCMV-C5p53 together with 1.5 μg of the CAT-plasmids pPRG(-589), pPRG(-589Δ$^{-284/-262}$), pPRG(-356), and pPRG(-186), respectively, and 1 μg pCMV-lacZ as control. Basal activity was determined by cotransfection of 2 μg pCMV-lacZ and the pPRG-CAT plasmids. CAT expression was quantified by a CAT-ELISA and normalized to β-galactosidase activity. Data are shown as mean ± SD, from four independent duplicate experiments. (**B**) Dose-dependent effect of p53 on p22/PRG1 promoter activity. HeLa cells were cotransfected with 1.5 μg pPRG(-356), the indicated amounts of pCMV-wtp53 or pCMV-C5p53 and pCMV-lacZ for adjusting the total plasmid content to 7.5 μg. CAT expression was determined as described above. Data are shown as mean ± SD from three independent experiments performed in duplicate.

cells cotransfected with pCMV-p53 or pCMV-lacZ and the p22/PRG1 promoter fragments lacking the p53 binding site.

To confirm the specificity for wild-type p53, HeLa cells were also cotransfected with various amounts (0.2, 0.5, and 1.5 μg) of a pCMV expression plasmid for a transcriptionally inactive mutant p53 (pCMV-c5p53).[14] In contrast to cells cotransfected with increasing amounts (0.2, 0.5, and 1.5 μg) of the wild-type p53 expression plasmid, no increasing effect on CAT expression could be observed (FIG. 3B). Therefore, it can be concluded that the p53 binding site in the p22/PRG1 promoter represents a functional site that mediates wild-type p53-dependent transactivation. Interestingly, the inducing effect of wild-type p53 on CAT expression was maximal at a plasmid amount below 1 μg, whereas at higher amounts the inducing effect was lower. This dose-dependent biphasic effect of wt p53 on target genes has been similarly described for PCNA[15] and could reflect two distinct actions of wt p53: when expressed at moderate levels it acts as a site-specific transcription factor[16] but at higher levels it negatively interferes with the basic transcription machinery.[17]

Induced Expression of p22/PRG1 and Its Relation to Activated wt p53

A close relation of p53 expression and the induction of p22/PRG1 *in vivo* was shown in Clone 6 cells, a rat fibroblast cell line stably transfected with a dominant negative temperature-sensitive mutant p53.[18] When grown at the non-permissive temperature of 39°C, which keeps the mutant p53 in its inactive state, only a moderate p22/PRG1 expression was detectable, as shown by Western blotting (FIG. 4). A shift of Clone 6 cells to the permissive temperature of 31°C promotes p53 to adopt a wild type–like active conformation with a decreased half-life (FIG. 4, upper panel). At this permissive temperature, the expression of p22/PRG1 was strongly elevated within 3 hours and remained at a high level for at least 24 hours. A comparable time course was observed for the expression of the established p53 target gene p21/Waf1, which exhibits a similar increase in Clone 6 cells upon the temperature shift. By contrast, p22/PRG1 and p21/Waf1 expression did not increase at 31°C in control cells expressing wild-type p53 (data not shown), thus supporting the view that in Clone 6 cells the temperature-dependent increase of p22/PRG1 and p21/Waf1 expression depends on active p53.

Using γ-irradiated rat splenocytes, which undergo p53-dependent apoptosis, we checked the possibility that p22/PRG1 is involved in this process. Increased binding of annexin to cell membranes within 24 h of irradiation indicated the onset of apoptosis in 40% of the splenocytes, preceded by an increase of p53 protein levels between 3–12 hours. Following the elevated p53 expression, mRNA levels of p22/PRG1 continuously increased between 6 and 24 h, as shown by RT-PCR (FIG. 5). A similar time course of transcriptional induction was exhibited by p21/Waf1, whereas the mRNA content of β-actin declined after 12 hours. Such a parallel increase of p22/PRG1 and p21/Waf1 mRNA levels was also observed in splenocytes treated with the genotoxic agent doxorubicin (data not shown). By contrast, dexamethasone-treated thymocytes, which undergo p53-independent apoptosis,

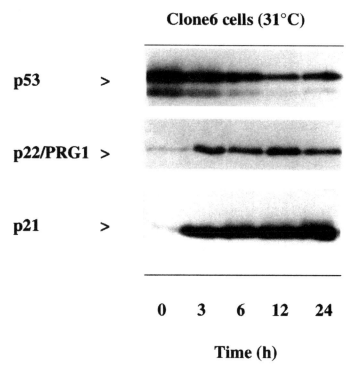

FIGURE 4. p53 status and induction of the expression of p22/PRG1 as well as of p21/Waf1 in Clone 6 cells. Clone 6 cells were precultured at the non-permissive (39°C) temperature before incubation at the permissive temperature (31°C) for the indicated periods. Expression of p53, p22/PRG1, and p21/Waf1 was determined by immunoblotting. All panels show a representative result of three experiments.

or γ-irradiated p53-*null* Saos2 cells did not exhibit the delayed and continuous increase of p22/PRG1 expression distinct from the still detectable early and transient increase in response to growth stimuli or stress factors (data not shown). Therefore, it can be assumed that the late induction of p22/PRG1 expression during apoptosis in DNA-damaged cells depends on wild-type p53.

A Putative Role of p22/PRG1 as p53 Target Gene

How does one explain the involvement of a potentially growth-associated early response gene, like p22/PRG1, in p53-mediated apoptosis and growth control? Although growth suppression via p21/Waf1 mediated G1-arrest and/or via apoptosis represents the most prominent function of p53, it seems plausible that its activation is not strictly mandatory for the cell to undergo apoptosis or to remain in cell cycle arrest. When successfully overcoming DNA damage or any other genotoxic insult, a cell may reenter the cell cycle and replenish the pool of cells depleted by p53-dependent apoptosis. By providing genes that bypass the cell cycle arrest, i.e., the MDM2 gene,[19,20] mediate growth induction, i.e., TGFα,[21] or support cell cycle entry, i.e., p22/PRG1, p53 might initiate a signaling program leading to cellular recovery and survival. However, these growth-promoting genes keep the potential to elicit the apoptotic program when facing prevailing growth-inhibiting signals,

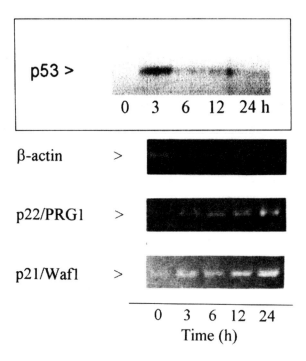

FIGURE 5. p22/PRG1 expression in p53-dependent apoptosis. Rat splenocytes were γ-irradiated (8 Gy) and analyzed after the indicated periods for expression of of p22/PRG1, p21/Waf1, and β-actin by RT-PCR. A representative Et-Br stained gel from three experiments is shown.

i.e., irreparable DNA damage or an antiproliferative environment. Such conflicting signals have been described to be responsible for c-myc–dependent apoptosis.[22,23] Accordingly, the recruitment of some growth-related genes by p53 provides a dual mechanism by which p53 determines the cellular fate: survival and growth on the one hand or apoptosis on the other hand. The exact mechanisms underlying p53-dependent survival or non-survival responses and the identity of cooperating genes are still unknown, but p22/PRG1 might be a novel candidate gene by which p53 can fulfill this dual function.

REFERENCES

1. HOLLSTEIN, M., D. SIDRANSKY, B. VOGELSTEIN & C. C. HARRIS. 1991. p53 mutations in human cancer. Science **253**: 49–53.
2. CROOK, T., N. J. MARSTON, E. A. SARA & K. H. VOUSDEN. 1994. Transcriptional activation by p53 correlates with suppression of growth but not transformation. Cell **79**: 817–827.
3. EL-DEIRY, W. S., J. W. HARPER, P. M. O'CONNOR, V. E. VELCULESCU, C. E. CANMAN, J. JACKMAN, J. A. PIETENPOL, M. BURRELL, D. E. HILL, Y. WANG *et al.* 1994. WAF1/CIP1 is induced in p53-mediated G1 arrest and apoptosis. Cancer Res. **54**: 1169–1174.
4. MIYASHITA, T. & J. C. REED. 1995. Tumor suppressor p53 is a direct transcriptional activator of the human bax gene. Cell **80**: 293–299.
5. DI-LEONARDO, A., S. P. LINKE, K. CLARKIN & G. M. WAHL. 1994. DNA damage triggers a prolonged p53-dependent G1 arrest and long-term induction of Cip1 in normal human fibroblasts. Genes & Dev. **8**: 2540–2551.
6. SIEGEL, J., M. FRITSCHE, S. MAI, G. BRANDNER & R. D. HESS. 1995. Enhanced p53 activity and accumulation in response to DNA damage upon DNA transfection. Oncogene **11**: 1363–1370.
7. MORRIS, G. F., J. R. BISCHOFF & M. B. MATHEWS. 1996. Transcriptional activation of the human proliferating-cell nuclear antigen promoter by p53. Proc. Natl. Acad. Sci. USA **93**: 895–899.
8. KASTAN, M. B., Q. ZHAN, W. S. EL-DEIRY, F. CARRIER, T. JACKS, W. V. WALSH, B. S. PLUNKETT, B. VOGELSTEIN & A. J. FORNACE, JR. 1992. A mammalian cell cycle checkpoint pathway utilizing p53 and GADD45 is defective in ataxia-telangiectasia. Cell **71**: 587–597.
9. SCHÄFER, H., A. TRAUZOLD, E. G. SIEGEL, U. R. FÖLSCH & W. E. SCHMIDT. 1996. p22/PRG1: A novel early-response gene transcriptionally induced by pituitary adenylate cyclase activating polypeptide in a pancreatic carcinoma cell line. Cancer Res. **56**: 2641–2648.
10. SCHÄFER, H., J. ZHENG, F. GUNDLACH, R. GÜNTHER, E. G. SIEGEL, U. R. FÖLSCH & W. E. SCHMIDT. 1996. Pituitary adenylate cyclase activating polypeptide stimulates proto oncogene expression and activates the AP 1 (c Fos/c Jun) transcription factor in AR4-2J pancreatic carcinoma cells. Eur. J. Biochem. **242**: 467–476.
11. ZAMBETTI, G. P., J. BARGONETTI, K. WALKER, C. PRIVES & A. J. LEVINE. 1992. Wild-type p53 mediates positive regulation of gene expression through a specific DNA sequence element. Genes & Dev. **6**: 1143–1152.
12. EL-DEIRY, W. S., T. TOKINO, V. E. VELCULESCU, D. B. LEVY, R. PARSONS, J. M. TRENT, D. LIN, W. E. MERCER, K. W. KINZLER & B. VOGELSTEIN. 1995. WAF1, a potential mediator of p53 tumor suppression. Cell **75**: 817–825.
13. HUPP, T. R. & D. P. LANE. 1994. Regulation of the cryptic sequence-specific DNA-binding function of p53 by protein kinases. Cold Spring Harbor Symp. Quant. Biol. **59**: 195–206.
14. ELIYAHU, D., N. GOLDFINGER, O. PINHASI-KIMHI, G. SHAULSKY, Y. SKURNIK, N. ARAI, V. ROTTER & M. OREN. 1988. Meth A fibrosarcoma cells express two transforming mutant p53 species. Oncogene **3**: 313–321.
15. SHIVAKUMAR, C. V., D. R. BROWN, S. DEB & S. P. DEB. 1995. Wild-type human p53 transactivates the human proliferating cell nuclear antigen promoter. Mol. Cell. Biol. **15**: 6785–6793.
16. UNGER, T., M. M. NAU, S. SEGAL & J. D. MINNA. 1992. p53: a transdominant regulator of transcription whose function is ablated by mutations occurring in human cancer. EMBO J. **11**: 1383–1390.
17. MACK, D. H., J. VARTIKAR, J. M. PIPAS & L. A. LAIMINS. 1993. Specific repression of TATA-mediated but not initiator-mediated transcription by wild-type p53. Nature **363**: 281–283.
18. MICHALOVITZ, D., O. HALEVY & M. OREN. 1990. Conditional inhibition of transformation and of cell proliferation by a temperature-sensitive mutant of p53. Cell **62**: 671–680.

19. WU, X., J. H. BAYLE, D. OLSON & A. J. LEVINE. 1993. The p53-MDM-2 autoregulatory feedback
     loop. Genes & Dev. **7**: 1126–1132.
20. XIAO, Z. X., J. CHEN, A. J. LEVINE, N. MODJTAHEDI, J. XING, W. R. SELLERS & D. M. LIVINGSTON.
     1995. Interaction between the retinoblastoma protein and the oncoprotein MDM2. Nature **375**:
     694–698.
21. SHIN, T. H., A. J. PATERSON & J. E. KUDLOW. 1995. p53 stimulates transcription from the human
     transforming growth factor alpha promoter: a potential growth-stimulatory role for p53. Mol.
     Cell. Biol. **15**: 4694–4701.
22. SAKAMURO, D., V. EVINER, K. J. ELLIOTT, L. SHOWE, E. WHITE & G. C. PRENDERGAST. 1995. c-
     Myc induces apoptosis in epithelial cells by both p53-dependent and p53-independent mech-
     anisms. Oncogene **11**: 2411–2418.
23. JIANG, M. C., H. F. YANG-YEN, J. K. LIN & J. Y. YEN. 1996. Differential regulation of p53, c-Myc,
     Bcl-2 and Bax protein expression during apoptosis induced by widely divergent stimuli in
     human hepatoblastoma cells. Oncogene **13**: 609–616.

Solution Structure Comparison of the VIP/PACAP Family of Peptides by NMR Spectroscopy

VICTOR WRAY,[a,d] KIYOSHI NOKIHARA,[b] AND SATORU NARUSE[c]

[a]Gesellschaft für Biotechnologische Forschung, Mascheroder Weg 1, D-38124 Braunschweig, Germany

[b]Shimadzu Scientific Research, Tokyo, Japan

[c]Nagoya University School of Medicine, Nagoya, Japan

ABSTRACT: The current status of structural studies of the VIP/PACAP family of peptides in solution by NMR spectroscopy is briefly reviewed. The structural elucidation methodology is described with examples from recent work and finally general structural conclusions are drawn from data from the now extensive literature.

The solution structures of a significant number of members of the VIP/PACAP family of neuropeptides/hormones are now available.[1–14] All are small linear peptides, 27 to 39 residues in length, that show significant sequence homology (FIG. 1). They initiate a variety of responses in the respiratory, genital-urinary, and gastrointestinal systems through interaction with membrane-bound G-protein coupled receptors in those cases investigated. In our own ongoing physiology/structure studies, we have recently investigated the structure of four members (PACAP27, PACAP38, PACAP-related peptide, helodermin)[9,11,14] under various solution conditions using 2D ^{1}H NMR spectroscopy. Although one of these (helodermin) originates in reptile venom, it reacts with mammalian receptors and helodermin-like peptides are probably present in mammalian tissues. In all the reported studies, interest lies in determining the change in peptide conformation on going from the free to the bound state. Ideally one would like to study the structure of the peptide bound or in equilibrium with the membrane-bound receptor. Unfortunately such studies are not possible: insufficient quantities of the receptor and its high molecular weight (57,000–70,000) make NMR studies impossible at present. Consequently one must compromise and study the free hormones under solution conditions that mimic as much as possible the hydrophobic environment of the peptide/receptor interface. This has been performed in a number of ways either by the use of organic solvents or micelles. In our own studies we have used aqueous trifluoroethanol. In the following sections the methodology is summarized using examples from our own work and then general conclusions applicable to all systems are enumerated.

[d]Corresponding author: Dr. Victor Wray, Gesellschaft für Biotechnologische Forschung, Mascheroder Weg 1, D-38124 Braunschweig, Germany, Tel.: 6181-362; Fax: 6181-355; E-mail: vwr@gbf-braunschweig.de

```
Peptide        Sequences

VIP            H-S-D-A-V-F-T-D-N-Y-T-R-L-R-K-Q-M-A-V-K-K-Y-L-N-S-I-L-N*
PACAP-38       H-S-D-G-I-F-T-D-S-Y-S-R-Y-R-K-Q-M-A-V-K-K-Y-L-A-A-V-L-G-K-R-Y-K-Q-R-V-K-N-K*
Secretin       H-S-D-G-T-F-T-S-E-L-S-R-L-R-E-G-A-R-L-Q-R-L-Q-G-L-V*
Glucagon       H-S-Q-G-T-F-T-S-D-Y-S-K-Y-L-D-S-R-R-A-Q-D-F-V-Q-W-L-M-N-T
Helodermin     H-S-D-A-I-F-T-E-E-Y-S-K-L-L-A-K-L-A-L-Q-K-Y-L-A-S-I-L-G-S-R-T-S-P-P-P*

Identity of helodermin with

VIP            + + + +   + + +     +       + + + +   + + + +
PACAP-38       + + + +   + + +     + +   + + + + +   + + + +
Secretin       + + +     + + +   +   + + +   + +   + + +     +
Glucagon       + +       + +     + + + + +   +   + +       +
```

Footnote:
§ All the sequences are of the human peptides, apart from helodermin which is from *Heloderma suspectum*
* Amidated peptide

FIGURE 1. Representative sequences of peptides from the secretin/VIP family.

PHASE I	Identification of amino acid spin systems starting at the amide protons with COSY and TOCSY experiments that provide through-bond correlations
PHASE II	Sequence specific assignments of amino acid spin systems using NOESY experiments that provide through-space correlations
PHASE III	Use of chemical shift data and qualitative NOE data to give secondary structure elements
PHASE IV	Calculation of complete structure from quantitative NOE data used as distance restraints in molecular dynamic/energy minimization calculations

FIGURE 2. Strategy for the structure elucidation of peptides in solution by ^{1}H NMR spectroscopy.

STRUCTURE ELUCIDATION METHODOLOGY

In general after initial CD studies, a standard strategy was followed using homonuclear 2D ^{1}H NMR techniques (COSY, TOCSY, NOESY), under various solution conditions, to identify amino acid spin systems and give sequence specific assignments (FIG. 2). Secondary structure regions were initially deduced from the difference in Hα chemical shifts from their random coil values[15] and qualitative NOE data (this is exemplified in FIG. 3 with the data for helodermin).

Detailed structures at the atomic level were then calculated from the quantitative NOE data, which were used as distance restraints in molecular dynamics and energy minimization calculations. In the most recent work, regions of stable secondary structure were defined from the resulting final peptide conformations using a new fitting program that uses a consecutive segment approach that takes into account the summed RMS differences between all structures for short segments of 2 to 5 residues in length.[14] This procedure allows a reasonably objective method of defining the edges of stable structure. Subsequently these limits were used for the alignment of all the final restrained structures (FIG. 4).

GENERAL STRUCTURAL CONCLUSIONS

In this final section we attempt to make several general statements that encompass all the studies that have been reported in the literature. Space does not allow us to go into the exact details of each structure, although these are found in the original studies, and only global features are considered here.

(1) The most striking feature of all the structures determined under limiting solution conditions, i.e., in the presence of organic solvents or micelles, is that all members of the family adopt regular secondary structure exhibiting the same general features in having a disordered N-terminal region of 6 to 8 residues followed by an extended helical region of 18 ± 2 residues, which begins at or near the conserved residues Phe-6 or Thr-7 (FIG. 5). Hence the greatest structural similarities are shown by regions of the peptides that have the least residues in common, while the N-terminus, which has the most residues in common remains flexible under all solution conditions. In several cases (VIP and PACAP)[7,9] there

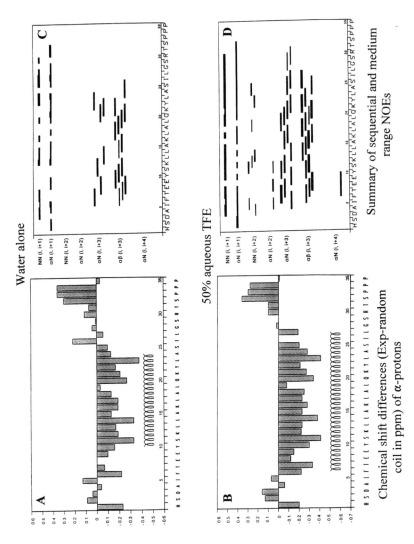

FIGURE 3. Hα shift differences and medium-range NOE's for helodermin in water alone (**A** and **C**, respectively) and in 50% TFE (**B** and **D**).

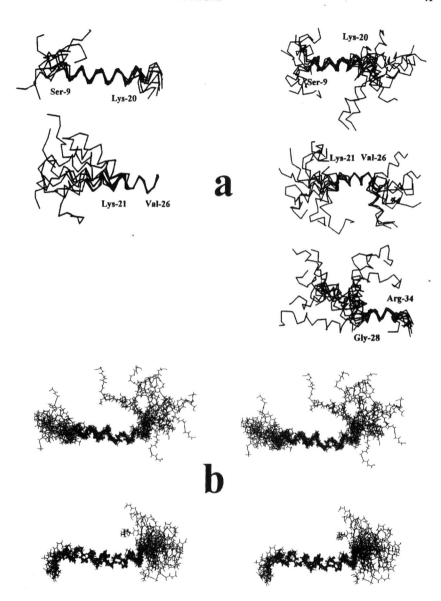

FIGURE 4. (a) Superposition of all final restrained structures for PACAP27 (*left*) and PACAP35 (*right*), after alignment of the backbone atoms, in 50% TFE. (b) Stereoview superposition of all final restrained structures for helodermin, after alignment of the backbone atoms, in water alone (*upper*) and in 50% TFE (*lower*), from Glu-9 to Leu-23 and Phe-6 to Ala-24, respectively.

is evidence of a central interruption within the helix, although in helodermin this is not apparent as the helix appears to be significantly more stable than in other members of the family.[14] The initial N-terminal section of the helix is amphipathic in character with

	1	5	10	15	20	25	30	35	Homology
PACAP38:	H S D G I F T D S Y S R Y R K Q M A V K K Y L A A V L G K R Y K Q R V K N K NH2								
	H H H H H H H H H H H H H H H H H H H H								
PACAP27:	H S D G I F T D S Y S R Y R K Q M A V K K Y L A A V L NH2								100
	H H H H H H H H H H H H H H H H								
VIP:	H S D A V F T D N Y T R L R K Q M A V K K Y L N S I L N								68
	H H H H H H H H H H H H H H H								
Glucagon:	H S Q G T F T S D Y S L Y L D S R R A Q D K V Q W L M N T								28
	H H H H H H H H H H H H H H H H H H H H H H H								
Secretin:	H S D G T F T S E L S R L R D S A R L Q R L L Q G L V								37
	H H H H H H H H H H H H H H H H								
GHRF29:	Y A D A I F T N S Y R K V L G Q L S A R K L L Q D I M S R NH2								31
	H H H H H H H H H H H H H H H H H H H H H H H H								
Helo-dermin:	H S D A I F T Q Q Y S K L L A K L A L Q K Y L A S I L G S R T S P P P NH2								46
	H H H H H H H H H H H H H H H H H H H H H H H H H H H								

FIGURE 5. Schematic of the positions of helical secondary structure found in members of the VIP/PACAP family of peptides.

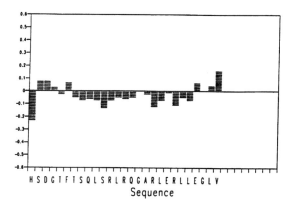

FIGURE 6. Chemical shift differences (Exp - random coil in ppm) of the Hα of a secretin mutant in water at pH 2.7.

a surface of hydrophobic residues opposed by a surface of charged residues, a feature that may suggest an orientation of the membrane-bound molecules.

(2) Most members of the secretin/VIP family of peptides show little evidence of stable structure in water alone. Only on exposure to a more hydrophobic environment, achieved by using organic solvents or micelles, could stable secondary structure be detected by NMR spectroscopy. Indeed in most cases ^{1}H NMR data show little or no evidence of medium range NOEs and little deviation of the Hα chemical shifts from their random coil values (an example of the latter is shown in FIG. 6).

(3) In our experience only helodermin shows stable structure without the necessity of using solvent mixtures[14] (compare FIGURES 3 and 6). Significantly addition of TFE only causes stabilization of helical regions that already have a tendency for such structures in water alone. Consequently there is no evidence of an induction of or change in secondary structure on adding TFE.

(4) Early spectroscopic evidence of global structure suggested that a number of residues in many members of the family may indeed be involved in α helices under certain aqueous conditions.[16,17] In combination with the observations for helodermin, this may now be rationalized by a transient involvement of a number of rapidly interconverting helix-containing species.

(5) The relationship of the solution structure of these conformationally mobile peptides to their receptor-bound structure is problematic. However, as an α-helical structure already exists almost complete (helodermin) or in part (others), it now appears most likely that they are bound in this conformation in the hydrophobic environment at or near to the membrane-bound receptor, especially as a more hydrophobic environment has a stabilizing rather than destabilizing influence

(6) The structural domains appear to be of biological importance. Space precludes a discussion here but details are to be found in Wray and colleagues[9] and Blankenfeldt and colleagues.[14]

REFERENCES

1. BRAUN, W., G. WIDER, K. H. LEE & K. WÜTHRICH. 1983. Conformation of glucagon in a lipid-water interphase by ^{1}H nuclear magnetic resonance. Mol. Biol. **169:** 921–948.
2. FOURNIER, A., J. K. SAUNDERS & S. ST-PIERRE. 1984. Synthesis, conformational studies and biological activities of VIP and related fragments. Peptides **5:** 169–177.

3. CLORE, G. M., S. R. MARTIN & A. M. GRONENBORN. 1986. Solution structure of human growth hormone releasing factor. J. Mol. Biol. **191:** 553–561.

4. GRONENBORN, A. M., G. BOVERMANN & G. M. CLORE. 1987. A ^{1}H-NMR study of the solution conformation of secretin. FEBS Lett. **215:** 88–94.

5. CLORE, G. M., M. NILGES, A. BRÜNGER & A. M. GRONENBORN. 1988. Determination of the backbone conformation of secretin by restraint molecular dynamics on the basis of interproton distance data. Eur. J. Biochem. **171:** 479–484.

6. FRY, D. C., V. S. MADISON, D. R. BOLIN, D. N. GREELEY, V. TOOME & B. B. WEGRZYNSKI. 1989. Solution structure of an analogue of vasoactive intestinal peptide as determined by two-dimensional NMR and circular dichroism spectroscopies and constrained molecular dynamics. Biochemistry **28:** 2399–2409.

7. THERIAULT, Y., Y. BOULANGER & S. ST-PIERRE. 1991. Structural determination of the vasoactive intestinal peptide by two-dimensional ^{1}H-NMR spectroscopy. Biopolymers **31:** 459–464.

8. INOOKA, H., S. ENDO, C. KITADA, E. MIZUTA & M. FUJINO. 1992. Pituitary adenylate cyclase activation polypeptide (PACAP) with 27 residues. Int. J. Peptide Protein Res. **40:** 456–464.

9. WRAY, V., C. KAKOSCHKE, K. NOKIHARA & S. NARUSE. 1993. Solution structure of pituitary adenylate cyclase activation polypeptide by nuclear magnetic resonance spectroscopy. Biochemistry **32** (No. 22): 5832–5841.

10. THORNTON, K. & D. G. GORENSTEIN. 1994. Structure of glucagon-like peptide(7-36) amide in a dodecylphosphocholine micelle as determined by 2D NMR. Biochemistry **33:** 3532–3539.

11. WRAY, V., K. NOKIHARA, S. NARUSE, E. ANDO, C. KAKOSCHKE & M. WEI. 1995. Synthesis, solution structure and biological action of PACAP-related peptide. Biomed. Peptides Proteins Nucl. Acids **1:** 77–82.

12. HAGHJOO, K., P. W. CASH, R. S. FARID, B. R. KOMISARUK, F. JORDAN & S. S. POCHAPSKY. 1996. Solution structure of vasoactive intestinal polypeptide (11-28)-NH$_2$, a fragment with analgesic properties. Peptide Res. **9:** 327–331.

13. GOOSSENS, J.-F., P. COTELLE, P. CHAVATTE & J.-P. HÉNICHART. 1996. NMR study of five N-terminal peptide fragments of the vasoactive intestinal peptide: Crucial role of aromatic residues. Peptide Res. **9:** 322–326.

14. BLANKENFELDT, W., K. NOKIHARA, S. NARUSE, U. LESSEL, D. SCHOMBURG & V. WRAY. 1996. NMR spectroscopic evidence that helodermin, unlike other members of the secretin/VIP family of peptides, is substantially structured in water. Biochemistry **35:** 5955–5962.

15. WISHART, D. S., B. D. SYKES & F. M. RICHARDS. 1992. The chemical shift index: A fast and simple method for the assignment of protein secondary structure through NMR spectroscopy. Biochemistry **31:** 1647–1651.

16. WAGMAN, M. E., C. M. DOBSON & M. KARPLUS. 1980. Proton NMR studies of the association and folding of glucagon in solution. FEBS Lett. **119:** 265–270.

17. BODANSZKY, A., M. A. ONDETTI, V. MUTT & M. BODANSZKY. 1969. Synthesis of secretin. IV. Secondary structure in a miniature protein. J. Amer. Chem. Soc. **91:** 944–949.

Cloning and Functional Characterization of PACAP-Specific Receptors in Zebrafish

Y. WEI,[a] S. C. MARTIN,[b] G. HEINRICH,[b,c] AND S. MOJSOV[a,d]

[a]The Rockefeller University, New York, New York USA 10021

[b]Boston University Medical Center, Boston, Massachusetts USA 02118

ABSTRACT: Two PACAP receptors were isolated from total zebrafish cDNA library prepared from 6-day old fish by a homology-based cloning strategy. The two zebrafish PACAP receptors have the same topology as the one found in other members of this class of seven membrane-spanning G-protein-coupled receptors. Each of the two zebrafish PACAP receptors shares about 70% sequence identity at the amino acid level with the human PACAP-type 1 receptor, and about 50% amino acid identity with PACAP/VIP R-1 and PACAP/VIP R-2 receptors. One of these zebrafish receptors contains the hop2 configuration found in the human and rat PACAP-type 1 receptors. On the basis of these structural characteristics the zebrafish PACAP receptors were classified as PACAP-type 1 and PACAP-type 2 receptors. In competitive binding experiments zebrafish PACAP-type 1 and PACAP-type 2 receptors showed similar binding specificity for zebrafish and human PACAP-38 and PACAP-27. Furthermore, the specificity of PACAP-type 1 and PACAP-type 2 receptors for zebrafish and human PACAPs is about 1,000-fold higher than for human VIP. These results demonstrate that zebrafish PACAP-type 1 receptor is a structural and pharmacological homolog of the mammalian PACAP-type 1 receptor. Additional pharmacological characterization is needed in order to classify the zebrafish PACAP-type 2 receptor.

Pituitary adenyláte cyclase activating polypeptide (PACAP)[1] exerts diverse metabolic, neuroendocrine, neurotransmitter and immunological effects by interacting with distinct G-protein–coupled receptors that are widely distributed throughout the peripheral and central nervous systems.[2] In mammals, three types of PACAP receptors with different pharmacological properties have been identified.[3,4] Two of these receptor types have similar affinities for PACAP and the structurally related vasoactive intestinal polypeptide (VIP). In addition to these common receptors for PACAP and VIP, a third, PACAP-specific receptor (also called PACAP type 1 receptor) has been characterized. PACAP-specific type 1 receptor, in contrast to the common PACAP/VIP receptors, recognizes with high affinity only the two molecular forms of PACAP peptides (PACAP-38 and PACAP-27) and interacts with VIP at high, non-physiological concentrations of 1 μM. The three different PACAP receptors share about 50% amino acid sequence homologies among each other.

The alternative splicing of mammalian PACAP-type 1 receptor gene leads to several splice variants of this PACAP receptor type. The short[5] or null[6] variant of the receptor shows similar topography of the third intracellular loop as the one found in the other two

[c]Current address: Department of Medicine, VA Northern California Health Care System, Martinez, CA 94553.

[d]Corresponding author: Laboratory of Cellular Physiology and Immunology, The Rockefeller University, 1230 York Avenue, New York, NY 10021; Tel.: 212-327-8108; Fax: 212-327-8875; E-mail: mojsov@rockvax.rockefeller.edu

45

common PACAP/VIP receptors. The other splice variants of the PACAP-type 1 receptor contain an insertion in the third intracellular loop of the receptor. This insertion can contain one of the two distinct 28 amino acid long cassettes, termed hip and hop1. Alternatively, the insertion can contain a single 56 amino acid–long cassette, consisting of the hip and hop1 cassettes arranged in tandem with each other. An additional splice variant of the hop configuration, termed hop2, has also been identified, where the amino terminal serine residue of the hop1 cassette is missing.

To study the role of PACAP in vertebrate development using the zebrafish as a model, we used a homology-based cloning strategy to isolate PACAP receptors from total zebrafish cDNA library prepared from 6-day-old fish. Using this strategy we isolated two putative zebrafish PACAP receptors. They share with each other about 85% sequence identity at the amino acid level and contain the same topology as the one found in this class of G–protein–coupled receptors. All the cysteine residues in the two putative zebrafish receptors are located in the same positions of their extracellular domains as the ones found in the sequence of the extracellular domain of mammalian PACAP-type 1 receptors.[6,7] In addition, each of these two putative zebrafish PACAP receptors showed approximately 70% amino acid sequence identity with the human PACAP-type 1 receptor and about 50% amino acid sequence identity with the other two human PACAP/VIP receptors. One of these putative zebrafish receptors contained the hop2 configuration of the mammalian PACAP-type 1 receptor. The 27 amino acid–long hop2 cassette of the putative zebrafish PACAP-type 1 receptor shares 19 identical amino acid residues with the human hop2 sequence. On the basis of these structural homologies, we classified the two putative zebrafish receptors as zebrafish PACAP type 1 (hop2) and zebrafish PACAP type 2 receptors.

To characterize the ligand specificity of the two putative zebrafish PACAP receptors, we expressed their cDNAs in a Chinese hamster lung (CHL) cell line and performed competitive binding experiments. We postulated that ^{125}I-labeled human PACAP-27 will be able to bind to the two putative zebrafish PACAP receptors because there are high structural homologies that exist between (1) the human and zebrafish PACAPs (approximately 80%)(Wei and colleagues, manuscript in preparation) and (2) human PACAP type 1 receptor and the two putative zebrafish PACAP receptors. In multiple experiments we found that the specific binding of ^{125}I-labeled human PACAP-27 (Amersham) to both zebrafish PACAP type 1 and PACAP type 2 receptors was between 4% and 6%. This level of specific binding is in the same range as that measured with the human PACAP/VIP R-2 receptor expressed in the CHL cells.[2]

The specific binding of radioiodinated human PACAP-27 (44 pM) to the zebrafish PACAP type 1 and PACAP type 2 receptors was displaced with increasing concentrations (1 pM to 1 μM) of zebrafish PACAP-38, zebrafish PACAP-27, human PACAP-38, human PACAP-27, human VIP (FIG. 1, A and B), and rat PHI (FIG. 1, B). As seen in FIGURE 1, both zebrafish PACAP receptors displayed about 1,000-fold higher ligand binding specificity towards zebrafish and human PACAP-38 and PACAP-27, respectively, than towards human VIP. Thus, the binding specificities of zebrafish PACAP type 1 and zebrafish PACAP type 2 receptors are similar to the binding specificity of the mammalian (human and rat) PACAP type 1 receptors.[7,8] Inhibition concentrations (IC$_{50}$) for zebrafish PACAP-38 and zebrafish PACAP-27, calculated from the displacement curves to represent the concentration of peptides that gives 50% inhibition of the specific binding, were similar for zebrafish PACAP type 1 and zebrafish PACAP type 2 receptors, respectively (7.5 nM vs. 6 nM for zebrafish PACAP-38; and 20 nM vs. 15 nM for zebrafish PACAP-27). The rat PHI displaced the binding of the ^{125}I-labeled human PACAP-27 to zebrafish PACAP type 2 receptor only at concentrations of 1 μM (FIG. 1, B).

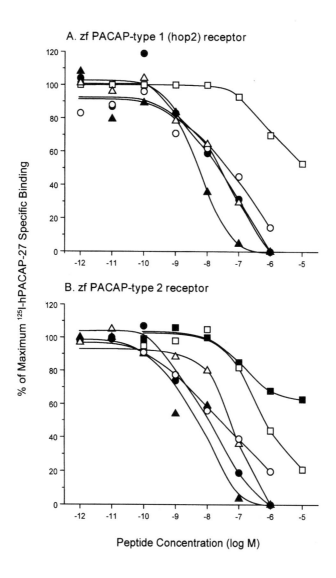

FIGURE 1. Binding specificity of zebrafish PACAP type 1 (hop2) receptor (**A**) and zebrafish PACAP type 2 (**B**) receptors expressed in CHL cells. Displacement of the binding of ^{125}I-labeled human PACAP-27 (44 pM, Amersham) with increasing concentrations of zebrafish PACAP-38 (*closed triangles*), zebrafish PACAP-27 (*open triangles*), human PACAP-38 (*closed circles*), human PACAP-27 (*open circles*), human VIP (*open squares*), and rat PHI (*closed squares*). The experimental conditions for the competitive binding experiments were the same as the ones described in Wei and Mojsov.[2] For each peptide concentration the individual data points represent an average of three independent measurements.

CONCLUSIONS

(1) We have identified the zebrafish homolog of the human and rat PACAP type 1 receptors. Our conclusion is based on the following evidence:

(*i*) High conservation of amino acid residues that are found in the sequences of the human PACAP type 1 and the zebrafish PACAP type 1 receptors. This homology includes the presence of the hop2 cassette in the third intracellular loop of the zebrafish PACAP type 1 receptor.

(*ii*) Similar binding specificities of the zebrafish PACAP type 1 receptor for zebrafish PACAP-38, zebrafish PACAP-27, human PACAP-38, human PACAP-27, and human VIP, as the binding specificity of the human and rat PACAP-type 1 receptors for human and rat PACAP-38, PACAP-27, and VIP, respectively.

(2) We have identified a second zebrafish PACAP type 2 receptor that shows higher structural similarities to the human and rat PACAP type 1 receptors than to the human and rat PACAP/VIP receptors. The binding specificity of the zebrafish PACAP type 2 receptor for zebrafish and human PACAP-38, PACAP-27 and human VIP is similar to the binding specificities of the human and rat PACAP type 1 receptors. However, we do not know whether zebrafish PACAP-type 2 receptor will have the same ligand specificity for zebrafish VIP, as the one that we measured with human VIP. The sequence of zebrafish VIP is not available yet. Therefore, we still can not conclude whether the zebrafish PACAP type 2 receptor is a novel type of a PACAP-specific receptor or is a zebrafish homolog of the mammalian PACAP/VIP receptors.

REFERENCES

1. ARIMURA, A. 1992. A pituitary adenylate cyclase activating polypeptide (PACAP): Discovery and current status of research. Regul. Pept. **37**: 287–303.
2. WEI, Y. & S. MOJSOV. 1996. Tissue specific expression of different human receptor types for pituitary adenylate cyclase activating polypeptide and vasoactive intestinal polypeptide: Implications for their role in human physiology. J. Neuroendocrinol. **8**: 811–817.
3. RAWLINGS, R. S. & M. HEZAREH. 1996. Pituitary adenylate cyclase-activating polypeptide (PACAP) and PACAP/Vasoactive intestinal polypeptide receptors: Actions on the anterior pituitary gland. Endocrine Rev. **17**: 4–29.
4. HARMAR, T. & E. LUTZ. 1994. Multiple receptors for PACAP and VIP. Trends. Pharm. Sci. **15**: 97–99.
5. SPENGLER, D. *et al*. 1993. Differential signal transduction by five splice variants of the PACAP receptor. Nature **365**: 170–175.
6. PISEGNA, J. R. & S. A. WANK. 1996. Cloning and characterization of the signal transduction of four splice variants of the human pituitary adenylate cyclase activating polypeptide receptor. J. Biol. Chem. **271**: 17267–17274.
7. OGI, K. *et al*. 1993. Molecular cloning and functional expression of a cDNA encoding a human pituitary adenylate cyclase activating polypeptide receptor. Biochem. Biophys. Res. Commun. **196**: 1511–1521.
8. HASHIMOTO, H. *et al*. 1993. Molecular cloning and tissue distribution of a receptor for pituitary adenylate cyclase-activating polypeptide. Neuron **11**: 333–342.

Induction of Type I PACAP Receptor Expression by the New Zinc Finger Protein Zac1 and p53

ANKE HOFFMANN,[a] ELISABETTA CIANI,[a] SOUHEIR
HOUSSAMI,[b] PHILIPPE BRABET,[b] LAURENT JOURNOT,[b] AND
DIETMAR SPENGLER[a,b]

[a]*Max-Planck Institute of Psychiatry, Molecular Neurobiology and
Neuropathology, Kraepelinstrasse 2-10, 80804 Munich, Germany*

[b]*Centre National de la Recherche Scientifique, Unité Propre de
Recherche 9023 du CNRS, CCIPE, 141, rue de la cardonille, 34094
Montpellier Cedex 05, France*

ABSTRACT: We reported recently the cloning of the type I PACAP receptor by a functional expression cloning technique. Unexpectedly, we observed additional PACAP-positive pools that turned out to encode the wild-type form of the tumor suppressor gene p53 and the novel zinc finger protein Zac1, which regulates apoptosis and cell cycle arrest.[1] Both Zac1 and p53 caused, under transient or stably regulated expression, induction of the type I PACAP receptor by transcriptional mechanisms. Transactivation of the type I PACAP receptor gene by Zac1 and p53 points to a subtle balance between death promoting and protective mechanisms. The control of these processes is central to various physiological conditions ranging from development to senescence, whereas dysregulation may lead to overt pathological outcomes, notably cancer, immune deficiency syndromes, and neurodegenerative disorders.

Pituitary adenylate cyclase activating polypeptide (PACAP) was first isolated from ovine hypothalamic tissues during an attempt to discover novel hypothalamic hypophysiotropic factors.[2] Studies on the ontogeny of PACAP and PACAP receptor expression[3–8] and *in vitro* functional studies[9–14] support the view that PACAP plays an important regulatory role in the development of the brain.

We reported recently the cloning of a novel gene designated Zac1, which encodes a protein with seven zinc fingers of the C_2H_2 type and is only distantly related to previously characterized zinc finger proteins, namely the Krüppel family, which is known to play an important role in the development of *Drosophila*.[15] Unexpectedly, we isolated Zac1 by a functional expression cloning technique, which also resulted in the isolation of the wild-type form of p53. This technique is based on cotransfection of pools of an expression library with a cAMP-responsive reporter gene.[1,16] We show here that Zac1 and p53 induce expression of the gene encoding the type I PACAP receptor through transcriptional mechanisms.

RESULTS

Functional Expression Cloning of the Type I PACAP Receptor by Zac1 and p53

We used a recently described expression cloning method[16] to screen simultaneously for different receptors positively coupled to adenylate cyclase (FIG. 1). This method is based on

49

transcriptional induction of a cAMP-responsive luciferase reporter gene by stimulation of adenylate cyclase through activated target receptors. Pools of clones from a newborn rat colliculi cDNA library and from a corticotroph tumor cell line (AtT-20) cDNA library were cotransfected with the cAMP-responsive reporter construct pΔMC16-LUC into the host cell line LLC-PK1. Separate aliquots of cells were incubated with various peptides, including PACAP-38, 20 h after electroporation. One pool of clones from the rat colliculi library consistently stimulated luciferase activity in the presence of PACAP and a functional clone encoding the type I PACAP receptor was isolated by successive subdivisions.[16] Several other pools displayed the same phenotype, namely a PACAP-dependent stimulation of the reporter gene and the corresponding active clones were isolated by the same subdivision process. Sequencing of one of these PACAP-positive clones from the rat library revealed that this cDNA was identical to the wild-type form of the tumor suppressor gene p53. Partial sequencing of two PACAP-positive clones from the AtT-20 library also proved the presence of the wild-type form of mouse p53. In contrast, two clones (p2195 and p1270) inducing type I PACAP receptor expression turned out to encode the same protein, which we named Zac1. This new seven zinc finger protein causes apoptosis and cell cycle arrest under constitutive and regulated expression.[1] The cloning of p53 and Zac1 by this functional expression cloning method prompted us to hypothesize that overexpression of Zac1 and p53 in LLC-PK1 cells induces, by direct or indirect means, the expression of the endogenous type I PACAP receptor gene. A schematic outline of this hypothesis is presented in FIGURE 2. According to this model, stimulation of type I PACAP receptor gene transcription results in increased synthesis of the receptor protein, which will be targeted to the cell membrane. Binding of the ligand PACAP and coupling to G-proteins lead to increased levels of intracellular cAMP and consequent induction of the cAMP-responsive reporter gene, which is monitored by luciferase measurements of the cell extracts.

Cotransfection of Zac1 and p53 Induces Dose-Dependent Activation of the cAMP-Responsive Reporter by PACAP-38 and PACAP-27 in a Panel of Unrelated Cell Lines

Since neither Zac1 nor p53 induced the cAMP-responsive reporter in the absence of PACAP, a direct regulation of the reporter plasmid could be ruled out. In addition, PACAP by itself did not stimulate significantly the reporter plasmid, arguing against the presence of type I PACAP receptors in the host cell line LLC-PK1[16] (data not shown). Conclusively, overexpression of Zac1 and p53 in LLC-PK1 cells seemed to induce the expression of the

FIGURE 1. Schematic representation of the expression cloning technique. Poly(A)$^+$ RNA was prepared from newborn rat colliculi and the murine corticotroph cell line AtT-20. Reverse transcription was performed with a random primer-NotI adapter and BstXI adapters were added after second-strand synthesis. The cDNAs were digested with NotI, size selected on a polyacrylamide gel, and cloned in the BstXI/NotI sites of the CMV-promoter driven expression vector pRK8. After electrotransformation of DH5α, pools about 2,000 clones were amplified and DNA (10 µg) was cotransfected into porcine renal epithelial LLC-PK1 cells with the luciferase reporter plasmid pΔMC16-LUC (1 µg) and the plasmid pCH110 (1 µg), containing a SV40-promoter driven β-galactosidase gene. The construct pΔMC16-LUC contains a modified mouse mammary tumor virus promoter in which the glucocorticoid-responsive region was replaced by 16 copies of a cAMP-responsive element derived from the promoter of the corticotropin-releasing hormone gene. Aliquots of the electroporated cells were plated in 12-well clusters and hormones were added 20 h later for 4 h. The ratio of luciferase to β-galactosidase activities in the cell lysates in the presence or absence of hormone was used as an index for the expression of the respective receptor in the transfected pool. Endogenous V$_2$ receptors were used as a positive control in each transfection experiment. A further internal control involved addition of 10 ng of the β$_1$-adrenergic receptor cDNA expressed from pRK8 into the pool DNA.

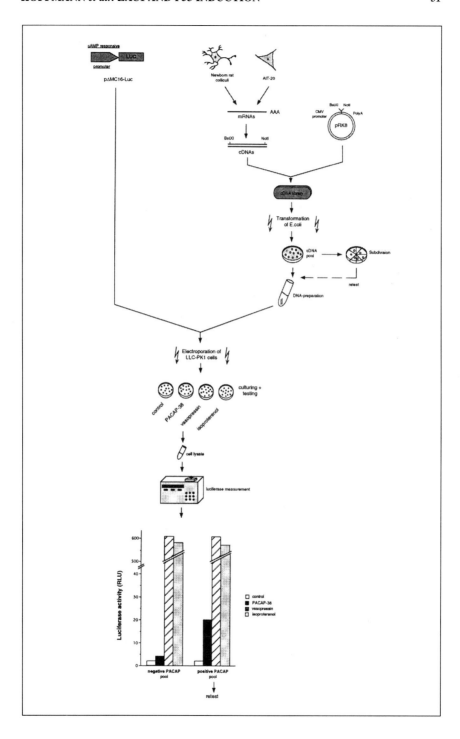

endogenous type I PACAP receptor gene (FIG. 2). To investigate this hypothesis the isolated Zac1 and p53 cDNAs were cotransfected into the LLC-PK1 cell line in order to characterize their response on the cAMP-responsive reporter gene in more detail (FIG. 3). Cotransfection of Zac1 and p53 caused a strong stimulation of the reporter plasmid. This response was dose-dependent with a maximum 100-fold induction at 30 ng of the cotransfected cDNAs under the described conditions. Stimulation by PACAP-38 was slightly more potent than PACAP-27 with EC_{50} values for both peptides in the nanomolar range (PACAP-38 EC_{50}= 2 nM and PACAP-27 EC_{50}= 5 nM; data not shown), whereas VIP required substantially higher concentrations (>1 μM) and led to a weak stimulation of the reporter system (data not shown). Induction of the cAMP-responsive reporter plasmid by PACAP due to transfection of p53 and Zac1 was less compared to the expression of the type I PACAP receptor cDNA (200–300-fold)(data not shown), which reflects the onset of apoptosis under higher levels of Zac1 and p53 expression as previously described.[1] Of note, induction of the reporter plasmid by cotransfection of Zac1 and p53 was not observed in the presence of other peptide ligands tested so far, including CRH, secretin, and glucagon, emphasizing the specificity of the observed response[16] (data not shown). Therefore, we wondered whether induction of the type I PACAP receptor gene is strictly cell-type specific or can be extended to various cellular models of different origin and derived from different species. Accordingly, we tested Zac1 and p53 cDNAs for induction of the cAMP-responsive reporter by PACAP-38 in a panel of unrelated cell lines of human, rat, and mouse origin. Due to apparent differences between the cell lines, such as transfection efficiency, expression levels from the employed CMV promoter, stability of the Zac1 and p53 proteins, and responsiveness of the cAMP system, the observed induction ratios served merely as a qualitative assessment as shown by their respective scores in TABLE 1. However, we observed a strong induction in at least two more cellular models tested, namely the rat pheochromocytoma cell line PC-12 and the mouse neuroblastoma cell line N2A. In addition, Zac1 appeared to be more effective than p53 in these transfection experiments in respect to a positive PACAP response. Importantly again, induction of the type I PACAP receptor gene by Zac1 and p53 was observed in the majority of the investigated cases, though to varying degrees, emphasizing the biological significance of this finding.

Induction of the Type I PACAP Receptor Gene Can Be Transferred to Heterologous Transactivators by the Zac1 DNA-Binding Domain

Previous experiments using stable cell clones of Zac1 and p53 under the control of the tetracycline-responsive expression system[1] provided evidence that expression of Zac1 and p53 leads to a 30- to 40-fold increase in type I PACAP receptor mRNA as demonstrated

TABLE 1. Zac1 and p53 induction of the type I PACAP receptor in different cell lines

Cell Line	Tissue	Species	Zac1	p53
SaOs-2	Osteosarcoma	Human	++++	++++
LLC-PK1	Epithelial kidney	Pig	++++	++++
C6	Glioma	Rat	+	–
CV-1	Kidney fibroblast	Monkey	–	–
PC12	Pheochromocytoma	Rat	+++	++
N2A	Neuroblastoma	Mouse	+++	++

Note: Cotransfections were carried out as described above. Scores indicate induction ratios obtained for the reporter plasmid by PACAP-38. (++++) 80–100-fold; (+++) 60–80-fold; (++) 40–60-fold; (+) 10–40-fold; (–) < 10-fold or not detectable.

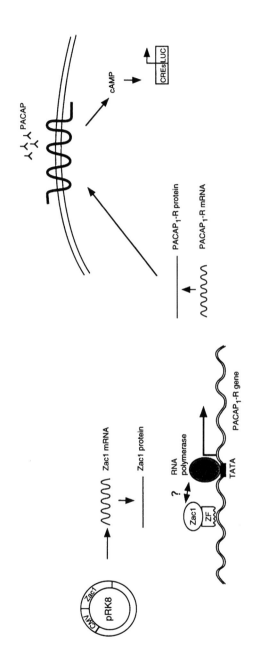

FIGURE 2. Schematic model for transactivation of the cAMP-responsive reporter plasmid by Zac1 and p53 through induction of the type I PACAP receptor gene. Zac1 or p53 expressed from the CMV promoter of pRK8 was cotransfected with the cAMP-responsive reporter plasmid pΔMC16-LUC (1 μg) and the β-galactosidase plasmid pCH110 (1 μg) into LLC-PK1 cells. Expression of Zac1 or p53 leads to DNA binding via their respective DNA-binding domains in the promoter region of the type I PACAP receptor gene and/or of other at present unknown target genes, whose regulation in turn contributes to or accounts for expression of the type I PACAP receptor gene. The type I PACAP receptor mRNA is translated into the receptor protein, which will be targeted to the cell membrane. Binding of the ligand PACAP and subsequent interaction with G-proteins positively coupled to adenylate cyclase will cause increases in intracellular levels of cAMP, which in turn lead to induction of the cAMP-responsive reporter gene, whose activity can be measured by the luciferase activity in the cell extracts.

p53/Zac1

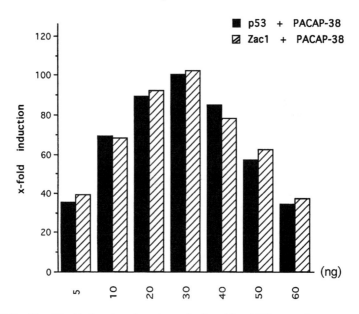

FIGURE 3. p53 and Zac1 induce dose-dependent activation of the cAMP-responsive reporter gene by PACAP-38. p53 and Zac1 cDNA expression vectors were cotransfected with the cAMP-responsive reporter pΔMC16-LUC into LLC-PK1 cells (2×10^6). PACAP-38 (10^{-9} M) was added after 20 h on the next day for 4 h to half of the samples before harvesting the cells. To calculate induction ratios, luciferase activity was standardized on β-galactosidase values. The cAMP-responsive reporter pΔMC16-LUC (1 μg) was cotransfected with the plasmid pGEM4 as carrier DNA and the amount of CMV-expression vector was kept constant with pCMVCAT. Cotransfected pCH110 (1 μg) was used for standardization. p53 and Zac1 were transfected at the indicated concentrations. Results are the averages from three independent experiments performed in triplicate.

FIGURE 4. Induction of the type I PACAP receptor gene can be tranferred to a heterologous trans-activator by the Zac1 DNA-binding domain. (**A**) Schematic representation of Zac1/steroid receptor hybrids. Structural analysis of Zac1 demonstrated features compatible with a transcription factor composed of an N-terminal seven zinc finger DNA-binding domain (AA1-208) and a putative C-terminal transactivation domain (grey box) (AA 275-668). Abbreviations used are for human glucocorticoid (GR) and mineralocorticoid receptor (MR) domains. The GR- and MR-transactivation domain are shown as a hatched box, the MR hormone binding domain as a black box; Zac1, GR, and MR zinc finger domains as grey boxes with the numbers of zinc fingers indicated. Amino acids (AA) are numbered. (**B**) Zac1 zinc finger domain confers type I PACAP receptor gene regulation. The construct $GZ_{ZF}M$ was cotransfected at the indicated amounts with the cAMP-responsive reporter pΔMC16-LUC and plated in the absence or presence of aldosterone (10^{-9} M) or spironolactone (10^{-7} M). PACAP-38 (10^{-9} M) was added after 20 h to half of the samples for 4 h. Luciferase activity of each sample was standardized to the activity in control wells. Results are the averages from three independent experiments performed in triplicate. The GR_{NX} and MR_{NX} constructs were previously described.[29] Primers used to create $GZ_{ZF}M$ were: 5'-gtgatggcggccgCCATTCCGCTGTCAAAAATGTG-3' (+7 bp to +27 bp) and 5'-ccgcgcctcgagGGTCTTCTTGGTGTGACG-3' (+618 bp to +601 bp). PCR products were sequenced to check integrity. The different constructs were subcloned into pRK5PUR. Transfected LLC-PK1 cells (2×10^6) were replated in charcoal-treated serum. The cAMP-responsive reporter pΔMC16-LUC (1 μg) was cotransfected with the plasmid pGEM4 as carrier DNA and the amount of CMV-expression vector was kept constant with pCMVCAT. Cotransfected pCH110 (1μg) was used for standardization.

by actin standardized RT-PCR experiments.[17] These observations prompted us to test the possibility that induction of the type I PACAP receptor gene depends on the DNA-binding domain of Zac1. The Zac1 protein displays at its N-terminus seven zinc fingers of the C_2H_2 type, suggesting a function in DNA binding and eventual transcriptional regulation of target genes. In support of this view, a construct of Zac1 deleted for the zinc finger region was devoid of type I PACAP receptor gene regulation in both LLC-PK1 and SaOs-2 cells under transient transfection.[17] This observation supports a crucial role for this region of Zac1 in DNA binding and led us to ask whether these regulatory properties could be transferred to a heterologous transactivator. Therefore, we created a hybrid fusion protein encompassing at its N-terminus the transactivation domain of the glucocorticoid receptor,

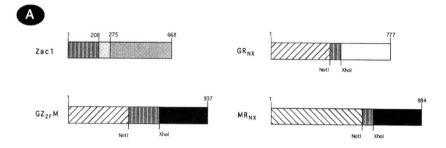

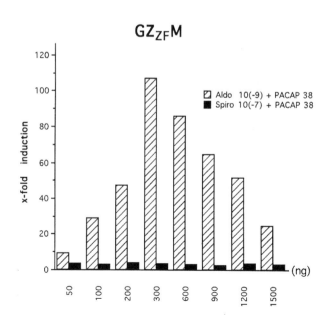

in its central part the DNA-binding domain of Zac1, and at the C-terminus the hormone-binding domain of the mineralocorticoid receptor to allow controlled expression (FIG.4, A). The hormone-binding domain of the mineralocorticoid receptor was used to prevent pleiotropic effects known for glucocorticoid hormones. We cotransfected the cAMP-responsive reporter gene with increasing amounts of the hybrid fusion protein $GZ_{ZF}M$ into the SaOs-2 cell line to measure luciferase activities in the absence and presence of PACAP-38. Increasing amounts of $GZ_{ZF}M$ led to a stepwise maximal 100-fold increase in reporter activity. Induction of the type I PACAP receptor gene by $GZ_{ZF}M$ required higher amounts of transfected DNA than in the case of the wild-type form of Zac1 (FIG. 2 and FIG. 4). This finding could indicate eventual differences in transactivation efficiency, differences in protein stabilities, or both. Importantly, however, induction was strictly dependent on the presence of the agonist aldosterone and PACAP-38, whereas the presence of the mineralocorticoid receptor antagonist spironolactone completely prevented induction by PACAP-38. We conclude that transactivation of the PACAP-receptor gene by Zac1 depends on its DNA-binding domain and can be successfully transferred to a heterologous activator. Interestingly again, cotransfection of higher amounts of the hybrid fusion protein $GZ_{ZF}M$ into the SaOs-2 was followed by decreasing induction ratios, as was observed in the case of the wild-type Zac1 protein, indicating the occurrence of apoptosis. Thus, transfer of the DNA-binding domain of Zac1 to a heterologous transactivator conferred regulation of both type I PACAP receptor gene expression and apoptosis, underlining the requirement for Zac1 DNA binding to regulate these processes.

DISCUSSION

We report here that wild-type p53 and the novel zinc finger protein Zac1 regulate expression of the type I PACAP receptor gene. Both proteins were isolated in a functional expression system based on cotransfection of a cAMP-responsive reporter gene and subsequent stimulation by PACAP-38. Transfection of the isolated cDNAs p53 and Zac1 allowed efficient stimulation of the reporter gene by PACAP-38 and PACAP-27 at nanomolar concentrations. Both p53 and Zac1 revealed at higher doses increasing signs of apoptosis, which might explain the decrease in this response at higher amounts of cotransfected cDNAs as compared to expression of the recombinant type I PACAP receptor cDNA. This behavior was mimicked by the transfer of the DNA-binding domain of Zac1 to a heterologous transactivator. The peptide VIP was considerably less effective in this test system, compatible with the idea of a *de novo* expression of the type I PACAP receptor in these cellular models.

De novo induction of the type I PACAP receptor gene by p53 and Zac1 was not confined to the LLC-PK1 cell line but was also observed in a panel of cell lines of various origins, from human, mouse, and rat, emphasizing the general biological importance of this finding. The observation that Zac1 and p53 are capable of inducing expression of the type I PACAP receptor gene in tumor cell lines devoid of endogenous type I PACAP receptors deserves particular attention. In addition, for several of these cellular models the normal nontransformed tissue counterpart is presumably not expressing the type I PACAP receptor gene. Loss of differentiation markers and reexpression of embryonic features appear to be key events in tumorigenesis, underlying logarithmic growth as opposed to the linear growth observed for cell renewal of stem cell populations in the mature organism. Accordingly, we speculate that induction of the type I PACAP receptor gene by p53 and Zac1 could mirror a step in early development of nontransformed tissues, which becomes paradoxically disclosed under the dedifferentiated state of tumor cells.

Cell proliferation is tightly regulated through molecular pathways controlling cell division, differentiation, growth arrest, and apoptosis. A strict control of these events is fundamental to the maintainance of homeostasis from development to senescence. A network of genes, including cell cycle regulatory genes, protooncogenes, and tumor suppressor genes, has emerged that plays a major role in normal physiological processes, such as development[18,19] and aging,[20] as well as under various pathological states,[21] such as neoplasia[22,23] and neurodegenerative disorders.[24,25]

At this time, the functional significance for type I PACAP receptor regulation by Zac1 and p53 *in vivo* can be only tentatively assigned. Proteins controlling cell cycle arrest and apoptosis coordinate cell cycle regulation and cell cycle exit during central nervous system formation and neuronal differentiation.[14,25,26] Due to its dual role, Zac1 could fulfill an additional role as caretaker in elimination of damaged cells, as has recently been reported for p53.[27] Therefore, in light of the neurotrophic functions of PACAP,[28] transactivation of the type I PACAP receptor gene by Zac1 and p53 could point to a subtle balance between death-promoting and protective mechanisms.

REFERENCES

1. SPENGLER, D., M. VILLALBA, A. HOFFMANN, C. PANTALONI, S. HOUSSAMI, J. BOCKAERT & L. JOURNOT. 1997. Regulation of apopotosis and cell cycle arrest by Zac1, a novel zinc finger protein expressed in the pituitary gland and the brain. EMBO J. **16**: 2814–2825.

2. MIYATA, A., A. ARIMURA, R. R. DAHL, N. MINAMINO, A. UEHARA, L. JIANG, M. D. CULLER & D. H. COY. 1989. Isolation of a novel 38 residue-hypothalamic polypeptide which stimulates adenylate cyclase in pituitary cells. Biochem. Biophys. Res. Commun. **164**: 567–574.

3. KÖVES, K., A. ARIMURA, T. G. GORCS & A. SOMOGYVARI-VIGH. 1991. Comparative distribution of immunoreactive pituitary adenylate cyclase activating polypeptide (PACAP) and vasoactive intestinale polypeptide (VIP) in rat forebrain. Neuroendocrinology **54**: 159–169.

4. KÖVES, K., A. ARIMURA, A. SOMOGYVARI-VIGH, S. VIGH & J. MILLER. 1993. Immunohisto-chemical demonstration of a novel hypothalamic peptide, pituitary adenylate cyclase polypeptide, in the ovine hypothalamus. Endocrinology **127**: 264–271.

5. ARIMURA, A., A. SOMOGYVARI-VIGH, R. C. FIORE & C. WEIL. 1993. Prevention of natural death of motorneurons by pituitary adenylate cyclase activating polypeptide (PACAP) in chick embryo. Presented at the 75th Meeting of the Endocrine Society. San Diego, CA, 1993. 897B [abstract].

6. ARIMURA, A., A. SOMOGYVARI-VIGH, A. MIYATA, K. MIZUNO, D. H. COY & C. KITADA. 1991. Tissue distribution of PACAP as determined by RIA: highly abundant in the rat brain and testes. Endocrinology **129**: 2787–2789.

7. ARIMURA, A. 1992. Pituitary adenylate cyclase activating polypeptide (PACAP): discovery and current status of research. Regul. Pept. **37**: 287–303.

8. SHUTO, Y., D. UCHIDA, H. ONDA & A. ARIMURA. 1995. Ontogeny of pituitary adenylate cyclase activating polypeptide and its receptor mRNA in the mouse brain. Regul. Pept. **67**: 79–83.

9. BASILLE, M., B. J. GONZALEZ, P. LEROUX, L. JEANDEL, A. FOURNIER & H. VAUDRY. 1993. Localization and characterization of PACAP receptors in the rat cerebellum during development: evidence for a stimulatory effect of PACAP on immature cerebellar granule cells. Neuroscience **57**: 329–338.

10. BASILLE, M., B. J. GONZALEZ, L. DESRUES, M. DEMAS, A. FOURNIER & H. VAUDRY. 1995. Pituitary adenylate cyclase-activating polypeptide (PACAP) stimulates adenylyl cyclase and phospho-lipase c activity in rat cerebellar neuroblasts. J. Neurochem. **65**: 1318–1324.

11. DICICCO-BLOOM, E. & P. J. DEUTSCH. 1992. Pituitary adenylate cyclase activating polypeptide (PACAP) potently stimulates mitosis, neuritogenesis and survival in cultured rat sympathetic neuroblasts. Regul. Pept. **37**: 319–325.

12. CHANG, J. Y. & V. V. KOROLEV. 1997. Cyclic AMP and sympathetic neuronal programmed cell death. Neurochem. Int. **31**: 161–167.

13. VILLALBA, M., J. BOCKAERT & L. JOURNOT. 1997. Pituitary adenylate cyclase-activating polypep-tide (PACAP-38) protects cerebellar granulae neurons from apoptosis by activating the mito-gen-activated protein kinase (MAP kinase) pathway. J. Neurosci. **17**: 83–90.

14. Lu, N. R. E. & E. DiCicco-Bloom. 1997. Pituitary adenylate cyclase-activating polypeptide is an autocrine inhibitor of mitosis in cultured cortical precursor cells. Proc. Natl. Acad. Sci. USA **94:** 3357–3362.
15. Schuh, R., W. Aicher, U. Gaul, S. Cote, A. Preiss, D. Maier, E. Siefert, U. Nauber, C. Schröder, R. Kemler & H. Jäckle. 1986. A conserved family of nuclear proteins containing structural elements of the zinc finger protein encoded by Krüppel, a Drosophila segmentation gene. Cell **47:** 1025–1032.
16. Spengler, D., C. Waeber, C. Pantaloni, F. Holsboer, J. Bockaert, P. H. Seeburg & L. Journot. 1993. Differential signal transduction by five splice variants of the PACAP receptor. Nature **365:** 170–175.
17. Hoffmann, A., E. Ciani, S. Houssami, P. Brabet, L. Journot & D. Spengler. (Unpublished observations.)
18. Raff, M. C., B. A. Barres, J. F. Burne, H. S. Coles, Y. Ishizaki & M. D. Jacobson. 1993. Programmed cell death and the control of cell survival: Lessons from the nervous system. Science **262:** 695–700.
19. Steller, H. 1995. Mechanisms and genes of cellular suicide. Science **267:** 1145–1149.
20. Jazwinski, S. M. 1996. Longevity, genes and ageing. Science **273:** 54–59.
21. Thompson, C. B. 1995. Apoptosis in the pathogenesis and treatment of disease. Science **267:** 1456–1462.
22. Hartwell, L. H. & M. B. Kastan. 1994. Cell cycle control and cancer. Science **266:** 1821–1828.
23. Karp, J. E. & S. Broder. 1995. Molecular foundations of cancer: new targets for intervention. Nature Med. **1:** 309–320.
24. Heintz, N. 1993. Cell death and the cell cycle: a relationship between transformation and neurodegeneration? Trends Biochem. Sci. **18:** 157–159.
25. Ross, M. E. 1996. Cell division and the nervous system: regulating the cycle from neural differentiation to death. Trends Neurosci. **19:** 62–68.
26. Schmid, P., A. Lorenz, H. Hameister & M. Montenarh. 1991. Expression of p53 during mouse embryogenesis. Development **113:** 857–865.
27. Eizenberg, O., A. Faber-Elman, E. Gottlieb, M. Oren, V. Rotter & M. Schwartz. 1996. p53 plays a regulatory role in differentiation and apoptosis of central nervous system-associated cells. Mol. Cell. Biol. **16:** 5178–5185.
28. Arimura, A., A. Somogyvari-Vigh, C. Weill, R. C. Fiore, I. Tatsuno, V. Bay & D. E. Brenneman. 1994. PACAP functions as a neurotrophic factor. Ann. N.Y. Acad. Sci. **739:** 228–243.
29. Rupprecht, R. et al. 1993. Mol. Endocrinol. **7:** 597–603.

Cloning and Functional Characterization of the Human VIP1/PACAP Receptor Promoter

ALAIN COUVINEAU,[a] JEAN JOSE MAORET, CHRISTIANE ROUYER-FESSARD, ISABEL CARRERO, AND MARC LABURTHE

Laboratoire de Neuroendocrinologie et Biologie Cellulaire Digestives, Institut National de la Santé et de la Recherche Médicale, INSERM U-410, Faculté de Médecine Xavier Bichat, BP 416, 75870 Paris Cedex 18, France

ABSTRACT: The 5'-flanking region (1.5 kb) of the gene coding for the human VIP1/PACAP receptor was isolated, sequenced, and characterized. Transient expression of constructs containing sequentially deleted 5'-flanking sequences of the VIP1/PACAP receptor fused to a luciferase reporter gene showed that this sequence was active as a promoter in the intestinal cancer cell line, HT-29, expressing endogenous VIP1/PACAP receptor. The shortest DNA fragment with significant promoter activity encompassed the region from -205 to +76 bp. Deletion of a CCAAT-box sequence in the construction corresponding to -173 to +76 bp dramatically reduced the promoter activity. The promoter -205 to +76 bp has a housekeeping gene structure without TATA-box. It contains GC-rich regions characterized by potential Sp1 and AP2 sites and some potential regulatory elements, such as CRE and ATF, and a CCAAT-box sequence (-182 to -178) crucial for gene transcription.

The VIP1/PACAP receptor is widely distributed in the central and peripheral nervous system.[1] In particular, VIP1/PACAP receptors are expressed along the digestive tract where VIP plays an important role in controlling ionic exchange by epithelial cells[2] and motility.[2] Cloning of the human intestinal VIP1/PACAP receptor showed that it is a glycoprotein of 457 amino acids with seven transmembrane domains belonging to the new class II subfamily of G protein–coupled receptors.[3] The gene of the VIP1/PACAP receptor spans about 22 kb and is composed of 13 exons and 12 introns.[4] Recently, structure-function relationship studies of the human VIP1/PACAP receptor have demonstrated that (1) several residues in the N-terminal domain and first and second extracellular loops are crucial for VIP binding;[5,6] (2) two glycosylation sites present in the N-terminal domain are necessary for correct delivery of the receptor to the plasma membrane;[7] (3) a structural determinant for peptide selectivity is made of three nonadjacent amino acid residues in the first extracellular loop and third transmembrane domain.[8]

In the present work, we have cloned and sequenced the 5' flanking region of human VIP1/PACAP receptor gene in order to characterize its promoter. Functional characterization of the 5' flanking region revealed that a 281 bp fragment was necessary and sufficient to have maximal promoter activity and showed the crucial role of a CCAAT-box.

[a]Corresponding author: Alain Couvineau; Tel.: 33 01 44 85 61 35; Fax: 33 01 44 85 61 24; E-mail: coucou@bichat.inserm.fr

MATERIALS AND METHODS

Cloning and Sequencing of the 5′-Flanking Region of the VIP1/PACAP Receptor Gene

The human placental genomic library in pWE 15 cosmid vector (Stratagene, LaJolla, CA, USA) was screened under standard conditions of plaque hybridization[3] with ^{32}P-labeled cDNA probe coding to human VIP1/PACAP receptor N-terminal region (Bases 1-440). One clone (L1) was selected, characterized by Southern blotting, sequenced, and subcloned upstream of the luciferase gene (*Firefly*) reporter vector (p1424Luc) at BamH I-Sst I sites.

Plasmid Constructions and Functional Characterization of the Promoter

The p1424Luc plasmid was used to perform deletion constructions. Briefly, each 5′ or 3′ deletion of the promoter was obtained by restriction fragments of p1424Luc and the resulting plasmids were named according to the size of their respective promoter fragments. Human colon cancer cells (HT-29) were transiently transfected with 15 μg of pSNeoLuc or promoter-luciferase gene construct and 15 μg of salmon sperm carrier DNA by calcium phosphate-DNA coprecipitation as previously described.[3] In order to estimate the transfection efficiency, pRL plasmid (Promega, Madison, WI, USA) containing *Renilla* luciferase reporter gene driven by cytomegalovirus (CMV) ubiquitous promoter was cotransfected with each above-described constructions. After 48 h to 72 h of culture, luciferase activities were independently measured using a dual-luciferase reporter assay system kit (Promega). Relative luciferase activity was calculated as the ratio between *Firefly* luciferase and *Renilla* luciferase activities for each transfection and then reported as a percentage of luciferase activity of p1424Luc construction.

RESULTS AND DISCUSSION

We screened 400,000 clones of the human placental genomic library with ^{32}P-labeled cDNA probe coding for the N-terminal region (bases 1 to 440) of the human VIP1/PACAP receptor. Six positive clones were further characterized by restriction mapping and Southern blotting analysis using ^{32}P-oligonucleotide probe corresponding to bases 1 to 63 of the human VIP1/PACAP receptor. A BamH I-Sst I restriction fragment of one of these clones corresponding to 1.5 kb upstream of the initiation codon was characterized by restriction mapping, Southern blotting, fully sequenced, and subcloned upstream luciferase reporter gene (p1424Luc) in order to characterize promoter activity. After transient transfection in the intestinal cancer cells HT-29, which express endogenous VIP1/PACAP receptor,[9] HT-29 cells showed a high level of luciferase expression (FIG. 1). This activity was about about 33 times higher than the background level determined by transfection of the pSNeoLuc promoterless plasmid. To determine the minimal promoter sequence having the maximal luciferase activity, 5′ deletion constructs were transfected in HT-29 cells (FIG. 1). Deletion of the -1348 to -254 region resulted in an increased luciferase activity (about 49 times the background level) strongly suggesting that the 330 bp minimal promoter was sufficient to have maximal activity. Moreover, inclusion of additional upstream sequences (-1348 to -1124) in conjunction with the 330 bp promoter, as seen with p224Δ330Luc construct, repressed promoter activity (FIG. 1), suggesting the presence of silencer element(s) in the -1348 to -1124 sequence. It should be noted that transfection of the biggest fragment cloned (-4024 to +76), corresponding to the p4100

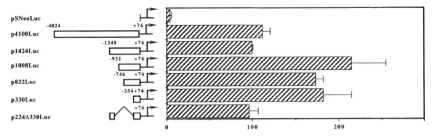

FIGURE 1. Expression of luciferase gene in transfected HT-29 cells. Deletion constructs containing various lengths of the 5'-flanking sequence of the gene of the human VIP1/PACAP receptor were transiently transfected in HT-29 cells. Luciferase activities (*Firefly* and *Renilla*) were measured in cell homogenates and results were normalized to *Renilla* luciferase activity (see *Materials and Methods*) and expressed as the percentage of activity obtained with the p1424Luc construct (100%).

construct, resulted in a luciferase activity similar to that observed with the p330Luc construct supporting that upstream sequences did not exhibit further regulatory sequences, at least when transfected in HT-29 cells. Analysis of the p330Luc sequence revealed the presence of some putative regulatory elements, such as a TATA box (-216 to -211), two CCAAT boxes (-182 to -178; -107 to -103), and some GC-rich regions. In order to determine the role of these regulatory elements, further 5' deletion constructions were transiently expressed in HT-29 cells. Deletion of the -254 to -205 sequence resulted in maximal promoter activity as shown in FIGURE 2, suggesting that the minimal promoter region encompasses the -205/+76 sequence and that the TATA box is not important for promoter activity. In contrast, deletion of the -254 to -173 sequence strongly reduced promoter activity, suggesting that the CCAAT box sequence (-182 to -178) is crucial for promoter activity. Similarly, a very low promoter activity was detected when this CCAAT box sequence was mutated (data not shown). Moreover, partial deletion of the GC sequence (-68 to -38) and of the transcription initiation site in the p330Δ130Luc construct abolished promoter activity. Finally, the construction containing GC sequence and transcription initiation site corresponding to the p130Luc construct was not sufficient to have promoter activity (FIG. 2).

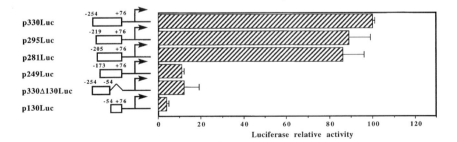

FIGURE 2. Expression of luciferase gene in transfected HT-29 cells. Complementary 5' and 3' deletion constructs were transiently transfected in HT-29 cells and luciferase activities were determined in cell homogenates, as described in the legend to FIGURE 1. Results are expressed as the percentage of activity obtained with the p330Luc construct (100%).

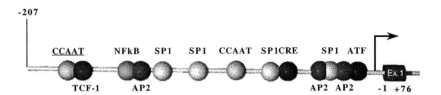

FIGURE 3. Schematic representation of putative nuclear protein binding to the regulatory elements of the human VIP1/PACAP receptor promoter. Computer analysis of the -207 to +76 fragment of the VIP1/PACAP receptor promoter determined the position of each putative regulatory element on the 281 bp sequence. The crucial CCAAT box sequence for promoter activity is underlined. Ex. 1 represents the +1 to +76 fragment of the first exon before the initiation codon of the VIP1/PACAP receptor gene and the arrow displays the trancriptional initiation site as determined in Sreedharan and coworkers.[4]

Taken together these results showed that a 281 bp fragment (FIG. 3) was crucial for promoter activity of the VIP1/PACAP receptor gene. Sequence analysis of this fragment revealed a housekeeping gene structure characterized by the absence of active TATA-box sequence[10] but having several potential regulatory elements (FIG. 3): (1) GC-rich region characterized by potential Sp1 and AP2 sites; (2) two CCAAT-box sequences, one of which (-182 to -178) was a crucial element for transactivation of VIP1/PACAP gene, whereas the other CCAAT box sequence (-107 to -103) was completely unnecessary (data not shown); and (3) some potential regulatory elements such as NFkB, CRE, and ATF. The promoter structure of the human VIP1/PACAP gene is similar to that of the rat VIP1/PACAP receptor promoter characterized by the absence of TATA-box sequence and the presence of various Sp1 sites.[11] In contrast, the rat VIP1/PACAP receptor promoter does not contain a CCAAT box sequence and has a GRE sequence involved in the regulation of receptor expression by glucocorticoid hormone.[12] The absence of an active TATA-box sequence seems to be a common feature of promoters for other members of the class II family of G protein–coupled receptors, such as the PTH receptor[13] or the calcitonin receptor.[14]

In conclusion, identification and functional characterization of the human VIP1/PACAP receptor promoter establish a foundation for the use of molecular genetic approaches to further study the regulation of its transcription during physiopathologic processes. In that respect, we have previously demonstrated that the expression of the VIP1/PACAP receptor is dependent upon the enterocytic differentiation of the human colon cancer cell line, Caco-2.[15] The role of the receptor promoter in this differentiation-dependent expression is currently under investigation in our group.

REFERENCES

1. FAHRENKRUG, J. 1989 *In* Handbook of Physiology. Section 6: The gastrointestinal system. G. M. Makhlouf, Ed. **2**: 611–629. American Physiological Society. Bethesda, MD.
2. LABURTHE, M. & B. AMIRANOFF. 1989 *In* Handbook of Physiology. Section 6: The gastrointestinal system. G. M. Makhlouf, Ed. **2**: 615–243. American Physiological Society. Bethesda, MD.
3. COUVINEAU, A., C. ROUYER-FESSARD, D. DARMOUL, J. J. MAORET, I. CARRERO, E. OGIER-DENIS & M. LABURTHE. 1994. Biochem. Biophys. Res. Commun. **200**: 769–776.
4. SREEDHARAN, S., J. X. HUANG, M. C. CHEUNG & E. GOETZL. 1995. Proc. Natl. Acad. Sci. USA **92**: 2939-2943.
5. COUVINEAU, A., P. GAUDIN, J. J. MAORET, C. ROUYER-FESSARD, P. NICOLE & M. LABURTHE. 1995. Biochem. Biophys. Res. Commun. **206**: 246–252.
6. GAUDIN, P., A. COUVINEAU, J. J. MAORET, C. ROUYER-FESSARD & M. LABURTHE. 1995. Biochem. Biophys. Res. Commun. **211**: 901–908.

7. COUVINEAU, A., C. FABRE, P. GAIDIN, J. J. MAORET & M. LABURTHE. 1996. Biochemistry **35**: 1745–1752.

8. COUVINEAU, A., C. ROUYER-FESSARD, J. J. MAORET, P. GAUDIN, P. NICOLE & M. LABURTHE. 1996. J. Biol. Chem. **271**: 12795–12800.

9. COUVINEAU, A., M. ROUSSET & M. LABURTHE. 1985. Biochem. J. **231**: 139–143.

10. SEHGAL, A., N. PATIL & M. CHAO. 1988. Mol. Cell. Biol. **8**: 3160–3167.

11. PEI, L. & S. MELMED. 1995. Biochem. J. **308**: 719–723.

12. PEI, L. 1996. J. Biol. Chem. **271**: 20879–20884.

13. MCCUAIG, K., J. CLARKE & J. WHITE. 1994. Proc. Natl. Acad. Sci. USA **91**: 5051–5055.

14. ZOLNIEROWICZ, S., P. CRON, S. SOLINAS-TOLDO, R. FRIES, H. LIN & B. HEMMINGS. 1994. J. Biol. Chem. **269**: 19530–19538.

15. LABURTHE, M., M. ROUSSET, C. ROUYER-FESSARD, A. COUVINEA, I. CHANTRET, G. CHEVALIER & A. ZWEIBAUM. 1987. J. Biol. Chem. **262**: 10180–10184.

Desensitization of the Human Vasoactive Intestinal Peptide Receptor (hVIP₂/PACAP R): Evidence for Agonist-Induced Receptor Phosphorylation and Internalization[a]

TERENCE P. MCDONALD,[b] DIANE M. DINNIS, CHRISTINE F. MORRISON, AND ANTHONY J. HARMAR

MRC Brain Metabolism Unit, Royal Edinburgh Hospital, Morningside Park, Morningside, Edinburgh EH10 5HF, United Kingdom

ABSTRACT: To investigate the role of phosphorylation and internalization in the desensitization of the hVIP₂/PACAP receptor, we expressed a C-terminal epitope–tagged (hemagglutinin; YPYDVPDYASL) receptor in COS7 and HEK293 cell lines. Radiolabeling experiments demonstrated that exposure to agonist induced receptor phosphorylation significantly above basal levels. This receptor phosphorylation was greater than that induced by receptor-independent activation of PKA with forskolin and that induced by co-application of forskolin and agonist. This suggests that receptor occupancy promotes phosphorylation and also that receptor phosphorylation may involve a specific G protein–coupled receptor kinase in addition to PKA. Immunocytochemical analysis showed that the receptor was internalized in response to agonist to a single site of accumulation within the cell and this was dependent on temperature, agonist concentration, and time. Further studies will focus on identifying phosphorylation sites and endocytic signals within the hVIP₂/PACAP R.

The neuropeptide vasoactive intestinal peptide (VIP) is a member of a family of regulatory peptides that includes pituitary cyclase-activating polypeptide, secretin, glucagon and growth hormone-releasing hormone. Receptors for these peptides belong to a family of seven-transmembrane-spanning G protein–coupled receptors (GPCRs) and are distinct from the rhodopsin superfamily.[1] All members of this family couple to adenylyl cyclase, probably mediated by G$_s$. Many are also capable of stimulating inositol phosphate production.[2-6] Two distinct receptors for VIP (VIP₁/PACAP R and VIP₂/PACAP R) have been cloned[7-8] that display similar pharmacological properties but are differentially distributed within the brain.[9] Exposure of GPCRs to ligands triggers not only their activation but also a chain of events, termed desensitization, that results in a reduction of cellular response to the agonist. This desensitization provides a means by which agonist-stimulated signal can be regulated. Two major components have been identified for G protein–coupled receptor desensitization: internalization of the occupied receptor (probably into vesicles physically separated from the plasma membrane) and phosphorylation of the intracellular domains of the receptor.

[a]*Note added in proof:* The IUPHAR Subcommittee on Nomenclature for Receptors for VIP and PACAP has proposed that the VIP₂/PACAP receptor be called VPAC₂.[12]

[b]Corresponding author: Tel.: 44 131 537 6530; Fax: 44 131 537 6110; E-mail: tmcdonald@srv1.bmu.mrc.ac.uk

Subtypes of a given receptor family can differ markedly in their desensitization properties and this may be one of the main reasons for the existence of receptor subtypes. We decided therefore to characterize the relative contributions of phosphorylation and internalization in mediating desensitization of the human VIP_2/PACAP R ($hVIP_2$/PACAP R). We have used a hemagglutinin epitope tag fused to the C-terminus of the $hVIP_2$/PACAP R to immunoprecipitate the receptor from cell lysates and localize it using immunofluorescence. To carry out these studies we expressed the tagged receptor ($hVIP_2$/PACAP R-HA) in both COS7 and HEK293 cells.

MATERIALS AND METHODS

Construction of Hemagglutinin-Tagged hVIP₂/PACAP R

The last codon of the $hVIP_2$/PACAP R cDNA contained in the expression vector pcDNA3 was changed by polymerase chain reaction from ATC to CTC, thus creating a *Xho*I restriction site. This restriction site together with that of *Xba*I in the plasmid multicloning site were digested with *Xho*I and *Xba*I and the 6.99-kb fragment isolated. The oligonucleotides 5′-TCGAGTACCCATACGATGTTCCAGATTACGCCTCCCTCTAGT-3′ and 5′-CTAGACTAGAGGGAGGCGTAATCTGGAACATCGTATGGGTAC-3′ were annealed together and ligated with the plasmid DNA recreating the *Xho*I and *Xba*I restriction sites. The resultant plasmid contained $hVIP_2$/PACAP R cDNA with the last C-terminal amino acid changed from Leu to Ileu followed directly by Glu and the 11–amino acid hemagglutinin epitope (HA; YPYDVPDYASL) and a translational stop signal (TAC).

Cell Culture and Transfection

Cells were transfected by way of electroporation; details of transfection are described elsewhere.[11] HEK293 and COS7 cells were grown in Dulbecco's modified Eagle medium supplemented with 10% fetal calf serum, 100 U/ml penicillin, and 100 µg/ml streptomycin. CHO cells were grown in nutrient mixture Ham's F-12 supplemented with 10% fetal calf serum, 100 U/ml penicillin, and 100 µg/ml streptomycin. Stable clones were selected with 400 µg/ml geneticin and maintained in 200 µg/ml geneticin. Cell cultures were maintained at 37°C in a humidified atmosphere of 95% air/5% CO_2.

Phosphorylation of the hVIP₂/PACAP R

Cells (60–80% confluent) were washed in phosphate-buffered saline (PBS) and incubated for 2 h in phosphate-free minimal essential medium containing 20 mM HEPES, pH 7.4, 0.5% BSA, and 200 µCi/ml [β-^{32}P]orthophosphate. After 2 h at 37°C, the cells were exposed to VIP or other agents for the times or concentrations described in the text. Following treatment, cells were washed once in PBS at room temperature and twice at 4°C. The cells were then resuspended into 50 mM Tris-HCl, 1 mM EGTA, pH 7.4, with protease inhibitors (2 µg/ml pepstatin A, 2 µg/ml aprotinin, and 4 µg/ml leupeptin) and phosphatase inhibitors (10 mM $Na_4P_2O_7$, 10 mM NaF, and 0.1 mM Na_3VO_4). The cells were then disrupted by homogenization and the membranes were collected by centrifugation. The membranes were solubilized in Nonidet buffer (0.5% Nonidet P-40, 50 mM Tris-HCl pH 7.4, 150 mM NaCl, 5 mM $MgCl_2$, and 1 mM EGTA, pH 7.4) containing protease and phosphatase inhibitors as above. A mouse anti-HA epitope monoclonal antibody (Autogen Bioclear, Wiltshire, UK) was added to the lysate at 1/1,000-fold dilution for 1 h (4°C) and unsolubilized membranes then removed by centrifugation. The receptor-antibody complex was precipitated (1 h/4°C) using protein G-sepharose (Autogen Bioclear) then washed five times in Nonidet buffer. De-glycosylation was carried out with PNGase (New England Bio-labs, Beverly, MA, USA). The proteins were separated by SDS-PAGE and the ^{32}P incorporated into the proteins detected by autoradiography. Precipitated receptor was detected using the anti HA-epitope antibody.

Immunocytochemistry

Cells grown overnight on glass coverslips were fixed with 4% paraformaldehyde, permeabilized with 0.2% Triton X-100 and non-specific binding was blocked with 10% goat serum. Fixed cells were then incubated sequentially (1 h at room temperature) with each of the following: anti-HA antibody, anti-mouse IgG biotinylated antibody, and avidin-fluorescein isothiocyanate. Coverslips were mounted in PBS/glycerol and viewed on a Zeiss inverted microscope (Axiovert 135 M).

RESULTS

Characterization of hVIP2/PACAP R-HA Bearing HEK293 Cell Lines

To facilitate the investigation of phosphorylation and internalization of the hVIP$_2$/PACAP receptor, the hemagglutinin epitope (HA; YPYDVPDYASL) was fused to the C-terminus (see *Methods*). The hVIP$_2$/PACAP R-HA protein elicited VIP-stimulated cAMP production when expressed stably in HEK293 cells. No differences were detected in the extent of stimulation of adenylyl cyclase activity or in the EC$_{50}$ values between the tagged (hVIP$_2$/PACAP-HA) and untagged receptors [EC$_{50}$ for untagged, 0.97 ± 0.17 (N=7) and for tagged, 1.00 ± 0.2 (N=8)].

Desensitization

To explore desensitization of the hVIP$_2$/PACAP R, HEK293 cells stably expressing the receptor were exposed to 10 nM VIP for 20 min (FIG. 1). After such exposure and extensive washing, stimulation with increasing concentrations of VIP elicited a reduction in the extent of cAMP production to 74.5 ± 6.0% (N = 4) of the maximal responses observed in controls not preincubated with VIP. In contrast, the EC$_{50}$ values were very similar [0 nM, 0.76 ± 0.26 nM (N = 5) and 10 nM, 0.75 ± 0.38 nM (N = 4)].

Effects of Phosphokinase Inhibitors

The role of the effector kinases PKA and PKC in desensitization of the hVIP2/PACAP R was investigated. COS7 cells expressing the receptor were pretreated with the specific inhibitors H-89 (PKA) and bisindolylmaleimide (PKC) and cAMP production in response to VIP was measured. cAMP production in response to a range of VIP concentrations was enhanced to a maximum of 31.1% after pretreatment with 20 μM H-89, although there was little variation in EC$_{50}$ values (FIG. 2). Pretreatment with bisindolylmaleimide caused a small reduction in cAMP levels at high concentrations. These data suggest that PKA (but not PKC) plays a role in the desensitization of the hVIP$_2$/PACAP R.

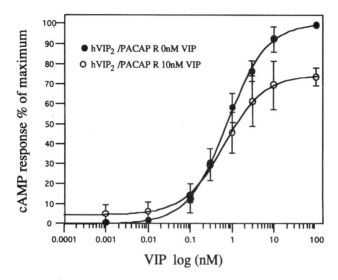

FIGURE 1. Desensitization of a hVIP₂/PACAP receptor bearing HEK293 cell line. The graph shows cAMP responses to increasing concentrations of VIP in the presence of phosphodiesterase inhibitor isobutylmethylxanthine (IBMX) with or without the pretreatment with 10 nM VIP for 20 min. Values are expressed as means (standard deviations from these means are indicated by the error bars) of five and four experiments in the untreated and treated experiments, respectively. The values are expressed as a percentage of the maximum response of the control and fitted by least-squares analysis to an equation for an asymmetric sigmoid. The maximum cAMP response in the VIP treated is $74.5 \pm 6.0\%$ ($N=4$) of that of the untreated. The maximum cAMP responses were $1,070 \pm 233$ pmol/0.5×10^6 cells for the VIP treated and 687 ± 410 pmol/0.5×10^6 cells for the untreated. Cellular cAMP levels were measured by radioimmunoassay as described elsewhere.[10]

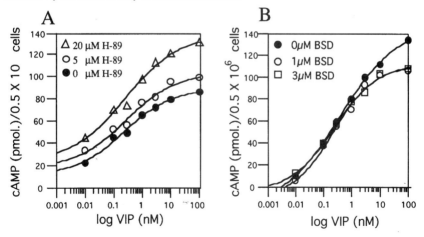

FIGURE 2. The PKA inhibitor H-89 but not the PKC inhibitor BSD enhanced VIP-stimulated cAMP production in COS7 cells expressing the hVIP₂/PACAP receptor. COS7 cells transiently transfected with hVIP₂/PACAP R cDNA were exposed to either H-89 (**A**) or to BSD (**B**) for 5 min before exposure to increasing amounts of VIP for 20 min in the presence of IBMX. Cellular cAMP levels were measured by radioimmunoassay as described elsewhere.[10]

Characterization of the hVIP$_2$/PACAP R by Western Blot Analysis

Western blot analysis of membranes from COS7 and CHO cells stably expressing the hVIP$_2$/PACAP R-HA revealed that two major immunoreactive proteins of 44 and 66 kD were expressed in COS7 cells and a single 66 kD protein in CHO cells (FIG. 3). No immunoreactive proteins were detected in cells not expressing the tagged receptor. Digestion with N-glycosidase F (PNGase F), which cleaves all of the N-linked sugars, revealed a single immunoreactive protein of 44 kD in both cell types, indicating that both forms of receptor have identical polypetide chains. However, exposure to endoglycosidase H (Endo H) did not modify the electrophoretic pattern, suggesting the absence of significant amounts of mannose-rich, N-linked carbohydrates.

Phosphorylation of the hVIP$_2$/PACAP Receptor

We investigated the ability of VIP to induce phosphorylation of the hVIP$_2$/PACAP R-HA expressed in both COS7 and HEK293 cells. Cells were incubated in media containing 200 μCi/ml [β-^{32}P]orthophosphate at 37°C before treatment with 100 nM VIP for 20 min. The receptor was immunoprecipitated and deglycosylated as described in *Methods*. Western blotting and autoradiography were used to analyze receptor phosphorylation. A small amount of basal phosphorylation of the receptor was observed. Treatment with VIP resulted in a significant increase in the incorporation of radioactivity into the 44 kD band identified as the deglycosylated receptor (FIG. 4). The VIP-induced incorporation of

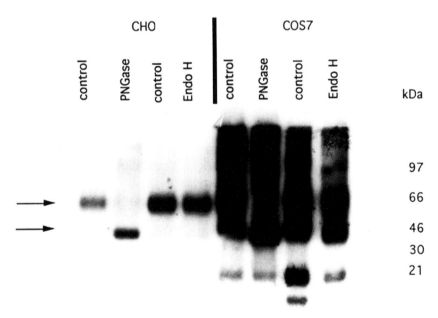

FIGURE 3. Characterization of the hVIP$_2$/PACAP R by Western blot analysis. COS7 and CHO membranes containing hVIP$_2$/PACAP R-HA were solubilized and treated with PNGase F or Endo H. The receptor was then detected after SDS-PAGE and Western blotting by antibodies specific to the HA epitope. The upper arrow indicates the mature hVIP$_2$/PACAP R-HA and the lower arrow the deglycosylated receptor.

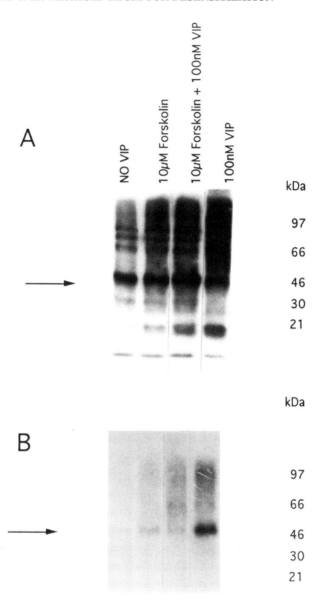

FIGURE 4. Phosphorylation of the human VIP$_2$ receptor is induced by agonist occupancy. HEK293 cells stably transfected with hVIP$_2$/PACAP R-HA cDNA were labeled in media containing 200 μCi/ml [β-^{32}P]orthophosphate for 2 h. Forskolin and VIP were added 20 min before the end of this incubation. The labeled receptor was immunoprecipitated with antibodies specific to the HA epitope and deglycosylated. The precipitated receptor was then detected after SDS-PAGE by immunoblotting (**A**) and the phosphorylated receptor by autoradiography (**B**). The arrows indicate the position of the immunoprecipitated and phosphorylated receptor. The results shown are representative of three independent experiments. Similar results were obtained with COS7 cells transiently transfected with hVIP$_2$/PACAP R-HA cDNA.

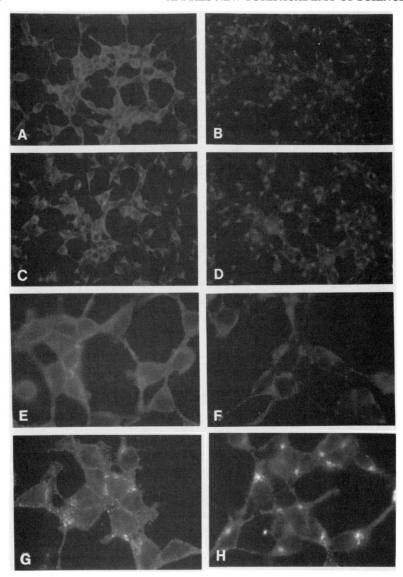

FIGURE 5. Internalization of the hVIP$_2$/PACAP R. Localization of the hVIP$_2$/PACAP R-HA in untreated HEK293 cells at × 40 magnification (**A**) and × 100 magnification (**E**). Cells were pretreated for 30 min at 37°C with VIP at 0.1 μM (**B**), 1 μM (**C**), and 10 μM (**D**) (× 40 magnification). The time course of agonist-induced internalization was investigated by pretreatment of cells at 37°C with 10 μM VIP for 5 min (**F**), 10 min (**G**), and 30 min (**H**) (× 100 magnification).

radioactive phosphate into the receptor protein was detectable after 2 min, the shortest time analyzed, and reached apparent saturation in less than 10 min (data not shown). Receptor-independent activation of PKA via stimulation of adenylyl cyclase with forskolin (100 μM), induced phosphorylation above basal levels, but to a lesser degree than VIP. These data indicate that receptor occupancy promotes phosphorylation. The addition of both VIP and forskolin together induced receptor phosphorylation at a level lower than that of VIP alone. These data suggest that there is more than one mechanism of phosphorylation. This gives rise to the possibility that receptor phosphorylation involves a specific G protein–coupled receptor kinase as well as PKA. G protein–coupled receptor kinases phosphorylate only agonist-occupied, active receptors and therefore promote homologous desensitization in contrast to the heterologous desensitization demonstrated by effector kinases such as PKA.

Internalization of the hVIP$_2$/PACAP Receptor

Internalization of the hVIP$_2$/PACAP R was investigated in HEK293 cells using immunocytochemistry and conventional fluorescence microscopy. In HEK293 cells stably expressing the hVIP$_2$/PACAP R-HA, the anti-HA antibody revealed intense staining at the cell membrane. This was reduced following exposure to VIP for 30 min at 37°C and fluorescent staining was predominant at an intracellular perinuclear site (FIG. 5). This internalization of receptor was dependent on the concentration of VIP, incubation time, and temperature (data not shown). Staining was not observed in untransfected HEK293 cells or cells expressing the untagged hVIP$_2$/PACAP receptor.

CONCLUSIONS

Our data show that the hVIP2/PACAP receptor is desensitized in response to exposure to VIP. Furthermore, we have shown that the receptor undergoes both agonist-induced phosphorylation and internalization and it is likely that both these processes are involved in its desensitization. The observation that agonist-induced phosphorylation is inhibited by the receptor-independent activation of PKA, via forskolin, also suggests the possible involvement of a G protein–coupled receptor kinase as well as PKA. Further studies are being carried out to delineate the roles and mechanisms of phosphorylation and internalization in the desensitization of the hVIP$_2$/PACAP R and to ascertain their relative contribution to the process. Experiments using a series of C-terminal truncated receptors will focus on identifying phosphorylation sites and endocytic signals. Other experiments are being carried out to determine the precise intracellular location of the internalized receptor.

REFERENCES

1. SEGRE, G. V. & S. R. GOLDRING. 1993. Trends Endocrinol. Metab. **4**: 309–314.
2. CHABRE, O., B. R. CONKLIN, H. Y. LIN, H. F. LODISH, E. WILSON, H. E. IVES, L. CATANZARITI, B. A. HEMMINGS & H. R. BOURNE. 1992. Mol. Endocrinol. **6**: 551–556.
3. ABOU-SAMRA, A.-B. B., H. JUEPPNER, T. FORCE, M. W. FREEMAN, X.-F. KONG, E. SCHIPANI, P. URENA, J. RICHARDS, J. V. BONVENTRE, J. T. POTTS, H. M. KRONENBERG & G. V. SEGRE. 1992. Proc. Natl. Acad. Sci. USA **89**: 2732–2736.
4. FORCE, T., J. V. BONVENTRE, M. R. FLANNERY, A. H. GORN, M. YAMIN & S. R. GOLDRING. 1992. Am. J. Physiol. **262**: 1110–1115.
5. WHEELER, M. B., M. LU, J. S. DILLON, X.-H. LENG, C. CHEN & A. E. BOYD III. 1993. Endocrinology **133**: 57–62.
6. SPENGLER, D., C. WAEBER, C. PANTALONI, F. HOLSBOER, J. BOCKAERT, P. H. SEEBURG & L. JOURNOT. 1993. Nature **365**: 170–175.
7. COUVINEAU, A., C. ROUYER-FESSARD, T. VOISIN & M. LABURTHE. 1990. Eur. J. Biochem. **187**: 605–609.

8. LUTZ, E. M., W. J. SHEWARD, K. M. WEST, A. J. MORROW, G. FINK & A. J. HARMAR. 1993. FEBS Lett. **334**: 3–8.
9. SHEWARD, W. J., E. M. LUTZ & A. J. HARMAR. 1995. Neuroscience **67**: 409–418.
10. MORROW, J. A., E. M. LUTZ, K. M. WEST, G. FINK & A. J. HARMAR. 1993 FEBS Lett. **329**: 99–105.
11. MENDELSON, S. C., C. F. MORRISON, J. MCALLISTER, J. M. PATERSON, S. P. DOBSON, P. K. MULDERRY & J. P. QUINN. 1995. Neuroscience **65**: 837–847.
12. HARMAR, A. J. *et al*. 1998. Pharmacol. Rev. **50**: 265–270.

Rat Aortic Smooth-Muscle Cell Proliferation Is Bidirectionally Regulated in a Cell Cycle–Dependent Manner via PACAP/VIP Type 2 Receptor[a]

ATSURO MIYATA,[b] KUMI SATO, JUN HINO, HIROKI TAMAKAWA, HISAYUKI MATSUO, AND KENJI KANGAWA

Department of Biochemistry, National Cardiovascular Center Research Institute, Suita, Osaka 565, Japan

ABSTRACT: In the cardiovascular system, vasoactive intestinal polypeptide (VIP) and pituitary adenylate cyclase activating polypeptide (PACAP) have been well characterized as potent vasodepressors or vasodilators. However, their pathophysiological implication in proliferation of vascular smooth muscle cells has not yet been elucidated. In the present study, we have first identified PACAP/VIP type 2 receptor as a dominant type in rat vascular smooth muscle cell (VSMC) by RT-PCR. PACAP and VIP increased cyclic AMP accumulation with similar potency. In 24-h [³H]thymidine incorporation assay, PACAP or VIP exhibited a suppressive effect on the DNA synthesis of rat VSMC stimulated by serum when added at the late G_1 phase. In contrast, when added at G_0/G_1 phase of the cell cycle, PACAP or VIP enhanced the serum-induced DNA synthesis. In 24-h incubation, PACAP alone has little mitogenic activity. However, when incubated up to 48 h, PACAP stimulated significantly the DNA synthesis and the cell proliferation of rat VSMC. These results suggest that PACAP and VIP regulate the proliferation of rat VSMC by enhancing or suppressing in a cell cycle–dependent manner and induce delayed mitogenesis and cell proliferation.

Pituitary adenylate cyclase activating polypeptide (PACAP),[1,2] a novel peptide isolated from ovine hypothalamus, has structural similarities to vasoactive intestinal polypeptide (VIP). PACAP and VIP belong to the secretin/glucagon family and share three types of specific receptors—PACAP receptor, PACAP/VIP type 1 receptor, and PACAP/VIP type 2 receptor.[3–10] In the cardiovascular system, PACAP and VIP have been well characterized as potent vasodepressors or vasodilators.[11–13] As for other effects on vascular smooth muscle cells (VSMC), several groups reported the growth inhibitory activity of PACAP or VIP.[14–16] However there is also a contradictory report of growth-promoting activity.[17] Thus, their pathophysiological implication in proliferation of VSMC has not yet been elucidated. The proliferation of VSMC is thought to play an important role in the development of hypertension and atherosclerosis. In the present study, we investigated the effect of PACAP and

[a]This research was supported by Special Coordination Funds for Promoting Science and Technology (Encouragement System of COE) from the Science and Technology Agency of Japan and a Scientific Research Grant from the Ministry of Education, Science and Culture.

[b]Corresponding author: Atsuro Miyata, Department of Biochemistry, National Cardiovascular Center Research Institute, 5-7-1 Fujishiro-dai, Suita, Osaka 565, Japan; Tel.: (81)6-833-5012; Fax: (81)6-872-7485; E-mail: amiyata@ri.ncvc.go.jp

VIP on DNA synthesis and cell proliferation of rat VSMC. Herein, we show that PACAP and VIP regulate the proliferation of rat VSMC by enhancing or suppressing in a cell cycle–dependent manner and induce delayed mitogenesis and cellular proliferation.

MATERIALS AND METHODS

Cell Culture

Rat vascular smooth muscle cells (VSMC) were obtained from thoracic aorta of male Sprague-Dawley rats (8 weeks old) by the explantation method as described previously[18] and cultured in DMEM containing 10% fetal calf serum at 37°C in a humidified atmosphere at 5% CO_2–95% air. The cells were used between the fifth and tenth passage from three independent thoracic aortas in this study.

Identification of the Expressed PACAP/VIP Receptors

To examine how three different PACAP/VIP receptor subtypes are expressed in rat aortic tissue and vascular smooth muscle cells, we performed the RT-PCR. The procedure of RNA preparation and reverse transcription were performed in a manner similar to that described previously.[19] In brief, total RNA was extracted with TRIzol reagent (Life Technologies, Inc., Rockville, MD) and poly(A)$^+$ RNA was prepared with oligo(dT)30 latex (Nippon Roche, Tokyo). The first-strand cDNA syntheses were performed with random hexamer and RNase H$^-$ reverse transcriptase (Superscript II, Life Technologies, Inc.). cDNA produced from RT reactions using total RNA from rat aortic tissue and VSMC was amplified using PCR with pairs of primers specific for PACAP receptor, PACAP/VIP1 receptor, and PACAP/VIP2 receptor. For all three receptors we used primers to amplify the sequences between the putative third and seventh membrane-spanning domains of the receptors. For PACAP receptor, the primers used were PACF (5'-GTGGTGTCCAAC-TACTTCTG-3'), PACR (5'-TGGAGAGAAGGCGAATAC-3'), which would be expected to produce PCR product for the basic receptor, a single cassette insert (hip, hop1, or hop2), and a double insert (hiphop1 or hiphop2) as bands of 411 base pairs (bp), 495 bp, and 579 bp, respectively. For the PACAP/VIP1 receptor, the primers used were PV1F (5'-CCAACTTCTTCTGGCTGC-3') and PV1R (5'-CACGAAACCCTGGAAAGA-3'), which should give a PCR product of 470 bp. For PACAP/VIP2 receptor, the primers used were PV2F (5'-TGGCGAACTTCTACTGGC-3') and PV2R (5'-GGAAGGAACCAACA-CATAAC-3'), yielding a predicted PCR product 460 bp in length. Thermal cycle profile: annealing at 54–57°C for 1 min, extension at 72°C for 1.5 min, denaturation at 94°C for 50 sec, all for a total of 35 cycles. The PCR reaction mixture was electrophoresed on a 1.5% agarose gel and photographed.

Assay for cAMP

The cells were plated into 24-well plates, cultured until confluent, then washed twice and preincubated for 30 min at 37°C in DMEM containing 0.1% BSA, 50 mM HEPES and 1 mM IBMX. The cells were then stimulated by various doses of PACAP38, PACAP27, or VIP for 1 h. The reaction was terminated by the addition of 1 ml of ice-cold ethanol. The accumulated cAMP was measured by a radioimmunoassay kit.

Measurement of DNA Synthesis

The cells (2×10^4 cells/ml) were plated into 24-well plates in serum containing media. After attachment for 24 h, the cells became quiescent by serum deprivation for 2 days and were blocked in G_0/G_1 phase. The quiescent cells were treated with the indicated samples for 24 h. Six hours before harvesting, the cells were pulsed with 1 µCi of [methyl-^{3}H]thymidine (Amersham, Buckinghamshire, UK). The incubation was terminated by aspirating the medium, the cells were washed twice with PBS, detached using trypsin, and harvested to count the radioactivity. Accordingly, [^{3}H]thymidine incorporation into DNA during 24 h was measured as an indicator of DNA synthesis.

Cell Proliferation Assay

The cells were plated into a 96-well plate at a density of 3×10^3 cells/well. After attachment, the cells became quiescent through 2 days of serum deprivation and were blocked in G_0/G_1 phase. Then cells were incubated for 4 days with the indicated samples, which were added every 24 h. After 4 days, the cell proliferation assay was performed by a calorimetric method using CellTiter 96 (Promega, Madison, WI) according to the manufacturer's protocol.

RESULTS AND DISCUSSION

Characterization of the Type of PACAP/VIP Receptor To Be Expressed in Rat Vascular Smooth Muscle Cells

To characterize the receptors specific for PACAP and VIP to be expressed in VSMC, RT-PCR was performed using the pair of degenerate oligonucleotide primers corresponding to conserved regions in the third and seventh transmembrane domains of the receptors for secretin/glucagon family, as reported previously.[3–10] Amplified DNA was cloned with pBluescript (Stratagene, LaJolla, CA) and clones with the expected size were sequenced. More than 90% of the tested clones contained inserts identical to PACAP/VIP2 receptor cDNA and the rest were PTH receptor, glucagon receptor, and so on (data not shown). Then we compared the expression of the three receptor mRNAs (PACAP receptor, PACAP/VIP1 receptor, and PACAP/VIP2 receptor) by RT-PCR using the specific primers on total RNAs extracted from rat aortic tissue and VSMC. As shown in FIGURE 1, the aortic tissue preparation exhibited bands corresponding to all of the receptors. The size of the two bands for the PACAP receptor lane is consistent with expression of the short (no insert) form of the receptor and a receptor containing a single (hip or hop) cassette. The size of the bands for the PACAP/VIP1 receptor and PACAP/VIP2 receptor is in agreement with the predicted fragment size produced with the primers specific for the respective receptor. However, in the de-endothelialized aortic tissue and cultured VSMC, a clear band was observed for only PACAP/VIP2 receptor and no products corresponding to PACAP receptor and PACAP/VIP1 receptor were detected. Occasionally, we observed a very faintly stained band in the lane of PACAP receptor of de-endothelialized aortic tissue (not clearly visible in FIG. 1). The present study is an attempt to make a comparative screen for the expression of mRNAs corresponding to the three types of PACAP/VIP receptors in the cardiovascular system. It was first demonstrated that the PACAP/VIP2 receptor is expressed as a dominant type in rat VSMC, suggesting an important role for the PACAP/VIP2 receptor in the physiological function of vascular smooth muscle cells.

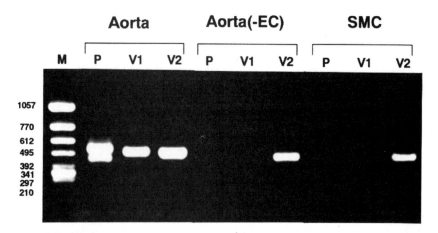

FIGURE 1. Expression of three PACAP/VIP receptor subtypes detected by RT-PCR in the vascular tissue and smooth muscle cell culture. This figure shows the ethidium bromide–stained PCR products resolved using agarose gel electrophoresis. cDNA produced from RT reactions using poly(A)⁺ RNA from normal rat aortic tissue and cultured smooth muscle cells was amplified using PCR with pairs of primers specific for PACAP receptor, PACAP/VIP1 receptor, PACAP/VIP2 receptor, as described in *Material and Methods*.

Effect of PACAP and VIP on cAMP Accumulation

We next examined the effects of PACAP and VIP on cAMP accumulation in rat VSMC. PACAP-27, PACAP-38, and VIP stimulated cAMP accumulation dose-dependently in the range of 0.01 nM and 100 nM, and exhibited almost equal potency (FIG. 2). This observation is consistent with the previous report of the PACAP/VIP2 receptor to be expressed in COS 7 cells.[10] On the other hand, neither PACAP nor VIP stimulated detectable changes in phosphoinositide hydrolysis or Ca^{2+} release from an intracellular store in rat VSMC, indicating that the PACAP/VIP2 receptor is probably not coupled to phospholipase C (PLC) in these cells (data not shown).

Time-Dependent Effects of PACAP on the Serum-Stimulated DNA Synthesis

When 10% fetal calf serum (FCS) was added at the G_0/G_1 phase of the cell cycle, DNA synthesis was markedly stimulated in rat VSMC. To determine the time course of this increase in DNA synthesis, cells in the G_0/G_1 phase were stimulated by FCS in the presence of [³H]thymidine for various lengths of time. As shown FIGURE 3, the DNA synthesis did not increase much in the first 10 h, increased slightly from 10 to 14 h, and increased most markedly from 14 to 20 h after FCS addition. These results indicate that about 14 h after FCS addition were required for the VSMC to move from the G_0 to the S phase. Then we examined the effect of PACAP on the time course (0–18 h) of DNA synthesis after the FCS stimulation, which was initiated at the G_0/G_1 phase of the cell cycle. After the start of the stimulation by FCS, the effect of PACAP changed from potentiation to inhibition in a time-dependent manner, namely, maximum potentiation was observed when PACAP was added at 0 h and maximum inhibition was observed when PACAP was added at 12 h (data not shown).

As shown in FIGURE 4A, PACAP potentiated the FCS-stimulated DNA synthesis in rat VSMC when added at the start of the stimulation (0 h). This potentiation by PACAP was

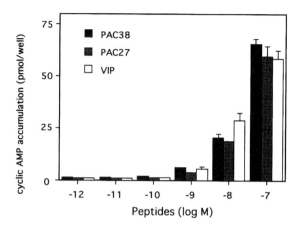

FIGURE 2. Dose-dependent effects of PACAPs and VIP on cAMP accumulation in rat VSMC. The cultured cells were stimulated by various doses of PACAP38, PACAP27, or VIP for 1 h, then cAMP accumulation was determined. Each value represents the mean ± SD of triplicate determinations.

also observed when stimulated by platelet-derived growth factor (PDGF) in FIGURE 4B. The potentiation by both PACAP and VIP was dose-dependent in the range of 0.1 nM to 100 nM with equal potency (data not shown). Further, this potentiation was mimicked by a stimulator of cAMP-dependent protein kinase (PKA), forskolin, or a cAMP analogue

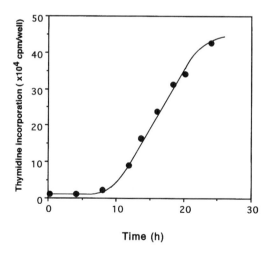

FIGURE 3. Time course of DNA synthesis in rat VSMC stimulated by FCS as a function of incubation time. Rat VSMC at subconfluency in the G_0 state were incubated with 0.5 ml of growth medium containing 10% fetal calf serum (FCS) and 1 μCi [^{3}H]thymidine for 0–24 h. The incorporated radioactivity of VSMC was measured at the indicated times of incubation. Values are averages for two separate cultures.

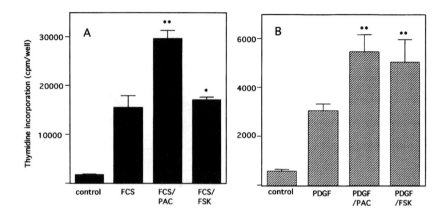

FIGURE 4. Potentiating effect of PACAP (PAC) or forskolin (FSK) on DNA synthesis of rat VSMC. Quiescent cells were stimulated by 5% FCS (**A**) or 10 ng/ml of PDGF (**B**) with PACAP38 (100 nM) or forskolin (10 μM) for 20 h and with 1 μCi [^{3}H]thymidine for an additional 4 h. Each value represents the mean ± S.D. ($N = 4$). *$p < 0.05$; **$p < 0.01$ compared with FCS or PDGF using Student's unpaired t-test.

(Bt$_2$cAMP or 8-bromo-cAMP) (data not shown). In addition, the PKA inhibitor, such as (Rp)-8-Br-cAMP or H-89, suppressed this activity (data not shown). These data indicate the involvement of the PKA pathway for potentiating the stimulated DNA synthesis at G$_0$/G$_1$ phase of the cell cycle in rat VSMC.

In contrast, when added 12 h after these stimulations, PACAP inhibited the FCS- or PDGF-stimulated DNA synthesis (FIG. 5). Thus, PACAP exhibited an inhibitory effect on the FCS-induced DNA synthesis in a cell cycle–dependent manner and the maximum effect was observed when added at the late G$_1$ phase of the cell cycle. Further, this inhibition was mimicked by forskolin, Bt$_2$cAMP, or 8-bromo-cAMP (data not shown). In addition, a PKA inhibitor, such as (Rp)-8-Br-cAMP, suppressed this inhibitory activity (data not shown), indicating the involvement of PKA pathway also for suppressing the stimulated DNA synthesis. These results are in agreement with the previous report in which PACAP inhibited the vasopressin-induced proliferation of rat aortic smooth muscle cell.[16]

Thus in the 24-h [^{3}H]thymidine incorporation assay, PACAP enhanced FCS-induced DNA synthesis when added at G$_0$/G$_1$ phase of the cell cycle. In contrast, PACAP exhibited a suppressive effect when added at the late G$_1$ phase. Interestingly the inhibitory action at the late G$_1$ phase was also reported in antiproliferative action of protein kinase C in rabbit smooth muscle cells.[18] Further, it was reported that vasopressin, which stimulates DNA synthesis when added at the G$_0$/G$_1$ phase, has a suppressing effect on DNA synthesis by inhibiting progression from late G$_1$ into the S phase of the cell cycle.[20] It should be interesting to see whether a common mechanism in such a dual regulation of cell proliferation is present or not.

The Sole Effect of PACAP in DNA Synthesis and Cellular Proliferation after 48 Hours

To evaluate the sole effect of PACAP on vascular cell growth, mitogenic assays for an extended incubation time were performed using VSMC (FIG. 6). Quiescent cells were

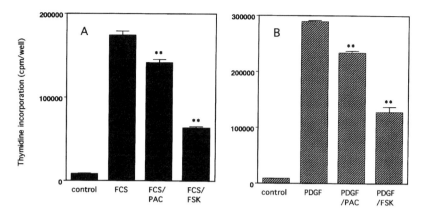

FIGURE 5. Suppressive effect of PACAP or forskolin on DNA synthesis of rat VSMC. Quiescent cells were incubated with 5% FCS (**A**) or 10 ng/ml of PDGF (**B**). At 12 h after the start of incubation, PACAP38 (100 nM) or forskolin (10 µM) was added for 8 h and with 1 µCi [³H]thymidine for an additional 4 h. Each value represents the mean ± SD ($N = 4$). *$p < 0.05$; **$p < 0.01$ compared with 5% FCS using Student's unpaired *t*-test.

incubated with 0.1 µM of PACAP or angiotensin II (AII). In these experiments, only AII stimulated increase in DNA synthesis when measured 24 h later. In the next effort to determine if PACAP effects were delayed in comparison to AII, in parallel experiments mitogenic assays were extended to 48 h. Relative levels of DNA synthesis stimulated by AII were similar to those observed after 24 h. However, by 48 h, PACAP (0.1 µM) now induced a consistent six- to eightfold stimulation of DNA synthesis in comparison to quiescent controls (FIG. 6). Thus, PACAP alone has little mitogenic activity in a 24-h incubation. However, when assays were carried out up to 48 h, PACAP induced a significant stimulation of DNA synthesis.

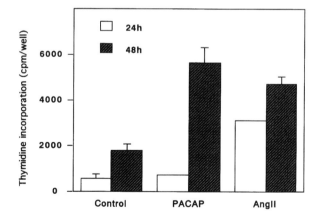

FIGURE 6. Effect of PACAP on delayed mitogenesis in cultured VSMC. Quiescent cells were incubated for 24 h *(open bars)* or 48 h *(shaded bars)* with [³H]thymidine in serum-free DMEM, PACAP (100 nM), or angiotensin II (100 nM). Each value represents the mean ± S.D. ($N = 4$).

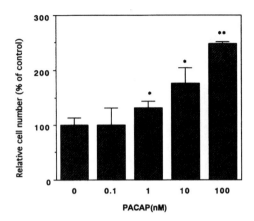

FIGURE 7. Dose-dependent effect of PACAP on proliferation of rat VSMC. Cells (3×10^3 per well) were incubated for 4 days with the indicated concentration of PACAP. Each value represents the mean ± S.D. ($N = 4$). *$p < 0.05$; **$p < 0.01$ compared with the control using Student's unpaired t-test.

We then investigated the effect of PACAP and VIP on cell proliferation of rat VSMC. After a 4-day incubation, PACAP induced dose-dependent cellular proliferation of VSMC (FIG. 7). Since delayed mitogenesis requires the increased expression of the new gene product before initiation of DNA synthesis, there is a possibility that PACAP might stimulate the expression of some endogenous growth factors.

In summary, we demonstrated that the PACAP/VIP type 2 receptor gene is dominantly expressed in rat VSMC by RT-PCR. PACAP and VIP regulate the proliferation of rat VSMC by enhancing or suppressing in a cell cycle–dependent manner and induce delayed mitogenesis and cellular proliferation. Further investigations concerning the intracellular signaling mechanism involved in such a dual regulation and the mechanism of delayed mitogenesis are in progress in our laboratory.

REFERENCES

1. MIYATA, A., L. JIANG, R. R. DAHL, C. KITADA, K. KUBO, M. FUJINO, N. MINAMINO & A. ARIMURA. 1990. Isolation of a neuropeptide corresponding to the N-terminal 27 residues of the pituitary adenylate cyclase activating polypeptide with 38 residues (PACAP38). Biochem. Biophys. Res. Commun. **170**: 643–648.
2. ARIMURA, A. 1992. Pituitary adenylate cyclase activating polypeptide (PACAP)–Discovery and current status of research. Regul. Pept. **37**: 287–303.
3. ISHIHARA, T., R. SHIGEMOTO, K. MORI, K. TAKAHASHI & S. NAGATA. 1992. Functional expression and tissue distribution of a novel receptor for vasoactive intestinal polypeptide. Neuron **8**: 811–819.
4. HASHIMOTO, H., T. ISHIHARA, R. SHIGEMOTO, K. MORI & S. NAGATA. 1993. Molecular cloning and tissue distribution of a receptor for pituitary adenylate cyclase-activating polypeptide. Neuron **11**: 333–342.
5. SPENGLER, D., C. WAEBER, C. PANTALONI, F. HOLSBOER, J. BOCKAERT, P. H. SEEBURG & L. JOURNOT. 1993. Differential signal transduction by five splice variants of the PACAP receptor. Nature **365**: 170–175.
6. HOSOYA, M., H. ONDA, K. OGI, Y. MASUDA, Y. MIYAMOTO, T. OHTAKI, H. OKAZAKI, A. ARIMURA & M. FUJINO. 1993. Molecular cloning and functional expression of rat cDNAs encoding the receptor for pituitary adenylate cyclase activating polypeptide (PACAP). Biochem. Biophys. Res. Commun. **194**: 133–143.

7. PISEGNA, J. R. & S. A. WANK. 1993. Molecular cloning and functional expression of the pituitary adenylate cyclase-activating polypeptide type I receptor. Proc. Natl. Acad. Sci. USA **90**: 6345–6349.
8. MORROW, J. A., E. M. LUTZ, K. M. WEST, G. FINK & A. J. HARMAR. 1993. Molecular cloning and expression of a cDNA encoding a receptor for pituitary adenylate cyclase activating polypeptide (PACAP). FEBS Lett. **329**: 99–105.
9. INAGAKI, N., H. YOSHIDA, M. MIZUTA, N. MIZUNO, Y. FUJII, T. GONOI, J. MIYAZAKI & S. SEINO. 1994. Cloning and functional characterization of a third pituitary adenylate cyclase-activating polypeptide receptor subtype expressed in insulin-secreting cells. Proc. Natl. Acad. Sci. USA **91**: 2679–2683.
10. LUTZ, E. M., W. J. SHEWARD, K. M. WEST, J. A. MORROW, G. FINK & A. J. HARMAR. 1993. The VIP2 receptor: molecular characterization of a cDNA encoding a novel receptor for vasoactive intestinal peptide. FEBS Lett. **334**: 3–8.
11. WARREN, J. B., L. E. DONNELLEY, S. CULLEN, B. E. ROBERTSON, M. A. GHATEI, S. R. BLOOM & J. MACDERMOT. 1991. Pituitary adenylate cyclase-activating polypeptide: a novel, long-lasting, endothelium-independent vasorelaxant. Eur. J. Pharmacol. **197**: 131–134.
12. GANZ, P., A. W. SANDROCK, S. C. LANDIS, J. LEOPOLD, M. A. GIMBRONE & R. W. ALEXANDER. 1986. Vasoactive intestinal peptide: vasodilatation and cyclic AMP generation. Am. J. Physiol. **250**: H755–H760.
13. MURTHY, K. S. & G. M. MAKHLOUF. 1994. Vasoactive intestinal peptide/pituitary adenylate cyclase activating peptide-dependent activation of membrane-bound NO synthase in smooth muscle mediated by pertussis toxin-sensitive Gil-2. J. Biol. Chem. **269**: 15977–15980.
14. ASSENDER, J. W., K. M. SOUTHGATE, M. B. HALLETT & A. C. NEWBY. 1992. Inhibition of proliferation, but not of Ca^{2+} mobilization, by cyclic AMP and GMP in rabbit aortic smooth muscle cells. Biochem. J. **288**: 527–532.
15. HULTGARDH-NILSSON, A., J. NILSSON, B. JONZON & C. J. DALSGAARD. 1988. Growth-inhibitory properties of vasoactive intestinal polypeptide. Regul. Pept. **22**: 267–274.
16. OISO Y., J. KOTOYORI, T. MURASE, Y. ITO & O. KOZAWA. 1993. Effect of pituitary adenylate cyclase activating polypeptide on vasopressin-induced proliferation of aortic smooth muscle cells: comparison with vasoactive intestinal polypeptide. Biochem. Cell Biol. **71**: 156–161.
17. MITSUHASHI, M. & D. G. PAYAN. 1987. The mitogenic effects of vasoactive neuropeptides on cultured smooth muscle cell lines. Life Sci. **40**: 853–861.
18. KARIYA, K., Y. FUKUMOTO, T. TSUDA, T. YAMAMOTO, Y. KAWAHARA, H. FUKUZAKI & Y. TAKAI. 1987. Antiproliferative action of protein kinase C in cultured rabbit aortic smooth muscle cells. Exp. Cell. Res. **173**: 504–514.
19. MIYATA, A. S. HARA, C. YOKOYAMA, H. INOUE, V. ULLRICH & T. TANABE. 1994. Molecular cloning and expression of human prostacyclin synthase. Biochem. Biophys. Res. Commun. **200**: 1728–1734.
20. MURASE, N., O. KOZAWA, M. MIWA, H. TOKUDA, J. KOTOYORI, K. KONDO & Y. OISO. 1992. Regulation of proliferation by vasopressin in aortic smooth muscle cells: function of protein kinase C. J. Hypertens. **10**: 1505–1511.

Identification of Binding Domains of Pituitary Adenylate Cyclase Activating Polypeptide (PACAP) for its Type 1 Receptor by Photoaffinity Labeling[a]

YONG-JIANG CAO,[b] ELZBIETA KOJRO,[b] MAREK JASIONOWSKI,[c] LESZEK LANKIEWICZ,[c] ZBIGNIEW GRZONKA,[c] AND FALK FAHRENHOLZ[b,d]

[b]Max-Planck-Institut für Biophysik, Kennedyallee 70, 60596 Frankfurt am Main, Germany

[c]Department of Chemistry, University of Gdansk, Sobieskiego 18, 80-952 Gdansk, Poland

ABSTRACT: Structure-function studies and photoaffinity labeling experiments were performed to identify residues and domains of PACAP involved in the interaction with PACAP receptors. For this purpose, a series of photoreactive analogues of PACAP(1-27) containing a photoreactive benzophenone (BP) residue in different peptide structural domains were utilized to analyze the interaction of PACAP(1-27) with pig PACAP type 1 receptors. Five PACAP derivatives were created with a photoreactive amino acid in the following peptide domains: either the disordered N-terminal or the helical C-terminal domain or a short loop region within the C-terminal helical domain of the peptide. Their receptor binding properties and efficiencies were tested on pig brain PACAP receptors. The results indicate the importance of the helical C-terminal domain of PACAP(1-27) for receptor binding affinity. Monoiodination of the photoreactive analogues did not change their binding affinities. Experiments with pig brain membranes demonstrated that the ^{125}I-labeled photoreactive analogues specifically label a protein band of M_r66,000. The efficiency of photoreactive labeling differed for the various analogues. These findings suggest that Tyr^{22} and Lys^{15} in PACAP(1-27) are located in or close to the hormone binding site of the PACAP type 1 receptor. The results provide evidence that the α-helical C-terminal region of PACAP is directly involved in receptor binding.

Pituitary adenylate cyclase activating polypeptide (PACAP) is found in two forms: PACAP(1-38) with 38 amino acids and PACAP(1-27) corresponding to the N-terminal 27 residues of PACAP(1-38). They belong to the secretin/glucagon/vasoactive intestinal peptide (VIP) family and show the highest homology (68%) to VIP. PACAP is widely distributed in mammals with the highest concentration in the central nervous system.[1,2] It is a hypophysiotropic hormone and functions also as neurotransmitter, neuromodulator, and neurotrophic factor in the central nervous system.[3,4]

[a]This work was supported by Deutsche Forschungsgemeinschaft (SFB 169) and by the Foundation of German-Polish cooperation (grant 1384/94/IS).

[d]Corresponding author: Dr. Falk Fahrenholz, Institut für Biochemie, Johannes Gutenberg-Universität Mainz, Becherweg 30, 55099 Mainz, Germany; Tel.: 0049-6131-39-5833; Fax: 0049-6131-39-5348.

At least two PACAP receptor types have been pharmacologically characterized. The type 1 receptor is distributed mainly in the central nervous system and has a high affinity for PACAP(1-38) and PACAP(1-27), but a 1,000-fold lower affinity for VIP.[5] In contrast, the type 2 receptor, which is widely distributed in peripheral tissues, recognizes PACAP and VIP with similar affinities and appears to be identical to the VIP receptor. Molecular cloning of both PACAP receptor types demonstrated that they belong to a new G protein–coupled receptor subfamily with seven transmembrane domains.[6–11] This subfamily includes receptors for secretin, glucagon, glucagon-like peptide 1 (GLP-1), VIP, PACAP, growth hormone–releasing hormone, and gastric inhibitory peptide and for corticotrophin-releasing factor, parathyroid hormone, and calcitonin.[12] These receptors share 25–50% amino acid identity among themselves and little primary sequence homology with other G protein–coupled receptors. A characteristic feature of these receptors is a relatively large amino-terminal extracellular region consisting of about 150 amino acids with several highly conserved cysteine residues. The amino-termini of rat PACAP type 1 receptor and rat GLP-1 receptor have been shown to function as a binding domain for their ligands.[13,14]

Structure-activity relations of PACAP have been extensively studied by means of amino acid substitutions and deletions in the PACAP sequence.[15–17] These investigations delineated the structural requirements for the occupancy of the PACAP receptors and adenylate cyclase activation. Therefore, highly active hormone antagonists and partial agonists for PACAP receptors were developed. However, these hormone structure-activity studies provide indirect information about the nature of the bimolecular interaction between ligands and their receptor. In the present study, we used photoaffinity labeling, which provides a method to directly identify contact regions of a receptor and its ligand. For this purpose photoreactive benzophenone (BP)–containing amino acids offer a promising approach.[18] Since amino acids of a peptide can be replaced by BP–containing moiety by chemical synthesis, well-defined photoreactive peptide ligands can be created that permit analysis of point-to-point interactions between a peptide and its macromolecular target protein.[19] This strategy has been employed successfully for photoaffinity labeling of several other receptor proteins.[20–25]

In order to define those structural elements involved in the interaction of PACAP with its type 1 receptor, we synthesized a series of photoreactive analogues of PACAP(1-27) that contain a photoreactive BP residue in different peptide structural domains: either the disordered N-terminal, the helical C-terminal domain, or a short loop region within the C-terminal helical domain of the peptide. The ligand binding properties of these analogues and their efficiencies as photoaffinity labels were tested for PACAP type 1 receptors from pig brain. The results show that the α-helical C-terminal region of PACAP is critical for receptor binding and that Lys[15] and Tyr[22] are probably in or close to the hormone binding site.

METHODS

Radioiodination of PACAP(1-27) and Its Photoreactive Analogues

The photoreactive analogues of PACAP(1-27) were synthesized by solid phase methodology using the Fmoc chemistry on the 9050 Plus Millipore Peptide Synthesizer (continuous flow). PACAP(1-27) and the photoreactive analogues were radioiodinated employing Iodogen® and was purified by HPLC as previously described for PACAP(1-27).[26] Briefly, peptide analogues dissolved in 20 μl sodium phosphate buffer (150 mM) was added to the Iodogen® reaction vessel followed by 1 mCi Na^{125}I in 10 μl of diluted NaOH solution (pH 7–11). The radioiodination was allowed to proceed for 10–15 min. The reaction was stopped by addition of 70 μl 0.1% TFA. The reaction mixture was immediately applied on a HPLC reversed-phase column (Vydac C 18 column; 250 × 4.5 mm, 5 μm). Monoiodinated ligands were separated with a linear gradient of acetonitrile from 27% to 41% for iodinated [Phe^6(pBz)]PACAP(1-27), from 30% to 36% for iodinated [Phe^{22}(pBz)]PACAP(1-27), and from 33% to 36% for iodinated [Lys^{15}(ε-pBz$_2$)]PACAP(1-27), [Lys^{20}(ε-pBz$_2$)]PACAP(1-27) and [Lys^{21}(ε-pBz$_2$)]PACAP(1-27) in 0.1% CF$_3$CO$_2$H for 50 min at a flow rate of 1 ml/min. Fractions of 0.5 ml each were collected and the radioactivity of the aliquots (10 µl of each fraction) was measured with a gamma spectrometer (Packard 5260). The tracers were neutralized and stored at -70°C. All isolated radioactive ligands had a specific activity of about 2,000 Ci/mmol.

Membrane Preparation

Plasma membranes from pig brain were prepared as previously described.[26] Briefly, the tissue was homogenized in 10 vol of buffer A (20 mM HEPES, pH 7.4, 0.32 M sucrose, 2 mM EDTA, 0.5 mg/ml bacitracin, 0.04 mg/ml soybean trypsin inhibitor, 1 mM PMSF) by using a Polytron PT-10 (setting 8) for 30 sec and then a Potter Elvjehm homogenizer. The homogenate was centrifuged at 1,000 × g for 10 min. The supernatant was recentrifuged at 50,000 × g for 50 min. The pellet was resuspended in buffer B (buffer A without sucrose) and recentrifuged. The membrane pellet was resuspended at a protein concentration of approximately 10 mg/ml in buffer B. The designated membrane fragments were frozen in aliquots and stored at –70°C. Protein content was quantified by the Bradford method using BSA as a standard.[27]

Receptor Binding Studies

Binding experiments with membrane-bound receptors were carried out at 25°C for 40 min with 0.1 nM iodinated PACAP(1-27) or its photoreactive analogues in binding buffer A (20 mM HEPES, pH 7.4, 5 mM MgCl$_2$, 1 mM EGTA, 0.5% BSA, 0.5 mg/ml bacitracin, 0.04 mg/ml soybean trypsin inhibitor, 1 mM PMSF). The ligand binding reaction was terminated by addition of 4 ml ice-cold washing buffer (20 mM HEPES, pH 7.4, 5 mM MgCl$_2$, 0.15 M NaCl) and rapidly filtered through GF/B (Whatman) filters that were presoaked in 0.3% polyethylenimine for about 4 h. The filters were washed three times and dried. The radioactivity was counted in the gamma spectrometer.

For saturation analysis, membranes were incubated for 60 min at 25°C with increasing concentrations of iodinated PACAP(1-27) or iodinated photoreactive analogues. Nonspecific binding was determined by incubation in the presence of a 1,000-fold excess of unlabeled PACAP(1-27). Competitive binding assays were carried out on membranes with 50 pM ^{125}I-labeled PACAP(1-27) and various amounts of the ligands (PACAP(1-27), [Phe^6(pBz)]PACAP(1-27), [Phe^{22}(pBz)]PACAP(1-27), [Lys^{15}(ε-pBz$_2$)]PACAP(1-27), [Lys^{20}(ε- pBz$_2$)]PACAP(1-27), and [Lys^{21}(ε-pBz$_2$)]PACAP(1-27). Results are expressed as means ± S.E. All experiments were done at least twice and each determination was performed in triplicate. The LIGAND computer program was used for the mathematical analysis of the binding data.[28]

Photoaffinity Labeling of PACAP Receptors

Photoaffinity labeling of the PACAP receptor was performed as described previously.[29] Membranes (100 µg) were incubated with 0.5 nM each of the iodinated photoreactive analogues (^{125}I-[Phe^6(pBz)]PACAP(1-27), ^{125}I-[Phe^{22}(pBz)]PACAP(1-27), ^{125}I-[Lys^{15}(ε-pBz$_2$)]PACAP(1-27), ^{125}I-[Lys^{20}(ε-pBz$_2$)]PACAP(1-27), and ^{125}I-[Lys^{21}(ε-pBz$_2$)]PACAP(1-27)) in a total volume of 200 µl binding buffer B (buffer A without BSA) for 60 min at 25°C. The reaction was then quenched by either diluting the mixture in 2 ml binding buffer B or by pelleting the membrane and resuspending it in 1 ml of the same buffer. Then the samples were photolysed. As a light source, an HBO 200 high-pressure mercury lamp (Osram) was used, which was built in a system with a reflector and a focusing lens (E. Leitz, Wetzlar). The sample solution was irradiated for 45 min in a quartz cuvette, which was placed in a cooled metal block open to the light source. A glass filter (Schott, Mainz) was used to cut off light with wavelengths shorter than 320 nm. The labeled membranes were then obtained by centrifugation. Unspecific labeling was determined in the presence of 0.25 µM PACAP(1-27). The samples were solubilized in sample buffer and applied to SDS-PAGE slab gel with a 3% stacking and a 10% polyacrylamide resolving gel. After visualizing the proteins with Coomassie Blue R-250 the gels were destained and dried. Autoradiography was performed for 1–2 days at –70°C on Hyperfilm MP films (Amersham) using one intensifying screen. For quantitative analysis of photoincorporation of the iodinated photoreactive ligands into the receptor, the gels were cut into 2-mm slices and counted in a gamma spectrometer.

RESULTS

Receptor Binding Analysis of the Photoreactive Analogues of PACAP(1-27)

A series of photoreactive analogues of PACAP(1-27), each containing a photoreactive benzophenone residue in different peptide structural domains, was utilized to analyze the interaction of PACAP(1-27) with pig PACAP type 1 receptor. FIGURE 1 shows the primary structure of the photoreactive analogues of PACAP(1-27). These photoreactive ligands were radioiodinated following the Iodogen® procedure and the reaction mixtures were separated by RP-HPLC.

The receptor binding properties of these photoreactive analogues were initially assessed by competitive displacement of ^{125}I-PACAP(1-27) binding to the PACAP type 1 receptor in pig brain membranes. As shown in FIGURE 2, binding of ^{125}I-PACAP(1-27) to porcine PACAP type 1 receptor was effectively displaced by the photoreactive analogues of PACAP(1-27). [Phe^6(pBz)]PACAP(1-27) (K_d = 1.3 nM) retained the high binding affinity of PACAP(1-27) (K_d = 0.57 nM), whereas substitutions in the C-terminal helical domain of PACAP(1-27) reduced the affinity 10- to 20-fold. The apparent dissociation constants are 4.4 nM for [Phe^{22}(pBz)]PACAP(1-27), 7.2 nM for [Lys^{15} (ϵ-pBz$_2$)]PACAP(1-27), 9.4 nM for [Lys^{20}(ϵ-pBz$_2$)]PACAP(1-27), and 7.7 nM for [Lys^{21}(ϵ-pBz$_2$)]PACAP(1-27) (TABLE 1). The results demonstrate the importance of the helical C-terminal domain of PACAP(1-27) for receptor binding affinity. The equilibrium binding of the iodinated photoreactive analogues to pig brain membranes showed that these monoiodinated photoreactive analogues retain the same high binding affinity as the non-iodinated photoreactive analogues for the porcine PACAP receptor.

Photoaffinity Labeling of PACAP Type 1 Receptors in Pig Brain Membranes

Pig brain membranes were incubated with 0.5 nM iodinated photoreactive analogue before photoactivation. The photolabeled membranes were analyzed by SDS-PAGE and autoradiography. A protein band with an apparent molecular mass of 66,000 was photolabeled by ^{125}I-[Phe^6(pBz)]PACAP(1-27) (FIG. 3, **A**) and ^{125}I-[Phe^{22}(pBz)]PACAP(1-27) (FIG. 3, **B**). The molecular mass of this labeled protein is identical to the mass of the porcine PACAP type 1 receptor determined by cross-linking with disuccinimidyl suberate, as previously described.[26,30] Complete inhibition of the labeled products with 500-fold excess of

PACAP(1-27)

His-Ser-Asp-Gly-Ile-**Phe**-Thr-Asp-Ser-Tyr-Ser-Arg-Tyr-Arg-**Lys**-Gln-Met-Ala-Val-**Lys**-**Lys**-**Tyr**-Leu-Ala-Ala-Val-Leu-NH$_2$

[Phe^6(pBz)]PACAP(1-27)

His-Ser-Asp-Gly-Ile-**Phe(pBz)**-Thr-Asp-Ser-Tyr-Ser-Arg-Tyr-Arg-Lys-Gln-Met-Ala-Val-Lys-Lys-Tyr-Leu-Ala-Ala-Val-Leu-NH$_2$

[Lys^{15}(ε-pBz$_2$)]PACAP(1-27)

His-Ser-Asp-Gly-Ile-Phe-Thr-Asp-Ser-Tyr-Ser-Arg-Tyr-Arg-**Lys(ε-Bz$_2$)**-Gln-Met-Ala-Val-Lys-Lys-Tyr-Leu-Ala-Ala-Val-Leu-NH$_2$

[Lys^{20}(ε-pBz$_2$)]PACAP(1-27)

His-Ser-Asp-Gly-Ile-Phe-Thr-Asp-Ser-Tyr-Ser-Arg-Tyr-Arg-Lys-Gln-Met-Ala-Val-**Lys(ε-Bz$_2$)**-Lys-Tyr-Leu-Ala-Ala-Val-Leu-NH$_2$

[Lys^{21}(ε-pBz$_2$)]PACAP(1-27)

His-Ser-Asp-Gly-Ile-Phe-Thr-Asp-Ser-Tyr-Ser-Arg-Tyr-Arg-Lys-Gln-Met-Ala-Val-Lys-**Lys(ε-Bz$_2$)**-Tyr-Leu-Ala-Ala-Val-Leu-NH$_2$

[Phe^{22}(pBz)]PACAP(1-27)

His-Ser-Asp-Gly-Ile-Phe-Thr-Asp-Ser-Tyr-Ser-Arg-Tyr-Arg-Lys-Gln-Met-Ala-Val-Lys-Lys-**Phe(pBz)**-Leu-Ala-Ala-Val-Leu-NH$_2$

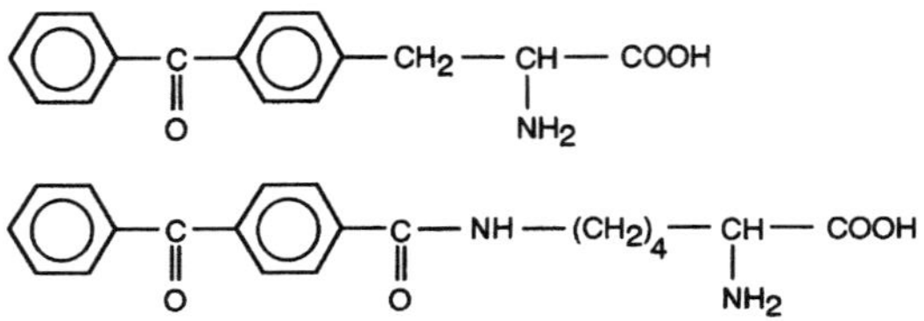

FIGURE 1. Structure of the photoreactive analogues of PACAP(1-27). Phe^6, Tyr^{22}, Lys^{15}, Lys^{20}, and Lys^{21} in PACAP(1-27) are replaced by photoreactive amino acids Phe(pBz) or Lys(ε-pBz$_2$) (**bold characters**) to give different photoreactive analogues.

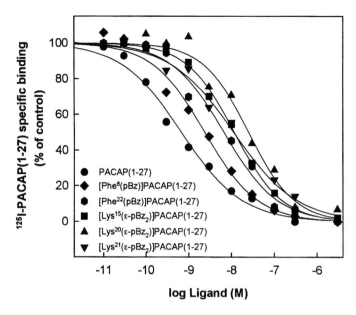

FIGURE 2. Competitive inhibition of specific ^{125}I-PACAP(1-27) binding to pig brain membranes by unlabeled PACAP(1-27) and its photoreactive analogues. Membranes from pig brain (0.1 mg/ml) were incubated with 50 pM ^{125}I-PACAP(1-27) and increasing concentrations of PACAP(1-27), [Phe^6(pBz)]PACAP(1-27), [Phe^{22}(pBz)]PACAP(1-27), [Lys^{15}(ε-pBz$_2$)]PACAP(1-27), [Lys^{20}(ε-pBz$_2$)]PACAP(1-27), and [Lys^{21}(ε-pBz$_2$)]PACAP(1-27)). Binding is expressed as percentage of specific binding obtained in the presence of ^{125}I-PACAP(1-27) alone.

PACAP(1-27) demonstrated the specificity of the photoaffinity labeling. These molecular masses include covalently bound photoreactive analogues of PACAP(1-27) with M_r 3,000. Therefore, for the native receptor itself a value of M_r 63,000 can be calculated. Similar results were obtained with PACAP analogues containing photoreactive benzophenone groups coupled to the Lys residue in position 15, 20, or 21 (^{125}I-[Lys^{15}(ε-pBz$_2$)]PACAP(1-27), ^{125}I-[Lys^{20}(ε-pBz$_2$)]PACAP(1-27), and ^{125}I-[Lys^{21}(ε-pBz$_2$)]PACAP(1-27)).

TABLE 1. Binding affinities and photolabeling efficiencies of the photoreactive analogues of PACAP(1-27) for the PACAP type 1 receptor in pig brain membranes

Ligands	K_d (nM)[a]	Efficiency (%)[b]
PACAP(1-27)	0.57 ± 0.12	-
[Phe^6(pBz)]PACAP(1-27)	1.32 ± 0.24	2
[Lys^{15}(ε-pBz$_2$)]PACAP(1-27)	7.16 ± 1.90	27
[Lys^{20}(ε-pBz$_2$)]PACAP(1-27)	9.40 ± 2.80	5
[Lys^{21}(ε-pBz$_2$)]PACAP(1-27)	7.73 ± 2.82	8
[Phe^{22}(pBz)]PACAP(1-27)	4.42 ± 0.35	45

[a]The apparent dissociation constants (K_d) values for the binding of the nonlabeled photoreactive analogues on pig brain membranes were determined by competitive binding experiments with ^{125}I-PACAP(1-27).

[b]Photoaffinity labeling efficiency refers to the percentage of ligand receptor complex that formed a covalent adduct after photoactivation of the ligand.

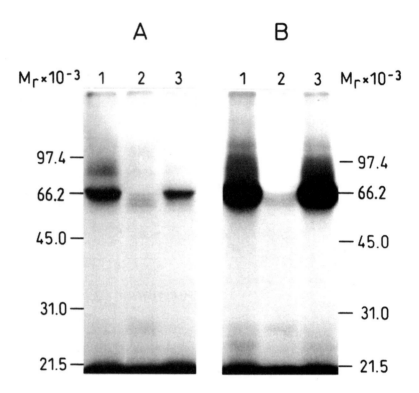

FIGURE 3. Photoaffinity labeling of PACAP receptors from pig brain membranes with [125]I-[Phe⁶(pBz)]PACAP(1-27) (**A**) or ¹²⁵I-[Phe²²(pBz)]PACAP(1-27) (**B**). Membranes from pig brain (0.1 mg) were incubated with 0.5 nM ¹²⁵I-[Phe⁶(pBz)]PACAP(1-27) or ¹²⁵I-[Phe²²(pBz)]PACAP(1-27) in the absence (lane 1, 3) and presence (lane 2) of 0.25 μM PACAP(1-27). After dilution (lane 1, 2) or centrifugation to remove the free ligand following pellet resuspension (lane 3) the samples were photolysed at wavelength > 320 nm. The samples were subjected to SDS-PAGE and autoradiography.

Quantitative measurements of photoincorporation of the iodinated photoreactive analogues into the receptor showed that the these photoreactive analogues labeled the PACAP receptor from pig brain membranes with different efficiency: [Phe²²(pBz)]PACAP(1-27) labeled the receptor with a yield of 45%, [Lys¹⁵(ε-pBz₂)]PACAP(1-27) with 27%, [Lys²¹(ε-pBz₂)]PACAP(1-27) with 8%, [Lys²⁰(ε-pBz₂)]PACAP(1-27) with about 5%, and [Phe⁶(pBz)]-PACAP(1-27) with 2% (TABLE 1). These results indicate that these photoreactive analogues interact distinctly with the PACAP receptor because of the different position of the benzophenone residue.

DISCUSSION

In the present study the photoaffinity labeling approach was used to analyze in more detail the interaction of PACAP with its type 1 receptor. The conformation of PACAP in solution has been determined by circular dichroism spectroscopy and 2D ¹H nuclear magnetic resonance spectroscopy.[31] Accordingly, PACAP(1-27) has a disordered N-terminal domain consisting of eight amino acids, followed by an α-helical structure stretching from Ser[9] to Val[26], which contains a discontinuity between Lys[20] and Lys[21]. In order to identify domains of the PACAP molecule that are involved in receptor binding, a series of photoreactive analogues of PACAP(1-27) were synthesized. Two different modes of incorporating the photoactivatable benzophenone (BP) residue were employed: Phe[6] or Tyr[22] within the PACAP sequence were replaced by *p*-benzoyl-L-phenylalanine (Phe(pBz)) and Lys[15] or Lys[20] or Lys[21] were acylated at their ε-amino group by *p*-benzoylbenzoic acid, thus creating five PACAP derivatives with a photoreactive amino acid in the following peptide domains: either the disordered N-terminal or the helical C-terminal domain or a short loop region within the C-terminal helical domain of the peptide. The BP moiety in Phe(pBz) is closer to the peptide backbone, as compared to *p*-benzoylbenzoyl-L-lysine (Lys(ε-pBz$_2$)) with its flexible aliphatic chain.[32] These photoreactive analogues, with their BP moiety in different positions of the peptide chain and with different distances to the peptide backbone, should allow a detailed analysis of ligand-receptor contact points.

The ligand binding properties and the yields of photoincorporation of these peptides into the PACAP type 1 receptors were determined. Substitution of Phe[6] in the disordered N-terminal domain had only a small effect on the binding affinity of the peptide while substitution of one of three Lys residues and Tyr[22] in the α-helical domain reduced the binding affinity about 10- to 20-fold. However, potential ligands for photoaffinity scanning have been estimated to have affinities in the 1–50 nM range.[32] Therefore, all of the photoreactive analogues of PACAP(1-27) are suitable candidates for the photoaffinity scanning studies. Despite of their lower receptor affinity [Phe[22](pBz)]PACAP(1-27) and [Lys[15](ε-pBz$_2$)]PACAP(1-27) labeled the PACAP type 1 receptor in pig brain membranes with much higher efficiency than [Phe[6](pBz)]PACAP(1-27), while [Lys[20](ε-pBz$_2$)]PACAP(1-27) and [Lys[21](ε-pBz$_2$)]PACAP(1-27), which have the photoreactive BP residue in the short loop region within C-terminal helical domain, showed a moderate efficiency. These findings provide evidence that Lys[15] and Tyr[22] in PACAP are important for the high affinity binding to the receptor and are probably located in or close to the receptor binding site. The importance of the residue in position 22 for receptor interaction has been shown for the related hormone VIP, where substitution of Tyr[22] by Ala resulted in a drastic loss of binding affinity.[33] Thus it appears most likely that the C-terminal α-helical conformation is retained in the hydrophobic environment at or near to the membrane-bound receptor. The involvement of the C-terminus for receptor binding is also supported by the fact that even the highly truncated PACAP fragment 14-27 is able to bind and initiate the signal transduction pathway.[16]

SUMMARY

Structure-activity studies and photoaffinity labeling experiments were performed to identify residues and domains of PACAP involved in the interaction with PACAP receptors. For this purpose, a series of photoreactive analogues of PACAP(1-27) containing a photoreactive benzophenone (BP) residue in different peptide structural domains were utilized to analyze the interaction of PACAP(1-27) with pig PACAP type 1 receptor. Two different modes of incorporating the BP residue were employed: Phe[6] or Tyr[22] within the PACAP sequence was replaced by *p*-benzoyl-L-phenylalanine (Phe(pBz)) and Lys[15] or Lys[20] or Lys[21] was acylated at their ε-amino group by *p*-benzoylbenzoic acid, thus creating five PACAP derivatives with a photoreactive amino acid in the following peptide domains: either the disordered N-terminal or the helical C-terminal domain or a short loop region within the C-terminal helical domain of the peptide.The receptor binding properties and the efficiencies of these peptide analogues as photolabels were tested for pig brain PACAP receptors. The results indicate the importance of the helical C-terminal domain of

PACAP(1-27) for receptor binding affinity. Monoiodination of the photoreactive analogues did not change the binding affinity of the photoreactive analogues. Results from photolabeling experiments with pig brain membranes demonstrated that the ^{125}I-labeled photoreactive analogues specifically label a protein band of M_r 66,000. The efficiency of photoreactive labeling differed from the various photoreactive analogues. Despite the lower receptor affinity, [Phe^{22}(pBz)]PACAP(1-27) and [Lys^{15}(ε-pBz$_2$)]PACAP(1-27) labeled the PACAP type 1 receptor in pig brain membranes with much higher efficiency than [Phe^6(pBz)]PACAP(1-27), while [Lys^{20}(ε-pBz$_2$)]PACAP(1-27) and [Lys^{21}(ε-pBz$_2$)]PACAP(1-27) showed a moderate efficiency. These findings suggest that Tyr^{22} and Lys^{15} in PACAP(1-27) are located in or close to the hormone binding site of the PACAP type 1 receptor. The results provide evidence that the α-helical C-terminal region of PACAP is directly involved in receptor binding.

REFERENCES

1. KOVES, K., A. ARIMURA, A. SOMOGYVARI-VIGH, S. VIGH & J. MILLER. 1990. Immunohistochemical demonstration of a novel hypothalamic peptide, pituitary adenylate cyclase activating polypeptide, in the ovine hypothalamus. Endocrinology **127:** 264–271.
2. ARIMURA, A., A. SOMOGYVARI-VIGH, A. MIYATA, K. MIZUNO, D. H. COY & C. KITADA. 1991. Tissue distribution of PACAP as determined by RIA: highly abundant in the rat brain and testes. Endocrinology **129:** 2787–2789.
3. ARIMURA, A., A. SOMOGYVARI-VIGH, C. WEILL, R. C. FIORE, I. TATSUNO, V. BAY & D. E. BRENNEMAN. 1994. PACAP functions as a neurotrophic factor. Ann. N. Y. Acad. Sci. **739:** 228–243.
4. ARIMURA, A. & S. SHIODA. 1995. Pituitary adenylate cyclase activating polypeptide (PACAP) and its receptors: neuroendocrine and endocrine interaction. Front. Neuroendocrinol. **16:** 53–88.
5. SHIVERS, B. D., T. J. GORCS, P. E. GOTTSCHALL & A. ARIMURA. 1991. Two high affinity binding sites for pituitary adenylate cyclase activating polypeptide have different tissue distributions. Endocrinology **128:** 3055–3065.
6. SPENGLER, D., C. WAEBER, C. PANTALONI, F. HOLSBOER, J. BOCKAERT, P. H. SEEBURG & L. JOURNOT. 1993. Differential signal transduction by five splice variants of the PACAP receptor. Nature **365:** 170–175.
7. HOSOYA, M., H. ONDA, K. OGI, Y. MASUDA, Y. MIYAMOTO, T. OHTAKI, H. OKAZAKI, A. ARIMURA & M. FUJINO. 1993. Molecular cloning and functional expression of rat cDNAs encoding the receptor for pituitary adenylate cyclase activating polypeptide (PACAP). Bichem. Biophys. Res. Commun. **194:** 133–143.
8. PISEGNA, J. R. & S. A. WANK. 1993. Molecular cloning and functional expression of the pituitary adenylate cyclase-activating polypeptide type 1 receptor. Proc. Natl. Acad. Sci. USA **90:** 6345–6349.
9. ISHIHARA, T., R. SHIGEMOTO, K. MORI, K. TAKAHASHI & S. NAGATA. 1992. Functional expression and tissue distribution of a novel receptor for vasoactive intestinal polypeptide. Neuron **8:** 811–819.
10. SREEDHARAN, S. P., D. R. PATEL, J.-X. HUANG & E. J. GOETZL. 1993. Cloning and functional expression of a human neuroendocrine vasoactive intestinal peptide. Biochem. Biophys. Res. Commun. **193:** 546–553.
11. USDIN, T. B., T. I. BONNER & E. MEZEY. 1994. Two receptors for vasoactive intestinal polypeptide with similar specificity and complementary distributions. Endocrinology **135:** 2662–2680.
12. SEGRE, G. V. & S. R. GOLDRING. 1993. Receptors for secretin, calcitonin, parathyroid hormone (PTH)/PTH-related peptide, vasoactive intestinal peptide, glucagonlike peptide 1, growth hormone-releasing hormone, and glucagon belong to a newly discovered G-protein-linked receptor family. Trends Endocrinol. Metab. **4:** 309–314.
13. CAO, Y.-J., G. GIMPL & F. FAHRENHOLZ. 1995. The amino-terminal fragment of the the pituitary adenylate cyclase activating polypeptide (PACAP) receptor functions as a high affinity PACAP binding domain. Biochem. Biophys. Res. Commun. **212:** 673–680.
14. WILMEN, A., B. GÖKE & R. GÖKE. 1996. The isolated N-terminal extracellular domain of the glucagon-like peptide-1 (GLP)-1 receptor has intrinsic binding activity. FEBS Lett. **398:** 43–47.

15. ROBBERECHT, P., P. GOURLET, P. DE NEEF, M.-C. WOUSSEN-COLLE, M.-C. VANDERMEERS-PIRET, A. VANDERMEERS & J. CHRISTOPHE. 1992. Structural requirements for the occupancy of pituitary adenylate-cyclase-activating-peptide (PACAP) receptors and adenylate-cyclase activation in human neuroblastoma NB-OK-1 cell membranes. Eur. J. Biochem. **207:** 239–246.

16. VANDERMEERS, A., S. VANDENBORRE, X. HOU, P. DE NEEF, P. ROBBERECHT, M.-C. VANDERMEERS-PIRET & J. CHRISTOPHE. 1992. Antagonistic properties are shifted back to agonistic properties by further N-terminal shortening of pituitary adenylate-cyclase-activating peptides in human neuroblastoma NB-OK-1 cell membranes. Eur. J. Biochem. **208:** 815–819.

17. BITAR, K. G., A. SOMOGYVARI-VIGH & D. H. COY. 1994. Cyclic lactam analogues of ovine pituitary adenylate cyclase activating polypeptide (PACAP): discovery of potent type II receptor antagonists. Peptide **15:** 461–466.

18. DORMAN, G. & G. D. PRESTWICH. 1994. Benzophenone photophores in biochemistry. Biochemistry **33:** 5661–5673.

19. WILLIANS, K. P. & S. E. SHOELSON. 1993. A photoaffinity scan maps regions of the p85 SH2 domain involved in phosphoprotein binding. J. Biol. Chem. **268:** 5361–5364.

20. BOYD, N. D., C. F. WHITE, R. CERPA, E. T. KAISER & S. E. LEEMAN. 1991. Photoaffinity labeling the substance P receptor using a derivative of substance P containing *p*-benzoylphenylalanine. Biochemistry **30:** 336–342.

21. THIELE, C. & F. FAHRENHOLZ. 1993. Photoaffinity labeling of central cholecystokinin receptors with high efficiency. Biochemistry **32:** 2741–2746.

22. SHOELSON, S. E., J. LEE, C. S. LYNCH, J. M. BACKER & P. F. PILCH. 1993. Bpa[B25] insulins, photoactivable analogues that quantitatively cross-link, radiolabel, and activate the insulin receptor. J. Biol. Chem. **268:** 4085–4091.

23. SERVANT, G., G. BOULAY, R. BOSSE, E. ESCHER & G. GUILLEMENTTE. 1993. Photoaffinity labeling of subtype 2 angiotensin receptor of human myometrium. Mol. Pharmacol. **43:** 677–683.

24. ADAMS, A. E., M. PINES, C. NAKAMOTO, V. BEHAR, Q. M. YANG, R. BESSALLE, M. CHOREV, M. ROSENBLATT, M. A. LEVINE & L. J. SUVA. 1995. Probing the bimolecular interactions of parathyroid hormone and the human parathyroid hormone/parathyroid hormone-related protein receptor. 2. Cloning, characterization, and photoaffinity labeling of the recombinant human receptor. Biochemistry **34:** 10553–10559.

25. HAMPE, W., R. W. FRANK, C. SCHULZE, I. DEHNING & H. C. SCHALLER. 1996. Photoaffinity labeling of the head-activator receptor from hydra. Eur. J. Biochem. **235:** 814–820.

26. CAO, Y.-J., G. GIMPL & F. FAHRENHOLZ. 1994. Molecular structure analysis of the pituitary adenylate cyclase activating polypeptide type 1 receptor from pig brain. Biochim. Biophys. Acta **1222:** 432–440.

27. BRADFORD, M. M. 1976. A rapid and sensitive method for the quantitation of microgram quantities of protein utilizing the principle of protein-dye binding. Anal. Biochem. **72:** 248–254.

28. MUNSON, P. J. & D. RODBARD. 1980. LIGAND: a versatile computerized approach for characterization of ligand-binding systems. Anal. Biochem. **107:** 220–239.

29. CAO, Y.-J., E. KOJRO, G. GIMPL, M. JASIONOWSKI, L. LANKIEWICZ & F. FAHRENHOLZ. 1997. Photoaffinity labeling analysis of the interaction of pituitary adenylate-cyclase-activating polypeptide (PACAP) with PACAP type 1 receptor. Eur. J. Biochem. **244:** 400–406.

30. SCHÄFER, H. & W. E. SCHMIDT. 1993. Characterization and purification of the solubilized pituitary adenylate-cyclase-activating polypeptide-1 receptor from porcine brain using a biotinylated ligand. Eur. J. Biochem. **217:** 823–830.

31. WRAY, V., C. KAKOSCHKE, K. NOKIHARA & S. NARUSE. 1993. Solution structure of pituitary adenylate cyclase activating polypeptide by nuclear magnetic resonance spectroscopy. Biochemistry **32:** 5832–5841.

32. NAKAMOTO, C., V. BEHAR, K. R. CHIN, A. E. ADAMS, L. J. SUVA, M. ROSENBLATT & M. CHOREV. 1995. Probing the bimolecular interactions of parathyroid hormone and the human parathyroid hormone/parathyroid hormone-related protein receptor. 1. Design, synthesis and characterization of photoreactive benzophenone-containing analogs of parathyroid hormone. Biochemistry **34:** 10546–10552.

33. O'DONNELL, M., R. J. GARIPPA, N. C. O'NEILL, D. R. BOLIN & J. M. COTTRELL. 1991. Structure-activity studies of vasoactive intestinal polypeptide. J. Biol. Chem. **266:** 6389–6392.

The Neurotrophic Activity of PACAP on Rat Cerebellar Granule Cells Is Associated with Activation of the Protein Kinase A Pathway and *c-fos* Gene Expression[a]

D. VAUDRY, M. BASILLE, Y. ANOUAR, A. FOURNIER,[b] H. VAUDRY,[c] AND B. J. GONZALEZ

European Institute for Peptide Research (IFRMP 23), Laboratory of Cellular and Molecular Neuroendocrinology, Institut National de la Santé et de la Recherche Médicale U413, UA Centre National de la Recherche, University of Rouen, 76821 Mont-Saint-Aignan, France

Institut National de la Recherche Scientifique-Santé, University of Québec, Pointe-Claire, Québec, Canada H9R 1G6

ABSTRACT: *In vitro* studies have shown that PACAP promotes cell survival and neurite outgrowth in immature cerebellar granule cells. In the present study, we have examined the transduction pathways involved in the neurotrophic activity of PACAP. Incubation of cultured granule cells with graded concentrations of PACAP produced a dose-dependent increase in *c-fos* mRNA level. The effects of PACAP on *c-fos* gene expression and granule cell survival were both mimicked by dbcAMP but not by PMA. The maximum effect of PACAP on *c-fos* gene expression was observed after 1 h of treatment. Similar effects of the peptide on granule cell survival were observed whether the cells were continuously incubated with PACAP for 48 h or only exposed to PACAP during 1 h. The PKA inhibitor H89 significantly reduced the effect of PACAP on *c-fos* mRNA level, whereas the specific PKC inhibitor chelerytrine had no effect. These data indicate that the action of PACAP on cerebellar granule cell survival and *c-fos* gene expression are both mediated through the adenylyl cyclase/PKA pathway.

Pituitary adenylate cyclase-activating polypeptide (PACAP) is widely distributed in the central nervous system.[1] Notably, high concentrations of PACAP are found in the rat cerebellum between postnatal day 4 (P4) and P20, a period that corresponds to intense neurogenesis.[2,3] In 8-day-old rats, PACAP-binding sites are also highly expressed in the external granule cell layer (EGL),[4,5] a neuroepithelium that generates the future cerebellar granule cells.[3] PACAP-binding sites, borne by cultured immature granule cells from the EGL, correspond to functional PVR1 receptors coupled to both adenylyl cyclase (AC) and phospholipase C (PLC).[6,7] Concurrently, a number of studies conducted in various tumoral cell lines or normal cells have shown that PACAP modulates cell proliferation, outgrowth of cell processes, and protein synthesis.[8–11] Altogether, these observations suggest that PACAP may act as a neurotrophic factor involved in the control of multiplication, differentiation, and/or migration of granule cells during cerebellar development.

[a]This work was supported by grants from Institut National de la Santé et de la Recherche Médicale (U413) and the Conseil Régional de Haute-Normandie.

[c]Corresponding author: Tel.: (33) 235.14.66.24; Fax: (33) 235.14.69.46; E-mail: hubert.vaudry@univ-rouen.fr

This review will focus on the neurotrophic activity of PACAP on cultured cerebellar granule cells and on the molecular mechanisms underlying this effect.

NEUROTROPHIC EFFECT OF PACAP ON CULTURED GRANULE CELLS

Cell Survival

The ability of PACAP to promote cell survival has been studied on cultured granule cells incubated in a chemically defined culture medium that favors apoptosis.[12,13] After 48 h, most cells incubated in control medium died and very few neurites had appeared (FIG. 1A). Exposure of cultured cells to 10^{-8} M PACAP38 resulted in a marked increase in the number of living cells and the formation of a dense neurite network (FIG. 1B).[14] Quantification of living cells was performed after 2, 24, and 48 h of culture by means of a computer-assisted image analyzer. Incubation of granule cells with graded concentrations of PACAP38 (10^{-6} to 10^{-10} M) for 2 h did not cause any modification in the number of living cells. After 24 h of culture in control conditions, the number of living cells was reduced by 58% as compared to 2-h cultures. Exposure of the cells to PACAP38 induced a dose-dependent increase in the number of living cells, the maximum effect being observed at a concentration of 10^{-7} M PACAP38 (FIG. 2A). When granule cells were cultured for 48 h in control conditions, only 36% of the cells survived. Addition of PACAP38 to the incubation medium significantly increased the number of living cells but the maximum effect was achieved at a concentration as low as 10^{-9} M PACAP38. At higher doses (10^{-8} to 10^{-6} M), PACAP38 was less efficient in promoting cell survival in 48-h- than in 24-h-old cells. This effect may be ascribed to the existence of two granule cell populations whose survival would be differentially regulated by PACAP. In agreement with this hypothesis, recent studies have shown that, in serum-free medium, granule cells die through both apoptosis and necrosis, suggesting the occurrence of at least two distinct cellular pathways leading to different types of cell death.[15,16]

Incubation of cultured cells with vasoactive intestinal polypeptide (VIP; 10^{-10} to 10^{-6} M) did not significantly affect the number of living cells.[14] The fact that VIP was at least 1,000 times less potent than PACAP38 in promoting cerebellar granule cell survival indicates that the effect of PACAP is not attributable to supplementation of the minimum culture medium with amino acids. The observation that the effect of PACAP38 on cell survival was significantly attenuated in the presence of the antagonist PACAP(6-38) confirmed that PACAP exerts a receptor-mediated trophic action on granule cells.[14]

Neurite Outgrowth

In order to investigate the effect of PACAP on neurite outgrowth, the length of individual neurites borne by granule cells was measured by means of a confocal laser scanning microscope. Exposure of cultured cells to various concentrations of PACAP38 for 2 or 24 h did not significantly affect neurite outgrowth. In contrast, incubation of granule cells in graded concentrations of PACAP38 for 48 h induced a dose-dependent increase in the cumulative length of neurites, the effect of PACAP38 being significant for doses ranging from 10^{-10} to 10^{-6} M (Fig. 2B).[14] Exposure of cells to PACAP38 provoked both an increase of the length of the processes and an augmentation of the number of neurites borne by each granule cell.

Transduction Mechanisms

PACAP receptors expressed by cultured rat cerebellar granule cells are positively coupled to both AC and PLC,[6,7] two transduction pathways usually associated with activation

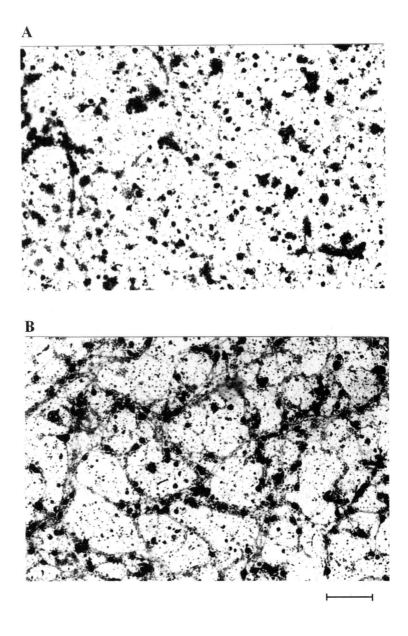

FIGURE 1. Microphotographs illustrating the effect of PACAP38 on survival and neurite outgrowth in cultured cerebellar granule cells. (**A**) Appearance of granule cells cultured in normal medium for 48 h. (**B**) Appearance of granule cells exposed to 10^{-8} M PACAP38 for 48 h. The PACAP38 solution was renewed after 24 h. Scale bar represents 50 μm.

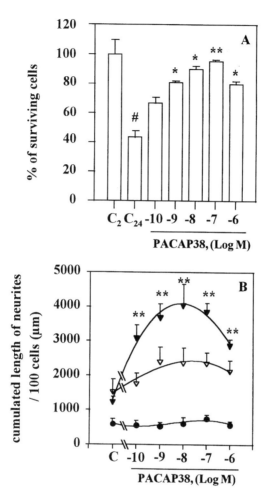

FIGURE 2. Effect of graded concentrations of PACAP38 on survival (**A**) and neurite outgrowth (**B**) of cultured cerebellar granule cells. (**A**)The cells were exposed for 24 h to 10^{-8} M PACAP38. The values are expressed as percentage of cultured cells surviving 2 h after incubation in normal medium (C_2; 100%). C_{24} represents the percentages of surviving cells after 24 h of culture in normal medium. # $p < 0.05$ *vs.* C_2;* $p < 0.05$ *vs.* C_{24}; ** $p < 0.01$ *vs.* C_{24}. (**B**) Effect of graded concentrations of PACAP38 on the cumulative length of neurites in cultured cerebellar granule cells. Granule cells were incubated with PACAP38 (10^{-10} to 10^{-6} M) during 2 h (●—●), 24 h (∇—∇), or 48 h (▼—▼). Each value represents the mean (± SEM) of at least three measurements made by two independent investigators. Representative data of three independent experiments. **$p < 0.01$.

of PVR1 receptors. To investigate the intracellular mechanisms mediating the effect of PACAP on granule cell survival, we have compared the capacity of PACAP38, the PKC activator PMA, and the PKA activator dbcAMP to promote cell survival. Incubation of granule cells with PMA did not affect cell survival and did not modify the morphological appearance of surviving cells. In contrast, the permeant cAMP analog dbcAMP mimicked the stimulatory effect of PACAP38 on cell survival (FIG. 3). These results, in agreement

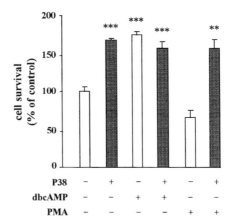

FIGURE 3. Effect of a 48-h treatment with PACAP38 (10^{-8} M), dbcAMP (10^{-3} M), and/or PMA (10^{-7} M) on granule cell survival. Each value is the mean (± SEM) of at least three measurements made by two independent investigators. Representative data of three independent experiments. $**p < 0.01$; $***p < 0.001$.

with recent data that revealed that PACAP blocks apoptosis through a PKA-dependent mechanism,[17] indicate that PACAP promotes cell survival via the adenylyl cyclase pathway and demonstrate that PKC is not implicated in the neurotrophic action of PACAP.

Exposure of granule cells to PACAP38 (10^{-8} M) caused a time-dependent increase in cell survival, the maximum effect being observed after a 1-h treatment (FIG. 4). Longer incubations with PACAP38 (2 h to 48 h) did not further increase cell survival, indicating that a short term stimulation of granule cells with PACAP38 causes activation of a specific biochemical pathway that results in a long-lasting survival of immature granule

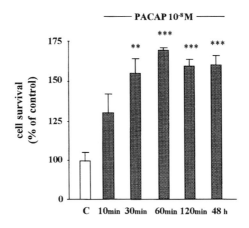

FIGURE 4. Time-course of the effect of PACAP38 on survival of cultured cerebellar granule cells. Granule cells were exposed to PACAP for durations ranging from 10 min to 48 h and cell quantification was performed 48 h after the beginning of the treatment. Each value is the mean (± SEM) of at least three measurements made by two independent investigators. Representative data of three independent experiments. $** p < 0.01$; $***p < 0.001$.

cells. This observation suggests that PACAP may stimulate early gene expression, known to mediate neurotrophic responses such as cell multiplication, cell differentiation, or programmed cell death.[18]

EFFECT OF PACAP ON *C-FOS* GENE EXPRESSION

Previous studies have shown that PACAP stimulates expression of the early gene *c-fos* in various cell types including cortical neurons and rat pancreatic carcinoma cells.[19,20] Since *c-fos* is known to interfere with cell growth,[21–23] it was conceivable that the effect of PACAP on cell survival could involve activation of *c-fos* gene expression. Incubation of cultured granule cells with graded concentrations of PACAP38 (10^{-10} to 10^{-6} M) induced a dose-dependent increase in *c-fos* mRNA level with an ED_{50} of 1.03 ± 0.08 nM, while VIP, in the same range of concentrations, had little effect on *c-fos* mRNA level.[24] Time-course experiments revealed that a 30-min incubation of granule cells with PACAP38 was sufficient to induce a significant increase in *c-fos* mRNA level with a maximum effect observed after 1-h exposure to the peptide (FIG. 5). The increase in *c-fos* mRNA evoked by PACAP38 was not blocked by cycloheximide, indicating that the effect of PACAP on *c-fos* gene expression does not require *de novo* protein biosynthesis.

The stimulatory effect of PACAP on *c-fos* mRNA levels was mimicked by incubation of the cells with dbcAMP, whereas PMA was devoid of effect. In addition, the selective protein kinase A inhibitor H89 abrogated the PACAP-evoked stimulation of *c-fos* mRNA level, whereas the PKC inhibitor chelerytrine did not affect the increase in *c-fos* mRNA provoked by PACAP. Taken together, these data demonstrate that PACAP stimulates *c-fos* gene expression through activation of the cAMP-dependent PKA pathway. Consistent with this notion, it has been demonstrated that, in transfected granule cells, PACAP induces the expression of a construct containing the full-length *c-fos* promoter by activating the PKA pathway.[25]

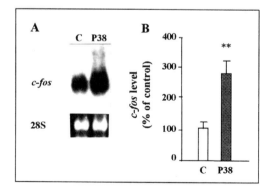

FIGURE 5. Effect of PACAP38 on *c-fos* mRNA levels. (**A**) Typical Northern blots illustrating the effect of PACAP38 (10^{-8} M). (**B**) Quantification of *c-fos* mRNA levels. Ethidium bromide–labeled 28S RNAs were used as internal standards and the mRNA levels were expressed as percentages of the control. Each value is the mean ($\pm$ SEM) of three independent experiments performed in triplicate. ** $p < 0.01$.

CONCLUSION

Both PACAP and its receptors are expressed in the developing rat brain. PACAP promotes survival, neurite outgrowth, and *c-fos* gene expression in cultured cerebellar granule cells through activation of the adenylyl cyclase pathway. Since *c-fos* is implicated in various trophic activities, the present data strongly suggest that activation of the early gene *c-fos* is involved in the neurotrophic action of PACAP in the developing cerebellum.

REFERENCES

 1. ARIMURA, A. 1992. Pituitary adenylate cyclase-activating polypeptide (PACAP): discovery and current status of research. Regul. Pept. **37**: 287–303.
 2. MASUO, Y. *et al.* 1994. Ontogeny of pituitary adenylate cyclase-activating polypeptide (PACAP) and its binding sites in the rat brain. Neurosci. Lett. **170**: 43–46.
 3. JACOBSON, M. 1991. Histogenesis and morphogenesis of cortical structures. *In* Developmental Neurobiology. M. Jacobson, Ed.: 401–452. Plenum Press. New York.
 4. BASILLE, M. *et al.* 1994. Ontogeny of pituitary adenylate cyclase-activating polypeptide (PACAP) receptors in the rat cerebellum: a quantitative autoradiographic study. Dev. Brain Res. **82**: 81–89.
 5. BASILLE, M. *et al.* 1993. Localization and characterization of PACAP receptors in the rat cerebellum during development: evidence for a stimulatory effect of PACAP on immature cerebellar granule cells. Neuroscience **57**: 329–338.
 6. BASILLE, M. *et al.* 1995. Pituitary adenylate cyclase-activating polypeptide (PACAP) stimulates adenylyl cyclase and phospholipase C activity in rat cerebellar neuroblasts. J. Neurochem. **65**: 1318–1324.
 7. FAVIT, A., U. SCAPAGNINI & P. L. CANONICO. 1995. Pituitary adenylate cyclase-activating polypeptide activates different signal transducing mechanisms in cultured cerebellar granule cells. Neuroendocrinology **61**: 377–382.
 8. HERNANDEZ, A. *et al.* 1995. Pituitary adenylate cyclase-activating peptide stimulates neurite growth in PC12 cells. Peptides **5**: 927–932.
 9. LU, N. & E. DICICCO-BLOOM. 1997. Pituitary adenylate cyclase-activating polypeptide is an autocrine inhibitor of mitosis in cultured cortical precursor cells. Proc. Natl. Acad. Sci. USA **94**: 3357–3362.
10. PESCE, M. *et al.* 1996. Pituitary adenylate cyclase-activating polypeptide (PACAP) stimulates adenylate cyclase and promotes proliferation of mouse primordial germ cells. Development **122**: 215–221.
11. WEST, A. P. *et al.* 1995. Pituitary adenylate cyclase-activating polypeptide can regulate testicular germ cell protein synthesis in vitro. J. Endocrinol. **144**: 215–223.
12. D'MELLO, S. R. *et al.* 1993. Induction of apoptosis in cerebellar granule neurons by low potassium: inhibition of death by insulin-like growth factor I and cAMP. Proc. Natl. Acad. Sci. USA **90**: 10989–10993.
13. MILLER, T. M. & E. M. JOHNSON. 1996. Metabolic and genetic analyses of apoptosis in potassium/serum-deprived rat cerebellar granule cells. J. Neurosci. **16**: 7487–7495.
14. GONZALEZ, B. J. *et al.* 1997. Pituitary adenylate cyclase-activating polypeptide promotes cell survival and neurite outgrowth in rat cerebellar neuroblasts. Neuroscience **78**: 419–430.
15. VILLALBA, M., J. BOCKAERT & L. JOURNOT. 1997. Concomitant induction of apoptosis and necrosis in cerebellar granule cells following serum and potassium withdrawal. Neuroreport **8**: 981–985.
16. WOOD, K. A., B. DIPASQUALE & R. J. YOULE. 1993. In situ labeling of granule cells for apoptosis-associated DNA fragmentation reveals different mechanisms of cell loss in developing cerebellum. Neuron **11**: 621–632.
17. VILLALBA, M., J. BOCKAERT & L. JOURNOT. 1997. Pituitary adenylate cyclase-activating polypeptide (PACAP-38) protects cerebellar granule neurons from apoptosis by activating the mitogen-activated protein kinase (MAP kinase) pathway. J. Neurosci. **17**: 83–90.
18. SHENG, M. & M. E. GREENBERG. 1990. The regulation and function of *c-fos* and other immediate early genes in the nervous system. Neuron **4**: 477–485.

19. MARTIN, J. L., D.GASSER & P. J. MAGISTRETTI. 1995. Vasoactive intestinal peptide and pituitary adenylate cyclase-activating polypeptide potentiate *c-fos* expression induced by glutamate in cultured cortical neurons. J. Neurochem. **65**: 1–9.

20. SCHÄFER, H. *et al.* 1996. PACAP stimulates transcription of *c-fos* and *c-jun* and activates the AP-1 transcription factor in rat pancreatic carcinoma cells. Biochem. Biophys. Res. Commun. **221**: 111–116.

21. HOLT, J. T. *et al.* 1986. Inducible production of *c-fos* antisense RNA inhibits 3T3 cell proliferation. Proc. Natl. Acad. Sci. USA **83**: 4794–4798.

22. MÜLLER, R. & E. F. WAGNER. 1984. Differentiation of F9 teratocarcinoma stem cells after transfer of *c-fos* proto-oncogenes. Nature **311**: 438–442.

23. RÜTHER, U. *et al.* 1987. Deregulated *c-fos* expression interferes with normal bone development in transgenic mice. Nature **325**: 412–416.

24. VAUDRY, D. *et al.* 1997. PACAP stimulates both *c-fos* gene expression and cell survival in rat cerebellar granule neurons through activation of the protein kinase A pathway. Neuroscience (Submitted for publication.)

25. CAMPARD, P. K. *et al.* 1997. PACAP type I receptor activation promotes cerebellar neuron survival through the cAMP/PKA signaling pathway. DNA Cell Biol. **16**: 323–333.

PACAP-38 Protects Cerebellar Granule Cells from Apoptosis

LAURENT JOURNOT,[a] MARTIN VILLALBA, AND JOËL
BOCKAERT

*Centre National de la Recherche Scientifique, Unité Propre de
Recherche, 9023, Mécanismes Moléculaires des Communications
Cellulaires, Centre CNRS-INSERM de Pharmacologie-Endocrinologie,
141, rue de la cardonille, 34094 Montpellier Cedex 05, France*

ABSTRACT: Pituitary adenylate cyclase–activating polypeptides (PACAP-27 and -38)
are neuropeptides of the vasoactive intestinal polypeptide (VIP)/secretin/glucagon
family. PACAP receptors are expressed in different brain regions including the cerebellum. We used primary culture of rat cerebellar granule neurons to study the effect
of PACAP-38 on apoptosis induced by potassium deprivation. We demonstrated that
serum and potassium withdrawal induces a mixture of apoptosis and necrosis rather
than apoptosis only. We showed that PACAP-38 increased survival of cerebellar neurons in a dose-dependent manner by specifically decreasing the extent of apoptosis
estimated by DNA fragmentation. PACAP-38 induced activation of the extracellular
signal–regulated kinase (ERK)-type of MAP kinase through a cAMP-dependent
pathway. PD98059, an inhibitor of MEK (MAP kinase kinase), completely abolished
the anti-apoptotic effect of PACAP-38, suggesting that MAP kinase pathway activation is necessary for PACAP-38 effect.

Cerebellar granule cells are among the most abundant neuronal population in the mammalian CNS. During the first few weeks of postnatal life, there is a well-documented
cell loss in the maturing granule cell layer of the cerebellum.[1] Cerebellar granule cells
undergo apoptosis between postnatal days 5 and 9 whereas cell loss between the third and
fifth postnatal weeks is not associated with DNA fragmentation.[2] *In vitro* culture of newborn rat cerebellar neurons provided a good model to study neuronal apoptosis due to the
high degree of cellular homogeneity.[3] Cerebellar granule cells survive and differentiate *in
vitro* in the presence of depolarizing concentrations of KCl (25–30 mM) without additional
need for neurotrophic factors.[4] In the presence of normal concentration of KCl (5–10 mM),
cerebellar granule cells undergo cell death, which is inhibited by different categories of
molecules: (1) forskolin[5] and cholera toxin,[6] which raise cAMP levels; (2) IGF-1,[5] which
activates a tyrosine kinase receptor; and (3) agonists of muscarinic cholinergic receptors[7]
and metabotropic glutamate receptors,[8] which stimulate phospholipase C. The effect of
cAMP is of particular interest since it was also demonstrated in other neuronal systems
such as sympathetic and sensory neurons,[9] dopamine neurons,[10] and developing septal
cholinergic neurons.[11] The mechanism underlying the cAMP survival effect is not well
understood, however it was suggested that the MAP kinase pathway is involved.[11] Though
cAMP inhibits the MAP kinase cascade in some cell lines[12–16] and has no effect in rat sympathetic neurons,[17] it stimulates MAP kinase cascade in other cell lines, including PC12.[18,19]

*[a]Corresponding author: Tel.: (33) (0)467 14 29 32; Fax: (33) (0)467 54 24 32; E-mail:
journot@ccipe.montp.inserm.fr*

100

MAP kinase activation has also been involved in protection of PC12 cells from NGF withdrawal-induced apoptosis.[20]

Modulation of granule cell loss by physiological agents has not been carefully described. Evidence for the presence of PACAP and PACAP receptor in the cerebellum is compelling and suggests a physiological role for PACAP in cerebellum development.[21-27] PACAPs are neuropeptides of the VIP/secretin/glucagon family and are named according to their amino acid number. PACAP-27 corresponds to the 27 N-terminal amino acids of PACAP-38 and displays 68% homology with VIP. Two classes of PACAP receptors have been described with respect to their pharmacological properties: type 1 PACAP receptors bind PACAP-27 and -38 two orders of magnitude more potently than VIP, whereas type 2 PACAP receptors do not discriminate between PACAP-27, PACAP-38, and VIP. At present three genes encoding PACAP/VIP receptors have been cloned. $PACAP_1$-R corresponds to type 1 binding sites whereas VIP_1/PACAP-R and VIP_2/PACAP-R correspond to type 2 PACAP receptors. No VIP-specific receptor has yet been cloned. PACAP-38 modulates the release of several pituitary hormones[28] and of catecholamines from the adrenal gland.[29,30] In addition, PACAP-38 promotes neurite outgrowth in PC12 cells[31,32] and NB-OK neuroblastoma,[33] stimulates neuritogenesis and survival of cultured rat sympathetic neuroblasts,[34,35] and prevents natural neuronal cell death in chick embryo and HIV gp120-induced cell death in hippocampal cultures.[36] Because of the demonstrated presence of PACAP-38 in the cerebellum, the PACAP neurotrophic and neuroprotective activity in other systems, and the PACAP stimulation of cAMP production, we tested PACAP-38 as a modulator of apoptosis in primary culture of cerebellar granule cells.

SERUM AND POTASSIUM WITHDRAWAL INDUCED A MIXTURE OF APOTOSIS AND NECROSIS

Development of the nervous system is controlled by neurotrophic factors that regulate survival and differentiation of neuronal precursors.[37,38] Neuroblasts are initially produced in large numbers and only those appropriately stimulated by neurotrophic factors will finally survive and differentiate. Elimination of unstimulated precursors is achieved through activation of genetic programs aimed at cell suicide, named programmed cell death or apoptosis.[39] It is therefore of importance to establish well-defined models of neuronal apoptosis to understand mechanisms underlying this process. Three models of choice have emerged and were used to generate an abundant literature. PC12 cells are pheochromocytoma-derived cells that survive and differentiate in the absence of serum and in the presence of nerve growth factor (NGF), whereas NGF withdrawal induces apoptosis.[20,40] The second model also involves NGF action but on a different neuronal population, namely culture of sympathetic neurons.[41] Finally, the third model consists of culture of cerebellar granule cells, the interneurons of the cerebellum, which survive and differentiate *in vitro* in the presence of serum and depolarizing concentrations of KCl (25–30 mM).[42] If the medium is changed to serum-free medium containing normal concentration of KCl (5–10 mM), cerebellar granule cells undergo cell death. In most studies, it was demonstrated (1) that serum and potassium withdrawal induced apoptosis and (2) that different compounds such as forskolin, cholera toxin, IGF-1, and agonists of the metabotropic glutamate or muscarinic receptors protected neurons from cell death. However, it was not demonstrated that apoptosis is the unique death process that takes place in this system yet it was assumed that protection from cell death occurs by protection from apoptosis.[5,6,43-45]

We compared the protective effects of PACAP-38 and IGF-1 by measuring (1) total cell death (necrosis+apoptosis) by the fluorescein diacetate conversion method and (2) apoptosis by quantifying the extent of DNA fragmentation. As shown in FIGURE 1, serum (FCS) and potassium withdrawal induced a dramatic decrease in neuronal survival, at least in part by

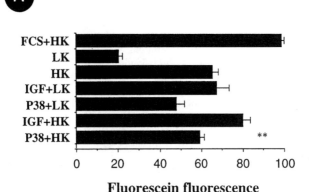

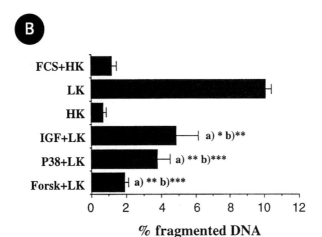

FIGURE 1. Serum and KCl withdrawal induces apoptosis and necrosis in cerebellar granule cells. (**A**) Survival following serum (FCS) or FCS/KCl withdrawal. Seven-day *in vitro* (DIV) cultures were deprived of serum or serum/KCl and maintained for 48 h in high potassium (HK) or low potassium (LK) medium with or without IGF-1 (IGF, 25 ng/ml) or PACAP-38 (P38, 100 nM). Survival was determined using the FDA conversion method. Results were expressed as the percentage of total fluorescein production in sister cultures rinsed and fed again with original medium. Data are mean ± S.E.M. of at least three experiments performed in triplicate. (**B**) Percentage of fragmented DNA after FCS or FCS/KCl withdrawal in the presence of different compounds. Seven DIV neurons were deprived of FCS and KCl and maintained for 24 h in HK or LK medium in the presence of IGF-1 (IGF, 25 ng/ml), or PACAP-38 (P38, 100 nM). Soluble and non-soluble DNA were isolated and quantified using Hoechst 33,258. Data are mean ± S.E.M. of at least three independent experiments. *$p < 0.01$; **$p < 0.0025$; ***$p < 0.0005$ as compared with (**a**) HK or (**b**) LK.

extensive apoptosis (FCS+HK vs. LK). Interestingly, FCS withdrawal alone did not induce significant apoptosis, whereas some cell death is still observed (FCS+HK vs. HK) indicating that FCS withdrawal leads mainly to necrosis and that high potassium concentrations fully prevented apoptosis. Although IGF-1 induced the same neuronal survival as high potassium concentrations, apoptosis could be observed in the presence of IGF-1 alone (HK vs. IGF1), suggesting that IGF-1 only partially protected against apoptosis but efficiently prevented necrosis. Accordingly, IGF-1 partially prevented FCS withdrawal–induced necrosis (HK vs. IGF-1+HK). On the other hand, activators of the cAMP pathway, such as PACAP-38, partially prevented apoptosis (HK vs. LK+P38) but not necrosis (HK vs. HK+P38).

Our results indicate that (1) serum and potassium withdrawal induced a mixture of apoptosis and necrosis, which were differentially prevented by different compounds, and (2) assessment of the anti-apoptotic effect of a given compound should be performed with a technique that specifically measures apoptosis rather than total cell death. Our results are in good agreement with a recent report by Miller and Johnson, who presented evidence of the existence of two neuronal populations in dying cerebellar granule cells cultures. Analysis of the time course and extent of death after removal of either serum or potassium alone demonstrated that a fast-dying ($T_{1/2}$ = 4 h) population (20%) responded to serum deprivation, whereas a slow-dying ($T_{1/2}$ = 25 h) population (80%) died in response to potassium deprivation.[46]

PACAP-38 PREVENTED POTASSIUM DEPRIVATION-INDUCED APOPTOSIS IN A DOSE-DEPENDENT MANNER THROUGH ACTIVATION OF PACAP-SPECIFIC RECEPTORS

We specifically assessed apoptosis by quantifying genomic DNA fragmentation and showed that protection by PACAP-38 was dose-dependent with a maximal effect at 100 nM and an EC_{50} of 5 nM (FIG. 2). Arimura and coworkers reported a neurotrophic biphasic effect of low PACAP-38 concentrations on gp120-induced apoptosis in hippocampal cultures.[36] At concentrations above 1 nM, PACAP-38 was not effective in their system. The effect of PACAP-38 on cerebellar granule neurons is therefore likely to involve mechanisms different from those recruited in hippocampal cultures.

To identify the PACAP receptor(s) involved, we performed RT-PCR using primers specific for the different PACAP/VIP receptor subtypes and splice variants. We demonstrated the expression of $PACAP_1$-R s, hop, and $VIP_1/PACAP$-R (data not shown). In addition, we performed pharmacological characterization of the expressed PACAP/VIP receptor(s) (FIG. 3). Stimulation of cAMP production by PACAP and VIP indicated the presence of type 1 PACAP receptor, which is compatible with RT-PCR experiments. On the other hand, the potency of VIP to stimulate cAMP production was low, indicating that no type 2 PACAP receptor protein ($VIP_1/PACAP$-R or $VIP_2/PACAP$-R) was significantly expressed in contrast to what was anticipated from RT-PCR experiments. The effect of PACAP-38 on cerebellar granule cells was therefore mainly mediated by $PACAP_1$-R activation.

Basille and coworkers documented the presence of PACAP receptors on cells of the proliferating external granule cell layer (EGL) at P8.[24] At that time, granule cells undergo both maximal proliferation and massive DNA fragmentation, indicating that apoptosis occurs in the EGL very soon after neurogenesis, before maximal migration to the internal granule cell layer (IGL) and synaptogenesis with Purkinje cells occur around P10. This indicates that factors other than synaptogenesis must regulate the number of granule cells that survive.[2] The present work suggests that PACAP-38 might be one of these factors.

PACAP-38 STIMULATES MAP-KINASE ACTIVITY

PACAP-38 was shown to display neurotrophic properties in several systems, namely PC12,[31] sympathetic neurons,[34,35] chick embryo, and hippocampal cultures.[36] In PC12 cells,

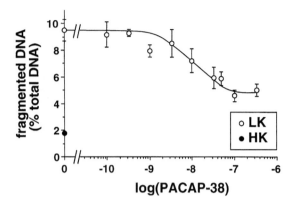

FIGURE 2. Dose-response curve of PACAP-38–induced decrease in DNA fragmentation. Seven-day-old neurons were deprived of serum and maintained for 24 h in LK medium supplemented with different PACAP-38 concentrations. Soluble and non-soluble DNA were isolated and quantified as described in FIGURE 1. Data are mean ± S.E.M. of three independent experiments.

it was demonstrated that MAP kinase activation is necessary and sufficient for differentiation[47] and that blockade of the MAP kinase pathway by PD98059, a specific MEK inhibitor,[48,49] prevented differentiation of PC12 cells by NGF.[50] Interestingly, it was also shown that cAMP-induced differentiation was accompanied by activation of the MAP kinase pathway,[18,51] and that PACAP-38 stimulates ERK1 activity.[18] These results suggested a possible mechanism for PACAP-38 action on cerebellar granule cells. Conversely, Edwards and coworkers[52] demonstrated that cAMP protected sympathetic neurons from

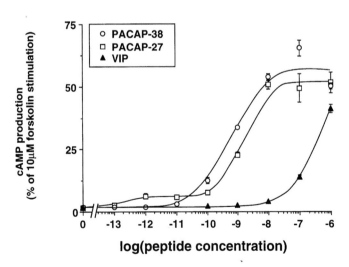

FIGURE 3. PACAP- or VIP-stimulated cAMP production in cerebellar granule cells. Neurons were incubated for 15 min at 37°C in LK medium containing the indicated concentrations of PACAP-38, PACAP-27, or VIP. Data are expressed as the percentage of cAMP production induced by 20 μM forskolin. Data are mean ± S.E.M. of three independent experiments performed in triplicate.

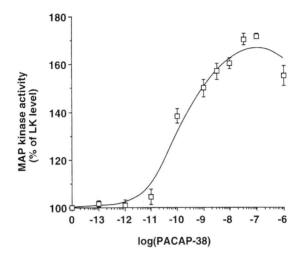

FIGURE 4. PACAP-38 stimulated MAP kinase activity. Neurons were incubated with the indicated concentrations of PACAP-38 for 10 min at 37°C. ERK activity was determined using the Biotrak p42/p44 MAP kinase enzyme assay kit as recommended by the manufacturer (Amersham). Data are expressed as the percentage of MAP kinase activity in LK medium alone. Data are mean ± S.E.M. of three independent experiments performed in triplicate.

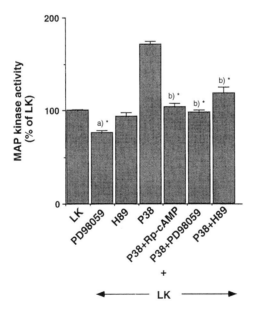

FIGURE 5. PACAP stimulates ERK activity through a PKA- and MEK-dependent mechanism. Neurons were incubated for 1 h in the presence or absence of different inhibitors (25 μM PD98059, 20 μM H89, 200 μM Rp-cAMP), washed twice, and incubated at 37°C for 10 min with 100 nM PACAP-38 in the presence or absence of the inhibitors. Data are mean ± S.E.M. of three independent experiments performed in triplicate. *$p < 0.0005$ with Student's *t*-test as compared with (**a**) LK or (**b**) PACAP-38.

NGF withdrawal-induced apoptosis without activating ERK.[17] PACAP-38 stimulated ERK activity in a dose-dependent manner with a maximal effect at 100 nM (FIG. 4). We demonstrated that tetrodotoxin (TTX) did not block PACAP-induced stimulation of MAP kinase activity (data not shown), suggesting that the effect of PACAP was not mediated by the release of neurotransmitters or neurotrophic factors. Rp-cAMP and H89, two inhibitors of PKA, and PD98059, an inhibitor of MEK (MAP Kinase kinase), blocked stimulation of ERK activity by PACAP-38 (FIG. 5). We extended our analysis by performing Western blots with anti-phosphoERKs antibodies. We demonstrated that ERK2 was more abundant than ERK1 and that PACAP-38 induced phosphorylation of both kinases (data not shown). Modulation of the phosphorylation state of ERK1 and ERK2 by the different treatments was in agreement with results obtained by the measurement of ERK activity. We also measured the activity of the stress-activated protein kinases (SAPK) p38 and JNK (c-Jun N-terminal kinase). We quantified the extent of *in vitro* phosphorylation of GST-ATF2 after immunoprecipitation of cellular extracts with anti-p38 or anti-JNK antisera. The activity of neither of the stress kinases was significantly modified by PACAP-38 (data not shown).

ACTIVATION OF PKA AND MAP-KINASE IS NECESSARY FOR PACAP-38 PROTECTIVE EFFECT

To test whether PACAP-induced ERK and PKA activation was involved in the anti-apoptotic effect of PACAP-38, we measured DNA fragmentation in the presence of

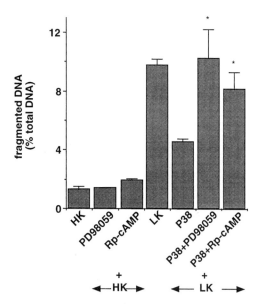

FIGURE 6. PACAP-38 decreased DNA fragmentation through a PKA- and MEK-dependent mechanism. Neurons were incubated with either 25 μM PD98059, 20 μM H89, or 200 μM Rp-cAMP for 1 h before addition of 100 nM PACAP-38. Cells were washed twice with HK medium, and incubated with different drugs and inhibitors for 24 h. Soluble and non-soluble DNA were isolated and quantified as indicated in FIGURE 1. Data are mean ± S.E.M. of at least three independent experiments. * $p < 0.005$ with Student's *t*-test as compared to PACAP-38 alone.

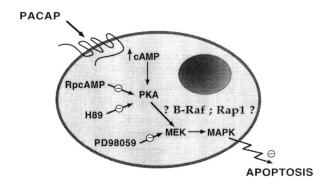

FIGURE 7. Possible mechanism of action of PACAP-38 on cerebellar granule neurons. PACAP-38 binding to PACAP$_1$-R induces cAMP formation, which results in stimulation of cAMP-dependent protein kinase (PKA) activity. PKA activates the MAP kinase pathway upstream of MAP kinase kinase (MEK) through possible activation of B-Raf and Rap1. Activation of MAP kinase leads to inhibition of potassium deprivation–induced apoptosis.

PD98059 and Rp-cAMP. Both compounds did not affect the protection induced by high KCl concentrations, excluding toxic or nonspecific effect (FIG. 6). Interestingly, both PD98059 and Rp-cAMP blocked the effect of PACAP-38 on DNA fragmentation (FIG. 6).

Conclusively, protection of cerebellar granule cells by PACAP-38 likely involves the same mechanism as the one suggested in PC12 cells for cAMP-induced differentiation, namely activation of PKA, which stimulates MEK activity resulting in activation of ERK (FIG. 7). A recent report by Vossler and coworkers using PC12 cells suggests a possible pathway to link PKA and MAP kinase pathways.[53]

Interestingly, we also demonstrated that activation of the MAP kinase pathway is not the exclusive way to protect cerebellar granule neurons from KCl deprivation-induced cell death. For instance, IGF-1 or high KCl concentration protected neurons (FIG. 1) but weakly stimulated ERK activity (FIG. 4). Furthermore, the protective effect of KCl was not affected by PD98059 (FIG. 5). This suggests that other pathways that work independently of ERK activation are possibly involved in protection from apoptosis. Xia and coworkers[20] recently proposed that NGF withdrawal–induced apoptosis of PC12 cells requires concurrent activation of the stress kinases [C-Jun N-terminal protein kinase (JNK) and p38] and inhibition of ERK kinases. Hence, either stimulation of ERK activity or inhibition of the JNK/p38 pathway could result in the same protection from apoptosis.

REFERENCES

1. LANDIS, D. M. D. & R. L. SIDMAN. 1978. Electron microscopic analysis of postnatal histogenesis in the cerebellar cortex of staggerer mutant mice J. Comp. Neurol. **179**: 831–864.
2. WOOD, K. A., B. DIPASQUALE & R. J. YOULE. 1993. In situ labelling of granule cells for apoptosis-associated DNA fragmentation reveals different mechanisms of cell loss in developing cerebellum. Neuron **11**: 621–632.
3. MARINI, A. M. & S. M. PAUL. 1992. N-methyl-D-aspartate receptor-mediated neuroprotection in cerebellar granule cells requires new RNA and protein synthesis Proc. Natl. Acad. Sci. USA **89**: 6555–6559.
4. GALLO, V., C. GIOVANINI & G. LEVI. 1990. Modulation of non-N-methyl-D-aspartate receptors in cultured cerebellar granule cells. J. Neurochem. **54**: 1619–1625.

5. D'MELLO, S. R., C. GALLI, T. CIOTTI & P. CALISSANO. 1993. Induction of apoptosis in cerebellar granule neurons by low potassium: inhibition of death by insulin-like growth factor I and cAMP. Proc. Natl. Acad. Sci. USA **90**: 10989–10993.

6. YAN, G.-M., S.-Z. LIN, R. P. IRWIN & S. M. PAUL. 1995. Activation of G proteins bidirectionally affects apoptosis of cultured cerebellar granule neurons. J. Neurochem. **65**: 2425–2431.

7. YAN, G.-M., S.-Z. LIN, R. P. IRWIN & S. M. PAUL. 1995. Activation of muscarinic cholinergic receptors blocks apoptosis of cultured cerebellar granule neurons. Mol. Pharmacol. **47**: 248–257.

8. COPANI, A., V. M. G. BRUNO, V. BARRESI, G. BATTAGLIA, D. F. CONDORELLI & F. NICOLETTI. 1995. Activation of metabotropic glutamate receptors prevents neuronal apoptosis in culture. J. Neurochem. **64**: 101–108.

9. RYDEL, R. E. & L. A. GREENE. 1988. cAMP analogs promote survival and neurite outgrowth in cultures of rat sympathetic and sensory neurons independently of nerve growth factor. Proc. Natl. Acad. Sci. USA **85**: 1257–1261.

10. MENA, M. A., M. J. CASAREJOS, A. BONIN, J. A. RAMOS & J. GARCIA DE YÉBENES. 1995. Effects of dibutyryl cyclic AMP and retinoic acid on the differentiation of dopamine neurons: prevention of cell death by dibutyryl cyclic AMP. J. Neurochem. **65**: 2612–2620.

11. KEW, J. N., D. W. SMITH & M. V. SOFRONIEW. 1996. Nerve growth factor withdrawal induces the apoptotic death of developing septal cholinergic neurons in vitro: protection by cyclic AMP analogue and high potassium. Neuroscience **70**: 329–339.

12. COOK, S. & F. MCCORMICK. 1993. Inhibition by cAMP of Ras-dependent activation of Raf. Science **262**: 1069–1072.

13. WU, J., P. DENT, T. JELINEK, A. WOLFMAN, M. J. WEBER & T. W. STURGILL. 1993. Inhibition of the EGF-activated MAP kinase signaling pathway by adenosine 3′,5′-monophosphate. Science **262**: 1065–1069.

14. BURGERING, B. M., G. J. PRONK, P. C. VAN WEEREN, P. CHARDIN & L. J. BOS. 1993. cAMP antagonizes p21ras-directed activation of extracellular signal-regulated kinase 2 and phosphorylation of mSos nucleotide exchange factor. EMBO J. **12**: 4211–4220.

15. GRAVES, L. M., K. E. BORNFELDT, E. W. RAINES, B. C. POTTS, S. G. MCDONALD, R. ROSS & E. G. KREBS. 1993. Protein kinase A antagonizes platelet-derived growth factor-induced signaling by mitogen-activated protein kinase in human arterial smooth muscle cells. Proc. Natl. Acad. Sci. USA **90**: 10300–10304.

16. SEVETSON, B. R., X. KONG & J. C. LAWRENCE, JR. 1993. Increasing cAMP attenuates activation of mitogen-activated protein kinase. Proc. Natl. Acad. Sci. USA **90**: 10305–10309.

17. VIRDEE, K. & A. M. TOLKOVSKY. 1995. Activation of p44 and p42 MAP kinases is not essential for the survival of rat sympathetic neurons. Eur. J. Neurosci. **7**: 2159–2169.

18. FRÖDIN, M., P. PERALDI & E. VAN OBBERGHEN. 1994. Cyclic AMP activates the mitogen-activated protein kinase cascade in PC12 cells. J. Biol. Chem. **269**: 6207–6214.

19. FAURE, M., T. A. VOYNO-YASENETSKAYA & H. BOURNE. 1994. cAMP and βγ subunits of heterotrimeric G proteins stimulate the mitogen-activated protein kinase pathway in COS-7 cells. J. Biol. Chem. **269**: 7851–7854.

20. XIA, Z., M. DICKENS, J. RAINGEAUD, R. J. DAVIS & M. E. GREENBERG. 1995. Opposing effects of ERK and JNK-p38 MAP kinases on apoptosis. Science **270**: 1326–1331.

21. HASHIMOTO, H., T. ISHIHARA, R. SHIGEMOTO, K. MORI & S. NAGATA. 1993. Molecular cloning and tissue distribution of a receptor for pituitary adenylate cyclase activating polypeptide. Neuron **11**: 333–342.

22. SPENGLER, D., C. WAEBER, C. PANTALONI, F. HOLSBOER, J. BOCKAERT, P. H. SEEBURG & L. JOURNOT. 1993. Differential signal transduction by five splice variants of the PACAP receptor. Nature **365**: 170–175.

23. FAVIT, A., U. SCAPAGNINI & P. L. CANONICO. 1995. Pituitary adenylate cyclase-activating polypeptide activates different signal transducing mechanisms in cultured cerebellar granule cells. Neuroendocrinology **61**: 377–382.

24. BASILLE, M., B. J. GONZALEZ, P. LEROUX, L. JEANDEL, A. FOURNIER & H. VAUDRY. 1993. Localization and characterization of PACAP receptors in the rat cerebellum during development: evidence for a stimulatory effect of PACAP on immature cerebellar granule cells. Neuroscience **57**: 329–338.

25. BASILLE, M., B. J. GONZALEZ, L. DESRUES, M. DEMAS, A. FOURNIER & H. VAUDRY. 1995. Pituitary adenylate cyclase-activating polypeptide (PACAP) stimulates adenylyl cyclase and phospholipase c activity in rat cerebellar neuroblasts. J. Neurochem. **65**: 1318–1324.
26. LAM, D.-C., K. TAKAHASHI, M. A. GHATEI, S. M. KANSE, J. M. POLAK & S. R. BLOOM. 1990. Binding sites of a novel neuropeptide pituitary adenylate cyclase-activating polypeptide in the rat brain and lung. Eur. J. Biochem. **193**: 725–729.
27. CAUVIN, A., P. ROBBERECHT, P. DE NEEF, P. GOURLET, A. VANDERMEERS, M.-C. VANDERMEERS-PIRET & J. CHRISTOPHE. 1991. Properties and distribution of receptors for pituitary adenylate cyclase activating peptide (PACAP) in rat brain and spinal cord. Regul. Pept. **35**: 161–173.
28. MIYATA, A., A. ARIMURA, R. R. DAHL, N. MINAMINO, A. UEHARA, L. JIANG, M. D. CULLER & D. H. COY. 1989. Isolation of a novel 38 residue-hypothalamic polypeptide which stimulates adenylate cyclase in pituitary cells. Biochem. Biophys. Res. Commun. **164**: 567–574.
29. WATANABE, T., Y. MASUO, H. MATSUMOTO, N. SUZUKI, T. OHTAKI, Y. MASUDA, C. KITADA, M. TSUDA & M. FUJINO. 1992. Pituitary adenylate cyclase activating polypeptide provokes cultured rat chromaffin cells to secrete adrenaline. Biochem. Biophys. Res. Commun. **182**: 403–411.
30. ISOKOBE, K., T. NAKAI & Y. TAKUWA. 1993. Ca^{2+}-dependent stimulatory effect of pituitary adenylate cyclase-activating polypeptide on catecholamine secretion from cultured porcine adrenal medullary chromaffin cells. Endocrinology **132**: 1757–1765.
31. DEUTSCH, P. J. & Y. SUN. 1992. The 38-amino acid form of pituitary adenylate cyclase-activating polypeptide stimulates dual signaling cascades in PC12 cells and promotes neurite outgrowth. J. Biol. Chem. **267**: 5108–5113.
32. HERNANDEZ, A., B. KIMBALL, G. ROMANCHUK & M. W. MULHOLLAND. 1995. Pituitary adenylate cyclase-activating peptide stimulates neurite growth in PC12 cells. Peptides **16**: 927–932.
33. DEUTSCH, P. J., V. C. SCHADLOW & N. BARZILAI. 1993. 38-amino acid form of pituitary adenylate cyclase activating peptide induces process outgrowth in human neuroblastoma cells. J. Neurosci. Res. **35**: 312–320.
34. PINCUS, D. W., E. M. DICICCO-BLOOM & I. B. BLACK. 1990. Vasoactive intestinal peptide regulates mitosis, differentiation and survival of cultured sympathetic neuroblasts. Nature **343**: 564–567.
35. DICICCO-BLOOM, E. & P. J. DEUTSCH. 1992. Pituitary adenylate cyclase activating polypeptide (PACAP) potently stimulates mitosis, neuritogenesis and survival in cultured rat sympathetic neuroblasts. Regul. Pept. **37**: 319–325.
36. ARIMURA, A., A. SOMOGYVARI-VIGH, C. WEILL, R. C. FIORE, I. TATSUNO, V. BAY & D. E. BRENNEMAN. 1994. PACAP functions as a neurotrophic factor. Ann. N. Y. Acad. Sci. **739**: 228–243.
37. LEVI-MONTALCINI, R. 1987. The nerve growth factor: thirty five years later. EMBO J. **6**: 1145–1154.
38. BARDE, Y. A. 1989. Trophic factors and neuronal survival. Neuron. **2**: 1525–1534.
39. RAFF, M. C., B. A. BARRES, J. F. BURNE, H. S. COLES, Y. ISHIZAKI & M. D. JACOBSON. 1993. Programmed cell death and the control of cell survival: Lessons from the nervous system. Science **262**: 695–700.
40. MESNER, P. W., C. L. EPTING, J. L. HEGARTY & S. H. GREEN. 1995. A timetable of events during programmed cell death induced by trophic factor withdrawal from neuronal PC12 cells. J. Neurosci. **15**: 7357–7366.
41. FARINELLI, S. E. & L. A. GREENE. 1996. Cell cycle blockers mimosine, ciclopirox, and deferoxamine prevent the death of PC12 cells and postmitotic sympathetic neurons after removal of trophic support. J. Neurosci. **16**: 1150–1162.
42. GALLO, V., A. KINGSBURY, R. BALASZ & O. S. JORGENSEN. 1987. The role of depolarization in the survival and differentiation of cerebellar granule cells in culture. J. Neurosci. **7**: 2203–2213.
43. YAN, G. M., B. NI, M. WELLER, K. A. WOOD & S. M. PAUL. 1994. Depolarization or glutamate receptor activation blocks apoptotic cell death of cultured cerebellar granule neurons. Brain Res. **656**: 43–51.
44. GALLI, C., O. MEUCCI, A. SCORZIELLO, T. M. WERGE, P. CALISSANO & G. SCHETTINI. 1995. Apoptosis in cerebellar granule cells is blocked by high KCl, forskolin, and IGF-1 through distinct mechanisms of action: the involvement of intracellular calcium and RNA synthesis. J. Neurosci. **15**: 1172–1179.

45. CHANG, J. Y., V. V. KOROLEV & J. Z. WANG. 1996. Cyclic AMP and pituitary adenylate cyclase-activating polypeptide (PACAP) prevent programmed cell death of cultured rat cerebellar granule cells. Neurosci. Lett. **206**: 181–184.

46. MILLER, T. M. & E. M. JOHNSON, JR. 1996. Metabolic and genetic analyses of apoptosis in potassium/serum-deprived rat cerebellar granule cells. J. Neurosci. **16**: 7487–7495.

47. COWLEY, S., H. PATERSON, P. KEMP & C. J. MARSHALL. 1994. Activation of MAP kinase kinase is necessary and sufficient for PC12 differentiation and for transformation of NIH 3T3 cells. Cell **77**: 841–852.

48. ALESSI, D. R., A. CUENDA, P. COHEN, D. T. DUDLEY & A. R. SALTIEL. 1995. PD 098059 is a specific inhibitor of the activation of mitogen-activated protein kinase kinase in vitro and in vivo. J. Biol. Chem. **270**: 27489–27494.

49. DUDLEY, D. T., L. PANG, S. J. DECKER, A. J. BRIDGES & A. R. SALTIEL. 1995. A synthetic inhibitor of the mitogen-activated protein kinase cascade. Proc. Natl. Acad. Sci. USA **92**: 7686–7689.

50. PANG, L., T. SAWADA, S. J. DECKER & A. R. SALTIEL. 1995. Inhibition of MAP kinase kinase blocks the differentiation of PC-12 cells induced by Nerve Growth Factor. J. Biol. Chem. **270**: 13585–13588.

51. YOUNG, S. W., M. DICKENS & J. M. TAVARÉ. 1994. Differentiation of PC12 cells in response to a cAMP analogue is accompanied by sustained activation of mitogen-activated protein kinase. FEBS Lett. **338**: 212–216.

52. EDWARDS, S. N., A. E. BUCKMASTER & A. M. TOLKOVSKY. 1991. The death programme in cultured sympathetic neurones can be suppressed at the posttranslational level by Nerve Growth Factor, cyclic AMP, and depolarization. J. Neurochem. **57**: 2140–2143.

53. VOSSLER, M. R., H. YAO, R. D. YORK, M.-G. PAN, C. S. RIM & P. J. S. STORK. 1997. cAMP activates MAP Kinase and Elk-1 through a B-Raf- and Rap1-dependent pathway. Cell **89**: 73–82.

PACAP Protects Hippocampal Neurons against Apoptosis: Involvement of JNK/SAPK Signaling Pathway[a]

SEIJI SHIODA,[b,g] HIROSHI OZAWA,[d] KENJI DOHI,[d] HIDEKATSU
MIZUSHIMA,[d] KIYOSHI MATSUMOTO,[d] SHIGEO NAKAJO,[e]
ATSUSHI TAKAKI,[f] CHENG JI ZHOU,[b] YASUMITSU NAKAI,[b] AND
AKIRA ARIMURA[c]

[b]Department of Anatomy, [d] Department of Neurosurgery, Showa
University School of Medicine, Tokyo 142-8555, Japan

[c] U.S.-Japan Biomedical Research Laboratories, Tulane University
Hebert Center, Belle Chasse, Louisiana 70037, USA

[e]Laboratory of Biological Chemistry, Showa University School of
Pharmaceutical Sciences, Tokyo 142-8555, Japan

[f]Department of Physiology, Kyushu University Faculty of Medicine,
Fukuoka 812-82, Japan

ABSTRACT: We have demonstrated that the ischemia-induced apoptosis of neurons in
the CA1 region of the rat hippocampus was prevented by either intracerebroventricular or intravenous infusion of pituitary adenylate cyclase–activating polypeptide
(PACAP). However, the molecular mechanisms underlying the anti-apoptotic effect
of PACAP remain to be determined. Within 3–6 h after ischemia, the activities of
members of the mitogen–activated protein (MAP) kinase family, including extracellular signal–regulated kinase (ERK), Jun N-terminal kinase (JNK)/stress-activated
protein kinase (SAPK), and p38 were increased in the hippocampus. The ischemic
stress had a potent influence on the MAP kinase family, especially on JNK/SAPK.
PACAP inhibited the activation of JNK/SAPK after ischemic stress. Secretion of
interleukin-6 (IL-6) into the cerebrospinal fluid was intensely stimulated after
PACAP infusion. IL-6 inhibited the activation of JNK/SAPK, while it activated ERK.
These observations suggest that PACAP and IL-6 act to inhibit the JNK/SAPK signaling pathway, thereby protecting neurons against apoptosis.

Cerebral ischemia causes various degrees of brain damage depending on its intensity
and duration.[1] It is believed that an increase in intracellular Ca^{2+} concentration, stimulated by several bioactive substances, induces apoptosis in neurons of the CA1 region of
the hippocampus.[2] Rat CA1 pyramidal cells are selectively vulnerable to transient

[a]This study was supported in part by grants (08458249 and 09558098) from the Ministry of
Education, Science, Sports, and Culture of Japan to (S.S.).

[g]Corresponding author: Dr. Seiji Shioda, Department of Anatomy, Showa University School of
Medicine, 1-5-8 Hatanodai, Shinagawa-ku, Tokyo 142, Japan; Tel.: 81-3-3784-8104; Fax: 81-3-3784-
6815; E-mail: shioda@med.showa-u.ac.jp

111

ischemic insult, and ischemia causes severe neuronal damage to the CA1 neurons, which start dying two or three days after the reperfusion (FIG. 1).[3]

The signaling pathways that lead to apoptosis are beginning to be defined and a number of proteins have been identified that induce or prevent apoptosis. There are molecular mechanisms that regulate apoptosis. Members of the mitogen-activated protein (MAP) kinase family, including extracellular signal-regulated kinase (ERK), c-Jun NH_2-terminal protein kinase (JNK) also termed stress-activated protein kinase (SAPK), and p38[4] (FIG. 2), are suggested to be contributors to cell death. It has been known that the activation of JNK/SAPK and p38, and concurrent inhibition of ERK are critical for the induction of apoptosis in PC12 cells.[6]

A number of neurotropic factors and nerve growth factors prevent the ischemia-induced degeneration of CA1 neurons. Among these factors, PACAP is a potent anti-apoptotic neurotropic factor. PACAP prevents the programmed death of dorsal root ganglion cells and lumbar motoneurons in chicken embryos.[7] PACAP also prevents apoptosis in cultured cerebellar granule cells.[8,9] In addition, hippocampal neuronal cell death induced by gp120 *in vitro* was completely prevented by PACAP at much lower concentrations than for vasoactive intestinal peptide (VIP).[7] Unlike the response to VIP, the dose-response curve was bimodal, with the greatest cytoprotective activity at 0.1 pM and 0.1 nM, suggesting that PACAP acts at two different receptors: type I (PACAP) and type II (VIP/PACAP) receptors. Recently, we have shown that PACAP also prevents the ischemia-induced death of hippocampal neurons *in vivo* (FIG. 3).[10] The results indicate that the neuroprotective effects of PACAP are mediated through the PACAP receptors (PACAPR) on astrocytes. Interleukin-6 (IL-6) has several effects on the central nervous system, including protection against ischemia and trauma, and stimulation of neuronal cell growth. It has been shown that neuronal cell death in the CA1 region was significantly prevented after intracerebroventricular (i.c.v.) infusion of IL-6[11] and PACAP stimulates IL-6 secretion in primary cultures of rat astrocytes.[12] These results suggest that PACAP and IL-6 may be secreted after ischemia *in vivo*. Although PACAP reduced the ischemia-induced loss of CA1 neurons, the precise molecular mechanism of neuronal apoptosis in the CA1 region and the contribution of PACAP and IL-6 to the regulation of MAP kinase family members need to be clarified.

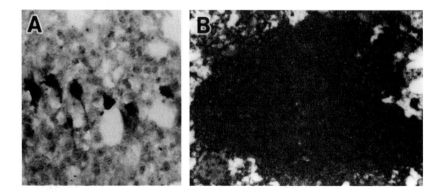

FIGURE 1. (A) Fragmentation of nuclear DNA into oligonucleosomal fragments is visualized with the TUNEL method in the CA1 neurons after ischemia. (B) DNA fragmentation is clearly visible in the CA1 neurons at the ultrastructural level.

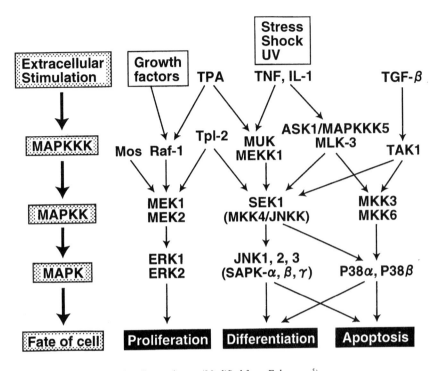

FIGURE 2. MAP kinase signaling pathway. *(Modified from Fukunaga.[5])*

MATERIALS AND METHODS

Ischemia Model

Global ischemia was induced in conscious rats for 5 min as described by Kawai and coworkers.[13] For i.c.v. administration of PACAP, PACAP was infused through a cannula at 1 pmol/h using an osmotic minipump (ALZA Corp., Palo Alto, CA). For i.c.v. administration of IL-6, IL-6 was infused through a cannula at 18 ng/h using an osmotic minipump. The animals were infused with PACAP or IL-6 for 2 days before the ischemia-reperfusion experiment.

Measurement of MAP Kinase Activity

Cell extracts from the hippocampus were assayed for MAP kinase activity using a synthetic peptide corresponding to the MAP kinase phosphorylation site of myelin basic protein as described by Fiore and coworkers[14] and Murphy and coworkers.[15] After incubation at 30°C for 15–20 min, the reaction was stopped by the addition of trichloroacetic acid, and 25 µl of the reaction mixture was spotted onto Whatman P81 phosphocellulose paper and washed with 0.5% phosphoric acid. The incorporated radioactivity was determined using liquid scintillation counting and normalized to the total protein concentration. Immunocomplex assays were also performed by using anti-MAP kinase, anti-JNK, and anti-p38 kinase antibodies as described by Rosen and coworkers.[16]

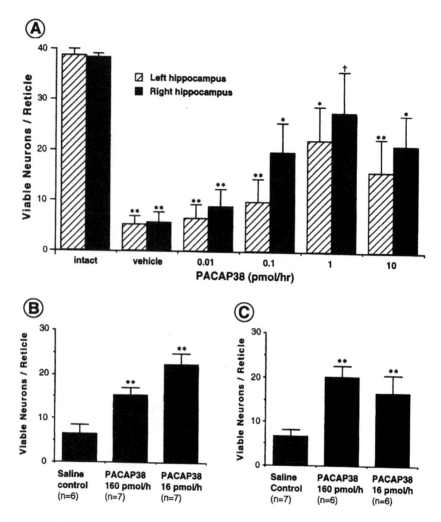

FIGURE 3. Effects of i.c.v. (**A**) or i.v. (**B, C**) infusion of PACAP38 on ischemia-induced neuronal death in CA1. (**A**) Vehicle-infused rats showed marked neuronal death on both sides of the hippocampus one week after ischemia. The cytoprotective effect of i.c.v. PACAP38 was maximal at 1 pmol/h. *$p < 0.05$ vs. untreated rats; ** $p < 0.01$ vs. untreated rats. Intravenous infusion of PACAP38 following an i.v. bolus (5 nmol/kg) begun immediately (**B**) or 24 h (**C**) after ischemia prevented neuronal death in CA1 at 7 days post-ischemia. $p < 0.01$ vs. saline-infused controls. (*Redrawn from Uchida and coworkers[10] with permission.*)

B9 Bioassay for IL-6

The bioactivity of IL-6 in each sample of cerebrospinal fluid (CSF) was measured using the proliferation of the murine IL-6–dependent hybridoma subclone B9 as described by Gottchall and coworkers.[17] Briefly, samples and standards were pipetted into 96-well plates in six consecutive one-to-one dilutions. B9 cells were added to the wells and the plates were incubated for 72 h. At the end of incubation, proliferation was determined colorimetrically using MTT. The cells were solubilized with acidic sodium dodecyl sulfate and the plates were read at 570 nm with a microplate reader.

Immunohistochemistry for IL-6

The animals were perfused with 4% paraformaldehyde in phosphate buffer for 12 h, or 1, 2, 4, or 7 days after the ischemia. Free-floating 20-μm frozen sections were incubated with a 1:2,000 dilution of a monoclonal mouse anti-human IL-6 antibody (Genzyme, Boston, MA), followed by immunostaining using the avidin-biotin complex method. Subsequently, they were developed with 3,3'-diaminobenzidine tetrahydrochloride. The primary antibody recognized both rat and guinea pig IL-6.

RESULTS AND DISCUSSION

In the ischemic hippocampal tissues, including the CA1 region, both in-gel and immunocomplex kinase assays indicated that both JNK and ERK activities increased during the first 6 h after ischemia-reperfusion. During the first 6 h, p38 was also stimulated after ischemia-reperfusion. Taken together, these observations suggest that the activation of JNK or p38 MAP kinase or both contributes to the induction of apoptosis in CA1 neurons. We found that this ischemic stress had a strong influence on MAP kinases, especially on JNK. In contrast, PACAP-treated animals showed no significant increase in JNK activity within the first 6 h. This indicates that PACAP inhibited the activation of JNK after ischemia-reperfusion stress. To identify the cytoprotective functions of PACAP precisely, we first mapped the distribution of PACAP and its receptors in the rat brain. The PACAPR was expressed in the CA1-CA3 regions and dentate gyrus of the rat hippocampus, but very few PACAP-containing perikarya and fibers were detected.[18] *In situ* hybridization for PACAPR mRNA in the rat brain showed that it is strongly expressed in the olfactory bulb, dentate gyrus, and cerebellar cortex, and moderately expressed in the CA1-CA3 regions of the hippocampus.[19] In emulsion-dipped tissue sections, PACAPR mRNA expression is visible in pyramidal cells in the CA1 region of untreated adult animals.[19]

In untreated animals, there were no cells that show IL-6–like immunoreactivity while a small number of IL-6-positive cells were visible 12 h after ischemia. A peak of IL-6–like immunoreactivity in the CA1 region was found at 1–2 days after ischemia and this intense immunoreactivity was found mostly in the reactive astrocytes. IL-6–positive cells were decreased in number at 4 and 7 days after ischemia. IL-6–like immunoreactivity was not detected in microglial cells during ischemia-reperfusion. Very high concentrations of IL-6 (pg range) in the CSF were detected at 6 h and 24 h after ischemia. We measured IL-6 concentrations in the CSF of the ischemic animals with or without infusion of PACAP (1 pmol/h). The secretion of IL-6 into the CSF was stimulated a little by the vehicle. Unexpectedly, IL-6 secretion into the CSF was stimulated intensely (ng range) after PACAP infusion (1 pmol/h). These findings suggest that PACAP stimulates IL-6 secretion from astrocytes in the rat brain.

The contribution of MAP kinase family members, including ERK, JNK/SAPK, and p38, was examined in the ischemic hippocampus with or without IL-6 infusion. No significant increase in JNK/SAPK activity was detected within the first 6 h in the IL-6–treated (2 ng/h) animals. Instead, ERK activity was increased gradually with time after IL-6 infusion. This indicates that IL-6 not only inhibits the activation of JNK, but also activates ERK after ischemia-reperfusion stress. These findings suggest that IL-6, released from the astrocytes immediately after the ischemic stress, prevents the apoptosis through

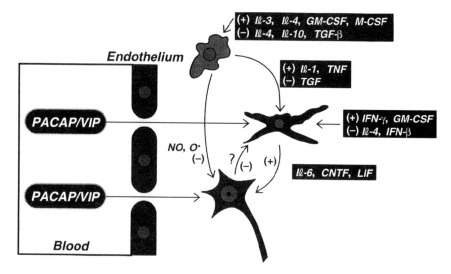

FIGURE 4. Proposed status of PACAP as a neuroprotective regulator. CNTF: ciliary neurotrophic factor; GM-CSF: granulocyte macrophage colony stimulating factor; IFN: interferon; IL: interleukin; LIF: leukemia inhibitory factor; M-CSF: macrophage colony stimulating factor; TGF: transforming growth factor; TNF: tumor necrosis factor.

the MAP kinase-signaling pathway. It may be considered that PACAP itself and IL-6–stimulated secretion by PACAP both inhibit the JNK signaling pathway, thereby protecting neurons against cell death.

From the results shown here, PACAP may not only suppress JNK and p38 MAP kinase activation, but also concurrently trigger the activity of ERKs directly on neurons or indirectly through astrocytes. PACAP has been shown to protect against apoptosis in rat primary cultures of cerebellar granule cells through the MAP kinase pathway.[8] There exist many mechanisms for the regulation of the relevant MAP kinase pathways and various forms of cross-talk between these signal transduction pathways (FIG. 2). PACAP may have a very important role on the decision for cellular life or death through the integration of multiple signals. Although both neurons and astrocytes respond to PACAP in a dose-related manner, no increase in cAMP levels for any CNS culture system has been observed at the lowest concentrations of the peptide that increase neuronal survival (0.1 pM PACAP).[7] These data strongly suggest that the cytoprotective action of low concentrations of PACAP is mediated by a second messenger other than cAMP.

In conclusion, very low concentrations of PACAP, after passing through the blood-brain barrier, may stimulate neurons directly and indirectly by stimulating astrocytes to secrete neuroprotective factors that affect neurons (FIG. 4). During the process of cytoprotection by PACAP, PACAP may regulate the dynamic balance between growth factor–activated ERK and stress-activated JNK-p38 pathways. The significant neuroprotection with intravenous infusions of PACAP begun 24 h after the ischemic insult suggests a potential clinical importance for PACAP.

REFERENCES

1. PULSINELLI, W. A., J. B. BRIERLEY & F. PLUM. 1982. Temporal profile of neuronal damage in a model of transient forebrain ischemia. Ann. Neurol. **11:** 491–498.

2. DESHPANDE, J. K., B. K. SIESJO & T. WIELOCH. 1987. Calcium accumulation and neuronal damage in the hippocampus following cerebral ischemia. J. Cereb. Blood Flow Metab. **7**: 89–95.

3. KIRINO, T. 1982. Delayed neuronal death in the gerbil hippocampus following ischemia. Brain Res. **239**: 57–69.

4. OPPENHEIM, R. W. 1991. Cell death during development of the nervous system. Annu. Rev. Neurosci. **14**: 453–501.

5. FUKUNAGA, R. 1997. Target proteins of MAP kinases. Transcription factors and MAP kinases. Jikkenigaku (in Japanese) **15**: 1733–1738.

6. XIA, Z., M. DICKENS, J. RAINGEAUD, R. J. DAVIS & M. E. GREENBERG. 1995. Opposing effects of ERK and JNK-p38 MAP kinases on apoptosis. Science **270**: 1326–1331.

7. ARIMURA, A., A. SOMOGYVARI-VIGH, C. WEILL, R. C. FIORE, I. TATSUNO, V. BAY, & D. E. BRENNEMAN. 1994. PACAP functions as a neurotrophic factor. Ann. N.Y. Acad. Sci. **739**: 228–243.

8. CAVALLARO, S., A. COPANI, V. D'AGATA, S. MUSCO, S. PETRALLIA, C. VENTRA, F. STIVALA, S. TRAVALI & P. L. CANONOCO. 1996. Pituitary adenylate cyclase activating polypeptide prevents apoptosis in cultured cerebellar granule neurons. Mol. Pharmacol. **50**: 60–66.

9. VILLABA, M., J. BOCKAERT & L. JOURNOT. 1997. Pituitary adenylate cyclase-activating polypeptide (PACAP-38) protects cerebellar granule neurons from apoptosis by activating the mitogen-activated protein kinase (MAP kinase) pathway. J. Neurosci. **17**: 83–90.

10. UCHIDA, D., A. ARIMURA, A. SOMOGYVARI-VIGH, S. SHIODA & B. BANKS. 1996. Prevention of ischemia-induced death of hippocampal neurons by pituitary adenylate cyclase activating polypeptide. Brain Res. **736**: 280–286.

11. MATSUDA, S., T-C. WEN, F. MORITA, H. OTSUKA, K. IGASE, H. YOSHIMURA & M. SAKANAKA. 1996. Interleukin-6 prevents ischemia-induced learning disability and neuronal and synaptic loss in gerbils. Neurosci. Lett. **204**: 109–112.

12. GOTTSCHALL, P. E., I. TATSUNO & A. ARIMURA. 1994. Regulation of interleukin-6 (IL-6) secretion in primary cultured rat astrocytes: synergism of interleukin-1 (IL-1) and pituitary adenylate cyclase activating polypeptide (PACAP). Brain Res. **637**:197–203.

13. KAWAI, K., L. NITECKA & C. A. RUETZLER. 1992. Global ischemia associated with cardiac arrest in the rat: I. Dynamics of early neuronal damage. J. Cereb. Blood Flow Metab. **12**: 238–249.

14. FIORE, R., T. H. MURPHY, J. S. SANGHERA, S. PELECH & J. M. BARABAN. 1993. Activation of p42 mitogen-activated protein kinase by glutamate receptor stimulation in rat primary cortical cultures. J. Neurochem. **61**: 1626–1633.

15. MURPHY, T. H., L. A. BLATTER, R. V. BHAT, R. S. FIORE, W. G. WIER & J. M. BARABAN. 1994. Differential regulation of calcium/calmodulin-dependent protein kinase II and p42 MAP kinase activity by synaptic transmission. J. Neurosci. **14**: 1320–1331.

16. ROSEN, L. B., D. D. GINTY, M. J. WEBER & M. E. GREENBERG. 1994. Membrane depolarization and calcium influx stimulate MEK and MAP kinase via activation of RAS. Neuron **12**: 1207–1221.

17. GOTTSCHALL, P. E., I. TATSUNO & A. ARIMURA. 1992. Increased sensitivity of glioblastoma cells to interleukin-1 and long-term incubation with dexamethasone. Mol. Cell Neurosci. **3**: 49–55.

18. ARIMURA, A. & S. SHIODA. 1995. Pituitary adenylate cyclase activating polypeptide (PACAP) and its receptors: neuroendocrine and endocrine interaction. Front. Endocrinol. **16**: 53–88.

19. SHIODA, S., Y. SHUTO, A. SOMOGYVARI, G. LEGRADI, H. ONDA, D. H. COY, S. NAKAJO & A. ARIMURA. 1997. Localization and gene expression of the receptor for pituitary adenylate cyclase-activating polypeptide in the rat brain. Neurosci. Res. **28**: 345–354.

Stimulatory Transducing Systems in Pancreatic Islet Cells[a]

SHAHIN EMAMI,[b] KARINE REGNAULD, NATHALIE FERRAND,
ANY ASTESANO, MARCIA PESSAH, HUAN PHAN, CLAUDINE
BOISSARD, JEAN-MICHEL GAREL, AND GABRIEL ROSSELIN

*Institut National de la Santé et de la Recherche Médicale, Centre de
Recherche Paris Saint-Antoine, Paris Cedex 12, France*

ABSTRACT: We have determined the cellular distribution of different alpha subtypes
of G proteins and adenylyl cyclase (AC) isoforms in endocrine, exocrine, and established pancreatic cell lines. VIP, PACAP, and tGLP-1 receptor proteins are expressed
to varying extents in A and B cells, whereas the expression of $G\alpha$ subunits is cell specific. Thus, $G_{olf}\alpha$ is detected in normal rodent B cells and immortalized pancreatic B
cell lines, whereas $G_s\alpha$ is more ubiquitously expressed. The cellular density of AC isoforms labeling (I, II, III, IV, V/VI) is also islet cell–specific and their distribution is
age- and species-dependent. The identification of numerous signaling molecule subtypes, together with the discovery of their specific subcellular distribution, will help
the functional characterization of their intraregulatory pathways, leading to the
extrusion of insulin or glucagon secretory granules, and those leading to differentiation and apoptosis of islet cells.

Glucose effects on insulin and glucagon secretion in B and A cells are regulated by neuropeptides and hormones according to the need of the organism and its adaptation to the environment.[66] The initial action of the ligands is the activation of the seven
transmembrane or serpentine receptors (SR). The message is then transduced to the effectors through the heterotrimeric G proteins.[62,70] We will focus on the islet cell signaling systems involving (1) vasointestinal peptide (VIP),[17,35] pituitary adenylyl cyclase activating
polypeptide (PACAP),[3,58,69] and truncated glucagon-like peptide (tGLP-1) receptors;[71]
(2) the alpha subunits of G_s,[62,70] and G_{olf};[38,31] and (3) the adenylyl cyclases (AC) as prevalent effectors.[29,60] Classically, these systems regulate the intracellular functions of hormones and neuromediators by catalyzing the formation of cyclic AMP. However, different
effects originating from each subunit can not be excluded.[11,52] Cyclic AMP production is
regulated by both AC and phosphodiesterase isoforms.[60] The activation of the cyclic
AMP–dependent protein kinases (cyclic-A-PK), including protein kinase A, results in
phosphorylation cascade of a variety of enzymes and molecules leading to the control of
metabolic pathways and gene transcription. It is well known in B cells, that glucose triggers a Ca^{2+} influx. Such an effect could be potentiated by cyclic AMP either in phospho-
rylating the B cell voltage-dependent Ca^{2+} channel,[36] as shown for the cardiac L-type Ca^{2+}
channel,[13] or through the opening of nonselective cationic channels.[32] Furthermore, the
activated PKA can phosphorylate cyclic AMP–responsive element binding protein
(CREB) involved in insulin gene transcription. Cloning of the signaling components has

[a]This work was supported in part by Fondation de France, N° de Project 96002060.
[b]Corresponding author: Tel.: 33 1 4928 4660; Fax: 33 1 4928 4669; E-mail: Emami@adr.st
antoine.inserm.fr

revealed a large number of isotypes, characterized by their chronological appearance during development, by their cellular distribution, and by their interaction with the other cell regulatory components.

SIGNALIZATION AND LANGERHANS ISLETS

Signalization is present in fetal pancreas leading to the differentiation of islets, and during that period the adult pattern is progressively acquired.[63] The abundance and the diversity of the SR in pancreatic islets determine the adaptation of insulin and glucagon biosynthesis and secretion to a number of life circumstances responsible for changes in the energy requirements of the organism[66]: modification of physiological states, development, fasting/feeding alternance, pregnancy, and occurrence of stress. The SR target islet-specific neuroregulators and hormones, which through the cell surface receptors modify the islet functions appropriately. Glucose-dependent insulinotropic polypeptide (GIP),[46] gut glucagon,[7] oxyntomodulin,[8] tGLP-1,[20,51,53] VIP,[7] and CGRP[66] prominently activate AC. Cholecystokinin and acetylcholine stimulate phospholipase C (PLC),[73] and PACAP activates both AC and PLC.[54,76] These regulatory peptides stimulate glucose-induced insulin secretion, and with the exception of glucagon and tGLP-1, for the most part, potentiate the glucagon secretion induced by amino acids or low glucose concentrations. Somatostatin, which has receptors in islet cells,[47] generally inhibits both AC and ion channels, whereas galanin[1,16] is species specific in inhibiting insulin and stimulating glucagon release.[64,65] Here we have attempted to further characterize by using specific anti-receptor antibodies the presence and the cellular distribution of VIP, PACAP, and tGLP-1 receptors in different islet preparations.

VIP, PACAP, AND tGLP-1 RECEPTORS IN B AND A CELLS

Previously, the presence of specific VIP binding sites and receptor-mediated active endocytosis was demonstrated on islet pancreatic B cells by Anteunis and coworkers.[2] Using electron microscopic autoradiography after perfusion of the pancreas with ^{125}I-labeled VIP, they found that the VIP-induced insulin release[7] was promoted directly.[2] Using the same approach, the presence of VIP receptors (VIP R) on glucagon cells was observed, (Fig. 1), indicating that the effect of VIP in stimulating the glucagon release[7] is direct. This finding must now be interpreted in relation to the discovery of PACAP receptors (PACAP-R). The PACAP-R type I specifically binds PACAP and, as obtained from a rat brain cDNA library, two PACAP-R type I subtypes originating from mRNA alternative splicing. One, encoded by the clone pRPACAPR12, has an additional 28–amino acid sequence as compared to that of the clone pRPACAPR46-5,[3] and both their sequences differ from the VIP-R sequence. The PACAP-R type I was shown to be present in rat B cells.[76] The VIP-R of the B and A cells, which bind the ^{125}I-labeled VIP, is likely able to equally bind PACAP since the VIP-R is identical to the PACAP type II receptor (also named VIP/PACAP-R). VIP was also shown to bind to the PACAP type III receptor of 437 amino acids found in the MIN 6 mouse B cell line, which shares 50% identity with VIP-R and PACAP-R type I. In this cell, the PACAP-R type III is positively coupled with AC.[34] We have further studied the distribution of the VIP-R and PACAP-R type I in islets and diverse cell lines, using a rat VIP-R antibody directed against the internal part of the VIP-R (generously given by J. Fahrenkrug, Bispebjerg Hospital, Copenhagen, Denmark) (Fig. 1), and the human antibody 93093-2 directed against the C-terminal 25–amino acid peptide fragment (K411-A435 / K339-A463) common to the two isotypes of the PACAP-R type I, but different from the VIP-R and the PACAP-3–R sequences. PACAP-R is detected, in pancreatic A and B cells and the corresponding cell lines. Cells are labeled by the PACAP-R

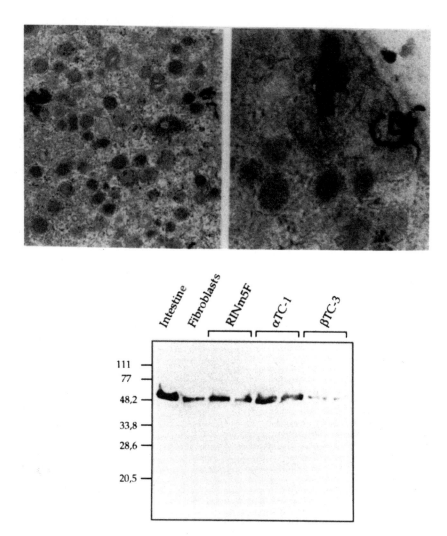

FIGURE 1. (Top) Autoradiography of pancreatic A cells after perfusion of the whole pancreas with ^{125}I-labeled VIP and processing of islet sections (80 nm) for ultrastructural examination according to a previously described method.[2] Briefly, 4.5 nM ^{125}I-labeled VIP, was injected for 3 min at 10°C to restrict endocytosis. The reaction was stopped by the fixation perfusion with 2.7% glutaraldehyde plus 0.8% formaldehyde in 80 mM sodium cacodylate buffer, and the islet cells stained for 3 min at 20°C with 0.12 mM dithizone. The islets were isolated by microdissection and prepared for electron microscopy. The autoradiographs are observed at 30,000 × magnification. **(Bottom)** Western blot of VIP receptors proteins in 300 μg protein obtained from whole cell lysate of the B cell lines RINm5F, αTC-1, βTC-3 rat intestinal epithelial cells, and embryonic mouse BP-A31 fibroblasts and detected by VIP receptor antisera (1:10,000). Blots were revealed according to the ECL procedure (Amersham, Life Sciences International, England).

type I antibody (FIG 2. c and d), mainly at the glucagon-labeled (FIG. 2, b) peripheral part (Glu 001, Novo-Nordisk) and to a lesser extent at the insulin-labeled (FIG. 2, a) part (Hui 018, Novo). The labeling was distributed inside the cell, excluding the nuclei. Acinar cells were poorly labeled (FIG. 2, d). The PACAP-R expression level is detected according to the following order: glucagon A cells>> insulin B cells > acinar cells. Some differences with the results of Yada, working with the same antibody, and a uniform labeling of the whole islet after avidin-biotin complex immunostaining[76] might be due to differences in the methodology and/or the dilution of the antibody. Using a PACAP-R type I antibody issued from another rabbit (93094-2), the order was: insulin B cells ≥ glucagon A cells > acinar cells with a definite labeling of the latter (data not shown). However the labeling is less intense than with 93093-2 antibody. The presence of PACAP receptor in acinar cells has already been reported in binding studies of calf pancreas.[42] Ultrastructural studies indicated that PACAP-R is present in subcellular structures from the endoplasmic reticulum to secretory granules and the plasma membrane (FIG. 3). The presence of specific PACAP-R on glucagon-producing cells strengthened the data obtained by stimulation of glucagon release after PACAP perfusion in rat,[77] mouse,[27] and human.[26] We have also detected the presence of PACAP-R type I by immunoblotting in different B cell lines (FIG. 3).

The insulinotropic effect of tGLP-1[51] was shown on B cells of different species including rat, mouse, hamster, and human. The tGLP-1 R was detected by immunoblotting or Northern analysis in rodents[10, 12, 24] as in different cell lines including RINm5F,[71] HIT T15,[24,48] β TC-1 cells, and human insulinoma. FIGURE 3(a) shows the immunofluorescence study of tGLP-1 R on RINm5F and HIT cells, using a specific rabbit polyclonal antibody raised against the entire extracellular domain of the rat tGLP-1 R sequence (up to the first membrane spanning domain) fused to the glutathione S-transferase (a generous gift of Dr. S. Mojsov, Rockefeller University, New York).[74] Thus, the presence of immunoreactive tGLP-1-R cells is corroborated in these two cell lines. Since the existence of the tGLP-R on A cells is still a controversial, we performed an immunoblot of tGLP-1-R in βTC-3 and αTC-1 cells (FIG. 3, b). The tGLP-1 receptors are clearly detected in the βTC-3 cell line as shown by the presence of bands of ~ 46 kD and ~ 70 kD. These molecular weights are identical to those found in RINm5F corresponding to the protein and its glycosylated form.[10] In αTC-1 cells, a 46 kD band, corresponding to the nonglycosylated tGLP-1 receptor, was observed but the 70 kD band was not found under this condition (FIG. 4). In other reports, the tGLP-1-R expression was either present or absent[23,50] in pancreatic A cells. In the first study, the presence of tGLP-1-R was assessed by immunofluorescence, using an antibody raised against a 18–amino acid sequence from the GLP-1-R N-terminal sequence, or by RT-PCR on a single rat islet cell containing proglucagon mRNA.[30] Furthermore, it was found that tGLP-1-R was functional in isolated rat pancreatic A cells.[19] However, these discrepancies extend to the effect of tGLP-1, which stimulated,[19] inhibited,[53] or did not modify the glucagon secretion. Cyclic AMP–producing agents are known to enhance glucagon secretion in perfused neonatal pancreas.[37] Since tGLP-1 is a cyclic AMP–producing agent, one could expect that it stimulates at least glucagon release, independently of its other secondary effects. Indeed, the inhibitory effect of tGLP-1 on glucagon secretion might occur through a tGLP-1–induced insulin[28] or somatostatin release,[23] which in turn inhibits glucagon secretion by a paracrine mechanism. Another explanation could be that the GLP-1-R is not processed regularly in normal A cells and that tGLP-1-R might remain immature and consequently not functional. Further investigations are necessary to define the mechanism of action of tGLP-1 on glucagon processing and secretion. Classically, the specificity of the signaling transduction systems was due to the receptor selectivity. But it now appears that the specificity of the ligand effects also occurs beyond the cell surface receptor and that each receptor could be able to discriminate the pattern of different Gα protein subtypes.[11] We have therefore initiated a study to characterize the heterotrimeric G proteins and the AC isoforms present in the pancreatic islet cells.

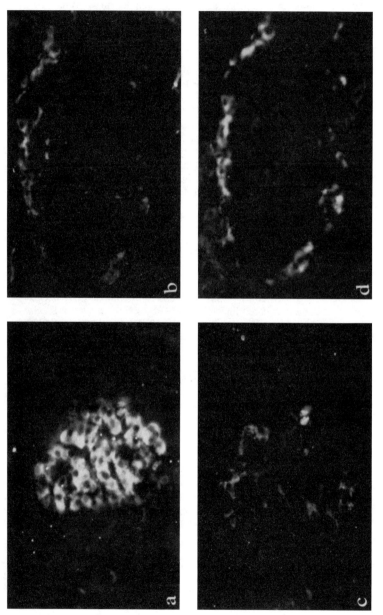

FIGURE 2. Localization of PACAP type I receptors (**c** and **d**) in Wistar rat pancreas. Immunohistochemical double-labeling immunofluorescence using monoclonal antibodies (Novo Nordisk) against insulin (HUI-108) (**a**) or glucagon (GLU-001) (**b**) tagged with DTAF (**a** and **c**) and with TRSC (**b** and **d**). Studies were performed on fixed sections in 4% paraformaldehyde, 100 mM phosphate buffer, pH 7.4 (1 h). Semithin sections (0.5–1.0 μm) were incubated with the primary antibody (1/100) followed by cross-absorbed DTAF-conjugated donkey antirabbit F(ab')2 of TRSC-conjugated donkey antimouse antirabbit F(ab')]. PACAP-R and glucagon labeling are colocalized. PACAP labeling is also observed in B cell.

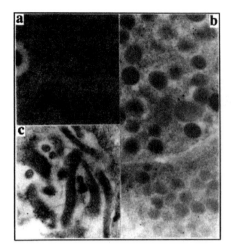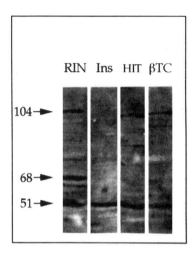

FIGURE 3. (Left) Subcellular distribution of the PACAP-R type I labeling in islet cells. Paralysine periodate–treated sections of Sprague-Dawley rat pancreas usually processed for EM (generously given by Prof. Bendayan, Départmente de Pathologie, et de la Biologie Cellulaire, Université de Montreal), were incubated with the specific PACAP-R type I antibody (1/100) overnight at 4°C, rinsed with PBS, incubated in PBS-ovalbumin, and then protein A–gold or IgG-gold complexes,[9] gold particles 5 nm for 30 min at room temperature. The grids were then washed with PBS, rinsed with distilled water, dried and stained with uranyl acetate before examination. Labeling is shown in the endoplasmic reticulum of a B cell (**a**), the glucagon secretory granules and very slighly on a somatostatin cells (**b**), and the microvilli of B cells (**c**). **(Right)** Western blot of the PACAP receptor in RINm5F, INS-1, HIT-T15, and β-TC3 cell lines. Whole cell lysates (100 μg) from cells proteins were separated by SDS-PAGE on a 12.5% acrylamide gel and transferred to a nitrocellulose membrane and analyzed by Western blotting using the antibody against the PACAP sequence KRKWRSWKVNRYFTMD-FKHRHPSLA (Dr. Arimura), also used for immunocytochemistry. A major band at about 51 kD is observed in all the cell lines. Several bands between 51 kD and 68 kD are particularly visible in the RINm5F cells, which might correspond to different degree of glycosylation. The bands at 104–106 kD are variable according to the cell lines and remain unexplained.

ISLET CELLS AND G PROTEINS

Gα subunits are responsible for the specificity in the coupling of the ligand/receptors to the intracellular effectors,[11] though the βγ subunits might also be effective in this respect.[52,61] The potential role of the G subtypes in the B cell has been recently reviewed.[46,67,78] Briefly, Gα subunits belong to a family of 39–52 kD proteins and are issued from at least 16 genes. They attach to the cellular membranes after specific post-translational acylation, myristoylation, or palmitoylation. They are classically responsible for the specificity of the signal transduction. Furthermore, the 36 kD Gβ subunits are tightly complexed with the 6–9 kD γ subunits. Isoprenylation and, in some instances, methylation of Gγ, are necessary for G protein function. The Gβγ complex is need for the Gα receptor interactions since only the membrane-inserted heterotrimeric Gαβγ protein is able to interact with the receptor proteins.[11] It is still unknown to what extent G subunits are linked to a specific AC isoform. The crystal structures of some G-protein subunits and heterotrimers[41] helped to identify the surface of interactions within the subunits and between the subunits with their receptors or their effectors. The successive interactions of ligands to their receptors, to G protein subtypes, and to AC isoforms finally result in the conformational changes underlying the receptor-catalyzed guanine nucleotide exchange.

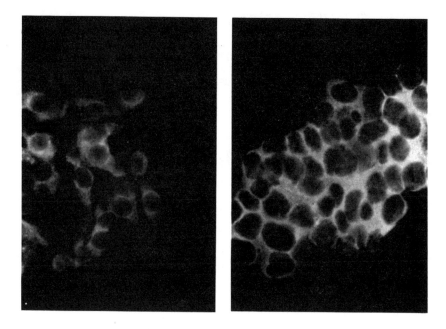

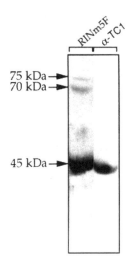

FIGURE 4. (**Top**) Immunocytochemical analysis of tGLP-1 receptor in RINm5F (**left**) and HIT-T15 (**right**). (Magnification 40×) (**Bottom**) Detection of tGLP-1 receptor in RINm5F and aTC-1. Proteins from whole cell lysates (300 μg) from cells were separated by SDS-PAGE on a 12.5% acrylamide gel and transferred to a nitrocellulose membrane and were analyzed by Western blotting using polyclonal antibody for rat tGLP-1 receptors and ECL system.

The presence of a number of distinct patterns in the molecular assembly amplifies the diversity of how the signal is transduced and accounts for the cell-specific diversity of the ligand-induced biological or pharmacological effects. The $G\alpha$ proteins are classified mainly on the basis of their interactions with the effectors. $G_s\alpha$ and probably $G_{olf}\alpha$ couple to AC stimulation and are sensitive to cholera toxin. $G_s\alpha$ is expressed ubiquitously in various tissues, while others like $G_{olf}\alpha$ are more or less cell specific. Gi/o are often responsible for AC inhibition and are sensitive to pertussis toxin (PTX). Gq/11 are coupling molecules activating PLC and are insensitive to PTX. In addition, transduced effects on ionic channels are observed with $G_q\alpha$, $G_{11}\alpha$, and $G_{16}\alpha$. On the other hand, this system includes the possibility of cross-talk between different $G\alpha$ subunits and the $\beta\gamma$ complex issued from another G protein upon receptor activation.[11] This explains the final cellular effect, which is subtype specific and depends on the overall pattern of the G subunits targeted to the plasma membrane and their effector isotypes. For example, $G_s\alpha$, which classically couples with the stimulating serpentine receptors, may also couple with EGF receptors.[69] In another study, $G_s\alpha$ couples the VIP receptor to AC to stimulate lacrimal secretion. This effect is suppressed by $G_s\alpha$ not by $G_{q/11}\alpha$ antibody. However, the effect of the cholinergic agent carbachol is blocked by $G_{q/11}\alpha$ antibody but also, although to a lesser extent, by Gs antibody.[49] Furthermore, the predominant action of $G_q\alpha$ is to transduce the CCK signal to PLC, but $G_q\alpha$ might also be responsible for protein kinase activation in rat pancreatic acinar cells.[18] $G_{11}\alpha$ was shown to increase intracellular inositol phosphate, though its final role has not yet been elucidated.[75] Type II AC might integrate coincident signals not only coming from Gs and Gi, but also from Gq in a kidney cell line.[43] Distinct PLC-beta isoenzymes are activated by Gq/11 after stimulation by CCK, carbachol, or bombesin in rat pancreatic acinar membranes.[56] In a recent study, it was shown that VIP or PACAP activate PI-3 kinase activity in islets,[68] an unidentified effect which might be mediated through an heterotrimeric G. The effects of this enzyme are likely to be involved in exocytosis since wortmannin, a known inhibitor of PI-3 kinase, inhibits the PACAP- and VIP-stimulated insulin secretion. This is a new function, since PI-3 kinase is known for its role in the transduction of mitogenic signals issued from growth factor receptors. This further substantiated the cross-talk between the proliferative and secretory functions of the signal molecules operating in the cell. What subtypes of G proteins are involved in this action remains to be explored. Due to their importance in regulating insulin and glucagon secretion,[66] as well as other general functions, it is of interest to provide direct information on the stimulatory GTP-binding proteins in B and A pancreatic islet cells of normal animals, and to identify and determine the cell specificity and function. The difficulty in studying the G proteins belonging to the cyclic AMP or the PLC stimulatory family is due to the strong subunit analogy between each other. For example $G_{olf}\alpha$ and $G_s\alpha$ or $G_q\alpha$ and $G_{11}\alpha$ share 88% and 88.5% amino acid identities, compelling the use of several specific antibodies for their characterization. From our data it is possible to prove the distribution of the G protein subunits in the islets. Immunoblotting studies of rat and hamster islet pancreas extracts show the presence of $G_{olf}\alpha$ and $G_s\alpha$.[55] The $G_{olf}\alpha$ is predominantly, if not exclusively, localized in B cells as detected by immunohistological, immunocytological, and *in situ* hybridization methods, whereas $G_s\alpha$ is present in B cells as well as in glucagon A cells and acinar cells.[4] Comparative expression of G proteins in the βTC-1 and αTC-1 cell lines shows the specific 46 kD band of $G_{olf}\alpha$ that is not expressed in A cells, whereas $G_s\alpha$ is expressed in both cell types. Furthermore, islet $G_s\alpha$ and $G_{olf}\alpha$ proteins labeled by nicotinamide adenine dinucleotide[adenylate-^{32}P] are ADP-ribosylated by cholera toxin, indicating their association with the $\beta\gamma$ subunits in islets, and consequently, demonstrate their functional activity.[55] Other approaches are used to determine the G protein isotype function. Subcellular localization might indicate whether their effects are not restricted to the plasma membrane. Ultrastructural data in the normal islet B cells show the presence of $G_{olf}\alpha$, $G_s\alpha$, and $G_q\alpha$ subunits from the endoplasmic reticulum to the plasma membrane, though the Golgi network and the secretory granules are likely to be the main intracellular pathway. These α subunits have a specific pattern of distribution in the subcellular compartments, $G_{olf}\alpha$ and $G_s\alpha$ being prevalent around the secretory granules and $G_q\alpha$ localized in the endoplasmic reticulum.[5] The intracellular roles of these α subunits are possible, since the βγ complex also has been detected in the insulin secretion granule,[39] permitting the assembly of a functional heterotrimer G protein. The function of the G protein isotypes was also determined with the creation of transgenic mice in which oligonucleotide antisense expression is achieved in neonatal adipocytes showing insulin resistance.[45] Our preliminary experiments indicate that transfection of HIT-T15 cells with $G_{olf}\alpha$ antisense produced major change in cell morphology accompanied by a decrease of $G_{olf}\alpha$ expression in these cells (FIG. 5). Overexpression of $G_s\alpha$ was attempted in islets of transgenic mice, where the increase in cyclic AMP was unfortunately attenuated by a simultaneous rise in phosphodiesterase activity.[44] In another cellular system, the overexpression of $G_s\alpha$ was shown to modulate myoblast differentiation.[72]

ADENYLYL CYCLASE

At least nine isoforms of AC have been identified in the past few years[15,29,59] that exhibit quite a different pattern of expression throughout the brain. AC isoforms are regulated independently by calcium, Gβγ subunits, and protein kinase C. To what extent the expression of AC isoforms is found to be cell specific in islets is still unknown. Presently, studies are undertaken in the laboratory, using antibodies from Santa Cruz Biotechnology, Inc. (Santa Cruz, CA) directed against the different AC isoforms, I (human), II (human), III (rat), IV (rat), and V/VI (human), to determine whether the AC expression is islet-cell specific. Immunohistological studies of AC isoforms in pancreatic endocrine cells were performed by double-labeling with either insulin or glucagon in different species including

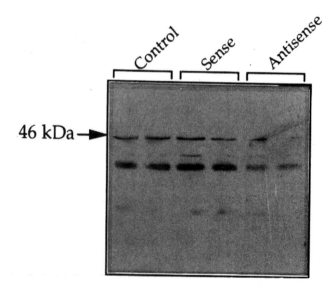

FIGURE 5. Western blot of $G_{olf}\alpha$ in the whole lysate of HIT-T15 cells (150 μg protein). Duplicate of control cells and of cells transfected with sense or antisense of Golf, kindly given by Dr. Shimizu (Otsuka Gen Research Institute, Takushima, Japan), were treated with the K19 antibody (1/500) from Santa-Cruz and bands were as indicated in the legend of FIGURE 1.

mouse, rat, hamster, and human[25] and are substantiated by ultrastructural analysis of rat pancreas.[5,25] Western blots were also performed on RINm5F cells,[25] and βTC-3 and αTC-1 cell lines. Preliminary results indicated that B and A cells were stained persistently by AC antibodies from Santa Cruz, and that the expression of AC isoforms was higher in neonatal rat than in the adult pancreas. The distribution of the AC immunoreactivity was cell dependent, the AC II–like labeling was more marked in rat and mouse A cells than in B islet cells, and was exclusively seen in A cells of human and hamster adult pancreas. Ultrastructural studies with the AC-II antibody, performed in adult rats, showed a discrete labeling at the plasma membrane level, while the largest part of the labeling was distributed within different cellular compartments, probably in relation to the processing of this peptide. Immunoblot studies showed the presence of bands with a molecular weight lower than that expected for AC, which suggested that some epitopes of these polyclonal antibodies may be shown by and likely to be related to protease digestion of the cyclase during the post-translational process. *In situ* hybridization of AC is under investigation in our laboratory to further document the nature of the AC detected by immunocytochemistry and Western blot. The processes under which the AC isoforms become functionally activatable are unknown. It is generally admitted that all mammalian AC isoforms ought to be inserted into a membrane to become activated, with the exception of a soluble form of AC found in testis stimulated by forskolin and $G_s\alpha$.[70] The proteolytic system leading to a soluble form of AC of low molecular weight depends on the isoform and therefore might play a important role for intracellular signaling pathway.[21] AC types were classified by their regulation, i.e., AC1, AC3, and AC8 are stimulated by Ca^{2+}/calmodulin complex but this appear extraphysiological; AC5 and AC6 are inhibited by elevating of calcium concentrations within the physiological range; AC2, AC4, and AC7 are insensitive to calcium, but activated by βγ subunits; and AC9 is insensitive to both calcium or βγ complex subunit. AC2 has also been reported to be activated by PKC. The nature of interactions between AC and other signaling proteins is far from clear. For example, the co-expression of the AC III isoform with $G_{olf}\alpha$ and the olfactive receptors in neuro-olfactive epithelium suggests that both of these molecules might have prevalent interactions and that AC III is probably stimulated by Ca^{2+}. In this respect, our finding that the AC III was present in both A and B cells associated with $G_{olf}\alpha$ expression in B cells[25] is a matter of interest, since Ca^{2+} cellular fluxes are of importance in insulin or glucagon release. However, further work is required since calcium regulation of AC isoforms remains controversial.[14] Using intact cell and *in vivo* studies, type III AC might be either inhibited or insensitive to calcium, so results obtained from a cell preparation cannot be extrapolated to another tissues.

The identification of a number of signaling molecule subtypes together with the discovery of their cellular distribution might well help in the definition of the regulatory pathways leading to the extrusion of insulin or glucagon secretion granules after appropriate ligand stimulation. It will also help clarify the link between these pathways and those involved in islet cells differentiation and apoptosis. The cellular specificity of the signaling protein localization might, for example, account for the differences observed in the stimulation of insulin or glucagon release by B and A cells, despite of the identity of this stimulating ligand. Insulin exocytosis is potentiated only in the presence of appropriate glucose concentrations, whereas that of glucagon is enhanced only at low glucose levels. Furthermore, the presence of a G protein on secretory granules leaves open the question of the role in driving the granules toward the plasma membrane once secretion is stimulated. Finally, the involvement of G protein and AC isoforms has been reported in some diseases.[14,33,57] It cannot be excluded that functional or genetic modifications of these proteins are involved in diabetes at the level of the islet-cell differentiation and function. For example, it has been recently demonstrated that the glucose stimulatory effect on B cells induces post-translational modifications of G protein γ subunit via a non receptor-mediated mechanism.[40]

REFERENCES

1. AMIRANOFF, B., A. M. LORINET & M. LABURTHE. 1989. Galanin receptor in the rat pancreatic beta cell line RINm5F. Molecular characterization by chemical cross-linking. J. Biol. Chem. **264**: 20714–20717.
2. ANTEUNIS, A. *et al.* 1989. Ultrastructural analysis of VIP internalization in rat β- and acinar cells in situ. Am. J. Physiol. **256**: G689–G697.
3. ARIMURA, A. & H. ONDA. 1993. Biosynthesis and processing of PACAP and characterization of its receptor. *In* International Symposium on Vasoactive Intestinal Peptide Pituitary Adenylate Cyclase Activating Polypeptide & Related Regulatory Peptides. G. Rosselin, Ed.:101–112. World Scientific Publ.
4. ASTESANO, A. *et al.* 1995. Subcellular localization of G-protein alpha subunits in adult rat islets. Ann. NY Acad. Sci. **808**: 549–554.
5. ASTESANO, A. *et al.* 1997. Diversity of the transduction system expression in pancreatic A cells. Diabetologia **40**(suppl. 1): A137–532.
6. BATAILLE, D., P. FREYCHET & G. ROSSELIN. 1974. Interactions of glucagon, gut glucagon, vasoactive intestinal polypeptide and secretion with liver and fat cell plasma membranes: binding to specific sites and stimulation of adenylate cyclase. Endocrinology **95**: 713–721.
7. BATAILLE, D. *et al.* 1977. Effect of vasoactive intestinal peptide 5 (VIP) and gastric inhibitory peptide (GIP) on insulin and glucagon release by perfused newborn rat pancreas. *In* Glucagon: Its Role in Physiology and Clinical Medicine. P. P. Foa, J. S. Bajaj & N. R. Foa, Eds.: 255–269. Springer Verlag. New York.
8. BATAILLE, D. *et al.* 1981. Bioreactive enteroglucagon (oxyntomodulin): present knowledge on its chemical structure and its biological activities. Peptides **2**: 41–44.
9. BENDAYAN, M. 1995. Colloïdal gold post-embedding immunocytochemistry. Prog. Histochem. Cytochem. **29**: 1–163.
10. BOISSARD, C. *et al.* 1997. Immunocharacterization and function of glucagon-like peptide-1 islet receptor in Syrian hamster. Diabetologia **40**(suppl. 1): A129–496.
11. BOURNE, H. R. 1997. How receptors talk to trimeric G proteins. Curr. Opin. Cell. Biol. **9**: 134–142.
12. CAMPOS, R. V., Y. C. LEE & D. J. DRUCKER. 1994. Divergent tissue-specific and developmental expression of receptors for glucagon and glucagon-like peptide-1 in the mouse. Endocrinology **134**: 2156–2164.
13. CHATTERJEE, T. K., R.V. SHARMA & R. A. FISHER. 1996. Molecular cloning of a novel variant of the pituitary adenylate cyclase-activating polypeptide (PACAP) receptor that stimulates calcium influx by activation of L-type calcium channels. J. Biol. Chem. **271**: 32226–32232.
14. CHETHAM, P. M. *et al.* Ca^{2+} -inhibitable adenylyl cyclase and pulmonary microvascular permeability. Am. J. Physiol. **273**: L22–L30.
15. COOPER, D. M., N. MONS & J. W. KARPEN. 1995. Adenylyl cyclases and the interaction between calcium and cyclic-AMP signaling. Nature **374**: 421–424.
16. CORMONT, M. *et al.* 1991. Identification of G protein α-subunits in RINm5F cells and their selective interaction with galanin receptor. Diabetes **40**: 1170–1176.
17. COUVINEAU, A. *et al.* 1994. Human intestinal VIP receptor: cloning and functional expression of two cDNA encoding proteins with different N-terminal domains. Biochem. Biophys. Res. Commun. **200**: 769–776.
18. DABROWSKI, A. *et al.* 1996. Cholecystokinin stimulates formation of shr-grb2 complex in rat pancreatic acinar cells through a protein kinase C-dependent mechanism. J. Biol. Chem. **271**: 27125–27129.
19. DING, W. G. *et al.* 1997. Glucagon-like peptide I glucose-dependent insulinotropic polypeptide stimulate Ca^{2+}-induced secretion in rat α-cells by a protein kinase A-mediated mechanism. Diabetes **46**: 792–800.
20. DRUCKER, D. J. *et al.* 1987. Glucagon-like peptide-1 stimulates insulin gene expression and increases cyclic-AMP levels in a rat islet cell line. Proc. Natl. Acad. Sci. USA **84**: 3434–3438.
21. EBINA, T. *et al.* 1997. Isoform-dependent activation of adenylyl cyclase by proteolysis. FEBS Lett. **401**: 223–226.
22. EFRAT, S. 1988. Beta-cell lines derived from transgenic mice expressing a hybrid insulin gene-oncogene. Proc. Natl. Acad. Sci. USA **85**: 9037–9041.

23. FEHMANN, H. C. & J. F. HABENER. 1991. Functional receptors for the insulinotropic hormone glucagon-like peptide-1 (7-37) on a somatostatin secreting cell line. FEBS Lett. **279**: 335–340.
24. FEHMANN, H. C. & J. F. HABENER. 1991. Homologous desensitization of the insulinotropic glucagon-like peptide-1-(7-37) receptor on insulinoma (HIT-T15) cells. Endocrinology **128**: 2880–2888.
25. FERRAND, N. *et al.* 1997. Specificity of adenylylcyclase isoforms expression in pancreatic islets and acini. Diabetologia **40**(suppl. 1): A113–438.
26. FILIPSSON, K. 1997. Pituitary adenylate cyclase-activating polypeptide stimulates insulin and glucagon secretion in humans. J. Clin. Endocrinol. Metab. **82**: 3093–3098.
27. FRIDOLF, T. 1992. Pituitary adenylate cyclase-activating polypeptide (PACAP) occurrence in rodent pancreas and effects on insulin and glucagon secretion in the mouse. Cell Tissue Res. **269**: 275–279.
28. GREENBAUM, C. J. *et al.* 1991. Intra-islet insulin permits glucose to directly suppress pancreatic A cell function. J. Clin. Invest. **88**: 767–773.
29. HANOUNE, J. *et al.* 1997. Adenylyl cyclases: structure, regulation and function in an enzyme superfamily. Mol. Cell. Endocrinol. **128**: 179–194.
30. HELLER, R. S., T. J. KIEFFER & J. F. HABENER. 1997. Insulinotropic glucagon-like peptide I receptor expression in glucagon-producing α-cells of the rat endocrine pancreas. Diabetes **46**: 785–791.
31. HERVE, D., M. ROGARD & M. LEVI-STRAUSS. 1995. Molecular analysis of the multiple $G_{olf}\alpha$ subunit mRNAs in the rat brain. Mol. Brain Res. **32**: 125–134.
32. HOLZ, C. G., C. A. LEECH & J. F. HABENER. 1995. Activation of a cyclic-AMP–regulated Ca(2+) signaling pathway in pancreatic beta-cells by the insulinotropic hormone glucagon-like peptide-1. J. Biol. Chem. **270**: 17749–17757.
33. IIRI, T., Z. FARFEL & H. R. BOURNE. 1997. Conditional activation defect of a human $G_s\alpha$ mutant. Proc. Natl. Acad. Sci. USA **94**: 5656–5661.
34. INAGAKI, N. 1994. Cloning and functional characterization of a third pituitary adenylate cyclase-activating polypeptide receptor subtype expressed in insulin-secreting cells. Proc. Natl. Acad. Sci. USA **91**: 2679–2683.
35. ISHIHARA, T. *et al.* 1992. Functional expression and tissue distribution of a novel receptor for vasoactive intestinal polypeptide. Neuron **8**: 811–819.
36. ISLAM, M. S. 1995. Effects of caffeine on cytoplasmic free Ca^2+ concentration in pancreatic beta-cells are mediated by interaction with ATP-sensitive K^+ channels and L-type voltage-gated Ca^{2+} channels but not the ryanodine receptor. Biochem. J. **306**: 679–686.
37. JARROUSSE, C. & G. ROSSELIN. 1975. Interaction of amino acids and cyclic-AMP on the release of insulin and glucagon by newborn rat pancreas. Endocrinology **96**: 168–177.
38. JONES, D. T. & R. R. REED. 1989. G_{olf}: an olfactory neuron specific-G protein involved in odorant signal transduction. Science **244**: 790–795.
39. KOWLURU, A. *et al.* 1996. A novel regulatory mechanism for trimeric GTP-binding proteins in the membrane and secretory granule fractions of human and rodent β cells. Biochem. J. **313**: 97–107.
40. KOWLURU, A. G. LI & S. A. METZ. 1997. Glucose activates the carboxyl methylation of subunits of trimeric GTP-binding proteins in pancreatic β cells. J. Clin. Invest. **100**: 1596–1610.
41. LAMBRIGHT, D. G. *et al.* 1996. The 2.0 Å crystal structure of a heterotrimeric G protein. Nature **379**: 311–319.
42. LE MEUTH, V. *et al.* 1991. Characterization of binding sites for VIP-related peptides and activation of adenylate cyclase in developing pancreas. Am. J. Physiol. **260**: G265–G274.
43. LUSTIG, K. D. *et al.* 1993. Type II adenylylcyclase integrates coincident signals from Gs, Gi, and Gq. J. Biol. Chem. **268**: 13900–13905.
44. MA, Y. H. *et al.* 1994. Constitutively active stimulatory G-protein alpha s in beta-cells of transgenic mice causes counterregulation of the increased adenosine 3′,5′-monophosphate and insulin secretion. Endocrinology **134**: 42–47.
45. MALBON, C. C. 1997. Heterotrimeric G-proteins and development. Biochem. Pharmacol. **53**: 1–4.
46. MALETTI, M. *et al.* 1987. Evidence of functional gastric inhibitory polypeptide (GIP) receptors in human insulinoma. Diabetes **36**: 1336–1340.

47. MALETTI, M. *et al.* 1992. Solubilization and partial purification of somatostatin-28 preferring receptors from hamster pancreatic beta cells. J. Biol. Chem. **267**:15620–15625.
48. MARIE, J. C. *et al.* 1996. Glucagon acts through its own receptors in the presence of functional glucagon-like peptide-1 receptors on hamster insulinoma. Endocrinology **137**: 4108–4114.
49. MENERAY, M. A., T. Y. FIELDS & D. J. BENNETT. 1997. Gs and Gq/11 couple vasoactive intestinal peptide and cholinergic stimulation to lacrimal secretion. Invest. Ophthalmol. Vis. Sci. **38**: 1261–1270.
50. MOENS, K. *et al.* 1996. Expression and functional activity of glucagon, glucagon-like peptide-1 and glucose-dependent insulinotropic peptide receptors in rat pancreatic islet cells. Diabetes **45**: 257–261.
51. MOJSOV, S., G. C. WEIR & J. F. HABENER. 1986. Insulinotropin: glucagon-like peptide-1(7-37) co-encoded in the glucagon gene is a potent stimulator of insulin release in the perfused rat pancreas. J. Clin. Invest. **79**: 616–619.
52. NEER, J. E. 1995. Heterotrimeric G proteins: organizers of transmembrane signals. Cell **80**: 249–257.
53. ORSKOV C, HOLST J. J. & O. V. NIELSEN. 1988. Effect of truncated glucagon-like peptide-1 [proglucagon-(78-107) amide] on endocrine secretion from pig pancreas, antrum, and nonantral stomach. Endocrinology **123**: 2009–2013.
54. PANTALONI, C. *et al.* 1996. Alternative splicing in the N-terminal extracellular domain of the pituitary adenylate cyclase activating polypeptide (PACAP) receptor modulates receptor selectivity and relative potencies of PACAP-27 and PACAP-38 in phospholipase C activation. J. Biol. Chem. **271**: 22146–22151.
55. PHAN, H. *et al.* 1997. Functional and immunohistological species specificity of $G_{olf}\alpha$ and $G_i\alpha$ expression in pancreas, islets and B-cells. Diabetologia **40**(suppl. 1, A 106): 410.
56. PIIPER, A. *et al.* 1997. CCK carbachol and bombesin activate distinct PLC-beta isoenzymes via Gq/11 in rat pancreatic acinar membranes. Am. J. Physiol. **272**: G135–G140.
57. PING P. *et al.* 1997. Adenylyl cyclase and G protein receptor kinase expression during development of heart failure. Am. J. Physiol. **273**: H707–H717.
58. PISEGNA, J. R. & S. A. WANK. 1993. Molecular cloning and functional expression of the pituitary adenylate cyclase-activating polypeptide type I receptor. Proc. Natl. Acad. Sci. USA **90**: 6345–6349.
59. PREMONT, R. *et al.* 1996. Identification and characterization of a novel and widely expressed isoform of adenylyl cyclase. J. Biol. Chem. **271**: 13900–1396.
60. PRENTKI, M. & R. A. RIUS. 1987. Ca^{2+}, cyclic-AMP, and phospholipid-derived messengers in coupling mechanisms of insulin secretion. Physiol. Rev. **67**: 1185–1248.
61. REHM, A. & H. L. PLOEGH. 1997. Assembly and intracellular targeting of the $\beta\gamma$ subunits of heterotrimeric G proteins. J. Cell. Biol. **137**: 305–317.
62. RODBELL, M. 1993. G-proteins and membrane transduction: a new model. *In* International Symposium on Vasoactive Intestinal Peptide Pituitary Adenylate Cyclase Activating Polypeptide & Related Regulatory Peptides. G. Rosselin, Ed.: 101–112. World Scientific.
63. ROSSELIN, G. & S. EMAMI. 1997. Growth and differentiation of the islet cells in neonates. *In* Pancreatic Growth and Regeneration. Chapter 3, Nora Sarvetnick, Ed. : 44–97. Karger Landes Systems. Basel.
64. SEAQUIST, E. R. *et al.* 1992. G-proteins and hormonal inhibition of insulin secretion from HIT-T15 cells and isolated rat islets. Diabetes **41**: 1390–1399.
65. SHARP, G. W. G. 1996. Mechanisms of inhibition of insulin release. Am. J. Physiol. **271**: C1781–C1799.
66. SKOGLUND, G. & G. ROSSELIN. 1993. Receptor coupled information processes in the regulation of pancreatic β-cell functions by the alimentary tract. Biomed. Res. **14**: 21–51.
67. Spengler, D. *et al.* 1993. Differential signal transduction by five splice variants of the PACAP receptor. Nature **365**: 170–175.
68. STRAUB, S. G. & G. W. G. SHARP. 1996. A wortmannin-sensitive signal transduction pathway is involved in the stimulation of insulin release by vasoactive intestinal polypeptide and pituitary adenylate cyclase-activating polypeptide. J. Biol. Chem. **271**: 1660–1668.
69. SUN, H. *et al.* 1997. The juxtamembrane, cytosolic region of the epidermal growth factor receptor is involved in association with α-subunit of Gs. J. Biol. Chem. **272**: 5413–5420.

70. SUNAHARA, R. K. *et al.* 1997. Interaction of Gsα with the cytosolic domains of mammalian adenylyl cyclase. J Biol Chem. **272**: 22265–22271.
71. THORENS, B. 1992. Expression cloning of the pancreatic β cell receptor for the gluco-incretin hormone glucagon-like peptide 1. Proc. Natl. Acad. Sci. USA **89**: 8641–8645.
72. TSAI, C. C., J. E. SAFFITZ & J. J. BILLADELLO. 1997. Expression of the Gs protein α-subunit disrupts in the normal program of differentiation in cultured murine myogenic cells. Clin. Invest. **99**: 67–76.
73. VERSPOHL, E. J. & K. HERRMANN. 1996. Involvement of G proteins in the effect of carbachol and cholecystokinin in rat pancreatic islets. Am. J. Physiol. **271**: E65–E72.
74. WEI, Y. & S. MOJSOV. 1995. Tissue-specific expression of the human receptor for glucagon-like peptide-1: brain, heart and pancreatic forms have the same deduced amino acid sequences. FEBS Lett. **358**: 219–224.
75. WISE, A., PARENTI, M. & G. MILLIGAN. 1997. Interaction of the G-protein G11alpha with receptors and phosphoinositidase C: the contribution of G-protein palmitoylation and membrane association. FEBS Lett. **407**: 257-260.
76. YADA, T. *et al.* 1994. Pituitary adenylate cyclase activating polypeptide is an extraordinarily potent intrapancreatic regulator of insulin secretion from islet β-cells. J. Biol. Chem. **2**: 1290–1293.
77. YOKOTA, C. *et al.* 1993. Stimulatory effects of pituitary adenylate cyclase-activating polypeptide (PACAP) on insulin and glucagon release from the isolated perfused rat pancreas. Acta Endocrinol. **129**: 473–479.
78. ZIGMAN, J. M. *et al.* 1993. Human G$_{olf}$α: complementary deoxyribonucleic acid structure and expression in pancreatic islets and other tissues outside the olfactory neuroepithelium and central nervous system. Endocrinology **133**: 2508–2514.

Miniglucagon: A Local Regulator of Islet Physiology

STÉPHANE DALLE, PHILIPPE BLACHE, DUNG LE-NGUYEN,
LAURENCE LE BRIGAND, AND DOMINIQUE BATAILLE[a]

*Institut National de la Santé et de la Recherche Médicale INSERM
U376, CHU Arnaud-de-Villeneuve, 34295 Montpellier Cedex, France*

ABSTRACT: Miniglucagon, or glucagon-[19-29], is partially processed from glucagon in its target tissues where it modulates the glucagon action. In the islets of Langerhans, the glucagon-producing A cells contain miniglucagon at a significant level (2–5% of the glucagon content). We studied a possible control of insulin release by miniglucagon using as a model the MIN6 cell line. Miniglucagon, in the 10^{-14} to 10^{-9} M range, inhibited insulin release induced by glucose, glucagon, tGLP-1, or glibenclamide by 85–100% with an IC_{50} close to 1 pM. While no change in the cyclic AMP content was noted, Ca^{2+} influx was reduced in parallel with the inhibition of insulin release. Use of pharmacological modulators of L-type voltage-sensitive Ca^{2+} channels and bacterial toxins indicates that miniglucagon blocks insulin release by closing this type of channel via a pertussis toxin–sensitive G protein. Miniglucagon is a novel, possibly physiologically relevant, local regulator of islet function.

Proglucagon processing, like proopiomelanocortin processing,[1] represents a complex post-translational event, leading to a variety of regulatory peptides.[2] Indeed, proglucagon, the processing of which is tissue specific, leads in the pancreatic alpha cells to glucagon (a regulator of blood sugar opposite that of insulin). In the intestinal L cells, proglucagon leads to several peptides with different biological roles: oxyntomodulin and glicentin, which contain the glucagon sequence plus a C-terminal extension, which provides their biological specificity directed towards the gut physiology.[3] Proglucagon also leads to the glucagon-like peptides GLP-1 and GLP-2, which display specific activities. The question of the post-translational processing of proglucagon is rendered even more complex by the fact that C-terminal fragments of the mature forms (in particular, glucagon, oxyntomodulin, and glicentin) have their own spectrum of activity. This is true for oxyntomodulin-(19-37)[3,4] and for glucagon-(19-29), both products of a cleavage at a basic doublet (Arg_{17}-Arg_{18}) present in the original hormones. The latter peptide, also referred to as "miniglucagon," displays biological activities that differ from that of glucagon, the parent hormone.[5–7] In several instances, the miniglucagon action is antagonistic to that of glucagon. This is true, for instance, in the cardiac myoblasts where glucagon displays positive inotropic action whereas low doses of miniglucagon exert a negative control on the same parameter.[8]

The presence of miniglucagon in pancreas and in an alpha cell line[9] led us to investigate whether miniglucagon might control insulin release. In a preliminary report, we showed that miniglucagon is able to reduce the glucose- and glucagon-stimulated insulin release from the insulin-secreting β-TC cell line.[10]

[a]Corresponding author: Tel.: 33 4 67 41 52 20/30; Fax: 33 4 67 41 52 22; E-mail: bataille@u376.montp.inserm.fr

We show here that miniglucagon is a very potent and efficient inhibitor of insulin release from the MIN6 β cell line by closing, in a pertussis toxin–sensitive pathway, the β cell plasma membrane voltage-sensitive calcium channel.

MATERIALS AND METHODS

Materials

[NorLeu[27]] miniglucagon was synthesized in our laboratory.[11] Synthetic glucagon-like peptide-1(7-36) amide [GLP-1(7-36) amide] was obtained from Peninsula Laboratories (San Carlos, CA, USA), glucagon from Novo Research Institute (Bagsvaerd, Denmark), and glibenclamide from Guidotti Laboratory (Pisa, Italy). $^{45}CaCl_2$ was obtained from NEN (Dupont de Nemours, France). Bay K-8644 was purchased from Calbiochem-Novabiochem (La Jolla, CA, USA).

Cell Culture

MIN6 cells[12] were obtained from Dr. H. Ishihara (Tokyo, Japan). The cells were grown in Dulbecco's modified Eagle's medium containing 25 mmol/l glucose (DMEM, Gibco, USA) equilibrated with 5% CO_2/95% air at 37°C. The medium was supplemented with 15% fetal calf serum (FCS, Gibco), 100 μg/ml streptomycin (Gibco), 100 U/ml penicillin sulfate (Gibco), and 75 nM β-mercapto-ethanol (Sigma). MIN 6 cells used in this present study were harvested at passages 18 to 25.

Insulin Release

MIN6 cells were plated in 1 ml DMEM in 24-well plates at a density of 10^6 cells per well for 3–5 days. Eighteen hours before the experiments, the culture medium was changed. Insulin release was determined using a static incubation method in 5% CO_2/95% air at 37°C using cells in exponential growth and performed in HEPES-balanced Krebs-Ringer bicarbonate buffer (119 mmol/l NaCl; 4 mmol/l KCl; 1.2 mmol/l KH_2PO_4; 1.2 mmol/l $MgSO_4$; 2.5 mmol/l $CaCl_2$; 20 mmol/l HEPES, pH 7.5) containing 0.5% BSA (KRB buffer). The day of the experiment, after washing twice with 500 μl KRB buffer, cells were preincubated for 1 h in 500 μl of KRB buffer containing 1 mmol/l glucose in 5% CO_2/95% air at 37°C. MIN6 cells were then preincubated for 2 h in 500 μl KRB buffer containing varying concentrations of glucose and other test agents. Media were then collected and assayed for immunoreactive insulin by radioimmunoassay as previously described, using rat insulin as standard.[13]

Measurement of $^{45}Ca^{2+}$ Influx

MIN6 cells were grown in 24-well plates for 3–5 days under the same conditions as for insulin release., The culture medium was changed 24 h before the experiment. On the day of the experiment, after washing twice with 500 μl of KRB buffer, MIN6 cells were preincubated for 30 min at 37°C in 250 μl KRB buffer containing 1 mmol/l glucose in 5% CO_2/95% air. The preincubation solution was then replaced by 250 μl KRB containing 8 μCi/ml $^{45}CaCl_2$ and the test agents (glucose, KCl, and Bay K-8644) for 10 min, miniglucagon was then added for 3 minutes. The reaction, developed at 37°C, was stopped by aspiration of the medium. The cells were rapidly washed four times with ice-cold buffer (135 mM NaCl, 5 mM KCl, 2.5 mM CaCl$_2$, 1 mM LaCl$_2$, 10 mM HEPES). The cells were then solubilized in 1 ml KRB containing 0.1% Triton for 1 h at room temperature. An aliquot of the solution (100 μl) was then assayed for ^{45}Ca^{2+} content in a beta counter after addition of a liquid scintillation medium (PCS, Amersham).

RESULTS

Modulation by Miniglucagon of Glucose- or Sulfonylurea-Stimulated Insulin Release

As shown in FIGURE 1, miniglucagon was able, at concentrations ranging from 0.01 pM to 1,000 pM, to reduce, in a dose-dependent manner, the 10 mM glucose-induced and the 20 nM glibenclamide-induced insulin release. At 1 nM, miniglucagon inhibited by 85% and 95% the stimulated release, while a half-maximal effect was observed at about 1 pM and 0.5 pM when glucose and glibenclamide were used as the stimulus, respectively.

Modulation by Miniglucagon of Glucagon or tGLP-1-Stimulated Insulin Release

Since both glucose and sulfonylureas use mainly the same pathway to stimulate insulin release, that is by closing the ATP-dependent potassium channels (K-ATP channels) present in the β cell plasma membrane, triggering a membrane depolarization that, in turn, opens the voltage-sensitive calcium channels,[14] we investigated whether miniglucagon was also able to modulate insulin release when stimulated by molecules that use a cyclic

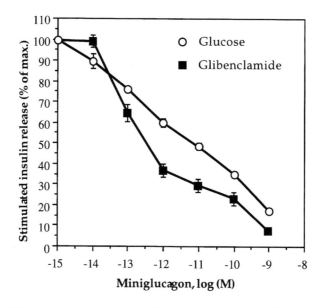

FIGURE 1. Effects of various concentrations of miniglucagon on the 10 mM glucose-stimulated or 20 nM glibenclamide-stimulated insulin release from MIN6 cells. Data are expressed as % of insulin release stimulated by the respective stimulus (glucose or glibenclamide), the basal release measured at 1 mM glucose representing zero in the vertical scale. Mean ± S.E.M. of 12 determinations.

AMP/protein kinase A pathway. For that purpose, we stimulated insulin release by either glucagon or tGLP-1, both known to act on β cell via cyclic AMP production,[15] and at the same time applied various concentrations of miniglucagon.

As shown in FIGURE 2, miniglucagon, within the same range of concentration as in the glucose and glibenclamide experiments (0.01 pM to 1,000 pM, see FIG. 1) was able to inhibit glucagon- or tGLP-1-stimulated release by 85 to 95%. Again, half-maximal inhibition was observed at about 1 pM miniglucagon.

Modulation of Ca^{2+} Entry by Miniglucagon

Since miniglucagon was a potent and efficient inhibitor of insulin release triggered by stimuli acting mainly either through the K-ATP channels (glucose or sulfonylurea) or through a cyclic AMP–dependent pathway (glucagon or tGLP-1), we hypothesized that miniglucagon was active at a common step in the action of all these stimuli. Because Ca^{2+} entry via the voltage-sensitive L-type calcium channel was a good candidate for being this common step,[16,17] we investigated whether miniglucagon was able to modulate Ca^{2+} entry into the MIN6 cells.

As shown in FIGURE 3, miniglucagon, in the concentration range where it was active on insulin release (FIGS. 1 and 2), completely suppressed the 10 mM glucose-induced Ca^{2+} entry. Again, half-maximal effect of miniglucagon was observed at about 1 pM. However, it must be noted that at concentrations from 10 to 1,000 pM the amount of Ca^{2+} entering the cell during the time of the experiment was even lower than that observed under basal (1 mM glucose) conditions. Very similar data were observed when the other stimuli were applied (sulfonylurea, glucagon, or tGLP-1): the Ca^{2+} entry triggered by these stimuli was totally suppressed by miniglucagon (data not shown).

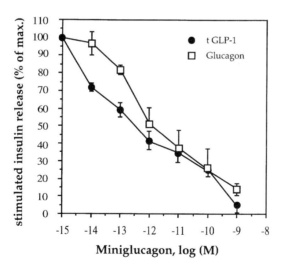

FIGURE 2. Effects of various concentrations of miniglucagon on the 10 nM glucagon-stimulated or 10 nM tGLP-1-stimulated insulin release from MIN6 cells. Data are expressed as % of insulin release stimulated by the respective stimulus (glucagon or tGLP-1), the basal release, measured at 1 mM glucose representing zero in the vertical scale. Mean ± S.E.M. of 9–12 determinations.

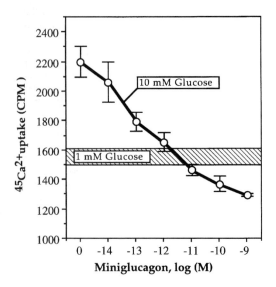

FIGURE 3. Effects of various concentrations of miniglucagon on the 10 mM glucose-stimulated Ca^{2+} uptake by MIN6 cells. Data are expressed as counts/min of $^{45}Ca^{2+}$ accumulated during the incubation time with miniglucagon and $^{45}Ca^{2+}$ (3 min), which followed a 10-min preincubation time with glucose and are mean ± S.E.M. of 9 determinations.

Modulation by Miniglucagon of Ca^{2+} Entry via Voltage-Sensitive Channels

The preceding data strongly suggested that miniglucagon inhibits insulin release by closing a Ca^{2+} channel. However, there was still no proof that this channel is the voltage-sensitive, L-type channel known to exist in the β cell membrane. To get further insights into this question, we designed the following two experiments.

(1) We induced a membrane depolarization of the MIN6 cells by increasing the extracellular potassium concentration to a level that mimicked that observed under stimulation by physiological or pharmacological stimuli (glucose, sulfonylurea, glucagon, or tGLP-1). Under these conditions, miniglucagon, in the concentration range shown to be active in the preceding experiments, totally suppressed the potassium-induced Ca^{2+} entry. Again, the highest miniglucagon concentrations induced a Ca^{2+} entry that was lower than the basal values (FIG. 4).

(2) We opened the L-type voltage-sensitive Ca^{2+} channel by means of a pharmacological compound of the dihydropyridine family (Bay K-8644) known to stimulate directly this type of channel.[18] FIGURE 5 shows that miniglucagon, in the 0.01–1,000 pM range, completely inhibited the Bay K-8644-induced Ca^{2+} entry. Thus, it was clearly demonstrated by our data that miniglucagon inhibits secretagogue-stimulated insulin release via closure of L-type voltage-sensitive Ca^{2+} channels.

Involvement of a Pertussis Toxin-Sensitive Pathway in the Miniglucagon Action

Next, we wanted to analyze the intracellular pathway through which miniglucagon acts on this channel. In view of preceding data showing that at least one G protein exists

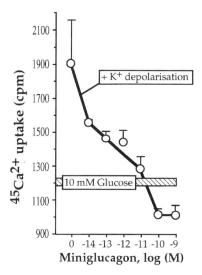

FIGURE 4. Effects of various concentrations of miniglucagon on the 15 mM KCl (K+ depolarization) -stimulated Ca^{2+} uptake by MIN6 cells. Data are expressed as counts/min of ^{45}Ca^{2+} accumulated during the incubation time with miniglucagon and ^{45}Ca^{2+} (3 min), which followed a 10-min preincubation with glucose and KCl and are mean ± S.E.M. of 6 determinations.

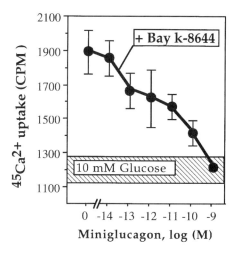

FIGURE 5. Effects of various concentrations of miniglucagon on the 1 µM Bay K-8644-stimulated Ca^{2+} uptake by MIN6 cells. Data are expressed as counts/min of ^{45}Ca^{2+} accumulated during the incubation time with miniglucagon and ^{45}Ca^{2+} (3 min), which followed a 10-min preincubation with glucose and Bay K-8644 and are mean ± S.E.M. of 9 determinations.

between the miniglucagon receptor and its effectors,[6] we compared the miniglucagon effectiveness on cells pretreated with bacterial toxins known to act on G proteins. While cholera toxin pretreatment of MIN6 cells did not modify significantly the action of miniglucagon on insulin release (data not shown), pertussis toxin pretreatment of the cells entirely suppressed it (TABLE 1). For comparison, pertussis toxin pretreatment also suppressed the ability of somatostatin, known to act on insulin release via a pertussis toxin–sensitive G protein,[19] to inhibit insulin release from MIN6 cells (TABLE 1).

DISCUSSION

Miniglucagon, the C-terminal(19-29) fragment of glucagon, is the product of a secondary processing. This fragment is produced (1) locally at the level of the glucagon target tissue, thanks to the presence at the plasma membrane outer surface of a specific protease, modulating the action of the mother hormone before being quickly destroyed and (2) in a unique endocrine cell, the islet A cell, which produces glucagon.

The possibility that miniglucagon acts as a physiological regulator depends on two factors: (1) the local concentration of miniglucagon at the immediate vicinity of its target cells and (2) the affinity of the peptide for its receptor. If we compute the amount of miniglucagon likely to be present within the glucagon target tissues according to its production and degradation rates (obtained, for obvious reasons, indirectly from *in vitro* experiments), we obtain figures that are close to one percent of the mother hormone (glucagon) concentrations. This concentration, measured by radioimmunoassay, is in the range of 10^{-11} to 10^{-9} M; one percent of that amount is 10^{-13} to 10^{-11} M. This is precisely the affinity range of miniglucagon for the biological phenomena in which it is implicated.[2,5–10] It is of special interest to point out that the same calculation concerning a possible regulatory role of miniglucagon within the islet of Langerhans gives similar figures: the miniglucagon concentrations present in the pancreatic A cells are a few[2–5] percent of that of glucagon,[9] while, as clearly indicated by the present data, miniglucagon is active on the pancreatic β cells in a dose range that is 1,000 to 100 times lower than the active concentrations of glucagon. All these data argue for a physiological role of miniglucagon both in the glucagon target tissues and within the islet of Langerhans. More investigations are necessary, however, to support definitely this physiological role, particularly by measuring the amounts of miniglucagon released from the A cells under various conditions.

The mode of action of miniglucagon, as evidenced by our data, implies: (1) a G-protein linked receptor, certainly of the seven transmembrane domains family, which remains to be characterized; (2) a pertussis toxin–sensitive G protein that is, thus, likely to be of the Gi or Go type; and (3) the dihydropyridine L-type plasma membrane Ca^{2+} channel.

All these features are reminiscent of the pathway used by other peptides that display an inhibitory action on insulin release, such as galanin or somatostatin.[19] Two specific features, however, set the miniglucagon apart: first, the extremely high affinity of the peptide for the regulatory processes that it controls and, second, the fact that the peptide derives

TABLE 1. Effects of pertussis toxin pretreatment

Peptide	Insulin Release (%) (no PT treatment)	Insulin Release (%) (PT treatment)	p
Somatostatin	100 ± 7.3	2.72 ± 3.18	<0.01
Miniglucagon	100 ± 22.7	1.81 ± 2.27	<0.01

Note: Effects of pertussis toxin pretreatment (PT, 200 ng/ml overnight) of the MIN6 cells on the ability of somatostatin and miniglucagon to inhibit insulin release. Data are mean ± S.E.M. of 9 determinations.

from another peptide that displays the opposite effect. This antagonistic action of a peptide fragment towards the effects of the hormone from which it derives is unique. We may think that both characteristics are linked together, since the miniglucagon production is partial and represents only a small part of the mother hormone concentration and thus requires that the peptide fragment act at much lower concentrations than the mother hormone. On the other hand, miniglucagon does not block just glucagon action, but also that of most (if not all) the insulino-secretagogues. This characteristic is linked to the pathway used (closing of the L-type Ca^{2+} channels), which is of major importance for the secretagogue action.

It remains to be established under what precise physiological conditions miniglucagon exerts its inhibitory action and whether a dysfunction of this regulatory system may be implicated in diabetic states. Together with the search for the specific miniglucagon receptor, this will be the subject of future investigations.

ACKNOWLEDGMENTS

We are highly indebted to Dr. Hisamitsu Ishihara (Third Department of Internal Medicine, Faculty of Medicine, University of Tokyo, Japan) for providing us with several batches of MIN6 cells.

REFERENCES

1. LAZURE, C., N. SEIDAH, D. PELAPRAT & M. CHRÉTIEN. 1983. Can. J. Biochem. Cell. Biol. **61**: 501–515.
2. BATAILLE, D. 1996. Preproglucagon and its processing. Handb. Exp. Pharmacol. **123**: 31–51.
3. BATAILLE, D. 1996. Oxyntomodulin and its related peptides. Handb. Exp. Pharmacol. **123**: 327–340.
4. JARROUSSE, C., C. CARLES-BONNET, H. NIEL, R. SABATIER, M.-P. AUDOUSSET-PUECH, P. BLACHE, A. KERVRAN, J. MARTINEZ & D. BATAILLE. 1993. Inhibition of gastric acid secretion by oxyntomodulin and its [19-37] fragment in the conscious rat. Am. J. Physiol. **264**: G816–G823.
5. MALLAT, A., C. PAVOINE, M. DUFOUR, S. LOTERSZTAJN, D. BATAILLE & F. PECKER. 1987. A glucagon fragment is responsible for the inhibition of the liver Ca^{2+} pump by glucagon. Nature **325**: 620–622.
6. LOTERSZTAJN, S., C. PAVOINE, V. BRECHLER, B. ROCHE, M. DUFOUR, D. LE-NGUYEN, D. BATAILLE & F. PECKER. 1990. Glucagon (19-29) exerts a biphasic action on the liver plasma membrane Ca^{2+} pump which is mediated by G proteins. J. Biol. Chem. **265**: 9876–9880.
7. BLACHE, P., A. KERVRAN, M. DUFOUR, J. MARTINEZ, D. LE-NGUYEN, S. LOTERSZTAJN, C. PAVOINE, F. PECKER & D. BATAILLE. 1990. Glucagon (19-29), a Ca^{2+} pump inhibitory peptide, is processed from glucagon in the rat liver plasma membrane by a thiol endopeptidase. J. Biol. Chem. **265**: 21514–21519.
8. PAVOINE, C., V. BRECHLER, A. KERVRAN, P. BLACHE, D. LE-NGUYEN, S. LAURENT, D. BATAILLE & F. PECKER. 1991. Miniglucagon <Glucagon-(19-29)> is a component of the positive inotropic effect of glucagon. Am. J. Physiol. **260**: C993–C999.
9. BLACHE, P., A. KERVRAN, D. LE-NGUYEN, J. LAUR, A. COHEN-SOLAL, J. DEVILLIERS, P. MANGEAT, J. MARTINEZ & D. BATAILLE. 1989. Glucagon-related peptides deriving from Proglucagon in the gastroenteropancreatic and central nervous systems. Biomed. Res. **9** (Suppl.3): 19–28.
10. BATAILLE D., S. KUROKI, P. BLACHE, A. KERVRAN, D. LE-NGUYEN, M. DUFOUR, S. LOTERSZTAJN, A. MALLAT, V. BRECHLER & F. PECKER. 1992. Glucagon [19-29]: A daughter-molecule which modulated the action of the mother-molecule. Biomed. Res. **13** (Suppl.2): 137–142.
11. LE-NGUYEN, D., M. DUFOUR, A. MALLAT, S. LOTERSZTAJN, C. PAVOINE, F. PECKER & D. BATAILLE. 1988. Synthesis of [Nle²⁷]-Glucagon-(19-29) and effectiveness on hepatic calcium pump. *In* Peptide Chemistry 1987. T. Shiba & S. Sakakibara, Eds: 391–395. Protein Research Foundation. Osaka.

12. ISHIHARA, H., T. ASANO, K. TSUKUDA, H. KATAGIRI, K. INUKAI, M. ANAI, M. KIKUCHI, Y. YAZAKI, J. I. MIYAZAKI & Y. OKA. 1993. Pancreatic beta cell line MIN 6 exhibits characteristics of glucose metabolism and glucose-stimulated insulin secretion similar to those of normal islets. Diabetologia **36**: 1139–1145.

13. KERVRAN, A., M. RIEUTORT & M. GUILLAUME. 1976. A simultaneous radioimmunoassay for growth hormone and insulin in the plasma of rats and rabbits. Diab. Métabol. **2**: 67–72.

14. ASHCROFT, F. M,. S. J. H. ASHCROFT & D. E. HARRISON. 1988. Properties of single potassium channels modulated by glucose in rat pancreatic β-cells. J. Physiol. (Lond) **400**: 501–527.

15. MOENS, K., H. HEIMBERG, D. FLAMEZ, P. HUYPENS, E. QUARTIER, Z. LING, D. PIPELEERS, S. GREMLICH, B. THORENS & F. SCHUIT. 1996. Expression and functional activity of glucagon, glucagon-like peptide I, and glucose-dependent insulinotropic peptide receptors in rat pancreatic islet cells. Diabetes **45**: 257–261.

16. MALAISSE-LAGAE, F., P. C. F. MATHIAS & W. J. MALAISSE. 1984. Gating and blocking channels by dihydropyridines in the pancreatic B-cell. Biochem. Biophys. Res. Commun. **123**: 1062–1068.

17. HENQUIN, J. C., W. SCHEEMER, M. NENQUIN & H. P. MEISSNER. 1985. Effects of a calcium channel agonist on the electrical, ionic and secretory events in mouse pancreatic B-cells. Biochem. Biophys. Res. Commun. **131**: 980–986.

18. BROWN, A. M., D. L. KUNZE & A. YATANI. 1984. The agonist effect of dihydropyridines on Ca channels. Nature (Lond) **311**: 570–572.

19. NILSSON, T., P. ARKHAMMAR, P. RORSMAN & P.-O. BERGGREN. 1989. Suppression of insulin release by galanin and somatostatin is mediated by a G-protein. An effect involving repolarisation and reduction in cytoplasmic free Ca^{2+} concentration. J. Biol. Chem. **264**: 973–980.

Pituitary Adenylate Cyclase Activating Polypeptide Induces Degranulation of Rat Peritoneal Mast Cells via High-Affinity PACAP Receptor-Independent Activation of G Proteins

JÖRG SEEBECK,[a,d] MARIE-LUISE KRUSE,[b] ANJONA SCHMIDT-CHOUDHURY,[c] AND WOLFGANG E. SCHMIDT[b]

[a]Department of Pharmacology, [c]Department of Pediatrics, [b]First Department of Internal Medicine, Christian-Albrechts University of Kiel, 24105 Kiel, Germany

ABSTRACT: In this study, the secretory effects of PACAP and PACAP analogues on [³H]serotonin-loaded purified rat peritoneal mast cells (RPMCs) were investigated. PACAP(1-27) and PACAP(6-27) stimulated [³H]serotonin release with low potency (ED_{50}: 2×10^{-6} M) but high efficacy. The N-terminally truncated PACAP form, PACAP(6-27), stimulated tracer release with an ED_{50} of 0.2×10^{-6} M, indicating a high-affinity PACAP receptor–independent mechanism of action. The secretory response to PACAP(1-27) could be inhibited by 60-min preincubation with pertussis toxin (ptx), which inhibits G proteins. U73122, a cell-permeable phospholipase C inhibitor, dose-dependently inhibited the secretory effect of 5 μM PACAP(1-27) with an IC_{50} value of 4 μM ($N = 4$; $p < 0.006$). We conclude that PACAP exerts a secretory effect in RPMCs by high-affinity PACAP receptor–independent direct activation of one or more G proteins, which may then activate the PLC-dependent signal-transduction pathway.

Recently, PACAP(1-27), PACAP-(1-38), and N-terminally truncated PACAP(16-38) and PACAP(28-38) have all been shown to dose-dependently stimulate histamine release from rat peritoneal mast cells (RPMCs) with low potency and high efficacy.[1] Since the N-terminus of PACAP is essential for the interaction with high-affinity PACAP receptors,[2,3] these findings suggest a different mechanism of action. PACAP receptors belong to the large family of G protein–coupled receptors with seven transmembranous segments.[4,5] In numerous studies with other neuropeptides and peptide hormones, similar, seemingly atypical (high-affinity receptor independent), secretory responses have been observed in RPMCs.[6] General features of these peptide responses were a non-cytolytic mechanism of action, an absolute requirement of positively charged amino acid residues, and the inhibition by pretreatment with G protein inhibitors like pertussis toxin (ptx). These peptide effects are mimicked by the non-peptide substance compound 48/80, a known direct activator of heterotrimeric G proteins and most frequently used as a reference substance in

[d]Corresponding author: Dr. Jörg Seebeck, Institut für Pharmakologie der CAU Kiel, Hospitalstr. 4, 24105 Germany; Tel.: 49-431-597-3508, Fax: 49-431-597-3522; E-mail: department@pharmakologie.uni-kiel.de

141

investigations of IgE-independent secretion in mast cells.[7,8] A challenge of RPMCs with compound 48/80 is associated with an increase of cyclic AMP (cAMP), inositol 1,4,5-trisphosphate (IP_3), and arachidonic acid.[9] The two latter ones have been shown to promote exocytosis, while the action of cAMP remains a matter of controversy.[10]

In the present investigation, inhibitors of different signal transduction pathways were tested to assess which of the signal transduction pathways activated by compound 48/80 are also involved in PACAP-induced degranulation of RPMCs.

MATERIAL AND METHODS

Preparation of Rat Peritoneal Mast Cells (RPMCs)

Peritoneal mast cells were obtained by peritoneal lavage from spontaneous hypertensive rats (weight 150–250 g) used for breeding purposes at the Institute of Pharmacology, University of Kiel. These cells showed no morphological or functional differences when compared with RPMCs from normal Wistar rats. RPMCs were purified using a BSA gradient [0.2/40% BSA (wt/vol)] centrifugation procedure as recently described.[6] This method yields mast cells at an average purity of more than 90% (FIG. 1,a). In general, lavages from three rats were combined for one experiment of 20–40 specimen. The buffer used for lavage and secretion experiments contained 0.2% BSA (γ-globulin and protease-free; Sigma A-3059) and [mM]: 137 NaCl, 2.7 KCl, 0.3 $CaCl_2$, 1 $MgCl_2$, 0.4 NaH_2PO_4, 10 HEPES, and 5.6 glucose. RPMCs could be easily identified by their characteristic small nucleus-to-plasma ratio and their highly contrasted, granular-filled three-dimensional appearance when observed by phase-contrast microscopy. These features were enhanced by alcian blue staining (FIG. 1,b). For this purpose, cells were fixed on coverslips by addition of ice-cold methanol for at least 20 min. Subsequently, coverslips were transferred into acetic acid for another 10 min. Alcian blue staining was carried out using a 1% aqueous solution. After a brief wash in water, cells were transferred to 1% alkaline ethanol for 10 min (color fixation). Finally, cells were counter-stained with 1% aqueous Kernechtrot, dehydrated, and mounted.

Secretion Experiments

Purified mast cells were incubated for 2 h at 37°C with 1 μCi/ml hydroxytryptamine creatine sulfate,5-[1,2-^{3}H(N)] ([^{3}H]5HT). Consecutively, cells were washed by centrifugation (5 min, 4°C, 100 × g) and resuspended in buffer (final volume: 300 μl/vial). Incubations were stopped on ice followed by centrifugation. The activity of tracer released into the supernatant divided by the total tracer activity in each individual sample was taken as measure of secretion. All concentrations were determined in duplicate. Results are given as mean ± SEM. Statistical evaluations were made by means of the paired t test (*$p < 0.05$, **$p < 0.005$, ***$p < 0.0005$).

RESULTS

Secretory Effect of PACAP(1-27), PACAP(6-27), and Compound 48/80 in RPMCs

RPMCs incubated for 10 min with PACAP(1-27) (10^{-7}–10^{-5} M) showed a dose-dependent release of [^{3}H]5HT (EC_{50}: 2×10^{-6} M, basal: 1.1 ± 0.8, maximum: 72 ± 2.8% of total;

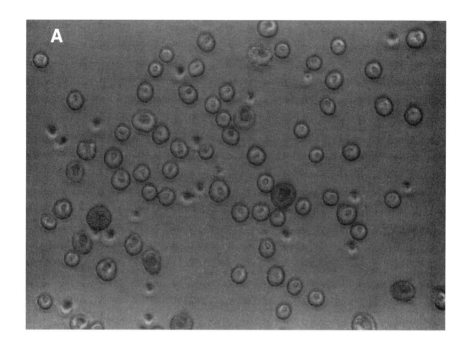

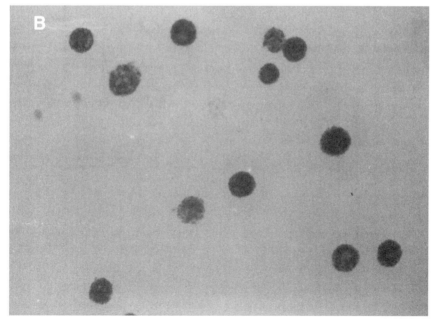

FIGURE 1. Purified rat peritoneal mast cells (RPMC). (**A**) Native purified mast cells. (**B**) RPMCs stained with alcian blue.

FIG. 2a). In the same series of experiments, compound 48/80 (0.03–1 µg/ml) stimulated [³H]5HT release with a maximum effect of 86.7% of total. In comparison to PACAP(1-27), PACAP(6-27) stimulated exocytosis with a somewhat higher potency (EC$_{50}$: 0.2 µM; $N = 4$; FIG. 2b).

Effects of Three Different Signal-Transduction Inhibitors on PACAP-Induced Secretion

Pertussis Toxin

A 60-min preincubation of RPMCs with 1 µg/ml pertussis toxin (ptx) significantly ($p < 0.005$) reduced the secretory response to 5 µM PACAP(1-27) [PA(1-27): 55 ± 7.6; basal: 3.6 ± 1.9; PA(1-27) + ptx: 15.0 ± 7.5; basal: 4.3 ± 0.7; $N = 4$; FIG. 2c].

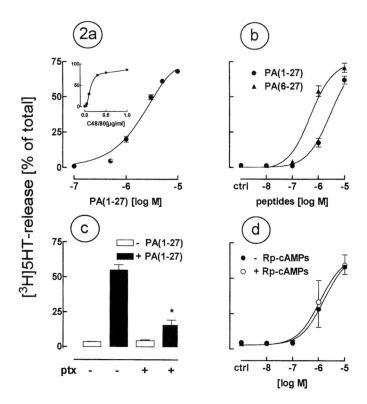

FIGURE 2. [³H]serotonin release [% of total; mean ± SEM] in rat peritoneal mast cells induced by PACAP(1-27), PACAP(6-27), or compound 48/80. (**a**) Secretory effects of compound 48/80- (*upper left*) and PACAP(1-27) within a single experiment (incubation time = 10 min). (**b**) Dose-dependent stimulation of [³H]serotonin release by PACAP(1-27) and PACAP(6-27) (incubation time = 10 min, $N = 4$). (**c**) Effect of a 60-min preincubation with 1 µg/ml pertussis toxin (ptx) on 5 µM PACAP-induced degranulation (*$p \leq 0.005$; $N = 4$). (**d**) Effect of 0.1 M Rp-cAMPs on PACAP-induced tracer release ($N = 4$). Cells were incubated for 10 min in the presence of 0.1 M Rp-cAMPs.

Rp-cAMPs

The dose-response curve (10^{-8}–10^{-5} M) of PACAP(1-27)–induced secretion in RPMCs was not changed by a 10-min preincubation period with 0.1 mM of the competitive cAMP-antagonist Rp-cAMPs [EC_{50}(– RpcAMPS): 1.6 µM; EC_{50}(+ RpcAMPs): 1.2 µM; $N = 4$; FIG. 2d].

U73122

Preincubation of RPMCs for 10 min with the phospholipase C inhibitor U73122 dose-dependently reduced the secretory effect of 5 µM PACAP(1-27) with an IC_{50} of 4 µM ($N = 4$). A maximum inhibition was observed at 10 µM (10.9 ± 5.5 vs. 53.3 ± 8.5; $p \leq 0.006$; FIG. 3). At higher concentrations, U73122 alone induced exocytosis (data not shown).

DISCUSSION

In this study we investigated the mechanism of PACAP-induced tracer release (exocytotic vs. cytolytic) in [³H]5HT–loaded peritoneal mast cells (RPMCs). Additionally, the influence of several pharmacological signal-transduction inhibitors was assessed to identify signal-transduction pathways underlying this PACAP response.

Pertussis toxin (ptx), an inhibitor of G proteins at a concentration of 1 µg/ml significantly reduced the effect of PACAP. This finding indicates that PACAP-induced degranulation is indeed due to an exocytotic and not a cytolytic process. Additionally, it can be concluded that one or more G proteins are involved in PACAP-induced stimulus-secretion coupling.

Previously, it has been shown that substance P- and compound 48/80-induced secretion in RPMCs is accompanied by an increase of inositol trisphosphate and intracellular Ca^{2+}-levels.[8,9] Dialysis of single rat peritoneal mast cells via the patchpipette with the phospholipase C inhibitor neomycin has been shown to completely prevent the exocytotic effect of compound 48/80.[10] These results suggest an important role of the PLC-associated signal-transduction

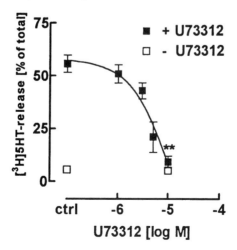

FIGURE 3. Effect of the phospholipase C inhibitor U73122 (preincubation time = 10 min), on 5 µM PACAP(1-27)–induced [³H]serotonin release in rat peritoneal mast cell (** $p \leq 0.005$; $N = 4$).

cascade in RPMCs. We used the cell-permeable PLC-inhibitor U73122[11] to assess the influence of PLC inhibition on PACAP-induced exocytosis. U73122 dose-dependently diminished the secretory effect of PACAP(1-27). The IC_{50} value of U73122 was somewhat higher (4 μM) in RPMCs as originally described for granule exocytosis in polymorphonuclear neutrophils (0.12 μM).[11] U73343, a derivative of U73122, previously classified as an inactive control,[11] showed no effect on PACAP-induced secretion at 10 μM. This finding may support the specificity of the U73122 effect in our system and the likely mediation of PLC.

SUMMARY

Our data show that PACAP peptides exert their secretory effect in rat peritoneal mast cells independently of high affinity PACAP receptors. The PACAP receptor molecule in RPMCs may be a highly promiscuous low affinity receptor, since it accepts several positively charged peptides. Alternatively, PACAP may directly interact with one or more G proteins. At least one of the heterotrimeric G proteins may activate the PLC-associated signal-transduction cascade and thereby activate downstream signaling cascades leading to exocytosis. Our findings further indicate that biological responses to PACAP at high concentrations, especially in whole animals or isolated organs, are not necessarily due to activation of high-affinity PACAP receptors and should be therefore interpreted with care.

ACKNOWLEDGEMENT

The authors gratefully acknowledge expert technical assistance by W. Gloyna.

REFERENCES

1. MORI, T. et al. 1994. Histamine release induced by pituitary adenylate cyclase activating polypeptide from rat peritoneal mast cells. Arzneim.-Forsch./Drug Res. **44** (II): 1044–1046.
2. BUSCAIL, L. et al. 1990. Presence of highly selective receptors for PACAP (pituitary adenylate cyclase activating peptide) in membranes from the rat pancreatic acinar cell line AR 4-2J. FEBS Lett. **262**: 77–81.
3. SCHMIDT, W. E. et al. 1993. PACAP and vip stimulate enzyme secretion in rat pancreatic acini via interaction with VIP/PACAP-2 receptors: Additive augmentation of CCK/carbachol-induced enzyme release. Pancreas **8**: 476–487.
4. SPENGLER, D. et al. 1993. Different signal transduction by five splice variants of the PACAP receptor. Nature **365**: 170–175.
5. PISEGNA, J. R. & S. A. WANK. 1993. Molecular cloning and functional expression of the pituitary adenylate cyclase activating polypeptide type 1 receptor. Proc. Natl. Acad. Sci. USA **90**: 6345–6349.
6. CHAHL, L. A. 1991. Role of histamine in the action of neuropeptides and local hormones. Handb. Pharmacol. **97**: 521–540.
7. ARIDOR, M. et al. 1993. Activation of exocytosis by the heterotrimeric G protein Gi3. Science **262**: 1569–1573.
8. MOUSLI, M. et al. 1989. Activation of rat peritoneal mast cells by substance P and mastoparan. J. Pharmacol. Exp. Ther. **250**: 329–335.
9. NAKAMURA, T. & M. UI. 1985. Simultaneous inhibitions of inositol phospholipid breakdown, arachidonic acid release, and histamine secretion in mast cells by islet-activating protein, pertussis toxin. J. Biol. Chem. **260**: 3584–3593.
10. PENNER, R. 1988. Multiple signalling pathways control stimulus-secretion coupling in rat peritoneal mast cells. Proc. Natl. Acad. Sci. USA **85**: 9856–9860.
11. SMITH, R. J. 1990. Receptor-coupled signal transduction in human polymorphonuclear neutrophils: Effects of a novel inhibitor of phospholipase c-dependent processes on cell responsiveness. J. Pharmacol. Exp. Ther. **253**: 688–697.

The Pituitary Adenylate Cyclase Activating Polypeptide Type 1 Receptor (PAC$_1$-R) Is Expressed on Gastric ECL Cells: Evidence by Immunocytochemistry and RT-PCR

NINGXIN ZENG, TAO KANG, RONG-MING LYU, HELEN WONG, YI WEN, JOHN H. WALSH, GEORGE SACHS, AND JOSEPH R. PISEGNA[a]

CURE: VA/UCLA Digestive Diseases Research Center, West Los Angeles Veterans Administration Medical Center and the University of California, Los Angeles, Los Angeles, CA 90073

ABSTRACT: The current study was undertaken to determine the presence and distribution of PAC$_1$-Rs within the gastric mucosa. Polyclonal antibodies to the carboxyl terminus of the rat PAC$_1$-R were generated and shown to be specific against the PAC$_1$-R expressed in NIH 3T3 cells. Western blot analysis using isolated ($\approx 85\%$ pure) ECL cell membranes identified a 48 kD protein consistent with the calculated molecular mass of the cloned PAC$_1$-R. RT/PCR performed using specific primers for the PAC$_1$-R confirmed the presence of splice variants of the rat PAC$_1$-R, but not VPAC$_1$-R or VPAC$_2$-R. These data provide the first direct evidence for the existence of functional PACAP Type I receptors on ECL cells of the gastric mucosa and suggest a potential role for PACAP in the stimulation of gastric acid secretion and in the regulation of the growth of ECL cells.

Pituitary adenylate cyclase activating polypeptide (PACAP) is the most recently discovered neuropeptide in the vasoactive intestinal polypeptide (VIP)/ secretin/glucagon family of peptide hormones that was isolated in 1989 from ovine hypothalamus. PACAP has potent adenylate cyclase–stimulating activity and occurs as two amidated forms: PACAP-38, a 38–amino acid peptide; and PACAP-27, a 27–amino acid peptide.[1] Radioligand binding and cloning studies have identified three major types of PACAP receptors (PACAP-R) distinguished pharmacologically by their relative affinities for PACAP, VIP, and helodermin.[2-6] The PACAP-R Type 1 (PAC$_1$) has high affinity for only PACAP; the Type 2 PACAP-R (classical VIP receptor, VPAC$_1$) has high affinity for PACAP-38, PACAP-27, and VIP; and the Type 3 PACAP-R (VAPC$_2$) shows high affinity for PACAP, VIP, and helodermin.[5-6] The PAC$_1$ in both rats and humans has been shown to be alternatively spliced and expressed as one of four major splice variant transcripts in a tissue-specific manner.[7,8] In native cell systems, PAC$_1$-Rs have been shown to exhibit coupling to both adenylate cyclase and phospholipase C signal transduction pathways, such as in the rat pheochromocytoma cell line, PC-12.[9]

[a]Corresponding author: Dr. Joseph R. Pisegna, CURE:VA/UCLA DDRC, Bldg 115, Room 315 West LA VA Medical Center, 11301 Wilshire Blvd, Los Angeles, CA 90073; Tel.: 310-478-3711, ext. 41940; Fax 310-268-4963.

Since its discovery, PACAP hormone expression has been identified in numerous tissues including the brain, gastrointestinal tract, adrenal gland, and testis by immunohistochemistry and radioimmunoassay.[3,4] In the gastrointestinal tract, PACAP has been shown to inhibit pentagastrin-stimulated gastric acid secretion, reduce basal smooth muscle contraction, and stimulate intestinal secretion and pancreatic exocrine secretion.[10-13]

The enterochromaffin-like (ECL) cells of the gastric mucosa are responsible for the paracrine secretion of histamine in response to gastrin and other ligands and play an important role in peripheral regulation of gastric acid secretion.[14-16] The exact role of PACAP and PAC_1-R in the gastric mucosa have not been clearly defined. Furthermore, the existence of PAC_1-R on select cells of the gastric mucosa as well as the physiological role of PACAP on these cells, such as for the gastric ECL cells, have not been previously determined. In the current study we isolated ECL cells in primary culture that were 85% pure.[14,15] These cellular preparations were used for both functional assays and for immunocytochemical and molecular assays. The present study demonstrates that rat gastric ECL cells express high levels of the rat PAC_1-R and PACAP is coupled to the release of intracellular calcium and to the release of histamine, suggesting a potentially important mechanism in the regulation of gastric acid secretion and ECL growth.

MATERIAL AND METHODS

Materials

Peptides corresponding to amino acids 509–523 (YQLRMSSLPADNLAT) of the carboxyl tail of the rat PAC_1-R were developed by G. Poy (National Institute of Diabetes, Digestive and Kidney Diseases, National Institutes of Health). This peptide was used to sensitize rabbits to develop polyclonal antibodies against the rat PAC_1-R by the Antibody Core (CURE, Los Angeles, CA). Oligonucleotides for the rat PAC_1-R were synthesized using a DNA synthesizer (Model 391 PCR-MATE, ABI, Perkin Elmer, Foster City, CA). All other reagents were analytical grade and were purchased from the indicated sources: *Taq* DNA polymerase (Stratagene, La Jolla, CA); PCR primers for β-actin (Clontech, Palo Alto, CA); BSA, acridine orange, DME-F12 medium, ITS, hydrocortisone, trypan blue, FSK, dbcAMP, gastrin, somatostatin, goat anti-rabbit fluorescein-conjugated IgG, ionomycin, TEA, EGTA (Sigma Chemical Co., St. Louis, MO); Cell-Tak (Collaborative Research); and Fura-2 (Molecular Probes, Oregon).

Methods

Isolation and Purification of ECL Cells

Rat gastric ECL cells were isolated by a combination of elutriation and density gradient centrifugation as described previously.[14,15] Briefly, approximately five rat stomachs were digested to yield approximately 1×10^6 cells in the low density layer. The viability of cells was determined using trypan blue exclusion. Cellular purity was determined by immunostaining with anti-HDC antibody and anti-histamine antibody and by using the fluorescent dye acridine orange. Using these techniques, approximately 65–75% of cells in the isolated cell population were ECL cells. Isolated ECL cells were rinsed by gentle centrifugation in growth medium containing DME/F_{12} supplemented with 2 mg/ml bovine serum albumin (BSA), 2.5% fetal calf serum (FCS), 100 μM hydrocortisone, 1% penicillin, 1 mg/100 ml streptomycin and 5 mg/ml insulin, 5 mg/ml transferrin, and 5 μg/l sodium selenite. Enrichment of the primary culture was performed by plating 15 μl of the cell suspension onto glass coverslips pre-coated with Cell-Tak® and incubated at 37°C for 45 min. Using this methodology, the ECL cell population was enriched to more than 85%.

Immunohistochemistry

Rat stomach was incised longitudinally, fixed in 4% formaldehyde in phosphate-buffered saline (PBS) for 12 h, dehydrated gradually in 70% and 100% ethanol, and embedded in paraffin. Paraffin-embedded tissue blocks were cut into 4-mm sections, deparaffinized, rehydrated, and treated with 3% hydrogen peroxide for 5 minutes. After blocking with 20% swine serum, sections were incubated with antibodies to the receptor at a 1:1,000 dilution overnight at room temperature, followed by biotinylated secondary antibodies (20 min). Streptavidin conjugated to peroxidase (20 min) and amino-ethyl carbazole-H_2O_2 or diaminobenzidine using the LSAB-2 kit were performed subsequently (DAKO Corp, Carpinteria, CA). All washes were done with 0.02 M Tris-HCl pH 7.2 containing 0.1% Tween-20 and 0.3 M NaCl. The slides were counter-stained with Mayer's hematoxylin and observed under Nikon Optiphot II light microscope. For absorption of the antibodies with corresponding peptides used for immunization, the peptides were added to prediluted antibodies to give a final concentration of 50 mg/ml of the peptide. The antibodies were allowed to react with the peptide for 8 h prior to use.

Immunocytochemistry

Pellets of isolated cells were fixed in 4% formaldehyde in PBS for 12 h, dehydrated gradually in 70% and 100% ethanol, and then embedded in paraffin. Following blocking with 20% swine serum, the cells were incubated with antibodies to PAC_1-R at a 1:1,000 dilution overnight at room temperature, followed by biotinylated secondary antibodies (20 min). Streptavidin conjugated to peroxidase (20 min) and amino-ethyl carbazole-H_2O_2 or diaminobenzidine using the LSAB-2 Kit were used for detection (DAKO Corp., Carpinteria, CA). All washes were performed using 0.02 M Tris-HCl, pH 7.2, containing 0.1% Tween-20 and 0.3 M NaCl. Slides were counter-stained with Mayer's hematoxylin and observed under a Nikon Optiphot II light microscope. For absorption studies, antibodies were incubated with corresponding peptides that were used for immunization (final concentration of 50 mg/ml). The antibodies were allowed to react with the peptide for 8 h prior to use.

Western Blot Analysis

ECL cells used for Western blot analysis were isolated as described above. Following isolation, the cells were washed twice in ice-cold PBS, centrifuged ($200 \times g$), and resuspended in homogenization buffer (HEPES 50 mM, pH 6.5, NaCl 0.1 M, $MgCl_2$ 5 mM, benzamidine 0.2 mg/ml, leupeptin 1 mg/ml, pepstatin 0.7 mg/ml, EGTA 1 mM) using a sonicator (50% power for 1 min) and subjected to a low speed centrifugation ($500 \times g$). The supernatant was centrifuged at $25,000 \times g$ for 20 min and the pellet resuspended in Tris-Glycine-SDS buffer (Tris HCl, pH 6.8, 63 mM glycerol 10%, SDS 2%). Following the determination of protein concentration, 10 µg of isolated protein was loaded per lane on a 10% Tris-Glycine Gel (Novex) and run at 125 V ($\approx$ 90 min). Transfer of protein to Hybond® (Millipore) was performed using a Millipore Gel Transfer apparatus. Blotting was performed using the anit-PAC_1-R antibody (1:1,000 dilution) at room temperature, and an anti-rabbit second antibody (1:10,000 dilution) at room temperature. Detection was performed using the ECL Detection Method (Amersham).

cDNA Synthesis and RT-PCR Analysis

RNA was extracted from 1 g of purified ECL cell preparation using the Fast Track RNA Purification kit (Invitrogen, La Jolla, CA). Total RNA (5 μg) from enriched rat ECL cells was used to synthesize cDNA by reverse transcriptase with oligo-dT$_{15}$ primers (Boerhinger-Mannheim). PCR was performed in low salt Taq$^+$ DNA polymerase buffer and 5 U Taq$^+$ DNA polymerase (Stratagene, La Jolla, CA) in the presence of oligonucleotide primers under the following conditions: the initial step (one cycle) was 94°C for 2 min, 57°C for 1 min, and 72°C for 2 min; followed by 94°C for 1 min, 57°C for 1 min, and 72°C for 2 min (30 cycles); and a final extension step (one cycle) at 94°C for 1 min, 57°C for 1 min, and 72°C for 15 min. The sense primers used were: (SENSE 1) 5′ CGAGTGGACAGTGGCAGGCGGTGA 3′ (52–77); and (SENSE 2): 5′GCTCTCCCT-GACTGCTCTCCTGCTG 3′ (145–170). The antisense primers used were: (ANTISENSE 1) 5′ CAGTAGTGAGGGTGGCGAGGGAAGT 3′ (611–636); and (ANTISENSE 2) 5′ CAGTAGGTGTCCCCCAGCCGATGAT 3′ (935–960). To standardize the amount of total RNA used for RT-PCR, duplicate samples were analyzed for β-actin as controls. The sense primer for β-actin was 5′ TTGTAACCAACTGGGACGATATGG (1552–1575) and antisense primer sequence was 5′ GATCTTGATCTTCATGGTGCTAGG (2991–2844). The rat PAC$_1$-R PCR products amplified from ECL cell cDNA were purified from 0.7% agarose gels using the Quiex Gel Extraction Kit, and subcloned into plasmid pCR-Script Amp SK(+)(Stratagene, La Jolla, CA). DNA sequence analysis was performed on 500 fmol of extracted DNA product using a DNA Autoanalyzer (ABT).

RESULTS

Validation of the Rat PAC$_1$-R Antibody

Polyclonal antibodies were generated against specific amino acids of the deduced amino acid sequence of the rat PAC$_1$-R as described in the *Methods*.[5] Specificity of this antibody was determined using both Western blot analysis and immunocytochemistry in NIH/3T3 cells stably expressing the rat PAC$_1$-R (FIG. 1A). Immunocytochemistry confirmed the expression of PACAP receptors on these NIH/3T3 cells expressing rPAC$_1$-Rs but not on the NIH/3T3 cells expressing the VPAC$_1$-R (FIG. 1B) or VPAC$_2$-Rs (data not shown). These results demonstrated that the polyclonal antibodies to the PACAP Type 1 receptor are specific.

Western Blot Analysis

To confirm the expression of PAC$_1$-Rs on ECL cells, membrane preparations from ECL cells and Western blot analysis using the polyclonal antibodies to the PAC$_1$-Rs (described above) were performed as shown in FIGURE 2. Western blot analysis demonstrated a single protein band with a molecular weight of ≈ 46 kD (indicated by the arrow) expressed on ECL cells. As a positive control, membranes from AR4-2J cells were used and showed a similar protein size (FIG. 2). No expression of the PACAP receptor was observed in the HEK293 cells (data not shown).

Reverse Transcriptase/PCR Confirmation

To further demonstrate that rat ECL cells express the PAC$_1$-R, reverse transcriptase–PCR, was performed using RNA extracted from isolated ECL cells. Amplification of the cDNA was performed with two different rat PAC$_1$-R primers selected because they span intron/exon junctions of the rat PACAP receptor gene (unpublished data). As demonstrated in FIGURE 3, specific DNA products were observed using separate PCR primer combinations.

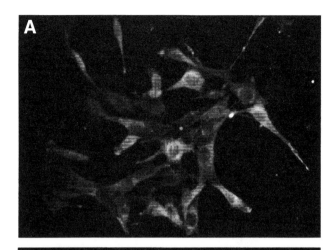

Figure 1 A
hPAC$_1$-R

Figure 1 B
rVPAC$_1$-R

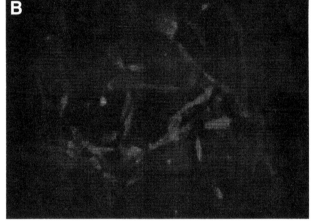

FIGURE 1. Immunocytochemistry using the anti-hPAC$_1$-R Ab on NIH/3T3 cells expressing the hPAC$_1$-R (**A**) or VPAC$_1$-R (**B**). Positive staining is present against the human PAC$_1$-R expressing cells. Preincubation of cells with the immunization peptides was negative (data not shown).

One primer combination, using Sense 1 and Antisense 2 primers, yielded a band at approximately 900 kbp with a predicted size of 883 kbp. A second primer combination using S1 and AS1 yielded a PCR product of approximately 600 kbp with a predicted size of 559 kbp. Amplification of reverse-transcribed RNA from ECL cells using primers specific for either the VPAC$_1$ or VPAC$_2$ receptor was negative (data not shown). Experiments using rat null, hop, hip, and hiphop primers based on the data by Spengler and coworkers[7] confirmed the expression of all four splice variants (data not shown).

Tissue Localization by Immunohistochemistry

To further confirm the results obtained by Western blot and RT–PCR, immunohistochemisty was performed using the anti-rat PAC$_1$-R antibody (dilution of 1:1,000). At low magnifications, the majority of the immunostaining is noted at the base of the oxynic gland

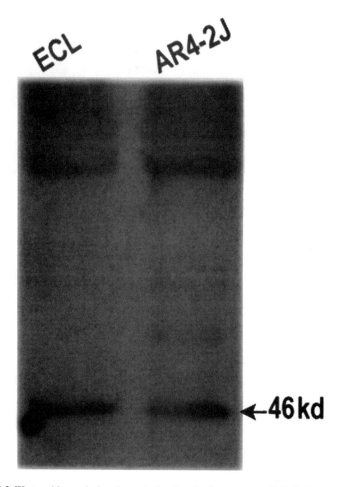

FIGURE 2. Western blot analysis using polyclonal antibodies to the rat PAC$_1$-R. Receptor expression is detected in membrane preparations from gastric enterochromaffin-like cells (ECL) (*left lane*). The antibody detects receptor expression corresponding to 46 kD (arrow). Similar expression was detected in the rat pancreatic cancer cell line, AR4-2J (*right lane*).

with little or no receptor expression observed at the isthmus or luminal regions of the gland. At higher magnifications, staining is present on ECL cells at the base of the gastric gland (data not shown).

Immunocytochemistry

The specific polyclonal antibodies generated against carboxl-terminal amino acids of the rat PAC$_1$-R were used for immunocytochemistry (FIG. 4) using isolated cell preparations of ECL cells isolated from the rat gastric mucosa. The cell composition was determined using antibodies to either histidine decarboxylase (ECL cells), gastrin (G cells), or somatostatin (D cells) in order to confirm that the cells were pure (data not shown). Intense

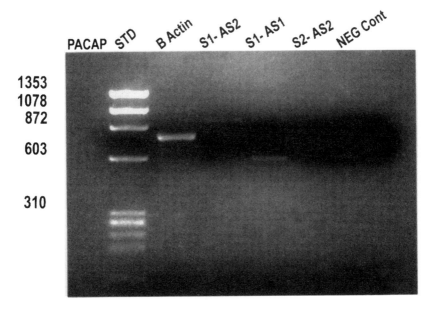

FIGURE 3. Reverse transcriptase/polymerase chain reaction (RT-PCR) demonstrating rat PAC_1-R expression in isolated rat gastric ECL cells. Rat gastric ECL cells were isolated as described previously.[14-16] Lane 1 indicates DNA standards at the sizes indicated on the left margin. Lane 2 demonstrates β-actin expression as a positive control. Lanes 3–5 indicate the expression of rat PAC_1-Rs using different primer pairs derived from the cloned $rPAC_1$-R cDNA.[5] Specific bands are identified for all three primer combinations and are the sizes expected from the primers used. The last lane demonstrates that no bands are detected when the enzme, reverse transcriptase, is omitted from the reaction mixture (negative control).

staining on the surface of the ECL cells but not of the G cells was observed using the anti PAC_1-R antibody (FIG. 4) and was specific compared to cells stained in the presence of pre-immune serum (data not shown).

DISCUSSION

It has been demonstrated by both pharmacological and molecular studies that three classes of PACAP receptors exist: PAC_1 (PACAP Type 1), $VPAC_1$ (VIP1/PACAP), and $VPAC_2$ (VIP2/PACAP).[5,6] However, much confusion exists as to their localization and physiological significance. Although PACAP hormone expression has been shown to occur in numerous peripheral tissues, their role has not been completely understood because PACAP-27 and PACAP-38 have high affinity for all three types of receptors. The current study is significant because it is the first study to identify the specific expression of PAC_1-Rs in the stomach. The current study also demonstrates that PAC_1-Rs are expressed on gastric ECL cells, suggesting that PACAP may regulate gastric acid secretion.

The cloning of the rat PAC_1-R afforded the generation of molecular probes and the construction of immunogenic peptides for the localization of the PAC_1-R. The rat PAC_1-R shares several structural features that are similar to the other members of the VIP/secretin/glucagon receptor family, such as a long amino terminus (>120 amino acids), which may be

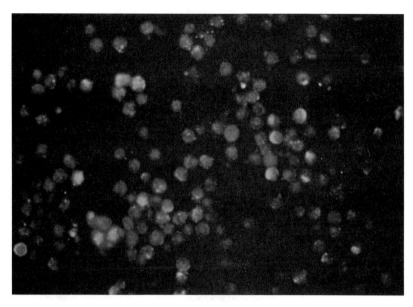

FIGURE 4. Localization of PACAP-Rs in isolated ECL cells by immunocytochemistry (50 × magnification). Immunocytochemistry was preformed on elutriated ECL cells (≈90% pure) with the specific anti-PACAP-R antibody showing the expression of PACAP-Rs on the cell surface. Immunocytochemistry on isolated ECL cells with pre-incubation of the anti-PACAP-R antibody with preimmune rabbit serum revealed no staining (data not shown).

important for ligand recognition; five potential N-linked glycosylation sites (ASN-X-SER/THR) in the amino terminus, which may be important for high affinity ligand; and seven cysteine residues, which may be necessary for agonist binding conformation.[5] The rat PACAP receptor expressed on ECL cells (FIG. 2) with a molecular mass of ≈ 46 kD is similar in size to the calculated molecular mass of 51 kD and is in close agreement to the 55 kD size previously published for the rat PACAP-R using crosslinking studies.[5]

PACAP hormone expression has been identified in numerous tissues including the brain, gastrointestinal tract, adrenal gland, and testis by immunohistochemistry and radioimmunoassay. The highest density of PACAP binding sites occurs in the brain cortex, olfactory bulb, hippocampus, and hypothalamus.[19,20] Therefore, PACAP appears to play an important role in the growth and development of the brain. In the gastrointestinal tract, PACAP-immunoreactive nerve fibers have been previously identified in the esophagus, stomach, duodenum, and small and large intestines.[21–23] In addition, PACAP receptors have been identified in the liver and pancreatic islets and on the rat pancreatic carcinoma cell line, AR-42J.[24] On smooth muscle cells of the gastrointestinal tract, PACAP acts as a potent non-adrenergic, non-cholinergic inhibitory neurotransmitter that directly relaxes intestinal smooth muscle cells, an action that is probably mediated by apamin-sensitive potassium channels or by the activation of nitric oxide.[25–27] Characterization of PACAP or VIP receptor subtypes mediating these effects on smooth muscles has not been determined previously. Immunohistochemistry performed in the current study shows that the PAC_1-R is expressed in high densities in the gastric mucosa. Examination of the gastric fundus demonstrates expression on ECL cell using ECL cell isolation techniques. No receptor expression was demonstrated on the mucous neck cells or on parietal cells of the stomach. Similarly, no receptor expression was demonstrated on the circular or longitudinal smooth muscles of the stomach. This is surprising given the potential role of PACAP on gastrointestinal motility previously reported.[25] These results also suggest that the role of PACAP in gastric motility may be mediated by another receptor subtype as has been shown recently in gallbladder smooth muscle.[27]

As previously shown in native cells such as in somatotrophs and gonadotrophs, PACAP stimulates intracellular Ca^{2+} and hormone secretion.[28] This study demonstrates that, like the pituitary gland, the PAC_1-R expressed on ECL cells may also be coupled to the stimulation of intracellular Ca^{2+} and the release of hormone. The specificity of this response is demonstrated by localization of PAC_1-Rs on ECL cells by immunohistochemical staining and RT-PCR. The dose response of intracellular calcium release following PACAP-38 or PACAP-27 is similar to that previously demonstrated for the cloned rat and human PAC_1-R (data not shown).[5,7,30] No comparable stimulation was observed following exposure to VIP suggesting the specificity of this response (data not shown). Because PAC_1-Rs have been shown to be expressed as splice variants in both rats and humans, studies were conducted to determine whether particular splice variants were expressed on the ECL cells. Using primer combinations based on published cDNA sequences,[7] RT-PCR demonstrated the expression of all of the rPACAP-R splice variants. Although the RT-PCR was performed on cells that were over 80% homogeneous, the possibility that contamination from other cells in the preparation occurred cannot be overlooked. The expression of rPACAP-R splice variants is important to understand in light of recent data demonstrating that particular splice variants (i.e., hop) in human cells are associated with an increased efficacy for phospholipase C coupling, immediate-early gene expression, and influence growth.[8,30]

The development of ECL cell neoplasia in both humans and rats has become of significant interest for the understanding of the development of tumors in general. Profound and long-term inhibition of gastric acid secretion has been considered to be one of the major mechanisms underlying the development of ECL tumors in rodents and humans. The mechanisms responsible for the development of ECL tumors are unclear. The suggestion that PACAP stimulates intracellular calcium and histamine release by ECL cells and the specific expression of PAC_1 on ECL cells may be of importance in understanding those mechanisms regulating ECL tumor growth in addition to the potential effects on gastric acid secretion. Evidence exists for suggesting that histamine secreted by ECL cells influences growth of ECL cells by an autocrine mechanism.[31] These observations and the observations that PACAP stimulates the growth of certain established tumor cell lines suggest a potential role for PACAP in influencing the growth of native cells.[32] It is unclear at the present time whether PACAP stimulation of histamine secretion or perhaps a direct growth effect of PACAP on ECL growth, as has been shown on certain lung cancer cell lines,[30] may be important for the genesis of gastric carcinoid tumors. Studies aimed at elucidating these complex interactions in the gastric mucosa deserve careful investigation.

REFERENCES

1. MIYATA, A., A. ARIMURA, R. R. DAHL, N. MINAMINO, A. VEHARA, L. JIANG, M. D. CULLER & D. H. COY. 1989. Biochem. Biophys. Res. Commun. **164**: 567–574.
2. LAM, H-C, K. TAKAHASHI, M. A. GHATEI, S. M. KANSE, J. M. POLAK & S. R. BLOOM. 1990. Eur. J. Biochem. **193**: 725–729.
3. SHIVERS, B. D., T. J. GORCS, P. E. GOTTSCHALL & A. ARIMURA. 1991. Endocrinology **128**: 3055–3065.
4. ISHIHARA, T., R. SHIGEMOTO, K. MORI, K. TAKAHASHI & S. NAGATA. 1992. Neuron **8**: 811–819.
5. PISEGNA, J. R. & S. A. WANK. 1993. Proc. Natl. Acad. Sci. USA **90**: 6345–6349.
6. LUTZ, E. M., W. J. SHEWARD, K. M. WEST, J. A. MORROW, G. FINK & A. J. HARMAR. 1993. FEBS Lett. **334**: 3–8.
7. SPENGLER, D., C. WAEBER, C. PANTALONI, F. HOLSBOER, J. BOCKAERT, P. H. SEEBURG & L. JOURNOT. 1993. Nature **365**: 170–175.

8. PISEGNA, J. R. & S. A. WANK. 1996. J. Biol. Chem. **271**:17267–17271.
9. DEUTSCH, P. J. & Y. SUN. 1992. J. Biol. Chem. **267**: 5108–5113.
10. SUDA, K., D. M. SMITH, M. A. GHATEI, J. K. MURPHY & S. R. BLOOM. 1991. J. Clin. Endocrinol. Methodol. **72**: 958–964.
11. CHRISTOPHE, J. 1993. Biochim. Biophys. Acta **1154**: 183–199.
12. SHEN, Z., L. T. LARSSON, G. MALMFORS, A. ABSOOD, R. HAKANSON & F. SUNDLER. 1992. Cell Tissue Res. **269**: 369–374.
13. MOODY, T. W., F. ZIA & L. MAKHEJA. 1993. Peptides **14**: 241–246.
14. PRINZ, C., M. KAJIMURA, D. R. SCOTT *et al.* 1993. Gastroenterology **105**: 449–461.
15. PRINZ, C., D. R. SCOTT, D. HURWITZ *et al.* 1994. Am. J. Physiol. **267**: G663–G675.
16. MODLIN, I. M., G. P. LAWTON, L. H. TANG, J. GEIBEL, R. ABRAHAM & U. DARR. 1994. Digestion **55**:31–37.
17. SANDVIK, A., E. BRENNA, H. ALDUM. 1993. Am. J. Physiol. **27**: G51–G56.
18. ZENG, N., J. H. WALSH, T. KANG, K. HELANDER, H. F. HELANDER & G. SACHS. 1996. Gastroenterology **110**: 1835–1846.
19. ARIMURA, A., A. SOMOGYVARI-VIGH, A. MIYATA, K. MIZUNO, D. H. COY & C. KITADA. 1991. Endocrinology **129**: 2787–2789.
20. TATSUNO, I., A. SOMOGYVARI-VIGH & A. ARIMURA. 1994. Peptides **15**: 55–60.
21. NGUYEN, T. D., G. G. HEINTZ & J. A. COHN. 1992 Gastroenterology **103**: 539–544.
22. MUNGAN, Z., A. ARIMURA, A. ERTAN, W. J. ROSSOWSKI & D. H. COY. 1992. Scand. J. Gastroenterol. **27**: 375–380.
23. COX, H. M. 1992. Br. J. Pharmacol. **106**: 498–502.
24. BUSCAIL, L., P. GOURLET, P. CAUVIN *et al.* 1990. FEBS Lett. **262**: 77–81.
25. SCHWORER, H., S. KATSOULIS, W. CREUTZFELDT & W. E. SCHMIDT. 1992. Naunyn-Schmiedeberg's Arch. Pharmacol. **346**: 511–515.
26. YADA, T. *et al.* 1994. J. Biol. Chem. **269**: 1290–1293.
27. MURTHY, K. S. & G. M. MAKHLOUF. 1994. J. Biol. Chem. **269**: 15977–15980.
28. SCHOMERUS, E., A. POCH, R. BUNTING, W. T. MASON & C. A. MCARDLE. 1994. Endocrinology **134**: 315–323.
29. SHEN, Z., L. T. LARSSON, G. MALMFORS, A. ABSOOD, R. HAKANSON & F. SUNDLER. 1992. Cell Tissue Res. **269**: 369–374.
30. PISEGNA, J. R, J. LEYTON, COELHO *et al.* 1997. Life Sci. **61**: 631–639.
31. MODLIN, I. M., R. R. KUMAR, C. J. SOROKA *et al.* 1994. Dig. Dis. Sci. **39**: 1446–53.
32. ZIA, F., M. FAGARASAN, K. BITAR, D. H. COY, J. R. PISEGNA, S. A. WANK & T. W. MOODY. 1996. Cancer Res. **55**: 4886–4891.

A Critical View of the Methods for Characterization of the VIP/PACAP Receptor Subclasses[a]

PATRICK ROBBERECHT[b] AND MAGALI WAELBROECK

Department of Biochemistry and Nutrition, School of Medicine,
Université Libre de Bruxelles, Bât. G/E, CP 611, 808 Route de Lennik,
B-1070 Brussels, Belgium

ABSTRACT: The binding properties of the three cloned VIP/PACAP receptors and their coupling to G proteins and effectors can be studied in cells expressing each recombinant protein. The data obtained in these models must be critically evaluated: the expression of a high receptor density may reveal irrelevant receptors states and coupling to non-cognate G protein, and entail a marked amplification of the response as well as distortions in the selectivity profile of full and partial agonists. These models are, however, of great interest in the design of selective agonists and antagonists for each receptor subtype. The availability of selective ligands will facilitate the identification of the receptor subtype responsible for PACAP and VIP actions in cells and tissues.

Three PACAP receptors have been cloned in different animal species: the PACAP I receptor that exists under several molecular forms resulting from alternative splicing,[1–3] and the VIP$_1$/PACAP and the VIP$_2$/PACAP receptors.[4–6] The PACAP I receptors are selective for PACAP-27 and PACAP-38 and do not recognize with a high affinity VIP or any other parent peptide.[7] The different splice variants do not differ markedly.[3,8] The VIP$_1$/PACAP and the VIP$_2$/PACAP receptors recognize with similar affinities PACAP and VIP.[4,5] However, they can be differentiated by secretin and growth hormone releasing factor (GRF), which have a very low affinity for the VIP$_2$ receptor.[9]

The properties of each receptor have been established on model cell lines expressing each recombinant receptor, by binding studies, and by evaluation of the second messenger's pathways.[8,10,11] These experiments are easy to perform and—theoretically—easy to interpret. They nevertheless have limitations (see below).

PROPERTIES OF THE RECOMBINANT RECEPTORS

Binding of Radiolabeled Ligands

The only labeled compounds available for identification of VIP/PACAP receptors are iodinated agonists. This is of paramount importance as the affinity of an agonist for a

---

[a]Supported by grant No. 3.4507.98 from the Fonds de la Recherche Scientifique Médicale and by an "Action de Recherche Concertée" from the Communauté Française de Belgique and by a "Interuniversity Poles of Attraction Programme—Belgian State, Prime Minister's Office—Federal Office for Scientific, Technical and Cultural Affairs."

[b]Corresponding author: Tel.: 32.2.555.62.15; Fax: 32.2.555.62.30; E-mail: probbe@ulb.ac.be

receptor typically depends on the receptor state, which means on its interaction with a G protein occupied or not by guanosyl diphosphate (GDP) or guanosyl triphosphate (GTP). By definition, if the labeled receptor state is relevant, its binding properties must be in agreement with functional data or with the second messenger pathway activation. Several reports indicate that some binding data may be difficult to reconcile with the functional data. Binding of ^{125}I-PACAP-27 to membranes of cells transfected with the PACAP I receptors is inhibited by nanomolar PACAP-27 and PACAP-38 concentrations. Both peptides stimulate the adenylate cyclase activity in the same concentration range. However, in the transfected porcine renal epithelial cells LCC PK$_1$,[2,12] PACAP-38 stimulates inositol phosphate turnover at low concentrations but PACAP-27 is effective only at very high concentrations. The observation that PACAP-27 activates phospholipase C at concentrations higher than those required for the apparent occupancy of all the receptors suggests that a second receptor state, with low affinity for PACAP-27 and not detected in ^{125}I-PACAP-27 binding studies, exists in LLC PK$_1$ cells. It must be noted that such a discrepancy between PACAP-27 and PACAP-38 induced phospholipase C activation was not observed when the rat PACAP receptor was transfected in Chinese hamster ovary (CHO) cells[13] nor when the human PACAP receptor was expressed in NIH/3T3 cells.[3]

In CHO cells expressing the PACAP receptor, ^{125}I-PACAP-27 binding is inhibited by nanomolar concentrations of PACAP-27 and PACAP-38 and by micromolar VIP concentrations. This agrees perfectly with the capacity of the three peptides to stimulate the adenylate cyclase activity. However, ^{125}I-VIP also labels PACAP receptors—an unexpected finding considering the low affinity of VIP for that receptor. Furthermore, ^{125}I-VIP binding is inhibited by nanomolar concentrations of not only PACAP-27 and PACAP-38 but also VIP.[14] This binding profile cannot be correlated with any biological activity: it is likely that ^{125}I-VIP labels a receptor state with no biological function.

PACAP I receptor expressed in transfected CHO cells has been characterized with two different tracers: ^{125}I-PACAP-27 and ^{125}I-PACAP-29.[15] The following observations have been described: (1) ^{125}I-PACAP-29 identifies more binding sites than ^{125}I-PACAP-27; (2) ^{125}I-PACAP-29 binding is GTP insensitive; (3) PACAP-27 and PACAP-38 (IC$_{50}$ values of 3 nM) were equipotent, and VIP (IC$_{50}$ value: 3 μM) poorly active for ^{125}I-PACAP-27 binding inhibition. In contrast, ^{125}I-PACAP-29 binding is inhibited with IC$_{50}$ values of 3 and 30 nM by PACAP-38 and PACAP-27, respectively, and VIP was completely inactive. A study of the binding properties of a chimeric receptor made of the N-terminal part of the PACAP I receptor grafted on the rat VIP$_1$/PACAP receptor C-terminal sequence was also performed with three different tracers (^{125}I-VIP, ^{125}I-PACAP-27, and ^{125}I-PACAP-29). The unlabeled peptide inhibition curves were different with each tracer and none of the observed profiles were strictly comparable to that of adenylate cyclase activation.[15]

From these studies it appears that the VIP/PACAP receptors exist in different states that can be labeled, more or less efficiently, by different tracers.

It is possible that the existence—or our ability to identify—these receptor states is due to the high receptors density expressed by the cell lines used. For instance, the number of G proteins may become limiting, generating receptor states uncoupled from the G proteins.

These results indicate that binding studies performed on recombinant systems must be interpreted with extreme caution.

Second Messengers Studies

The study of adenylate cyclase activation on membranes, cyclic AMP levels, phosphatidyl inositol turnover, or intracellular calcium concentrations in intact cells may be useful for receptors identification. This is of a particular interest when studying models expressing a very low level of receptors or mutant receptors with a low affinity for the tracer. In these cases, binding studies are not feasible and the study of the second messenger pathway activation is the only available procedure to evaluate the ligand's recognition. How is the EC_{50} value (concentration required for half maximal activation of the second messenger pathway) related to the receptor dissociation constant (ligand concentration required for the occupancy of 50% of the receptors)? When a high number of receptors is expressed (and this is often the case in cells transiently or permanently transfected with the cDNA coding for a recombinant receptor), the agonist dose-response curves are rarely in perfect agreement with the binding data. Indeed, the maximal stimulation may occur when a limited number of receptors are occupied and the existence of spare or reserve receptors may generate erroneous conclusions. In amplified systems, the two major findings are that (1) the full agonists have low EC_{50} values (as compared to the expected, predicted, or measured receptor occupancy K_d values and (2) partial agonists may stimulate maximally the effector activity, and their EC_{50} values are, usually, in good agreement with the K_d value of receptor occupancy. As a consequence, receptor selectivity, judged by the relative potencies of agonist ligands, may depend on receptor density and on the efficacy of the receptors to the G protein coupling or on the efficacy of the receptor–G protein complex for effector activation. These considerations are important when mutant receptors are created to identify the receptors and ligands areas responsible for receptor occupancy and second messenger activation, but also when screening new ligands designed for their capacity to recognize selectively one receptor subtype.

When functional data are considered in absence of binding studies, it is necessary to ensure that the system studied does not involve spare receptors. This can be achieved, for instance, by pretreating the cells expressing the studied receptor with an agonist for 18 to 24 hours.[16] This should induce a downregulation of the receptors and a decreased number of receptors. The comparison of the EC_{50} values and of the maximal stimulatory effect of full and partial agonists, in control and "downregulated" cells, may be used as a test for the existence of an amplification process in the cell line studied.

FUNCTIONAL RECEPTORS IN TISSUE OR CELL LINES

In tissue or cell lines that respond to VIP and PACAP, it is of interest to identify the receptor subtype(s) involved.

Expression of the mRNA Coding for the Receptor

It is now rather easy to detect the presence of the mRNA coding for a defined receptor. mRNA extraction can be performed on small amounts of tissues. A DNase treatment is often required to eliminate contaminating genomic DNA. The reverse transcriptase (RT) transforms the mRNA in a stable complementary DNA (cDNA) that can be amplified with selective primers by the polymerase chain reaction (PCR). If the primers used are selected appropriately, it is possible to exclude a possible contamination by genomic DNA by determination of the size of the amplified segment.[17,18]

The RT-PCR technique is highly sensitive but in most cases non quantitative. Northern blots or RNase protection assays require higher amounts of mRNA but are more appropriate for mRNA quantification.

The results obtained by RT-PCR detect the presence of the mRNA coding for the receptor, not the presence of the receptor. As the stability and the rate of translation of a given mRNA depend on the 3′ and 5′ non-translated sequences and probably also on cellular factors, there is only a poor correlation between the level of mRNA and the receptor expression. Furthermore, when studying tissue preparations, it must be kept in mind that the mRNA present in a minority of cells (that do not necessarily participate to the biological response of the tissues under study) can be detected by RT-PCR. In our experience, the search for the mRNA coding for a defined receptor is nevertheless of interest for tissue analysis after cell dissociation and preparation of fractions enriched in a defined cell type. Under these conditions, and even if the dissociation process, which often requires a pretreatment with proteolytic enzymes, modifies the receptor-mediated response, it is possible to establish that the cell has the potential to synthesize the receptor.[19]

Expression of the Receptors

Binding studies with radioligands and selected unlabeled ligands as well as functional studies may be performed for receptor subtype identification.

The limitations are, however, obvious: (1) The success of binding studies depends essentially on the number of receptors expressed (considering that ligand's affinity is not tissue dependent). A low receptor density may be compensated for by using a large number of cells or membranes in the incubation medium, but the ligand's stability must then be carefully checked. (2) If non-selective ligands are used (like ^{125}I-PACAP that labels PACAP I, VIP$_1$ and VIP$_2$ receptors, or ^{125}I-VIP that labels VIP$_1$, and VIP$_2$ receptors), the contribution of different receptor subtypes can be evaluated by competition curves using selective unlabeled ligands. For instance, the comparison of the inhibition curves of ^{125}I-PACAP binding by PACAP and VIP may indicate the presence of both PACAP and VIP receptors but does not allow the discrimination of VIP$_1$ or VIP$_2$ receptors. The inhibition curves of ^{125}I-VIP binding by VIP help to detect the presence of VIP$_1$ and VIP$_2$ receptors, secretin being even less potent for VIP$_2$ than VIP$_1$ receptor recognition. However, the competition curve analysis is often difficult, particularly if one receptor subtype is predominant.

The determination of second messenger accumulation of adenylate cyclase activation has the advantage of being still possible even when the receptor density is low. The comparison of PACAP and VIP dose-effect curves[20] may indicate the possible contribution of both VIP and PACAP receptors to the response, but does not allow the measurement of the relative number of each receptor subclass. The maximal response indeed depends on not-yet-established parameters, such as the "intrinsic" activity of each receptor subtype, and may depend on each cell type considered.

The development of selective, high affinity ligands for each receptor subclass should help to overcome the problem of identification, quantification, and relative contribution of each receptor to the biological effect of VIP and PACAP.

Maxadilan, a peptide extracted from a sand fly, is the only described ligand selective for the PACAP I receptor.[21-23] Its primary amino acid sequence has no obvious similarity with PACAP sequence; the published studies on maxidalan are rare.

We were the first to discover that the long-acting VIP analogue RO 25-1553, a potent antiinflammatory and broncho-relaxant agent in various models of asthma,[24,25] is a highly selective VIP$_2$ receptor subtype agonist.[26] Its affinity for the VIP$_2$ receptor was even better than that of VIP itself and was 300- to 1,000-fold higher than its affinity for the VIP$_1$ receptor. Furthermore, at least on human receptors, RO 25-1553 behaves like a full agonist on the VIP$_2$ receptors and as a partial agonist only on the VIP$_1$ receptor.[26] ^{125}I-RO 25-1553 labels almost exclusively the VIP$_2$ receptors and is an ideal ligand for receptor identification on both cell membranes and histological slices.[27]

The strategy for the design of selective VIP$_1$ receptor ligands was based on the clear preference of porcine secretin and human GRF for the VIP$_1$ receptor subtype (despite their rather low affinity). Thus, secretin and GRF were considered as lead compounds, that needed to be modified in order to reduce their affinity for their cognate receptor (the secretin and the GRF receptors) and to increase their affinity for VIP receptors without changing their preference for the VIP_1 receptor subclass. To design new secretin derivatives, we first compared porcine, rabbit, and avian sequences. Rabbit secretin[28] had a higher affinity for the VIP receptors than porcine secretin; this was due to the introduction of an arginine residue in position 16 instead of a glutamine residue.[29] Avian secretin,[30] which differs from mammalian secretin in 13 residues, had a comparable affinity for the secretin and the VIP_1 receptor but retained a low affinity for the VIP_2 receptor. Combining these two observations led to the synthesis of [Arg[16]]chicken (or avian) secretin that had the expected properties: a high affinity for the VIP_1 receptor and a low affinity for the VIP_2 and PACAP receptors. This compound did retain a non-negligible affinity for the secretin receptor and had a higher affinity for the rat VIP_1 than for the human VIP_1 receptor.[31] Nevertheless, [125]I-[Arg[16]]chicken secretin was an excellent tracer for selective labeling of the rat VIP_1 receptor on both membranes and histological slices, in tissues that do not express secretin receptors.[27]

For GRF derivatives, we created a chimeric ligand consisting in the N-terminal sequence of VIP (VIP(1-7)) and the carboxy-terminal sequence of human GRF, and shortened the GRF sequence to reduce the new peptide's interaction with the GRF receptor.[32] The VIP(1-7)/GRF(8-27) hybrid retained the GRF preference for the VIP_1 receptor, did not recognize the PACAP or the secretin receptor, and did not reproduce the effects of GRF on pituitary membranes.[31] It had, however, a rather low affinity for the VIP_1 receptor. The introduction of an arginine and a lysine residue in positions 16 and 15, respectively, led to a high affinity VIP_1 agonist without any significant affinity for the VIP_2, GRF, secretin, and PACAP receptors. Its affinity and selectivity were comparable in rat and human. Based on previous studies,[33] the synthesis of the acetylated peptide with a D-Phe residue in position 2 led to a potent and selective antagonist of the VIP_1 receptor.[34]

CONCLUSIONS

There are three practical consequences of the recent cloning of the VIP/PACAP receptors. (1) The availability of model cell lines expressing one defined receptor subtype allows a precise evaluation of the receptor properties and an easy screening of new analogues in the search of selective agonists and antagonists. (2) The design of mutated receptors may help to elucidate the binding domain structure and the structural changes responsible for G-protein activation, a prerequisite for the rational synthesis of new ligands. (3) The detection in small tissue samples of the mRNA coding for a receptor subtype is now easy and sensitive.

These powerful models and techniques must, however, be critically evaluated. When studying recombinant receptors, it is tempting to use preparations expressing a large number of receptors. This may lead to an excess of receptors over the number of the cognate G protein with the appearance of unusual receptor states, to an increased agonists potency with a distortion of the selectivity profile, and to receptor coupling to G proteins not normally used as transducers.[35]

REFERENCES

1. SVOBODA, M., M. TASTENOY, E. CICCARELLI, M. STIEVENART & J. CHRISTOPHE. 1993. Cloning of a splice variant of the pituitary adenylate cyclase-activating polypeptide (PACAP) type I receptor. Biochem. Biophys. Res. Commun. 195: 881–888.
2. SPENGLER, D., C. WAEBER, C. PANTALONI, F. HOLSBOER, J. BOCKAERT, P. H. SEEBURG & L. JOURNOT. 1993. Differential signal transduction by five splice variants of the PACAP receptor. Nature 365: 170–175.
3. PISEGNA, J. R. & S. A. WANK. 1996. Cloning and characterization of the signal transduction of four splice variants of the human pituitary adenylate cyclase activating polypeptide receptor.

Evidence for dual coupling adenylate cyclase and phospholipase C. J. Biol. Chem. **271**: 17267–17274.

4. ISHIHARA, T., R. SHIGEMOTO, K. MORI, K. TAKAHASHI & S. NAGATA. 1992. Functional expression and tissue distribution of a novel receptor for vasoactive intestinal polypeptide. Neuron **8**: 811–819.

5. LUTZ, E. M., W. J. SHEWARD, K. M. WEST, J. A. MORROW, G. FINK & A. J. HARMAR. 1993. The VIP₂ receptor: molecular characterisation of a cDNA encoding a novel receptor for vasoactive intestinal peptide. FEBS Lett. **334**: 3–8.

6. SVOBODA, M., M. TASTENOY, J. VAN RAMPELBERGH, J.-F. GOOSSENS, P. DE NEEF, M. WAELBROECK & P. ROBBERECHT. 1994. Molecular cloning and functional characterization of a human VIP receptor from SUP-T1 lymphoblasts. Biochem. Biophys. Res. Commun. **205**: 1617–1624.

7. CHRISTOPHE, J. 1993. Type I receptors for PACAP (a neuropeptide even more important than VIP?). Biochim. Biophys. Acta **1154**: 183–199.

8. CICCARELLI, E., M. SVOBODA, P. DE NEEF, E. DI PAOLO, A. BOLLEN, C. DUBEAUX, J.-P. VILARDAGA, M. WAELBROECK & P. ROBBERECHT. 1995. Pharmacological properties of two recombinant splice variants of the PACAP type I receptor transfected and stably expressed in CHO cells. Eur. J. Pharmacol. **288**: 259–267.

9. ROBBERECHT, P., P. GOURLET, P. VERTONGEN & M. SVOBODA. 1996. Characterization of the VIP receptor from SUP T1 lymphoblasts. Ad. Neuroimmunol. **6**: 49–57.

10. CICCARELLI, E., J.-P. VILARDAGA, P. DE NEEF, E. DI PAOLO, M. WAELBROECK, A. BOLLEN & P. ROBBERECHT. 1994. Properties of the VIP-PACAP type II receptor stably expressed in CHO cells. Regul. Pept. **54**: 397–407.

11. GAUDIN, P., A. COUVINEAU, J.-J. MAORET, C. ROUYER-FESSARD & A. LABURTHE. 1996. Stable expression of the recombinant human VIP₁ receptor in clonal Chinese hamster ovary cells: pharmacological, functional and molecular properties. Eur. J. Pharmacol. **302**: 207–214.

12. PANTALONI, C., P. BRABET, B. BILANGES, A. DUMUIS, S. HOUSSAMI, D. SPENGLER, J. BOCKAERT & L. JOURNOT. 1996. Alternative splicing in the N-terminal extracellular domain of the pituitary adenylate cyclase-activating polypeptide (PACAP) receptor modulates receptor selectivity and relative potencies of PACAP-27 and PACAP-38 in phospholipase C activation. J. Biol. Chem. **271**: 22146–22151.

13. DELPORTE, C., P. POLOCZEK, P. DE NEEF, P. VERTONGEN, E. CICCARELLI, M. SVOBODA, A. HERCHUELZ, J. WINAND & P. ROBBERECHT. 1995. Pituitary adenylate cyclase activating polypeptide (PACAP) and vasoactive intestinal peptide stimulate two signaling pathways in CHO cells stably transfected with the selective type I PACAP receptor. Mol. Cell. Endocrinol. **107**: 71–76.

14. HASHIMOTO, H., N. OGAWA, N. HAGIHARA, K. YAMAMOTO, K. IMANISHI, H. NOGI, A. NISHINO, T. FUJITA, S. NAGATA & A. BABA. 1997. Vasoactive intestinal polypeptide and pituitary adenylate-cyclase activating polypeptide receptor chimeras reveal domains that determine specificity of vasoactive intestinal polypeptide binding and activation. Mol. Pharmacol. **52**: 128–135.

15. VAN RAMPELBERGH, J., P. GOURLET, P. DE NEEF, P. ROBBERECHT & M. WAELBROECK. 1996. Properties of the PACAP I, PACAP II VIP₁ and chimeric N-terminal PACAP/VIP₁ receptors: Evidence for multiple receptor states. Mol. Pharmacol. **50**: 1596–1605.

16. VILARDAGA, J.-P., P. DE NEEF, E. DI PAOLO, A. BOLLEN, M. WAELBROECK & P. ROBBERECHT. 1995. Properties of chimeric and VIP receptor proteins indicate the importance of the N-terminal domain for ligand discrimination. Biochem. Biophys. Res. Commun. **211**: 885–891.

17. VERTONGEN, P., J. D'HAENS, A. MICHOTTE, B. VELKENIERS, J. VAN RAMPELBERGH, M. SVOBODA & P. ROBBERECHT. 1995. Expression of Pituitary Adenylate Cyclase Activating Polypeptide and receptors in human brain tumors. Peptides **16**: 713–719.

18. VERTONGEN, P., C. DEVALCK, E. SARIBAN, M.-H. DE LAET, H. MARTELLI, F. PARAF, P. HELARDOT & P. ROBBERECHT. 1996. Pituitary Adenylate Cyclase Activating Peptide and its receptors are expressed in human neuroblastomas. J. Cell. Physiol. **167**: 36–46.

19. VERTONGEN, P., B. VELKENIERS, E. HOOGHE-PETERS & P. ROBBERECHT. 1995. Differential alternative splicing of PACAP receptor in pituitary cell subpopulations. Mol. Cell. Endocrinol. **113**: 131–135.

20. VERTONGEN, P., M. DELHASE, F. RAJAS, J. TROUILLAS, E. HOOGHE-PETERS, M. SVOBODA & P. ROBBERECHT. 1996. Pituitary adenylate cyclase-activating polypeptide/vasoactive intestinal

polypeptide receptor subtypes are differently expressed in rat transplanted pituitary tumours (SMtTW) and in the normal gland. J. Mol. Endocrinol. **16**: 239–248.

21. LERNER, E. A., J. M. C. RIBEIRO, R. J. NELSON & M. R. LERNER. 1991. Isolation of Maxadilan, a potent vasodilatory peptide from the salivary glands of the sand fly *Lutzomyia longipalpis*. J. Biol. Chem. **266**: 11234–11236.

22. GREVELINK, S. A., J. OSBORNE, J. LOSCALZO & E. A. LERNER. 1995. Vasorelaxant and second messenger effects of Maxadilan. J. Pharmacol. Exper. Ther. **272**: 33–37.

23. MORO, E. & E. A. LERNER. 1997. Maxadilan, the vasodilator from sand flies, is a specific pituitary adenylate cyclase activating type I receptor agonist. J. Biol. Chem. **272**: 966–970.

24. O'DONNELL, M., R. J. GARIPPA, N. RINALDI, W. M. SELIG, B. SIMKO, L. RENZETTI, S. A. TANNU, M.A. WASSERMAN, A. WELTON & D.R. BOLIN. 1994. RO 25-1553: A novel, long-acting vasoactive intestinal peptide agonist. Part I: *In vitro* and *in vivo* bronchodilator studies. J. Pharmacol. Exp. Ther. **270**: 1282–1288.

25. O'DONNELL, M., R. J. GARIPPA, N. RINALDI, W. M. SELIG, J.E. TOCKER, S. A. TANNU, M. A. WASSERMAN, A. WELTON & D.R. BOLIN. 1994. RO 25-1553: a novel; long-acting vasoactive intestinal peptide agonist. Part II: Effect on in vitro and in vivo models of pulmonary anaphylaxis. J. Pharmacol. Exp. Ther. **270**: 1289–1294.

26. GOURLET, P., P. VERTONGEN, A. VANDERMEERS, M.-C. VANDERMEERS-PIRET, J. RATHE, P. DE NEEF & P. ROBBERECHT. 1997. The long-acting vasoactive intestinal polypeptide agonist RO 25-1553 is highly selective of the VIP$_2$ receptor subclass. Peptides **18**: 403–408.

27. VERTONGEN, P., S. N. SCHIFFMANN, P. GOURLET & P. ROBBERECHT. 1997. Autoradiographic visualization of the receptor subclasses for Vasoactive Intestinal Polypeptide (VIP) in rat brain. Peptides **18**: 1547–1554.

28. GOSSEN, D., L. BUSCAIL, A. CAUVIN, P. GOURLET, J. DE NEEF, J. RATHE, P. ROBBERECHT, M.-C. VANDERMEERS-PIRET, A. VANDERMEERS & J. CHRISTOPHE. 1990. Amino acid sequence of VIP, PHI and secretin from the rabbit small intestine. Peptides **11**: 123–128.

29. GOURLET, P., A. VANDERMEERS, M.-C. VANDERMEERS-PIRET, P. DE NEEF, M. WAELBROECK & P. ROBBERECHT. 1996. Effect of introduction of an Arginine[16] in VIP, PACAP and secretin on ligand affinity for the receptors. Biochim. Biophys. Acta **1314**: 267–273.

30. MUTT, V. 1988. Vasoactive intestinal polypeptide and related peptides. Isolation and chemistry. Ann. N.Y. Acad. Sci. **527**: 1–19.

31. GOURLET, P., A. VANDERMEERS, P. VERTONGEN, J. RATHE, P. DE NEEF, J. CNUDDE, M. WAELBROECK & P. ROBBERECHT. 1997. Development of high affinity selective VIP$_1$ receptor agonists. Peptides **18**: 1539–1545.

32. COY, D. H., W. A. MURPHY, V.A. LANCE & M. L. HEIMAN. 1987. Differential effects of N-terminal modifications on the biological potencies of growth hormone releasing factor analogues with varying chain lengths. J. Med. Chem. **30**: 219–222.

33. WAELBROECK, M., P. ROBBERECHT, D. H. COY, J. C. CAMUS, P. DE NEEF & J. CHRISTOPHE. 1985. Interaction of growth hormone-releasing factor (GRF) and 14 GRF analogs with vasoactive intestinal peptide (VIP) receptors of rat pancreas. Discovery of (N-Ac-Tyr^1,D-phe^2)-GRF(1-29)-NH^2 as a VIP antagonist. Endocrinology **116**: 2643–2649.

34. GOURLET, P., P. DE NEEF, J. CNUDDE, M. WAELBROECK & P. ROBBERECHT. 1997. *In vitro* properties of a high affinity selective antagonist of the VIP$_1$ receptor. Peptides **18**: 1555–1560.

35. VAN RAMPELBERGH, J., P. POLOCZEK, I. FRANOYS, C. DELPORTE, J. WINAND, P. ROBBERECHT & M. WAELBROECK. 1997. The pituitary adenylate cyclase activating polypeptide (PACAP I)- and VIP (PACAP II VIP$_1$) receptors stimulate inositol phosphate synthesis in transfected CHO cells through interaction with different G proteins. Biochim. Biophys. Acta (Mol. Cell Res.) **1357**: 259–255.

Mechanisms of Pituitary Adenylate Cyclase Activating Polypeptide (PACAP)-Induced Depolarization of Sympathetic Superior Cervical Ganglion (SCG) Neurons[a]

VICTOR MAY, MATTHEW M. BEAUDET, RODNEY L. PARSONS, JEAN C. HARDWICK, ERIC A. GAUTHIER, J. PETER DURDA, AND KAREN M. BRAAS

Department of Anatomy and Neurobiology, University of Vermont College of Medicine, Given Health Science Center, Burlington, Vermont, 05405 USA

Pituitary adenylate cyclase activating polypeptides (PACAP) demonstrate unique expression and regulatory functions in the peripheral nervous system. Not only do a variety of peripheral ganglia express PACAP immunoreactivity and mRNA, but many ganglia systems are potently regulated by PACAP27 or PACAP38 peptide. A population of sensory neurons in the dorsal root and trigeminal ganglia, for example, exhibit proPACAP mRNA and possess PACAP immunoreactivity, which can be upregulated by axotomy and neurogenic inflammation.[1-6] Dense PACAP immunoreactive fiber plexuses formed by sensory nerve projections have been found in the superficial layers of the substantia gelatinosa of the spinal cord. Furthermore, high levels of PACAP mRNA have been identified in lamina I and II neurons in the spinal cord dorsal horn suggesting that PACAP peptides participate at multiple sites of the primary sensory processing pathway.[7,8] PACAP peptides and mRNA have also been described in subpopulations of the parasympathetic sphenopalatine, otic, ciliary, and submandibular ganglion neurons,[1,9-11] implicating PACAP modulation of vascular smooth muscle tone and salivation. More recently, we have described PACAP and PACAP-selective receptor expression in guinea pig atrial parasympathetic cardiac ganglia.[12,13] PACAP immunoreactivity was identified in interganglionic fibers presumably of vagal derivation, which often formed pericellular baskets around the principal cardiac neurons, and in a small population of intrinsic neurons. Ligand activation of PACAP$_1$ receptors in the ganglia alters the membrane properties of the principal neurons and appears to modify the resulting cardiac output, suggesting that PACAP peptides may serve as an integrative neuromodulator in parasympathetic autonomic function.[12-14]

Given the well established effects of VIP in sympathoneuroblast survival and superior cervical ganglion (SCG) tyrosine hydroxylase activity, and the homology of PACAP to the VIP family of peptides, the possible roles of PACAP in the sympathetic system have generated considerable interest at several different levels. In concurrence with the observed neurotrophic properties of VIP, PACAP has been shown to promote potently sympathoneuroblast survival, neuronal division and differentiation, and neurite outgrowth.[15,16] These

[a]This work was supported by grants HD-27468 (V.M. and K.M.B.), NS-01636 (V.M.), and NS-23978 (R.L.P.) from the National Institutes of Health and grant 94015540 (K.M.B.) from the American Heart Association.

effects are in accord with related studies demonstrating the ability of PACAP peptides to prevent neuronal apoptosis upon growth factor withdrawal and ischemic insult.[17–19] Our laboratory has examined a different facet of PACAP regulation of sympathetic function and shown recently that PACAP27 and PACAP38 stimulated with high potency and efficacy primary SCG neuron catecholamine/neuropeptide Y(NPY) production and secretion[20,21]; 100- to 1,000-fold higher concentrations of VIP were required to modulate secretion, as would be predicted for neurons expressing the PACAP-selective PACAP$_1$ receptor. Both PACAP peptides elicited sustained increases in the rates of NPY and catecholamine release; the increase in total NPY and catecholamine production and secretion was in concert with elevated levels in proNPY and tyrosine hydroxylase mRNA, and consistent with other studies using cells derived from the sympathoadrenal lineage, including PC12 pheochromocytoma and adrenal medullary chromaffin cells.[22–26] PACAP$_1$ receptor mRNA expression in the SCG was identified by reverse transcription PCR, which was consistent with the cellular neuropharmacological response profiles to PACAP and VIP peptides.[21] Many neuropeptides have been localized to the intermediolateral cell column (IML) of the spinal cord and among several possibilities, PACAP in the IML sympathetic preganglionic neurons represented one potential extrinsic peptide source for SCG PACAP$_1$ receptor activation. In accord with that hypothesis, PACAP mRNA expression was identified in IML preganglionic projection neurons to the SCG.[7,8] Transection of the cervical sympathetic trunk diminished the PACAP-immunoreactive fibers in the SCG.[27] In sum, these results provided complementary evidence implicating PACAP peptides as candidate noncholinergic regulators in the sympathetic pathway.

But in addition to the extrinsic sources of PACAP peptides modulating the sympathetic effects, SCG neurons also demonstrated regulated intrinsic PACAP production and release.[1,27–30] Endogenous SCG PACAP38 was verified by reverse phase HPLC analysis. Reverse transcription PCR and sequence-specific hybridization revealed proPACAP mRNA in SCG neurons. Immunocytochemical and *in situ* hybridization histochemical studies localized PACAP peptides and proPACAP transcripts to a small subset of the SCG neuronal population, thereby increasing the neurophenotypic and functional diversity of sympathetic neurons. Depolarization stimulated not only endogenous SCG PACAP content, but also augmented PACAP secretion, suggesting that these peptides may function as signaling molecules. By Northern analysis, depolarizing conditions increased levels of the 2.2 kb form of proPACAP mRNA identified in most neuronal tissues, but also induced the novel expression of a shortened, 0.9 kb transcript.[28] The smaller transcript was not the previously reported short testicular form or the result of alternative splicing. Further studies demonstrated that the novel shortened form of sympathetic neuron proPACAP mRNA arose from alternative 3′ untranslated region cleavage and polyadenylation.[31–33] The shortened proPACAP transcripts were induced also in the SCG following axotomy[34]; accordingly, these posttranscriptional modifications may represent critical regulatory mechanisms with important consequences in PACAP expression during altered physiological states, especially neuronal injury.

To gain a better understanding of how extrinsic and intrinsic PACAP peptides impact and potentially regulate sympathetic neuron development and function, we have examined the prevalence of the PACAP$_1$ receptor protein on SCG neurons and the PACAP peptidergic effects on sympathetic neuronal membrane properties. These studies were intended to provide a broader appreciation of the functional relevance of PACAP peptides as neuromodulators and neurotrophic factors in the peripheral nervous system.

SUPERIOR CERVICAL GANGLION PACAP$_1$ RECEPTOR EXPRESSION

Based on current work, the receptor-mediated effects of PACAP have been suggested to be mediated by three G protein–coupled receptors. The PACAP$_1$ receptor is PACAP-selective;

PACAP and VIP exhibit equally high potency at the VIP$_1$/PACAP and VIP$_2$/PACAP receptors.[35-44] The PACAP$_1$ receptor belongs to group III of G protein receptors coupled to multiple intracellular signaling pathways, especially adenylyl cyclase and phospholipase C, whereas the VIP/PACAP receptors appear to be coupled only to adenylate cyclase. From neuropharmacological considerations, the sympathetic neurons appeared to express preferentially the PACAP$_1$ receptor than either of the nonselective receptor subtypes.[20,21] Multiple isoforms of the PACAP$_1$ receptor have been described. Alternative splicing of the 84 base pair HIP and/or HOP cassettes in the receptor sequence encoding the third cytoplasmic loop appears to discriminate second messenger coupling.[35] Splicing in the amino terminal extracellular domain (short and very short variants) has been suggested to determine PACAP27/PACAP38 potency.[45] By reverse transcription PCR, we demonstrated that adult and neonatal SCG and cultured SCG neurons express PACAP-selective receptors with one cassette insert in the third cytoplasmic loop. Based on correlative sequence-specific hybridization and diagnostic restriction analyses of the amplified products, the SCG receptor represented predominantly the HOP1 receptor variant. To examine potential alternative splicing in the amino terminal extracellular domain, we employed similar approaches using primers flanking the insert site (FIG. 1). Amplification of SCG cDNA templates with primers PACAPR3/PACAPR4 produced a 413 base pair product that represented the receptor with the short 63 base pair insert; digestion of the amplified material with AvaII produced the expected restriction fragments verifying the identity of the product (FIG. 1). From these results, the receptor expressed by sympathetic neurons of the SCG appeared predominantly to be the PACAP$_1$HOP1(short) isoform. These characteristics predicted a series of attributes in receptor second messenger coupling and, consistent with the responses associated with the HOP isoform, activation of the SCG PACAP$_1$ receptor stimulated both cyclic AMP and inositol phosphate production.[21,46] The

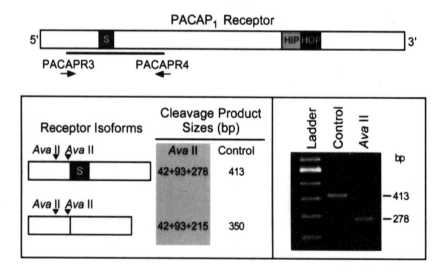

FIGURE 1. SCG neurons express the short variant of the PACAP$_1$ receptor. Reverse transcription PCR was used to identify the PACAP$_1$ receptor variants. Amplification of cDNA templates from rat SCG using primers PACAPR3 and PACAPR4 which differentiate the amino terminal extracellular domain splice variants, demonstrated that the sympathetic neurons express the short (s) receptor variant (413 bp product). The amplified product was analyzed by restriction enzyme digestion. Cleavage of the 413 bp amplified fragment with AvaII generated products anticipated for the short variant.

expression of the other PACAP$_1$ receptor variants was low by comparison; using the same methodologies, the VIP$_1$/PACAP and VIP$_2$/PACAP receptors did not appear to be expressed at appreciable levels in the adult SCG, which was in agreement with the results of VIP/PACAP receptor *in situ* hybridization histochemistry.[47]

DISTRIBUTION AND LOCALIZATION OF PACAP$_1$ RECEPTORS IN SCG SYMPATHETIC NEURONS

Although the reverse transcription PCR studies examining PACAP$_1$ receptor mRNA were a necessary requisite to implicate SCG PACAP$_1$ receptor expression, these results were not indicative of the abundance and distribution of the PACAP$_1$ receptor, which related directly to the relative significance of PACAP peptides in sympathetic regulation. The pertinent questions were: What is the relative proportion of SCG neurons expressing the PACAP$_1$ receptor? Are PACAP$_1$ receptors expressed solely in principal neurons or in the supporting cells? And if the receptors, as expected, are localized to neurons, are the PACAP receptor–bearing neurons distributed homogeneously within the tissue or segregated anatomically within the ganglion? PACAP$_1$ receptor expression in neurons would reinforce the roles of PACAP peptides as transmitters with neuroregulatory roles in the sympathetic pathways. Receptor expression in a majority of sympathetic neurons would imply a more global role for these peptides, while a more restricted distribution would suggest PACAP regulation of neurons of specific phenotypic determination that project to specific effector targets. Accordingly, localization and distribution studies of the PACAP receptor appeared central to evaluating the function and contributions of PACAP peptides in sympathetic regulation.

Using *in situ* hybridization histochemical techniques, several laboratories independently produced corroborating results demonstrating a striking pattern of PACAP$_1$ receptor mRNA expression in sympathetic neurons.[1,8,27,30,47] In our studies employing an antisense [^{35}S]radiolabeled riboprobe, which encoded the intracellular cytoplasmic tail of the receptor and did not discriminate the receptor variants, more than 90% of the principal cells of the SCG demonstrated dense autoradiographic grains for the PACAP$_1$ receptor mRNA. There was no apparent labeling in the small interstitial cells or intercellular matrix. From these results, the sympathetic neuronal PACAP effects appeared to be broad in scope.

But how prevalent was the PACAP$_1$ receptor protein and did the receptor proteins exhibit specific cellular localizations? While the demonstration of cellular mRNA expression is frequently used as necessary requisite for protein production, the corollary studies to demonstrate protein expression and distribution have become increasing essential. To begin addressing these issues, we have produced antisera to two synthetic peptides of the PACAP$_1$ receptor. Rabbit polyclonal antisera to the synthetic peptide acetyl-ERIQRANDLMGLNESSPGC-amide corresponding to an amino acid segment in the extracellular amino terminal of the PACAP$_1$ receptor and to acetyl-DFKHRHPSLASSGVNGGC amide corresponding to amino acids in the intracellular carboxyl terminal tail were produced and affinity-purified on homologous peptide columns. Both antisera demonstrated operative titers of 1:10,000 to 1:15,000 in solid phase enzyme linked immunoassays and recognized a cluster of 55 to 58 kD proteins corresponding to the receptor protein on Western blot analyses of detergent-solubilized brain membrane preparations.

Using the amino terminal antisera, immunocytochemical staining of cultured primary SCG neurons showed that more than 90% of the neurons were labeled in direct correlation with the *in situ* hybridization studies. The staining appeared over the cell soma; closer inspection revealed that the heaviest labeling appeared on the cell surface circumscribing the neuronal soma, which was compatible with functional localization and distribution of cellular receptors (FIG. 2). Some of the labeling appeared as punctate dots over the neuronal

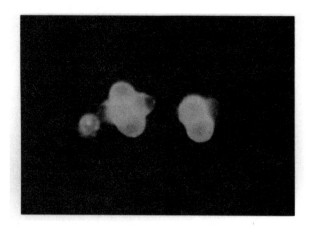

FIGURE 2. SCG neurons express PACAP$_1$ receptor protein. Primary cultured SCG neurons were fixed with 4% paraformaldehyde and immunocytochemically stained for PACAP$_1$ receptor using the amino-terminal extracellular domain–directed antibodies. Consistent with the neuropharmacological responses of sympathetic neurons, the majority of SCG neurons expressed the receptor protein. Neuronal labeling for the receptor was most prominent on the somal surface; areas of punctate staining may represent receptor clusters.

surface, which may represent receptor clusters. When intact adult SCG was examined, the staining patterns were recapitulated; nearly all of the principal neurons were labeled and, again, the immunofluorescent staining appeared heaviest at the membranes outlining the principal neurons. Whether the immunocytochemical patterns represented receptor staining on the neuronal soma or dendritic baskets is under investigation. Analogous to the distribution of VIP receptors in enteric neurons, the PACAP receptors may be localized to somal membranes and primary dendrites which may have significant functional implications with respect to PACAP peptide modulation of preganglionic neuroregulatory signals on postganglionic sympathetic neurons.[48] Nevertheless, the immunocytochemical staining patterns for the PACAP$_1$ receptor complemented the *in situ* hybridization histochemical studies; data from the two different approaches appeared identical, revealing a relatively homogeneous distribution of PACAP$_1$ receptor-expressing neurons in the SCG. These results implied that the majority of the SCG postganglionic neurons, regardless of neurophenotypic characteristics or final target tissue destinations, may be under PACAP peptide regulation supporting a physiologically relevant role for these peptides in sympathetic function.

PACAP-INDUCED ELECTROPHYSIOLOGICAL RESPONSES IN SCG NEURONS

We had described previously PACAP-induced sympathetic neuron neurotransmitter and neuropeptide production and release.[20,21] To assess more inclusively the rapid dynamic and direct effects of these peptides on SCG neurons, electrophysiological studies were undertaken to address whether PACAP peptides altered sympathetic neuronal membrane properties and if so, investigate the ionic conductances activated by PACAP. For congruity, a primary cultured SCG neuronal system was employed.[49] Neonatal ganglia were enzymatically dissociated, treated with cytosine arabinoside to remove supporting cell development, and the neurons were cultured on collagen-coated Aclar substrates under serum-free defined media conditions supplemented with 50 ng/ml nerve growth fac-

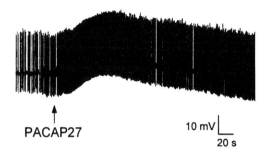

↑
PACAP27   10 mV |_____
                                    20 s

FIGURE 3. PACAP peptides depolarize and increase action potential frequency of sympathetic neurons. PACAP-induced depolarization of SCG neurons was measured using intracellular electrodes. Superfusion of cultures with 100 nM PACAP27 for 30 sec depolarized SCG neurons with an increase in action potential frequency. More than 90% of the cultured neurons responded to PACAP.

tor.[20,21,49] Under these conditions, in the presence of saturating NGF levels and in the absence of elevated cytokine levels from supporting cells, the sympathetic neuronal cultures no longer acquired characteristics induced during neuronal injury and reflected more closely neurons with properties found *in vivo*. The Aclar culture preparations were pinned to sylgard-lined dishes and experiments examining primary SCG neuronal responses to PACAP were conducted using standard intracellular microelectrode recording techniques. Under current clamp conditions, superfusion with 100 nM PACAP27 for 30 sec produced pronounced membrane depolarization from an average resting membrane potential of -50 mV. Depolarizations of approximately 10 mV were accompanied commonly by an increase in action potential firing frequency (FIG. 3). PACAP27 depolarized SCG neurons in a concentration-dependent manner with an apparent half-maximal response at 5 nM peptide; maximal depolarizations averaging 10 mV were achieved with 32 to 100 nM peptide concentrations (FIG. 4). PACAP38 elicited similar responses and concentration-dependence profiles; however, VIP had no apparent effects at the PACAP effective concentrations. In over 200 cells studied to date, more than 90% of the sympathetic neurons responded to PACAP27 or PACAP38, which was consistent with the *in situ* histochemical hybridization and immunocytochemical results.

Neuronal responses to PACAP27 treatment persisted in the presence of either 300 nM tetrodotoxin, a potent inhibitor of voltage-gated sodium channels, or 200 μM cadmium, an inhibitor of voltage-gated calcium channels (FIG. 5), demonstrating that the cellular responses to PACAP represented direct peptide actions on sympathetic neurons and were not mediated through interneuronal signaling events within the cell preparation. Additionally, the PACAP-induced depolarizations persisted in the presence of either 20 mM apamin or 1 mM tetraethylammonium (TEA; FIG. 5), obviating the closing of calcium-activated potassium channels as a possible mechanism for the observed effect of PACAP on membrane potential. Sympathetic neurons, however, have been well described to possess a non-inactivating barium-sensitive, outward potassium M-current, which is activated as neurons depolarize from resting potential.[50] Pressure application of PACAP27 (50 μM) reduced the M-current in the neurons by approximately 50%, suggesting one potential mechanism for PACAP-induced events.

In the presence of tetrodotoxin to eliminate action potential firing, pressure application of PACAP27 onto sympathetic neurons voltage-clamped to -50 mV produced inward currents of approximately 170 pA (FIG. 6). If M-current inhibition were the primary mechanism generating the PACAP-induced currents, then inhibition of the M-current prior to

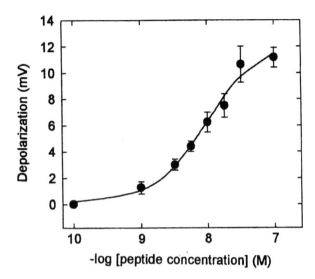

FIGURE 4. PACAP-induced SCG neuron depolarization is concentration dependent. SCG neurons were superfused with various concentrations of PACAP27 with the extent of depolarization measured with an intracellular electrode. Maximal depolarizations of approximately 10 mV were attained with 32 to 100 nM peptide; half-maximal responses were obtained with approximately 5 nM PACAP. Each point represents the mean of 3 to 7 neurons ± SEM.

peptide application should reduce substantially the size of the inward currents. In the presence of 1 mM barium, the PACAP-induced current was reduced by approximately 25%, indicating that the remaining current was produced by alternative mechanisms. To examine the possibility that the residual current could result from PACAP activation of a nonselective cationic conductance, neurons were bathed in 26 mM HEPES-buffered Krebs solution in which all of the sodium ions had been replaced with the non-permeant N-methyl-D-glucamine (NMG). In the sodium-deficient solution the PACAP-induced current was significantly reduced (FIG. 6). Accordingly, from these studies, the PACAP-induced inward current in sympathetic neurons appeared to be primarily a consequence of combined M-current inhibition and non-selective cationic conductance activation. These mechanisms for the excitatory physiological properties of PACAP on sympathetic neurons appear to be similar to those described for pituitary melanotropes, sympathetic preganglionic neurons, pancreatic beta cells, and HIT-T15 insulinoma cells,[51–54] and may be fundamentally important in understanding the acute and long term effects of PACAP peptides on sympathetic function.

SUMMARY

Our understanding of PACAP expression and regulation of sympathetic neuronal function has been augmented considerably over the last few years. Among the three major VIP/PACAP receptor subtypes, the SCG appears to express preferentially one particular variant of the PACAP-selective PACAP$_1$ receptor coupled to multiple intracellular signaling cascades. The *in situ* histochemical hybridization and immunocytochemical studies of PACAP$_1$ receptor mRNA and protein are in good agreement; nearly all of the SCG neu-

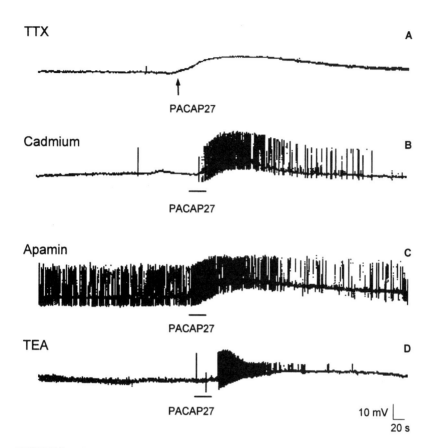

FIGURE 5. PACAP-induced depolarization represents a direct peptide effect on sympathetic neurons and does not result from inhibition of a calcium-activated potassium channel. PACAP-induced depolarization of SCG neurons was recorded in the presence of pharmacological agents that preferentially block specific ionic conductances. (A) Tetrodotoxin (TTX, 300 nM) was used to block voltage-gated sodium channels; (B) 200 μM cadmium was employed to block voltage-gated calcium channels; and (C) apamin (20 nM); and (D) tetraethylammonium (TEA, 1 mM) were used to inhibit the small and large conductance calcium-activated potassium channels, respectively. Depolarizations induced by local puffer application or superfusion of PACAP27 (100 nM) persisted under these paradigms, suggesting that modulation of these ionic conductances was not required to initiate the PACAP-induced depolarization in SCG neurons.

rons express the PACAP-selective receptor, suggesting that most of the sympathetic neurons are under PACAP neuromodulation. In accord with that possibility, several independent studies have now demonstrated PACAP peptide expression in the IML sympathetic preganglionic neurons and fibers, including those projecting to the SCG, further emphasizing the significance of PACAP peptides as a preganglionic noncholinergic mediator of sympathetic function.

Given the high potency of PACAP on any of a number of cellular responses, the functional relevance of PACAP peptides on SCG neurons is considerable. We have previously demonstrated the potency and efficacy of both PACAP27 and PACAP38 on sympathetic

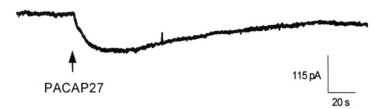

PACAP27   115 pA
                                                                    20 s

Treatment	Current Amplitude (pA)	n
TTX	176 ± 34	6
TTX + Barium	134 ± 25	6
TTX + NMG	53 ± 8	7

FIGURE 6. Inhibition of M-current and activation of a nonselective cationic conductance contribute to the PACAP-induced depolarization of SCG neurons. Local pressure application of PACAP27 (50 μM, 1 sec) induced a slowly developing inward current of approximately 150 pA in neurons clamped to -50 mV, which was not affected by the presence of tetrodotoxin (TTX, 300 nM; *table*). Inhibition of the M-current with 1 mM barium (Ba^{2+}) reduced the PACAP-induced current approximately 25%, whereas the inward current was inhibited approximately 70% upon replacement of sodium ions in the extracellular medium with the nonpermeant *N*-methyl-D-glucamine (NMG). The latter observation implicated the involvement of a nonselective cationic conductance.

neuron neurotransmitter/neuropeptide production and secretion; the ability of these peptides to stimulate neuronal second messenger activation was also in the nanomolar range. These results are congruous with our current electrophysiological studies, which were driven to further define the dynamic sympathetic responses to PACAP. In line with the morphological studies, for example, more than 90% of the sympathetic neurons responded to PACAP. In agreement with previous neuropharmacological data, the PACAP-induced depolarizations were elicited at physiologically relevant peptide concentrations at high affinity PACAP-selective receptors. The effects were direct and the alterations in postganglionic neuronal membrane properties appeared to be mediated by several ionic mechanisms. If these studies were analogous to pieces in a puzzle to understand the effects of PACAP in sympathetic development and function, the picture of late has been more completely assembled. But several important challenges still remain. What are the signal transduction mechanisms that mediate the PACAP-induced changes in sympathetic membrane properties? How do the resulting alterations impact the acute and more long-term responses of sympathetic neurons? Does the coupling of $PACAP_1$ receptors to intracellular signaling pathways differ during development, resulting in a transition from the neurotrophic properties of PACAP in neuroblasts to neuromodulatory roles of the peptides in postmitotic neurons? By looking at these issues in one distinct neuronal system, we enlarge our understanding and appreciation of peptides, and PACAP in particular, in the molecular and cellular events guiding neuronal development, function, and plasticity.

REFERENCES

1. SUNDLER, F., E. EKBLAD, J. HANNIBAL, K. MOLLER, Y-Z. ZHANG, H. MULDER, T. ELSAS, T. GRUNDITZ, N. DANIELSEN, J. FAHRENKRUG & R. UDDMAN. 1996. Pituitary adenylate cyclase activating polypeptide in sensory and autonomic ganglia: localization and regulation. Ann. N. Y. Acad. Sci. **805**: 410–428.

2. ZHANG, Y-Z., Q. ZHANG, T. J. SHI, R. R. JI, F. SUNDLER, J. HANNIBAL, J. FAHRENKRUG & T. HOKFELT. 1995. Expression of pituitary adenylate cyclase activating polypeptide in dorsal root ganglia following axotomy: time course and coexistence. Brain Res. **705**: 149–158.

3. ZHANG, Y-Z., J. HANNIBAL, Q. ZHAO, K. MOLLER, N. DANIELSEN, J. FAHRENKRUG & F. SUNDLER. 1996. Pituitary adenylate cyclase activating peptide (PACAP) expression in the rat dorsal root ganglia. Upregulation after peripheral nerve injury. Neuroscience **74**: 1099–1110.

4. MULDER, H., R. UDDMAN, K. MOLLER, Y-Z. ZHANG, E. EKBLAD, J. ALUMETS & F. SUNDLER. 1994. Pituitary adenylate cyclase activating polypeptide (PACAP) expression in sensory neurons. Neuroscience **63**: 307–312.

5. HANNIBAL, J., J. OMAR LARSEN, S. M. KNUDSEN & J. FAHRENKRUG. 1997. Expression of pituitary adenylate cyclase activating polypeptide (PACAP) in the mesencephalic trigeminal nucleus of the rat after transections of the masseteric nerve. Soc. Neurosci. Abstr. **23**: 1768.

6. ZHANG, Y-Z., H. MULDER, N. DANIELSEN & F. SUNDLER. 1997. Pituitary adenylate cyclase activating peptide (PACAP) expression is upregulated in inflammation. Soc. Neurosci. Abstr. **23**: 1768.

7. BEAUDET, M. M., K. M. BRAAS & V. MAY. 1998. Pituitary adenylate cyclase activating polypeptide (PACAP) expression in sympathetic preganglionic projection neurons to the superior cervical ganglion. J. Neurobiol. In press.

8. BEAUDET, M. M., K. M. BRAAS & V. MAY. 1996. Pituitary adenylate cyclase polypeptide (PACAP) expression in preganglionic neurons innervating the superior cervical ganglion. Soc. Neurosci. Abstr. **22**: 1998.

9. TOBIN, G., A. ASZTELY, V. EDWARDS, J. EKSTROM, R. HAKANSON & F. SUNDLER. 1995. Presence and effects of pituitary adenylate cyclase activating peptide in the submandibular gland of the ferret. Neuroscience **66**: 227–235.

10. ELSAS, T., R. UDDMAN, H. MULDER & F. SUNDLER. 1997. Pituitary adenylate cyclase activating polypeptide and nitric oxide synthase are expressed in the rat ciliary ganglion. Brit. J. Ophthalmol. **81**: 223–227.

11. MULDER, H., R. UDDMAN, K. MOLLER, T. ELSAS, E. EKBLAD, J. ALUMETS & F. SUNDLER. 1995. Pituitary adenylate cyclase activating polypeptide is expressed in autonomic neurons. Regul. Pept. **59**: 121–128.

12. HARDWICK, J. C., K. M. BRAAS, S. A. HARAKALL, V. MAY & R. L. PARSONS. 1997. Endogenous pituitary adenylate cyclase activating polypeptide (PACAP) regulates guinea pig cardiac neuron excitability through PACAP$_1$-selective receptors. Soc. Neurosci. Abstr. **23**: 428.

13. BRAAS, K. M., V. MAY, S. A. HARAKALL, J. C. HARDWICK & R. L PARSONS. 1998. Pituitary adenylate cyclase activating polypeptide (PACAP) modulation of guinea pig parasympathetic cardiac ganglion neuronal activity. J. Neurosci. In press.

14. SEEBECK, J., W. E. SCHMIDT, H. KILBINGER, J. NEUMANN, N. ZIMMERMAN & S. HERZIG. 1996. PACAP induces bradycardia in guinea pig heart by stimulation of atrial cholinergic neurones. Naunyn-Schmieddeberg's Arch. Pharmacol. **354**: 424–430.

15. DICICCO-BLOOM, E. & P. DEUTSCH. 1992. Pituitary adenylate cyclase activating polypeptide (PACAP) potently stimulates mitosis, neuritogenesis and survival in cultured rat sympathetic neuroblasts. Regul. Pept. **37**: 319.

16. DICICCO-BLOOM, E. 1996. Region-specific regulation of neurogenesis by VIP and PACAP: direct and indirect modes of action. Ann. N. Y. Acad. Sci. **805**: 244–258.

17. VILLALBA, M., J. BOCKAERT & L. JOURNOT. 1997. Pituitary adenylate cyclase activating polypeptide (PACAP38) protects cerebellar granule neurons from apoptosis by activating the mitogen-activated protein kinase (MAP kinase) pathway. J. Neurosci. **17**: 83–90.

18. CHANG, J. Y. & V. V. KOROLEV. 1997. Cyclic AMP and sympathetic neuronal programmed cell death. Neurochem. Int. **31**: 161–167.

19. ARIMURA, A., A. SOMOGYVARI-VIGH, C. WEILL, R. C. FIORE, I. TATSUNO, V. BAY & D. E. BRENNEMAN. 1994. PACAP functions as a neurotrophic factor. Ann. N. Y. Acad. Sci. **739**: 228–243.

20. MAY, V. & K. M. BRAAS. 1995. Pituitary adenylate cyclase activating polypeptide (PACAP) regulation of sympathetic neuron neuropeptide Y and catecholamine expression. J. Neurochem. **65**: 978–987.

21. BRAAS, K. M. & V. MAY. 1996. Pituitary adenylate cyclase activating polypeptides, PACAP38 and PACAP27, regulation of sympathetic neuron catecholamine sympathetic neuron and neuropeptide Y expression through activation of type I PACAP/VIP receptor isoforms. Ann. N. Y. Acad. Sci. **805**: 204–218.

22. DEUTSCH, P. J. & Y. SUN. 1992. The 38-amino acid form of pituitary adenylate cyclase activating polypeptide stimulates dual signaling cascades in PC12 cells and promotes neurite outgrowth. J. Biol. Chem. **267**: 5108–5113.

23. BARRIE, A. P., A. M. CLOHESSY, C. S. BUENSUCESO, M. V. ROGERS & J. M. ALLEN. 1997. Pituitary adenylate cyclase activating peptide stimulates extracellular signal-regulated kinase 1 or 2 (ERK 1/2) activity in a Ras-independent, mitogen-activated protein kinase/ERK kinase 1or 2-dependent manner in PC12 cells. J. Biol. Chem. **272**: 19666–19671.

24. COLBERT, R. A., D. BALBI, A. JOHNSON, J. A. BAILEY & J. M. ALLEN. 1994. Vasoactive intestinal peptide stimulates neuropeptide Y gene expression and causes neurite extension in PC12 cells through independent mechanisms. J. Neurosci. **14**: 7141–7147.

25. ISOBE, K., T. NAKAI & Y. TAKUWA. 1993. Ca^{2+} dependent stimulatory effect of pituitary adenylate cyclase activating polypeptide on catecholamine secretion from cultured porcine adrenal medullary chromaffin cells. Endocrinology **132**: 1757–1765.

26. TANAKA, K., I. SHIBUYA, T. NAGAMOTO, H. YAMSHITA & T. KANNO. 1996. Pituitary adenylate cyclase activating polypeptide causes rapid Ca^{2+} release from intracellular stores and long lasting Ca^{2+} influx mediated by Na^+ influx-dependent membrane depolarization in bovine adrenal chromaffin cells. Endocrinology **137**: 956–966.

27. MOLLER, K., M. REIMER, E. EKBLAD, J. HANNIBAL, J. FAHRENKRUG, M. KANJE & F. SUNDLER. 1997. The effects of axotomy and preganglionic denervation on the expression of pituitary adenylate cyclase activating peptide (PACAP), galanin and PACAP type I receptors in the rat superior cervical ganglion. Brain Res. **75**: 166–182.

28. BRANDENBURG, C. A., V. MAY & K. M. BRAAS. 1997. Identification of endogenous sympathetic neuron pituitary adenylate cyclase activating polypeptide (PACAP): depolarization regulates production and secretion through induction of multiple propeptide transcripts. J. Neurosci. **17**: 4045–4055.

29. KLIMASCHEWSKI, L., C. HAUSER & C. HEYM. 1996. PACAP immunoreactivity in the rat superior cervical ganglion in comparison to VIP. Neuroreport **7**: 2797–2801.

30. MOLLER, K., M. REIMER, J. HANNIBAL, J. FAHRENKRUG, F. SUNDLER & M. KANJE. 1997. Pituitary adenylate cyclase activating peptide (PACAP) and PACAP type I receptor expression in regenerating adult mouse and rat superior cervical ganglia in vitro. Brain Res. **75**: 156–165.

31. BRANDENBURG, C. A., V. MAY & K. M. BRAAS. 1996. Induction of novel pituitary adenylate cyclase activating polypeptide (PACAP) precursor messenger RNA in rat superior cervical ganglion. Soc. Neurosci. Abstr. **22**: 1764.

32. BRANDENBURG, C. A., S. A. HARAKALL, V. MAY & K. M. BRAAS. 1997. Novel forms of neuronal pituitary adenylate cyclase activating polypeptide (PACAP) mRNA are generated by alternative polyadenylation. Soc. Neurosci. Abstr. **23**: 2317.

33. HARAKALL, S. A., C. A. BRANDENBURG, G. A. GILMARTIN, V. MAY & K. M. BRAAS. 1998. Induction of multiple pituitary adenylate cyclase activating polypeptide (PACAP) transcripts through alternative cleavage and polyadenylation of proPACAP precursor mRNA. Ann. N. Y. Acad. Sci. This volume.

34. MAY, V., C. A. BRANDENBURG & K. M. BRAAS. 1996. Axotomy and decentralization regulate pituitary adenylate cyclase activating polypeptide (PACAP) expression in rat superior cervical ganglion. Soc. Neurosci. Abstr. **22**: 1764.

35. SPENGLER, D., C. WAEBER, C. PANTALONI, F. HOLSBOER, J. BOKAERT, P. H. SEEBURG & L. JOURNOT. 1993. Differential signal transduction by five splice variants of the PACAP receptor. Nature **365**: 170–175.

36. HOSOYA, M., H. ONDA, K. OGI, Y. MASUDA, Y. MIYAMOTO, T. OHTAKI, H. OKAZAKI, A. ARIMURA & M. FUJINO. 1993. Molecular cloning and functional expression of rat cDNAs encoding the receptor for pituitary adenylate cyclase activating polypeptide (PACAP). Biochem. Biophys. Res. Commun. **194**: 133–143.

37. PISEGNA, J. R. & S. A. WANK. 1993. Molecular cloning and functional expression of the pituitary adenylate cyclase activating polypeptide type I receptor. Proc. Natl. Acad. Sci. USA **90**: 6345–6349.

38. SVOBODA, M., M. TASTENOY, E. CICCARELLI, M. STIEVENART & J. CHRISTOPHE. 1993. Cloning of a splice variant of the pituitary adenylate cyclase activating polypeptide (PACAP) type I receptor. Biochem. Biophys. Res. Commun. **195**: 881–888.

39. MORROW, J. A., E. M. LUTZ, K. M. WEST, G. FINK & A. J. HARMAR. 1993. Molecular cloning and expression of a cDNA encoding a receptor for pituitary adenylate cyclase activating polypeptide (PACAP). FEBS Lett. **329**: 99–105.

40. RAWLINGS, S. R. & M. HEZAREH. 1996. Pituitary adenylate cyclase-activating polypeptide (PACAP) and PACAP/vasoactive intestinal polypeptide receptors: actions on the anterior pituitary gland. Endocrine Rev. **17**: 4–29.

41. ISHIHARA, T., R. SHIGEMOTO, K. TAKAHASI & S. NAGATA. 1992. Functional expression and tissue distribution of a novel receptor for vasoactive intestinal peptide. Neuron **8**: 811–819.

42. LUTZ, E. M., W. J. SHEWARD, K. M. WEST, J. A. MORROW, G. FINK & A. J. HARMAR. 1993. The VIP$_2$ receptor: molecular characterization of a cDNA encoding a novel receptor for vasoactive intestinal peptide. FEBS Lett. **334**: 3–8.

43. HASHIMOTO, H., T. ISHIHARA, R. SHIGEMOTO, K. MORI & S. NAGATA. 1993. Molecular cloning and tissue distribution of a receptor for pituitary adenylate cyclase activating polypeptide. Neuron **11**: 333–342.

44. CHATTERJEE, T. K., X. LIU, R. L. DAVISSON & R. A. FISHER. 1997. Genomic organization of the rat pituitary adenylate cyclase activating polypeptide receptor gene. J. Biol. Chem. **272**: 12122–12131.

45. PANTALONI, C., P. BRABET, B. BILANGES, A. DUMUIS, S. HOUSSAMI, D. SPENGLER, J. BOCKAERT & L. JOURNOT. 1996. Alternative splicing in the N-terminal extracellular domain of the pituitary adenylate cyclase-activating polypeptide (PACAP) receptor modulates receptor selectivity and relative potencies of PACAP27 and PACAP38 in phospholipase C activation. J. Biol. Chem. **271**: 22146–22151.

46. BRAAS, K. M. & V. MAY. 1995. Novel activation and interaction of sympathetic neuron signal transduction pathways by pituitary adenylate cyclase polypeptides. Soc. Neurosci. Abstr. **21**: 1845.

47. NOGI, H., H. HASHIMOTO, N. HAGIHARA, S. SHIMADA, K. YAMAMOTO, T. MATSUDA, M. TOHYAMA & A. BABA. 1998. Distribution of mRNAs for pituitary adenylate cyclase activating polypeptide (PACAP), PACAP receptor, vasoactive intestinal polypeptide (VIP) and VIP receptors in the rat superior cervical ganglion. Neurosci. Lett. **227**: 37–40.

48. SZURSZEWSKI, J. H. 1998. The functional significance of VIP-containing viscerofugal neurons. Ann. N.Y. Acad. Sci. This volume.

49. MAY, V. C., A. BRANDENBURG & K. M. BRAAS. 1995. Differential regulation of sympathetic neuron neuropeptide Y and catecholamine content and secretion. J. Neurosci. **15**: 4580–4591.

50. BROWN, D. A. & A. A. SELYANKO. 1985. Two components of muscarine-sensitive membrane current in rat sympathetic neurons. J. Physiol. **358**: 335–363.

51. TANAKA, K., I SHIBUYA, N. HARAYAMA, M. NOMURA, N. KABASHIMA, Y. UETA & H. YAMASHITA. 1997. Pituitary adenylate cyclase activating polypeptide potentiation of Ca^{2+} entry via protein kinase C and A pathways in melanotrophs of the pituitary pars intermedia of rats. Endocrinology **138**: 4086–4095.

52. LAI, C. C., S. Y. WU, H. H. LIN & N. J. DUN. 1997. Excitatory action of pituitary adenylate cyclase polypeptide on rat sympathetic preganglionic neurons in vivo and in vitro. Brain Res. **748**: 189–194.

53. LEECH, C. A., G. G. HOLTZ & J. F. HABENER. 1996. Signal transduction of PACAP and GLP-1 in pancreatic beta cells. Ann. N. Y. Acad. Sci. **805**: 81–92.

54. LEECH, C. A., G. G. HOLTZ & J. F. HABENER. 1995. Pituitary adenylate cyclase activating polypeptide induces the voltage-dependent activation of inward membrane currents and elevation of intracellular calcium in HIT-T15 insulinoma cells. Endocrinology **136**: 1530–1536.

Sympathetic Neurons of the Chick Embryo Are Rescued by PACAP from Apoptotic Death

ARUN R. WAKADE[a] AND DIMITRY LEONTIV

Department of Pharmacology, Wayne State University, School of Medicine, 540 E. Canfield Avenue, Detroit, Michigan 48201 USA

There is now convincing evidence that VIP/PACAP are the neurotransmitters in the rat adrenal medulla. These peptides have satisfied four out of five major criteria to play the role of neurotransmitters in controlling diverse functions of adrenal chromaffin cells. For example, (1) VIP/PACAP-containing nerve terminals innervating the chromaffin cells have been found in the adrenal medulla;[1,2] (2) VIP/PACAP are released in the medium after stimulation of neurons innervating the adrenal gland;[1,3] (3) specific receptors for these peptides are located on cell membranes of the chromaffin cells;[4] and (4) exogenous VIP/PACAP stimulate secretion of catecholamines (CA)–mimicking the effects of nerve stimulation.[5,6] Surprisingly, very little is known about the metabolic fate of VIP/PACAP in the adrenal medullary synapse or other synapses.

There is additional evidence that VIP/PACAP affect several other functional aspects of the chromaffin cells. For example, mRNA for proenkephalin was enhanced by VIP in bovine adrenal chromaffin cells.[7] Recently, it was demonstrated that PACAP increased chromogranin A expression in PC12 cells.[8] Furthermore, there are several other reports that tyrosine hydroxylase activity is enhanced by VIP/PACAP in chromaffin cells.[9,10] Taken together, evidence is emerging from different laboratories to imply that neuropeptides, VIP/PACAP, have multiple actions on chromaffin cells. It is well accepted that VIP/PACAP are now considered to be the noncholinergic transmitters in the adrenal medulla.

Sympathetic ganglia, which are embryologically related to the adrenal medulla, also receive peptidergic innervation in addition to classical cholinergic innervation.[11] VIP and its related members of the secretin family were found to modulate tyrosine hydroxylase activity of postsynaptic sympathetic neurons.[12] More recently, May and his colleagues have demonstrated that PACAP exerts a variety of effects on sympathetic neurons.[13] These results imply that PACAP and related peptides have multiple effects on the function of postsynaptic sympathetic neurons. Most important, these actions are different from those mediated by the cholinergic neurotransmitter and thereby play a unique role in autonomic synapses.

We have extended our studies from the adrenal medulla to autonomic ganglia to study the effects of PACAP on two different functions of sympathetic neurons derived from 10-day-old chick embryos maintained in culture. First, the effect of PACAP on transmitter release properties of neurons was monitored using ³H-norepinephrine (³H-NE). Second, survival and growth promoting effects of PACAP were examined in nerve growth factor–supported and

[a]Corresponding author: Arun R. Wakade, Ph.D., Department of Pharmacology, Wayne State University, School of Medicine, 540 E. Canfield Avenue, Detroit, MI 48201. Tel.: 313-577-2495; Fax: 313-577-6739; E-mail: aaade@med.wayne.edu

–deprived sympathetic neurons. Methods for studying these parameters have been published in the past.[14]

Effects of 100 nM PACAP on ^{3}H-NE release under nonstimulation and acetylcholine (ACh) stimulation conditions are shown in FIGURE 1. Spontaneous release of ^{3}H-NE was measured over four collection periods (2 min each). Addition of 3μM ACh caused more than twofold increase in ^{3}H-NE release over the background sample. ACh-evoked release gradually declined over three to four sample collections. When 30 nM PACAP was added to the medium there was a gradual increase in ^{3}H-NE release. Although not shown, PACAP caused about two- to threefold increase in the release, which lasted up to 10–15 minutes. When ACh was retested after PACAP the release of ^{3}H-NE was greatly potentiated. Potentiated response lasted for the next two samples as shown in FIGURE 1.

Since ACh acts through nicotinic and muscarinic receptors and only the nicotinic receptors are linked to secretory apparatus and muscarinic receptors are not,[15] we examined the effect of PACAP on nicotine-evoked ^{3}H-NE release. As shown in FIGURE 2, nicotine caused a significant increase in ^{3}H-NE release and this effect was potentiated by PACAP treatment. These experiments clearly show that PACAP has receptors on sympathetic neurons that are linked to the secretory mechanism, just as ACh nicotinic receptors are. The exact physiological significance and the mechanism of PACAP action await further investigations.

It is established that when well-developed sympathetic neurons in culture are deprived of NGF (washed with NGF-free medium and containing antibodies to NGF) they die as a result of apoptosis or program cell death. This model has been widely used to explore the mechanism of apoptosis in neuronal cells.[16] For example, Johnson and his colleagues have demonstrated that inhibitors of protein synthesis prevent apoptotic death induced by NGF deprivation. From these and other experiments it was proposed that new protein synthesis was necessary for the execution of the cell death program in sympathetic neurons.[17] There is also evidence that agents that increase cAMP are capable of interfering with sympathetic neuronal apoptosis.[18] Thus, forskolin, 8-bromo cAMP, etc., are neuroprotective drugs for neurons undergoing NGF-deprived cell death. Since PACAP is the potent activator of adenylate cyclase-cAMP-signaling system[19] and, most importantly, it is an endogenous peptide in

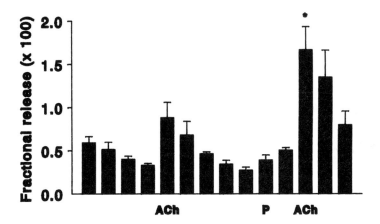

FIGURE 1. Potentiation of ACh-evoked release of ^{3}H-NE by PACAP. After four collections of samples to obtain spontaneous release of ^{3}H-NE, acetylcholine (ACh, 3 μM) was added for 2 min. After 8 min PACAP (P, 30 nM) was added for 2 min and ACh was added again after 6 min. Each column is the mean of three experiments. Vertical bars show S. E. of mean. $^*p < 0.01$.

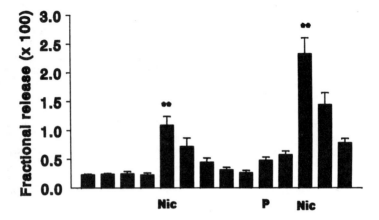

FIGURE 2. Potentiation of nicotine-evoked release of ³H-NE by PACAP. After four collections of samples (2 min each) nicotine (Nic, 10 μm) was added for 2 min. After 8 min PACAP (P, 30 nM) was added for 2 min and finally nicotine was added 6 min after PACAP. Each column represents the mean of six experiments. **$p < .001$.

autonomic ganglia (*see above*), we considered the possibility that PACAP should function as a neuroprotective presynaptic transmitter for postsynaptic sympathetic neurons.

In the second set of experiments, PACAP was tested for its neurotrophic action in chick embryo sympathetic neurons. It is well known that these neurons require NGF for their survival in culture.[20] When freshly plated neurons are not supplemented with NGF virtually all neurons die with two days (not shown). When 100 nM PACAP was added to the cultures at the time of plating the neurons there was no survival-promoting effect and all the neurons died (FIG. 3,a). This action cannot be considered toxic because when PACAP was added along with NGF there was no adverse peptide effect on neuronal

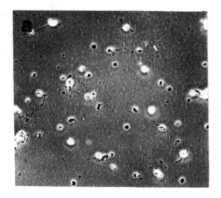
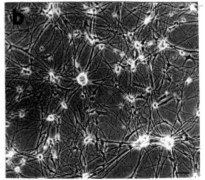

FIGURE 3. PACAP does not support the survival of freshly plated sympathetic neurons. PACAP (100 nM) added at the time of seeding the neurons in culture medium did not support survival (**a**) but those supported by NGF alone (not shown) or NGF plus PACAP showed survival and robust neuritic outgrowth (**b**). Bar represents 50 μm and the traces are representatives of five similar experiments.

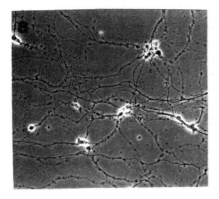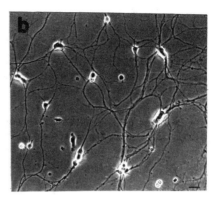

FIGURE 4. PACAP rescues NGF-supported sympathetic neurons from apoptotic death. Neurons were grown in NGF-supplemented medium for two days and then washed with NGF-free medium three times and maintained in NGF-free plus antibodies to NGF medium (**a**) or this medium plus 100 nM PACAP (**b**) for 30 h. Bar represents 50 μm and the photomicrographs are representative of seven similar experiments.

survival (FIG. 3, b). Neuronal growth was robust and comparable to that supported by NGF (not shown). These data clearly show that PACAP is not a neurotrophic molecule to support the survival of chick embryo sympathetic neurons in culture. Washout of NGF (addition of antibodies to NGF) from two-day-old neuronal cultures resulted in fragmentation of neurites and shrinkage of cell bodies (FIG. 4,a). However, 100 nM PACAP added to NGF-deprived neurons almost completely prevented the degenerative effects of NGF withdrawal (FIG. 4,b). This neuroprotective action of PACAP lasted for 24 to 48 hours. If NGF was added within a 24- to 48-hour period to PACAP-rescued neurons, most of them continued to grow for several days and were indistinguishable from neurons maintained in NGF all the time (not shown).

This brief report conclusively shows that PACAP is not a neurotrophic molecule in its own right as NGF and several other neurotrophic factors are for sympathetic neurons. Virtually all neurons died when PACAP was added to freshly plated neurons. However, the peptide proved to be very effective as a neuroprotective agent when added to neurons matured by NGF but deprived of NGF subsequently. This rescuing action of PACAP on NGF-deprived sympathetic neurons is exciting because of the newly discovered neurotransmitter role of the peptide in the autonomic ganglia and the adrenal medulla (*see above*). Based on our observations, it is tempting to postulate that PACAP could function as a temporary rescuing agent for sympathetic neurons in a situation where NGF supply to the neurons is transiently interrupted. Under this scenario release of PACAP from presynaptic nerve terminals on to postsynaptic cell bodies could serve as a rescue transmitter until NGF becomes available again. Thus, it seems that the sympathetic neuron is being supported from both ends, postsynaptically by NGF derived from the effector organs and presynaptically by PACAP released from nerve endings. It will be rewarding to determine if the neuropeptide could play a neuroprotective role in the rescue of sympathetic neurons *in vivo*.

REFERENCES

1. WAKADE, T. D., M. A. BLANK, R. K. MALHOTRA, R. POURCHO & A. R. WAKADE. 1991. The peptide VIP is a neurotransmitter in rat adrenal medulla: Physiological role in controlling catecholamine secretion. J. Physiol. (London) **444**: 349–362.

2. TABARIN, A., D. CHEN, R. HAKANSON & F. SUNDLER. 1994. Pituitary adenylate cyclase-activating peptide in the adrenal gland of mammals: distribution, characterization and responses to drugs. Neuroendocrinology **59**(2): 113–119.
3. WAKADE, A. R., X. GUO, R. STRONG, A. ARIMURA & J. HAYCOCK. 1992. Pituatory adenylate activating polypeptide (PACAP) as a neurotransmitter in rat adrenal medulla. Regul. Pept. **37**: 331.
4. MOLLER, K. & F. SUNDLER. 1996. Expression of pituitary adenylate cyclase activating peptide (PACAP) and PACAP type I receptors in the rat adrenal medulla. Regul. Pept. **63**(2–3): 129–139.
5. WATANABE, T., Y. MASUO, H. MATSUMOTO, N. SUZUKI, T. OHTAKI, Y. MASUDA, C. KITADA, M. TSUDA & M. FUJINO. 1992. Pituitary adenylate cyclase activating polypeptide provokes cultured rat chromaffn cells to secrete adrenaline. Biochem. Biophys. Res. Commun. **182**: 403–411.
6. CHOWDHURY, P. S., X. GUO, T. D. WAKADE, D. A. PRYWARA & A. R. WAKADE. 1994. Exocytosis from a single rat chromaffin cell by cholinergic and peptidergic neurotransmitters. Neurosci. **59**: 1–5.
7. WAN, D. C.-C. & B. G. LIVETT. 1989. Vasoactive intestinal peptide stimulates proenkephalin A mRNA expression in bovine adrenal chromaffin cells. Neurosci. Lett. **101**: 218–222.
8. TAUPENOT, L., S. K. MAHATA, H. WU & D. T. O'CONNOR. 1998. Peptidergic activation of transcription and secretion in chromaffin cells: CIS and TRANS signalling determinants of pituitary adenylate cyclase-activating polypeptide (PACAP). J. Clin. Invest. **101**: 863–876.
9. HAYCOCK, J. W. & A. R. WAKADE. 1992. Activation and multiple site phosphorylation of tyrosine hydroxylase in perfused rat adrenal gland. J. Neurochem. **58**: 57–64.
10. MARLEY, P. D., C. Y. CHEUNG, K. A. THOMSON & R. MURPHY. 1996. Activation of tyrosine hydroxylase by pituitary adenylate cyclase-activating polypeptide (PACAP-27) in bovine adrenal chromaffin cells. J. Auton. Nerv. Syst. **60**(3): 141–146.
11. SUNDLER, F., E. EKBLAD, J. HANNIBAL, K. MOLLER, Y-Z. ZHANG, H. MUDLER, T. ELSAS, T. GRUNDITZ, N. DANIELSEN, J. FAHRENKRUG & R. UDDMAN. 1996. Pituitary adenylated cyclase-activating peptide in sensory and autonomic ganglia: Localization and regulation. Ann. N.Y. Acad. Sci. **805**: 410–428.
12. BALDWIN, C., C. A. SASEK & R. E. ZIGMOND. 1991. Evidence that some preganglionic sympathetic neurons in the rat contain vasoactive intestinal peptide or peptide histidine isoleucine amidelike immunoreactives. Neuroscience **40**: 175–184.
13. MAY, V. & K. M. BRAAS. 1995. Pituitary adenylate cyclase-activating polypeptide (PACAP) regulation of sympathetic neuron neuropeptided Y and catecholamine expression. J. Neurochem. **65**: 978–987.
14. KULKARNI, J. S. & A. R. WAKADE. 1996. Quantitative analysis of similarities and differences in neurotoxicities caused by adenosine and 2-deoxyadenosine in sympathetic neurons. J. Neurochem. **67**: 778–786.
15. BHAVE, S. V., R. K. MALHATRA, T. D. WAKADE & A. R. WAKADE. 1988. Formation of inositol triphosphate by muscarinic agents does not stimulate transmitter release in cultured sympathetic neurons. Neurosci. Lett. **90**: 234–238.
16. DESHMUKH, M. & E. M. JOHNSON, JR. 1997. Programmed cell death in neurons: focus on the pathway of nerve growth factor deprivation-induced death of sympathetic neurons. Molec. Pharmacol. **51**: 897–906.
17. MARTIN, D. P., R. E. SCHMIDT, P. S. DISTEFANO, O. H. LOWRY, J. G. CARTER & E. M. JOHNSON, JR. 1988. Inhibitors of protein synthesis and RNA synthesis prevent neuronal death caused by nerve growth factor deprivation. J. Cell Biol. **106**: 829–844.
18. EDWARDS, S. N., A. E. BUCKMASTER & A. M. TOLKOVSKY. 1991 The death programme in cultured sympathetic neurones can be supressed at the posttranslational level by nerve growth factor, cyclic AMP, and depolarization. J. Neurochem. **57**: 2140–2143.
19. MARGIOTTA, J. F. & D. PARDI. 1995. Pituitary adenylate cyclase-activating polypeptide type I receptors mediate cyclic AMP-dependent enhancement of neuronal acetylcholine sensitivity. Molec. Pharmacol. **48**(1): 63–71.
20. EDGAR, D., Y. A. BARDE & H. THOENEN. 1981. Subpopulations of cultured chick sympathetic neurons differ in their requirements for survival factors. Nature **289**: 294–295.

Differential Display PCR Reveals Induction of Immediate Early Genes by Vasoactive Intestinal Peptide in PC12 Cells[a]

ANDREAS ESCHELBACH, ANDREAS HUNZIKER,[b] AND
LARS KLIMASCHEWSKI[c]

Institute of Anatomy and Cell Biology, University of Heidelberg, Im Neuenheimer Feld 307, 69120 Heidelberg, Germany

[b]*German Cancer Research Institute, Division of Cell Biology, Im Neuenheimer Feld 280, 69120 Heidelberg, Germany*

ABSTRACT: In order to identify genes regulated by vasoactive intestinal peptide, we performed differential display PCR as originally described by Liang and Pardee.[1] Messenger RNA of PC12 cells treated with vasoactive intestinal peptide or nerve growth factor for one hour was reverse transcribed and amplified using different sets of oligo-dT and random primers. Radioactively labeled PCR products were displayed on polyacrylamide gels and candidate cDNAs extracted from the gel, re-amplified by PCR, cloned, and sequenced. Differential expression was verified by RT-PCR applying sets of specific primers obtained from the sequence. The specificity of the PCR product was confirmed by Southern blotting using a radioactively labeled internal primer and semi-quantitative densitometric analysis. This rapid and sensitive protocol led to the isolation of two immediate early genes, pip92[2] and PC4,[3] known to be increased on mRNA level by nerve growth factor in PC 12 cells.

PC12 cells are derived from a rat pheochromocytoma and differentiate into sympathetic neuron-like cells in response to a variety of external stimuli, e.g., growth factors.[4,5] These act through specific cell surface receptors, which are either coupled to guanyl nucleotide binding proteins (e.g., receptors for vasoactive intestinal peptide, VIP) or exhibit intrinsic tyrosine kinase activity (e.g., TRK receptors activated by nerve growth factor, NGF). Following stimulation of multiple second messenger pathways, long processes resembling axons develop and prototypic growth cones become apparent. Therefore, PC12 cells provide a model system to study neuronal differentiation as well as axon formation.

In PC12 cells, VIP stimulates extension of neurites within minutes after addition of the peptide.[6,7] This effect is accompanied by tyrosine phosphorylation of several proteins[8] and by increased levels of proteins associated with neuronal differentiation (e.g., VGF or MAP kinases).[9,10] Besides VIP, other activators of adenylate cyclase (e.g., forskolin) also induce

[a]This work supported by The German Research Foundation (Zi 110/22-3) and Medical Faculty of Heidelberg (93/96).

[c]Corresponding author: Tel.: 0049/6221-548353; Fax: 0049/6221-544951; E-mail: lars. klimaschewski@urz.uni-heidelberg.de

neurite outgrowth. Therefore, this effect is most likely due to stimulation of cAMP-dependent pathways. Although the short-term induction of neurites mediated by an increase in cAMP does not appear to require ongoing RNA synthesis,[11] the continued extension and maintenance of processes induced by trophic factors involves major changes in gene expression.[12,13] The early activation of certain transcription factors within the first hour after addition of NGF is followed by an increase of mRNAs necessary for the production of cytoskeletal components and transmitter synthesizing enzymes, among others.

Until now, it has not been known whether VIP induces the same set of genes as NGF. VIP and NGF share an important intracellular signaling pathway, the MAP-kinase cascade in PC12 cells.[9,14] Therefore, changes in gene transcription induced by activation and nuclear translocation of MAP kinase are expected to occur in both VIP and NGF-differentiated PC12 cells. However, the activation of additional intracellular signaling pathways that also influence gene expression may account for differences in the morphological response observed after treatment with VIP and NGF, respectively. These include the time course of process formation, the percentage of differentiating cells, and the shape as well as length of neurites of PC12 cells.[6]

In order to understand the molecular mechanism of VIP-induced neurite outgrowth and to establish VIP as neurotrophic factor, it is important to identify genes that are regulated by VIP and to compare them with the gene expression profile induced by treatment with established trophic factors, e.g., NGF. Here, we outline a rapid and sensitive differential display PCR procedure developed by Liang and Pardee,[1] followed by confirmatory RT-PCR for analysis of gene expression in PC12 cells exposed to VIP and NGF, respectively. We focused on early changes in gene expression in order to identify immediate early genes (IEGs) that may be specific for either VIP or NGF.

METHODS

Cell Culture

PC12 cells were grown at low density in poly-L-lysine coated culture flasks (10^7 cells per 175 cm^2). RPMI 1640 medium supplemented with 10% horse serum, 5% fetal calf serum, and 1% PSN (5 mg/ml penicillin, 5 mg/ml streptomycin, 10 mg/ml neomycin) was changed two to three times per week. Cells were subcultured every four to five days and serum-starved 24 h before treatment with either VIP (1 μM; Bachem, Heidelberg, Germany) or NGF (10 ng/ml; Boehringer Mannheim, Germany). Cycloheximide is often added to the cultures to superinduce certain mRNA species. However, we did not use protein synthesis inhibitors because cycloheximide may downregulate or even prevent the induction of certain genes.[15,16]

mRNA Differential Display

Total RNA was extracted from PC12 cells using TRI reagent (Molecular Research Center, Cincinnati, OH, USA) according to the manufacturer's instructions. Chromosomal DNA contamination was removed from RNA using the MessageClean kit (GenHunter, Brookline, MA, USA) and mRNA differential display performed with RNAmap kits (Genhunter) as described.[1] Briefly, four reverse transcription reactions were prepared for each RNA sample using 0.2 μg total RNA in $1 \times$ RT buffer ($5 \times$ RT buffer = 125 mM Tris-Cl, pH 8.3, 188 mM KCl, 7.5 mM MgCl$_2$, and 25 mM dithiothreitol); 20 μM each dATP, dCTP, dGTP, and dTTP; and 1 μM of either $T_{12}MA$, $T_{12}MC$, $T_{12}MG$, or $T_{12}MT$ oligonucleotide (with M being a degenerate mixture of dA, dC, and dG). The tubes were transferred into a thermocycler (PerkinElmer GeneAmp PCR system 2400) programmed to 65°C for 5 min, followed by 37°C for 60 min, and 95°C for 5 min. After 10 min at 37°C, 200 units of MMLV reverse transcriptase was added to each tube.

PCR reaction mixtures contained 0.1 volume of cDNA, 1 × PCR buffer (10 × PCR buffer = 100 mM Tris-Cl, pH 8.4, 500 mM KCl, 15 mM MgCl$_2$, and 0.01% gelatin); 2 µM each dATP, dCTP, dGTP, and dTTP; 10 µCi of [^{35}S]dATP; 1 µM of 3′-primer used for reverse transcription (T$_{12}$MA, T$_{12}$MC, T$_{12}$MG, or T$_{12}$MT); 0.2 µM 5′-arbitrary 10-mer primer; and 1 unit of AmpliTaq DNA polymerase (PerkinElmer). Cycling parameters were as follows: 94°C for 30 sec (denaturation), 40°C for 2 min (primer annealing), 72°C for 30 sec (primer extension) for 40 cycles, followed by 72°C for 5 min (final extension). The amplified cDNAs were mixed with DNA sequencing stop buffer (USB) and incubated at 85°C for 2 min immediately before loading onto a 6% polyacrylamide sequencing gel. Gels were run at 70W constant current, blotted onto a piece of Whatman 3M paper, and dried under vacuum at 80°C for 1 h without prior fixation. Autoradiogram and dried gel were oriented with radioactive ink and exposed for 48 h at room temperature.

cDNA Recovery and Reamplification

After development of the film, cDNA bands clearly evident under one experimental condition and absent in the other were located by cutting through the film. The dried gel slice along with the Whatman 3M paper was placed in 200 µl TE buffer (10 mM TrisCl, 1 mM EDTA, pH 8.0) for 10 min followed by boiling the microfuge tube with cap tightly closed for 15 min. The supernatant was removed and cDNA recovered by ethanol precipitation in the presence of 0.3 M NaOAC and 5 µl 10 mg/ml glycogen as carrier. The pellet was resuspended in 10 µl TE following ethanol wash. Reamplification of 4 µl cDNA probe in a 40 µl reaction volume was performed using the same primer pairs and PCR conditions as used in the mRNA display except that dNTP concentration was 20 µM and no isotope was added.

Cloning and DNA Sequencing

Following reamplification, PCR products were excised from 1.5% agarose gels stained with ethidium bromide, eluted using QIAEX II extraction kit (Qiagen), and cloned into the PCRII vector using the TA cloning system from Invitrogen (San Diego, CA, USA). Plasmid DNA sequencing was performed on a DNA Sequencer 373A (ABI) using the PRISM Ready Reaction Dye Terminator Cycle Sequencing kit with AmpliTaq DNA polymerase FS (ABI) and M13 universal primers according to the instructions.

RT-PCR and Southern Hybridization

For verification of differential expression of cDNAs obtained by DD-PCR, RNA was extracted from new cultures and reverse transcribed in a volume of 20 µl containing 2 µg total RNA dissolved in diethylpyrocarbonate (DEPC)-treated H$_2$O, 1 × first-strand RT buffer (Life Technologies), 10 mM dithiothreitol, 0.5 mM dNTP, and 3 µg random primers (Life Technologies). Samples were heated to 70°C for 2 min, followed by 37°C for 60 min, and 65°C for 5 min. After 10 min at 37°C, 200 units of MMLV reverse transcriptase (Life Technologies) was added.

PCR was performed in triplicate in a volume of 50 µl using 1.5 µl of cDNA, 1 µM of each sense and antisense primer, 1 × PCR buffer containing 15 mM MgC1$_2$ (Perkin Elmer), 0.1 mM dNTP, and 1 unit AmpliTaq DNA polymerase (PerkinElmer). Primers for pip92 were 5′-CAC-CGATGGACTTTTATTTTCC-3′ (sense) and 5′-CCAGCGAATTTAAGAAAGTCTG-3′ (antisense), while those for PC4 were 5′-TTTGACCAAGCATCACAAAC-3′ (sense) and

mouse pip92 cDNA

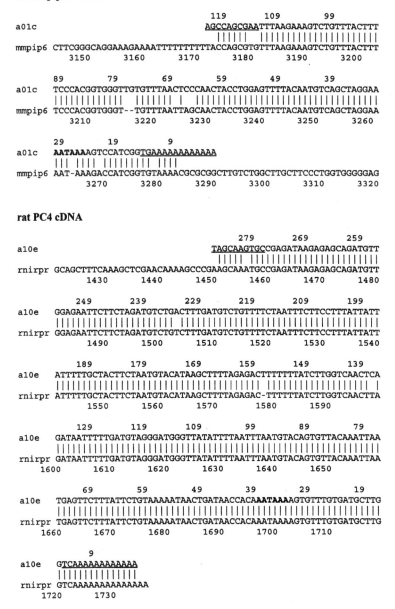

rat PC4 cDNA

FIGURE 1. Homology of nucleotide sequences of clones a01 c (corresponding to mouse pip92 cDNA, 93% identity in 98 bases) and al0e (corresponding to rat PC4 cDNA, 99% identity in 262 bases). Flanking primer sequences are underlined and putative polyadenylation sequences are in boldfaced letters. Numbering correlates with references given in the text. Note mispairings at the primer sites.

5'-TAGCAAGTGCCGAGATAAGAG-3' (antisense). PCR was performed within the linear range of amplification using the following cycle parameters: 93°C for 30 sec, 58°C for 1 min, 72°C for 1 min for 30 cycles, followed by 72°C for 10 min. Eight µl of PCR products were separated by agarose gel electrophoresis and blotted to porablot NY plus membrane (Macherey Nagel, Düren, Germany). Internal detection oligonucleotides (pip92: 5'-GCTGA-CATTGTAAAACTCCAGG-3'; PC4: 5'-GCAGATGTTGGAGAATTCTTCT-3') were labeled with [^{32}P]-γATP and subsequent Southern hybridization was performed according to Sambrook and colleagues.[17] Signal intensities were determined after exposure of membranes to PhosphoImager screens from BioRad and processed using Molecular Analyst software (BioRad). Experiments were repeated twice using cDNA derived from new cultures.

RESULTS AND DISCUSSION

In order to identify genes regulated by VIP in PC12 cells, we employed a novel technique, differential display PCR, developed by Liang and Pardee.[1] Only 2 µg of total RNA are needed for the complete display, which is about two orders of magnitude less than that required for subtractive hybridization. The latter method requires additional steps, e.g., plaque purification and rescreening, which are time consuming and may lead to the loss of genes of interest in the subtraction procedure. The greatest advantages of DD-PCR, however, are (1) the possibility of detecting low abundant mRNAs, e.g., those encoding transcriptional regulators; (2) the simultaneous identification of up- and down-regulated genes in the same experiment; and (3) the comparison of multiple samples in the same gel. The two major drawbacks of DD-PCR are (1) the high frequency of false positives and (2) the short sizes of differentially displayed fragments (for review see Debouck[18]). In the present experiment, the presumed differential expression of 20 bands excised from the gel was confirmed in 15 cases (75%) following subcloning and verification by specific RT-PCR and Southern hybridization using an internal oligonucleotide. This is an extremely sensitive method to confirm differential expression with the possibility of acquiring semi-quantitative data from densitometric analysis. Several groups have reported difficulties in carrying out Northern blot analysis, although this is regarded as the confirmatory method of choice since mRNA size can be determined and quantitative data obtained. However, Northern analysis is less sensitive and more labor-intensive than RT-PCR. In the present screening, 3 out of 15 differentially expressed genes (20%) were identified as expressed sequence tags (ESTs) while 4 cDNAs (27%) did not match with any sequence in the database. This may be due to the fact that the size of differentially displayed fragments ranges from only 43 to about 431 bases in length. These short fragments may correspond to 3' untranslated regions (UTRs) since the reverse DD-PCR primer is anchored next to the poly(A)-tail. 3'-UTRs of known genes are often not contained in sequence databases. Furthermore, they may differ considerably between homologous genes, making it difficult to identify even the gene family to which the fragment belongs.

In this study, 20 arbitrary 10-mer 5'-primers and four 3'-primers (T_{12}MA, T_{12}MC, T_{12}MG, T_{12}MT with M being a degenerate mixture of dA, dC, and dG) were used. The combination of these primers divided the mRNA population into 80 groups. Each primer combination yielded approximately 80 discrete bands per lane resulting in visualization of 6,000 mRNA molecules. These represent about 50% of the estimated 10,000 to 15,000 cellular mRNA species. Interestingly, the highest number of bands per lane was obtained with 3'-primer T_{12}MC, followed by T_{12}MG > T_{12}MA > T_{12}MT. About 75% of bands were reproducible in separate experiments. Only those bands were excised from the gel that appeared to be differentially expressed in at least two independent experiments. Performing this additional step may explain the low frequency of false positives (25%) in our study when compared with other reports applying the same technique.

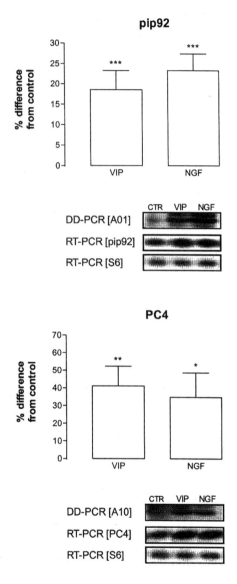

FIGURE 2. RT-PCR and Southern analysis of pip92 and PC4 cDNA as putatively regulated mRNAs identified by DD-PCR. The band pattern of the original display corresponds to the specific RT-PCR signals. Semi-quantitative analysis reveals the induction of both mRNAs compared to untreated control cells ($N = 9$, mean ± S.E.M., * $p < 0.05$, ** $p < 0.01$, *** $p < 0.001$ using unpaired t-test). PCR reactions were carried out in triplicate using three cDNAs derived from different RNA samples, and all measurements per group were pooled. cDNA of the ribosomal protein S6 was used as control template.

While 9 out of 15 differentially expressed genes (60%) are upregulated by both VIP and NGF, one clone (7%) has been found to be significantly downregulated by these factors. One cDNA was up-regulated by VIP only while NGF had no effect. Three clones (20%) were overexpressed and one underexpressed in NGF-treated cells with no influence of VIP on mRNA levels of these cDNAs.

The aim of this study was to identify IEGs induced by VIP in PC12 cells. Until now, two genes corresponding to known IEGs were identified by database searches, pip92 and PC4. The cDNA sequences of the respective DD-PCR clones are depicted in FIGURE 1. Primer sequences identical to those used in the differential display are indicated. In addition, degenerate binding of primers is confirmed by the mispairings at the primer sites of both clones. Confirmatory RT-PCR revealed that VIP and NGF, respectively, induced PC4 mRNA by 30–40%, while pip92 mRNA was upregulated by about 20% compared to untreated control cells (FIG. 2). Although the densitometric analysis was performed in the linear phase of the PCR, it is likely that these values underestimate the true difference in mRNA levels between control cells and differentiated cells.[19]

Pip92 is a protein of unknown function expressed during the G_0/G_1 transition of growth factor–stimulated fibroblasts.[2] In PC12 cells, pip92 mRNA accumulates within one hour after treatment with NGF, epidermal growth factor (EGF), or membrane depolarization with KCl.[2,20] The proline-rich protein is found in the cytoplasm and has an extremely short half-life. Interestingly, a 73-bp promoter fragment upstream of the transcription start site contains two binding sites for transcription factors of the Ets family,[21] which are primary targets of MAP kinases.[22]

The γ-interferon–related protein PC4[3] is induced in PC12 cells by VIP and NGF, respectively. Like pip92, PC4 is a cytoplasmic protein of unknown function transiently increased by NGF within three hours of treatment. Thereafter, the protein appears to be localized at the plasma membrane and, after longer time periods, is shifted to the nucleus. The presence of putative sequences for nuclear localization and for the addition of myristic acid within the 49 kD protein corroborates the immunohistochemical evidence for NGF-dependent subcellular redistribution.[3]

Since each of those genes is induced by VIP or NGF within one hour of treatment to a similar degree, it is likely that pip92 and PC4 are regulated in PC12 cells by the MAP kinase pathway. MAP kinase is activated by NGF via Ras, B-raf, and MAP kinase kinase. VIP feeds into this pathway at the level of B-raf, which is activated by the small G-protein Rapl. VIP stimulates Rapl via cAMP-dependent protein kinase (PKA).[14]

In summary, the direct comparison of gene expression profiles by DD-PCR followed by RT-PCR and Southern analysis provides a powerful tool for the identification of genes induced by VIP or other trophic factors. The method is sensitive enough to reveal changes in low abundant mRNAs, like IEG transcripts, and provides information on the intracellular signaling pathways possibly involved in induction of these genes.

ACKNOWLEDGMENTS

We thank C. Suter-Crazzolara, R. Witzgall, P. Mundel, and K. Krohn for valuable advice and K. Unsicker for providing laboratory facilities.

REFERENCES

1. LIANG, P. & A. B. PARDEE. 1992. Differential display of eukaryotic messenger RNA by means of the polymerase chain reaction. Science **257**: 967–971.

2. CHARLES, C. H., J. S. SIMSKE, T. P. O'BRIEN & L. F. LAU. 1990. Pip92: a short-lived, growth factor-inducible protein in BALB/c 3T3 and PC12 cells. Mol. Cell Biol. **10**: 6769–6774.
3. GUARDAVACCARO, D., A. MONTAGNOLI, M. T. CIOTTI, A. GATTI, L. LOTTI, C. DI LAZZARO, M. R. TORRISI & F. TIRONE. 1994. Nerve growth factor regulates the subcellular localization of the nerve growth factor-inducible protein PC4 in PC 12 cells. J. Neurosci. Res. **37**: 660–674.
4. GREENE, L. A. & A. S. TISCHLER. 1976. Establishment of a noradrenergic clonal line of the rat adrenal pheochromocytoma cells which respond to nerve growth factor. Proc. Natl. Acad. Sci. USA **73**: 2424–2428.
5. GREENE, L. A. & A. S. TISCHLER. 1982. PC12 pheochromocytoma cultures in neurobiological research. Adv. Cell Neurobiol. **3**: 373–414.
6. KLIMASCHEWSKI, L., K. UNSICKER & C. HEYM. 1995. Vasoactive intestinal peptide but not galanin promotes survival of neonatal rat sympathetic neurons and neurite outgrowth of PC12 cells. Neurosci. Lett. **195**: 133–136.
7. COLBERT, R. A., D. BALBI, A. JOHNSON, J. A. BAILEY & J. M. ALLEN. 1994. Vasoactive intestinal peptide stimulates neuropeptide Y gene expression and causes neurite extension in PC12 cells through independent mechanisms. J. Neurosci. **14**: 7141–7147.
8. OKUMURA, N., M. OKADA, K. NAGAI & H. NAKAGAWA. 1994. Vasoactive intestinal peptide induces tyrosine phosphorylation in PC12h cells. J. Biochem.(Tokyo) **116**: 341–345.
9. OKUMURA, N., Y. MIYATAKE, T. TAKAO, T. TAMARU, K. NAGAI, M. OKADA & H. NAKAGAWA. 1994. Vasoactive intestinal peptide induces differentiation and MAP kinase activation in PC12h cells. J. Biochem.(Tokyo) **115**: 304–308.
10. YOUNG, S. W., M. DICKENS & J. M. TAVARÉ. 1994. Differentiation of PC12 cells in response to a cAMP analogue is accompanied by sustained activation of mitogen-activated protein kinase: Comparison with the effects of insulin, growth factors and phorbol esters. FEBS Lett. **338**: 212–216.
11. GUNNING, P. W., G. E. LANDRETH, M. A. BOTHWELL & E. M. SHOOTER. 1981. Differential and synergistic actions of nerve growth factor and cyclic AMP in PC12 cells. J. Cell Biol. **89**: 240–245.
12. BATISTATOU, A., C. VOLONTE & L. A. GREENE. 1992. Nerve growth factor employs multiple pathways to induce primary response genes in PC12 cells. Mol. Biol. Cell **3**: 363–371.
13. LEONARD, D. G. B., E. B. ZIFF & L. A. GREENE. 1987. Identification and characterization of mRNAs regulated by nerve growth factor in PC12 cells. Mol. Cell. Biol. **7**: 3156–3167.
14. VOSSLER, M. R., H. YAO, R. D. YORK, M. G. PAN, C. S. RIM & P. J. STORK. 1997. cAMP activates MAP kinase and Elk-1 through a B-Raf- and Rap 1-dependent pathway. Cell **89**: 73–82.
15. CHO, K. O., W. C. SKARNES, B. MINSK, S. PALMIERI, L. JACKSON-GRUSBY & J. A. WAGNER. 1989. Nerve growth factor regulates gene expression by several distinct mechanisms. Mol. Cell. Biol. **9**: 135–143.
16. KAPLAN, M. D., J. A. OLSCHOWKA & M. K. O'BANION. 1997. Cyclooxygenase-1 behaves as a delayed response gene in PC12 cells differentiated by nerve growth factor. J. Biol. Chem. **272**: 18534–18537.
17. SAMBROOK, J., E. F. FRITSCH & T. MANIATIS, EDS. 1993. Molecular Cloning. Cold Spring Harbor Laboratory Press, New York.
18. DEBOUCK, C. 1995. Differential display or differential dismay. Curr. Opin. Biotechnol. **6**: 597–599.
19. SUTER-CRAZZOLARA, C. & K. UNSICKER. 1996. GDNF mRNA levels are induced by FGF-2 in rat C6 glioblastoma cells. Mol. Brain Res. **41**: 175–182.
20. BARTEL, D. P., M. SHENG, L. F. LAU & M. E. GREENBERG. 1989. Growth factors and membrane depolarization activate distinct programs of early response gene expression: dissociation of fos and jun induction. Genes Dev. **3**: 304–313.
21. LATINKIC, B.V. & L. F. LAU. 1994. Transcriptional activation of the immediate early gene pip92 by serum growth factors requires both Ets and CArG-like elements. J. Biol. Chem. **269**: 23163–23170.
22. MARAIS, R., J. WYNNE & R. TREISMAN. 1993. The SRF accessory protein Elk-1 contains a growth factor-regulated transcriptional activation domain. Cell **73**: 381–393.

Developmental Regulation of Pituitary Adenylate Cyclase Activating Polypeptide (PACAP) and Its Receptor 1 in Rat Brain: Function of PACAP as a Neurotrophic Factor[a]

DAN LINDHOLM,[b] YLVA SKOGLÖSA, AND NOBUYUKI TAKEI

Department of Developmental Neuroscience, Box 587 Biomedical Center, Uppsala University, 23 Uppsala, Sweden

ABSTRACT: To function as a trophic factor PACAP and PACAP-R must be expressed in the nervous system during early development. We report here on the distribution of PACAP mRNA in the developing nervous system of the rat and compare its expression with that of PACAP-R. We discuss primary neuron culture experiments that study the neurotrophic activity of PACAP. Experimental results that indicate the presence of PACAP and its receptor in the developing nervous system, together with the observed neuropeptide activity on various populations of neurons, support the view that PACAP exhibits important neurotrophic activities comparable to those of the classical neurotrophic factors.

Development of the nervous system is characterized by distinct phases of cellular proliferation, migration, differentiation, and cell death. Neurotrophic factors are crucial for the proper development of neurons and act through specific neuronal receptors.[1,2] Among the neurotrophic factors, the neurotrophins, consisting of nerve growth factor (NGF), brain-derived neurotrophic factor (BDNF), neurotrophin-3, and neurotrophin 4/5, are the best known molecules.[3] The neurotrophins support the survival and differentiation of specific populations of neurons both in the peripheral and central nervous system.[3] Besides the classical neurotrophic factors, neurotransmitters and some neuropeptides have been shown to exhibit neurotrophic activities.[5,6] The first neuropeptide to be studied for its neurotrophic actions was vasoactive intestinal peptide, which affects the development of sympathetic neuroblasts in culture.[7] In neurons, VIP increases cAMP, a key molecule for regulation of neuronal development. The neuropeptide PACAP is structurally related to VIP[8] and has a much stronger effect on cAMP levels in target cells.[8-10] PACAP specifically activates its receptor PACAP receptor 1 (PACAP-R) in a nanomolar range leading to various biological effects in target cells.[9,10] PACAP acts in many instances as a neuropeptide, having a neuromodulatory function in different systems.[8] In recent years the possibility that PACAP might have a trophic activity on neurons has received increased attention.[11-14] However, to function as a trophic factor PACAP and PACAP-R need to be expressed in the nervous system during early development.

[a]This work supported by grants from the Swedish Medical Research Council (MFR) and the Swedish Cancer Foundation (Cancerfonden). N. T. received scholarships from Wenner Grenska Samfundet and from Wenner Gren Stiftelsen.
[b]Corresponding author: Dan Lindholm, Fax: 18-559017 and E-mail: dan.lindholm@mun.uu.se

189

In this chapter we report on the distribution of PACAP mRNA in the developing nervous system of the rat and compare its expression with that of PACAP-R. In addition, we discuss some recent experiments studying the neurotrophic activity of PACAP in primary neuron cultures. The results obtained on the presence of PACAP and its receptor in the developing nervous system, together with the observed activity of the neuropeptide on various populations of neurons support the view that PACAP exhibits important neurotrophic activities that are comparable to those of the classical neurotrophic factors.

PACAP mRNA IS PRESENT IN DEVELOPING AND ADULT RAT BRAIN

In situ hybridization studies with a synthetic oligonucleotide probe complementary to rat PACAP mRNA[14] showed that the neuropeptide was already expressed in nervous tissue during embryonic development. At embryonic day 13 (E13), the earliest investigated time point, high expression of PACAP mRNA was found in the differentiating layers of the neuroepithelium of forebrain, midbrain, and hindbrain and in the pituitary (FIG. 1, A and B). In addition, there was a distinct expression of PACAP mRNA in the spinal cord and the dorsal root ganglia (FIG. 1, A and B). At E20, PACAP mRNA was expressed highly in the cortical plate, in several thalamic and hypothalamic nuclei, in the habenular nucleus, in developing hippocampus, in the trigeminal ganglion, and in the amygdala (FIG. 2, A–C). PACAP mRNA was also expressed in the olfactory bulbs, inferior colliculus, solitary nucleus, inferior olive, and other pontine nuclei, in the dorsal root ganglia, in the spinal cord, and also weakly in the cerebellum (data not shown). Postnatally, the expression of PACAP mRNA diminished towards adulthood. As shown in FIGURE 3A, PACAP mRNA was still moderately high in alternating layers of the cortex, within specific fields of the hippocampus, olfactory bulbs, hypothalamic and thalamic nuclei, and some pontine nuclei. In addition, we observed a weak signal in the granular layer of the cerebellum.

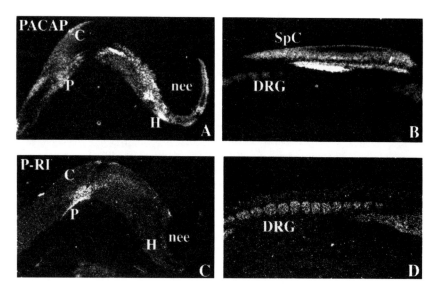

FIGURE 1. (**A** and **B**) Expression pattern of PACAP mRNA in saggital sections of E 13 rats. (**C** and **D**) Expression of PACAP-R mRNA in saggital sections of E 13 rats. C, cerebellum; P, pons; H, hypothalamus; nee, neuroepithelium; SpC, spinal cord; DRG, dorsal root ganglion.

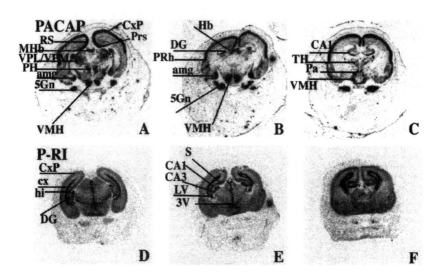

FIGURE 2. Expression of PACAP mRNA (**A–C**) and PACAP-R (**D–F**) in coronal sections of E 20 rats. RS, retrosplenial cortex; MHb, medial habenular nucleus; VPL, ventral posterolateral thalamic nucleus; VPM, ventral posteromedial thalamic nucleus; PH, posterior hypothalamic area; amg, amygdaloid neuroepithelium; 5Gn, trigeminal ganglion; VMH, ventromedial hypothalamic nucleus; CxP, cortical plate; Prs, presubiculum; Hb, habenular nucleus; DG, dentate gyrus; PRh, perirhinal cortex; CA1, CA1 field of the hippocampus; TH, thalamus; Pa, paraventricular hypothalamic nucleus; cx, cortical neuroepithelium; hi, hippocampal formation epithelium; S, subiculum; LV, lateral ventricle; 3V, third ventricle.

EXPRESSION OF mRNA ENCODING PACAP-R

In situ hybridization studies with a synthetic oligonucleotide probe complementary to rat PACAP-R mRNA, which recognizes all splice variants,[9] showed that PACAP-R mRNA was also expressed within the nervous system during embryonic development. In E13 rats the strongest expression of PACAP-R mRNA was in the mitotic layer of the neuroepithelium of forebrain, midbrain, and hindbrain (FIG. 1). PACAP-R mRNA was furthermore expressed moderately in the dorsal root ganglia and the spinal cord. Contrary to the expression pattern of PACAP, PACAP-R was expressed more evenly all over the brain during development, with exception to the very strong signals of the mitotic layers of the neuroepithelium (FIG. 2). Postnatal development of the rat central nevous system was accompanied by reduced PACAP receptor 1 mRNA expression. As shown in FIGURE 3C, in adult brain the strongest signals were observed in the olfactory bulbs, the dentate gyrus, the medial vestibular nuclei, and the granular layer of the cerebellum. In addition, a broad but moderate expression could be noticed in the thalamus, the hypothalamus, and in the pons.

DIFFERENTIAL EXPRESSION OF PACAP, PACAP-R, AND BDNF-mRNA IN THE ADULT HIPPOCAMPUS

We investigated the neurotrophic effect of PACAP in comparison to known neurotrophins, such as the brain-derived neurotrophic factor (BDNF), because we were interested in the different localization of them in the adult hippocampus.

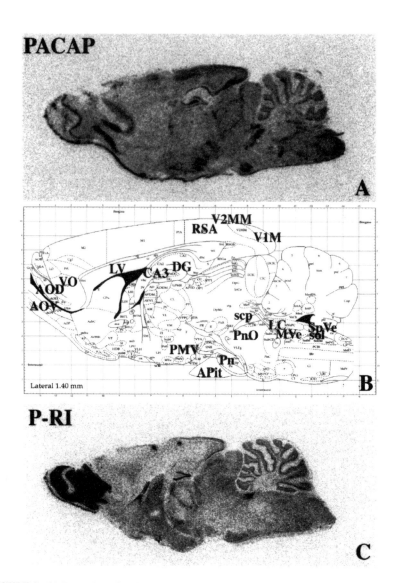

FIGURE 3. (**A**) Expression of PACAP mRNA in saggital section of adult rat. (**B**) Diagram of the respective saggital section at lateral 1.40 mm. (**C**) Expression of PACAP-R mRNA in saggital section of adult rat. AOD, anterior olfactory nucleus, dorsal part; AOV, anterior olfactory nucleus, ventral part; VO, ventral orbital cortex; LV, lateral ventricle; RSA, retrosplenial agranular cortex; V1M, primary visual cortex, monocular area; V2MM, secondary visual cortex, mediomedial area; CA3, field CA3 of hippocampus; DG, dentate gyrus; scp, superior cerebellar peduncle (brachium conjunctivum); PMV, premammillary nucleus, ventral part; Pn, pontine nuclei; APit, anterior lobe of pituitary; PnO, pontine reticular nucleus, oral part; MVe, medial vestibular nucleus; sol, solitary tract; LC, locus coeruleus; SpVe, spinal vestibular nucleus.

In situ hybridization studies with synthetic oligonucleotide probes complementary to PACAP, PACAP-R, and BDNF mRNA revealed a highly specific expression pattern of each mRNA. As shown in FIGURE 4 (A and D), PACAP mRNA was expressed in the CA1, CA4 regions, and the dentate gyrus, whereas the receptor was expressed exclusively in the dentate gyrus (FIG. 4, B and E). Also note the high expression of PACAP mRNA in the precursor cells in the hilus. In comparison, BDNF mRNA was expressed in all subfields of the hippocampus, such as in the CA1, CA2, CA3, and CA4, with a slightly enhanced signal in the dentate gyrus and in cells within hilus.

NEUROTROPHIC ACTIONS OF PACAP

In view of the expression of PACAP and PACAP-R mRNA during early development, we investigated whether PACAP has an effect on the development of various primary neurons in culture.

The results obtained showed that PACAP has a neurotrophic activity on a wide variety of neurons. TABLE 1 shows a list of target neurons that respond to PACAP. PACAP promoted the survival of cortical and hippocampal neurons isolated from embryonic rats and cultured under serum-free conditions. The effect on survival was revealed using the MTT assay for survival and by counting the number of neurons stained with MAP-2, which is specific for neurons. In hippocampal cultures, 10 nM PACAP promoted survival two- to threefold above controls. PACAP also promoted the survival of cholinergic neurons in cultures from both embryonic and postnatal rat brain. PACAP has been reported to increase the survival of cultured cerebellar granule neurons.[12,13] In addition to its survival-promoting effect, PACAP also enhanced the functional maturation of catecholaminergic neurons. Thus, PACAP increased the number of tyrosine hydroxylase neurons in mesencephalic cultures and enhanced neuronal uptake of dopamine. Besides brain neurons, PACAP acts on sympathetic[7] and dorsal root ganglion neurons.[14] PACAP supported the survival of DRG neurons and also induced the immunoreactivity of calcitonin gene–related peptide in these cultures.[14]

In addition to its function during neuronal development, PACAP has also a neuroprotective effect against certain toxic insults. It has been reported that PACAP suppresses cell death of cortical neurons induced by excessive glutamate.[15] This neuroprotective effect of PACAP against glutamate was also observed with hippocampal neurons in culture. Furthermore, PACAP counteracted apoptosis of cortical neurons induced by ionomycin, a potent Ca^{2+} ionophore.[16] Beside calcium-induced toxicity, PACAP also inhibited the action of 6-hydroxydopamine, which specifically causes cell death of dopaminergic neurons.

PACAP acts as a neuroprotective molecule not only *in vitro* but also *in vivo*. PACAP protected hippocampal neurons from hypoxic insult.[17] Recently, we have also observed

TABLE 1. Neurotrophic effects on PACAP

Target Neurons	Effects
Central nervous system	
Cortical neurons	Survival
Hippocampal neurons	Survival
Cerebellar granule neurons	Survival (prevention of apoptosis)
Septal cholinergic neurons	Survival
Mesencephalic dopaminergic neurons	Increase of TH neurons
	Enhancement of dopamine uptake
Peripheral nervous system	
DRG neurons	Survival
	Induction of CGRP

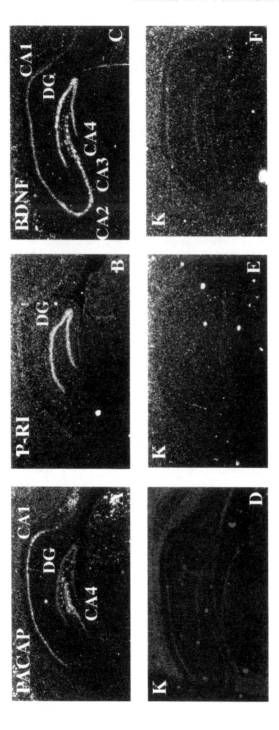

FIGURE 4. Expression of PACAP, PACAP-R, and BDNF mRNA in adult rat hippocampus. (**A**) PACAP mRNA, (**B**) PACAP-R mRNA, (**C**) BDNF mRNA, (**F**) control for BDNF hybridization. (**D-F**) Control hybridization of PACAP, PACAP-R, and BDNF, respectively using a 200-fold excess of unlabeled oligonucleotide. DG, dentate gyrus; CA1,2,3,4, field CA1,2,3,4 of hippocampus.

TABLE 2. Neuroprotective effects on PACAP

Target Neurons	Insults
Cortical neurons	Glutamate
	Ionomycin
Hippocampal neurons	Glutamate
	Hypoxia (*in vivo*)
Septal cholinergic neurons	Axotomy (*in vivo*)
Mesencephalic dopaminergic neurons	6-OHDA

that the infusion of PACAP partially rescues septal cholinergic neurons degenerating after a fimbria-fornix lesion (Takei *et al.*, unpublished observation).

The spectrum of the neurotrophic and neuroprotective effects of PACAP under normal and pathophysiological conditions is shown in TABLES 1 and 2.

CONCLUSION

The results in this study show the presence of PACAP and PACAP-R in various parts of developing rat brain and in the peripheral nervous system. Employing various populations of neurons in culture, PACAP was also found to support neuronal survival and to promote neuronal development. Taken together these findings support the view that PACAP is an important neurotrophic factor for developing neurons and that PACAP probably acts in conjunction with other known neurotrophic factors in promoting neuronal survival.

REFERENCES

1. OPPENHEIM, R. W. 1991. Cell death during development of the nervous system. Annu. Rev. Neurosci. **14**: 453–501.
2. LINDSAY, R. M., S. J. WIEGAND, C. A. ALTAR & P. S. DiSTEFANO. 1994. Neurotrophic factors: from molecule to man. Trends Neurosci. **19**: 182–190.
3. LEWIN, G. & Y.-A. BARDE. 1996. Physiology of the neurotrophins. Annu. Rev. Neurosci. **19**: 289–317.
4. DAVIES, A. M. 1994. The role of neurotrophins in the developing nervous system. J. Neurobiol. **25**: 1334–1348.
5. SCHWARTZ, J. P. 1992. Neurotransmitters as neurotrophic factors—a new set of functions. Int. Rev. Neurobiol. **34**: 123.
6. BRENNEMAN, D. E. & G. FOSTER. 1987. Structural specificity of peptides influencing neuronal survival during development. Peptides **8**: 687–694.
7. PINCUS, D. W., E. DiCicco-BLOOM & I. B. BLACK. 1990. Vasoactive-intestinal peptide regulates mitosis, differentiation and survival of cultured sympathetic neuroblasts. Nature (London*)* **343**: 564–567.
8. ARIMURA, A. & S. SHIODA. 1995. Pituitary adenylate cyclase activating polypeptide (PACAP) and its receptors: neuroendocrine and endocrine interaction. Front. Neuroendocrinol. **16**: 53–88.
9. SPENGLER, D., C. WAEBER, C. PANTALONI, F. HOLSBOER, J. BOCKAERT, P. H. SEEBURG & L. JOURNOT. 1993. Differential signal transduction by five splice variants of the PACAP receptor. Nature **365**: 170–175.
10. RAWLINGS, S. R. & M. HEZAREH. 1996. Pituitary adenylate cyclase-activating polypeptide (PACAP) and PACAP/VIP intestinal polypeptide receptors: Actions on the anterior pituitary gland. Endocrine Rev. **17**: 429.
11. ARIMURA, A., A. SOMOGYVARI-VIGH, C. WEILL, R. C. FIORE, I. TATSUNO, V. BAY & D. E. BRENNEMAN. 1994. PACAP functions as a neurotrophic factor. Ann. N. Y. Acad. Sci. **739**: 228–243.

12. CAVALLARO, S., A. COPANI, V. D'AGATA, S. MUSCO, S. PETRALIA, C. VENTRA, F. STIVALA, S. TRAVALI & P. L. CANONICO. 1996. Pituitary adenylate cyclase activating polypeptide prevents apoptosis in cultured cerebellar granule neurons. Mol. Pharmacol. **50**: 60–66.
13. VILLALBA, M., J. BOCKAERT & L. JOURNOT. 1997. Pituitary adenylate cyclase-activating polypeptide (PACAP-38) protects cerebellar granule neurons from apoptosis by activating the mitogen-activated protein kinase (MAP kinase) pathway. J. Neurosci. **17**: 83–90.
14. LIOUDYNO, M., Y. SKOGLÖSA, N. TAKEI & D. LINDHOLM. 1998. Pituitary adenylate cyclase-activating polypeptide (PACAP) protects dorsal root ganglion neurons from death and induces calcitonin gene-related peptide (CGRP) immunoreactivity in vitro. J. Neurosci. Res. **51**: 243–256.
15. MORIO, H., I. TATSUNO, A. HIRAI, Y. TAMURA & Y. SAITO. 1996. Pituitary adenylate cyclase-activating polypeptide protects rat cultured cortical neurons from glutamate-induced cytotoxicity. Brain Res. **741**: 8288.
16. TAKEI, N. & Y. ENDO. 1994. Ca^{2+} inophore-induced apoptosis of cultured rat cortical neurons. Brain Res. **652**: 6570.
17. BANKS, W. A., D. UCHIDA, A. ARIMURA, A. SOMOGYVARI-VIGH & S. SHIODA. 1996. Transport of pituitary adenylate cyclase-activating polypeptide across the blood-brain barrier and the prevention of ischemia-induced death of hippocampal neurons. Ann. N. Y. Acad. Sci. **805**: 270–277.

Pituitary Adenylate Cyclase Activating Peptide (PACAP) in the Retinohypothalamic Tract: A Daytime Regulator of the Biological Clock[a]

JENS HANNIBAL,[b,g] JIAN M. DING,[c] DONG CHEN,[d] JAN FAHRENKRUG,[b] PHILIP J. LARSEN,[e] MARTHA U. GILLETTE,[c,d] AND JENS D. MIKKELSEN[f]

[b]Department of Clinical Biochemistry, Bispebjerg Hospital, University of Copenhagen, DK-2400 Copenhagen NV, Denmark

[c]Department of Cell & Structural Biology and [d]Molecular and Integrative Physiology, University of Illinois, Urbana, Illinois USA

[e]Institute of Medical Anatomy, University of Copenhagen, Denmark

[f]Department of Neurobiology, Research and Development, H. Lundbeck A/S, Valby Copenhagen, Denmark

ABSTRACT: The retinohypothalamic tract (RHT) relays photic information from the eyes to the brain biological clock in the suprachiasmatic nucleus (SCN). Activation of this pathway by light plays a role in adjusting circadian timing to light exposure at night. Here we report a new signaling pathway by which the RHT regulates circadian timing in the daytime as well. Using dual-immunocytochemistry for PACAP and the in vivo tracer Cholera toxin subunit B (ChB), intense PACAP immunoreactivity (PACAP-IR) was observed in retinal afferents at the rat SCN as well as in the intergeniculate leaflet (IGL) of the thalamus. This PACAP-IR was nearly lost upon bilateral eye enucleation. PACAP afferents originated from ganglion cells distributed throughout the retina. The phase of circadian rhythm measured as SCN neuronal activity in vitro was significantly advanced by application of PACAP-38 during the subjective day, but not at night. The effect is channelled to the clock via a PACAP 1 receptor–cAMP signaling mechanism. Thus, in addition to its role in nocturnal regulation by glutamatergic neurotransmission, the RHT can adjust the biological clock by a PACAP-cAMP–dependent mechanism during the daytime.

Mammalian circadian rhythms are generated by an endogenous circadian clock in the suprachiasmatic nucleus (SCN) of the hypothalamus.[1–5] The timing of the circadian clock is adjusted daily by the environmental light:dark cycle via the retinohypothalamic tract (RHT), a direct neural projection from the retina to the SCN.[6,7] Exposure to light in their subjective night can reset the animal's phase of circadian rhythms.[8,9] These light-induced phase shifts involve the release of glutamate from the RHT terminals in the SCN and subsequent activation of NMDA receptor, calcium influx, and nitric oxide signaling pathway.[9–13]

[a]This study was supported by the DanishMRC (12-1642), Danish Biotechnology program for Cellular Communication, PHS (USA) grant NS22155 from NINDS (M. U. G.).

[g]Corresponding author: Jens Hannibal; Phone 45 35 31 26 46; Fax: 45 35 31 3955; E-mail: biochbbh@inet.uni2.dk

Arousal or exposure to a dark pulse in the subjective day can phase-shift the animal's circadian rhythm.[14,15] The dark pulse–induced phase shift in the daytime involves the release of neuropeptide Y (NPY) from the intergeniculate leaflet (IGL) of the thalamus through a secondary pathway known as the geniculohypothalamic tract (GHT).[16-18] It has long been known that daytime phase-shifting involves the activation of a cAMP signaling pathway within the SCN.[19] However, the primary neurotransmitter mediating daytime phase shifts through cAMP is not yet known.

Pituitary adenylate cyclase activating peptide (PACAP) is a new member of the vasoactive intestinal peptide/secretin/glucagon family. It exists in two forms, PACAP-27 and PACAP-38, and is a powerful stimulator of adenylate cyclase.[20,21] It has been demonstrated in a high concentration in neuroendocrine areas in mammals, suggesting a role as a hypothalamic regulatory peptide, but has also been found to be distributed throughout the brain though in smaller concentrations.[22,23] PACAP-38 is the dominant product of posttranscriptional processing of the PACAP precursor in the brain.[22]

The endogenous transmitter activating cAMP in the SCN has not been established. The present study provides evidence that PACAP may be such a factor.

MATERIALS AND METHODS

Immunocytochemistry

On the day of fixation, the animals were anesthetized with tribromoethanol (20 mg/100 g body weight) and perfused via the left ventricle with a room temperature solution of saline (0.9%) to which heparin (15,000 IU/l) was added (75–100 ml over 3 min). This perfusion was followed by 2% paraformaldehyde, 0.2% picric acid in 0.1 M sodium phosphate buffer, pH 7.2 (300 ml over 15 min). After fixation, the brains were rapidly removed and postfixed in the same fixative for 24 h. After postfixation, the brains were equilibrated in phosphate-buffered saline (PBS, 0.05 M, pH 7.4) containing 30% sucrose for 48 h at 4°C and then sectioned in a freezing microtome into 40 μm sections. Whole mounts of the retina were processed as the free-floating brain sections as described below. Immunocytochemical visualization of PACAP immunoreactivity (IR) was carried out as described previously using the avidin-biotin bridge method.[22] The sections and the entire retina whole-mounts were incubated for 24 h with a monoclonal anti-PACAP antibody at 4°C. The specificity of the monoclonal antibody (code MabJHH1) has been characterized previously and displays equal affinity for PACAP-38 and PACAP-27, recognizing an epitope between amino acid 6–16, but has no affinity for structurally related peptides such as VIP.[22] Control sections for single antigen immunocytochemistry were routinely processed by either omitting or replacing the primary antibody with an equivalent concentration of either goat or rabbit preimmune serum or with antibody preabsorbed with PACAP-38 and PACAP-27 (20 μg/ml). Using these procedures, all immunocytochemical staining was blocked. Immunocytochemical visualization of PACAP-IR and ChB was performed by the procedure described previously for visualization of two antigens[24] using a mixture of monoclonal PACAP-antibody (supernatant diluted 1:2) and goat anti-ChB antiserum (List Biologicals, Campbell, CA) (diluted 1:750) for 24 h at 4°C.

In Situ Hybridization Histochemistry

In situ hybridization was performed using a slight modification of the previously described procedure.[25] 12-μm sections from three rats were used. The ^{35}S-UTP-labeled antisense and sense RNA probes were prepared by in vitro transcription using T7 (antisense) and SP6 (sense) RNA polymerase. The template containing a cDNA encoding the whole PACAP type I receptor sequence (nucleotide 20–1546)[26] was kindly given by Dr. Steven A. Wank. The plasmid (pGEM-3Z) was linearized with *Hind*III for antisense probe and with *Eco*RI for the sense probe. Transcription was performed at 37°C for 2 h in 20 μl containing 5 × TB buffer (Boehringer Mannheim, Germany), 25 mM dithiothreitol (DTT), 20 U RNasin (Amersham, DK), 1.5 mM NTP-mix (Boehringer Mannheim, Germany), 40 U polymerase (T7; Stratagene, USA or SP6; Boehringer Mannheim, Germany), and 2 μM ^{35}S-UTP (3,000 mCi, Amersham, DK). After removal of the DNA template by adding 1 μl RNasin (30–40 units), 2 μl tRNA (10 μg/μl), and 1 μl DNase (Boehringer Mannheim, Germany), and incubation for further 15 min at 37°C, the probes were purified by water/phenol extraction followed by chloroform/isoamyl alcohol extraction, and finally, NH_4 acetate/ethanol precipitation. The labeled product was fragmented by incubation in hydrolysis buffer for 50 min at 60°C and used in a concentration of 1×10^7 cpm/ml. After hybridization overnight at 53°C, the sections were washed in 4 × saline sodium citrate (4 × SSC = 0.60 M NaCl, 0.060 M sodium citrate), 4 mM DTT for a few minutes at room temperature followed by RNase treatment for 30 min (RNase A buffer, Sigma, USA). After washing in 2 × SSC, 2 mM DDT at room temperature for 60 min followed by washing in 0.01 × SSC, 2 mM DDT at 60° C for 60 min and 1 × SSC, 2 mM DDT for 10 min at room temperature, the sections were dehydrated through a series of alcohols. The slides were finally exposed to Amersham Hyperfilm for three weeks. For control purposes, hybridization was performed in parallel using an antisense and a sense probe on consecutive sections.

SCN Brain Slice and Neurophysiological Methods

These methods have been described in detail previously.[9] Briefly, a 500 μm coronal hypothalamic slice containing the paired SCN was prepared at least 2 h before the onset of the dark phase from 6- to 9-week-old inbred Long-Evans rats housed in a 12-h light: 12-h dark lighting schedule. Brain slices survived for three days with continuous perfusion (34 ml/h) by Earle's balanced salt solution (EBSS), supplemented with 24.6 mM glucose, 26.2 mM sodium bicarbonate, and 5 mg/l of gentamicin and saturated with 95% O_2; 5% CO_2 at 37° C (pH 7.4). The single unit activity of SCN neurons was recorded extracellularly with a glass microelectrode, and running means were calculated to determine the peak of activity. Effects of 10^{-6} M PACAP-38 (Sigma, St. Louis, USA), the dominant product of posttranscriptional processing of the PACAP precursor in the rat brain,[22] were examined at circadian time (CT) 6, 14, and 19. For treatments, the perfusion was stopped and a 1.0 μl microdrop of test substance dissolved in EBSS was applied directly to the SCN. After 10 min, the SCN surface was washed with EBSS, perfusion was resumed, and the time of peak was assessed on the subsequent days. Extracellular single unit activities were sampled throughout the SCN in brain slice in 10 sec intervals over 2 min and grouped into a 2 h running average to determine the peak of firing activity.[9] Because a PACAP/VIP-R2 receptor has been demonstrated in the SCN,[27] we evaluated effects of VIP at the time of maximal SCN sensitivity to PACAP. Dose-response curves were generated at CT 6 by applying PACAP-38 at a dose from 10^{-10} M to 10^{-5} M or VIP from 10^{-7} M to 10^{-4} M for a 10-min pulse in a 1 μl droplet. For each dose, three to four experiments were performed. To investigate the specificity of the PACAP effect as well as the signaling pathway involved, a PACAP antagonist PACAP 6-38[28] (10 μM) and a competitive inhibitor for cAMP-dependent processes, Rp-cAMPS (Rp)[29] (10 μM), were added to EBSS 20 min before PACAP-38 was applied. Experiments were performed with the experimenter "blind" to the treatment protocol.

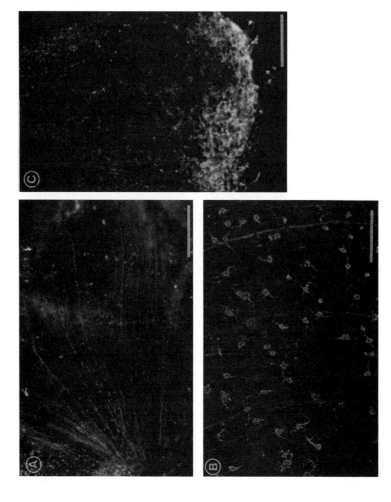

FIGURE 1. PACAP-IR is present in retinal ganglion cells and in fibers in the retinorecipient SCN. Texas red® streptavidin-biotin immunocytochemistry using a monclonal antibody against PACAP demonstrates an accumulation of fibers coursing into the retinal papil (**A**). PACAP-IR retinal ganglion cells are distributed throughout the retina (**B**) cells are mostly small with less extensive arborization. A dense accumulation of PACAP-IR nerve fibers is observed in the ventral part of the SCN, overlapping with the retinal innervation (**C**). Scale bars: **A**, 400 µm, **B** and **C**, 200 µm.

RESULTS AND DISCUSSION

Using immunocytochemistry we showed PACAP-IR fibers coursing towards the retinal papil (FIG. 1,A), in retinal ganglion cells (FIG. 1,B), and in nerve fibers and terminals in the ventrolateral part of the SCN (FIG. 1,C) of adult rats. The exact position of PACAP terminals in the SCN varied along the rostroventral axis of the SCN, but overlapped entirely with the retinorecipient area. In the rostral SCN, the PACAP-IR nerve fibers were located in the extreme ventral part of the nucleus, whereas in the middle and caudal SCN, the location changed to a more lateral and dorsal position. The retinal ganglion cells projecting to the circadian system originate from a distinct subset of neurons spread throughout the retina.[30,31] Ganglion cells expressing PACAP-IR were frequent (FIG. 1,B) and seemed to represent a population of small neurons with a few branching processes. They were widely distributed

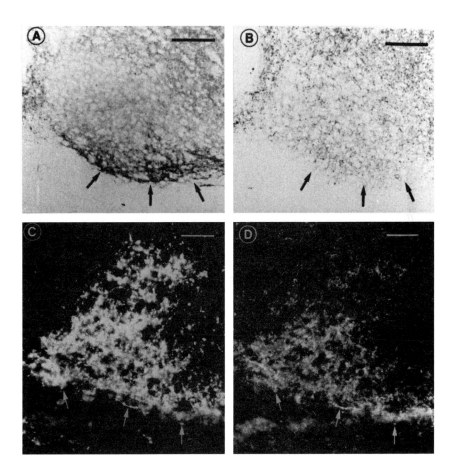

FIGURE 2. PACAP is present in the RHT. A high accumulation of PACAP-IR nerve fibers is present in the ventral SCN (**A**), but this is dramatically reduced by enucleation (**B**). Dual immunocytochemistry showing the distribution of ChB (**C**) and PACAP-IR (**D**). The arrows point to positive elements that contain both ChB- and PACAP-IR. Scale bars **A–D**, 100 μm.

in the retina and likely belong to the subset of ganglion cells of the W-type characterized by their capacity to transport viral tract tracers from the eye to the circadian system.[30,31] To determine the extent of labeling due to retinal innervation, PACAP distribution was studied in normal and bilaterally enucleated rats. In the enucleated animals PACAP-IR nerve fibers in the SCN were greatly diminished. In particular, prominent reduction was observed in the retinorecipient area of enucleated animals (FIG. 2,A and B). The nerve fibers originate from the ganglion cells, since intraocular injections of ChB and concomitant visualization of PACAP-IR and ChB in the SCN revealed that the majority of PACAP-IR nerve fibers also exhibited ChB-IR (FIG. 2,C and D). However, detectable levels of immunoreactivity were still found in the SCN after enucleation, indicating that a minor afferent system may originate from the brain. Notably, the IGL also contained a considerable plexus of PACAP-IR nerve fibers and varicose terminals, which overlapped extensively with the distribution of retinal afferents and disappeared after enucleation (not shown).

At present, three types of receptors with affinity for PACAP have been characterized and cloned,[27,32–34] some of which also have affinity for VIP. The PACAP type 1 receptor, which is specific for PACAP, activates cAMP and inositol triphosphate (IP-3) pathways, depending upon the splice variant.[34] The PACAP type 2 (VIP type 1 and VIP type 2) receptors show similar affinities for VIP and PACAP, and are coupled to cAMP. Since the PACAP/VIP type 2 receptor mRNA is confined to the dorsal SCN,[27,32,33] it is unlikely to be regulated directly by the retinal afferents. We found here that the PACAP-R1 mRNA is located in the ventral SCN (FIG. 3) indicating that the PACAP-R1 on the SCN cells could be primarily affected by the PACAP afferents from the retina.

In order to determine the functional implications of PACAP innervation of the SCN, the effect of PACAP-38 on the phasing of SCN rhythm of neuronal activity was assessed. Application of PACAP-38 could alter the phase of the circadian rhythm of neuronal activity (FIG. 4,A). The phase shift occurred as a prominent advance of the activity peak by 3.5 h ± 0.4 h when PACAP-38 was applied in a 1 µl drop at CT6, mid subjective day. The CT

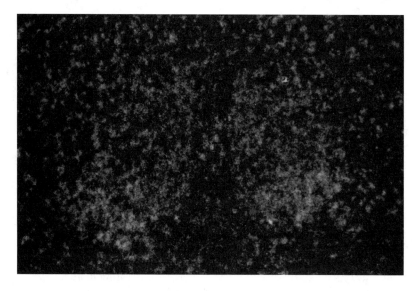

FIGURE 3. A high accumulation of PACAP-R1 mRNA is present in the SCN using a cRNA antisense probe.

0 is defined as the time when light comes on in the donor colony. This effect was dose dependent, with a half-maximum shift occurring in response to a microdrop of 5×10^{-9} M of PACAP-38. Notably, the 1 µl drop of PACAP-38 may be significantly diluted by diffusion into the SCN. Thus, the effective concentration would be in the range seen in different binding assays of PACAP receptors.[26] Interestingly, the effect could only be produced at CT6, but not at CT14 or CT19 (FIG. 4,C). This pattern is in antiphase to the timing of clock sensitivity to light, glutamate, NMDA receptor activation, NO donors, and the transcriptional factors CREB and Fos.[11,35] However, it is overlapping with SCN sensitivity to cAMP, serotonin, and NPY.[19,29,36–39] To examine whether the PACAP-selective type-1

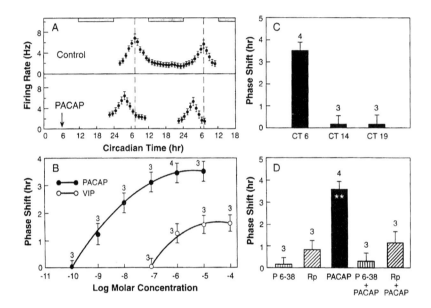

FIGURE 4. (A) PACAP directly resets the phase of the SCN circadian rhythm of neuronal activity. (*Top panel*) Circadian rhythms of neuronal activity of the SCN in brain slice recorded from 112 units over 38 h under constant conditions *in vitro*. The rhythm peaked in mid-subjective day at CT7, on both day 2 and 3 *in vitro*. (*Bottom panel*) Effect of PACAP applied at CT6 advanced the peak of the SCN activity rhythm by 3.5 h. A 1 µl droplet of 1×10^{-6} M PACAP-38 was directly applied to the SCN for 10 min, followed by rinsing in medium. Horizontal bars indicate subjective night. (B) Dose-response curve for a 10-min pulse of 1 µl of PACAP-38 (*closed circles*) and VIP (*open circles*) to the SCN *in vitro* at CT 6. Each data point represents the mean ± SD of three to four experiments, as indicated, measuring the time-of-peak as in FIG. 3 (A). Half-maximal response was achieved at 3×10^{-9} M PACAP and 7×10^{-7} M VIP. Experiments were performed with the experimenter "blind" to the treatment protocol. (C) Phase advance by PACAP depends on the circadian time of application to the SCN (dosage as in FIG. 4,A). Each data point represents three to four experiments as indicated. Phase advance is 3.5 ± 0.4 h at CT 6. No significant phase shift was detected at CT 14 or CT 19, points of maximal responsiveness to light and glutamate.[9] (D) The phase shift by PACAP was blocked by the PACAP receptor antagonist PACAP 6-38, and a competitive inhibitor for cAMP-dependent processes, Rp-cAMPS. Brain slices were incubated for 20 min with 10 µM PACAP 6-38 or 10 µM Rp-cAMPS before PACAP application in a microdrop onto the SCN for 10 min. Each data point represents the mean ± SD of three to four experiments as indicated. Significant difference was found between PACAP-treated vs. Rp-cAMPS-treated groups, and between groups treated with PACAP and PACAP 6-38 + PACAP and Rp-cAMPS + PACAP, respectively. No significant difference was detected between PACAP 6-38, Rp-cAMPS, and antagonist + PACAP-treated groups. (** $p \leq 0.01$)

receptor mediates phase resetting of the biological clock, we investigated the effects of VIP *in vitro*, since a PACAP/VIP-R2 receptor, which has equal affinities for both PACAP and VIP, had also been demonstrated in the SCN.[27] Therefore, we examined the response to VIP over a range of concentrations. VIP was 1,000-fold less potent than PACAP at altering the phasing of the SCN circadian rhythm. The half-maximal response to VIP was calculated as a 0.75 h phase advance to a microdrop containing 7×10^{-7} M VIP. As can be seen in FIGURE 4, B, a shift of this magnitude would be produced by 7×10^{-10} M PACAP. To investigate the second messenger pathway activated by PACAP, we tested the effect of PACAP together with a competitive inhibitor for cAMP-dependent processes, Rp-cAMPS. Application of Rp-cAMPS before PACAP application completely blocked the phase advance of PACAP at CT 6 (FIG. 4, D), confirming that the PACAP-R1 receptor stimulates a second messenger pathway involving cAMP/protein kinase A.

We have shown that PACAP is located in the RHT and that the peptide adjusts the phase of SCN through a PACAP-R1 receptor located in the ventral SCN. It is striking that SCN sensitivity to PACAP-38 lies in the subjective day, a time when light stimuli would normally activate the RHT. Yet, light by itself does not cause phase shifting in the daytime. Arousal stimuli, such as dark pulses and intense locomotor activity, do induce daytime phase advances, and NPY release from the GHT is essential for generation of this type of phase shift.[29,38,39] Additionally, serotonin (5-hydroxytryptamine; 5-HT) induces significant phase advances at the time PACAP is effective.[36] Activation of a 5-HT_7 receptor, which is coupled to cAMP stimulation, is required for this shift, but mRNA for this receptor has not been demonstrated in the SCN.[40] Based on the present findings it is possible that serotonin may act through presynaptic receptors on RHT terminals and thereby stimulate a selective release of PACAP, activate the PACAP-R1 receptor, and stimulate cAMP pathways within the SCN. Localization of PACAP-IR in the retinal afferents, where NPY and serotonin projections terminate, suggests that integration of the geniculate and raphe signals may occur presynaptically at the PACAP-containing boutons. This may provide the intense stimulation for mobilization of peptide release into the synapse and may underlie the lack of effect of light alone in the daytime. The localization of PACAP-R1 mRNA and the profound involvement of cAMP in SCN phase advances suggest that PACAP exerts its effect postsynaptically. Both cAMP and nonphotic stimuli have phase-shifting capacities during the subjective day, and this sensitivity is regulated downstream from cAMP activation within the cells of the SCN.[14,15] The phase shift exerted by PACAP during the subjective day, if associated with arousal, is not transmitted through expression of the transcriptional factor Fos. Another regulatory path could be through a cAMP-dependent cascade that involves expression of inducible cyclic AMP early repressor (ICER) mRNA.[41,42]

REFERENCES

1. TUREK, F. W. 1985. Circadian neural rhythms in mammals. [Review]. Ann. Rev. Physiol. **47**: 49–64.
2. HASTINGS, M. H. 1991. Neuroendocrine rhythms. Pharmac. Ther. **50**: 35–71.
3. KLEIN, D. C., R. Y. MOORE & S. M. REPPERT. 1991. Suprachiasmatic Nucleus. The Mind's Clock. Oxford University Press. New York.
4. HARRINGTON, M. E., B. RUSAK & R. E. MISTLBERGER. 1994. Anatomy and physiology of the mammalian circadian system. *In* Principles and Practice of Sleep Medicine. M. H. Kryger, T. Roth & W. C. Dement, Eds.: 286–301. W.B. Saunders Company. Philadelphia.
5. MORIN, L.P. 1994. The circadian visual system. [Review]. Brain Res. Rev. **19**: 102–127.
6. MOORE, R. Y. & N. J. LENN. 1972. A retinohypothalamic projection in the rat. J. Comp. Neurol. **146**: 1–14.
7. JOHNSON, R. F., R. Y. MOORE & L. P. MORIN. 1988. Loss of entrainment and anatomical plasticity after lesions of the hamster retinohypothalamic tract. Brain Res. **460**: 297–313.
8. TAKAHASHI, J. S., P. J. DECOURSEY, L. BAUMAN *et al.* 1984. Spectral sensitivity of a novel photoreceptive system mediating entrainment of mammalian circadian rhythms. Nature **308**: 186–188.

9. DING, J. M., D. CHEN, E. T. WEBER *et al.* 1994. Resetting the biological clock: mediation of nocturnal circadian shifts by glutamate and NO. Science **266**: 1713–1717.

10. CASTEL, M., M. BELENKY, S. COHEN *et al.* 1993. Glutamate-like immunoreactivity in retinal terminals of the mouse suprachiasmatic nucleus. Eur. J. Neurosci. **5**: 368–381.

11. REA, M. A., B. BUCKLEY & L. M. LUTTON. 1993. Local administration of EAA antagonists blocks light-induced phase shifts and c-fos expression in hamster SCN. Am. J. Physiol. **265**: R1191–R1198.

12. MIKKELSEN, J. D., P. J. LARSEN & F. J. EBLING. 1993. Distribution of N-methyl D-aspartate (NMDA) receptor mRNAs in the rat suprachiasmatic nucleus. Brain Res. **632**: 329–333.

13. MIKKELSEN, J. D., P. J. LARSEN, G. MICK *et al.* 1995. Gating of retinal inputs through the suprachiasmatic nucleus: role of excitatory neurotransmission. Neurochem. Int. **27**: 263–272.

14. MROSOVSKY, N. & P. A. SALOMON. 1987. A behavioral method for accelerating re-entrainment of rhythms to new light-day cycles. Nature **330**: 372–373.

15. SUMOVA, A., F. J. EBLING, E. S. MAYWOOD *et al.* 1994. Non-photic circadian entrainment in the Syrian hamster is not associated with phosphorylation of the transcriptional regulator CREB within the suprachiasmatic nucleus, but is associated with adrenocortical activation. Neuroendocrinol. **59**: 579–589.

16. HARRINGTON, M. E., D. M. NANCE & B. RUSAK. 1985. Neuropeptide Y immunoreactivity in the hamster geniculo-suprachiasmatic tract. Brain. Res. Bull. **15**: 465–472.

17. CARD, J. P. & R. Y. MOORE . 1989. Organization of lateral geniculate-hypothalamic connections in the rat. J. Comp. Neurol. **284**: 135–147.

18. MIKKELSEN, J. D. 1990. Projections from the lateral geniculate nucleus to the hypothalamus of the Mongolian gerbil (Meriones unguiculatus): an anterograde and retrograde tracing study. J. Comp. Neurol. **299**: 493–508.

19. PROSSER, R. A. & M. U. GILLETTE. 1989. The mammalian circadian clock in the suprachiasmatic nuclei is reset in vitro by cAMP. J. Neurosci. **9**: 1073–1081.

20. MIYATA, A., A. A. ARIMURA, R.R. DAHL *et al.* 1989. Isolation of a novel 38 residue-hypothalamic polypeptide which stimulates adenylate cyclase in pituitary cells. Biochem. Biophys. Res. Commun. **164**: 567–574.

21. ARIMURA, A. & S. SHIODA. 1995. Pituitary adenylate cyclase activating polypeptide (PACAP) and its receptors: neuroendocrine and endocrine interaction. Front. Neuroendocrinol. **16**: 53–88.

22. HANNIBAL, J., J. D. MIKKELSEN, H. CLAUSEN *et al.* 1995. Gene expression of pituitary adenylate cyclase activating polypeptide (PACAP) in the rat hypothalamus. Regul. Pept. **55**: 133–148.

23. MIKKELSEN, J. D., J. HANNIBAL, J. FAHRENKRUG *et al.* 1995. Pituitary adenylate cyclase activating peptide-38 (PACAP-38), PACAP-27, and PACAP related peptide (PRP) in the rat median eminence and pituitary. J. Neuroendocrinol. **7**: 47–55.

24. HANNIBAL J., J. D. MIKKELSEN, J. FAHRENKRUG *et al.* 1995. Pituitary adenylate cyclase-activating peptide gene expression in corticotropin-releasing factor-containing parvicellular neurons of the rat hypothalamic paraventricular nucleus is induced by colchicine, but not by adrenalectomy, acute osmotic, ether, or restraint stress. Endocrinology **136**: 4116–4124.

25. FAHRENKRUG, J. & J. HANNIBAL. 1996. Pituitary adenylate cyclase activating polypeptide innervation of the rat female reproductive tract and the associated paracervical ganglia: effect of capsaicin. Neuroscience **73**: 1049–1060.

26. PISEGNA, J. R. & S. A. WANK. 1993. Molecular cloning and functional expression of the pituitary adenylate cyclase-activating polypeptide type I receptor. Proc. Natl. Acad. Sci. USA **90**: 6345–6349.

27. MORROW, J. A., E. M. LUTZ, K. M. WEST *et al.* 1993. Molecular cloning and expression of a cDNA encoding a receptor for pituitary adenylate cyclase activating peptide (PACAP). FEBS Lett. **329**: 99–105.

28. ROBBERECHT, P., P. GOURLET, P. DE NEEF *et al.* 1992. Structural requirements for the occupancy of pituitary adenylate-cyclase-activating-peptide (PACAP) receptors and adenylate cyclase activation in human neuroblastoma NB-OK-1 cell membranes. Discovery of PACAP(6-38) as a potent antagonist. Eur. J. Biochem. **207**: 239–246.

29. PROSSER, R. A., H. C. HELLER & J. D. MILLER. 1994. Serotonergic phase advances of the mammalian circadian clock involve protein kinase A and K^+ channel opening. Brain Res. **644**: 67–73.

30. COOPER, H. M., M. HERBIN & E. NEVO. 1993. Ocular regression conceals adaptive progression of the visual system in a blind subterranean mammal. Nature **361**: 156–159.
31. MOORE, R. Y., J. C. SPEH & J. P. CARD. 1995. The retinohypothalamic tract originates from a distinct subset of retinal ganglion cells. J. Comp. Neurol. **352**: 351–366.
32. ISHIHARA, T., R. SHIGEMOTO, K. MORI *et al.* 1992. Functional expression and tissue distribution of a novel receptor for vasoactive intestinal polypeptide. Neuron **8**: 811–819.
33. LUTZ, E. M., W. J. SHEWARD, K. M. WEST *et al.* 1993. The VIP2 receptor: molecular characterisation of a cDNA encoding a novel receptor for vasoactive intestinal peptide. FEBS Lett. **334**: 3–8.
34. SPENGLER, D., C. WAEBER, C. PANTALONI *et al.* 1993. Differential signal transduction by five splice variants of the PACAP receptor. Nature **365**: 170–175.
35. GINTY, D. D., J. M. KORNHAUSER, M. A. THOMPSON *et al.* 1993. Regulation of CREB phosphorylation in the suprachiasmatic nucleus by light and a circadian clock. Science **260**: 238–241.
36. PROSSER, R. A., J. D. MILLER & H. C. HELLER. 1990. A serotonin agonist phase-shifts the circadian clock in the suprachiasmatic nuclei in vitro. Brain Res. **534**: 336–339.
37. MEDANIC, M. & M. U. GILLETTE. 1992. Serotonin regulates the phase of the rat suprachiasmatic circadian pacemaker in vitro only during the subjective day. J. Physiol. **450**: 629–642.
38. MEDANIC, M. & M. U. GILLETTE. 1993. Suprachiasmatic circadian pacemaker of rat shows two windows of sensitivity to neuropeptide Y in vitro. Brain Res. **620**: 281–286.
39. BIELLO, S. M., D. JANIK & N. MROSOVSKY. 1994. Neuropeptide Y and behaviorally induced phase shifts. Neuroscience **62**: 273–279.
40. LOVENBERG, T. W., B. M. BARON, L. DE LECEA *et al.* 1993. A novel adenylate cyclase-activating serotonin receptor (5-HT7) implicated in regulation of mammalian circadian rhythms. Neuron **11**: 449–458.
41. JANIK, D., J. D. MIKKELSEN & N. MROSOVSKY. 1995. Cellular colocalization of Fos and neuropeptide Y in the intergeniculate leaflet after nonphotic phase-shifting events. Brain Res. **698**: 137–145.
42. STEHLE, J. H., M. PFEFFER, R. KUHN & H. W. KORF. 1996. Light-induced expression of transcription factor ICER (inducible cAMP early repressor) in rat suprachiasmatic nucleus is phase-restricted. Neurosci. Lett. **217**: 169–172.

VIP Neurotrophism in the Central Nervous System: Multiple Effectors and Identification of a Femtomolar-Acting Neuroprotective Peptide

DOUGLAS E. BRENNEMAN,[a] GORDON GLAZNER,[a]
JOANNA M. HILL,[a] JANET HAUSER,[a] ARIANE DAVIDSON,[b]
AND ILLANA GOZES[b]

[a]Section on Developmental and Molecular Pharmacology, National
Institute of Child Health and Human Development, National Institutes
of Health, Bethesda, Maryland 20892 USA

[b]Department of Clinical Biochemistry, Sackler School of Medicine, Tel
Aviv University, Tel Aviv, Israel

ABSTRACT: Vasoactive intestinal peptide has neurotrophic and growth-regulating properties. As in the case of many neurotrophic molecules, VIP also has neuroprotective properties, including the prevention of cell death associated with excitotoxicity (NMDA), beta-amyloid peptide, and gp120, the neurotoxic envelope protein from the human immunodeficiency virus. The neurotrophic and neuroprotective properties are mediated in part through the action of glial-derived substances released by VIP. These substance include cytokines, protease nexin I, and ADNF, a novel neuroprotective protein with structural similarities to heat–shock protein 60. Antiserum against ADNF produced neuronal cell death and an increase in apoptotic neurons in cell culture. A 14 amino acid peptide (ADNF-14) derived from ADNF has been discovered that mimics the survival-promoting action of the parent protein. These studies support the conclusion that VIP, PACAP, and associated molecules are both important regulators of neurodevelopment and strong candidates for therapeutic development for the treatment of neurodegenerative disease.

V asoactive intestinal peptide (VIP), a 28-amino acid neuropeptide widely distributed in the mammalian nervous system, has potent neurotrophic actions that influence mitosis, neuronal survival, and neurodifferentiation.[1–4] In 1985, we first reported that VIP had survival-promoting properties in primary neuronal cultures derived from the embryonic mouse spinal cord.[5,6] This effect was also neuroprotective in that it was observed in a culture system co-treated with tetrodotoxin (TTX), a neurotoxin that blocks synaptic activity. Subsequent studies revealed that VIP produced neuroprotection in spinal cord cultures indirectly through its secretagogue actions on astroglia.[7] A complex array of neurotrophic substances was found to be released by subnanomolar amounts of VIP. High affinity receptors (K_d 30 pM) detected on astroglia were associated with the secretory effects.[8] Mechanistic investigations of this release suggested that increases in intracellular calcium,[9]

[c]Corresponding author: Dr. Douglas E. Brenneman, Chief, SDMP/LDN/NICHD, Building 49 Room 5A38, National Institutes of Health, Bethesda, Maryland 20892; Tel.: 301-496-7649; Fax: 301-480-5041.

stimulation of protein kinase C (PKC),[9] and the translocation of specific isozymes of PKC were correlated with the secretory and neurotrophic actions of VIP.[10] Cyclic AMP, the widely recognized second messenger of VIP actions, was not obviously involved in this effect on glia. Although cAMP may not regulate the secretion of survival-promoting substances from astroglia, the second mesenger is most likely cell specific.[11] The receptor that mediates this action of VIP is not yet known. Similar neurotrophic actions have been reported for PACAP38, with the biological response being a potent bimodal effect.[12]

Much of our effort has focused on the identification of glia-derived substances associated with VIP and the investigation of their potential role in development and neuroprotection. To date, the substances known to be released in response to VIP include cytokines,[13,14] a novel growth factor,[15] and a serine protease inhibitor, protease nexin I.[16] VIP has been shown to release at least eight different cytokines from cerebral cortical astrocytes. These include interleukin-1 alpha and beta, interleukin-3, interleukin-6, tumor necrosis factor alpha, interferon gamma, macrophage-colony stimulating factor, and granulocyte-colony stimulating factor. Many of these substances have been shown to regulate cell division, gene expression, and neuronal survival. The variety and demonstrated biological activity of cytokines that are secreted after stimulation with VIP strongly suggest that their release is an important element in the control of central nervous system development. VIP has also been shown to increase the release of protease nexin-1 (PN-1), a serine protease inhibitor. This is particularly significant in that PN-1 has marked effects on neurite extension [17] and neuronal survival.[16] The presumed mechanism of PN-1 involves inhibition of thrombin-like protease activity in the CNS. Indeed, thrombin has been shown to decrease neuronal survival in developing spinal cord cultures.[16]

A novel growth factor has recently been isolated from VIP-stimulated astroglia—activity-dependent neurotrophic factor (ADNF).[15] Whereas all other molecules shown thus far to be released by VIP were previously identified substances, ADNF is novel, not only in structure but also as a representative of a new concept in neuroprotection, extracellular stress proteins.[18] Here, we will emphasize the pharmacology of ADNF and highlight the identification of a 14 amino acid peptide derived from ADNF that shares its femtomolar-acting, survival-promoting action on CNS neurons.

ISOLATION OF ACTIVITY-DEPENDENT NEUROTROPHIC FACTOR

The discovery of ADNF originated from investigations of the trophic support for activity-dependent neurons, a class of cells that is dependent on electrical activity for their survival. Blockade of spontaneous synaptic activity by TTX results in the death of this population of neurons. The isolated protein was named activity-dependent neurotrophic factor as it protected neurons from death associated with electrical blockade. The strategy for the isolation of ADNF employed conditioned medium from VIP-stimulated astrocytes as the biological source of the activity. With this tactic, only secreted proteins were analyzed, lending physiological relevance to the scheme. Using a sequential chromatographic analysis of anion exchange, size exclusion, and hydrophobic interaction, a neuroprotective, glia-derived neurotrophic protein (14 kD and pI 8.3 ± 0.25) was isolated.[15] As shown in FIGURE 1, the survival-promoting activity of purified ADNF was evident with an EC_{50} of 2×10^{-16} M, making this substance among the most potent ever described. The pharmacology was also unusual in that the survival response attenuated at increasing concentrations of ADNF. The reason for this complex dose response is not yet clear. As revealed in FIGURE 1, the response of cerebral cortical cultures to co-treatment with ADNF and TTX produced cell counts that were greater than that of control cultures. This response above controls is interpreted as a prevention of apoptotic, neuronal cell death that occurs naturally in these cultures during the 5-day test period. This possibility is further supported experimentally in the next section on anti-ADNF.

NEURONAL DEATH PRODUCED BY ANTI-ADNF

A classical test of the biological importance of a molecule is to prepare neutralizing antibodies to it and then investigate the biological effects elicited by the antiserum. These experiments often provide important clues as to the function of the test substance. To assess the biological role of ADNF, antiserum was produced following sequential injections of purified ADNF into mice.[19] As shown in FIGURE 2, anti-ADNF ascites fluid (1:10,000) decreased neuronal survival by 35–50% in comparison to untreated cultures or cultures treated with control ascites. The neuronal death after anti-ADNF treatment was observed in cultures derived from the spinal cord, hippocampus, or cerebral cortex at similar IC_{50}'s. Neuronal cell death produced by the antiserum to ADNF was prevented in cultures co-treated with purified ADNF or ADNF-14, an active peptide derived from the parent ADNF.[19] These data strongly support the hypothesis that ADNF or ADNF-like substance is both endogenous to the CNS cultures and important to the survival of a subpopulation of neurons.

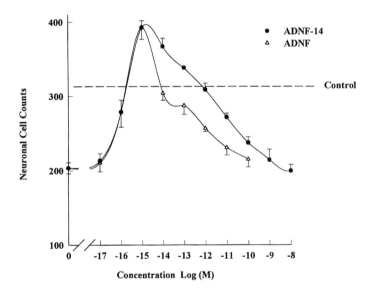

FIGURE 1. ADNF and ADNF-14 prevented neuronal cell death produced by 1 µM tetrodotoxin. Treatment of cerebral cortical cultures was started one week after plating on a confluent layer of cortical astrocytes. The duration of treatment was 5 days. At termination, cultures were fixed in glutaraldehyde. Cell counts were made at pre-determined coordinate locations. All counts were conducted without knowledge of the treatment group. Neuronal identity was determined immunocytochemically with antiserum to neuron-specific enolase.[24] Purified ADNF (*open triangles*) produced increases in neuronal cell counts at concentrations from 10^{-16} M to 10^{-12} M. ADNF-14 (*closed circles*), a peptide derived from ADNF, had identical potency and efficacy to that of ADNF. Each point is the mean ± standard error of three determinations. The dotted line is the mean of cell counts of control cultures at the conclusion of the experiment. Figure derived in part from previously described data.[15]

To test if the cell death associated with anti-ADNF might be apoptotic, a terminal deoxynucleotidyl transferase *in situ* assay was used to assess the cerebral cortical cultures.[19] Anti-ADNF was shown to produce a 70% increase in the number of labeled cells in comparison to controls (FIG. 3). The increase in apparent apoptosis produced by the antiserum was prevented by co-treatment with purified ADNF. Addition of ADNF alone to the cultures significantly decreased the number of labeled cells in comparison to controls. Treatment with control ascites produced no significant change in the number of labeled cells from control cultures. Together, these studies support the conclusion that electrical activity and ADNF are influencing apoptotic regulation of neuronal survival in these developing cerebral cortical cultures.

DISCOVERY OF ADNF-14

While studying the structural characteristics of ADNF, an active peptide fragment, ADNF-14,[15] was discovered. The amino acid sequence of this peptide is VLGGGSALLRSIPA.

This peptide had strong homology, but not identity, to an intracellular stress protein, heat-shock protein 60 (hsp60). ADNF-14, like ADNF, exhibited neuroprotection from neuronal death associated with TTX (FIG. 1). The EC_{50} was identical to that observed with purified ADNF. Similar to ADNF, the peptide also produced an attenuation of biological response at higher concentrations. Nevertheless, the peptide was more active over a broader range of concentrations than the parent protein. Subsequent studies demonstrated that ADNF-14 provided neuroprotection from a wide variety of neurotoxic substances,

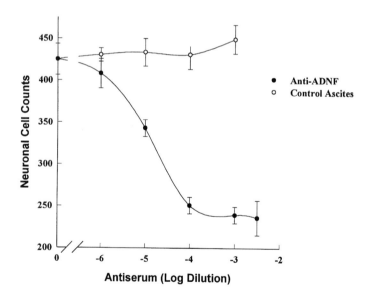

FIGURE 2. Antiserum to ADNF increased neuronal cell death in dissociated cerebral cortical cultures. Antiserum to ADNF was produced following sequential injections of purified ADNF (0.5–1.0 μg)/injection) into Balb/c mice.[19] Cells were treated with the antiserum for five days. Antiserum to ADNF (*closed circles*) decreased neuronal cell count in comparison to controls at all dilutions between 10^{-5} to 3×10^{-3}) . Control ascites (*open circles*) had no detectable effect on neuronal cell count in comparison to controls. Each point is the mean ± the standard error of three to four determinations. Figure based on previously described data.[19]

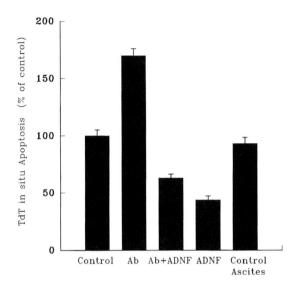

FIGURE 3. Anti-ADNF increased the number of labeled cells in a terminal deoxynucleotidyl transferase apoptosis assay. Cerebral cortical cultures were treated with anti-ADNF (Ab), purified ADNF, or control ascites for 5 days. The anti-ADNF and the control ascites was tested at 1:5,000. The concentration of the purified ADNF was 10^{-13} M, as determined from a bioassay.[15] A TACS *in situ* apoptosis detection kit was used to assess the number of apoptotic neurons. The number of labeled cells was counted in 20 fields at 400× magnification. Control counts were 18.2 + 0.9 cells/20 fields. Significant increases ($p < 0.001$) in the number of labeled cells were observed in cultures treated with anti-ADNF in comparison to controls. Significant decreases ($p < 0.001$) in the number of labeled cells were observed in cultures treated with ADNF alone or ADNF plus anti-ADNF in comparison to control cultures. Each value is the mean of 4–6 determinations. The error bar is the SEM. Figure based on previously described data.[19]

including the envelope protein from the human immunodeficiency virus, N-methyl D-aspartate (excitotoxicity), beta-amyloid peptide, all of which are toxic substances that have been implicated in neurodegenerative disease in humans.[20–22] Recent studies have extended the neuroprotection findings to a model of excitotoxicity *in vivo*.[23] Gressens and colleagues found that brain lesions produced by ibotenate were reduced by co-treatment with VIP or ADNF peptide. Neither forskolin or PACAP provided protection similar to that observed with VIP.

REFERENCES

1. WASCHEK, J. A. 1995. Vasoactive intestinal peptide: an important trophic factor and developmental regulator. Dev. Neurosci. **17**: 1–7 (abstract).
2. GOZES, I. & D. E. BRENNEMAN. 1989. VIP: molecular biology and neurobiological function. Mol. Neurobiol. **3**: 201–236.
3. GRESSENS, P., J. M. HILL, I. GOZES, M. FRIDKIN & D. E. BRENNEMAN. 1993. Growth factor function of vasoactive intestinal peptide in whole cultured mouse embryos. Nature **362**: 155–158.
4. PINCUS, D. W., E. DICICCO-BLOOM & I. B. BLACK. 1994. Trophic mechanisms regulate mitotic neuronal precursors: role of vasoactive intestinal peptide (VIP). Brain Res. **663**: 51–60.
5. BRENNEMAN, D. E., L. E. EIDEN & R. E. SIEGEL. 1985. Neurotrophic action of VIP on spinal cord cultures. Peptides **6** (Suppl 2): 35–39.

6. BRENNEMAN, D. E. & L. E. EIDEN. 1986. Vasoactive intestinal peptide and electrical activity influence neuronal survival. Proc. Natl. Acad. Sci. USA **83**: 1159–1162.
7. BRENNEMAN, D. E., E. A. NEALE, G. A. FOSTER, S. W. D'AUTREMONT & G. L. WESTBROOK. 1987. Nonneuronal cells mediate neurotrophic action of vasoactive intestinal peptide. J. Cell Biol. **104**: 1603–1610.
8. GOZES, I., S. K. MCCUNE, L. JACOBSON, D. WARREN, T. W. MOODY, M. FRIDKIN & D. E. BRENNEMAN. 1991. An antagonist to vasoactive intestinal peptide affects cellular functions in the central nervous system. J. Pharmacol. Exp. Ther. **257**: 959–966.
9. FATATIS, A., L. A. HOLTZCLAW, R. AVIDOR, D. E. BRENNEMAN & J. T. RUSSELL. 1994. Vasoactive intestinal peptide increases intracellular calcium in astroglia: synergism with alpha-adrenergic receptors. Proc. Natl. Acad. Sci. USA **91**: 2036–2040.
10. OLAH, Z., C. LEHEL, W. B. ANDERSON, D. E. BRENNEMAN & D. VAN AGOSTON. 1994. Subnanomolar concentration of VIP induces the nuclear translocation of protein kinase C in neonatal rat cortical astrocytes. J. Neurosci. Res. **39**: 355–363.
11. SCHETTINI, G., M. GRIMALDI, P. NAVARRA, G. POZZOLI, S. REICHLIN & P. PREZIOSI. 1994. Regulation of interleukin 6 production by cAMP-protein kinase-A pathway in rat cortical astrocytes. Pharmacol. Res. **30**: 13–24.
12. ARIMURA, A., A. SOMOGYVARI-VIGH, C. WEILL, R. C. FIORE, I. TATSUNO, V. BAY & D. E. BRENNEMAN. 1994. PACAP functions as a neurotrophic factor. Ann. N. Y. Acad. Sci. **739**: 228–243.
13. BRENNEMAN, D. E., J. M. HILL, G. W. GLAZNER, I. GOZES & T. W. PHILLIPS. 1995. Interleukin-1 alpha and vasoactive intestinal peptide: enigmatic regulation of neuronal survival. Int. J. Dev. Neurosci. **13**: 187–200.
14. BRENNEMAN, D. E., M. SCHULTZBERG, T. BARTFAI & I. GOZES. 1992. Cytokine regulation of neuronal survival. J. Neurochem. **58**: 454–460.
15. BRENNEMAN, D. E. & I. GOZES. 1996. A femtomolar-acting neuroprotective peptide [see comments]. J. Clin. Invest. **97**: 2299–2307.
16. FESTOFF, B. W., J. B. NELSON & D. E. BRENNEMAN. 1996. Prevention of activity-dependent neuronal death: vasoactive intestinal polypeptide stimulates astrocytes to secrete the thrombin-inhibiting neurotrophic serpin, protease nexin-1. J. Neurobiol. **30**: 255–266. (abstract)
17. MONARD, D., E. NIDAY, A. LIMAT & F. SOLOMON. 1983. Inhibition of protease activity can lead to neurite extension in neuroblastoma cells. Prog. Brain Res. **58**: 359–364.
18. GOZES, I. & D. E. BRENNEMAN. 1996. Activity-dependent neurotrophic factor (ADNF). An extracellular neuroprotective chaperonin? J. Mol. Neurosci. **7**: 235–244.
19. GOZES, I., A. DAVIDSON, Y. GOZES, R. MASCOLO, R. BARTH, D. WARREN, J. HAUSER & D. E. BRENNEMAN. 1997. Antiserum to activity-dependent neurotrophic factor produces neuronal cell death in CNS cultures: immunological and biological specificity. Dev. Brain Res. **99**: 167–175.
20. BRENNEMAN, D. E., G. L. WESTBROOK, S. P. FITZGERALD, D. L. ENNIST, K. L. ELKINS, M. R. RUFF & C. B. PERT. 1988. Neuronal cell killing by the envelope protein of HIV and its prevention by vasoactive intestinal peptide. Nature **335**: 639–642.
21. GOZES, I., A. BARDEA, A. RESHEF, R. ZAMOSTIANO, S. ZHUKOVSKY, S. RUBINRAUT, M. FRIDKIN, & D. E. BRENNEMAN. 1996. Neuroprotective strategy for Alzheimer disease: intranasal administration of a fatty neuropeptide. Proc. Natl. Acad. Sci. USA **93**: 427–432.
22. MARK, R. J., E. M. BLANC & M. P. MATTSON. 1996. Amyloid beta-peptide and oxidative cellular injury in Alzheimer's disease. Mol. Neurobiol. **12**: 211–224.
23. GRESSENS, P., S. MARRET, J. M. HILL, D. E. BRENNEMAN, I. GOZES, M. FRIDKIN & P. EVRARD. 1997. Vasoactive intestinal peptide prevents excitotoxic cell death in the murine developing brain. J. Clin. Invest. **100**: 390–397.
24. SCHMECHEL, D., P. J. MARANGOS, A. P. ZIS, M. BRIGHTMAN & F. K. GOODWIN. 1978. Brain endolases as specific markers of neuronal and glial cells. Science **199**: 313–315.

VIP and PACAP in the CNS: Regulators of Glial Energy Metabolism and Modulators of Glutamatergic Signaling[a]

P. J. MAGISTRETTI,[b] J.-R. CARDINAUX, AND J.-L. MARTIN

Laboratoire de Recherche Neurologique, Institut de Physiologie et Service de Neurologie du CHUV, Faculté de Médecine, Université de Lausanne, CH-1005 Lausanne, Switzerland

ABSTRACT: VIP neurons are a homogeneous population of intracortical bipolar cells. They receive excitatory synapses from afferent circuits to the cortex and exert effects on neurons, astrocytes, and capillaries. Effects on the two latter cell types imply that VIP neurons can translate incoming neuronal signals into local metabolic actions. Indeed, VIP tightly regulates glycogen metabolism in astrocytes. In this cell type VIP regulates the expression of a number of genes related to energy metabolism, such as glycogen synthase. These effects of VIP involve the transcription factor family C/EBP and result in the induction of at least seven new proteins by astrocytes. The actions of VIP on neurons appear to be of a modulatory nature: thus VIP enhances glutamate-mediated neurotransmission by potentiating the effects of glutamate on arachidonic acid formation and on the induction of *c-fos* and on BDNF expression. These effects indicate that VIP can actually increase the strength of glutamate-mediated neurotransmission.

MORPHOLOGY AND CONNECTIONS OF VIP NEURONS IN THE CEREBRAL CORTEX

In the cerebral cortex, VIP is contained in a homogeneous population of radially oriented, bipolar interneurons.[1] Because their dendritic arborization diverges only minimally from the main axis of the cell, these intracortical neurons exert very localized input-output functions within radial cortical "columns"[1] (FIG. 1). VIP neurons are distributed throughout the cortical mantle, with a slight rostro-caudal gradient. In addition, the density of VIP-containing neurons is such that the columnar ensembles that they define partially overlap, meaning that despite their radial nature, VIP neurons can "cover" the entire cerebral cortex.[1] VIP neurons receive predominantly asymmetric synapses,[2,3] implying that they are driven by excitatory afferents to cortical columnar ensembles. VIP neurons engage in synaptic contacts (mostly symmetric, i.e., inhibitory) on pyramidal neurons[4,5]; the inhibitory nature of the synaptic contacts that VIP-containing neurons make is consistent with the fact that GABA co-localizes with VIP in at least 30% of the cases.[4,5] At least 70% of VIP neurons also contain acetylcholine.[4,5] In addition to engaging in synaptic contacts with other neurons, VIP-containing processes are also in close apposition with astrocytes and intraparenchymal blood vessels.[6] Consistent with the existence of functional interactions

[a]This work was supported by grants No. 31-40'565.94 (P. J. M.) and No. 3100-050806.97 (J. L. M.) of the Swiss National Science Foundation.

[b]Corresponding author: Pierre J. Magistretti, M.D., Ph.D., Institut de Physiologie, 7 Rue du Bugnon, 1005 Lausanne, Switzerland. Tel.: 41-21-692.5542; Fax: 41-21-692.5595; E-mail: Pierre. Magistretti@iphysiol.unil.ch

213

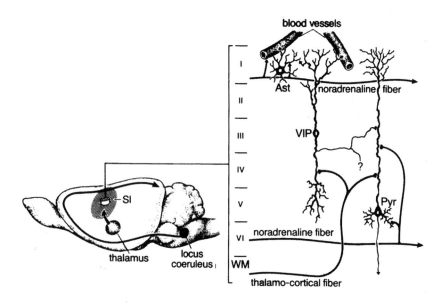

FIGURE 1. Anatomical organization and putative targets of the noradrenaline- and VIP-containing neuronal circuits in rat cerebral cortex. (*Left*) Noradrenergic fibers originate in locus coeruleus and project to cerebral cortex where they adopt horizontal trajectories parallel to pial surface. (*Right*) VIP neurons are intrinsic to the cerebral cortex and are oriented vertically perpendicular to pial surface. Astrocytes (Ast), intraparenchymal blood vessels, and neurons, such as certain pyramidal cells (Pyr), are potential target cells for VIP neurons. Roman numerals indicate cortical layers. VIP neurons can be activated by specific afferents (e.g., thalamo-cortical fibers). WM, white matter; SI, primary somatosensory cortex. (Reproduced with permission from Magistretti & Morrison.[1])

between VIP neurons and non-neuronal cells is the fact that receptors for VIP have been characterized on astrocytes[7] and intraparenchymal capillaries.[7,8] These receptors are functional, as they are coupled to second messenger systems, and in particular to the cAMP cascade.[7,8] Since synapses between neurons and astrocytes have not been described in the mammalian brain,[9] the interaction between VIP-containing neurons and non-neuronal cells is likely to occur at extrasynaptic sites.

These cytological characteristics indicate that VIP neurons are ideally positioned to translate excitatory inputs into diverging outputs to multiple cell types within a cortical column through synaptic contacts on neurons and through non-synaptic relationships with astrocytes and capillaries.[1] The pleiotropic nature of VIP neurons' targets is reflected in the action that VIP exerts on them. In this article, we review these actions, which have been revealed for the most part in purified preparations (primary cultures) of astrocytes and neurons.

CELLULAR DISTRIBUTION OF VIP RECEPTORS IN THE CEREBRAL CORTEX

Analysis of the cellular distribution of VIP receptors is consistent with the multiplicity of targets that the cytological studies suggest. Thus using purified preparations of mouse cerebral cortex consisting of primary astrocyte cultures, intraparenchymal microvessels

TABLE 1. Proposed classification of VIP receptor subtypes in cerebral cortex

Subtype	Localization	K_d (nM)	Competition by Secretin
VIP 1	Astrocytes (I,II)[a]	3.3; 3.6	-
	Microvessels	1.4	-
	Synaptosomes	4.9	-
VIP 2	Synaptosomes	42.8	+
VIP 3	Astrocytes (II)[a]	41.3	-
	Microvessels	30.3	-

[a]I, primary cultures; II, secondary cultures.

and synaptosomal membranes, respectively, three VIP receptor subtypes, with differential cellular localizations, were identified (TABLE 1).[7] The first subtype (VIP 1) is ubiquitous and of high affinity with K_ds of 3.3 nM (astrocytes), 1.4 nM (microvessels), and 4.9 nM (synaptosomes). Secretin does not interact with this site. The second receptor subtype (VIP 2) is exclusively present on synaptosomal membranes. It is a low affinity site, with a K_d of 42.8 nM. Secretin interacts with this site with an IC_{50} of 150 nM. The third subtype (VIP 3) is also of low affinity, with a K_d of 30.3 nM and is localized on microvessels. Secretin does not interact with this site. In addition to providing a classification for VIP receptor subtypes, these observations suggest that secretin may represent a useful tool to discriminate between neuronal and non-neuronal VIP binding sites.

VIP IS A KEY REGULATOR OF GLYCOGEN METABOLISM IN ASTROCYTES

Glycogen is the single largest energy reserve of the brain.[10] It is predominantly localized in astrocytes, to the point where this cell type can be positively identified at the ultrastructural level by the presence of glycogen granules.[10]

VIP promotes a concentration-dependent glycogenolysis with an EC_{50} of 3 nM (TABLE 2). Peptides sharing sequence homologies with VIP, such as PACAP, PHI, and secretin, are also glycogenolytic, while the two structurally unrelated peptides somatostatin and neuropeptide Y (NPY) are without effect.[11]

While several neurotransmitters, for which the presence of receptors has been demonstrated on astrocytes (acetylcholine, glutamate, GABA), do not promote glycogenolysis,[11] three other neurotransmitters are glycogenolytic, further illustrating the tight regulation of glycogenolysis in astrocytes. These glycogenolytic neurotransmitters are noradrenaline, adenosine, and ATP, which exert their action with EC_{50}s of 20 nM, 0.8 µM, and 1.3 µM, respectively (TABLE 2).

TABLE 2. Induction of glycogenolysis by neurotransmitters in primary astrocyte cultures of mouse cerebral cortex

Neurotransmitter	EC_{50} (nM)
VIP	3
PHI	6
Secretin	0.5
PACAP	0.08
Noradrenaline	20
Adenosine	800
ATP	1,300

PHI, Peptide histidine isoleucine; PACAP, pituitary adenylate cyclase-activating peptide.

The glycogenolytic action of VIP is rapid, with an initial rate of 9.1 nmol/mg protein/min.[11] Interestingly, this value is close to the rate of glucose utilization by the same culture[12] and even by cerebral cortex *in situ*.[13] These observations indicate that the glycosyl units released by the neurotransmitter-evoked glycogenolysis can provide energy substrates that match the energy demands of cortical gray matter. The precise metabolic fate of the released glycosyl units is not yet unequivocally determined; however, *in vitro* evidence suggests that lactate may in fact be produced from glycogen via glycogenolysis.[14] Evidence has accumulated over the years showing that lactate is an adequate metabolic substrate for sustaining synaptic activity.[15,16]

In view of the morphology and arborization pattern of VIP-containing neurons (FIG. 1), we have proposed that these cells could regulate the availability of energy substrates locally, within cortical columns.[1,17]

In addition to its glycogenolytic action discussed above,[11] which occurs within minutes, VIP also induces a temporally delayed resynthesis of glycogen, resulting, within nine hours, in glycogen levels that are 6 to 10 times higher than those measured before application of either neurotransmitter.[18] The continued presence of the neurotransmitter is not necessary for this long-term effect since pulses as short as one minute result in the doubling of glycogen levels nine hours later. The induction of glycogen resynthesis triggered by VIP is dependent on protein synthesis, since both cycloheximide and actinomycin D abolish it entirely. These results indicate that the same neurotransmitter, e.g., VIP or noradrenaline, can elicit two actions with different time-courses. Thus, by increasing cAMP levels, VIP simultaneously triggers a short-term effect, glycogenolysis, as well as a delayed one, transcriptionally regulated glycogen resynthesis. This longer-term effect ensures that sufficient substrate (glycogen) is available for the continued expression of the short-term action of VIP (glycogenolysis) (FIG. 2).

REGULATION BY VIP OF GENES CONTROLLING GLYCOGEN METABOLISM

The C/EBP Family of Transcription Factors

The massive glycogen resynthesis triggered by VIP involves the synthesis of new proteins, implying the induction of transcription of relevant genes. We have therefore engaged in the molecular dissection of the mechanisms that mediate this transcriptionally regulated physiological action of VIP. Since the resynthesis of glycogen appeared to be an exclusively cAMP-dependent phenomenon,[18] we searched for cAMP-inducible transcription factors for which evidence existed that they could regulate the expression of genes encoding for enzymes involved in energy metabolism in peripheral tissues. Such evidence existed for a family of transcription factors termed C/EBP (for CCAAT/enhancer binding protein); C/EBPs appear to be pleiotropic transactivators involved in metabolic gene transcription in liver and adipose tissue.[19,21] Of particular interest was the fact that some C/EBP isoforms could stimulate, in a cAMP-dependent fashion, the expression of genes encoding for enzymes involved in energy metabolism regulation. For this reason we tested the effect of VIP and PACAP on the level of expression of C/EBPβ and δ. Results obtained can be summarized as follows: VIP, PACAP, and noradrenaline (via β-adrenergic receptors coupled to cAMP-generating signal transduction pathways) increase C/EBPβ and δ isoforms at the mRNA (FIG. 3) and protein levels. Induction of C/EBPβ and δ mRNAs by VIP is rapid (FIG. 3) and occurs in the presence of a protein-synthesis inhibitor, indicating that the two members of the C/EBP transcription factor family behave as immediate-early genes. These results, in addition to identifying a key step in the cascade of molecular events that begin with activation by VIP of receptors at the membrane and lead to glycogen resynthesis, also

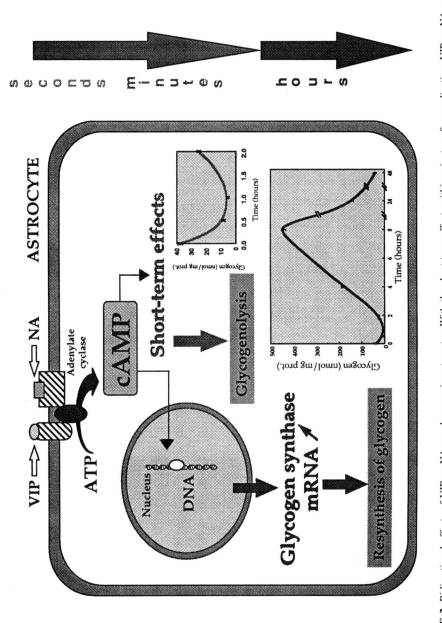

FIGURE 2. Bidirectional effects of VIP or NA on glycogen in astrocytes. With the short-term effect, within minutes after application, VIP or NA promote glycogenolysis. This effect is due to cAMP-dependent phosphorylation of preexisting proteins. With the long-term effect, within a few hours after application of VIP or NA, glycogen levels are increased 6–10 times above control levels. This effect is due to cAMP-dependent induction of new protein synthesis.

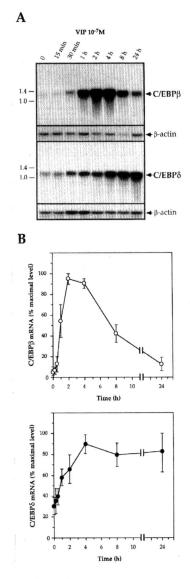

FIGURE 3. Induction of C/EBPβ and -δ mRNA expression by VIP in cultures of mouse cortical astrocytes. (**A**) Time course of C/EBPβ and -δ mRNA accumulation after exposure of cultures of cortical astrocytes to 10^{-7} M VIP. Equal amounts of total RNA (20 μg) were electrophoresed and analyzed by Northern blotting using ^{32}P riboprobes complementary to C/EBPβ and -δ mRNAs. Similar results were obtained in three independent experiments. RNA markers are indicated (in kb) at the left. Blots were hybridized sequentially with C/EBP and β-actin probes; the latter was used as a loading control. (**B**) Results in (**A**) were quantified by scanning densitometry and are expressed as mean ± SEM (N=3) compared with the maximal level of expression for each C/EBP mRNA. (Reproduced with permission from Cardinaux & Magistretti.[22])

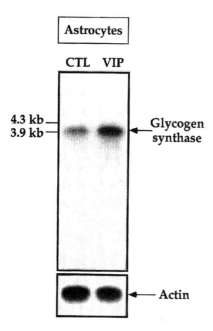

FIGURE 4. Induction of glycogen synthase mRNA expression by VIP in cultured cortical astrocytes. Cultures of cortical astrocytes were stimulated for 30 min in the presence of 1 μM VIP. Ten hours after the end of the pulse stimulation poly(A)⁺ mRNA was isolated. In each lane, 4 μg of poly(A)⁺ mRNA were loaded, electrophoresed, and hybridized with a 2.4 kb mouse brain glycogen synthase cDNA fragment random-primed labeled with [³²P]dCTP. Autoradiogram shows results from one experiment, repeated twice with similar results. (Reproduced with permission from Pellegri *et al.*[23])

demonstrate that VIP and PACAP regulate the level of expression of a family of transcription factors in astrocytes.

Glycogen Synthase

The most likely candidate gene for induction by VIP in relation to its effect on glycogen resynthesis is glycogen synthase. Mouse astrocyte glycogen synthase was cloned in our laboratory[23] and its distribution was characterized in the mouse brain by *in situ* hybridization. In addition, we examined whether its level of expression could be regulated by VIP in primary astrocyte cultures, where VIP induces a massive, transcriptionally regulated glycogen resynthesis.[18] As shown in FIGURE 4, VIP indeed stimulated the level of expression of mRNA encoding for glycogen synthase.

INDUCTION BY VIP OF ASTROCYTIC PROTEINS: ANALYSIS BY TWO-DIMENSIONAL POLYACRYLAMIDE GEL ELECTROPHORESIS

In view of the transcriptionally mediated effects of VIP on glycogen metabolism, which involve the regulation of specific genes related to energy metabolism, we decided to extend our analysis and to search for proteins that could potentially be regulated by VIP. A convenient way to screen for protein induction by a given manipulation is to compare their pattern of expression by two-dimensional polyacrylamide gel electrophoresis (2D-PAGE). This technique combines two electrophoretic techniques (isoelectric focusing and SDS-polyacrylamide gel electrophoresis) to provide a considerably greater resolution than either of the individual procedures. 2D-PAGE separates proteins by charge in the first dimension and by molecular size in the second dimension. This technique is therefore particularly useful to analyze complex mixtures of proteins such as total cellular extracts.

We have examined the induction of intracellular proteins by VIP and NA in primary cultures of astrocytes. Experiments were carried out according to the method described in detail in Cardinaux and colleagues.[24] High resolution 2D-PAGE was performed on the Oxford GlycoSystems' Investigator™ System. Isoelectric focusing was performed according to the method of Hochstrasser and colleagues.[25] Seven proteins were shown to be induced by exposure to VIP or NA (FIG. 5).

MODULATION BY VIP OF GLUTAMATE-MEDIATED NEUROTRANSMISSION IN CEREBRAL CORTICAL NEURONS

Over the past few years we have accumulated evidence that VIP and PACAP interact with glutamate to potentiate glutamate-mediated signaling in the cerebral cortex. Evidence supporting such an interaction has been observed at the second messenger level and at the transcription level.

VIP and PACAP Potentiate the Glutamate-Evoked Release of Arachidonic Acid

Glutamate is the main excitatory neurotransmitter in the cerebral cortex. Activation of glutamate receptors results both in rapid changes in ionic currents and in second messenger formation. Among the second messenger pathways stimulated by glutamate, the release of arachidonic acid (AA) has been shown in brain slices[26] and in primary cultures of neurons and astrocytes.[27] The glutamate-evoked release of AA could be of particular relevance to the physiological regulation of excitatory signaling, since AA appears to amplify glutamatergic neurotransmission. Indeed, AA enhances glutamate release from presynaptic terminals,[28] inhibits glutamate reuptake,[29] and increases NMDA currents.[30]

By exploring the potential interactions existing between VIP and glutamate, we found that VIP and PACAP potentiate the glutamate-evoked release of AA from primary cultures of cerebral cortical neurons devoid of glial cells.[31] The EC_{50} values for these potentiating effects correspond to 1 μM for VIP and 0.7 nM for PACAP, suggesting that these peptides might exert their effect by activating PACAP 1 receptors.[31] Interestingly, despite the clear mediation of this effect via PACAP 1 receptors, which are in principle associated with cAMP-generating signal transduction mechanisms, the cAMP/PKA pathway does not appear to be involved.[31]

VIP and PACAP Potentiate the Actions of Glutamate on BDNF and c-fos Expression

At the transcriptional level, interactions between VIP or PACAP and glutamate have been shown on the expression of two genes, namely BDNF and c-fos.[32,33] BDNF is a neurotrophic factor that has been demonstrated to promote the survival and differentiation of selective populations of neurons during development.[34] More recently its involvement in synaptic plasticity has been demonstrated.[35] Regulation of BDNF expression by VIP and PACAP was examined in primary cultures of mouse cerebral cortical neurons devoid of

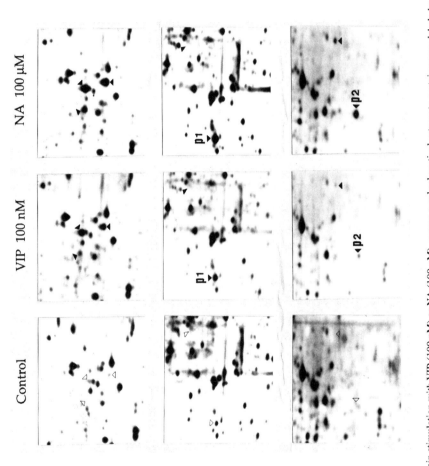

FIGURE 5. After a 30-min stimulation with VIP (100 nM) or NA (100 μM), mouse cerebral cortical astrocyte proteins were labeled with ^{35}S-methionine and ^{35}S-cysteine for 8 hours. Control astrocytes were labeled for the same time without stimulation. The same amount of protein (100 μg) was loaded on each 2D-PAGE. Induced proteins are indicated by a *dark arrow* for VIP or NA-stimulated astrocytes and by a *white arrow* for control astrocytes. The proteins p1 and p2 are specifically indicated since they are visible on a silver-stained gel, and are therefore susceptible to purification and microsequencing.

glial cells.[32,33] Stimulation of cultured cortical neurons with VIP or PACAP upregulates BDNF expression[32] in a time- and concentration-dependent manner. Increases in BDNF mRNA reach maximal levels between one and three hours after the beginning of the stimulation.[32] Both VIP and PACAP stimulate BDNF expression in a concentration-dependent manner, PACAP being more potent.[32] Since VIP receptors are also present on cortical astrocytes in culture,[7] we examined whether VIP and PACAP could stimulate BDNF mRNA in this preparation. Results indicate that VIP and PACAP upregulate BDNF expression although to a lesser degree than in cortical neurons.[32]

In cortical neurons, induction of BDNF mRNA by these two neuropeptides is completely inhibited by the NMDA receptor antagonists MK-801 and AP5,[32] indicating that VIP and PACAP do not stimulate BDNF expression directly but rather potentiate the effect of glutamate released by neurons (~ 1 μM) and acting at NMDA receptors.[32,33] This effect is postsynaptic in nature since neither VIP nor PACAP stimulates the release of glutamate in this preparation.[33]

A similar interaction between VIP or PACAP and glutamate has also been observed on the transcription of the immediate-early gene c-fos.[33] The Fos protein is thought to stimulate the transcription of other genes by binding, in heterodimeric association with members of the Jun family, to the AP-1 site.[36] Induction of c-fos mRNA by VIP and PACAP is completely inhibited by the noncompetitive NMDA receptor antagonist MK-801, therefore indicating that, similarly to BDNF, VIP and PACAP stimulate c-fos expression by potentiating the effect of glutamate.[33] Given that VIP and PACAP stimulate c-fos expression in a glutamate-dependent manner and that c-fos encodes a transcription factor, we raised the question of whether the induction of BDNF expression could be a consequence of the induction of c-fos expression. Results showed that although c-fos mRNA is induced earlier than BDNF mRNA, there is no sequential link since both genes behave as immediate-early genes.[32]

At the intracellular level, coactivation of protein kinases A and C mediates the glutamate-dependent stimulation of BDNF and c-fos expression evoked by VIP and PACAP.[32,33] Indeed, H-89 and staurosporin, which inhibit protein kinases A and C, respectively, totally abolish the induction of c-fos and BDNF expression evoked by these two neuropeptides.[32,33] The inhibitory effect of H-89 on the induction of c-fos and BDNF expression is consistent with the stimulatory effect of VIP and PACAP on cAMP formation observed in this preparation.[33] The inhibitory effect of staurosporin can be ascribed to the fact that induction of c-fos and BDNF expression by VIP and PACAP is mediated by glutamate acting on NMDA receptors. Indeed, activation of NMDA receptors by glutamate increases Ca^{2+} entry, which in turn triggers the translocation of protein kinase C from the cytoplasm to the membrane. Translocation of protein kinase C from the cytoplasm to the membrane is a necessary prerequisite for the activation of this enzyme by membrane-associated diacylglycerol. Thus, by blocking protein kinase C activation induced by endogenous glutamate, staurosporin indirectly inhibits induction of c-fos and BDNF expression evoked by VIP or PACAP.

The transduction mechanism underlying c-fos and BDNF expression points to a "biochemical AND gate" mechanism, which can be defined as "a regulatory mechanism that requires that two independent receptor-regulated processes act simultaneously to produce a full response,"[37] and that implies the obligatory activation of both protein kinases A and C in the induction of c-fos and BDNF expression by VIP and PACAP.[32,33]

Altogether these observations suggest that the potentiation of the glutamate response by VIP or PACAP, observed in cultured cortical neurons, could be a general mechanism that occurs at the second messenger level by stimulating the formation of arachidonic acid[31] and at the transcription level by enhancing the expression of c-fos and BDNF mRNAs.[32,33] As noted earlier, VIP intracortical neurons receive predominantly excitatory synaptic contacts, most likely glutamatergic.[2,3] In addition, glutamate has been shown to evoke VIP release in the cerebral cortex[38]; thus, the activation of excitatory afferents could result in synergistic interactions between VIP and glutamate on common target cells within discrete cortical domains (FIG. 6).

In conclusion, these observations strongly suggest that VIP and PACAP interact synergistically with glutamate to increase the "throughput" or "strength" of glutamate-mediated signaling in the cerebral cortex.

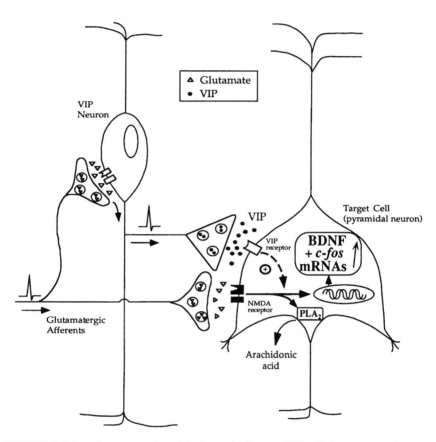

FIGURE 6. Schematic representation of the interaction between VIP and glutamate on *c-fos* and BDNF expression as well as on the release of arachidonic acid. Within the cerebral cortex, VIP is almost exclusively localized to a homogeneous population of bipolar, radially orientated neurons.[1] These morphological characteristics imply that when glutamate is released from activated afferent pathways to a given cortical area, VIP-containing bipolar neurons are ideally positioned to receive laminarly specified excitatory inputs from afferent pathways. In support of this hypothesis, frequent asymmetrical (excitatory) synapses have been identified on VIP-containing neurons[2,3] and glutamate has been shown to evoke VIP release.[38] Thus, activation of excitatory afferents can result in synergistic interactions between VIP and glutamate on potential common target cells, leading to the induction of *c-fos* and BDNF expression and to the release of arachidonic acid.

REFERENCES

1. MAGISTRETTI, P. J & J. H. MORRISON. 1988. Noradrenaline- and vasoactive intestinal peptide-containing neuronal systems in neocortex: functional convergence with contrasting morphology. Neurosci. **24**: 367–378.
2. HAJOS, F., K. ZILLES, A. SCHLEICHER & M. KALMAN. 1988. Types and spatial distribution of vasoactive intestinal polypeptide (VIP)-containing synapses in the rat visual cortex. Anat. Embryol. **178**: 207–217.
3. STAIGER, J. F., K. ZILLES & T. F. FREUND. 1996. Innervation of VIP-immunoreactive neurons by the ventroposteromedial thalamic nucleus in the barrel cortex of the rat. J. Comp. Neurol. **367**: 194–204.
4. PETERS, A. & K. HARRIMAN. 1988. Enigmatic bipolar cell of rat visual cortex. J. Comp. Neurol. **267**: 409–432.
5. BAYRAKTAR, T., J. F. STAIGER, L. ACSADY, C. COZZARI, T. F. FREUND & K. ZILLES. 1997. Co-localization of vasoactive intestinal polypeptide, gamma-aminobutyric acid and choline acetyltransferase in neocortical interneurons of the adult rat. Brain Res. **757**: 209–217.
6. CHEDOTAL, A., D. UMBRIACO, L. DESCARRIES, B. K. HARTMAN & E. HAMEL. 1994. Light and electron microscopic immunocytochemical analysis of the neurovascular relationships of choline acetyltransferase and vasoactive intestinal polypeptide nerve terminals in the rat cerebral cortex. J. Comp. Neurol. **343**: 57–71.
7. MARTIN, J.-L., D. L. FEINSTEIN, N. YU, O. SORG, C. ROSSIER & P. J. MAGISTRETTI. 1992. VIP receptor subtypes in mouse cerebral cortex: evidence for a differential localization in astrocytes, microvessels and synaptosomal membranes. Brain Res. **587**: 1–12.
8. HUANG, M. & O. P. RORSTAD. 1984. Cerebral vascular adenylate cyclase: evidence for coupling to receptors for vasoactive intestinal peptide and parathyroid hormone. J. Neurochem. **43**: 849–856.
9. PETERS, A., S. L. PALAY & H. V. DEF. WEBSTER. 1991. The Fine Structure of the Nervous System. Saunders. Philadelphia.
10. MAGISTRETTI, P. J., O. SORG & J.-L. MARTIN. 1993. Regulation of glycogen metabolism in astrocytes: physiological, pharmacological, and pathological aspects. *In* Astrocytes: Pharmacology and Function S. Murphy, Ed.: 243–265. Academic Press. San Diego.
11. SORG, O. & P. J. MAGISTRETTI. 1991. Characterization of the glycogenolysis elicited by vasoactive intestinal peptide, noradrenaline and adenosine in primary cultures of mouse cerebral cortical astrocytes. Brain Res. **563**: 227–233.
12. YU, N., J.-L. MARTIN, N. STELLA & P. J. MAGISTRETTI. 1993. Arachidonic acid stimulates glucose uptake in cerebral cortical astrocytes. Proc. Natl. Acad. Sci. USA **90**: 4042–4046.
13. SOKOLOFF, L. M. REIVICH, C. KENNEDY, M. H. DES ROSIERS, C. S. PATLAK, K. D. PETTIGREW, O. SAKURADA & M. SHINOHARA. 1997. The [^{14}C]deoxyglucose method for the measurement of local cerebral glucose utilization: theory, procedure, and normal values in the conscious and anesthetized albino rat. J. Neurochem. **28**: 897–916.
14. DRINGEN, R., R. GEBHARDT & B. HAMPRECHT. 1993. Glycogen in astrocytes: possible function as lactate supply for neighboring cells. Brain Res. **623**: 208–214.
15. ISUMI, Y., A. M. BENZ, H. KATSUKI & C. F. ZORUMSKI. 1997. Endogenous monocarboxylates sustain hippocampal synaptic function and morphological integrity during energy deprivation. J. Neurosci. **17**: 9448–9457.
16. MAGISTRETTI, P. J. & L. PELLERIN. 1997. The cellular bases of functional brain imaging: evidence for astrocyte-neuron metabolic coupling. Neuroscientist **3**: 361–365.
17. MAGISTRETTI, P. J., J. H. MORRISON, W. J. SHOEMAKER, V. SAPIN & F. E. BLOOM. 1981. Vasoactive intestinal polypeptide induces glycogenolysis in mouse cortical slices: a possible regulatory mechanism for the local control of energy metabolism. Proc. Natl. Acad. Sci. USA **78**: 6535–6539.
18. SORG O. & P. J. MAGISTRETTI. 1992. Vasoactive intestinal peptide and noradrenaline exert long-term control on glycogen levels in astrocytes: blockade by protein synthesis inhibition. J. Neurosci. **12**: 4923–4931.
19. YEH, W.-C., Z. CAO, M. CLASSON & S. L. MCKNIGHT. 1995. Cascade regulation of terminal adipocyte differentiation by three members of the C/EBP family of leucine zipper proteins. Genes Dev. **9**: 168–181.

20. WANG, N.-D., M. J. FINEGOLD, A. BRADLEY, C. N. OU, S. V. ABDELSAYED, M. D. WILDE, L. R. TAYLOR, D. R. WILSON & G. J. DARLINGTON. 1995. Impaired energy homeostatis in C/EBPα knockout mice. Science **269:** 1108–1112.
21. VALLEJO, M., D. RON, C. P. MILLER & J. F. HABENER. 1993. C/ATF, a member of the activating transcription factor family of DNA-binding proteins, dimerizes with CAAT/enhancer-binding proteins and directs their binding to cAMP response elements. Proc. Natl. Acad. Sci. USA **90:** 4679–4683.
22. CARDINAUX, J.-R. & P. J. MAGISTRETTI. 1996. Vasoactive intestinal peptide, pituitary adenylate cyclase-activating peptide, and noradrenaline induce the transcription factors CCAAT/enhancer binding protein (C/EBP)-β and C/EBP δ in mouse cortical astrocytes: involvement in cAMP-regulated glycogen metabolism. J. Neurosci. **16:** 919–929.
23. PELLEGRI, G., C. ROSSIER, P. J. MAGISTRETTI & J.-L. MARTIN. 1996. Cloning, localization and induction of mouse brain glycogen synthase. Mol. Brain Res. **38:** 191–199.
24. CARDINAUX, J.-R., P. J. MAGISTRETTI & J.-L. MARTIN. 1997. Brain-derived neurotrophic factor stimulates phosphorylation of stathmin in cortical neurons. Mol. Brain Res. **51:** 220–228.
25. HOCHSTRASSER, D. F., A. PATCHORNIK & C. R. MERRIL. 1988. Development of polyacrylamide gels that improve the separation of proteins and their detection by silver staining. Anal. Biochem. **173:** 412–423.
26. PELLERIN, L. & L. S. WOLFE. 1991. Release of arachidonic acid by NMDA-receptor activation in the rat hippocampus. Neurochem. Res. **16:** 983–989.
27. STELLA, N., M. TENCE, J. GLOWINSKI & J. PREMONT. 1994. Glutamate-evoked release of arachidonic acid from mouse brain astrocytes. J. Neurosci. **14:** 568–575.
28. HERRERO, I., M. T. MIRAS-PORTUGAL & J. SANCHEZ-PRIETO. 1992. Positive feedback of glutamate exocytosis by metabotropic presynaptic receptor stimulation. Nature **360:** 163–166.
29. YU, A. C., P. H. CHAN & R. A. FISHMAN. 1986. Effects of arachidonic acid on glutamate and gamma-aminobutyric acid uptake in primary cultures of rat cerebral cortical astrocytes and neurons. J. Neurochem. **47:** 1181–1189.
30. MILLER, B., M. SARANTIS, S. F. TRAYNELIS & D. ATTWELL. 1992. Potentiation of NMDA receptor currents by arachidonic acid. Nature **355:** 722–725.
31. STELLA, N. & P. J. MAGISTRETTI. 1996. Vasoactive intestinal peptide (VIP) and pituitary adenylate cyclase-activating polypeptide (PACAP) potentiate the glutamate-evoked release of arachidonic acid from mouse cortical neurons. J. Biol. Chem. **271:** 23705–23710.
32. PELLEGRI, G., P. J. MAGISTRETTI & J.-L. MARTIN. 1998. VIP and PACAP potentiate the action of glutamate on BDNF expression in mouse cortical neurones. Eur. J. Neurosci. **10:** 272–280.
33. MARTIN, J.-L., D. GASSER & P. J. MAGISTRETTI. 1995. Vasoactive intestinal peptide and pituitary adenylate cyclase-activating polypeptide potentiate c-fos expression induced by glutamate in cultured cortical neurons. J. Neurochem. **65:** 1–9.
34. DAVIES, A. M. 1994. The role of neurotrophins in the developing nervous system. J. Neurobiol. **25:** 1334–1348.
35. THOENEN, H. 1995. Neurotrophins and neuronal plasticity. Science **270:** 593–598.
36. SHENG, M. & M. E. GREENBERG. 1990. The regulation and function of c-fos and other immediate early genes in the nervous system. Neuron **4:** 477–485.
37. NIKODIJEVIC O. & D. C. KLEIN. 1989. Adenosine stimulates adenosine 3′,5′-monophosphate and guanosine 3′,5′-monophosphate accumulation in rat pinealocytes: evidence for a role for adenosine in pineal neurotransmission. Endocrinology **125:** 2150–2157.
38. WANG, J.-Y., T. L. YAKSH, G. J. HARTY & V. L. W. GO. 1986. Neurotransmitter modulation of VIP release from cat cerebral cortex. Am. J. Physiol. **250:** R104–R111.

Glutamate Toxicity in the Lung and Neuronal Cells: Prevention or Attenuation by VIP and PACAP

S. I. SAID,[b] K. DICKMAN, R. D. DEY,[a] A. BANDYOPADHYAY,
P. DE STEFANIS, S. RAZA, H. PAKBAZ, AND H. I. BERISHA

*V.A. Medical Center, Northport, New York and SUNY Stony Brook,
New York USA*

[a]*Department of Anatomy, School of Medicine, West Virginia University,
Morgantown, West Virginia 26506-9128*

ABSTRACT: VIP, which has been demonstrated to reduce or prevent oxidant injury in the lungs and other organs, is shown here to protect against excitotoxic injury of the lung and excitotoxic death of cortical neuronal cells in primary culture. Glutamate killing of neuron-like PC-12 cells, attributable to oxidant stress rather that to excitotoxicity, is also reduced or prevented by VIP and by the closely related peptide PACAP. The exact mechanisms of this protection remain to be determined, but appear to include antioxidant and anti-apoptotic actions, and suppression of glutamate-induced upregulation of its own receptor. Both VIP and PACAP offer the promise of novel and nontoxic means of defending against NMDA and glutamate toxicity.

Over the past two decades, we have investigated the mechanisms of acute lung injury and the modulation or prevention of this injury by various pharmacological agents, particularly the neuropeptide vasoactive intestinal peptide (VIP) and related peptides.[1-5]

We have recently concentrated on an experimental model in which lung injury, manifested by acute, high-permeability pulmonary edema, is produced by exogenous administration of the glutamate agonist *N*-methyl-D-aspartate (NMDA).[6,7] This model is well known in the field of neuroscience, where the excitatory amino acids glutamate and aspartate, the principal physiological neurotransmitters,[8] are also recognized as potential triggers of neuronal cell death in certain acute neurological disorders, e.g., cerebral ischemia and stroke,[9] and chronic neurodegenerative diseases, including Alzheimer's disease.[10]

Because of the favorable results we obtained with VIP in attenuating the severity, or delaying the onset, of lung injury due to NMDA, we turned our focus to glutamate neurotoxicity and the possible protection against this toxicity by VIP and the related peptide pituitary adenylate cyclase–activating peptide (PACAP). In this report we describe our findings on glutamate toxicity in the lungs and in neuronal and neuron-like cells, with emphasis on the mechanisms of toxicity and its prevention or attenuation by VIP, PACAP, and antioxidant compounds.

METHODS

Experiments in Isolated Lungs

Isolated rat lungs were ventilated at constant tidal volume with 95% O_2–5% O_2, and perfused at constant flow with Krebs solution containing 4% BSA, as described.[6,7] NMDA

[b]Corresponding author: 17 - 040 HSC; SUNY, Stony Brook, NY 11794-8172; Tel.: 516-444-1754; Fax: 516-444-7502; E-mail: SSAID@EPO.SOM.SUNYSB.EDU

was added to the perfusate at concentrations of 100–500 μM. The occurrence of acute lung injury was detected and its severity gauged by increases in airway inflation pressure, pulmonary artery perfusion pressure, wet-to-dry lung weight ratio, and protein leakage into the alveolar space.

In most experiments, these markers of high-permeability pulmonary edema were examined 60 min after the addition of NMDA. In other experiments, lung perfusion was continued until overt pulmonary edema occurred, as evidenced by the appearance of foam in the airway. VIP, or other modulators of injury, were added to the perfusate beginning 10 min before NMDA. Protection against NMDA toxicity was assessed by attenuation of signs of edematous pulmonary injury when experiments were terminated 60 min from the addition of NMDA, or by delay in the onset of overt lung edema (foam) when perfusion was continued until edema occurred.

EXPERIMENTS ON NEURONAL AND NEURONAL-LIKE CELLS

Cell Preparations

We examined glutamate toxicity in two types of neuronal and "neuronal-like" cells: adrenal chromaffin-like PC-12 cells, derived from a rat pheochromocytoma, and fetal rat cerebral cortical cells in primary cultures. We had earlier experimented with, but later discontinued, the use of neuroblastoma cell lines and NT-2 cells[11,12] because of inconsistent expression of glutamate receptors on these cells in our laboratory.

PC-12 Cells

Originally established in 1976 by Greene and Tischler[13] from a rat adrenal pheochromocytoma, the PC-12 clonal cell line has been extensively used as a neuronal model. PC-12 cells express a number of peptide and other neurotransmitter receptors.[14] Grown in the presence of NGF, they differentiate into a nonmitotic neuronal phenotype, expressing ion channels and fuctional transduction systems.[15] In our experiments, however, as in the work of major groups who have studied glutamate toxicity in PC-12 cells,[16–20] the cells were not pretreated with NGF.

The cells were obtained from Prof. Simon Halegoua, of our Department of Neurobiology and Behavior. As described by Schubert and colleagues and others,[16–20] the cells were cultured in DMEM supplemented with 10% horse serum, 5% fetal calf serum, 2 mM Glutamax, 100 μg/ml streptomycin, and 100 U/ml penicillin. The cells, kept in a humidified atmosphere of 95% air and 5% CO_2 at 37°C, were routinely split 1:6 on a weekly basis. For experimental use, 1-week-old cultures were split 1:10 into collagen-coated, 35-mm dishes in the same medium described above, except that dialyzed horse and fetal calf sera were used. Twenty-four hours after plating, the cultures were refed and exposed for 24 h to various treatments that may modulate glutamate toxicity.

Cortical Neurons in Primary Culture

Mixed glia and neuron primary cultures were prepared from fetal rat brains, as described by Choi and colleagues[21,22] and Dawson and colleagues.[23-25] The embryos (13 to 14 days gestation) were rapidly removed under sterile conditions and placed into ice-cold Hanks' balanced salt solution (HBSS). Neocortices were dissected under sterile conditions, incubated for 20 min in 0.0027% trypsin and saline solution, then transferred to modified Eagle's medium (MEM) supplemented with 10% each fetal calf and horse sera,

100 U/ml penicillin, 100 µg/ml streptomycin, 2 mM pyruvate, and 2 mM Glutamax. The cells were dissociated by trituration and plated onto poly-ornithine–coated, 24-well plates (3–4×10^5 ml cells/well). Glial overgrowth was inhibited by treatment on days 7–9 in culture with 10 µg/ml of 5-fluoro-2-deoxyuridine. Cells were maintained thereafter in the MEM medium described above except that fetal calf serum was omitted. Experiments were performed on cultures that were allowed to mature for 2–3 weeks after plating, to allow time for expression of NMDA receptors.

Injury Models

Cell injury and death were produced in PC-12 cells by incubation with glutamate (1–10 mM) or NMDA (500 µM), for 24 h, with and without VIP, PACAP, or other potential modulators of injury. Both molecular forms of PACAP, the 38– and the 27–amino acid residue peptides, were tested, because they act via different signal transduction pathways. Cortical neurons were treated with control salt solution (CSS), NMDA (300 or 500 µM), glutamate (5 mM). All drugs were applied in CSS for 5–10 min or for 20–24 h: Incubations were performed in Mg^{2+}-free MEM + 0.1% BSA and 100–200 µM glycine.

Cell death was evaluated by measurement of LDH release into the medium and by the MTT assay of mitochondrial enzyme activity.[17,18,20]

Chemicals and Reagents

6(5H) phenantridinone was from Aldrich, MK-801 was from Professor Leslie D. Iversen (then at Merck Sharp and Dohme), and VIP was from Professor Viktor Mutt (Karolinska Institute, Stockholm). All other chemicals including NMDA, L-NAME, D-NAME, L-nitroarginine, AP-5, AP-7, and benzamide, were from Sigma Chemical Co. (St. Louis, MO, USA).

Statistical Analysis

The influence of each intervention on functional and biochemical measurements was assessed by analysis of variance followed by Tukeys' procedures for intergroup comparisons. Repeated measurements of lung airway and vascular pressures over time were analyzed by multiple-paired t tests.

RESULTS AND DISCUSSION

Protection by VIP against Excitotoxic Injury of the Lung [6,7]

The addition of NMDA (1 mM) to the perfusate induced acute injury in isolated rat lungs perfused with Krebs/4%BSA. Within approximately 60 min, peak airway pressure, mean pulmonary vascular perfusion, wet/dry lung weight ratio, and protein content of broncho-alveolar lavage fluid had increased significantly ($p < 0.01$–0.05) (FIG. 1). These findings reflected the occurrence of acute lung injury with high-permeability edema. The injury was directly related to activation of NMDA receptors, as it was prevented by competitive NMDA receptor antagonists DL-2-amino-5-phosphonovaleric acid (AP-V or AP-5, 100 µM) or AP-7 (100 µM), and by channel blocker MK-801 (5–10 µM). The injury was

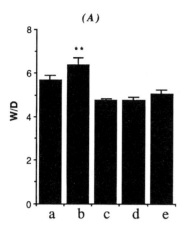
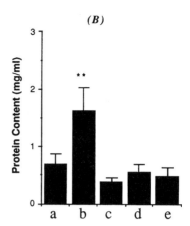

FIGURE 1. Excitotoxicity in the lung and its prevention by VIP. Induction of high-permeability edema in guinea pig lungs by NMDA and its prevention by NMDA receptor antagonist AP-5, by NOS inhibitor L-NAME, and by VIP. W/D lung ratios *(A)* BAL protein content and *(B)* in several experimental groups: (a) control, untreated lungs, $N = 8$; (b) NMDA (1 mM) + L-arginine (10 mM), $N = 16$; (c) NMDA + L-arginine +AP-5 (100 µM), $N = 3$; (d) NMDA + L-arginine + L-NAME (2 mM), $N = 4$; (e) NMDA + L-arginine + VIP 10 µM/kg·min^{-1}), $N = 5$. ** $p < 0.01$ versus control. (Reprinted from Said *et al.*[7] with permission)

prevented by competitive NMDA receptor antagonists or by channel-blocker MK-801 (1–10 µM), and was reduced in the presence of Mg^{2+} (FIG. 1).

As with NMDA toxicity to central neurons, the lung injury was nitric oxide (NO) dependent: it required L-nitroarginine, was associated with increased production of NO due to upregulation of neuronal NO synthase (NOS) (FIG. 2), and was attenuated by either of two NOS inhibitors, L-NAME (0.1 and 2 mM) or L-nitroarginine (1.0 mM). VIP (10 µM/kg·min^{-1}) and either of two selective inhibitors of poly(ADP-ribose) polymerase (PARP), benzamide (1 mM) or 6(5H) phenanthridinone (1 mM), also prevented this injury (FIG. 3), but without inhibiting NO synthesis.

The findings, while providing indirect evidence that NMDA receptors exist in the lung, demonstrate that excessive activation of these receptors may provoke acute edematous lung injury as seen in the "adult respiratory distress syndrome." This injury can be modulated by blocking one of three critical steps: NMDA receptor binding, inhibition of NO synthesis, or activation of PARP. The presence of NMDA receptors in the lung, suggested by these observations, has now been confirmed by the demonstration of NMDA binding sites on alveolar and bronchial epithelial cells, as well as pulmonary vessels.[26,27]

Upregulation of VIP Gene Expression by NMDA

In experiments on tracheal ganglia we observed that addition of NMDA (1.0 mM) stimulated VIP gene expression (mRNA, normalized to G3PDH mRNA) by 50%, and this stimulation was reversed by NMDA channel-blocker MK801 (10 µM). These findings are consistent with a role for VIP as an endogenous modulator of lung injury.[5,28]

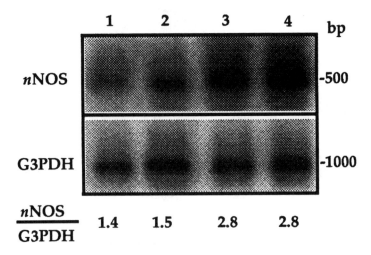

FIGURE 2. NMDA upregulates neuronal *n*NOS expression in rat lung. PCR amplification of *n*NOS and G3PDH mRNA in rat lung. (Lane 1) Control; (Lane 2) 2 mM NMDA for 10 min; (Lane 3) 10 mM NMDA for 10 min; and (Lane 4) 10 mM NMDA for 30 min.

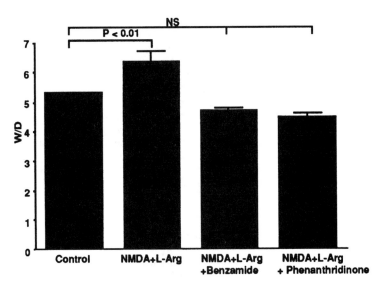

FIGURE 3. NMDA-induced lung injury and its modulation by PARP inhibitors. Either of two selective PARP inhibitors, benzamide or phenanthridinone, prevented the induction of pulmonary edema, measured as increased wet/dry (W/D) ratio in rat lungs (more details in text).

PROTECTION AGAINST GLUTAMATE TOXICITY IN CORTICAL NEURONAL CELLS

Cortical neurons in primary cultures exhibited normal morphologic features and neuronal properties, as demonstrated by immunochemical staining for neuron-specific enolase. Neuronal cell death was induced by acute (5–10 min) or chronic (20–24 h) incubation with either NMDA (500 μM) or glutamate (5 mM) (FIGS. 4 and 5). The cell death was mediated by NMDA-receptor activation and by NO excessive production, as it was markedly attenuated, respectively, by NMDA-receptor antagonist MK-801 and by NOS inhibitor L-NAME. VIP (3 μM) and PACAP (1 μM) were similarly protective against toxicity due to NMDA or glutamate. The protection by VIP or PACAP was mimicked by the cell-permeant cAMP analog 8-bromo-cAMP (2.5 mM), supporting the view that the protection by VIP in this preparation may be mediated in large part, though not exclusively, via the cAMP pathway.

DEFENSE AGAINST GLUTAMATE TOXICITY IN PC-12 CELLS

Mechanisms of Glutamate-Induced PC-12 Cell Death: Excitotoxicity versus Oxidant Stress

Glutamate, at high concentrations (10 mM), produced PC-12 cell death. Under the same experimental conditions, however, these cells were resistant to NMDA (500 μM) (FIG. 6A). This was in contrast to rat cortical neurons in primary cell culture, which were killed by

(A)   **(B)**

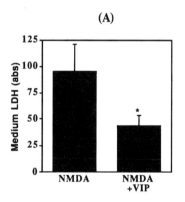
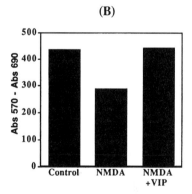

FIGURE 4. VIP protects cultured cortical neurons from excitotoxicity. (**A**) LDH release. Rat embryonic cortical neurons in culture were incubated with 500 μM NMDA for 20 h in MEM+0.1% BSA and 100 μM glycine. VIP exposure (3 μM) was initiated 30 min prior to NMDA treatment and continued throughout the exposure to NMDA. Toxicity was assessed by release of LDH into the culture medium (minus control levels). Values are mean+SEM ($N = 5$ culture preparations), *$p < 0.05$ vs. NMDA. (**B**) MTT assay for toxicity. Neurons were incubated in Mg^{2+}-free control saline solution with 10 μM glycine (CSS), CSS + 300 μM NMDA, for 10 min at room temperature, or CSS + NMDA + 3 μM VIP. Cultures were then washed twice, incubated in MEM+ 0.1% BSA for 20 h, then exposed to 0.5 mg/ml MTT for 15 min at 37°C. The reaction was stopped by acid isopropanol, samples were solubilized by sonication, and absorbance was read at 570 λ with background correction at 690 λ. When used, VIP was present 30 min before NMDA exposure, during exposure, and for the following 20 h period.

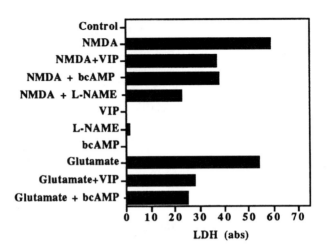

FIGURE 5. Protection from chronic excitotoxicity in cortical neurons. Viability of 3-week-old cortical neuron primary cultures was determined by release of LDH into the medium. LDH levels were measured after 24 h exposure to 5 mM glutamate or 500 μM NMDA with various interventions: VIP (3 μM), 8-bromo-cAMP (bcAMP, 2.5 mM), L-NAME (1 mM). When used, VIP and bcAMP were added 30 min prior to NMDA or glutamate and maintained throughout the 24 h period. All experiments were done in MEM with 0.1% BSA and 200 μM glycine. Values are means of duplicate culture wells.

either glutamate or NMDA. Further, the glutamate toxicity of PC-12 cells was not prevented by NMDA receptor channel-blocker MK-801 (FIG. 6B). Similar results were obtained by Schubert and colleagues and others in PC-12 cells,[17,18,20] and Murphy and colleagues in neuroblastoma × primary embryonic hybrid cell line N18-RE-105[29] and immature rat cortical neurons (1–3 days in culture).[30] These observations suggest that glutamate toxicity in PC-12 cells is not due to excitotoxicity, i.e., it is not mediated by NMDA receptors.

Murphy and colleagues proposed that glutamate toxicity in PC-12 (and immature cortical neurons) results from competitive inhibition of cysteine transport, leading to depletion of the cellular antioxidant glutathione (GSH) and the accumulation of cellular oxidants.[17,18] We have confirmed this view of the pathogenesis of glutamate toxicity in PC-12 cells by two lines of evidence: (1) PC-12 cell toxicity was induced by the neuronal cysteine uptake inhibitor L-homocysteate (1 mM), and by buthionine sulfoximime (BSO, 30 μM), a specific inhibitor of GSH synthesis (FIG. 7); and (2) the antioxidant butylated hydroxyanisole (BHA) totally prevented glutamate-induced killing of PC-12 cells (FIG. 7).

Role of NO in Glutamate Toxicity

While inhibition of NOS activity prevented NMDA-induced lung injury (FIG. 1) and reduced excitotoxic cortical cell death (FIG. 5), it did not protect against glutamate killing of PC-12 cells.

Prevention of Oxidative Glutamate Toxicity by VIP and PACAP

We have examined the protection by VIP and PACAP against oxidative glutamate neurotoxicity in PC-12 cells, and the mechanisms of such protection. Glutamate (10 mM) produced

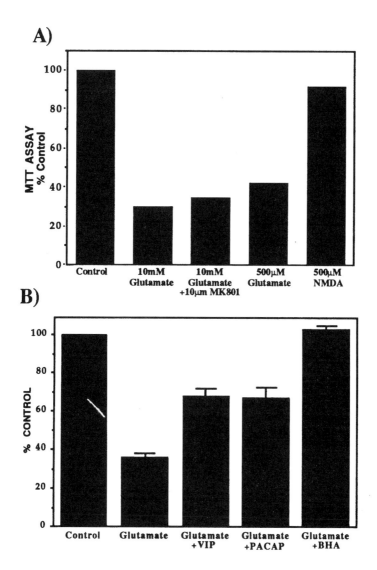

FIGURE 6. (**A**) Glutamate toxicity is not mediated by NMDA receptors in PC12 cells. PC12 cells were exposed to glutamate, NMDA, or glutamate plus the NMDA receptor antagonist MK801 for 24 h. Cell viability was measured by the MTT assay. Data are mean values, N=1–2. (**B**) Protection from glutamate toxicity in PC12 cells. MTT assay in PC12 cells exposed for 24 h to 10 mM glutamate, alone or with either VIP (5 μM), PACAP-38 (2 μM), or the antioxidant BHA (butylated hydrox-yanisole, 100 μM); N=3.

90% killing of the cells at 24 h, as judged by MTT assay (FIGS. 6B and 7). VIP (1–10 μM) and PACAP (2 μM) protected against this killing, in a concentration-dependent manner, with 10 μM VIP providing full protection (FIGS. 6B and 7). VIP also protected against toxicity due

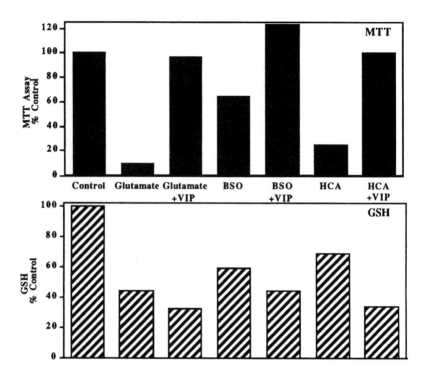

FIGURE 7. VIP prevents PC12 cell injury due to glutathione depletion. PC12 cells were exposed to glutamate (10 mM), homocysteic acid (HCA, 1 mM), or buthionine sulfoximine (BSO, 30 μM) in the presence or absence of VIP (10 μM). Glutathione (GSH) content was measured after 6 h treatment (*bottom*); cell viability was evaluated by the MTT assay after 24 h treatment (*top*). Data are % of control values.

to BSO (30 μM), and DL-homocysteic acid (1 mM) (FIG. 7). Despite this protection, VIP did not block the glutamate-induced depletion of GSH. These results suggest that VIP protects PC-12 cells against death due to oxidative stress by glutamate, but without preventing the associated decline in GSH levels. The neuroprotective effect of VIP in PC-12 cells, therefore, involves antioxidant mechanisms that are independent of GSH synthesis.

Relative Potency and Efficacy of VIP and PACAP

PACAP was highly effective in protecting PC-12 cells against glutamate toxicity. As reported by other investigators in other systems,[31] PACAP exhibited greater potency than VIP in these experiments, although by a smaller margin (FIG. 6B).

Upregulation of NMDA Receptor by Glutamate and Its Inhibition by VIP

Glutamate (10 mM) upregulated NMDA-receptor heteromeric subunit NMDAR1 protein in PC-12 cells (FIG. 8), and VIP (1 μM) prevented this upregulation (FIG. 8). Increased NMDAR1 gene expression is associated with neuronal toxicity due to severe ischemia.[32]

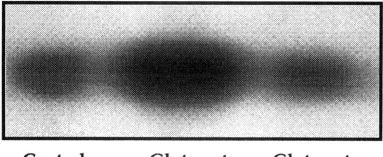

Control Glutamate Glutamate +VIP

FIGURE 8. NMDA upregulates NMDA receptor 1 protein in PC12 cells; VIP prevents this upregulation. Western blot of NMDA-R1 in PC12 cells incubated for 24 h with 10 mM glutamate, in the presence or absence of 1 μM VIP.

Even though glutamate toxicity in PC-12 cells is not mediated via NMDA receptors, the suppressant effect of VIP on this upregulation may be a neuroprotective mechanism.

Apoptosis in Glutamate Toxicity: VIP as an Anti-apoptotic Agent

Glutamate toxicity has been reported to involve apoptosis, as well as necrosis, as a mechanism of cell death.[33] We found that glutamate (10 mM) suppressed the expression of anti-apoptotic proto-oncogene *bcl-2* mRNA (by 43%) in PC-12 cells (FIG. 9). The addition of VIP (1 μM) abrogated this downregulation of *bcl-2* and preserved expression at neuronal basal levels (FIG. 9).

SUMMARY AND CONCLUSIONS

VIP, which has been demonstrated to reduce or prevent oxidant injury in the lungs and other organs, is shown here to protect against excitotoxic injury of the lung and excitotoxic

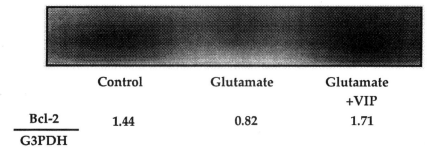

	Control	Glutamate	Glutamate +VIP
Bcl-2 / G3PDH	1.44	0.82	1.71

FIGURE 9. VIP preserves bcl-2 mRNA levels during glutamate toxicity in PC12 cells. PCR amplification of bcl-2 mRNA (normalized to G3PDH mRNA) from PC12 cells treated for 24 h with 10 mM glutamate, in the presence or absence of 1 μM VIP.

death of cortical neuronal cells in primary culture. Glutamate killing of neuron-like PC-12 cells, attributable to oxidant stress rather than to excitotoxicity, is also reduced or prevented by VIP and by the closely related peptide PACAP. The exact mechanisms of this protection remain to be determined, but appear to include antioxidant and anti-apoptotic actions and suppression of glutamate-induced upregulation of its own receptor. Both VIP and PACAP offer the promise of novel and nontoxic means of defending against NMDA and glutamate toxicity.

REFERENCES

1. BERISHA, H., H. FODA, H. SAKAKIBARA, M. TROTZ, H. PAKBAZ & S. I. SAID. 1990. VIP prevents lung injury due to xanthine/xanthine oxidase. Am. J. Physiol. **259:** L151–L155.
2. PAKBAZ, H., H. D. FODA, H. I. BERISHA, M. TROTZ & S. I. SAID. 1993. Paraquat-induced lung injury: Prevention by VIP and the related peptide helodermin. Am. J. Physiol. **265:** L369–L373.
3. BERISHA, H. I., H. PAKBAZ, A. ABSOOD & S. I. SAID. 1994. Nitric oxide as a mediator of oxidant lung injury due to paraquat. Proc. Natl. Acad. Sci. USA **91:** 7445–7449.
4. SAID, S. I. 1996. Vasoactive intestinal peptide and nitric oxide: Divergent roles in relation to tissue injury. Ann. N.Y. Acad. Sci. **805:** 379–388.
5. SAID, S. I. 1998. Anti-inflammatory actions of VIP in the lungs and airways. *In* Pro-Inflammatory and Anti-Inflammatory Peptides. Lung Biology in Health and Disease S. I. Said, Ed. **112:** 345–361. Marcel Dekker, Inc. New York.
6. SAID, S. I., H. I. BERISHA & H. PAKBAZ. 1995. NMDA receptors outside the CNS: activation causes acute lung injury that is mediated by nitric oxide synthesis and prevented by VIP. Neuroscience **65:** 943–946.
7. SAID, S. I., H. I. BERISHA & H. PAKBAZ. 1996. Excitotoxicity in lung: N-methyl-D-aspartate-induced, nitric oxide-dependent, pulmonary edema is attenuated by vasoactive intestinal peptide and by inhibitors of poly (ADP-ribose) polymerase. Proc. Natl. Acad. Sci. USA **93:** 4688–4692.
8. MAYER, M. L. & G. L. WESTBROOK. 1987. The physiology of excitatory amino acids in the vertebrate central nervous system. Prog. Neurobiol. **28:** 197–276.
9. CHOI, D. W. 1992. Excitotoxic cell death. J. Neurobiol. **23:** 1261–1276,
10. LIPTON, S. A. & P. A. ROSENBERG. 1994. Excitatory amino acids as a final common pathway for neurologic disorders. New Engl. J. Med. **330:** 613–622.
11. MUNIR, M., L. LU & P. MCGONIGLE. 1995. Excitotoxic cell death and delayed rescue in human neurons derived from NT2 cells. J. Neurosci. **15:** 7847–7860.
12. YOUNKIN, D. P., C. M. TANG, M. HARDY, U. R. REDDY, Q. Y. SHI, S. J. PLEASURE, V. M. LEE & D. PLEASURE. 1993. Inducible expression of neuronal glutamate receptor channels in the NT2 human cell line. Proc. Natl. Acad. Sci. USA **90:** 2174–2178.
13. GREENE, L. A. & A. S. TISCHLER. 1976. Establishment of a noradrenergic clonal line of rat adrenal pheochromocytoma cells which respond to nerve growth factor. Proc. Natl. Acad. Sci. USA **73:** 2424–2428.
14. ISOM, G. E. & J. L. BOROWITZ. 1993. PC-12 cells. Methods Toxicol. **1A:** 82–93.
15. CASADO, M., A. LOPEZ-GUAJARDO, B. MELLSTROM, J. R. NARANJO & J. LERMA. 1996. Functional N-methyl-D-aspartate receptors in clonal rat phaeochromocytoma cells. J. Physiol. **490:** 391–404.
16. BEHL, C., L. HOVEY III, S. KRAJEWSKI, D. SCHUBERT & J. C. REED. 1993. Bcl-2 prevents killing of neuronal cells by glutamate but not by amyloid beta protein. Biochem. Biophys. Res. Commun. **197:** 949–956.
17. FROISSARD, P., H. MONROCQ & D. DUVAL. 1997. Role of glutathione metabolism in glutamate-induced programmed cell death of neuronal-like PC12 cells. Eur. J. Pharmacol. **326:** 93–99.
18. PEREIRA, C. M. & C. R. OLIVEIRA. 1997. Glutamate toxicity on a PC12 cell line involves glutathione (GSH) depletion and oxidative stress. Free Radical Biol. Med. **23:** 637–647.
19. SCHUBERT, D. & C. BEHL. 1993. The expression of amyloid beta protein precursor protects nerve cells from β-amyloid and glutamate toxicity and alters their interaction with the extracellular matrix. Brain Res. **629:** 275–282.

20. SCHUBERT, D., H. KIMURA & P. MAHER. 1992. Growth factors and vitamin E modify neuronal glutamate toxicity. Proc. Natl. Acad. Sci. USA **89:** 8264–8267.
21. CHOI, D. W., J. Y. KOH & S. PETERS. 1988. Pharmacology of glutamate neurotoxicity in cortical cell culture: attenuation by NMDA antagonists. J. Neurosci. **8:** 185–196.
22. CHOI, D. W., M. MAULUCCI-GEDDE & A. R. KREIGSTEIN. 1987. Glutamate neurotoxicity in cortical cell culture. J. Neurosci. **7:** 357–368.
23. DAWSON, V.L. & T. M. DAWSON. 1996. Free radicals and neuronal cell death. Cell Death Differ. **3:** 71–78.
24. DAWSON, V. L., T. M. DAWSON, D. A. BARTLEY, G. R. UHL & S. H. SNYDER. 1993. Mechanisms of nitric oxide-mediated neurotoxicity in primary brain cultures. J. Neurosci. **13:** 2651–2661.
25. GONZALEZ-ZULUETA, M., L. M. ENSZ, G. MUKHINA, R. M. LEBOVITZ, R. M. ZWACKA, J. F. ENGELHARDT, L. W. OBERLEY, V. L. DAWSON & T. M. DAWSON. 1998. Manganese superoxide dismutase protects nNOS neurons from NMDA and nitric oxide-mediated neurotoxicity. J. Neurosci. **18:** 2040–2055.
26. ROBERTSON B., R. D. DEY, J. L. HUFFMAN, M. J. POLAK & S. I. SAID. 1997. Autoradiographic localization of NMDA receptors in the rat lung and implications in lung injury. Abstr. Soc. Neurosci. **23:** 931.
27. ROBERTSON, B., B. E. SATTERFIELD, S. I. SAID & R. D. DEY. 1998. NMDA-receptors are expressed by intrinsic neurons of rat larynx and esophagus. Neurosci. Lett. **244:** 77–80.
28. SAID, S. I. 1996. Molecules that protect: The defense of neurons and other cells (Editorial). J.Clin. Invest. **97:** 2163–2164.
29. MURPHY, T. H., M. MIYAMOTO, A. SASTRE, R. L. SCHNAAR & J. T. COYLE. 1989. Glutamate toxicity in a neuronal cell line involves inhibition of cystine transport leading to oxidative stress. Neuron **2:** 1547–1558.
30. MURPHY, T. H., R. L. SCHNAAR & J. T. COYLE. 1990. Immature cortical neurons are uniquely sensitive to glutamate toxicity by inhibition of cystine uptake. FASEB J. **4:** 1624–1633.
31. DICICCO-BLOOM, E. 1996. Region-specific regulation of neurogenesis by VIP and PACAP: Direct and indirect modes of action. Ann. N.Y. Acad. Sci. **805:** 244–256.
32. KOKAIA, Z., Q. ZHAO, M. KOKAIA, E. ELMER, M. L. METSIS, B. K. SIESJO & O. LINDVALL. 1995. Regulation of brain-derived neurotrophic factor gene expression after transient middle cerebral artery occlusion with and without brain damage. Exp. Neurol. **136:** 73–88.
33. BONFOCO, E., D. KRAINC, M. ANKARCRONA, P. NICOTERA & S. A. LIPTON. 1995. Apoptosis and necrosis: Two distinct events induced, respectively, by mild and intense insults with NMDA or nitric oxide/superoxide in cortical cell cultures. Proc. Natl. Acad. Sci. USA **92:** 7162–7166.

Autoantibody Catalysis: No Longer Hostage to Occam's Razor

SUDHIR PAUL[a]

Department of Pathology, University of Texas Medical School, Houston, Texas 77030

ABSTRACT: Autoantibody catalysis is now a well-established phenomenon, but the initial finding of autoantibody-catalyzed VIP cleavage was suspected to be an artefact by certain practitioners of designer antibody catalysis, mainly because of conceptual complexities not foreseen in our training about the immune system and about the mechanisms of biological catalysis. Confirmation that antibodies can acquire proteolytic activity by entirely natural means has emerged, ironically, in part from the field of designer catalytic antibodies. Recent studies have provided insight into the molecular strategies whereby antibodies can combine antigen binding with chemical catalysis, and the contributions of innate and adaptive immune mechanisms in the proteolytic activity. The history of this field illustrates the dangers of assuming that novel observations must fit into the simple confines of established theories. Scientific theories are changeable entities, dependent on empirical data and interpretations of the data, and their growth is better served by keeping an open mind.

OCCAM'S RAZOR

Webster's New Collegiate Dictionary states the meaning of Occam's razor as follows: "A scientific and philosophical rule that entities should not be multiplied unnecessarily, which is interpreted as requiring that the simplest of competing theories be preferred to the more complex, or that explanation of unknown phenomenon be sought first in terms of known quantities." The intent of the razor is reasonable, i.e., to promote an orderly advance of knowledge, by weighing each novel experimental observation within the confines of current dogma, and when possible, accommodating the new observation into accepted theories. Major advances in understanding of biological processes usually occur, however, via "paradigm shifts," as opposed to incremental revisions of established theories. This state of affairs is not unexpected, because biological knowledge, unlike physical and mathematical knowledge, is still imprecise, and biologists freely admit that current biological theories do not fully explain various life processes and their origins. In biological research, therefore, the value of the razor is offset substantially by its constraints, i.e., it discourages the positing of novel theories based on unexpected experimental observations, even though biologists recognize that new theories are needed to fully understand life.

The razor's dangers are exacerbated by the academic and commercial ambitions of competing scientists. Novel biological theories often bear the potential of changing the way we might cure diseases and our perception of ourselves and the world. Thus, the razor can be applied honorably to question new hypotheses by invoking the need for simplicity,

[a]Corresponding author: Sudhir Paul, Professor, Dept. Pathology and Lab Medicine, MSB 2.250, University of Texas Medical School, Houston, TX 77030; Tel.: 713-500-5347; Fax: 713-500-0574; E-mail: paul@casper.med.uth.tmc.edu or s.paul@worldnet.att.net

or it can be used for personal ends, when a new, competing theory threatens loss of academic power or assets.

CATALYTIC ANTIBODIES

The development of enzymatic activity in antibodies by entirely natural means was first proposed in 1989, based on the accidental observation of catalytic cleavage of VIP by human autoantibodies.[1] In studies designed to assess whether autoantibodies to VIP inhibit the binding of the tissue receptors for this neuropeptide,[2] control radiolabeled VIP incubated with the autoantibodies in the absence of receptor preparations was observed to undergo degradation, as determined by trichloroacetic acid precipitation of the intact peptide. The phenomenon of catalytic cleavage of VIP by autoantibodies has been independently confirmed.[3] Further, autoantibody catalysis is not restricted to VIP. Autoantibodies in Hashimoto's thyroiditis catalyze the cleavage of thyroglobulin,[4] autoantibodies found in lupus patients cleave DNA,[5,6] and light chains found in multiple myeloma patients cleave gpl20,[7] Arg-vasopressin, and synthetic protease substrates.[9,10]

Until recently, it was assumed that proteins capable of high affinity binding of stable ligands, such as antibodies and cell surface receptors, do not express catalytic activity, and that nature had evolved a mechanistically different class of proteins, the enzymes, to catalyze biochemical reactions. In accordance with the strictures of Occam's razor, the initial report of VIP cleavage by autoantibodies was met with disbelief by certain catalytic antibody researchers. Although evidence to the contrary had been presented, explanations such as trace contaminants and radio-induced damage were proposed as the cause of the proteolytic activity of the antibodies.[11] In the mid and late 1980s, heavy publicity was given to the idea that designer catalytic antibodies with any desired specificity and chemical reactivity can be generated by immunization of experimental animals with transition state analogs.[12] The discovery of the autoantibody proteolytic activity implied that the catalytic function can arise naturally, without coaxing of the immune system by unnatural transition state analogs. Somehow this discovery was perceived to diminish the idea of designer catalysts. With time, however, it has become evident that the natural origin of the catalytic function does not minimize the power of catalyst design, and conceptual and experimental syntheses of the two approaches will likely hasten further progress in this field. For instance, the development of therapeutically useful antibodies, such as catalytic antibodies to tumor associated and microbial antigens, may be facilitated by combining the tendency towards catalytic antibody synthesis in autoimmune disease and immunization with covalently reactive antigen analogs that recruit the germline proteolytic site for antibody synthesis.

A STRATEGY TO COMBINE GROUND STATE AND
TRANSITION STATE BINDING

The ability to combine strong binding to the transition state of the acyl-enzyme complex and to the stable form of the antigen can be anticipated to impart a high level of substrate specificity to the catalytic antibodies not ordinarily associated with conventional proteases. Strong binding of the stable form of the antigen via noncovalent and multiresidue interactions is a hallmark of antibody combining sites, and the expression of this activity by catalytic antibodies is unremarkable. Catalysis is mediated by catalyst binding to the transition state of chemical reactions,[13] i.e., the short-lived, high energy intermediate structure separating the reactants and products. To enable catalysis, proteolytic antibodies must bind the transition state even more strongly than the stable antigen. In the case of conventional serine proteases and certain newly discovered serine protease-like antibodies, the rate-limiting step in the peptide bond cleavage reaction appears to be the formation of an acyl-enzyme complex containing a serine residue of the enzyme linked covalently to the carbonyl group of the erstwhile peptide bond (FIG 1).

The key structural features responsible for stabilization of the transition state of peptide bond cleavage by serine protease-like catalytic antibodies likely include: (1) the tetrahedral geometry of the carbon atom formed in the transition state of the scissile peptide bond, which must be spatially complementary to the catalytic site; (2) the electrophilicity of the carbon atom in the developing transition state, which must be capable of binding nucleophilic serine residues in the catalyst; (3) the oxyanionic structure formed at the carbon atom, which can be stabilized by ion pairing with residues like Asn, Gln, or Arg in the catalyst (the so-called oxyanion hole); (4) the basic residue on the N-terminal side of the scissile peptide bond, recognition of which may occur by ion pairing with acidic residues such as Asp or Glu located within or close to the catalytic site in the Abs.

Recognition of the positively charged side chain of the flanking residue, although not directly involved in bond making and breaking processes during catalysis, may be important for the following reasons. First, the proteolytic site present in certain antibodies cleaves Arg-X and Lys- X bonds preferentially over other types of peptide bonds.[14,15] Second, the positively charged side chain may occupy a different spatial position in the transition state than in the ground state, because the partial double-bond character of the scissile peptide bond is lost upon formation of the transition state, permitting rotation around this bond, and consequent changes in the positions of remote groups. Thus, Arg/Lys binding to the antibody catalytic site may be stronger in the transition state compared to the ground state, permitting a reduction of the the energy barrier that must be transcended to reach the transition state (activation energy).

An objection to the notion of antigen-specific proteolysis by antibodies is that strong binding to the stable form of the antigen may exert an anti-catalytic effect, because activation energy will increase if stabilization of the ground state structure occurs without an

Factors Stabilizing theTransition State

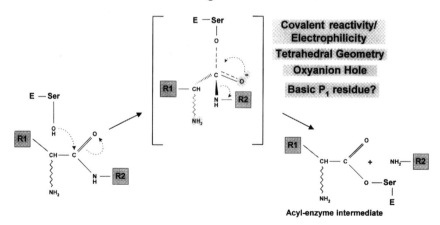

FIGURE 1. The putative transition state of acyl-enzyme formation during peptide bond cleavage by serine proteases (bracketed structure). In addition to recognizing the tetrahedral, electrophilic carbon atom formed in the transition state at the scissile peptide bond, serine protease-like catalytic antibodies may catalyze peptide bond cleavage by binding the basic residue on the N-terminal side of the scissile peptide bond as well as additional flanking residues (R1, R2).

equivalent or greater stabilization of the transition state.[16] This objection can be resolved by positing the development of physically distinct subsites in proteolytic antibodies mediating the high affinity binding of the reactant and the subsequent catalytic step (FIG 2). Evidence for this proposal has been gathered by studies on the interaction of various fragments of VIP with autoantibodies,[17] and more recently, by study of a light chain raised by immunization with VIP,[18] catalytically deficient mutants of which retain their reactant binding properties (FIG. 3). Menger[19] has recently shown on purely theoretical grounds that initial high affinity binding of the reactants can be pro-catalytic under certain conditions, i.e., the activation energy can be diminished by strong reactant binding when the catalyst contains distinct reactant and transition state binding subsites.

Another counterintuitive observation was the evident cleavage of polypeptide substrates at multiple peptide bonds by a recombinant light chain (FIG. 4)[15] and by autoantibodies to VIP[20] and thyroglobulin.[21] The focus of designer catalytic antibody research had been to target individual bonds for cleavage, and it was anticipated that a proteolytic antibody with specificity for a particular antigen ought to cleave a single peptide bond in that antigen. This expectation appears to be invalid in the "split-site" model of antibody catalysis. According to this model, in the ground state antigen-antibody complex there is no serious constraint on the conformational freedom of the catalytic subsite. Thus, the catalytic subsite can be positioned in register with different peptide bonds in alternate conformational states of the antigen-antibody complex, which can then proceed to form alternate transition state complexes, resulting in cleavage at multiple peptide bonds in the antigen.

The foregoing demonstrates the weaknesses of Occam's razor. The razor, if taken too seriously, can lead to non-constructive controversy and to delays in examining novel ideas that depart from dogma. It imposes an unduly harsh burden on novel hypotheses, and, often, the hypotheses are brought to the attention of the scientific community only when their proponents

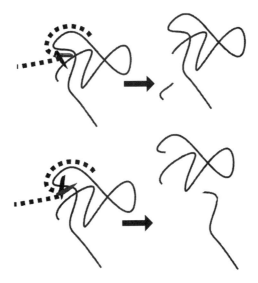

FIGURE 2. Conformational epitope cleavage—the role of GS and TS binding subsites. Alternate cleavage sites within an antigen epitope positioned in register with the catalytic subsite (scissor blades), which binds weakly in the ground state and strongly in the transition state (TS). The catalytic subsite is physically distinct from the antigen ground state (GS) binding site. Binding interactions at the flanks of the cleavage site are strong. The alternate cleavage sites are spatially close but distant in the linear sequence, resulting in product fragments with very different mass.

Undiminished VIP binding by catalytically deficient L chain mutants

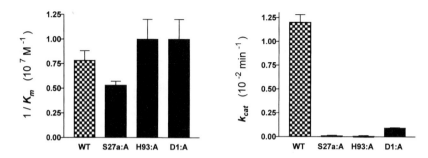

FIGURE 3. Maintenance of VIP binding activity by catalytically deficient mutants of the antibody light chain (clone c23.5). Substitution of light chain residues S27a, His 93, and Aspl by Ala resulted in reduced catalytic rate constants (kcat, *right*) but no sign)ficant change of $1/K_m$ values ($1/K_m$ approximates the equilibrium affinity constant). WT, wildtype. (Data are from Gao *et al.*[18])

are willing to take on the establishment with a conviction bordering on the fanatic. To reject complicated ideas out of hand because of an overwhelming desire for simplicity is clearly counter-progressive, particularly in biology, a discipline in which much remains to be learned. Note that if the new hypotheses are sufficiently important, their merit or specious nature will invariably emerge with time, because other investigators will subject the hypotheses to tests of experimental reproducibility and theoretical consistency.

INNATE VERSUS SOMATIC DEVELOPMENT OF THE PROTEOLYTIC ACTIVITY

The rate-accelerating capability of the enzymes is believed to have developed over the course of biological evolution because mutations and domain shuffling events imparted to the proteins an increasing capability to bind the transition state. Among the proteins that mediate life processes in higher organisms, antibodies are unique in that their structure undergoes diversification by somatic and adaptive mechanisms. Unlike other proteins, therefore, antibodies can potentially evolve new biological functions in short order in a matter of weeks, following stimulation of the immune system with the antigen. It was assumed, therefore, that adaptive sequence diversification mechanisms must be invoked to explain antibody catalysis. Like several other expectations in this field, however, experimental analysis has forced reexamination of this expectation.

The antigen combining site of antibodies is formed from the variable (V) regions of the light and heavy chain subunits. In the human, about fifty VL and VH germline genes are passed on to the offspring via sperms and eggs. The first indication that antibody proteolytic activity might be a heritable function encoded by the germline V genes emerged from observations that antibodies isolated from the sera of unimmunized humans and ani-

$$\text{HSDAVFTDNYTRLR} \uparrow \text{K} \uparrow \text{QMAVK} \uparrow \text{K} \uparrow \text{YLNSILN-NH}_2$$
1   28

FIGURE 4. Peptide bonds in VIP cleaved by antibody light chain clone c23.5. Large arrows denote the major cleavage sites and small arrows, the minor cleavage sites. (Data are from Sun *et al.*[15])

mals express a comparatively nonspecific proteolytic activity (designated polyreactive activity).[14] Because of the apparent inability of polyreactive proteolytic antibodies to discriminate between different polypeptide antigens, it was suggested that the antibodies had not been subjected to somatic sequence diversification, and the activity was probably encoded by a germline V gene(s). Antibody light chains isolated from multiple myeloma patients expressed a similar polyreactive proteolytic activity.[8–10] Previously, Erhan and Greller predicted immunoglobulins to possess peptidase activity based on sequence homology between the CDR1 of the light chains, including certain germline-encoded residues, and the region surrounding the active site Ser residue of serine proteases.[22] As revealed by molecular modeling and mutagenesis studies, the catalytic site of a VIP cleaving light chain is composed of a serine, a histidine, and an aspartate residue, resembling the constitution of the catalytic site of conventional serine proteases like trypsin and subtilisin.[18] The three catalytic residues (Ser27a-His93-Asp I) are also found in the VL germline gene from which the VIP cleaving light chain is derived (GenBank number Z72384).

The apparent germline origin of the proteolytic activity suggests it to be a component of heritable, innate immune responsiveness. Thus, the origins of the catalytic activity must be sought in phylogenetic evolutionary mechanisms, as opposed to the somatic sequence diversification mechanisms that underlie adaptive immunity. In this respect, examples of antibody proteases cited here are no different from conventional enzymes, which also arose over millions of years of evolution. The biological functions of the proteolytic antibodies remain to be defined. If it is accepted that only useful biological functions tend to be embedded in the heritable genes, it may be hypothesized that the activity might possess an important function(s). Even the esterase activity of a designer antibody isolated by immunization with the transition state analog has been traced to be due, in part, to the expression of catalytic activity by its germline-encoded counterpart.[23] Regardless of our original expectations, therefore, it appears that various types of catalytic activities have developed in phylogenetic evolution (as opposed to somatic evolution) of antibodies.

The germline origin of the antibody proteases described here does not exclude the possibility of their improvement by somatic means. An available mechanism for this purpose is the combinatorial variability arising from pairing of the different light and heavy chain subunits. FIGURE 5 compares the catalytic activities of single chain Fv constructs composed of a VIP- cleaving VL domain linked to its natural VH domain partner or an irrelevant VH

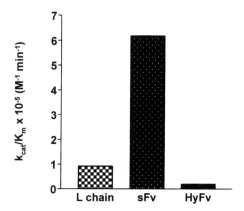

FIGURE 5. Improved cleavage of VIP by anti-VIP single chain Fv (sFv) compared to the light chain and hybrid Fv. The VL and VH domains in the anti-VIP Fv and the light chain are from clone c23.5. The hybrid Fv (HyFv) is composed of the anti-VIP VL domain linked to an anti-lysozyme VH domain. (Data are from Sun *et al.*[15])

domain from an anti-lysozyme antibody. VIP was bound by the anti-VIP Fv with an equilibrium dissociation constant (K_d) of 5 nM, a value close to that of the parent antibody.[15] The Michaelis-Menten constant for VIP cleavage (K_m) by the anti-VIP Fv was decreased substantially compared to the light chain, reflecting the beneficial contribution of the VH domain in VIP binding (the K_m value of a catalyst approximates its K_d value). There was also a significant improvement in the catalytic proficiency of the Fv relative to the L chain. On the other hand, the "hybrid" Fv composed of the anti-VIP VL domain linked to the irrelevant VH domain displayed considerably reduced VIP cleavage compared to the anti-VIP Fv and L chain constructs.

Somatic mutations may also improve the catalytic activity encoded by a germline VL gene. Certain VIP-cleaving L chains with high levels of catalytic efficiency have been observed to be highly mutated in comparison to their germline gene counterparts.[24] Further, polyclonal proteolytic antibodies isolated from patients with autoimmune disease display high affinities for their autoantigens,[1,4-6] which is a classical sign that the antibodies have been subjected to somatic mutations and clonal selection.

Implied in the foregoing discussion is the premise of a selection pressure(s) that might confer some type of survival advantage to the B cells responsible for synthesis of proteolytic antibodies. The existence of such a selection pressure(s) might also argue that *de novo* somatic generation of catalytic sites is possible. Note, however, that creating the catalytic site *de novo* from a noncatalytic site is a considerably more difficult task than improving a pre-existing catalytic site, because it may require sequence diversification rates beyond those offered by the antibody gene system.

AUTOIMMUNE DISEASE AND CATALYTIC ANTIBODIES

The association of catalytic antibodies with autoimmune disease has been confirmed in several studies,[1-6] but the reasons for the association are not known. Perhaps patients with autoimmune disease are genetically predisposed towards catalyst synthesis because of selective expression of particular germline V genes or increased formation of catalytic sites during somatic sequence diversification of antibody V domains. The hypothesis that autoimmune disease is associated with biased usage of different V genes is well-established in the literature.[25] Other genes relevant to antibody synthesis may also contribute to catalytic activity levels in autoimmune disease. An example of the power of autoantibody catalysis is provided by observations that mice with a genetic predisposition to autoimmune disease make catalytic antibodies more frequently upon immunization with the transition state analog of an esterase reaction compared to control non-autoimmune mice.[26] In this mouse strain, a mutation of the *Fas* apoptosis gene is believed to permit proliferation of T and B cells and expression of lupus- like disease.[27]

ACKNOWLEDGMENTS

The author is grateful for the support of the U.S. Public Health Service, the University of Nebraska and the University of Texas, and for generous collaborations extended by coworkers.

REFERENCES

1. PAUL, S., D. J. VOLLE, C. M. BEACH, D. R. JOHNSON, M. J. POWELL & R. J. MASSEY. 1989. Catalytic hydrolysis of vasoactive intestinal peptide by human autoantibody. Science **244**: 1158–1162.

2. PAUL, S., S. I. SAID, A. B. THOMPSON, D. J. VOLLE, D. K. AGRAWAL, H. FODA & S. DE LA ROCHA. 1989. Characterization of autoantibodies to vasoactive intestinal peptide in asthma. J Neuroimmunol. **23**: 133–142.

3. SUZUKI, H., H. IMANISHI, T. NAKAI & Y. K. KONISHI. 1992. Human autoantibodies that catalyze the hydrolysis of vasoactive intestinal polypeptide. Biochemistry (Life Sci. Adv.) **11**: 173–177.

4. LI, L., S. KAVERI, S. TYUTYULKOVA, M. KAZATCHKINE & S. PAUL. 1995. Catalytic activity of antithyroglobulin antibodies. J. Immunol. **154**: 3328–3332.

5. SHUSTER, A. M., G. V. GOLOLOBOV, O. A. KVASHUK, A. E. BOGOMOLOVA, I. V. SMIRNOV & A. G. GABIBOV. 1992. DNA hydrolyzing autoantibodies. Science **256**: 665–667.

6. GOLOLOBOV, G. V., E. A. CHERNOVA, D. V. SCHOUROV, I. V. SMIRNOV, I. A. KUDELINA & A. G. GABIBOV. 1995. Cleavage of supercoiled plasmid DNA by autoantibody Fab fragment: application of the flow linear dichroism technique. Proc. Natl. Acad. Sci. USA **92**: 254–257.

7. KALAGA, R., H. HUANG, F. J. STEVENS, A. SOLOMON & S. PAUL. 1995. gp120 hydrolysis by catalytic antibody light chain. Presented at the Ninth International Congress of Immunology. San Francisco, CA, July 23–29, 1995.

8. MATSUURA, K. & H. SINOHARA. 1996. Catalytic cleavage of vasopressin by human Bence Jones proteins at the arginylglycinamide bond. Biol. Chem. **377**: 587–589.

9. MATSUURA, K., K. YAMAMOTO & H. SINOHARA. 1994. Amidase activity of human Bence Jones proteins. Biochem. Biophys. Res. Commun. **204**: 57–62.

10. PAUL, S., L. LI, R. KALAGA, P. WILKINS-STEVENS, F. J. STEVENS & A. SOLOMON. 1995. Natural catalytic antibodies: Peptide hydrolyzing activities of Bence Jones proteins and V_L fragment. J Biol. Chem. **270**: 15257–15261.

11. CHADWICK, D. J. & J. MARSH, Eds. 1991. Catalytic Antibodies. Ciba Foundation Symposium. Vol.159. John Wiley & Sons. Chichester, England.

12. LEANER, R. A., S. J. BENKOVIC & P. G. SCHULTZ. 1991. At the crossroads of chemistry and immunology: Catalytic antibodies. Science **252**: 659–667.

13. PAULING, L. 1948. Chemical achievement and hope for the future. Am. Sci. **36**: 51.

14. KALAGA, R., L. LI, J. O'DELL & S. PAUL. 1995. Unexpected presence of polyreactive catalytic antibodies in IgG from unimmunized donors and decreased levels in rheumatoid arthritis. J. Immunol.**155**: 2695–2702.

15. SUN, M., Q.-S. GAO, L. KIRNARSKIY, A. REES & S. PAUL. 1997. Cleavage specificity of a proteolytic antibody light chain and effects of the heavy chain variable domain. J. Mol. Biol. **271**: 374–385.

16. MURPHY, D. J. 1995. Revisiting ground-state and transition-state effects, the split-site model, and the "fundamentalist position" of enzyme catalysis. Biochemistry **34**: 4507–4510.

17. PAUL, S., D. J. VOLLE, M. J. POWELL & R. J. MASSEY. 1990. Site specificity of a catalytic vasoactive intestinal peptide antibody: An inhibitory VIP subsequence distant from the scissile peptide bond. J. Biol. Chem. **265**: 11910–11913.

18. GAO, Q.-S., M. SUN, A. REES & S. PAUL. 1995. Site-directed mutagenesis of proteolytic antibody light chain. J. Mol. Biol. **253**: 658–664.

19. MENGER, F. M. 1992. Analysis of ground-state and transition-state effects in enzyme catalysis. Biochemistry **31**: 5368–5373

20. PAUL, S., M. SUN, R. MODY, S. H. EKLUND, C. M. BEACH, R. J. MASSEY & F. HAMEL. 1991. Cleavage of vasoactive intestinal peptide at multiple sites by autoantibodies. J. Biol. Chem. **256**: 16128–16134.

21. PAUL, S., L. LI, R. KALAGA, R. E. O'DELL, R. E. DANNENBRING, JR., S. SWINDELLS, S. HINRICHS, P. CATUREGLI & N. ROSE. 1997. Characterization of thyroglobulin-directed and polyreactive catalytic antibodies in autoimmune disease. J. Immunol. **159**: 1530–1536.

22. ERHAN, S. & L. D. GRELLER. 1974. Do immunoglobulins have proteolytic activity? Nature **251**: 353–355.

23. PATTEN, P. A., N. S. GRAY, P. L. YANG, C. B. MARKS, G. J. WEDEMAYER, J. J. BONIFACE, R. C. STEVENS & P.G. SCHULTZ. 1996. The immunological evolution of catalysis. Science **271**: 1086–1091.

24. TYUTYULKOVA, S., Q.-S. GAO, A. THOMPSON, A. RENNARD & S. PAUL. 1996. Efficient vasoactive intestinal polypeptide hydrolyzing antibody light chains selected from an asthma patient by phage display. Biochim. Biophys. Acta **1316**: 217–223.

25. SCHWARTZ, R.S. 1993. Autoimmunity and Autoimmune Diseases. Fundamental Immunology, Third Edition. W.E. Paul, Ed.: 1033–1097. Raven Press. New York, NY.

26. TAWFIK, D., R. CHAP, B. GREEN, M. SELA & Z. ESHHAR. 1995. Unexpectedly high occurrence of catalytic antibodies in MRL/lpr and SJL mice immunized with a transition state analog. Is there a linkage to autoimmunity? Proc. Natl. Acad. Sci. USA **92**: 2145–2149.

27. WATANABE-FUKUNAGA, R., C. I. BRANNAN, N. G. COPELAND, N. A. JENKINS & S. NAGATA. 1992. Lymphoproliferation disorder in mice explained by defects in Fas antigen that mediates apoptosis. Nature **356**: 314–317.

Analogues of VIP, Helodermin, and PACAP Discriminate between Rat and Human VIP₁ and VIP₂ Receptors[a]

P. GOURLET, A. VANDERMEERS, J. VAN RAMPELBERGH, P. DE NEEF, J. CNUDDE, M. WAELBROECK, AND P. ROBBERECHT[b]

Department of Biochemistry and Nutrition, School of Medicine, Université Libre de Bruxelles, Bât. G/E, CP 611, 808 Route de Lennik, B-1070 Brussels, Belgium

ABSTRACT: Vasoactive intestinal polypeptide (VIP) acts through interaction with two subclasses of seven transmembrane G protein–coupled receptors named VIP₁ and VIP₂ receptors. These receptors have been cloned in different species, such as rat and human. Considering the different distribution of both receptor subclasses, there is considerable interest in the development of selective agonists and antagonists. The present study compares the binding properties of VIP, PACAP, GRF, secretin, and helodermin analogues on recombinant rat and human VIP₁ and VIP₂ receptors. On both rat and human receptors, secretin and GRF had a higher affinity for the VIP₁ receptor subtypes. The amino-shortened VIP, and the carboxy terminal–shortened VIP and PACAP analogues also presented a higher affinity for the VIP₁ receptor. PHI, PHV, helodermin, and helospectin were selective for the human VIP₂ receptor subtypes. These results suggest that the helical structure of the carboxy terminal end is necessary for VIP₂ recognition. The differences between species were the following: PHI, PHV, helodermin, and helospectin had a higher affinity for the rat VIP₁ receptor than for the human VIP₁ receptor. On both rat and human receptors, D-Ala⁴ VIP and D-Phe⁴ VIP had a high affinity for the VIP₁ receptor and a low affinity for the VIP₂ receptor. Thus, three domains of the ligand involved in VIP₁/VIP₂ receptor discrimination were identified: the amino acid residue in position 4 ([D-Ala⁴], [D-Phe⁴]VIP), in positions 8 and 9 (the effects of helodermin and helospectin), and the carboxy terminal end (the effects of the shortened VIP and pituitary adenylate cyclase activating polypeptide analogues).

V asoactive intestinal polypeptide (VIP) acts through interaction with two subclasses of seven transmembrane G protein–coupled receptors that have been cloned: the VIP₁ receptor was cloned in rat[1] and human[2]; the VIP₂ receptor was cloned in rat[3], mouse[4], and human[5]. These receptors belong to a superfamily that also includes secretin[6], PACAP[7], glucagon[8], the glucagon-like peptide I[9], the GIP[10], the GRF[11], CRF[12], PTH[13], and calcitonin[14] receptors.

Before cloning, the binding properties of the rat and human VIP receptors were studied, for instance, on epithelial intestinal cells[15,16] that expressed mainly, if not exclusively, the VIP₁ receptor. The human VIP₂ receptor was studied on SUP-T1 lymphoblasts[17,18] that

[a]This work supported by grants Nos. 3.4513.95 from the F.R.S.M., by an "Action de Recherche Concertée" from the Communauté Française de Belgique and by a "Interuniversity Poles of Attraction Programme—Belgian State, Prime Minister's Office—Federal Office for Scientific, Technical and Cultural Affairs."

[b]Corresponding author: Tel.: 32.2.555.62.15; Fax: 32.2.555.62.30; E-mail : probbe@ulb.ac.be

express only that receptor subtype, but the rat VIP_2 receptor was not identified at that time. The properties of the recombinant rat and human VIP_1 and of human VIP_2 receptors were identical to those of the natural receptors.[5,19,20]

The distribution of the VIP_1 and VIP_2 receptor mRNA was performed in rat tissues by *in situ* hybridization.[21] VIP_1 and VIP_2 receptor mRNA were differently expressed: for instance, in the rat central nervous system, the VIP_1 receptor was present in cortex layers III and IV and the hippocampus, whereas the VIP_2 receptor was present in cortex layer VI, in the thalamus, in the nucleus accumbens, and in the hypothalamus.

We recently demonstrated using selective ligands an excellent correlation between mRNA and protein distribution.[22] Considering the different distribution of both receptor subclasses, there is considerable interest in the development of selective agonists and antagonists. The present study compares the binding properties of VIP, PACAP, GRF, secretin, and helodermin analogues on recombinant rat and human VIP_1 and VIP_2 receptors and gives indications of the amino acid sequences required for recognition of each receptor subclass.

MATERIAL AND METHODS

Cell Lines Used for Receptor Characterization

The DNA coding for rat VIP_1, human VIP_2, and human VIP_1 receptors were cloned into a mammalian expression vector containing the selectable neomycin phosphotransferase gene. The resulting recombinant plasmids were transfected into the CHO cell line DG44 by electroporation using a gene pulser.

The selection of the clones was performed on the basis of their resistance to geneticin and their capability to bind iodinated VIP. The CHO cells expressing the rat VIP_2 receptor were kindly provided by Dr. E.M. Lutz from the MRC Brain Metabolism Unit (Edinburgh, U.K.) Cells were maintained in α-minimal essential medium (α-MEM), supplemented with 10% fetal calf serum, 2 mM L-glutamine, 100 µg/ml penicillin, and 100 µg/ml streptomycin with an atmosphere of 95% air, 5% CO_2 at 37°C. Geneticin (0.4 mg/ml) was maintained in the culture medium of the stock culture. Subcultures prepared for membrane purification were done in a medium without geneticin.

Membrane Preparation, Receptor Identification, and Adenylate Cyclase Determination

Cells were harvested with a rubber policeman and pelleted by low speed centrifugation. The supernatant was discarded and the cells lysed in 1 mM $NaHCO_3$ and immediately frozen in liquid nitrogen. After thawing, the lysate was first centrifuged at 4°C for 10 min at $400 \times g$ and the supernatant was further centrifuged at $20,000 \times g$ for 10 min. The pellet resuspended in 1 mM $NaHCO_3$ was used immediately as a crude membrane preparation. [^{125}I]VIP (specific radioactivity of 0.7 mCi/mmol) was used as a tracer for the identification of the VIP_1 receptors and [^{125}I]RO 25-1553 (specific radioactivity of 0.8 mCi/mmol) was used as a tracer for VIP_2 receptors.[23] The binding was performed as described[19]; in all cases the non-specific binding was defined as the residual binding in the presence of 1 µM of VIP. Competition curves were performed by incubating membranes and tracer in the presence of increasing concentrations of unlabeled peptides. Peptide potency was assessed by the IC_{50} values, the peptide concentration required for half-maximal inhibition of tracer binding.

Peptide Synthesis

All peptides were synthesized as C-terminal amides by solid-phase methodology using an automated Applied Biosystems apparatus using the Fmoc ((9-fluorenylmethoxy carbonyl) strategy as described.[24] The peptides were cleaved and purified by reverse-phase chromatography in Mono S HR 5/5. The peptide purity was assessed (95%) by capillary electrophoresis and the sequence conformity was verified by direct sequencing and ion spray mass spectrometry.

Helospectin was prepared from heloderma venom and corresponded to the sequence presented in TABLE 1.

RESULTS

The primary sequences of the unmodified peptides tested were given in TABLE 1. The IC$_{50}$ values of binding on the rat and human VIP$_1$ and VIP$_2$ receptors were given in TABLE 2.

If we consider that a ratio of at least 30-fold in the IC$_{50}$ values of one peptide on two receptor subclasses defines a selective molecule, the following considerations emerged. (1) On rat VIP receptors, secretin; GRF; the carboxy terminal–shortened analogues VIP(1-25), PACAP (1-25), (1-24), (1-23); and the amino terminal–shortened analogues VIP(10-28), [D-Ala^4]VIP, [D-Phe^4]VIP, and [D-Phe^4]PHI were selective. Each had a preference for the VIP$_1$ receptor. (2) On human VIP receptors, GRF; VIP(1-25); PACAP(1-25), (1-24), (1-23), and VIP(10-28) were selective with a higher affinity for the VIP$_1$ receptor subtype. PHI, PHV, helodermin, and helospectin had a higher affinity for the VIP$_2$ receptor.

Based on the capacity of the ligands to stimulate or to inhibit the VIP-stimulated adenylate cyclase activity, VIP(6-28) and VIP(10-28) were found to be antagonists on the VIP receptors (data not shown). All other molecules were full or partial agonists.

DISCUSSION

Species Differences among the VIP$_1$ Receptors

It has been known since Laburthe's group papers[15] on epithelial digestive cell receptors that rat VIP$_1$ receptor recognized both secretin and PHI with a higher affinity than the human VIP$_1$ receptor. These observations were confirmed on recombinant receptors expressed in COS cells.[25] This occurs despite 80% similarity in amino acid sequences. The molecular basis for the selective recognition of PHI by the rat VIP$_1$ receptor was attributed to three non-adjacent amino acids of the receptor located in the first exoloop between transmembrane helix 2 and 3.[25] Which ligand amino acids are involved in this recognition has not been established and thus it is not possible to generalize the observation to the other peptides able to differentially interact with the rat and human receptors.

In the present work we confirmed the data on secretin and PHI and extended them to helodermin and helospectin, two VIP/PHI-related peptides extracted from lizard venom.[26] Helodermin preference for the rat receptor (125-fold) was higher than that of helospectin (25-fold). The main difference between helodermin and helospectin is the presence in

TABLE 1. Amino acid sequences of VIP and related peptides

VIP	H S D A V F T D N Y T R L R K Q M A V K K Y L N S I L N *
PHI	H A D G V F T S D Y S R L L G Q L S A K K Y L E S L I *
PHV	H A D G V F T S D Y S R L L G Q I S A K K Y L E S L I G K R I S S S I S E D P V P V
Secretin (Sn)	H S D G T F T S E L S R L R D S A R L Q R L L Q G L V *
GRF	Y A D A I F T N S Y R K V L G Q L S A R K L L Q D I L S R *
Helodermin (Hd)	H S D A I F T E E Y S K L L A K L A L Q K Y L A S I L G S R T S P P P *
Helospectin (Hsp)	H S D A T F T A E Y S K L L A K L A L Q K Y L E S I L G S S T S P R P P S S

* Indicates the presence of a carboxy terminal amide

TABLE 2. IC_{50} values of binding (nM)

	VIP_1 Rat	VIP_2 Rat	VIP_1 Human	VIP_2 Human
VIP	1 ± 1	4 ± 2	2 ± 1	3 ± 2
PHI	3 ± 1	30 ± 5	1,000 ± 100	30 ± 5
PHV	3 ± 1	8 ± 6	3,000 ± 200	10 ± 3
PACAP-27	1 ± 1	10 ± 2	3 ± 1	10 ± 3
Secretin	300 ± 10	30,000 ± 5,000	1,500 ± 300	5,000 ± 2,000
GRF (1-29)	80 ± 10	30,000 ± 5,000	100 ± 30	30,000 ± 5,000
Helodermin	8 ± 3	3 ± 2	1,000 ± 300	3 ± 1
Helospectin	20 ± 5	20 ± 5	500 ± 200	5 ± 3
VIP 1-27	5 ± 2	20 ± 5	2 ± 2	8 ± 3
VIP 1-26	50 ± 10	40 ± 15	100 ± 10	300 ± 100
VIP 1-25	100 ± 10	30,000 ± 1,000	200 ± 10	30,000 ± 1,000
PACAP 1-26	20 ± 5	3,000 ± 500	150 ± 10	3,000 ± 1,000
PACAP 1-25	50 ± 5	50,000 ± 10,000	150 ± 10	10,000 ± 1,000
PACAP 1-24	8 ± 3	800 ± 300	30 ± 20	800 ± 200
PACAP 1-23	10 ± 3	2,000 ± 200	80 ± 20	2,000 ± 1,000
VIP 2-28	100 ± 15	800 ± 100	150 ± 15	300 ± 20
VIP 3-28	800 ± 20	1,500 ± 100	300 ± 30	300 ± 20
VIP 4-28	300 ± 20	1,500 ± 100	500 ± 20	1,500 ± 20
VIP 5-28	300 ± 30	2,000 ± 100	300 ± 20	200 ± 30
VIP 6-28	300 ± 30	3,000 ± 200	300 ± 20	3,000 ± 1,000
VIP 10-28	1,500 ± 100	> 30,000	1,000 ± 300	> 30,000
D-Ala⁴ VIP	1 ± 1	300 ± 100	5 ± 2	60 ± 20
D-Phe⁴ VIP	1 ± 1	1,500 ± 100	20 ± 4	250 ± 100
D-Phe⁴ PHI	20 ± 3	10,000 ± 1,000	4,000 ± 1,000	1,000 ± 500

Values were given ± SEM and were the means of at least three determinations.

helodermin of two consecutive acidic residues in positions 8 and 9. The different behavior of helodermin and helospectin in rat and human is of particular pharmacological interest as lizard peptides are long-acting VIP analogues.[27]

Both carboxy- and amino-terminal–shortened VIP and PACAP show a similar affinity for the rat and the human VIP_1 receptors. The (10-28)VIP fragment, frequently used as a VIP receptor antagonist,[28] had the same affinity for rat and human receptors.

Species Differences among the VIP_2 Receptors

The ratio between the IC_{50} values for rat and human VIP_2 receptors occupancy never exceeded 10, suggesting quite similar pharmacological profiles. This is consistent with the high degree (86%) of similarities between both primary sequences of the receptors.[3,5] It must also be noticed that the mouse receptor also had 88 and 85% identities with the rat and human receptor, respectively. The selective and high affinity VIP_2 receptor agonist RO 25-1553 also had the same affinity for both rat and human VIP_2 receptor.[23]

Sequence of the Ligand Necessary for Discrimination of VIP_1 and VIP_2 Receptors

On both rat and human recombinant receptors, secretin and GRF have a lower affinity for the VIP_2 receptors than for the VIP_1 receptors, confirming previous findings on the SUP-T1 lymphoblastic cell line expressing the VIP_2 receptor.[18]

VIP and PACAP derivatives shortened at the carboxy terminus both have a clear preference for the VIP_1 receptor subtype. This suggests that the helical structure of the carboxy terminal end[29] is necessary for VIP₂ recognition. Along that line, it must be noticed that the selective VIP₂ receptor agonist RO 25-1553 is elongated at the carboxy terminal with a helical structure stabilized by a lactam ring between amino acid residues 21 and 25.

REFERENCES

1. ISHIHARA, T., R. SHIGEMOTO, K. MORI, K. TAKAHASHI & S. NAGATA. 1992. Functional expression and tissue distribution of a novel receptor for vasoactive intestinal polypeptide. Neuron **8**: 811–819.

2. COUVINEAU, A., C. ROUYER-FESSARD, D. DARMOUL, J.-J. MAORET, I. CARRERO, E. OGIER-DENIS & M. LABURTHE. 1994. Human intestinal VIP receptor: Cloning and functional expression of two cDNA encoding proteins with different N-terminal domains. Biochem. Biophys. Res. Commun. **200**: 769–776.

3. LUTZ, E. M., W. J. SHEWARD, K. M. WEST, J. A. MORROW, G. FINK & A. J. HARMAR. 1993. The VIP₂ receptor: molecular characterisation of a cDNA encoding a novel receptor for vasoactive intestinal peptide. FEBS Lett. **334**: 3–8.

4. INAGAKI, N., H. YOSHIDA, M. MIZUTA, N. MIZUNO, Y. FUJII, T. GONOI, J.-I. MIYAZAKI & S. SEINO. 1994. Cloning and functional characterization of a third pituitary adenylate cyclase-activating polypeptide receptor subtype expressed in insulin-secreting cells. Proc. Natl. Acad. Sci. USA **91**: 2679–2683.

5. SVOBODA, M., M. TASTENOY, J. VAN RAMPELBERGH, J.-F. GOOSSENS, P. DE NEEF, M. WAELBROECK & P. ROBBERECHT. 1994. Molecular cloning and functional characterization of a human VIP receptor from SUP-T1 lymphoblasts, Biochem. Biophys. Res. Commun. **205**: 1617–1624.

6. ISHIHARA, T., S. NAKAMURA, Y. KAZIRO, T. TAKAHASHI, K. TAKAHASHI & S. NAGATA. 1991. Molecular cloning and expression of a cDNA encoding the secretin receptor. EMBO J. **10**: 1635–1641.

7. PISEGNA, J. R. & S. A. WANK. 1993. Molecular cloning and functional expression of the pituitary adenylate cyclase-activating polypeptide type I receptor. Proc. Natl. Acad. Sci. USA **90**: 6345–6349.

8. SVOBODA, M., E. CICCARELLI, M. TASTENOY, P. ROBBERECHT & J. CHRISTOPHE. 1993. A cDNA construct allowing the expression of rat hepatic glucagon receptors. Biochem. Biophys. Res. Commun. **192**: 135–142.

9. THORENS, B. 1992. Expression cloning of the pancreatic β cell receptor for the gluco-incretin hormone glucagon-like peptide 1. Proc. Natl. Acad. Sci. USA **89**: 8641–8645.

10. USDIN, T. B., E. MEZEY, D. C. BUTTON, M. J. BROWNSTEIN & T. I. BONNER. 1993. Gastric inhibitory polypeptide receptor, a member of the secretin-vasoactive intestinal peptide receptor family, is widely distributed in peripheral organs and the brain. Endocrinology **133**: 2861–2870.

11. MAYO, K. E. 1992. Molecular cloning and expression for pituitary-specific receptor for growth hormone-releasing hormone. Mol. Endocrinol. **6**: 1737–1744.

12. CHALMERS, D. T., T. W. LOVENBERG, D. E. GRIGORIADIS, D. P. BEHAN & E. B. DE SOUZA. 1996. Corticotrophin-releasing factor receptors: from molecular biology to drug design. Trends Pharmacol. Sci. **17**: 166–172.

13. JÜPPNER, H., A.-B. ABOU-SAMRA, M. FREEMAN, X. F. KONG, E. SCHIPANI, J. RICHARDS, L. F. KOLAKOWSKI JR., J. HOCK, J. T. POTTS JR., H. M. KRONENBERG & G. V. SEGRE. 1991. A G protein-linked receptor for parathyroid hormone and parathyroid hormone-related peptide. Science **254**: 1024–1026.

14. LIN, H. Y., T. L. HARRIS, M. S. FLANNERY, A. ARUFFO, E. H. KAJI, A. GORN, L. F. KOLAKOWSKI JR., H. F. LODISH & S. R. GOLDRING. 1991. Expression cloning of an adenylate cyclase-coupled calcitonin receptor. Science **254**: 1022–1024.

15. LABURTHE, M., A. COUVINEAU, C. ROUYER-FESSARD & L. MORODER. 1985. Interaction of PHM, PHI and 24-glutamine PHI with human VIP receptors from colonic epithelium: comparison with rat intestinal receptors. Life Sci. **36**: 991–995.

16. LABURTHE, M., A. COUVINEAU & C. ROUYER-FESSARD. 1986. Study of species specificity in growth hormone-releasing factor (GRF) interaction with vasoactive intestinal peptide (VIP) receptors using GRF and intestinal VIP receptors from rat and human: evidence that Ac-Tyr¹ hGRF is a competitive VIP antagonist in the rat. Mol. Pharmacol. **29**: 23–27.

17. GOURLET, P., P. ROBBERECHT & J. CHRISTOPHE. 1991. Molecular characterization of helodermin-
     preferring VIP receptors in SUP T[1] lymphoma cells: Evidence for receptor glycosylation. J.
     Receptor Res. 11: 831–848.
18. GOURLET, P., P. DE NEEF, M.-C. WOUSSEN-COLLE, A. VANDERMEERS, M.-C. VANDERMEERS-PIRET,
     P. ROBBERECHT & J. CHRISTOPHE. 1991. The activation of adenylate cyclase by pituitary adeny-
     late cyclase activating polypeptide (PACAP) via helodermin-preferring VIP receptors in
     human SUP-T1 lymphoblastic membranes. Biochim. Biophys. Acta 1066: 245–251.
19. CICCARELLI, E., J.-P. VILARDAGA, P. DE NEEF, E. DI PAOLO, M. WAELBROECK, A. BOLLEN & P.
     ROBBERECHT. 1994. Properties of the VIP-PACAP type II receptor stably expressed in CHO
     cells. Regul. Pept. 54: 397–407.
20. GAUDIN, P., A. COUVINEAU, J.-J. MAORET, C. ROUYER-FESSARD & A. LABURTHE. 1996. Stable
     expression of the recombinant human VIP[1] receptor in clonal Chinese hamster ovary cells:
     pharmacological, functional and molecular properties. Eur. J. Pharmacol. 302: 207–214.
21. USDIN, T. B., T. I. BONNER & E. MEZEY. 1994. Two receptors for vasoactive intestinal polypep-
     tide with similar specificity and complementary distributions. Endocrinology 135: 2662–2680.
22. VERTONGEN, P., S. N. SCHIFFMANN, P. GOURLET & P. ROBBERECHT. 1997. Autoradiographic visu-
     alization of the receptor subclasses for Vasoactive Intestinal Polypeptide (VIP) in rat brain.
     Peptides. 18: 1547–1554.
23. GOURLET, P., P. VERTONGEN, A. VANDERMEERS, M.-C. VANDERMEERS-PIRET, J. RATHE, P. DE NEEF
     & P. ROBBERECHT. 1997. The long-acting vasoactive intestinal polypeptide agonist RO 25-
     1553 is highly selective of the VIP[2] receptor subclass. Peptides 18: 403–408.
24. ROBBERECHT, P., P. GOURLET, P. DE NEEF, M.-C. WOUSSEN-COLLE, M.-C. VANDERMEERS-PIRET,
     A. VANDERMEERS & J. CHRISTOPHE. 1992. Receptor occupancy and adenylate cyclase activa-
     tion in AR 4-2J rat pancreatic acinar cell membranes by analogs of Pituitary Adenylate
     Cyclase-Activating Peptides amino-terminally shortened or modified at position 1,2,3,20, or
     21. Mol. Pharmacol. 42: 347–355.
25. COUVINEAU, A., C. ROUYER-FESSARD, J.-J. MAORET, P. GAUDIN, P. NICOLE & M. LABURTHE. 1996.
     Vasoactive intestinal peptide (VIP)[1] receptor. Three nonadjacent amino acids are responsible
     for species selectivity with respect to recognition of peptide histidine isoleucineamide. J. Biol.
     Chem. 271: 12795-12800.
26. VANDERMEERS, A., P. GOURLET, M.-C. VANDERMEERS-PIRET, A. CAUVIN, P. DE NEEF, J. RATHE, M.
     SVOBODA, P. ROBBERECHT & J. CHRISTOPHE. 1987. Chemical, immunological and biological
     properties of peptides like vasoactive- intestinal-peptide and peptide-histidine-isoleucinamide
     extracted from the venom of two lizards (Heloderma horridum and Heloderma suspectum).
     Eur. J. Biochem. 164: 321–327.
27. FODA, H. D. & S. I. SAID. 1989. Helodermin, a C-terminally extended VIP like peptide, evokes
     long-lasting relaxation. Biomed. Res. 10: 107–110.
28. TURNER, J. T., S. B. JONES & D. B. BYLUND. 1986. A fragment of vasoactive intestinal peptide,
     VIP(10-28), is an antagonist of VIP in the colon carcinoma cell line, HT 29. Peptides 7:
     849–854.
29. THERIAULT, Y., Y. BOULANGER & S. ST-PIERRE. 1991. Structural determination of the vasoactive
     intestinal peptide by two-dimensional H-NMR spectroscopy. Biopolymers 31: 459–464.

Maxadilan Is a Specific Agonist and Its Deleted Peptide (M65) Is a Specific Antagonist for PACAP Type 1 Receptor

D. UCHIDA, I. TATSUNO,[b] T. TANAKA, A. HIRAI, Y. SAITO,
O. MORO,[a] AND M. TAJIMA[a]

*Second Deptartment of Internal Medicine, Chiba University School of
Medicine, China 260, Japan*

[a]*Shiseido Research Center, Shiseido Company, Japan*

ABSTRACT: Maxadilan is a potent vasodilator peptide isolated from salivary glands
extracts of the hematophagous sand fly. Recently, it was demonstrated that maxadilan binds to PACAP receptor type 1 in mammals, although maxadilan has no significant amino acid sequence homology with PACAP. In the present study, we
demonstrated that maxadilan is a specific agonist of PACAP type 1 receptor
(PACAP/VIP receptor 1; PVR1) as determined by the binding assay of [^{125}I]PACAP27
and cAMP accumulation using CHO cells stably expressing PVR1, VIP1 receptor
(PVR2), and VIP2 receptor (PVR3), and that the deleted peptide (#25–41) of maxadilan (termed as M65) is a specific antagonist of PVR1. In addition, maxadilan shares
the binding sites for PACAP and stimulates cAMP in cultured rat cortical neurons.
VIP stimulates cAMP accumulation probably through the binding to PVR1 since
M65 blocks the VIP-induced cAMP accumulation in cultured rat cortical neurons.

Maxadilan was isolated from sand fly salivary gland extracts as a vasodilator peptide in 1991.[1] The sand fly is the vector of the protozoan disease leishmaniasis, which occurs in many parts of the world including Central and South America, the Middle East, and Indian subcontinent. Maxadilan is produced as a 63 amino acid peptide that undergoes C-terminal cleavage and amidation to a 61 amino acid peptide.[1,2] It contains four cysteine residues that participate in the formation of disulfide bonds between positions 1–5 and 14–51.[3] Although maxadilan possesses both biological activity and specific binding sites, it is not present in mammals. As flies are evolutionarily distant from vertebrates and maxadilan does not share sequence homology with other known peptides, it was postulated that this peptide would act either at an orphan receptor whose endogenous ligand was unknown or at a receptor whose ligand had a structure distinct from maxadilan.[1,2,4] The receptor for maxadilan in mammals has been extensively explored for the past several years. Finally, it was clarified that maxadilan binds to the pituitary adenylate cyclase acti-vating polypeptide (PACAP) receptor in mammals[5] even though maxadilan has no signif-icant amino acid sequence homology with PACAP.

Pituitary adenylate cyclase activating polypeptide with 38 residues (PACAP38) and a shorter form of the peptide corresponding to the N-terminal 27 residues (PACAP27) were isolated from the ovine hypothalamus based on the activity of adenylate cyclase in the pituitary.[6,7] PACAP shows a 68% amino acid sequence homology with vasoactive

[b]Corresponding author: Ichiro Tatsuno, M.D., Second Department of Internal Medicine, Chiba University School of Medicine, 1-8-1 Inohana, Chuou-ku, Chiba-city, Japan 260-8670. Tel.: 81-43-226-2094; Fax: 81-43-226-2095; E-mail: ichico@intmed02.m.chiba-u.ac.jp

intestinal peptide (VIP). However, it is 1,000 times more potent than VIP in stimulating adenylate cyclase in pituitary cells.[6,7] Although PACAP was initially postulated to be a hypophysiotropic hormone,[8] it has been clarified that it has many important roles including a neurotrophic effect in CNS as reported by us.[9,10] These functions of PACAP are thought to be exerted through the activation of specific receptors and three subtypes of PACAP/VIP family receptors has been cloned.[11] One subtype is PACAP type 1 receptor (PACAP/VIP family receptor 1; PVR1), which prefers PACAP over VIP and possesses variant forms. Two other subtypes are VIP 1 receptor (PVR2) and VIP 2 receptor (PVR3), both of them equally prefer PACAP and VIP. However, the physiological meaning of the presence of PACAP/VIP family receptors remains obscure.

In the present paper, we demonstrated that (1) maxadilan specifically binds to PACAP type 1 receptor PVR1 to stimulate cAMP accumulation using CHO cells stably expressing PACAP/VIP family receptors; (2) deleted peptide (#25–41) of maxadilan (termed M65) is a specific antagonist of PVR1; and (3) maxadilan shares the binding sites with PACAP and stimulates cAMP in rat cultured cortical neurons.

MATERIALS AND METHODS

Preparation of Peptides

Recombinant maxadilan and its deleted peptide M65, produced in *Escherichia coli,* contained four additional amino acid residues (glycine, serine, isoleucine, and leucine) at the N terminus as a result of construction in the pGEX vector designed for cleavage with thrombin.[4] They were purified to homogeneity using reverse phase high performance liquid chromatography.[4] PACAP 27, PACAP 38, PACAP6-38, and VIP were obtained from Peptide Institute (Osaka, Japan). Iodination of PACAP27 and VIP was performed using the lactoperoxidase method as previously described.

CHO Cells Stably Expressing PACAP/VIP Receptors

The expression plasmids that carry full-length cDNAs encoding rat PACAP type 1 (basic)[12] and rat VIP 1 receptor[13] were cotransfected with pSV2neo into CHO cells using the Lipofectin reagent (GIBCO/BRL). Stable transformants were selected in a minimal essential medium containing 10% (vol/vol) fetal calf serum (Hyclone) and G418 at 400 μg/ml. The CHO cells stably expressing mouse VIP 2 receptor were kindly provided from Dr. S. Seino (Chiba University School of Medicine).[14]

Cultured Rat Cortical Neurons

The forebrain was removed from 16- to 18-day-old fetal SD rats (Charles River Breeding Labs., Tokyo, Japan) as previously reported by us.[10] Dissociated cortical neurons were seeded onto culture plates or dishes (Costar Scientific Corporation, Cambridge, MA, USA) precoated with 10 μg/ml poly-L-lysine (Sigma). Neurons were cultured with chemically defined medium (D-MEM/F-12 containing 2.5 mM L-glutamine, supplemented by 5 μg/ml insulin, 100 μg/ml human transferrin, 3×10^{-4} M selenium, 1×10^{-4} M putrescine, 2×10^{-8} M progesterone, 1×10^{-12} M 17β-estradiol, 37.5 μg/ml bovine serum albumin (fatty acid free), 1×10^{-8} M triiothyronine, and 1×10^{-7} M corticosterone) in an atmosphere of 95% air and 5% CO_2 at 37 °C for 11 days before experiments.

Binding Assay

Binding assay was performed as described.[15,16] Briefly, cortical neurons and CHO cells were scraped mechanically from the culture dishes using a rubber policeman. The cells were resuspended in ice-cold 50 mM Tris buffer (pH 7.4) containing 5 mM $MgCl_2$, 0.5 mg/ml bacitracin, and 0.5 mM *p*-amidinophenyl methanesulfonyl fluoride hydrochloride (p-APMSF) with protease inhibitor cocktail (10 μg/ml pepstatin A, 10 μg/ml antipain, 10 μg/ml chymostatin, 10 μg/ml leupeptin) (membrane buffer). The cells were homogenized using a glass Teflon homogenizer and the homogenate was centrifuged at $250 \times g$ for 10 min to collect the supernatant. The membrane preparations were prepared by the ultracentrifugation of supernatant (50,000 $g \times 30$ min, twice). The binding assay was performed in a total volume of 300 μl of the membrane buffer containing 1% BSA (binding buffer) at 22°C. [^{125}I]PACAP27 or [^{125}I]VIP, membrane preparations, and various concentrations of peptides were incubated. The assay was terminated by rapid filtration through Whatman GF/B glass paper filters, which were presoaked in 0.5% polyethylenimine using a cell harvester (Brandel Biomedical Research & Development Laboratories, Gaithersburg, MD). Each filter was counted for radioactivity by an automatic gamma counter. The binding data were analyzed with a computer program for Scatchard plot analysis called "Ligand." The program was kindly provided by Dr. Peter J. Munson (National Institutes of Health, Bethesda, MD).

cAMP Accumulation

Cortical neurons and CHO cells were washed twice with DMEM containing 0.1% BSA [DMEM(-)] and 500 μl of various concentration of peptides in DMEM(-) containing 0.5 mM IBMX were added to each well. The cultures were then incubated at 37°C in an atmosphere of 95% air and 5% CO_2 for the indicated time. At the end of the incubation, the conditioned medium was removed and stored at –20°C for determination of extracellular cAMP. cAMP was measured by radioimmunoassay using a commercial kit provided from Yamasa Corporation (Cyoshi, Chiba, Japan).

RT-PCR

Total cellular RNA was isolated by the guanidinium isothiocyanate/CsCl procedure from the cultured rat cortical neurons. RT-PCR was performed using RNA PCR Kit (TaKaRa, Ohtsu, Japan). Three μg RNA was used for each RT-PCR. The template produced from the RT reaction was amplified using one of three sets of primers dependent upon the PACAP/VIP receptor as reported by Rawlings.[17] For PVR1, the two primers used were PAC1-FL (5'-TTTCATCGGCATCATCATCATCATCCTT-3') and PAC1-VK (5'-CCTTCCAGCTCCTCCATTTCCTCTT-3'), which would be expected to produce PCR product sizes of 280 base pairs (bp) for the basic receptor, 364 bp for a single cassette insert (hip, hop1, or hop2), and 448 bp for a double insert (hiphop1 or hiphop2).[23] For PVR2, the primers used were VIP1-AI (5'-GCCCCCATCCTCCTCTCCATC-3') and VIP1-EL (5'-TCCGCCTGCACCTCACCATTG-3'), which should give a PCR product of 299 bp. The PVR3 primers used were VIP2-AE (5'-ATGGATAGCAACTCGC-CTTTCTTTAG-3') and VIP2-QL (5'-GGAAGGAACCAACACATAACTCAAACAG-3'), yielding a predicted PCR product 325 bp in length.

RESULTS AND DISCUSSION

Maxadilan and M65 in PACAP/VIP Family Receptors

The binding of maxadilan and its antagonist M65 on PACAP receptors was analyzed using [^{125}I]PACAP27 and [^{125}I]VIP in the membrane of CHO cells stably expressing PACAP/VIP family receptors. Although maxadilan and M65 did not displace the binding of both [^{125}I]PACAP27 and [^{125}I]VIP in CHO cells stably expressing VIP 1 receptor (PVR2-CHO) and VIP 2 receptor (PVR3-CHO), maxadilan and M65 displaced the [^{125}I]PACAP27 binding in a dose-dependent fashion in CHO cells stably expressing PACAP type 1 (PVR1-CHO), indicating that maxadilan specifically binds to PACAP type 1 receptor in PACAP/VIP family receptors. The Scatchard analysis of the displacement of [^{125}I]PACAP27 binding indicated the existence of a single class of binding sites and the dissociation constants (K_d) were calculated to 0.018 ± 0.021 nM for PACAP38, 0.055 ± 0.037 nM for maxadilan, 0.349 ± 0.458 nM for PACAP27, and 0.574 ± 0.501 nM for M65, respectively. The maximal binding capacity (B_{max}) was 0.854 ± 0.613 pmol/mg protein.

Maxadilan stimulated cAMP accumulation with the minimum effective concentration of 0.01 nM in a dose-dependent manner in PVR1-CHO, but not PVR2-CHO and PVR3-CHO. However, PACAP stimulated it in all three types of CHO cells. Therefore, the data of cAMP accumulation were very consistent with the binding data indicating that maxadilan specifically binds to PVR1 and stimulates cAMP accumulation. Interestingly, M65 itself did not stimulate cAMP accumulation in all three types of PACAP/VIP receptors and M65 inhibited the PACAP, maxadilan, and VIP-induced cAMP accumulation in a dose-dependent manner, although M65 specifically displaced the binding of [^{125}I]PACAP27 in PVR1-CHO indicating that M65 is a specific antagonist of PACAP type 1 receptor.

Maxadilan has an apparent vasodilating activity although it is not present in mammals.[1,2] It has been being quite interesting to investigate what kind of receptor maxadilan binds to in mammals. The present data clearly demonstrated that maxadilan is a specific agonist of PACAP type 1 receptor (PVR1), and that M65 is its specific antagonist. Maxadilan is produced as a 63 amino acid peptide that undergoes C-terminal cleavage and amidation to a 61 amino acid peptide.[1,2] It contains four cysteine residues that participate in the formation of disulfide bonds between positions 1–5 and 14–51.[3] The amino acid sequences of maxadilan, M65, and PACAP are shown in FIGURE 1. PACAP shows a 68% amino acid sequence homology with VIP. In contrast, maxadilan has no significant primary homology with PACAP/VIP family. As shown in FIGURE 2, the Chou and Fasman method of secondary structure predicts two α-helical domains in the positions 10–22 and 47–61.[10] The present data of maxadilan and M65 may suggest that (1) the sequence between the two helices, in the center of which the shared phenylalanine/threonine pair with PACAP/VIP family peptides are located, has a critical portion to activate PVR1 and (2) the two α-helical domains coupled by disulfide linkage determine the binding to PVR1.

Maxadilan	CDATCQFRKAIDDCQKQAHHSNVLQTSVQTTAT**FT** SMDTSQLPG N S**V**FKECMKQK**K**K**E**F K**AGK
M65	CDATCQFRKAIDDCQKQAHHSNVL LPG N S**V**FKECMKQK**K**K**E** F**K**AGK
PACAP 38	HSDGI **FT** DS YSRYRKQMA**VK**K**YLAA VLG**KR**YK**QRVKNK
PACAP 27	HSDGI **FT** DS YSRYRKQMA**VK**K**YLAA VL
VIP	HSDAV**FT** DNYTRLRKQMA**VK**K**YLN S I LN

FIGURE 1. The complete amino acid sequences of the peptides are presented. The sequences for maxadilan and the PACAP/VIP family of peptides have been arbitrarily arranged such that the shared phenylalanine/threonine pair and common C-terminal lysines and valine are aligned.

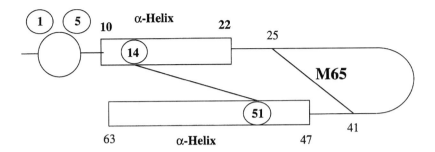

FIGURE 2. Amino acid sequence of maxadilan and M65 and their predicted structure.

PACAP/VIP Family Receptors in Cultured Rat Cortical Neurons: Effect of Maxadilan and M65

The binding of maxadilan and its antagonist M65 on PACAP receptors in the membrane of cultured rat cortical neurons was analyzed using [^{125}I]PACAP27. Maxadilan completely displaced the [^{125}I]PACAP27 binding in the same manner as PACAP38. The Scatchard analysis of it indicated the existence of a single class of binding sites and the affinities of maxadilan, PACAP38, and their related peptides were as follows: maxadilan = PACAP38 > PACAP27 = M65 >>> VIP. The K_ds were calculated as 0.059 ± 0.015 nM for PACAP38, 0.148 ± 0.065 nM for maxadilan, 0.928 ± 0.535 nM for PACAP27, and 3.92 ± 0.535 nM for M65, respectively. The B_{max} was 3.41 ± 1.30 pmol/mg protein. Maxadilan stimulated cAMP accumulation with the minimum effective concentration of 0.01 nM in a dose-dependent manner more potently than PACAP38, which stimulated it with the minimum effective concentration of 0.1 nM, although VIP stimulated it with the minimum effective concentration of 10 nM or higher in rat cultured cortical neurons. M65 (1 μM) completely blocked the cAMP accumulation stimulated by 100 nM of VIP, and partially inhibited the cAMP accumulation stimulated by 1 nM of maxadilan in rat cortical neurons. Analyses of expression of PACAP/VIP family receptors by RT-PCR demonstrated that the PCR products for PVR1 at both 280 bp of the basic receptor and 364 bp of a single cassette insert (hip, hop1, or hop2), not 448 bp for a double insert (hiphop1 or hiphop2), were equally present. In contrast, the PCR products for both PVR2 and PVR3 were not detected. These data indicated that (1) maxadilan shares binding sites with PACAP; (2) maxadilan stimulates cAMP accumulation through the activation of PVR1; (3) PVR1 is dominantly expressed in cultured rat cortical neurons; and (4) VIP may stimulate cAMP accumulation through the activation of PVR1 in cultured rat cortical neurons.

Maxadilan and its deleted peptide M65 seem to be two unique and important tools to clarify the physiological role of PVR1 among PACAP/VIP family receptors *in vitro* and *in vivo*.

ACKNOWLEDGMENT

We thank Drs. N. Inagaki and S. Seino (Chiba University School of Medicine, Chiba, Japan) for the gift of CHO cells stably expressing PVR3 and acknowledge Drs. K. Kashimoto, Y. Nagano, T. Yatogo (Ito Ham Foods Inc., Central Research Institute, Ibaragi, Japan), and Ms. H. Suto (Göttingen University, Germay) for their kind technical assistance.

REFERENCES

1. LERNER, E. A., J. M. RIBEIRO, R. J. NELSON & M. R. LERNER. 1991. Isolation of maxadilan, a potent vasodilatory peptide from the salivary glands of the sand fly Lutzomyia longipalpis. J. Biol. Chem. **266**: 11234–11236.
2. LERNER, E. A. & C. B. SHOEMAKER. 1992. Maxadilan. Cloning and functional expression of the gene encoding this potent vasodilator peptide. J. Biol. Chem. **267**: 1062–1066.
3. YOSHIDA, S., T. TAKAMATSU, S. DENDA, M. OHNUMA, M. TAJIMA, E. A. LERNER & F. KANDA. 1996. Structural characterization and location of disulphide linkages of a potent vasodilatory peptide, recombinant maxadilan, by a multiple mass spectrometric approach. Rapid Comm. Mass Spectrometry **10**: 641–648.
4. MORO, O., T. J. TSOMIDES, M. TAJIMA & E. A. LERNER. 1995. Maxadilan binds to membrane fractions of brain tissue. Biochem. Biophys. Res. Commun. **216**: 234–241.
5. MORO, O. & E. A. LERNER. 1997. Maxadilan, the vasodilator from sand flies, is a specific pituitary adenylate cyclase activating peptide type I receptor agonist. J. Biol. Chem. **272**: 966–970.
6. MIYATA, A., A. ARIMURA, R. R. DAHL, N. MINAMINO, A. UEHARA, L. JIANG, M. D. CULLER & D. H. COY. 1989. Isolation of a novel 38 residue-hypothalamic polypeptide which stimulates adenylate cyclase in pituitary cells. Biochem. Biophys. Res. Commun. **164**: 567–574.
7. MIYATA, A., L. JIANG, R. D. DAHL, C. KITADA, K. KUBO, M. FUJINO, N. MINAMINO & A. ARIMURA. 1990. Isolation of a neuropeptide corresponding to the N-terminal 27 residues of the pituitary adenylate cyclase activating polypeptide with 38 residues (PACAP38). Biochem. Biophys. Res. Commun. **170**: 643–648.
8. ARIMURA, A. 1992. Pituitary adenylate cyclase activating polypeptide (PACAP): discovery and current status of research. [Review]. Regul. Pept. **37**: 287–303.
9. UCHIDA, D., A. ARIMURA, V. A. SOMOGYVARI, S. SHIODA & W. A. BANKS. 1996. Prevention of ischemia-induced death of hippocampal neurons by pituitary adenylate cyclase activating polypeptide. Brain Res. **736**: 280–286.
10. MORIO, H., I. TATSUNO, A. HIRAI, Y. TAMURA & Y. SAITO. 1996. Pituitary adenylate cyclase-activating polypeptide protects rat-cultured cortical neurons from glutamate-induced cytotoxicity. Brain Res. **741**: 82–88.
11. RAWLINGS, S. R. & M. HEZAREH. 1996. Pituitary adenylate cyclase-activating polypeptide (PACAP) and PACAP/vasoactive intestinal polypeptide receptors: actions on the anterior pituitary gland. [Review]. Endocr. Rev. **17**: 4–29.
12. HOSOYA, M., H. ONDA, K. OGI, Y. MASUDA, Y. MIYAMOTO, T. OHTAKI, H. OKAZAKI, A. ARIMURA, & M. FUJINO. 1993. Molecular cloning and functional expression of rat cDNAs encoding the receptor for pituitary adenylate cyclase activating polypeptide (PACAP). Biochem. Biophys. Res. Comm. **194**(1): 133–43.
13. ISHIHARA, T., R. SHIGEMOTO, K. MORI, K. TAKAHASHI & S. NAGATA. 1992. Functional expression and tissue distribution of a novel receptor for vasoactive intestinal polypeptide. Neuron **8**(4): 811–819.
14. INAGAKI, N., H. YOSHIDA, M. MIZUTA, N. MIZUNO, Y. FUJII, T. GONOI, J. MIYAZAKI & S. SEINO. 1994. Cloning and functional characterization of a third pituitary adenylate cyclase-activating polypeptide receptor subtype expressed in insulin-secreting cells. Proc. Natl. Acad. Sci. USA **91**: 2679–2683.
15. TATSUNO, I., P. E. GOTTSCHALL, K. KOVES & A. ARIMURA. 1990. Demonstration of specific binding sites for pituitary adenylate cyclase activating polypeptide (PACAP) in rat astrocytes. Biochem. Biophys. Res. Commun. **168**: 1027–1033.
16. TATSUNO, I., P. E. GOTTSCHALL & A. ARIMURA. 1991. Specific binding sites for pituitary adenylate cyclase activating polypeptide (PACAP) in rat cultured astrocytes: molecular identification and interaction with vasoactive intestinal peptide (VIP). Peptides **12**: 617–621.
17. RAWLINGS, S. R., I. PIUZ, W. SCHLEGEL, J. BOCKAERT & L. JOURNOT. 1995. Differential expression of pituitary adenylate cyclase-activating polypeptide/vasoactive intestinal polypeptide receptor subtypes in clonal pituitary somatotrophs and gonadotrophs. Endocrinology **136**: 2088–2098.

Importance of Conserved Cysteines in the Extracellular Loops of Human PACAP/VIP₁ Receptor for Ligand Binding and Stimulation of cAMP Production[a]

SANNE MØLLER KNUDSEN,[b,d] JEPPE WEGENER TAMS,[b]
BIRGITTE S. WULFF,[c] AND JAN FAHRENKRUG[b]

[b]Department of Clinical Biochemistry, Bispebjerg Hospital, University
of Copenhagen, DK-2400 Copenhagen NV, Denmark

[c]Novo Nordisk Park, Måløv, Denmark

ABSTRACT: The importance of two highly conserved cysteines in the human pituitary adenylate cyclase activating polypeptide (PACAP)/vasoactive intestinal peptide 1 (VIP₁) receptor was examined. Using site-directed mutagenesis, each Cys residue was converted into Ala or Ser. The mutant and wildtype genes were transfected into HEK293 cells and tested for the ability to bind VIP and to activate cAMP production. Cys215Ala/Ser and Cys285Ala/Ser showed at least a tenfold decrease in binding affinity and receptor potency when compared to the wildtype. In contradiction to the wildtype receptor, both mutations were insensitive to dithiothreitol (DTT). The results indicate the existence of a disulfide bond between Cys215 and Cys285, which is important for stabilizing the receptor in the correct conformation for ligand binding and activation.

The human PACAP/VIP₁ receptor (hPV₁R) is a member of the secretin receptor family, which belongs to the group of G protein–coupled receptors (GPCR) having seven transmembrane helices (TM).[1–3] In addition to the sequence identity in this family of receptors, there are several common features: (1) a large N-terminal extracellular region; (2) the receptors all bind small peptide ligand and can activate Gs; (3) at the genomic level, all the receptors contain many introns in contrast to the major rhodopsin family; and (4) a conserved pair of cysteines in extracellular loops 1 and 2. Most of all GPCRs have a pair of conserved residues in the ultimate C-terminal end of extracellular loop 1 and in extracellular loop 2. In the rhodopsin family, which is the largest and most investigated GPCR family, the covalent linking of the first and second extracellular loop by a disulfide bond is considered to be important for the generation and stabilization of the receptor structure.[4] By sequence alignment analysis, there is no obvious sequence identity between hPV₁R/the secretin receptor family and the major rhodopsin family. However, there are a lot of similarities, including (1) comparable fold of the helices, (2) an anticlockwise arrangement of the seven transmembrane helices (seen from the extracellular space), and (3) a conserved

[a]This work supported by the Danish Biotechnology Centre for Cellular Communication and by grant No. 12-1640-1 from the Danish Medical Research Council.

[d]Corresponding author: Department of Clinical Biochemistry, Bispebjerg Hospital, DK-2400, Copenhagen NV, Denmark; Tel.: 45-35312646; Fax: 45-35315955; E-mail: smk@biobase.dk

259

pair of cysteines in extracellular loops 1 and 2.[3] Since all GPCR have the same property to bind ligands and activate an appropriate G protein in response to ligand binding, it is reasonable to postulate that the receptors have evolved to contain conserved structural features related to these shared functions.

The purpose of the present study was to examine the importance of conserved cysteines and disulfide bonding for the binding and activation of hPV₁R. In addition, the effect of a disulfide-reducing agent (dithiothreitol, DTT) on VIP binding was examined. Our findings suggest that a disulfide bond is present in the human PACAP/VIP₁ receptor between the extracellular loop 1 and loop 2, and that this linkage is critical for maintaining a functional conformation.

MATERIALS AND METHODS

Receptor Constructs and Site-Directed Mutagenesis

The cDNA encoding the human PACAP/VIP₁ receptor was kindly donated by Drs. A. Couvineau and M. Laburthe (INSERM U410, Paris, France). The coding region was subcloned into the expression vector pcDNA3 from Invitrogen. Mutagenesis was carried out by the overlap extension method using Pwo-polymerase (Boehringer Mannheim). Identification of all the mutants was confirmed by sequencing (Perkin Elmer).

Cell Culture and Transient Transfection

HEK293 cells were grown in MEM (Earl's salt), supplemented with 10% fetal bovine serum and 1% gentamicin. HEK293 cells were plated in 200-mm dishes (4×10^6 cells/dish) and transiently transfected with wildtype cDNA and mutant cDNA, using the $CaPO_4$ precipitation method. Cells were harvested 72 h after transfection for membrane preparation and VIP-stimulated cAMP production.

Preparation of Plasma Membranes

Cells from confluent 200-mm culture plates were rinsed with PBS buffer and scraped off in the same buffer. After centrifugation, the supernatant was resuspended in 15 ml binding buffer containing 20 mM HEPES, 2 mM $CaCl_2$, 150 mM NaCl, 5 mM EDTA, 1 mg/ml Bacitracin, and 200 µg/ml bovine serum albumin. The cell suspension was homogenized with a polytron for 30 sec, centrifuged at 20,000 g for 20 min, and resuspended in binding buffer. The procedure was repeated once and the membranes were resuspended in 10 ml binding buffer and stored until use at –80°C. Protein concentration was determined by the Bradford analysis (Bio-Rad).

Binding of ^{125}I-VIP to Plasma Membrane

The functional properties of wildtype and mutant PACAP/VIP₁ receptors were analyzed by ^{125}I-VIP binding to plasma membranes. VIP was labeled and purified as described by Martin and colleagues.[5] Binding reactions were carried out on 10 µg of membranes in a total volume of 150 µl with a constant amount of radioiodinated VIP (10^{-10} M) and increasing concentrations of unlabeled VIP (0–10^{-6} M) for 60 min at room temperature.

Bound and free radioligands were separated by centrifugation. To examine the effect of the reducing agent DTT on ^{125}I-VIP binding, wildtype and mutant receptors were incubated with 10 mM DTT. Data were analyzed by non-linear least square regression.[6]

Intracellular cAMP Assay

Intracellular cAMP levels were assayed with a [^{125}I]cAMP assay system from Amersham (Birkerød, Denmark). HEK293 cells were transfected and seeded to 2×10^5 cells/well 48 h after transfection. After further 24 h cells were washed with MEM-medium and incubated with 500 μl MEM-medium containing 0.1 mM isobutyl-methylxanthine (IBMX) for 20 min at 37°C and for further 25 min at 37°C with VIP in increasing concentrations (0 M–10^{-6} M). cAMP was extracted by incubating the cells with 50 μl trichloroacetic acid (10%) and neutralized by 50 μl 0.8 M Tris-base.

The generation of standard curves and the measurement of cAMP levels in supernatant of cell lysates were performed according to the manufacturer's instructions. Data were analyzed by non-linear least square regression.[6]

RESULTS AND DISCUSSION

Two residues, Cys215 in extracellular loop 1 and Cys285 in extracellular loop 2 of the human PACAP/VIP$_1$ receptor (FIG. 1), corresponding to the highly conserved positions in the rhodopsin family were mutated into alanine (A) or serine (S) using site-directed mutagenesis.[7,8] By an alanine substitution, the sidechain is removed and the normal stereochemistry is maintained. In the serine substitution, a thiol group is replaced by the analogue hydrophilic hydroxyl group. The wildtype and mutant receptors were transfected into HEK293 cells.

The expression level, binding affinity on purified membranes, and production of cAMP on cell lysates from C215A/S, C285A/S, and wildtype were examined. Using the equation of Akera and Cheng[9] to calculate the expression level (B_{max}), we obtained similar data for C215A/S, C285A/S, and the wildtype (TABLE 1). However, a marked decrease in binding affinity of C215A/S and C285A/S compared to the wildtype receptor was observed (FIG. 2). The effect of the disulfide-reducing agent DTT was examined on wildtype and mutant receptors. Incubation of the wildtype with 10 mM DTT caused, in agreement with the findings described by Robberecht and colleagues,[10] a loss in VIP binding and receptor number, whereas the binding characteristics of the mutant receptors were unaffected (data not shown). In accordance with our binding data, the dose-response curves for VIP-induced cAMP production in the mutants were shifted 30–300 fold to the right (FIG. 3). The difference between wildtype and mutants was most significant with the serine substitution, which probably depends on the hydrophilic nature of the unpaired hydrogen bond from serine.

Inside the major rhodopsin family, the disulfide bond between extracellular loops 1 and 2 is essential for the expression and stabilization of the high affinity state of the receptors.[11-15] In studies on the secretin receptor, Vilardaga and colleagues[16] found that the investigated disulfide bond facilitates the ligand recognition. In contrast, Gaudin and colleagues[17] concluded that a disulfide bridge between extracellular loops 1 and 2 is not present in hPV$_1$R. They have performed a number of point mutations in the hPV$_1$R including C215G and C285G. They examined the effect on VIP binding and found similar binding affinity of wildtype and C215G, whereas binding to C285G was not detectable. Their conclusion was, however, that a disulfide bridge is not present. The following explanations can account for the discrepancies between theirs and our results: (1) different methods of labeling and purifying the radioligand, (2) different buffer systems, (3) different preparations of plasma

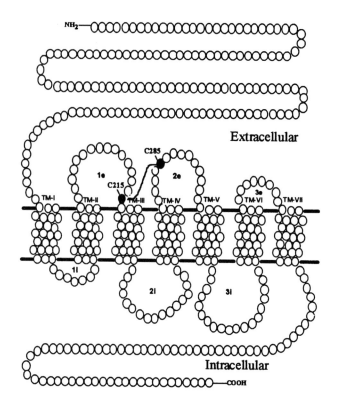

FIGURE 1. Schematic representation of the human PACAP/VIP$_1$ receptor, consisting of seven transmembrane helices (TMI-TMVII), an extracellular N-terminal domain, and a C-terminal domain located inside the cell. 1e-3e corresponds to the extracellular loops and 1i-3i to the intracellular loops. The investigated cysteine residues C215 and C285 are presented in black.

membranes, and (4) different cell lines. These variable conditions influence the following equilibrium between two interconvertible allosteric states of the receptor:

Active Receptor (R*)	⇔	Inactive Receptor (R)
+ ligand		+ligand
⇕		⇕
R*-ligand	⇔	R-ligand

TABLE 1. Binding affinity (IC_{50}), activation (EC_{50}), and receptor number (B_{max}) for human PACAP/VIP$_1$ receptor wildtype and Cys residue mutations, transiently expressed in HEK293 cells

Receptor Mutation	IC_{50} (nM) ± SEM	EC_{50} (nM) ± SEM	B_{max} (pmol/mg protein)± SEM
Wildtype (wt)	3.1 ± 0.6	0.6 ± 0.4	2.5 ± 0.5
C215A	42 ± 6	18 ± 3	1.5 ± 0.8
C215S	61 ± 28	40 ± 21	3.6 ± 0.8
C285A	60 ± 25	75 ± 35	2.4 ± 0.9
C285S	62 ± 31	232 ± 30	4.0 ± 0.8

Data represent the mean ± SEM of at least five experiments, each made in duplicate.

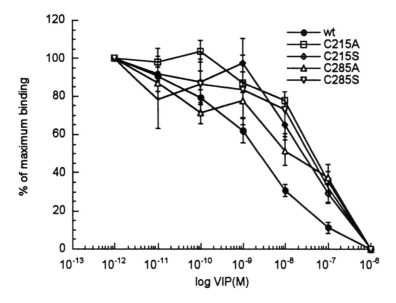

FIGURE 2. Competitive inhibition of ^{125}I-VIP binding to membrane preparation from HEK293 cells, transiently expressing hPV$_1$R (wildtype) and mutants, by unlabeled VIP. Data points represent the mean ± SEM of 8–13 separate experiments, each made in duplicate. The IC$_{50}$ values for wildtype and mutants are given in TABLE 1.

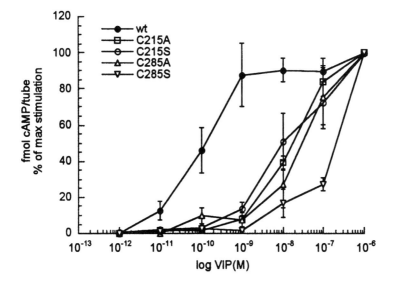

FIGURE 3. VIP induced cAMP production in HEK293 cells expressing wildtype and mutant receptors for hPV$_1$R. Data points represent the mean ± SEM of 4–6 separate experiments, each made in duplicate. The EC$_{50}$ values for wildtype and mutants are given in TABLE 1.

By site-directed mutagenesis, the normal equilibrium can be shifted either to the active conformation (R*) or to the inactive conformation (R). The disulfide bridge has a stabilizing effect on the active conformation, and by substitution of those cysteines with alanine or serine, the equilibrium is shifted towards the inactive conformation, which has an effect on the activity assay, where a higher concentration of ligand is needed in order to have full activation. If the binding assay is primarily measuring the low affinity binding, it will be difficult to detect a mutant, which shifts this equilibrium further to the right. However if the binding assay is measuring the high affinity binding, a change in binding affinity can be distinguished between wildtype and mutants. In the present study we were able to measure a change in binding affinity between wildtype and cys-mutants. This is most likely due to the use of HEK293 cells, which are known to contain a high amount of Gs proteins and contribute to stabilization of the active conformation.

In conclusion, the sequence alignment and our results suggest the existence of a disulfide bond between the two conserved extracellular cysteines (C215 and C285) in the human PACAP/VIP$_1$ receptor. This bond can contribute to secure extracellular loop 2 in a conformation, which makes it possible to send messages through the membrane, eventually by a movement of the helices. The arrangement of the seven transmembrane helices in hPV$_1$R shows that nearly 75% of the positions in the middle and in the intracellular part of helix V have sidechains that are not able to form hydrogen bonds (FIG. 4). This probably makes the helix more flexible, suggesting an involvement in receptor activation. It is likely that the cysteines and the disulfide bridge are important for the maintenance of the correct high affinity structure of the human PACAP/VIP$_1$ receptor.

ACKNOWLEDGMENTS

We thank Drs. A. Covineau and M. Laburthe for donating the human PACAP/VIP$_1$ receptor cDNA.

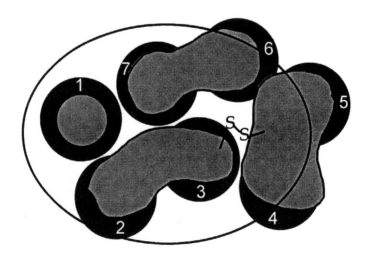

FIGURE 4. The arrangement of the seven transmembrane helices, viewed from the extracellular space. The spotted domains symbolize the extracellular loop 1, 2, and 3. Loop 1 and loop 2 are connected by the proposed disulfide bridge. The N-terminal extracellular domain is symbolized by an ellipse, which is anchored to TM1.

REFERENCES

1. COUVINEAU, M., C. ROUYER-FESSARD, D. DARMOUL, J. MAORET, I. CARRERO, E. OGIER-DENIS & M. LABURTHE. 1994. Human intestinal VIP receptor: Cloning and functional expression of two cDNA encoding proteins with different N-terminal domains. Biochem. Biophys. Res. Commun. **200**: 769–776.

2. LUTZ, E. M., W. J. SHEWARD, K. M. WEST, J. A. MORROW, G. FINK & A. J. HARMAR. 1993. The VIP$_2$ receptor: molecular characterisation of a cDNA encoding a novel receptor for vasoactive intestinal peptide. FEBS Lett. **334:** 3–8.

3. TAMS, J. W., S. M. KNUDSEN & J. FAHRENKRUG. 1997. Proposed arrangement of the seven transmembrane helices in the secretin receptor family. Receptors Channels **5**: 79–90.

4. STRADER, C. D., T. M. FONG, M. R. TOTA & D. UNDERWOOD. 1994. Structure and function of G protein-coupled receptors. Annu. Rev. Biochem. **63**: 101–132.

5. MARTIN, J-L., K. ROSE, G. HUGHES & P. J. MAGISTRETTI. 1986. [mono[^{125}I]iodo-Tyr^{10},MetO^{17}]-Vasoactive intestinal polypeptide. J. Biol. Chem. **261**: 5320–5327.

6. DE LEAN, A., A. A. HANCOCK & R. J. LEFKOWITZ. 1981. Validation and statistical analysis of a computer modelling method for quantitative analysis of radioligand binding data for mixtures of pharmacological receptor subtypes. Mol. Pharm. **21**: 5–16.

7. PROBST, W. C., L. A. SNYDER, D. I. SCHUSTER, J. BROSIUS & C. SEALFON. 1992. Sequence alignment of the G-protein coupled receptor superfamily. DNA Cell Biol. **11**: 1–20.

8. BALDWIN, J. M. 1993. The probable arrangement of the helices in G protein-coupled receptors. EMBO J. **12**: 1693–1703.

9. AKERA, T. & V-J. K. CHENG. 1977. A simple method for the determination of affinity and binding site concentration in receptor binding studies. Biochim. Biophys. Acta **470**: 412–423.

10. ROBBERECHT, P., M. WAELBROECK, L. CAMUS, P. DE NEEF & J. CHRISTOPHE. 1984. Importance of disulfide bonds in receptors for vasoactive intestinal peptide and secretin in rat pancreatic plasma membranes. Biochim. Biophys. Acta **773**: 271–278.

11. KARNIK, S. S., T. P. SAKMANN, H. B. CHEN & H. G. KHORANA. 1988. Cysteine residues 110 and 187 are essential for the formation of correct structure in bovine rhodopsin. Proc. Natl. Acad. Sci. USA **85**: 8459–8463.

12. DAVIDSON, F. F., P. C. LOEWEN & H. G. KHORANA. 1994. Structure and function in rhodopsin: Replacement by alanine of cysteine residues 110 and 187, components of the conserved disulfide bond in rhodopsin, affects the light-activated metarhodopsin II state. Proc. Natl. Acad. Sci. USA **91**: 4029–4033.

13. SAVARESE, T. M., C-D WANG & C. FRASER. 1992. Site-directed mutagenesis of the rat m$_1$ muscarinic acetylcholine receptor J. Biol. Chem. **267**: 11439–11448.

14. DOHLMAN, H. G., M. G. CARON, A. DEBLASI, T. FRIELLE & R. J. LEFKOWITZ. 1990. Role of extracellular disulfide-bonded cysteines in the ligand binding function of the β_2-adrenergic receptor. Biochemistry **29**: 2342–2348.

15. BOYD, N. D., R. KAGE, J. J. DUMAS, J. E. KRAUSE & S. E. LEEMAN. 1996. The peptide binding site of the substance P (NK-1) receptor localized by a photoreactive analogue of substance P: presence of a disulfide bond. Proc. Natl. Acad. Sci. USA **93**: 433–437.

16. VILARDAGA, J-P., E. DI PAOLO, C. BIALEK, P. DE NEEF, M. WAELBROECK, A. BOLLEN & P. ROBBERECHT. 1997. Mutational analysis of extracellular cysteine residues of rat secretin receptor shows that disulfide bridges are essential for receptor function. Eur. J. Biochem. **246**: 173–180.

17. GAUDIN, P., A. COUVINEAU, J. MAORET, C. ROUYER-FESSARD & M. LABURTHE. 1995. Mutational analysis of cysteine residues within the extracellular domains of the human vasoactive intestinal peptide (VIP) 1 receptor identifies seven mutants that are defective in VIP binding. Biochem. Biophys. Res. Commun. **211**: 901–908.

Multiple Actions of a Hybrid PACAP Antagonist: Neuronal Cell Killing and Inhibition of Sperm Motility[a]

ILLANA GOZES,[b,d] ORLY PERL,[b] RACHEL ZAMOSTIANO,[b]
SARA RUBINRAUT,[c] MATI FRIDKIN,[c] LEAH SHOCHAT,[b] AND
LAWRENCE M. LEWIN[b]

[b]Department of Clinical Biochemistry, Sackler School of Medicine, Tel Aviv University, Tel Aviv, Israel

[c]Department of Organic Chemistry, Weizmann Institute of Science, Rehovot, Israel

ABSTRACT: Pituitary stimulating adenylate cyclase (PACAP) is a major regulatory peptide with two active molecular forms: PACAP-27 and PACAP-38. Both molecular forms promote neuronal survival and protect against neurotoxicity. Based on our previous hybrid peptide strategy in designing vasoactive intestinal peptide (VIP) antagonists, novel PACAP analogues were synthesized (neurotensin$_{6-11}$PACAP$_{7-27}$ and neurotensin$_{6-11}$ PACAP$_{7-38}$). In addition to the hybrid modification, the methionine in position 17 was replaced by norleucine (Nle). Treatment of rat cerebral cortical cultures for five days with the putative PACAP antagonists (1 nM) resulted in a 35–45% reduction in neuronal cell counts as compared to controls. Neuronal cell death was already obtained at picomolar concentrations for the neurotensin$_{6-11}$PACAP$_{7-27}$ antagonist with 70% death at 10^{-8} M. Co-administration of the PACAP hybrid analogue with picomolar amounts of PACAP-27 or Nle^{17}-PACAP-27 attenuated the reduction in neuronal cell counts. While the protective effects of both analogues exhibited a peak at 1 pM concentrations, the Nle-containing agonist displayed a broader range of active concentrations (10^{-12} M–10^{-9} M)

The putative PACAP antagonist also inhibited sperm motility (golden hamster) in a dose-dependent manner as assessed *in vitro*. Complete inhibition was observed at 10 µM, suggesting a role for PACAP in sperm motility and sexual function. Thus, previous findings of a large number of PACAP and PACAP receptors in the nervous system and the reproductive system are now correlated with a function in neuronal survival and sperm motility. The structure-activity studies suggest that the methionine in position 17 and the first six amino acids are important in the determination of PACAP activity, knowledge that may facilitate PACAP-based drug design.

The availability of specific PACAP agonists and antagonists should help to clarify the physiological roles of this major neuropeptide and make PACAP-based drug design possible. The two molecular forms of PACAP, PACAP-27 and PACAP-38, have already

[a]This work was supported in part by Fujimoto Pharmaceutical Corporation (Osaka, Japan).

[d]Corresponding author: Prof. Illana Gozes, Department of Clinical Biochemistry, Sackler School of Medicine, Tel Aviv University, Tel Aviv, 69978 Israel; Tel.: 972-3-640-7240; Fax: 972-3-640-8541; E-mail: igozes@post.tau.ac.il

been shown to be involved in a multiplicity of functions,[1] including maintenance of neuronal survival.[2] PACAP has been localized in the brain as well as in the reproductive system, e.g. the ovaries.[3,4] It has been previously shown that PACAP6-38 can block some of the PACAP effects), for example, on muscle relaxation[5] and on gonadotropin-releasing hormone and somatostatin gene expression,[6] suggesting both neurotransmitter and hormone functions for PACAP.

We have previously designed and synthesized a VIP-neurotensin hybrid antagonist (neurotensin$_{6-11}$VIP$_{7-28}$). This receptor antagonist[7] inhibited VIP-stimulated sexual behavior, neuronal survival, learning and memory (when applied directly into the brain), and brain development.[8-15] Importantly, it also inhibited cancer growth (lung cancer,[16,17] neuroblastoma,[18] and breast cancer[19]).

In work based on hybrid peptide strategy for the design of VIP antagonists,[11] it was discovered that replacement of the methionine in position 17 by norleucine (Nle) may render the resulting molecule more stable and that this specific Nle^{17}-VIP hybrid antagonist inhibited VIP-mediated diurnal rhythms of motor behavior. Thus, novel PACAP hybrid antagonists were synthesized utilizing the VIP hybrid antagonists as prototypes.

Initially, a neurotensin$_{6-11}$PACAP$_{7-38}$ hybrid antagonist was shown to inhibit ^{125}I-PACAP-27 binding to NIH/3T3 cells expressing PACAP receptors.[20] Furthermore, this PACAP hybrid antagonized the ability of PACAP-27 to elevate formation of cAMP and increase cytosolic Ca^{2+} and c-fos mRNA.

This report introduces neurotensin$_{6-11}$PACAP$_{7-27}$ and Nle^{17}neurotensin$_{6-11}$PACAP$_{7-27}$ as novel PACAP antagonists. The results confirm the neuroprotective effects of PACAP that are now shown to be inhibited by the specific antagonists. In addition, a new role for PACAP as a chemotactic hormone in the reproductive tract is suggested.

EXPERIMENTAL DESIGN AND METHODS

Peptide Synthesis

Peptide synthesis was carried out via solid phase methodology as before.[21,22] Several peptides were synthesized as follows:

PACAP-27
H-S-D-G-I-F-T-D-S-Y-S-R-Y-R-K-Q-M-A-V-K-K-Y-L-A-A-V-L-NH$_2$
Nle^{17}-PACAP-27
H-S-D-G-I-F-T-D-S-Y-S-R-Y-R-K-Q-Nle-A-V-K-K-Y-L-A-A-V-L-NH$_2$

PACAP-38 and Nle-PACAP-38 were as above and contained the additional amino acids found in the original PACAP27 and PACAP38 molecules.[1,2,20] Hybrid peptides were neurotensin$_{6-11}$PACAP$_{7-27}$, Nle^{17}- neurotensin$_{6-11}$PACAP$_{7-27}$, neurotensin$_{6-11}$PACAP$_{7-38}$, and Nle^{17}-neurotensin$_{6-11}$PACAP$_{7-38}$.

Cell Culture

For measurements of neuroprotective actions, cerebral cortical cultures derived from newborn rats[22] were used. In this system, post-mitotic neurons were maintained on a confluent layer of cortical astrocytes. Neuronal identity was established by an immunocytochemical stain with antiserum against neuron-specific enolase.[23] Dishes were coded and counts conducted without knowledge of the treatment group. Statistical comparisons were made by ANOVA with a Student-Newman-Keuls multiple comparison of means test.

Sperm Motility

Experiments were performed utilizing sperm cells from golden hamsters that were at least two months old.[24,25] Breifly, mature sperm cells were removed (with care to insure the intactness of the sperm) and placed into phosphate-buffered saline containing 280 mM sucrose (1:1, vol/vol). Sperm cells were viewed under the light microscope to determine cell density. The test peptide was added to the sperm sample for a 5-min incubation period at room temperature. Motile sperm cells and immobile cells were counted and compared to untreated controls.

RESULTS

The New Neurotensin-PACAP Hybrids Produce Neuronal Cell Death

FIGURE 1 demonstrates that upon addition of increasing concentrations of the synthetic peptide, neurotensin$_{6-11}$PACAP$_{7-27}$, to cerebral cortical cells in culture, significant neuronal cell death occurred at concentrations of 10^{-12} M (about 40% death as compared to control cultures; $p<0.02$). Maximal cell death of about 70% was observed at peptide concentrations of 10^{-8} M. Co-administration of PACAP-27 together with neurotensin$_{6-11}$PACAP$_{7-27}$ (10^{-9} M) resulted in neuroprotection demonstrated at 10^{-12} M PACAP-27 (FIG. 2A). A similar results were observed with the analogue Nle^{17}-neurotensin$_{6-11}$PACAP$_{7-27}$. However, the neuroprotection obtained with Nle^{17}-PACAP-27 (FIG. 2B) displayed a broader range of

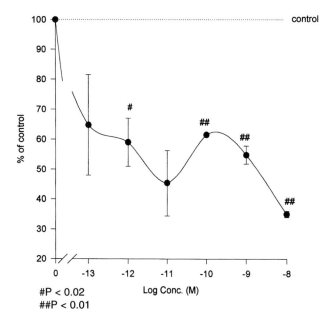

FIGURE 1. Neurotensin$_{6-11}$PACAP$_{7-27}$ produces neuronal cell killing in cerebral cortical cultures. Experiments were performed in duplicate and repeated at least two independent times. Student t-test was performed for each concentration point to evaluate the significance of the data in comparison to untreated control. (Control neurons = 353).

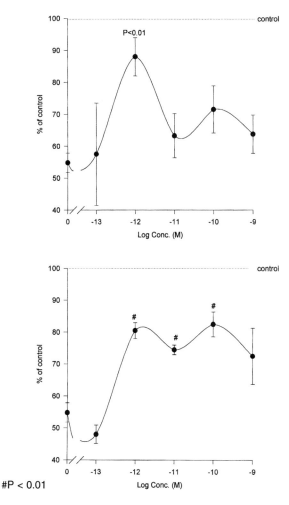

FIGURE 2. (A) PACAP-27 protects against neuronal cell killing caused by the hybrid antagonist neurotensin$_{6-11}$PACAP$_{7-27}$. Experiments were performed as above. All samples contained 10^{-9} M neurotensin$_{6-11}$PACAP$_{7-27}$. (Control neurons=353). (B) Nle^{17}PACAP-27 protects against neuronal cell killing caused by the hybrid antagonist Nle^{17}neurotensin$_{6-11}$PACAP$_{7-27}$. Experiments were performed as above. All samples contained 10^{-9} M Nle^{17}-neurotensin$_{6-11}$PACAP$_{7-27}$. (Control neurons= 260).

effective concentrations (10^{-12}–10^{-9} M) against the neuronal killing observed with 10^{-9} M Nle^{17}neurotensin$_{6-11}$PACAP$_{7-27}$.

While the short (27 amino acids) PACAP antagonists produced 45% neuronal cell death at 10^{-9} M (FIG. 1 and FIG. 2), the long neurotensin$_{6-11}$PACAP$_{7-38}$ hybrids were less efficacious, producing 35% death. Specificity was demonstrated in that the respective agonist protected against neuronal cell killing with the respective antagonist (FIG. 3). Thus, 10^{-11} M PACAP-38 protected against 10^{-9} M neurotensin$_{6-11}$PACAP$_{7-38}$ (*p*<0.01) and 10^{-12} M Nle^{17}-PACAP-38 provided some protection against Nle^{17}-neurotensin$_{6-11}$PACAP$_{7-38}$. The Nle^{17}-PACAP-38 seemed to be tenfold more potent (exhibiting maximal activity at tenfold

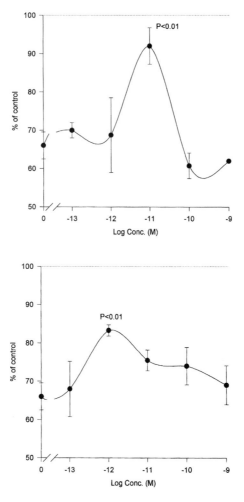

FIGURE 3. PACAP-38 protects against neurotensin$_{6-11}$PACAP$_{7-38}$ neuronal cell killing (A) (control cultures=315). (B) as in (A) but with the Nle^{17} analogues (control cultures=220). The results are the mean of three experiments with duplicates.

lower concentration) than the native analogue (containing methionine in position 17), yet it was less efficacious (fewer neurons were protected).

Antagonism of VIP and PACAP Inhibits Sperm Motility

Exogenous addition of PACAP or VIP to golden hamster sperm did not change their motility (84% sperm cells were motile in the control situation and 70% sperm cells were motile in the presence of 10^{-5} M VIP, N=10). However, addition of increasing concentrations of either the neurotensin$_{6-11}$VIP$_{7-28}$ hybrid antagonist or Nle^{17} neurotensin$_{6-11}$PACAP$_{7-27}$

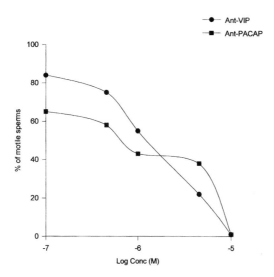

FIGURE 4. Increasing concentrations of neurotensin$_{6-11}$VIP$_{7-28}$ or neurotensin$_{6-11}$ PACAP$_{7-38}$ result in inhibition of sperm motility. Experiments were repeated ten times (independently). Motility was monitored under the microscope and was timed.

hybrid antagonist resulted in dose-dependent reduction in sperm motility with complete loss of motility at an antagonist concentration of 10^{-5} M (FIG. 4).

DISCUSSION

We have introduced here a family of PACAP antagonists and agonists. These hybrid antagonists display multiple sites of action, i.e., in the nervous system and in the reproductive system. Previous studies have shown that a PACAP hybrid antagonist belonging to this family (neurotensin$_{6-11}$PACAP$_{7-38}$) functions as a PACAP receptor antagonist. This PACAP hybrid was 2.5-fold more potent than PACAP6-38 with regard to receptor binding.[20] The higher affinity was attributed to electrostatic interactions of the neurotensin fragment (L-P-R-R-P-Y) with the receptor. However, this neurotensin moiety had no effect on neurotensin binding (Moody, unpublished observation mentioned by Pisegna and colleagues[20]).

The PACAP receptor has been cloned and shown to express six splice variants[26], acting through cAMP formation and Ca^{2+} mobilization, with some splice variants preferring one signal transduction mechanism over another. The neurotensin$_{6-11}$PACAP$_{7-38}$ hybrid recognized all receptors types tested, with the hop PACAP receptor (having a 28 amino acid insert at the third cytoplasmic domain and operating through both cAMP production and Ca^{2+} mobilization) being the major one tested.[20]

This report introduces PACAP agonists with a Nle replacement of the methionine in position 17. These analogues are similar to the ones previously described as the superactive VIP agonists.[11] Here, too, the Nle replacement resulted in a ten-fold increase in potency when the analogue tested was PACAP-38 (FIG. 3). The Nle replacement had a broader range of effective concentrations when the agonist tested was PACAP-27 (FIG. 2).

The effect of PACAP on neuronal survival is more complex than that of VIP. While VIP exhibits a bell-shaped dose-response curve of neuroprotection,[27] PACAP exhibits a bimodal type of response,[2] suggesting several mechanisms of neuroprotection and the involvement of more than one receptor subtype. Interestingly, apart from the six splice variants of the PACAP receptor, there are also two cloned VIP receptor that recognize both VIP and PACAP.[28] The novel PACAP antagonists together with the known VIP antagonists may help decipher the different activities of the various receptor molecules. Indeed, specificity was demonstrated for the neurotensin$_{6-11}$VIP$_{7-28}$ antagonist, in that it did not recognize the VIP receptor on lymphocytes.[10]

The recognition of sperm cell PACAP receptors and inhibition of sperm motility by the PACAP and VIP antagonists are interesting observations from the point of view of the existence of VIP and PACAP in the sexual organs. We have shown the existence of VIP mRNA in the ovaries,[29] and others have shown the existence of PACAP in the ovaries.[3,4,30] It is intriguing to hypothesize that PACAP may act as a chemotactic pheromone secreted in the female to attract the sperm and to increase its motility toward the ovum. Regardless of the mechanism, development of PACAP and/or VIP antagonists may present a novel approach to future contraceptive design.

ACKNOWLEDGMENTS

Professor Illana Gozes is the incumbent of the Lily and Avraham Gildor Chair for the Investigations of Growth Factors. We are very grateful to Mrs. S. Yossefi and A. Reshef for their help with the initial experiments.

REFERENCES

1. ARIMURA, A. 1992. Pituitary adenylate cyclase activating polypeptide (PACAP): discovery and current status of research. Regul. Pept. **37**: 287–303.
2. ARIMURA, A., A. SOMOGYVARI-VIGH, C. WEILL, R. C. FIORE, I. TATSUNO, V. BAY & D. E. BRENNEMAN. 1994. PACAP functions as a neurotrophic factor. Ann. N.Y. Acad. Sci. **739**: 228–243.
3. GOBBETTI, A., M. ZERANI, A. MIANO, M. BRAMUCCI, O. MURRI & D. AMICI. 1997. Presence of pituitary adenylate cyclase-activating polypeptide 38-immunolike material in the brain and ovary of the female crested newt, *Triturus carnifex*: its involvement in the ovarian synthesis of prostaglandins and steroids. J. Endocrinol. **152**: 141–146.
4. HRINDEL, J. J., J. SNEEDEN, C. J. POWELL, B. DAVIS & M. D. CULLER. 1996. A novel hypothalamic peptide, pituitary adenylate cyclase-activating peptide regulates the function of rat granulosa cells in vitro. Biol. Rep. **54**: 523–530.
5. KATSOULIS, S., W. E. SCHMIDT, R. SCHWAHOFF, U. R. FOLSCH, J. G. JIN, J. R. GRIDER & G. M. MAKHLOUF. 1996. Inhibitory transmission in guinea pig stomach mediated by distinct receptors for pituitary adenylate cyclase-activating peptide. J. Pharmacol. Exp. Ther. **278**: 199–204.
6. LI, S., V. GRINEVICH, A. FOURNIER & G. PELLETIER. 1996. Effects of pituitary adenylate-cyclase activating polypeptide (PACAP) on gonadotropin-releasing hormone and somatostatin gene expression in the rat brain. Mol. Brain Res. **41**: 157–162.
7. GOZES, I., M. FRIDKIN & D. E. BRENNEMAN. 1995. A VIP hybrid antagonist: from neurotrophism to clinical applications. Cell. Mol. Neurobiol. **15**: 675–687.
8. GOZES, I., E. MELTZER, S. RUBINRAUT, D. E. BRENNEMAN & M. FRIDKIN. 1989. Vasoactive intestinal peptide potentiates sexual behavior: Inhibition by novel antagonist. Endocrinology **125**: 2945–2949.
9. GOZES, I., S. K. MCCUNE, L. JACOBSON, D. WARREN, T. W. MOODY, M. FRIDKIN & D. E. BRENNEMAN. 1991. An antagonist to vasoactive intestinal peptide: effects on cellular functions in the central nervous system. J. Pharm. Exp. Ther. **257**: 959–966.
10. GOZES, Y., D. E. BRENNEMAN, M. FRIDKIN, R. ASOFSKY & I. GOZES. 1991. A VIP antagonist distinguishes VIP receptors on spinal cord cells and lymphocytes. Brain Res. **540**: 319–321.

11. GOZES, I., G. LILLING, R. GLAZER, A. TICHER, I. E. ASHKENAZI, A. DAVIDSON, S. RUBINRAUT, M. FRIDKIN & D. E. BRENNEMAN. 1995. Superactive lipophilic peptides discriminate multiple VIP receptors. J. Pharm. Exper. Ther. **273**: 161–167.

12. GLOWA, J. R., L. V. PANLILIO, D. E. BRENNEMAN, I. GOZES, M. FRIDKIN & J. M. HILL. 1992. Learning impairment following intracerebral administration of the HIV envelope protein gp120 or a VIP antagonist. Brain Res. **570**: 49–53.

13. HILL, J. M., I. GOZES, J. L. HILL, M. FRIDKIN & D. E. BRENNEMAN. 1991. Vasoactive intestinal peptide antagonist retards the development of neonatal behaviors in the rat. Peptides **12**: 187–192.

14. GRESSENS, P., J. M. HILL, I. GOZES, M. FRIDKIN & D. E. BRENNEMAN. 1993. Growth factor function of vasoactive intestinal peptide in whole cultured mouse embryos. Nature **362**: 155–158.

15. GRESSENS, P., J. M. HILL, B. PAINDAVEINE, I. GOZES, M. FRIDKIN & D. E. BRENNEMAN. 1994. Severe microcephaly induced by blockade of vasoactive intestinal peptide function in the primitive neuroepithelium of the mouse. J. Clin. Invest. **94**: 2020–2027.

16. MOODY, T. W., F. ZIA, A. L. GOLDSTEIN, P. H. NAYLOR, P. H., E. SARIN, D. E. BRENNEMAN, A. M. C. KOROS, J. C. REUBI, L. Y. KORMAN, M. FRIDKIN & I. GOZES. 1992. VIP analogues inhibit small cell lung cancer growth. Biomed. Res. **13** (2): 131–135.

17. MOODY, T. W., F. ZIA, M. DRAOUI, D. E. BRENNEMAN, M. FRIDKIN, A. DAVIDSON & I. GOZES. 1993. A novel VIP antagonist inhibits non-small cell lung cancer growth. Proc. Natl. Acad. Sci. USA **90**: 4345–4349.

18. LILLING, G., Y. WOLLMAN, M. N. GOLDSTEIN, S. RUBINRAUT, M. FRIDKIN D. E. BRENNEMAN & I. GOZES. 1995. Inhibition of human neuroblastoma growth by a specific VIP antagonist. J. Mol. Neurosci. **5**: 231–239.

19. ZIA, H., T. HIDA, S. JAKOWLEW, M. BIRRER, Y. GOZES, J. C. REUBI, M. FRIDKIN, I. GOZES & T. W. MOODY. 1996. Breast cancer growth in inhibition by vasoactive intestinal peptide (VIP) hybrid, a synthetic VIP receptor antagonist. Cancer Res. **56**: 3486–3489.

20. PISEGNA, J. R., J. LEYTON, T. COELHO, S. HIDA, S. JAKOWLEW, M. BIRRER, M. FRIDKIN, I. GOZES & T. W. MOODY. 1997. PACAP hybrid antagonizes PACAP splice variants. Life Sci. **61**: 631–639.

21. GOZES, I., A. RESHEF, D. SALAH, S. RUBINRAUT & M. FRIDKIN. 1994. Stearyl-Norleucine-VIP: a novel VIP analogue for noninvasive impotence treatment. Endocrinology **134**: 2121–2125.

22. GOZES, I., A. BARDEA, A. RESHEF, R. ZAMOSTIATNO, S. ZHUKOVSKY, S. RUBINRAUT, M. FRIDKIN & D. E. BRENNEMAN. 1996. Novel neuroprotective strategy for Alzheimer's disease: inhalation of a fatty neuropeptide. Proc. Natl. Acad. Sci. USA **93**: 427–432.

23. SCHMECHEL, D., P. J. MARANGOS, A. P. ZIS, M. BRIGHTMAN & F. K. GOODWIN. 1978. Brain enolases as specific markers of neuronal and glial cells. Science **199**: 313–315.

24. WEISSENBERG, R, S. YOSSEFI, R. BELLA & L. M. LEWIN. 1995. Changes during puberty in chromatin condensation, morphology, and fertilizing ability of epididymal spermatozoa of the golden hamster. Andrologia **27**: 341–344.

25. JEULIN, C., L. M. LEWIN, C. CHEVRIER & D. SCHOEVAERT. 1996. Changes in flagellar movement of rat spermatozoa along the length of the epididymis: manual and computer aided image analysis. Cell Motil. & Cytoskeleton **35**: 147–161.

26. JOURNOT, L., C. WAEBER, C. PANTALONI, F. HOLSBOER, P. H. SEEBURG, J. BOCKAERT & D. SPENGLER. 1995. Differential signal transduction by six splice variants of the PACAP receptor. Biochem. Soc. Trans. **23**: 133–137.

27. BRENNEMAN, D. E. & L. E. EIDEN. 1986. Vasoactive intestinal peptide and electrical activity influence neuronal survival. Proc. Natl. Acad. Sci. USA **83**: 1162.

28. HARMAR, T. & E. LUTZ. 1994. Multiple receptors for PACAP and VIP. Trends Pharmacol. Sci. **15**: 97–99.

29. GOZES, I. & A. TSAFRIRI. 1986. Detection of VIP-encoding mRNA in the rat ovaries. Endocrinology **119**: 2606–2610.

30. STEENTRUP, B. R., P. ALM, J. HANNIBAL, J. C. JORGENSEN, C. PALLE, J. JUNGE, H. B. CHRISTENSEN, B. OTTESEN & J. FAHRENKRUG. 1995. Pituitary adenylate cyclase-activating polypeptide: occurrence and relaxant effect in female genital tract. Am. J. Physiol. **269**: E108–E117.

The PACAP Ligand/Receptor System Regulates Cerebral Cortical Neurogenesis[a]

EMANUEL DICICCO-BLOOM,[b,c] NAIRU LU, JOHN E. PINTAR, AND JIWEN ZHANG

Department of Neuroscience and Cell Biology, Department of Pediatrics, University of Medicine and Dentistry of New Jersey/Robert Wood Johnson Medical School, 675 Hoes Lane, Room 338 CABM, Piscataway, New Jersey 08854 USA

ABSTRACT: The PACAP ligand/type I receptor system is expressed throughout the embryonic nervous system, suggesting roles in regulating neural patterning and neurogenesis. In the forebrain, precursors of the six-layered cerebral cortex cease dividing in a highly reproducible spatiotemporal sequence. The time of cell cycle exit in fact determines neuron laminar fate. Our studies indicate that PACAP signaling may elicit cortical precursor withdrawal from the cell cycle, antagonizing mitogenic stimulators. PACAP inhibited embryonic day 13.5 rat cortical precursor [^{3}H]thymidine incorporation, decreasing the proportion of mitotic cells. PACAP promoted morphological and biochemical differentiation, indicating that PACAP-induced cell cycle withdrawal was accompanied by neuronal differentiation.

In vivo, embryonic cortex contains PACAP. In culture, 85% of cells expressed PACAP while 64% exhibited receptor. Co-localization studies indicated that PACAP ligand and receptor were expressed by the mitotic precursors that divided in response to bFGF, suggesting that precursors integrate mitogenic and anti-mitogenic signals to determine the timing of cell cycle exit.

The expression of PACAP ligand and receptor in precursors raised the possibility of autocrine function. Indeed, peptide antagonists increased proliferation, suggesting that the PACAP system is expressed to elicit cell cycle exit. During ontogeny, an inhibitory signal, such as PACAP, may be required to counter the stimulatory activity of mitogenic bFGF and IGFI whose expression during cortical neurogenesis is sustained. The dynamic interplay of positive and negative regulators would regulate the timing of cell cycle withdrawal, and thus neuronal phenotype and laminar position.

Emerging evidence suggests that the PACAP ligand/receptor system plays a central role in early development of the nervous system. Indeed, recent studies demonstrating expression in the embryonic neural tube indicate that the signaling system is likely to interact with intrinsic cellular genetic programs and extracellular factors regulating proliferation, neural determination, and neuronal differentiation that underlie nervous system patterning. While much work will be needed to define the role of PACAP signaling during ontogeny, we will focus on the developing cerebral cortex, responsible for our most sophisticated perceptual and cognitive abilities, to illustrate potential PACAP functions.

[a]This work was supported by National Institutes of Health, NINDS grant NS32401 (E.D.-B.).
[b]Member, Cancer Institute of New Jersey
[c]Corresponding author: Tel.: 732-235-5381; Fax: 732-235-4990; E-mail: diciccem@umdnj.edu

Following a brief review of cortical development *in vivo*, we describe PACAP effects on proliferation and differentiation of cerebral cortical precursors in culture.

DEVELOPMENT OF THE CEREBRAL CORTEX

Evidence from human clinical studies as well as animal models indicates that different regions of the cerebral cortex subserve distinct cognitive abilities. For example, visual and somatosensory perception localize primarily to occipital and parietal cortices, while motor planning and movement depend on the prefrontal and frontal cortex, respectively.[1] However, while different regions contribute to different cognitive functions, the overall cortical cytoarchitecture is remarkably conserved among brain areas, revealing a pattern of six neuronal cell layers. Neurons in each layer exhibit distinct cellular morphologies and elaborate characteristic synaptic connections: pyramidal neurons in layer 3 establish synapses within and between cortical hemispheres, whereas deeper layer 5/6 neurons project primarily to subcortical nuclei.[1,2] Further, neurons exhibit layer-specific molecular diversity, including transcription factors, neurotransmitter biosynthetic enzymes, and calcium-binding proteins.[3–9] Significantly, evolutionary studies suggest that the emergence of the six-layered cortex from the primordial three-layered brain allowed for the advance in mammalian cognitive function.[10]

We have had a long-standing interest in how distinct populations of neurons in different regions of the nervous system arise during development.[11–22] In light of the cellular complexity of the cortex, the same approaches can be applied to this structure, since it is composed of diverse neuronal cell types, which emerge in highly reproducible spatiotemporal patterns. During early development, following closure of the neural plate to form the neural tube, the anterior nervous system divides into three vesicles, the forebrain, midbrain, and hindbrain. Subsequently, subdivision of the forebrain yields the diencephalon from which emerge the paired telencephalic vesicles, the primordia of the cerebral hemispheres.[2,10] Classical studies have extensively described the process of neuronal generation that produces the six cortical layers.[23,24] Precursors proliferate in the ventricular zone, a pseudostratified columnar neuroepithelium that surrounds the lateral ventricles (FIG. 1). The day on which a precursor leaves the cell cycle, detected by nuclear incorporation and retention of a DNA precursor base, such as tritiated thymidine or bromodeoxyuridine, is termed the cell's "birthdate." Such studies indicate that the earliest born cells migrate outward radially to populate the deepest cortical layers, whereas later-born neurons migrate past deep layer cells to occupy the outermost layers (FIG. 1). Significantly, recent studies indicate that the time at which a precursor leaves the cell cycle determines the future laminar fate of the cell, and thus its ultimate phenotype.[25] That is, environmental signals apparently assign the laminar fate while the cell is engaged in synthesizing DNA during S-phase. However, if the cell re-enters the cell cycle, fate specification is lost, allowing for a new determination.[26] Consequently, signals regulating cell cycle progress may be critical for neural determination and neuronal differentiation. In particular, signals that elicit cell cycle exit may be required to stabilize fate specification.

Previous studies indicate that basic fibroblast growth factor (bFGF) and insulin-like growth factor I (IGF-I) ligands and receptors are expressed in the cortical ventricular zone and are likely to stimulate precursors mitosis.[27-34] Further, these endogenous mitogens persist throughout the period of cortical neurogenesis, from embryonic day 12 (E12) to E20 in the rat, suggesting that an additional signal to induce cell cycle exit is required. Our evidence indicates that the PACAP ligand/receptor system elicits withdrawal of cortical precursors from the cell cycle, suggesting that the pathway plays a role in fate determination. We expect that precursor proliferation depends on a balance of stimulatory mitogens and inhibitory anti-mitogens to generate the complement of cortical neurons during ontogeny.

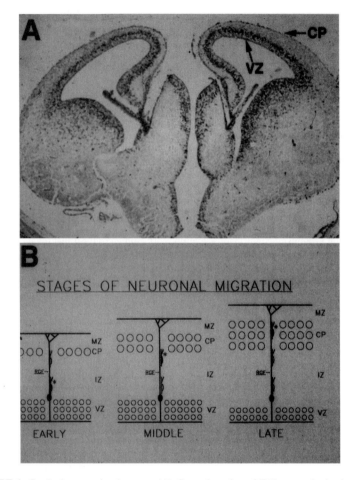

FIGURE 1. Cerebral cortex development. (**A**) Coronal section of E14 mouse brain showing the paired lateral ventricular spaces surrounded by the developing cortical hemispheres. The proliferative ventricular zone (VZ) precursors appear black after injection and immunocytochemical processing to detect bromodeoxyuridine labeling of mitotic cells. (**B**) After cells leave the mitotic cycle, newly born neurons migrate outward radially along radial glial fibers (RGF) to take up position in the cortical plate (CP). Cells born at progressively later times localize to more superficial cortical layers. (Pictures courtesy of R. Nowakowski, UMDNJ/RWJMS.)

PACAP EXPRESSION IN DEVELOPING FOREBRAIN

Early studies of PACAP protein levels and receptor binding indicated expression as early as E14 in the rat forebrain.[35] More recent studies demonstrate ligand and receptor mRNA expression,[36] as well as specific localization to the developing neural tube at E9,[37] indicating that the peptide system is positioned to interact with cell intrinsic and extrinsic factors regulating neural patterning and neurogenesis. We have begun to localize PACAP immunoreactivity in developing rat cortex at E14, identifying strong peptide signal in the emerging cortical plate region, which is composed of the earliest born, postmitotic neuronal precursors

(FIG. 2). However, in addition, we found PACAP expressed in the cytoplasm of proliferating ventricular zone cells, suggesting that the peptide may directly influence cell division. Moreover, our recent *in situ* hybridization studies indicate that receptor mRNA is expressed throughout the cortex, while the ligand message apparently localized to the postmitotic cells alone (FIG. 3). These preliminary *in situ* hybridization studies suggest that PACAP is produced by postmitotic, differentiating neurons and that released peptide participates in a feedback loop to influence ventricular zone production. While this may be one mechanism active at E14, it seems likely that the ventricular zone cells will produce PACAP themselves at earlier times when no postmitotic neurons have yet been generated. In sum, PACAP ligand and receptor are expressed in the developing cortex during the period of neuronal determination and differentiation, as precursors cease proliferating and begin differentiation. Further, co-localization of the peptide system with other signaling systems, such as bFGF and IGF-I,[27-34] suggests that multiple factors interact to regulate cortical neurogenesis.

PACAP REGULATES THE TRANSITION FROM PROLIFERATION TO DIFFERENTIATION IN CEREBRAL CORTICAL PRECURSORS IN CULTURE

Characterization of the Cortical Precursor Culture Model

To begin defining the role of PACAP in cortical development, we documented the presence of PACAP protein using radioimmunoassay, which revealed significant peptide levels in freshly dissected cortex tissue (TABLE 1).[38] In turn, we established fully defined, serum-free, low-density cultures of E13.5 rat cortical precursors to examine activities. While approximately 50% of cells expressed two neuronal markers, neuron-specific enolase and microtubule-associated protein-2 (MAP-2), at 3 hours, 90% of cells exhibited these traits at 24 hours and virtually all were positive at 48 hours (FIG. 4), indicating that we had obtained precursors that underwent neuronal differentiation *in vitro*.

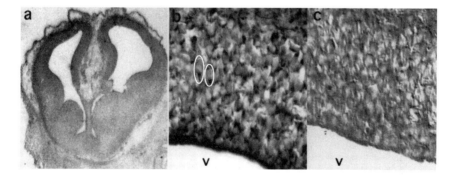

FIGURE 2. Immunocytochemical detection of PACAP-38 protein expression in the E14 rat cerebral cortex. (a) Coronal section of the cerebral hemispheres indicates staining in the medial walls (hippocampus) and dorsolateral walls (cortical plates), with minimal signal in the ventrally located ganglionic eminences (basal ganglia). In the cortex, most intense staining is observed in the outermost region, the emerging cortical plate (preplate). (b) At higher magnification of the VZ, intense signal is observed in triangular crescents, the cytoplasm surrounding oval nuclei of the spindle-shaped proliferative precursor cells. The ovals identify two single VZ cells with PACAP staining in apical cytoplasm. (c) Preincubation of PACAP antiserum with PACAP, but not VIP (not shown), markedly reduces peptide staining. v=ventricle.

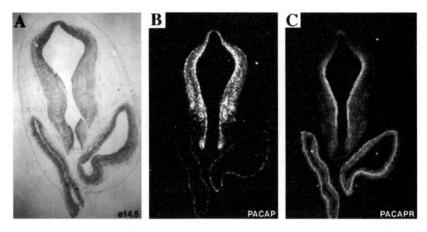

FIGURE 3. PACAP and PACAP-R mRNA expression in E14.5 rat cerebral cortex. The cerebral hemispheres are oriented downward. (**A**) Light microscopic image reveals the typical intense cellularity of the VZ within each hemisphere, surrounding the ventricular spaces. (**B**) PACAP mRNA is localized to the outer regions of the cortex, in which lie the first-born postmitotic cells. (**C**) PACAP-R appears more diffuse, present in all layers, including VZ and CP, suggesting peptide actions on both proliferation and neuronal differentiation.

In vivo, precursors cease proliferating prior to initiating neuronal differentiation. To examine relationships *in vitro*, precursors were exposed for 3 hours to BrdU, a marker for cells synthesizing DNA, and then fixed and analyzed immediately or after 24 hours incubation. As observed *in vivo*, cells actively engaged in DNA synthesis (BrdU-positive cells at 3 hours) were not positive for MAP-2. Conversely, at 24 hours, when cell division was complete, newly born precursors expressed MAP-2, indicating that following exit from the mitotic cycle, precursors underwent neuronal differentiation (FIG. 5). In turn, our system allowed analysis of signals potentially involved in neurogenetic regulation. Significantly, with time in low density culture, precursors progressively left the cell cycle, raising the possibility that signals inducing mitotic inhibition were active *in vitro*.

Effect of PACAP on Cortical Precursor Proliferation

As a first step in defining PACAP effects, we examined precursors for functional receptors. Indeed, at 24 hours, PACAP elicited a fivefold increase in cAMP content after 15 minutes exposure, with a peak at 0.1 nM peptide, indicating that the cells possessed active receptors (FIG. 6). Since similar effects were elicited by 1,000-fold higher concentrations of VIP, the data suggest the presence of a PACAP type I receptor.

To define PACAP effects on precursors proliferation, cultures were incubated for 24 hours in the absence and presence of 10 nM PACAP, and exposed to [^{3}H]thymidine ([^{3}H]dT)

TABLE 1. PACAP expression in neural tissues

Tissue	Protein (pg/mg)
Cerebral cortex (E14.5)	480 ± 15
Superior cervical ganglion (E15.5)	1,101 ± 117
Liver (adult)	N.D.

Note: Liver serves as a negative control. N.D.=not detected. Data from Lu and DiCicco-Bloom.[38]

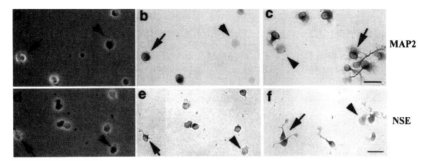

FIGURE 4. Characterization of E13.5 cortical precursor cultures. Cells were incubated for 3 h (**a,b,d,e**) or 24 h (**c,f**) and examined by phase (**a,d**) or brightfield microscopy to detect neuronal markers, immunoreactive MAP2 (**b,c**) and NSE (**e,f**). Neuronal markers increase from ~50% at 3 h (**b,e**) to ~90% at 24 h (**c,f**) (Positive staining indicated by *arrows*; negative cells = *arrowhead*). Cells were unstained when primary antibody was omitted. Bar = 50 μm. (*Reprinted with permission from Lu and DiCicco-Bloom.*[38])

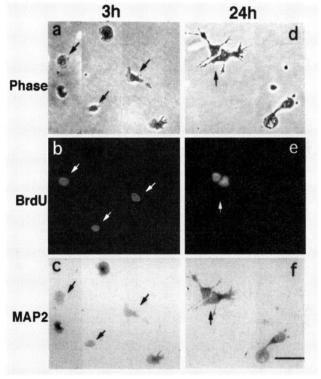

FIGURE 5. Cortical precursors differentiate following proliferation. Cells were exposed to BrdU for 3 h after plating and were fixed either immediately (**a,b,c**) or after further incubation in BrdU-free medium until 24 h (**d,e,f**). The fixed cultures were processed for double immunostaining for BrdU and MAP2. Cells were examined under phase (**a,d**), epifluorescence (**b,e**), and brightfield (**c,f**) microscopy. (**a,b,c**) BrdU-labeled precursors (*arrows*) fixed at 3 h did not express MAP2, suggesting that mitosis and differentiation do not occur at the same time. (**d,e,f**) BrdU-positive cells subsequently expressed MAP2 at 24 h (*arrow*), indicating that following mitosis, cortical precursors differentiated. Bar=50 μm. (*Reprinted with permission from Lu and DiCicco-Bloom.*[38])

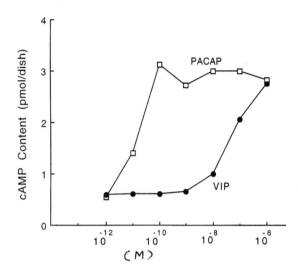

FIGURE 6. PACAP elicits intracellular signaling in precursors. 24 h cultures were treated with increasing concentrations of PACAP or VIP for 15 min. PACAP increased cAMP levels in a dose-dependent manner, indicating that precursors express functional receptors. VIP was 1,000-fold less potent, consistent with action via PACAP-R.

during the final 4 hours prior to assay for incorporation. PACAP elicited a 43% reduction in [^{3}H]dT incorporation, suggesting that the peptide inhibited DNA synthesis (FIG. 7). However, to examine possible peptide toxicity, we examined cell numbers over 3 days in culture. PACAP did not reduce cell number compared to controls at any time point, suggesting that the peptide was not a general cell toxin (FIG. 7). To further reduce possible confounding effects of changes in cell numbers, we developed a 6-hour culture paradigm,[39] in which cell numbers at termination were not different from those at initial plating. While addition of PACAP elicited no change in cell number at 6 hours, the peptide reduced incorporation by 30%, suggesting that the peptide inhibited DNA synthesis (FIG. 7). Moreover, PACAP reduced the percent of cells engaged in mitotic S-phase (the labeling index) by 35%, suggesting that the peptide induced cells to exit the cell cycle (FIG. 7). Finally, as PACAP elicits increased cAMP in cortical precursors, we examined effects of this pathway. Indeed, activation of the cAMP pathway also inhibited DNA synthesis without inducing cell death (FIG. 7). Thus the PACAP ligand/receptor system may play a role in inducing precursors to withdraw from the proliferative cycle in the developing nervous system.

As discussed above, PACAP signaling is likely to interact with other factors involved in cortical neurogenesis. Since both bFGF and IGF-I stimulate cortical precursor mitosis, we examined potential interactions. In fact, PACAP not only inhibited ongoing DNA synthesis, it also blocked mitogenic stimulation elicited by bFGF and IGF-I in culture (FIG. 8). While further studies will be required, PACAP may inhibit mitosis by several possible mechanisms including direct interaction with cell cycle machinery, interruption of growth factor signaling via tyrosine kinase receptors, or by inducing cell lineage decisions that indirectly determine cell responsiveness to bFGF and IGF-I. All of these pathways are currently under examination.

Effect of PACAP on Cortical Precu   Differentiation

As cortical precursors initiate differentiation following exit from the cell cycle, we examined PACAP effects on two markers of the neuronal phenotype, neuritic process outgrowth and

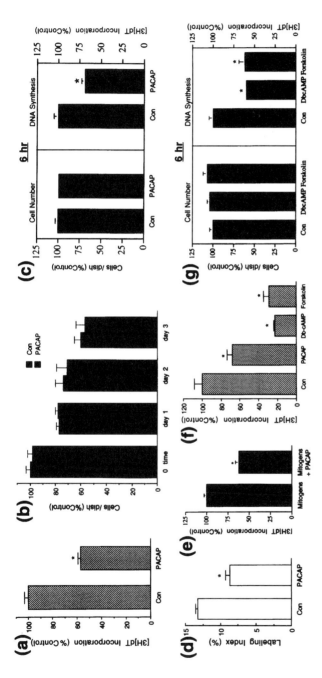

FIGURE 7. PACAP inhibits precursor mitosis. (a) At 24 h, PACAP (10 nM) inhibited [³H]dT incorporation by 43%. Data were from seven experiments, four culture wells/group, N=28. (Con cpm=534–1508) (*) Differs from control at p<0.02. (b) PACAP did not elicit decreased cell numbers compared to respective controls over 3 days (3 experiments). (c) At 6 h, while PACAP did not elicit decreased cell number, the peptide inhibited [³H]dT incorporation, indicating that the inhibitory effect was not due to toxicity. Data expressed as % control±SEM (*) p<0.02. (d) PACAP decreased the proportion of cells engaged in the mitotic cycle at 24 h (labeling index, LI=BrdU-positive cells/total). Data expressed as mean%±SEM. (*) p<0.001. (e) PACAP inhibited [³H]dT incorporation in the presence of endogenous mitogens at 24 h [insulin (10 μg/ml) or IGF1 (10 nM), bFGF (20 ng/ml), and EGF (20 ng/ml) (3 experiments). (Mitogens cpm=3,204–4,700) (*) p<0.01. (f) Db-cAMP (1 mM) and Forskolin (30 μM) inhibited [³H]dT incorporation at 24 h, producing similar though greater effects than PACAP (2 experiments). (*) p<0.001. (g) At 6 h, while Db-cAMP and Forskolin did not elicit decreased cell numbers, the drugs inhibited [³H]dT incorporation, indicating that the inhibitory effects were not due to toxicity. (*) p<0.0004. (Reprinted with permission from Lu and DiCicco-Bloom.[38])

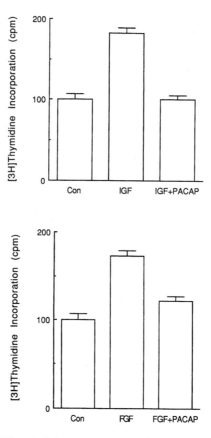

FIGURE 8. PACAP inhibits mitosis induced by cortical mitogens. Cultures were incubated in control medium or in medium containing PACAP (10 nM), IGF-I (10 ng/ml), or bFGF (10 ng/ml). [³H]dT incorporation was assayed at 24 h. PACAP blocked stimulation elicited by both IGF-I and bFGF, suggesting that stimulatory and inhibitory signals interact in mitogenic regulation. Data expressed as % control±SEM.

expression of the neurotrophin receptor for BDNF, trkB. PACAP increased the proportion of precursors expressing trkB immunoreactivity by 30%, suggesting that the peptide may influence cell response to this long-term survival factor (FIG. 9). Moreover, PACAP doubled the proportion of precursors exhibiting neurites in culture, suggesting that the peptide enhanced both morphological and biochemical differentiation (FIG. 9). Finally, increases in differentiation were also induced by cAMP agonists (FIG. 9). Thus we tentatively conclude that PACAP elicits the transition from proliferation to differentiation during cerebral cortex ontogeny.

Origin and Autocrine Function of PACAP Signaling

In light of the ontogenetic functions of exogenous PACAP and our detection of peptide in freshly dissected tissue, we were particularly interested in ligand and receptor expression

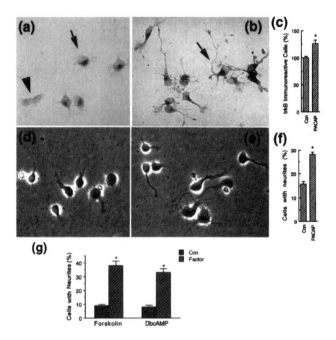

FIGURE 9. PACAP enhances cortical precursor differentiation. Precursors were incubated in control (plus vehicle) or PACAP-containing medium, and analyzed for *trk*B expression (**a,b,c**) and neurite outgrowth (**d,e,f**). (**a**) Brightfield photomicrograph of *trk*B immunoreactive precursors at 24 h: only a subset express *trk*B, localized to cytoplasm (*arrows*; *arrowhead*=negative cell; Santa Cruz antiserum) (**b**) Similar staining was observed with a different antiserum (Dr. D. Kaplan, Montreal University). (**c**) PACAP increased *trk*B-positive cells by 30%. Data expressed as % control ± SEM, range of positive cells: Con=36–67%; PACAP=57–84%. (four expertiments using both antisera), N=14, (*) $p < 0.01$. (**d,e**) Phase photomicrographs of control (**d**) and PACAP-treated cultures at 48 h (**e**). (**f**) PACAP doubled the percent of cells with neurites that were ≥2 cell soma diameters. Data expressed as mean % SEM, $N = 15$, (*) $p < 0.007$. (**g**) DbcAMP and Forskolin elicited a fourfold increase in neurite-bearing cells. (*) $p < 0.001$. Bar=50 μm. (*Reprinted with permission from Lu and DiCicco-Bloom.*[38])

in cortical precursors. That is, do the dividing cells possess receptor and/or peptide, as suggested from immunohistochemical studies? Potentially, PACAP may act on postmitotic cells that indirectly influence proliferation. To define expression, we initially examined PACAP and type I receptor immunoreactivity in cultured precursors at 24 hours. We found that 85% of precursors exhibited cytoplasmic PACAP signal, whereas 64% exhibited receptor immunoreactivity (Fig. 10). Further, to determine whether receptor expression correlated with function, we exposed precursors to PACAP for 15 minutes, and then used immunocytochemistry to detect nuclear expression of phosphorylated cAMP response element binding protein, P-CREB, a transcription factor downstream of cAMP/PKA pathway activation. Remarkably, 62% of cells exhibited nuclear P-CREB signal after PACAP, whereas only 2% of control cells responded, indicating that approximately two thirds of the population expressed functional PACAP receptors (Fig. 10).

The high proportions of cells exhibiting PACAP and type I receptor expression indicated that at least 50% expressed both components, raising the possibility that some cells may release PACAP and respond to the peptide, in an autocrine fashion. In this model, PACAP may be tonically released and active *in vitro*, providing an ongoing signal to leave

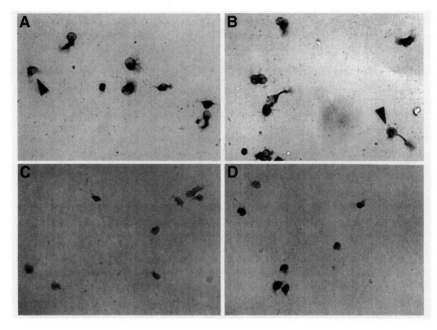

FIGURE 10. Precursors express both PACAP and PACAP receptor. (**A**) At 24 h, 85% of precursors expressed intense PACAP immunoreactivity localized to the cytoplasm. A sole negative cell is indicated by the arrowhead. Staining was blocked by antibody preincubation with PACAP, but not related peptides, secretin, and PHI. (**B**) At 24 h, 64% of cells exhibited immunoreactivity with a PACAP type I receptor antiserum. *Arrowhead* indicates negative cell. (**C, D**) Detection of PACAP-responsive precursors by examining P-CREB immunoreactivity in 24 h cultures. Fifteen minutes after PACAP exposure, nuclear P-CREB staining was detected in 62$\pm$3% of precursors (**D**), whereas only 2.3$\pm$0.3% of cells exhibited signal in the vehicle-treated control (**C**). (*Reprinted with permission from Lu and DiCicco-Bloom.*[38])

the cell cycle. If this is the case, addition of a PACAP antagonist would block ongoing inhibition, resulting in increased proliferation. Indeed, PACAP(6-38), a selective PACAP antagonist, increased both [3H]thymidine incorporation, by 30%, and the labeling index, by 60%, indicating that more precursors entered the mitotic cycle after blocking endogenous peptide function (FIG. 11). Moreover, similar increases were elicited by inhibiting peptide activity using a different reagent, PACAP neutralizing antibody, confirming that the peptide serves as an autocrine regulator of mitosis (FIG. 12). Finally, when cultured in an enriched serum-containing environment, blockade of PACAP activity resulted in a net 16% increase in cell number, whereas proliferation was prevented by peptide addition (FIG. 13). Thus ongoing expression and function of PACAP in culture inhibits precursor proliferation, potentially accounting for the previously documented withdrawal of cells from the cycle under these low density culture conditions. These observations suggest that the peptide signaling system plays a similar role during neurogenesis *in vivo*.

To directly define relationships of peptide signaling to mitosis, we subsequently employed double immunostaining. Cortical cells were plated for only 3 hours in BrdU, and then fixed and analyzed. We expect that peptide and receptor expression 3 hours after dissection likely reflects patterns in the embryo. We found that 95% of mitotic, BrdU-positive precursors expressed both PACAP peptide and receptor, suggesting that the peptide acts directly on the mitotic cells. Moreover, since PACAP is contained in the cytoplasm of cells in S-phase, we hypothesize that ventricular zone precursors initiated PACAP expression to

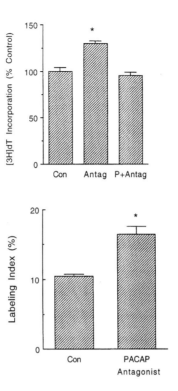

FIGURE 11. Blockade of PACAP function increased [³H]dT incorporation. (*Top*) Cells were incubated for 24 h in control medium or in medium containing peptide antagonist, PACAP$_{6-38}$ (100 nM). Antagonist-induced stimulation was reversed by coincubation with PACAP. (*) $p<0.004$. (*Bottom*) Blockade of PACAP function by antagonist elicited a 60% increase in the proportion of mitotic precursors (LI). (*) $p<0.001$.

induce their own cell cycle exit in autocrine fashion, though a feedback mechanism may also operate at later developmental times. Finally, it should be noted that PACAP may not be the only signal inducing cell cycle exit. As postmitotic neurons accumulate in the forming cortical plate, developing neurotransmitter systems, such as GABA and glutamate, may also play a feedback role in neurogenetic regulation.[40-42]

PACAP Receptor Signaling in Cortical Precursors

Since the PACAP receptor may exist in several splice isoforms that couple differentially to cAMP and phosphatidyl inositol (PI) pathways,[43] we examined cortical precursors and compared expression to embryonic sympathetic neuroblasts. As described above, PACAP elicited cAMP (FIG. 6) and P-CREB activation (FIG. 10) in 15 minutes and mitotic inhibition in 6 hours (FIG. 7), indicating the importance of this transduction pathway. However, PACAP did not elicit changes in PI turnover in cortical precursors, while the peptide increased sympathetic inositol phosphates fivefold.

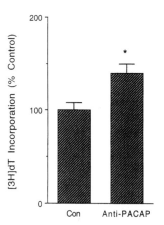

FIGURE 12. PACAP neutralizing antiserum increased [³H]dT incorporation. Precursors were incubated in control medium containing nonimmune rabbit serum or medium containing PACAP antiserum (1:3,000) for 24 h. (*) $p < 0.016$.

Northern analysis indicated that cortical precursors expressed easily detectable levels of PACAP receptor mRNA. Further analysis, using published primers,[43,44] indicated that precursors expressed primarily the "short" receptor isoform, containing no insert in the third intracellular loop, while much lower levels of a one 84 bp insert isoform were detected. This contrasts with sympathetic neuroblasts in which we only found a one insert receptor form. Further RT-PCR analysis revealed that the minor insert receptor isoform in cortex was the "hop" isoform, whereas sympathetics apparently expressed this form as its major mRNA, as shown also for newborn sympathetic neurons.[45,48] Thus these studies suggest that PACAP inhibits cortical precursor mitosis via the "short" (non-insert) receptor

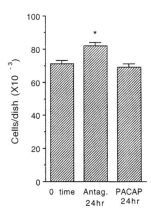

FIGURE 13. Blockade of PACAP function elicits precursor proliferation. In serum containing medium, cell growth was observed at 24 h only in the presence of PACAP antagonist (100 nM), but not PACAP peptide (10 nM), suggesting that endogenous PACAP inhibits proliferation. Data expressed as mean cells/dish±SEM, (2 experiments), N=6, (*) $p < 0.009$. (*Reprinted with permission from Lu and DiCicco-Bloom.*[38])

isoform, while the peptide stimulates sympathetic neuroblasts by activating both cAMP and PI through the "hop" receptor subtype.

IMPLICATIONS OF PACAP SIGNALING IN CORTICAL DEVELOPMENT

Our studies suggest that the PACAP ligand/receptor system elicits the transition from proliferation to differentiation in cortical precursors. Potentially, cells express the peptide to initiate this ontogenetic sequence. A not entirely exclusive alternative is that as postmitotic neurons accumulate in the emerging cortical plate, peptide is synthesized there and released to influence mitosis in the underlying proliferative ventricular zone. Ongoing *in vivo* studies will begin to address these issues. However, since the PACAP system is expressed widely throughout the anterior and posterior developing neural tube and neural crest–derived peripheral nervous system,[37,49] the peptide may play a critical role in multiple sites. While PACAP may simply elicit cell cycle exit in all mitotic precursors, studies in culture, in which PACAP receptor activation elicits increased sympathetic mitogenesis,[14,15,46] suggest that activity may be cell-type specific.[47] Significantly, the PACAP system is expressed at a time when proliferation, neural determination, and neuronal differentiation are actively underway. Thus it will be interesting to investigate PACAP interactions with the array of intrinsic genetic programs and extracellular factors currently known to participate in these patterning events.

REFERENCES

1. KANDEL, E. R., J. H. SCHWARTZ & T. M. JESSELL. 1991. Principles of Neural Science. Elsevier. New York.
2. JACOBSON, M. 1991. Developmental Neurobiology. Plenum Publishing Corp. New York.
3. FRANTZ, G. D., A. P. BOHNER, R. M. AKERS & S. K. MCCONNELL. 1994. Regulation of the POU gene SCIP during cerebral cortical development. J. Neurosci. **14:** 472–485.
4. FRANTZ, G. D., J. M. WEIMANN, M. E. LEVIN & S. K. MCCONNELL. 1994. Otx1 and Otx2 define layers and regions in developing cerebral cortex and cerebellum. J. Neurosci. **14:** 5725–5740.
5. BULFONE, A., S. M. SMIGA, K. SHIMAMURA, A. PETERSON, L. PUELLES & J. L. R. RUBINSTEIN. 1995. T-brain-1: A Homolog of Brachyury whose expression defines molecularly distinct domains within the cerebral cortex. Neuron **15:** 63–78.
6. BREDT, D. S., C. E. GLATT, P. M. HWANG, M. FOTUHI, T. M. DAWSON & S. H. SNYDER. 1991. Nitric oxide synthase protein and mRNA are discretely localized in neuronal populations of the mammalian CNS together with NADPH diaphorase. Neuron **7:** 615–624.
7. REINHARD, E., H. S. SUIDAN, A. PAVLIK & D. MONARD. 1994. Glia-derived nexin/protease nexin-1 is expressed by a subset of neurons in the rat brain. J. Neurosci. **37:** 256–270.
8. STANFIELD, B. B. & D. M. JACOBOWITZ. 1990. Antibody to a soluble protein purified from brain selectively labels layer V corticofugal projection neurons in rat neocortex. Brain Res. **532:** 219–224.
9. DEL RIO, J. A., A. MARTINEZ, M. FONSECA, C. AULADELL & E. SORIANO. 1995. Glutamate-like immunoreactivity and fate of Cajal-Retzius cells in the murine cortex as identified with calretinin antibody. Cerebral Cortex **5:** 13–21.
10. PETERS, A. & E. G. JONES, Eds. 1988. Cerebral Cortex: Development and Maturation of Cerebral Cortex. Plenum Press. New York, NY.
11. DICICCO-BLOOM, E. & I. B. BLACK. 1988. Insulin growth factors regulate the mitotic cycle in cultured rat sympathetic neuroblasts. Proc. Natl. Acad. Sci. USA **85:** 4066–4070.
12. DICICCO-BLOOM, E. & I. B. BLACK. 1989. Depolarization and insulin-like growth factor-I (IGF-I) differentially regulate the mitotic cycle in cultured rat sympathetic neuroblasts. Brain Res. **491:** 403–406.
13. DICICCO-BLOOM, E., E. TOWNES-ANDERSON & I. B. BLACK. 1990. Neuroblast mitosis in dissociated culture: Regulation and relationship to differentiation. J. Cell Biol. **110:** 2073–2086.

14. PINCUS, D. W., E. DICICCO-BLOOM & I. B. BLACK. 1990. Vasoactive intestinal peptide regulates mitosis, differentiation and survival of cultured sympathetic neuroblasts. Nature **343:** 564–567.
15. PINCUS, D. W., E. DICICCO-BLOOM & I. B. BLACK. 1990. Vasoative intestinal peptide regulation of neuroblast mitosis and survival: Role of cAMP. Brain Res. **514:** 355–357.
16. PINCUS, D. W., E. DICICCO-BLOOM & I. B. BLACK. 1991. Role of voltage-sensitive calcium channels in mitogenic stimulation of neuroblasts. Brain Res. **553:** 211–214.
17. PINCUS, D. W., E. DICICCO-BLOOM & I. B. BLACK. 1994. Trophic mechanisms regulate mitotic neuronal precursors: Role of vasoactive intestinal peptide (VIP). Brain Res. **663:** 51–60.
18. DICICCO-BLOOM, E. M., W. J. FRIEDMAN & I. B. BLACK. 1993. NT3 stimulates sympathetic neuroblast proliferation by promoting precursor survival. Neuron **11:** 1101–1111.
19. DUNCAN, M., E. M. DICICCO-BLOOM, X. XIANG, R. BENEZRA & K. CHADA. 1992. The gene for the helix-loop-helix protein, Id, is specifically expressed in neural precursors. Dev. Biol. **154:** 1–10.
20. DUNCAN, M. K., L. BORDAS, E. DICICCO-BLOOM & K. K. CHADA. 1997. The expression of the helix-loop-helix genes Id-1 and NSCL-1 during cerebellar development. Dev. Dynamics **208:** 107–114.
21. TAO, Y., I. B. BLACK & E. DICICCO-BLOOM. 1996. Neurogenesis in neonatal rat brain is regulated by peripheral injection of basic fibroblast growth factor (bFGF). J. Comp. Neurol. **376:** 653–663.
22. TAO, Y., I. B. BLACK & E. DICICCO-BLOOM. 1997. *In vivo* neurogenesis is inhibited by neutralizing antibodies to basic fibroblast growth factor (bFGF). J. Neurobiol. **33:** 289–296.
23. ANGEVINE, J. B. & R. L. SIDMAN. 1961. Autoradiographic study of cell migration during histogenesis of cerebral cortex in the mouse. Nature **192:** 766–768.
24. BERRY, M. & A. W. ROGERS. 1965. The migration of neuroblasts in the developing cerebral cortex. J. Anat. **99:** 691–709
25. MCCONNELL, S. K. & C. E. KAZNOWSKI. 1991. Cell cycle dependence of laminar determination in developing neocortex. Science **254:** 282–285.
26. MCCONNELL, M. K. 1991. The generation of neuronal diversity in the central nervous system. Annu. Rev. Neurosci. **14:** 269–300.
27. NURCOMBE, V., M. D. FORD, J. A. WILDSCHUT & P. F. BARTLETT. 1993. Developmental regulation of neural response to FGF-1 and FGF-2 by heparan sulfate proteoglycan. Science **260:** 103.
28. FAYEIN, N., Y. COURTOIS & J. JEANNY. 1992. Basic fibroblast growth factor high and low affinity binding sites in developing mouse brain, hippocampus and cerebellum. J. Biol. Cell. **76:** 1–13.
29. YAZAKI, N. *et al.* 1994. Differential expression patterns of mRNAs for members of the fibroblast growth factor receptor family, FGFR-1 - FGFR-4, in rat brain. J. Neurosci. Res. **37:** 445–452.
30. ORR-URTREGAR, A., D. GIVOL, A. YAYON, Y. YARDEN & P. LONAI. 1991. Developmental expression of two murine fibroblast growth factor receptors, flg and bek. Development **113:** 1419–1434.
31. ORR-URTREGAR, A., M. T. BEDFORD, T. BURAKOVA, E. ARMAN. Y. ZIMMER, A. YAYON, D. GIVOL & P. LONAI. 1993. Developmental localization of the splicing alternatives of fibroblast growth factor receptor-2 (FGFR2). Dev. Biol. **158:** 475–486.
32. DRAGO, J., M. MURPHY, S. M. CARROLL, R. P. HARVEY & P. F. BARTLETT. 1991. Fibroblast growth factor-mediated proliferation of central nervous system precursors depends on endogenous production of insulin-like growth factor I. Proc. Natl. Acad. Sci. USA **88:** 2199–2203.
33. GHOSH, A. & M. E. GREENBERG. 1995. Distinct roles for bFGF and NT-3 in the regulation of cortical neurogenesis. Neuron **15:** 89–103.
34. VICARIO-ABEJON, C., K. K. JOHE, T. G. HAZEL, D. COLLAZO & R. D. G. MCKAY. 1995. Functions of basic-fibroblast growth factor and neurotrophins in the differentiation of hippocampal neurons. Neuron **15:** 104–114.
35. TATSUNO, I., A. SOMOGYVARI-VIGH & A. ARIMURA. 1994. Developmental changes of pituitary adenylate cyclase activating polypeptide (PACAP) and its receptor in the rat brain. Peptides **15:** 55–60.
36. SHUTO, Y., D. UCHIDA, H. ONDA & A. ARIMURA. 1996. Ontogeny of pituitary adenylate cyclase activating polypeptide and its receptor mRNA in the mouse brain. Regul. Pep. **67:** 79–83.

37. SHEWARD, W. J., E. M. LUTZ & A. J. HARMAR. 1996. Expression of pituitary adenylate cyclase activating polypeptide receptors in the early mouse embryo as assessed by reverse transcription polymerase chain reaction and in situ hybridisation. Neurosci. Lett. **216:** 45–48.
38. LU, N. & E. DICICCO-BLOOM. 1997. Pituitary adenylate cyclase activating polypeptide is an autocrine inhibitor of mitosis in cultured cortical precursor cells. Proc. Natl. Acad. Sci. USA **94:** 3357–3362.
39. LU, N., I. B. BLACK & E. DICICCO-BLOOM. 1996. A paradigm for distinguishing the roles of mitogenesis and trophism in neuronal precursor proliferation. Dev. Brain Res. **94:** 31–36.
40. LOTURCO, J., D. OWENS, M. J. S. HEATH, M. B. E. DAVIS & A. R. KRIEGSTEIN. 1995. GABA and glutamate depolarize cortical progenitor cells and inhibit DNA synthesis. Neuron **15:** 1287–1298.
41. ANTONOPOULOS, J., I. S. PAPPAS & J. G. PARNAVELAS. 1997. Activation of the GABA$_A$ receptor inhibits the proliferative effects of bFGF in cortical progenitor cells. Eur. J. Neurosci. **9:** 291–298.
42. LAMANTIA, A. S. 1995. The usual suspects: GABA and glutamate may regulate proliferation in the neocortex. Neuron **15:** 1223–1225.
43. SPENGLER, D., C. WAEBER, C. PANTALONI, F. HOISBOER, J. BOCKERT, P. SEEBURG & L. JOURNOT. 1993. Differential signal transductions by five splice variants of the PACAP receptor. Nature **360:** 170–175.
44. HOSOYA, M., H. ONDA, K. OGI, Y. MASUDA, Y. MIYAMOTO, T. OHTAKI, H. OKAZAKI, A. ARIMURA & M. FUJINO. 1993. Biochem. Biophys. Res. Commun. **194:** 133–143.
45. MAY, V. & K. M. BRAAS. 1995. Pituitary adenylate cyclase-activating polypeptide (PACAP) regulation of sympathetic neuron neuropeptide Y and catecholamine expression. J. Neurochem. **65:** 978–987.
46. DICICCO-BLOOM, E. & P. DEUTSCH. 1992. Pituitary adenylate cyclase activating peptide (PACAP) potently stimulates mitosis, neuritogenesis and survival in cultured rat sympathetic neuroblasts. Regul. Pep. **37:** 319.
47. DICICCO-BLOOM, E. 1996. Region-specific regulation of neurogenesis by VIP and PACAP: Direct and indirect modes of action. Ann. N.Y. Acad. Sci. **805:** 244–258.
48. MAY, V., M. M. BEUDEF, R. L. PARSONS, J. C. HARDWICK, E. A. GAUTHIER, P. J. DURDA & K. M. BRAAS. 1998. Mechanisms of Pituitary Adenylate Cyclase Activating polypeptide (PACAP)-induced depolarization of superior cervical ganglion (SGG) neurons. Ann. N.Y. Acad. Sci. This volume.
49. LINDHOLM, D., Y. SKOGLÖSA & N. TAKEI. 1998. Developmental regulation of pituitary adenylate cyclase activating polypeptide (PACAP) and its receptor 1 in rat brain: Function of PACAP as a neurotrophic factor. Ann. N.Y. Acad. Sci. This volume.

VIP and Breast Cancer

T. W. MOODY,[a,d] J. LEYTON,[a] I. GOZES,[b] L. LANG, AND
W. C. ECKELMAN[c]

[a]Medicine Branch, National Cancer Institute, Rockville, Maryland, USA

[b]Department of Pathological Chemistry, University of Tel Aviv, Tel Aviv, Israel

[c]Positron Emission Tomography Department, Clinical Center, Bethesda, Maryland, USA

ABSTRACT: VIP_1 receptors are present in breast cancer cells. VIP elevates the cAMP and stimulates nuclear oncogene expression in MCF-7 cells. VIPhybrid is a VIP receptor antagonist that inhibits breast cancer proliferation. A VIP analog has been developed for imaging breast tumors. Therefore VIP_1 receptors may be utilized for the early detection and treatment of breast cancer.

B reast cancer kills approximately 40,000 women in the United States annually.[1] Traditionally it is treated with surgical resection and/or chemotherapy. Approximately half of the breast cancers are estrogen receptor–positive and hence the antiestrogen, tamoxifen, is utilized for treatment.[2] Better survival rates for breast cancer patients may be obtained with early detection and treatment.

Several growth factors have been identified in breast cancer including epidermal growth factor (EGF), erbB2, insulin-like growth factor (IGF), and platelet-derived growth factor (PDGF).[3–6] EGF binds with high affinity to breast cancer cells and can cause signal transduction through the RAS, phosphatidylinositol (PI), and Janus kinase pathways.[7] Of these, the RAS pathway appears to be most important and RAS activates RAF, which subsequently phosphorylates MEK; activated MEK phosphorylates MAPK. Activated MAPK can enter the nucleus altering expression of nuclear oncogenes such as c-fos, c-jun, and/or c-myc.[8] C-myc and c-erbB2 are overexpressed in approximately 30% of the breast cancer cases.[9,10]

High densities of vasoactive intestinal peptide (VIP) receptors have been detected in breast cancer.[11] Recently, we found that the receptor antagonist VIPhybrid inhibited the growth of breast cancer cells.[12] In particular VIPhybrid inhibited binding of ^{125}I-VIP, elevation of cAMP caused by VIP, elevation of nuclear oncogenes caused by VIP, and proliferation. Here the effects of VIP agonists on breast cancer were investigated.

VIP RECEPTORS

Three types of VIP receptors have been cloned including VIP_1, VIP_2, and pituitary adenylate cyclase activating polypeptide (PACAP) receptor.[13–16] Each of these G protein–coupled receptors stimulates adenylate cyclase but PACAP receptors stimulate both adenylate cyclase

[d]Corresponding author: Dr. Terry W. Moody, Medicine Branch, Bldg. KWC, Rm. 300, 9610 Medical Center Drive, Rockville, MD 20850 USA; Tel.: 301-402-3128; Fax: 301-402-4422

and PI pathways. The pharmacology of these receptors is distinct in that VIP_1 receptors prefer VIP and PACAP relative to helodermin (Hd), VIP_2 receptors prefer Hd, and PACAP receptors prefer PACAP.

TABLE 1 shows that VIP binds with high affinity to breast cancer cells and specific ^{125}I-VIP binding is inhibited with high affinity by VIP and PACAP-27 (IC_{50} values of 10 and 5 nM, respectively) and moderate affinity by VIPhybrid (IC_{50} value of 360 nM). Specific ^{125}I-VIP binding is inhibited with moderate affinity by peptide histidine isoleucine (PHI), Hd, and VIP^{10-28} (IC_{50} values of 200, 100, and 2,000 nM respectively, TABLE 1). These data suggest that ^{125}I-VIP is binding with high affinity to VIP_1 receptors present in MCF-7 cells. Previously we found that ^{125}I- VIP binds with high affinity to breast cancer biopsy specimens.[12]

VIP_1, VIP_2, and PACAP receptors contain 459, 437, and 495 amino acids, respectively, and cross the plasma membrane seven times. Each of the receptors has approximately 50% homology at the protein level. Using molecular biology techniques, the receptors can be detected at the mRNA level. By using primers for VIP_1 receptors and RT-PCR techniques, it can be shown that VIP_1 receptor mRNA is present in breast cancer cells (FIG. 1). In particular, a major 324 bp band was detected using cDNA derived from MCF-7 and T47D cells; similar results were obtained using cDNA derived from MDA-MB231 and SKBr3 cells. In contrast, using primers for the human VIP_2 receptor no bands were detected. These data indicate that breast cancer cell lines have VIP_1 but not VIP_2 receptor mRNA. Also, VIP_1 receptor mRNA was detected in all breast biopsy specimens examined.

ADENYLATE CYCLASE

VIP_1 receptors interact with a stimulatory guanine nucleotide binding protein activating adenylate cyclase. TABLE 2 shows that VIP and PACAP strongly elevated intracellular cAMP as determined by radioimmunoassay. In contrast, helodermin slightly stimulated cAMP, whereas glucagon had no effect. These data suggest that breast cancer VIP_1 receptors stimulate adenylate cyclase activity. Previously we showed that VIPhybrid had no effect on basal cAMP but inhibited the increase in cAMP caused by VIP.[12]

NUCLEAR ONCOGENES

The elevated cAMP stimulates protein kinase A. Protein kinase A may phosphorylate RAF, which will stimulate MEK and MAP kinase.[20] MAP kinase may phosphorylate sis-inducing factor (SIS) causing nuclear translocation and interaction with the sis-inducing

TABLE 1. Specificity of VIP binding

Peptide	IC_{50}(nM)
VIP	10
PACAP-27	5
(RR)VIP	15
Helodermin	100
PHI	200
VIPhybrid	360
VIP^{10-28}	2,000

The ability of peptides to half-maximally inhibit specific ^{125}I-VIP binding to MCF7 cells was determined using the radioreceptor assay described previously.[17] The structures of the peptides are shown below and amino acid homologies relative to VIP are underlined.

VIP	HSDAVFTDNYTRLRKQMAVKKYLNSILN*
VIPhyb	KPRRPYTDNYTRLRKQMAVKKYLNSILN*
(R^{15},R^{21})VIP	HSDAVFTDNYTRLRRQMAVKRYLNSILN*
PACAP-27	HSDGIFTDSYSRYRKQMAVKKYLAAVL*

 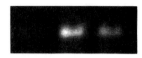

FIGURE 1. VIP_1 receptor mRNA. A 324 bp PCR product was observed for MCF-7 (**b**) and T47D (**c**) cells, but not a control lacking Taq polymerase (**a**). Total RNA was isolated from breast cancer cells using the guanidinium isothiocyanate (GIT) method. The cDNA was made from the RNA by reverse transcriptase. Specific primers for human VIP_1 receptor were 5′- ATGTGCAGATGATC-GAGGTG-3′ (sense, 127–146); 5′- TGTAGCCGGTCTTCACAGAA-3′ (antisense, 431–450). PCR was performed on a Perkin-Elmer/Cetus thermal cycler using the following conditions: 94°C for 45 see; 60°C for 30 sec, and 72°C for 60 sec. After 35 cycles of amplification using Taq polymerase, the PCR products were separated using a 2% agarose gel. Ethidium bromide fluorescence was used to assess the PCR products.

element (SIE) resulting in increased nuclear oncogene expression.[21,22] After 1 hour, VIP increased c-myc mRNA in a concentration-dependent manner with little increase at 0.1 or 1 nM and a maximal increase at 100 nM (FIG. 2). Ten nM VIP or PACAP-27 half-maximally stimulated c-myc mRNA. FIGURE 3 shows that VIP transiently stimulated c-myc gene expression in a time-dependent manner with a maximum increase at 1 h. Previously we showed that the increase in c-myc or c-fos mRNA stimulated by VIP was inhibited by VIPhybrid.[12] Similarly, VIP stimulated c-fos and c-jun mRNA and the increase caused by VIP was reversed by VIPhybrid.

PROLIFERATION

c-Fos and c-jun proteins form heterodimers and bind to the API recognition sequence TGACTCA altering the expression of downstream genes. VIP may increase expression of growth stimulatory proteins, such as cyclin D, which in turn phosphorylate Rb stimulating the G1 to S transition.[2] TABLE 3 shows that VIPhybrid decreased the proliferation of breast cancer cells using a clonogenic assay. Similarly, using nude mice bearing breast cancer xenografts, VIPhybrid (0.4 mg/kg; s.c.) slowed the growth of MDA-MB231 and T47D xenografts, but caused regression of MCF7 xenografts. Currently, we are investigating if VIPhybrid causes apoptosis of MCF7 cells.

EARLY DETECTION

Traditionally breast cancer is detected by mammography. It was recently shown that using imaging techniques and ^{123}I-VIP, colon tumors could be detected.[23] Here a ^{18}F-VIP analog was generated and its effects investigated *in vitro* and *in vivo*. (Arg^{15}, Arg^{21})VIP

TABLE 2. VIP and cAMP

Addition	fmol
None	2.7 ± 0.6
VIP, 10 nM	70.5 ± 4.2
PACAP, 10 nM	64.6 ± 6.4
Helodermin, 10 nM	7.5 ± 0.5
Glucagon, 1,000 nM	2.3 ± 0.2

The cAMP was determined by radioimmunoassay using MCF-7 cells.[18] The mean value ± S.E. of four determinations is indicated.

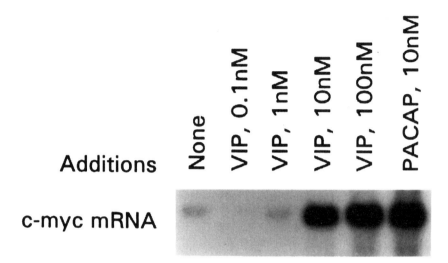

FIGURE 2. Nuclear oncogene mRNA. The ability of various concentrations of VIP and PACAP-27 to elevate c-myc mRNA was determined 1 h after addition to MCF-7 cells. For the c-myc experiments, breast cancer cells were cultured in SIT medium containing 0.5% fetal bovine serum for 4 h. The cells were treated with VIP peptides for 45 min and the RNA isolated using guanidine isothiocyanate (GIT). The RNA was fractionated using an agarose gel, transferred to Nytran and hybridized using a ^{32}P-labeled c-myc cDNA probe. After washing, the membrane was exposed to Kodak XAR-2 film at -80°C for 1 day and the autoradiogram developed. The autoradiograms were analyzed using a Molecular Dynamics densitometer.[19]

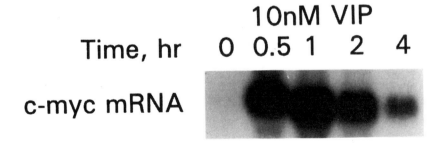

FIGURE 3. c-myc mRNA. The ability of 10 nM VIP to elevate c-myc was determined as a function of time after addition to MCF-7 cells. This experiment is representative of two others.

TABLE 3. Proliferation of breast cancer cell lines

Addition	MCF7	MDAMB231	T47D
None	24 ± 4	93 ± 10	44 ± 6
VIPhybrid, 1 μM	7 ± 2	77 ± 8	18 ± 4

The clonogenic assay was described previously.[12] The mean colony number ± S.E. of three determinations is indicated.

[(RR)VIP)]was a VIP agonist with a single lysine at position 20. (RR)VIP was conjugated to ^{18}F-succinimidylbenzoate and its properties determined.[24] TABLE 1 shows that (RR)VIP binds with high affinity to breast cancer cells.

[18]F-(RR)VIP was internalized by MDF-7 cells at 37°C. Because [18]F-(RR)VIP bound with high affinity to breast cancer cells *in vitro* its ability to bind to breast tumors was investigated *in vivo*. Using nude mice bearing breast cancer tumors [18]F-(RR)VIP was injected intravenously into the tail vein and its distribution determined after 2 hours. TABLE 4 shows that the density of [18]F-(RR)VIP was high in the tumor, kidney, lung, and spleen; moderate in the liver, heart, breast, and bone; and low in the brain (VIP does not readily cross the blood-brain barrier). Also, high levels of [18]F-(RR)VIP were detected in the body fluids, such as the blood and urine. These data suggest that [18]F-(RR)VIP, which is initially present in the blood, is taken up by organs enriched in VIP receptors, but is then concentrated in the kidney and excreted into the urine. Because the density of [18]F-(RR)VIP is higher in malignant than in normal tissues it may be utilized for early detection of breast cancer.

SUMMARY

VIP$_1$ receptors are abundant in lung cancer.[25] The density of ^{125}I-VIP binding sites is an order of magnitude greater than other peptide receptors such as somatostatin. Somatostatin analogs such as octreotide were labeled with ^{125}I or ^{123}I and their distribution determined. ^{125}I-octreotide localized well to neuroendocrine tumors, such as small cell lung cancer, but poorly to epithelial tumors, such as non-small cell lung or breast cancer.[26] ^{125}I-octreotide bound with high affinity to approximately half of the breast cancer biopsy specimens examined. In contrast, ^{125}I-VIP bound with high affinity to all breast cancer biopsy specimens examined.[27] Therefore VIP receptors may be utilized for imaging breast cancer tumors.

^{18}F-VIP localized preferentially to breast tumors as apposed to normal organs. A problem, however, was that ^{18}F-VIP was highly concentrated in the urine. The ^{18}F in the urine appeared to be due to a VIP metabolite, however, and was not authentic VIP. Current

TABLE 4. Biodistribution of [18]F-(RR)VIP

Organ	Density (cpm/mg wet tissue)
Blood	20.2 ± 2.7
Bone	4.9 ± 1.6
Brain	1.3 ± 0.3
Breast	4.7 ± 1.3
Heart	6.3 ± 0.7
Kidney	26.1 ± 11.0
Liver	9.3 ± 1.0
Lung	17.2 ± 1.4
Spleen	13.4 ± 2.0
Tumor	24.6 ± 1.7
Urine	986 ± 226

The mean value ± S.E. of four determinations is indicated after intravenously injecting 1,000,000 cpm of [18]F-(Arg[15],Arg[21])VIP into nude mice bearing T47D xenografts.

efforts focus on reducing the accumulation of product in the urine by prolonging the half-life of VIP in the blood. Because breast cancer cells have approximately 100,000 VIP receptors/cell, the VIP receptor represents a good target for imaging breast tumors.

A goal in cancer research is early detection and early treatment. Previously we identified that a VIP receptor antagonist, VIPhybrid, slows breast cancer growth *in vitro* and *in vivo*. Recently we identified that VIPhybrid is a chemopreventive agent for cancer.[28] It remains to be determined if VIPhybrid can be used to treat breast cancer patients.

ACKNOWLEDGMENTS

The authors thank Drs. M. Birrer and S. Jakowlew for helpful discussions.

REFERENCES

1. HARRIS, H., M. MORROW & L. NORTON. 1997. Malignant tumors of the breast. *In* Cancer: Principles and Practice of Oncology. V. T. Devita Jr., S. Hellman & S. A. Rosenberg, Eds.: 1557–1616. Lippincott-Raven Press. Philadelphia, PA.
2. DICKSON, R. B. & M. E. LIPPMAN. 1997. Molecular biology of breast cancer. *In* Cancer: Principles and Practice of Oncology. V. T. Devita Jr., S. Hellman & S. A. Rosenberg, Eds.: 1541–1557. Lippincott-Raven Press. Philadelphia, PA.
3. DICKSON, R. B. & M. E. LIPPMAN 1995. Growth factors in breast cancer. Endocr. Rev. **16**: 559–589.
4. RAM, T. G., K. E. KOKENY, C. A. DILTS & S. P. ETHIER. 1995. Mitogenic activity of neu differentiation factor/heregulin mimics that of epidermal growth factor and insulin-like growth factor I in human mammary epithelial cells. J. Cell Physiol. **163**: 589–596.
5. CHRYSOGELOS, S. A. & R. B. DICKSON. 1994. EGF receptor expression, regulation and function in breast cancer. Breast Cancer Res. Treat. **29**: 29–40.
6. COLEMAN-KRNACIK, S. & J. M ROSEN. 1994. Differential temporal and spatial gene expression of fibroblast growth factor family members during mouse mammary gland development. Mol. Endocrinol. **8**: 218–229.
7. ROBERTS, T. 1992. A signal chain of events. Nature **360**: 534–535.
8. BEN-LEVY, R., H. F. PATERSON, C. J. MARSHALL & Y. YARDEN. 1994. A single autophosphorylation site confers oncogenicity to the neu/erbB- 2 receptor and enables coupling to the MAP kinase pathway. EMBO J. **13**: 3302–3311.
9. EISENMAN, R. N. 1994. Myc, Max and Mad: A regulatory network. Adv. Oncol. **10**: 7.
10. HYNES, N. E. & D. F. STERN. 1995. The biology of erbB-2/neu/her2 and its role in cancer. Biochim. Biophys. Acta **1198**: 165.
11. GESPACH, C., W. BAWAB, P. DECREMOUX & F. CALVO. 1988. Pharmacology, molecular identification and functional characteristics of vasoactive intestinal peptide receptors in human breast cancer cells. Cancer Res. **48**: 5079–5083.
12. ZIA, H., T. HIDA, S. JAKOWLEW, M. BIRRER, Y. GOZES, J. C. REUBI, M. FRIDKIN, I. GOZES & T. W. MOODY. 1996. Breast cancer growth is inhibited by VIPhybrid, a synthetic VIP receptor antagonist. Cancer Res. **56**: 3486–3489.
13. HARMAR, T. & E. LUTZ. 1994. Multiple receptors for PACAP and VIP. Trends Pharmacol. Sci. **15**: 97–99.
14. ISHIHARA, T., R. SHIGEMOTO, K. MORI, K. TAKAHASHI & S. NAGATA. 1992. Functional expression and tissue distribution of a novel receptor for vasoactive intestinal polypeptide. Neuron **8**: 811–819.
15. PISEGNA, J. R. & S. A. WANK. 1993. Molecular cloning and functional expression of the pituitary adenylate cyclase activating polypeptide type I receptor. Proc. Natl. Acad. Sci. USA **90**: 6345–6349.
16. SPENGLER, D., C. WAEBER, C. PANTALONI, G. HOLSBOER, J. BOCKAERT, P. H. SEEBERG & L. JOURNOT. 1993. Differential signal transduction by five splice variants of the PACAP receptor. Nature **365**: 170–175.

17. MOODY, T. W., F. ZIA, D. BRENNEMAN, M. FRIDKIN & I. GOZES. 1993. A VIP antagonist inhibits the growth of non-small cell lung cancer. Proc. Natl. Acad. Sci. USA **90**: 4345–4349.

18. KORMAN, L. Y., D. N. CARNEY, M. CITRON & T. W. MOODY. 1986. Secretin/VIP stimulated secretion of bombesin-like peptides from human small cell lung cancer. Cancer Res. **46**: 1214–1218.

19. DRAOUI, M., T. HIDA, S. JAKOWLEW, M. BIRRER, F. ZIA & T. W. MOODY. 1996. PACAP stimulates c-fos mRNAs in small cell lung cancer cells. Life Sci. **59**: 307–313.

20. ZHONG, Y. 1995. Mutation of PACAP-like neuropeptide transmission by coactivation of Ras/Raf and cAMP signal transduction pathways in Drosophila. Nature **375**: 588–592.

21. BHAT, G. J., T. THEKKUMKARA, W. THOMAS, K. M. CONRAD & K. BAKER. 1994. Angiotensin II stimulates sis-inducing factor-like DNA binding activity. J. Biol. Chem. **269**: 31443–31449.

22. ROSEN, L. B., D. D. GINTY, M. J. WEBER & M. E. GREENBERG. 1994. Membrane depolarization and calcium influx stimulate MEK and MAP kinase via activation of Ras. Neuron **12**: 1207–1221.

23. ANGELBERGER, P., B. BANYAI, S. BANYAI, A.DURTARAN, S. LI, J. PIDLICH, M. RADERER, I. VIRGOLINI, Q. YANG, B. NIEDERLE, W. SCHEITHAUER & P. BALENT. 1994. Vasoactive intestinal peptide in receptor imaging for the localization of intestinal adenocarcinomas and endocrine tumors. New Engl. J. Med. **331**: 1116–1121.

24. LANG, L. & W. C. ECKELMAN. 1994. One step synthesis of ^{18}F labeled (^{18}F)-N-succinimidyl-4-(fluoromethyl)benzoate for protein labeling. Appl. Radiation Isot. **45**: 1155–1163.

25. LEE, M., R. T. JENSEN, S. C. HUANG, G. BEPLER, L. Y. KORMAN & T. W. MOODY. 1990. Vasoactive intestinal polypeptide binds with high affinity to non-small cell lung cancer cells and elevates cAMP levels. Peptides **11**: 1205–1209.

26. REUBI, J. C. 1995. Neuropeptide receptors in health and disease: The molecular basis for in vivo imaging. J. Nucl. Med. **36**: 1825–1835.

27. REUBI, J. C. 1995 In vitro identification of vasoactive intestinal peptide receptors in human tumors: Implications for tumor imaging. J. Nucl. Med. **36**: 1846–1853.

28. MOODY, T. W., L. YOU, S. JAKOWLEW, P. RICE & A. MALKINSON. 1997. Chemoprevention of lung adenomas by a VIP receptor antagonist. 1997. Proc. Am. Assoc. Cancer Res. **38**: 364.

Coordinated Role of Vasoactive Intestinal Peptide and Nitric Oxide in Cardioprotection[a]

DIPAK K. DAS,[b] RENI KALFIN,[c] NILANJANA MAULIK, AND RICHARD M. ENGELMAN

Division of Cardiovascular Research, Department of Surgery, University of Connecticut School of Medicine, Farmington, Connecticut, USA

[c]*University of Siena, Siena, Italy*

ABSTRACT: The present study sought to examine the interrelationship between nitric oxide (NO) and vasoactive intestinal peptide (VIP) in myocardial protection. Isolated rat hearts were perfused for 15 min with buffer only (Group I); 0.3 mM VIP (Group II); 3 mM L-arginine (a precursor of NO) (Group III); VIP and aminoguanidine (iNOS blocker) (Group IV); or L-arginine plus VIP 10-28 (VIP inhibitor) (Group V). Each heart was then made globally ischemic for 30 min followed by 2 h reperfusion. Both VIP and NO were found to provide cardioprotection during ischemia and reperfusion. However, the beneficial effects of VIP and NO were reduced by inhibition of NO and VIP, respectively, suggesting that cardioprotection by VIP is modulated by NO and *vice versa*. The results of this study suggested a coordinated regulation by cardioprotection by NO and VIP.

Previous studies from our laboratory demonstrated the release of vasoactive intestinal peptide (VIP) and nitric oxide (NO) from the heart during reperfusion of ischemic myocardium.[1,2] Replenishing the hearts with lost VIP or NO reduced the ischemic reperfusion injury as evidenced by improved postischemic ventricular functions and reduced creatine kinase release.[3,4] The results of these studies suggest that both VIP and NO play cardioprotective roles.

VIP was initially postulated to function as a local neurotransmitter, being localized primarily in perivascular nerves surrounding coronary arteries and within the vascular wall.[5] Consequently, it was found that VIP is a potent vasodilator of many vascular beds including coronary arteries. More recent studies demonstrated the presence of a VIP-like immunoreactive substance in the nerves of mammalian (including human) coronary arteries.[6] VIP has also been implicated in the pathogenesis of hypertension, hemorrhagic shock, heart failure, and ischemic reperfusion injury.[7]

Like VIP, NO has also been found to play a crucial role in the pathophysiology of a number of cardiovascular diseases. There is little doubt that NO plays a significant role in the intracellular signaling in the cardiovascular system. Several recent reports have demonstrated that myocardial ischemia and reperfusion are associated with decreased

[a]This study was supported by National Institutes of Health grants HL 34360, HL 22559, HL 56803, and HL 56322, and a Grant-in-Aid from the American Heart Association.

[b]Corresponding author: Dipak K. Das, Ph.D., University of Connecticut, School of Medicine, Farmington, CT 06030-1110; Tel.: 860-679-3687; Fax: 860-679-4606; e-mail: DDAS@NEURON.UCHC.EDU

vasorelaxation of isolated coronary arteries and that the impaired vasorelaxation can be restored by the NO precursor, L-arginine.[8] It has been suggested that NO may reduce ischemic reperfusion injury by blocking the formation of hydroxyl radicals (OH·) generated during the reperfusion of ischemic myocardium, augmenting coronary flow, and reducing leukocyte interactions with vascular endothelium.[9,10] Studies from this laboratory have demonstrated that NO exerts direct cardioprotective effect by modulating signal transduction and by acting as an intracellular antioxidant.[11]

The intention of this review is to elaborate the coordinated role of VIP and NO in cardioprotection and to discuss the results of a few recent experiments demonstrating the similarities of their mechanism of action.

SIGNAL TRANSDUCTION

A growing body of evidence supports the role of VIP in signal transduction. VIP is one of the most effective neuromodulators in promoting the production of intracellular cAMP in many biological systems, including the central neuronal and glial cells[12] and the retinal pigment epithelium (RPE).[13] The VIP-stimulated cAMP-dependent protein kinases have been demonstrated in many systems.[14] Another study indicated that VIP receptors might substitute functionally for β-adrenoceptors, because both activate adenylate cyclase.[15,16] Combined treatment with VIP and α1-adrenergic agonists increased cAMP accumulation 10–20-fold and cGMP accumulation 60–150-fold.[17] The role of VIP in intracellular signal transduction receives further support from the evidence that VIP increases inositol phospholipid breakdown in rat superior cervical ganglion.[18]

Recent data suggest that NO may act as an intracellular messenger molecule. The NO signaling appears to be extremely complex. In a physiologic system such as the heart, once NO is produced it immediately diffuses within the cell and interacts with the heme prosthetic group of cytosolic guanylate cyclase (GMP), activating guanylyl cyclase or other heme proteins. Activation of guanylate cyclase by NO was first reported during 1970s.[19] In this report, the authors also demonstrated that such activation was heme dependent. Subsequently, it was found that cytosolic guanylate cyclase is a hemoprotein containing 1 mol heme/mol of holoenzyme dimer, and heme-containing guanylate cyclase can be activated up to 100-fold by NO.[20] In addition, NO can readily react with other cellular hemoproteins such as hemoglobin and myoglobin to produce corresponding NO-heme adducts, which can rapidly activate guanylate cyclase.[21] Activation of guanylate cyclase causes cGMP accumulation. Only 10 nM of nitric oxide is necessary to enhance cGMP *in vitro* and cause smooth muscle relaxation. The physiologic effect is then produced by the stimulated cGMP, which acts as an intracellular messenger. Many of the biologic effects of NO, including relaxation of vascular smooth muscle and inhibition of platelet aggregation, are believed to be mediated by cGMP. The enhanced cGMP is likely to regulate the phosphodiesterases leading to the modulation of the diacylglycerol and protein C kinase. Thus, this NO-cGMP signal transduction system is believed to be the major pathway for both intracellular and intercellular communication and represents an unique system by which an extracellular signal gains access to an intracellular enzyme.

A study from our laboratory revealed that cGMP remained unaffected by ischemia and reperfusion, but was stimulated significantly after L-arginine treatment.[22] The cGMP level persisted up to 10 min of reperfusion and then dropped slightly. Reperfusion of ischemic myocardium resulted in significant accumulation of radiolabeled inositol phosphate, inositol bisphosphate, and inositol triphosphate. Isotopic incorporation of [³H]inositol into phosphatidylinositol, phosphatidylinositol-4-phosphate, and phosphatidylinositol-4, 5-bisphosphate was increased significantly during reperfusion. Reperfusion of the ischemic heart prelabeled with [^{14}C]arachidonic acid resulted in modest increases in [^{14}C]diacylglycerol and [^{14}C]phosphatidic acid. Pretreatment of the heart with L-arginine significantly reversed this enhanced phosphodiesteratic breakdown during ischemia and early reperfusion. However, at the end of the reperfusion the inhibitory effect of L-arginine on the phosphodiesterases seems to be reduced. In L-arginine–treated hearts, SOD activity was progressively decreased with the duration of reperfusion time. The results suggest that NO plays a significant role in transmembrane signaling in the ischemic myocardium. The signaling seems to be transmitted via cGMP and opposes the effects of phosphodiesterases by inhibiting the ischemia/reperfusion-induced phosphodiesteratic breakdown. This signaling effect appears to be reduced as reperfusion progresses.

VASORELAXATION

Among neuropeptides that may be involved in the regulation of myocardial function, VIP probably possesses the most marked influence on coronary circulation in the regulation of vascular tone.[23] This peptide is located in various functionally important regions of the heart as well as in the perivascular neuroplexus of the coronary vessels. Coronary arteries, atrial myocardium, and nodal regions of the heart contain the highest density of VIP-immunoreactive cardiac nerve fibers.[24] VIP relaxes isolated coronary arteries and increases coronary blood flow *in vivo*.[25] VIP possesses positive inotropic effects and improves left ventricular function.[26]

Coronary arteries isolated from cats relaxed after administration of VIP (10–100 nM) . VIP (0.1 µM)-induced relaxation was nearly in the same order of magnitude as that of noradrenaline (1 µM)-induced relaxation. However, in the presence of propranolol (1 µM), noradrenaline reverted its action from inhibitory to excitatory, while VIP produced the same relaxation as that observed in the absence of propranolol, suggesting that VIP induces relaxation of the cat coronary arteries *in vitro*, which is not mediated through β-adrenergic pathways. Thus, it appears that the presence of this vasoactive neuropeptide in the coronary arteries may have potential significance in the pathophysiology of coronary heart disease, especially diseases related to ischemia and reperfusion. It has been suggested that VIP has a role in maintaining a state of relaxation of the epicardial coronary arteries, and any change in the vasoactive intestinal polypeptidergic innervation may result in focal coronary spasm.[27]

Similar to VIP, NO also acts as a vasodilator to the cardiovascular system. Several recent reports have demonstrated that myocardial ischemia and reperfusion are associated with decreased vasorelaxation of isolated coronary arteries and that the impaired vasorelaxation can be restored by the NO precursor, L-arginine.[28] Numerous evidence exists to support the vasodilatory properties of NO.[29]

FREE RADICAL SCAVENGERS

VIP possesses a potent antioxidant property. Although VIP does not have significant O_2^-, OH·, or H_2O_2 scavenging ability, it can inhibit, in a dose-dependent manner, the $^1O_2$-dependent 2,2,6,6-tetramethylpiperidine N-oxyl (TEMPO) formation.[30] To demonstrate the ability of VIP to quench $^1O_2$, the free radical was generated in photosensitizing systems using rose bengal or methylene blue as sensitizers. Electron paramagnetic resonance (EPR) spectroscopy was used to detect the TEMP-$^1O_2$ product, TEMPO. In this study, VIP inhibited the TEMPO formation in a dose-dependent manner. The amount of VIP needed to cause 50% inhibition of the rate of $^1O_2$ quenching was found to be 37 µM.

VIP also possesses free radical scavenging properties in *in vivo* systems. One of the biological sources of the reactive oxygen species is the xanthine/xanthine oxidase system. The addition of this oxygen free radical generating system to perfused rat lungs increased both peak airway pressure and perfusion pressure simultaneously resulting in pulmonary edema and increased protein content in bronchoalveolar lavage fluid.[31] Treatment with 1–10 mg/kg/min of VIP significantly inhibited or completely abolished all signs of injury and reduced or abolished the generation of arachidonic acid products suggesting that VIP may function as a physiological modulator of inflammatory tissue damage resulting from toxic oxygen species. Another related study demonstrated that VIP does not cause generation of oxygen free radicals in any form and that the vasodilatory actions of VIP are not mediated through the generation of reactive oxygen species.[32]

By definition, NO is a free radical because, one atom of nitrogen combines with one atom of oxygen to produce NO with an unpaired electron. Unlike highly detrimental reactive oxygen species, hydroxyl radical, OH· which attacks a large number of biomolecules, NO can only react with a small range of compounds. Compared to OH·, it is virtually harmless. Nevertheless, NO can rapidly react at near diffusion-limited rate (6.7×10^9 $M^{-1}s^{-1}$) with superoxide anion (O_2^-) to form highly reactive peroxynitrite radical ($ONOO^-$) which is subsequently protonated to form OH· according to following scheme:

$$NO· + O_2^- \longrightarrow ONOO^- \text{ [i]}$$

$$ONOO^- + H^+ \longrightarrow ONOOH \text{ [ii]}$$

$$ONOOH \longrightarrow OH· + NO_2 \text{ [iii]}$$

Although this scheme occurs in an *in vitro* system, there is little evidence that such a system can occur in a physiological system. To the contrary, $ONOO^-$ radical should preferentially react with -SH group and ascorbate, which are present in most biological systems including heart. In this regard, NO can be considered an antioxidant. In addition, the affinity of NO towards O_2^- is greater than that of SOD for O_2^-. In fact, NO may compete with SOD for O_2^-, thereby removing O_2^- and preserving superoxide dismutase (SOD), further supporting its antioxidant role.

The results of our study indeed support the antioxidant role of NO.[33,34] Significant reduction of MDA formation was observed in L-arginine–treated hearts. MDA is considered to be the presumptive marker for free radical generation and development of oxidative stress, thus reduction of MDA formation implicates removal of oxidative stress from the ischemic reperfused myocardium.

The antioxidant activity of NO· may be realized through the reduction of reactive oxygen species produced by myoglobin in the presence of H_2O_2 and organic hydroperoxide. In particular, our *in vitro* study has demonstrated that oxoferrylmyoglobin and related free radical species formed from metmyoglobin (metMb) and tert-butyl hydroperoxide (t-BuOOH) can be directly reduced by NO· to ferric metMb. As a result of this, NO· can protect against oxidative damage produced by oxoferrylmyoglobin (oxoferryl Mb) as evidenced by the ability of NO· to inhibit oxoferrylMb radical–catalyzed oxidation of *cis*-parinaric acid. This effect of NO· is likely to contribute to protect against ischemia/reperfusion injury of the heart where high concentrations of myoglobin Mb and peroxidation products (viz., lipid hydroperoxides) set the stage for the enhancement of oxidative damage through the formation of oxoferrylMb.

Myoglobin is the major heme-containing protein of cardiomyocytes. It is involved in the intracellular transport and storage of molecular oxygen. Under conditions of oxidative stress in the heart, such as during ischemia and reperfusion, a potential damaging role of myoglobin has been proposed to be associated with its ability to form a potent oxidant, oxoferrylMb, through the interactions with H_2O_2 or lipid peroxidation products (e.g., lipid hydroperoxides). In particular oxoferrylMb has been implicated to be one of the important pathogenetic mechanisms of ischemia/reperfusion-induced oxidative injury to the heart. The results of our experiments demonstrate that NO· may counteract the damaging effect of oxoferrylMb. NO can act directly as a reductant to quench oxoferrylMb-derived free radical species and protect the heart against oxoferrylMb-induced peroxidations similar to the ischemia/reperfusion scenario.

The chemical reactions responsible for the biological activities of NO· appear to be its coordinate covalent binding to Fe and Cu in metalloproteins and its covalent modification of sulfhydryl groups in proteins. In biological systems, such as during myocardial ischemia and reperfusion, NO· is likely to undergo numerous redox reactions and simultaneously acts as either a weak oxidizing agent or a potent reducing agent. As discussed earlier, one of the most important reactions of NO· is its interaction with superoxide to form peroxynitrite (-OONO), which can decompose to yield highly reactive OH· radicals. The reaction is considered to be responsible for the oxidative damage associated with NO. On the other hand, strong radical scavenging effects of NO· towards peroxyl radicals and OH· radicals, as well as towards oxoferryl-hemoproteins,[31] as shown in the present study, strongly support the notion that NO acts as an antioxidant rather than a prooxidant. Indeed, our results have demonstrated decreased oxidative stress by NO as evidenced by the reduced MDA formation.

INTRACELLULAR Ca²⁺ REGULATION

Many peptides including VIP exert potent positive contractile responses directly in ventricular cardiomyocytes. In a recent study, the involvement of L-type calcium channels in the contractile responses elicited by VIP was investigated using selective antagonists at L-type calcium channels, verapamil and diltiazem.[35] The results of this study indicated that positive contractile responses to VIP in ventricular cardiomyocytes involve the influx of Ca^{2+} via L-type calcium channels. Crude membrane fractions prepared from rabbit gastric fundic muscle degraded VIP.[36] The VIP degradation was inhibited by EDTA and enhanced by Ca^{2+} in the concentration of 0.3–1.0 mM range. In another related study, VIP stimulated cAMP production and its interaction with protein kinase C activation and elevation of intracellular Ca^{2+} in NIE-115 neuroblastoma cells.[37] The results of this study indicated that VIP stimulates cAMP accumulation in NIE-115 cells, and that although activation of protein kinase C inhibits the VIP-stimulated cAMP response, elevation of intracellular Ca^{2+} potentiates this signaling pathway. VIP can induce glycogenolysis through a Ca^{2+}-dependent mechanism, and the restriction of cAMP accumulation during the infusion of high concentrations of VIP is caused by Ca^{2+}-induced phosphodiesterase activation.[38] In another study, Ca^{2+}-mobilizing agonists were found to potentiate VIP-stimulated cAMP production in human colonic cell line, HT29-cl.19A.[39] Calcium regulation of VIP mRNA abundance in SH-SY5Y human neuroblastoma cells has been described.[40]

Several recent studies indicate that the NO effect in heart is modulated by Ca^{2+}. For example, a recent clinical trial demonstrated a favorable effect of amlodipine, a Ca^{2+} channel blocker, on the survival of patients with heart failure resulting from nonischemic dilated cardiomyopathy.[41] Immunohistochemistry revealed that the number of cells stained with antibody against an inducible NO synthase decreased significantly in the amlodipine-treated group compared with that in the control group, suggesting that therapeutic effect of amlodipine may in part result from inhibition of NO overproduction. Another recent study showed that nifedipine attenuated systemic and renal vasoconstriction during nitric oxide inhibition in humans.[42] Voltage-gated calcium channel currents in human coronary myocytes were found to be regulated by cGMP and NO.[43] NO can regulate the calcium current in isolated human atrial myocytes.[44]

CO-LOCALIZATION OF VIP AND NO

Nerves from different organs have been found to contain VIP and NOS simultaneously. In the submucous plexus of the guinea pig ileum, VIP and NOS immunoreactive terminals are found predominantly in association with VIP-immunoreactive nerve cell bodies.[45] Double-label immunohistochemistry demonstrated co-localization of NOS and VIP immunoreactivity in axon terminals in submucous ganglia.[46] Co-localization of NOS and VIP has also been reported in human colon.[47]

RELEASE OF VIP AND NO FROM CORONARY CIRCULATION

As mentioned earlier, VIP relaxes isolated coronary arteries and increases coronary blood flow. In a recent study, we measured the release of VIP in the coronary effluents of rats undergoing ischemia and reperfusion.[1] The basal level of VIP in the effluent was 3.19 ± 0.05 pg/ml. The amount of VIP released from the heart increased progressively with the duration of reperfusion (15 min R = 5.16 ± 0.17 pg/ml; 30 min R = 6.11 ± 0.11 pg/ml; 45 min R = 7.69 ± 0.58 pg/ml, $p < 0.001$ in each case). Maximal VIP level (11.66 ± 2.34 pg/ml, $p < 0.007$) was noted at the end of the 60-min reperfusion period.

Similar to VIP, NO is also released from the ischemic reperfused myocardium.[48] However, unlike VIP, the amount of NO released from the coronary sinus decreased such that at the time of reperfusion the amount of NO was significantly lower as compared to the baseline values. For example, after 5 min of reperfusion, the amount of NO was 139 ± 15 nmol/g compared to the baseline value of 225 ± 18 nmol/g. NO release was further diminished as reperfusion progressed: 93 ± 13 and 93 ± 16 nmol/g after 20 and 30 min of reperfusion, respectively.

CARDIOPROTECTION BY VIP AND NO

Both VIP and NO were found to provide cardioprotection during ischemia and reperfusion. In an attempt to examine the effects of VIP on ischemia/reperfusion injury, isolated rat hearts were pre-perfused with three different doses of VIP for 15 min followed by 30 min of ischemia and 60 min of reperfusion.[1] Administration of VIP significantly increased the LV contractility before ischemia ($p < 0.001$). A 0.3 μM dose of VIP was found to be optimal (TABLE 1). Less, 0.1 μM, was much less effective, while more, 1 μM, increased the heart rate to over 70 beats/min and showed extremely potent positive inotropic action. The left ventricular functions were lowered in the post-ischemic hearts as expected, but significantly better recoveries in LVDP and LVdp/dt were observed in the VIP-treated group (0.3 and 1 μM) ($p < 0.05$). For example, after 30 min of reperfusion, LVDP in the control group was 77 ± 1, while in the treated group the value was 107 ± 9 (0.3 μM) and 114 ± 11 (1 μM). LVdp/dt in the control and treated groups were $1,637 \pm 114$ and $2,667 \pm 566$, respectively ($p < 0.05$). At 60-min reperfusion after the ischemic insult, the VIP treated group again showed much better recovery of LV developed pressure (81 ± 3 [0.3 μM], and 78 ± 5 [1 μM] compared to only 54 ± 7 in the controls, $p < 0.01$) and its maximum first derivative LVdp/dt max ($1,847 \pm 152$ [0.3 μM] and $1,675 \pm 166$ [1 μM] compared to $1,410 \pm 142$ in the controls, $p < 0.01$). Administration of VIP in pre-ischemic hearts increased the coronary flow ($p < 0.05$) compared with control. Throughout the reperfusion, recovery of coronary flow for the VIP-treated group was significantly better compared to the control group. After 5-min reperfusion, coronary flow in the VIP treated group was 12.38 ± 0.47 ml/min (0.3 μM) and 13.41 ± 0.68 ml/min (1 μM) compared to only 9.25 ± 0.25 ml/min ($p < 0.001$) in the

TABLE 1. Effects of intracoronary perfusion of VIP and NO and their inhibitors on myocardial contractile functions

		Control	VIP	NO	VIP + AMG	NO + VIP10-28
Coronary flow (ml/min)	Baseline	24.1 ± 0.6	24.2 ± 0.5	23.9 ± 0.4	24.0 ± 0.3	24.2 ± 0.4
	30-min Reperfusion	23.2 ± 1.5	25.3 ± 1.0	24.7 ± 0.8	23.7 ± 1.1	23.8 ± 0.7
	60-min Reperfusion	21.1 ± 1.0	24.5 ± 0.7*	25.0 ± 0.7*	23.1 ± 0.5	24.2 ± 0.5
	120-min Reperfusion	18.1 ± 0.4	22.6 ± 0.5*	23.7 ± 0.4*	20.6 ± 0.3*†	20.0 ± 0.4*†
Developed pressure (mm Hg)	Baseline	73 ± 1.5	74 ± 2.0	71.7 ± 1.9	75.1 ± 3.2	73.8 ± 2.7
	30-min Reperfusion	67 ± 6.5	70.8 ± 3.3	68.1 ± 1.2	69.0 ± 2.5	68.5 ± 1.8
	60-min Reperfusion	47.1 ± 2.0	62.4 ± 2.4*	64.4 ± 1.1*	55.8 ± 2.0*	57.1 ± 1.5*
	120-min Reperfusion	35.4 ± 1.8	50.7 ± 2.7*	53.1 ± 2.0*	44.0 ± 1.2*†	45.1 ± 0.9*†
Aortic flow (ml/min)	Baseline	43.5 ± 0.9	42.8 ± 0.7	44.0 ± 1.0	42.2 ± 0.9	43.8 ± 1.1
	30-min Reperfusion	30.3 ± 2.2	34.5 ± 0.8	36.0 ± 0.6	32.4 ± 1.0	32.1 ± 0.8
	60-min Reperfusion	24.3 ± 1.2	33.2 ± 1.1*	34.5 ± 1.5*	27.7 ± 0.4*	29.2 ± 0.4*
	120-min Reperfusion	14.4 ± 0.7	25.0 ± 0.5*	27.3 ± 0.7*	18.5 ± 0.3*†	20.2 ± 0.5*†
Creatine kinase (U/L)	Baseline	0.5 ± 0.4	0.9 ± 0.3	0.8 ± 0.2	0.9 ± 0.1	0.7 ± 0.3
	30-min Reperfusion	11.8 ± 1.5	7.7 ± 1.2*	6.8 ± 0.3*	8.2 ± 0.7	9.0 ± 0.4
	60-min Reperfusion	13.0 ± 1.9	8.0 ± 0.2*	7.2 ± 0.4*	11.5 ± 0.4	10.2 ± 0.5
	120-min Reperfusion	25.8 ± 3.7	15.0 ± 1.7*	14.1 ± 1.1*	20.3 ± 1.1*†	19.3 ± 0.4*†

*$p < 0.05$ compared to control; †$p < 0.05$ compared to either VIP or NO.

control group. In the control group, coronary flow was further reduced during the reperfusion so that at the end of reperfusion coronary flow in the treated group was 10.15 ± 0.45 ml/min ($0.3~\mu M$) and 10.0 ± 0.5 ml/min ($1~\mu M$), as compared to 7.65 ± 0.75 ml/min ($p < 0.05$) in the untreated group.

The myocardial CK release is a sensitive indicator of cellular injury. During the stabilization period, the release of CK in all groups was negligible. Reperfusion following 30 min of ischemia enhanced CK release. However in the $0.3~\mu M$ VIP-treated group, CK release from post-ischemic hearts was significantly lower compared to the untreated group. For example, after 30- and 60-min reperfusions, CK release from the heart in the control group was 11.79 ± 1.5 and 13.01 ± 1.9 IU/L, respectively, as compared to only 7.68 ± 1.2 and 8.8 ± 0.2 ($p < 0.05$ in each case), respectively, in the $0.3~\mu M$ VIP-treated group. A VIP concentration of $0.1~\mu M$ was totally ineffective, and $1~\mu M$ reduced the CK release minimally but not significantly.

To examine the effects of NO on ischemic myocardium, isolated perfused rat hearts were treated with L-arginine for 15 min. Hearts were then made ischemic for 30 min followed by 1 h of reperfusion.[2,3] Coronary flow and aortic flow were decreased in all the groups following ischemia and reperfusion as compared to the baseline values. However, the level of these parameters in the control group was significantly lower compared to that in the L-arginine group. For example, at the end of 30-min reperfusion, the aortic flow was reduced by 79% of the baseline value in the control group vs. 46% of the baseline level in the L-arginine–treated group. These values were 85% and 65% in the protoporphyrin and in the protoporphyrin plus L-arginine supplemented group, respectively, indicating that inhibition of heme oxygenase retarded the NO-induced improvement of these parameters. Similar results were obtained with methylene blue–treated rats, the aortic flow was reduced by 69% in this group compared to the baseline value. However when methylene blue was added in the L-arginine group the reduction was by 58%. Coronary flow was also reduced in all the experimental groups as well as in the control group of animals after 30 min of reperfusion. In the control group, this reduction was 24% whereas in the L-arginine–treated animals the reduction was 8% only when compared to the baseline value. The protoporphyrin-treated group showed 32% inhibition, whereas when protoporphyrin was added in combination with L-arginine to the animal the reduction was 20% compared to the baseline value. Similarly, the developed pressures in the control and L-arginine–treated groups were reduced by 51% and 24% of the baseline values, respectively, after 30-min reperfusions. During the same time period, i.e., after 30-min reperfusion, the developed pressure in the control group was reduced by 50%, whereas that for the L-arginine group was 24%. For the protoporphyrin group the inhibition was 61%, but when protoporphyrin was added in combination with L-arginine, the inhibition was 40%. Methylene blue and methylene blue and L-arginine group also showed 42% and 28% inhibition, respectively. The results dp/dtmax followed a similar pattern.

COORDINATED CARDIOPROTECTION BY VIP AND NO

In a recent study, a role of VIP was shown in the regulation of NO inhibition of intestinal function.[49] In this study, an addition of L-NAME to the perfusate obtained from perfused canine ileal segments causes, after a delay, a concentration-dependent persistent increase in tonic and phasic activity of circular muscle with corresponding reduction in tonic VIP output. Removal of Ca^{2+} to the perfusate markedly reduced this response. VIP has been described as the primary inhibitory transmitter that stimulates the production of

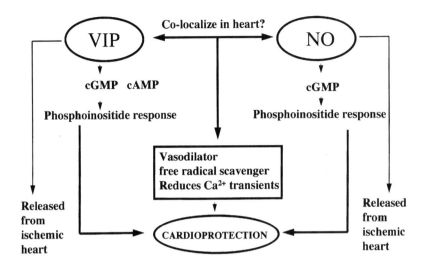

FIGURE 1. Schematic diagram showing cardioprotection by VIP and NO

NO. VIP-mediated G protein–coupled Ca^{2+} influx was found to activate the constitutive NOS in dispersed gastric muscle cells.[50]

In a recent study, isolated rat hearts were randomly divided into five groups. Hearts were perfused for 15 min with buffer only (Group I); 0.3 μM VIP (Group II); 3 mM L-arginine (a precursor of NO) (Group III); VIP and aminoguanidine (AMG, an iNOS blocker) (Group IV); and L-arginine plus VIP10-28 (a VIP inhibitor) (Group V). Each heart was then made globally ischemic by terminating the coronary flow for 30 min followed by 2 h of reperfusion. The results shown in TABLE 1 clearly indicate that beneficial effects of VIP and NO are blunted by inhibition of NO and VIP, respectively.

SUMMARY AND CONCLUSION

It has become increasingly apparent that VIP and NO share many properties that play a crucial role in cardioprotection (FIG. 1). VIP and NO, which are co-localized, are released from the ischemic/reperfused myocardium. Intracoronary delivery of these compounds reduces ischemic/reperfusion injury. Both cause vasorelaxation to the coronary arteries and tend to preserve coronary and arterial flow during ischemia and reperfusion. NO and VIP take part in intracellular signal transduction through the activation of cGMP and cAMP (for VIP) leading to the phosphoinositide response and reduction in intracellular Ca^{2+} overloading. Both act as intracellular antioxidants by scavenging the oxygen free radicals produced during the reperfusion of ischemic myocardium. Finally, our recent results demonstrate that cardioprotection by VIP can be modulated by inhibiting NO and, *vice versa*, suggesting a coordinated regulation of vascular tone by these compounds.

REFERENCES

1. KALFIN, R., N. MAULIK, R. M. ENGELMAN, G. A. CORDIS, K. MILENOV, L. KASAKOV & D. K. DAS. 1994. Protective role of intracoronary vasoactive intestinal peptide in ischemic and reperfused myocardium. J. Pharmacol. Exp. Therap. **268**: 952–958.

2. ENGELMAN, D. T., M. WATANABE, N. MAULIK, G. A. CORDIS, R. M. ENGELMAN, J. A. ROUSOU, J. E. FLACK, D. W. DEATON & D. K. DAS. 1995. L-arginine reduces endothelial inflammation and myocardial stunning during ischemia/reperfusion. Ann. Thoracic Surg. **60**: 1275–1281.
3. ENGELMAN, D. T., M. WATANABE, R. M. ENGELMAN, J. A. ROUSOU, D. W. DEATON & J. E. FLACK. 1995. Constitutive nitric oxide release is impaired following ischemia and reperfusion. J. Thoracic Cardiovasc. Surg. **110**: 1047–1053.
4. DAS, D. K., N. MAULIK & R. M. ENGELMAN. 1997. Role of vasoactive intestinal peptide in myocardial ischemia reperfusion injury. In Pro-Inflammatory and Anti-Inflammatory Peptides. S. Said, Ed.: 323–344. Marcel Dekker Inc. New York.
5. SAID, S.I., Ed. 1982. Vasoactive Intestinal Peptide: Advances in Peptide Hormone Research Series. Raven Press, New York.
6. DELLA, N. G., R. E. PAPKA, J. B. FURNESS & M. COSTA. 1983. Vasoactive intestinal peptide-like immunoreactivity in nerves associated with the cardiovascular system of the guinea pigs. Neuroscience **9**: 605–619.
7. BRUM, J. M., A. A. BOVE, Q. SUFAN, W. REILLY & V. L. W. GO. 1986. Action and localization of vasoactive intestinal peptide in the coronary circulation: evidence for nonadrenergic, non-cholinergic coronary regulation. J. Am. Coll. Cardiol. **7**: 406–413.
8. KUROSE, I., R. WOLF, M. B. GRISHAM & D. N. GRANGER. 1994. Modulation of ischemia/reperfusion-induced microvascular dysfunction by nitric oxide. Circ. Res. **74**: 376–382.
9. NAKANISHI, K., J. VINTEN-JOHANSEN, D. J. LEFER, Z. ZHAO, W. C. FOWLER, S. McGEE & W. E. JOHNSTON. 1992. Intracoronary L-arginine during reperfusion improves endothelial function and reduces infarct size. Am. J. Physiol. **263**: H1650–1658.
10. RUBANYI, G. M., E. H. HO, E. H. CANTOR, W. C. LUMMA & L. H. PARKER-BOTELHO. 1991. Cytoprotective function of nitric oxide: inactivation of superoxide radicals produced by human leukocytes. Biochem. Biophys. Res. Commun. **181**: 1392–1397.
11. MAULIK, N., D. T. ENGELMAN, M. WATANABE, R. M. ENGELMAN, G. MAULIK, G. A. CORDIS & D. K. DAS. 1995. Nitric oxide signaling in ischemic heart . Cardiovasc. Res. **30**: 593–601.
12. CHNEIWEISS, J., H. GLOWINSKI & J. PREMONT. 1985. Vasoactive intestinal polypeptide receptors linked to an adenylate cyclase, and their relationship with biogenic amine and somatostatin-sensitive adenylate cyclases on central neuronal and glial cells in primary cultures. J. Neurochem. **44**: 779–786.
13. KOH, S. W. M. & G. J. CHADER. 1984. Elevation of intracellular cyclic AMP and stimulation of adenylate cyclase activity by vasoactive intestinal peptide and glucagon in the retinal pigment epithelium. J. Neurochem. **43**: 1522–1526.
14. GUERRERO, J. M., J. C. PRIETO, J. R. CALVO & R. GOBERNA. 1984. Activation of cyclic AMP dependent protein kinase by VIP in blood mononuclear cells. Peptides **5**: 371–373.
15. CHIK, C. L. & A. K. HO. 1991. See-Saw signal processing: Reciprocal effects of stimulus deprivation on VIP-stimulated AMP and GMP accumulation in rat pinealocytes. Endocrinology **128**: 850–856.
16. KANEKO, T., P. Y. CHENG, H. OKA, T. ODA, N. YANAIHARA & C. YANAIHARA. 1980. VIP stimulate adenylate cyclase and serotonin N-acetyltransferase activities in rat pineal in vitro. Biomed. Res.1: 84–87.
17. CHIK, C. L., A. K. HO & D. C. KLEIN. 1988. Dual receptor regulation of cyclic nucleotides: a1-adrenergic potentiation of VIP stimulation of pinealocyte cAMP. Endocrinology **122**: 1646–1651.
18. AUDIGIER, S., C. BARBERIS & J. S. JARD. 1988. Vasoactive intestinal polypeptide increases inositol phospholipid breakdown in rat superior cervical ganglion. Ann. N.Y. Acad. Sci. **527**: 578–581.
19. KATSUKI, S., W. ARNOLD, C. MITTAL & F. MURAD. 1977. Stimulation of guanylate cyclase by sodium nitroprusside, nitroglycerine and nitric oxide in various tissue preparations and comparison to the effects of sodium azide and hydroxylamine. J. Cyclic Nucl. Res. **3**: 23–35.
20. CRAVEN, P. A. & F. R. DERUBERTIS. 1978. Restoration of the responsiveness of purified guanylate cyclase to nitrosoguanidine, nitric oxide, and related activators by heme and heme proteins: Evidence for the involvement of the paramagnetic nitrosyl-heme complex in enzyme activation. J. Biol. Chem. **253**: 8433–8443.
21. MARLETTA, M. A. 1989. Nitric oxide: Biosynthesis and biologic significance. Trends Biol. Sci. **14**: 488–492.

22. MAULIK, N., D. T. ENGELMAN, M. WATANABE, R. M. ENGELMAN, J. A. ROUSOU, J. E. FLACK, D. W. DEATON, N. V. GORBUNOV, N. M. ELSAYED, V. E. KAGAN & D. K. DAS. 1996. Nitric oxide/Carbon monoxide. A molecular switch for myocardial preservation during ischemia. Circulation **94**: 398–406.

23. BITAR, K. N. & G. M. MAKHLOUF. 1982. Relaxation of isolated gastric smooth muscle cells by vasoactive intestinal peptide. Science **216**: 531–533.

24. WEIHE, E., M. REINECKE & W. G. FORSSMANN. 1984. Distribution of vasoactive intestinal polypeptide-like immunoreactivity in the mammalian heart. Interrelation with neurotensin and substance P-like immunoreactive nerves. Cell Tissue Res. **236**: 527–540.

25. SMITHERMAN, T. C., H. SAKIO, A. M. GEUMEI, T. YOSHIDA, M. OYAMADA & S. I. SAID. 1982. Coronary vasodilator action of VIP. *In* Vasoactive Intestinal Peptide: 169–176. Raven Press. New York.

26. ANDERSON, F. L., A. C. KRALIOS, R. HERSHBERGER & M. R. BRISTOW. 1988. Effect of vasoactive intestinal peptide on myocardial contractility and coronary blood flow in the dog: Comparison with isoproterenol and forskolin. J. Cardiovasc. Pharmacol. **12**: 365–371.

27. BRUM, J. M., A. A. BOVE, Q. SUFAN, W. REILLY & V. L. W. GO. 1986. Action and localization of vasoactive intestinal peptide in the coronary circulation: evidence for nonadrenergic, noncholinergic coronary regulation. J. Am. Coll. Cardiol. **7**: 406–413.

28. ARCHER, S. L., J. M. HUANG, V. HAMPL, D. P. NELSON, P. J. SHULTZ & E. K. WEIR. 1994. Nitric oxide and cGMP cause vasorelaxation by activation of acharybdotoxin-sensitive K channel by cGMP dependent protein kinase. Proc. Natl. Acad. Sci. USA **91**: 7583–7587.

29. IGNARRO, L. J. 1993. Nitric oxide mediated vasorelaxation. Thromb. Haemost. **70**:148–151.

30. MISRA, B. R & H. P. MISRA. 1990. Vasoactive intestinal peptide, a singlet oxygen quencher. J. Biol. Chem. **265**: 15371–15274.

31. BERISHA, H., H. FODA, H. SAKAKIBARA, M. TROTZ, H. PAKBAZ & S. I. SAID. 1990. Vasoactive intestinal peptide prevents lung injury due to xanthine/xanthine oxidase. Am. J. Physiol. **259**: L151–L155.

32. BALLON, B. J., E. P. WEI & H. A. KONTOS. 1986. Superoxide anion radical does not mediate vasodilation of cerebral arterioles by vasoactive intestinal polypeptide. Stroke **17**: 1287–1290.

33. MAULIK, N., M. WATANABE, D. T. ENGELMAN, R. M. ENGELMAN, J. A. ROUSOU, J. E. FLACK, D. DEATON & D. K. DAS. 1996. Nitric oxide attenutes myocardial ischemia-reperfusion injury by functioning as synchronizing messenger with carbon monoxide and intracellular antioxidant. Surgical Forum **XL VII**: 230–233.

34. MAULIK, N., D. T. ENGELMAN, M. WATANABE, R. M. ENGELMAN & D. K. DAS. 1996. Nitric oxide—a retrograde messenger for carbon monoxide signaling in ischemic heart. Mol. Cell. Biochem. **157**: 75–86.

35. BELL, D. & B. J. MCDERMOTT. 1995. Inhibition by verapamil and diltiazem of agonist-stimulated contractile responses in mammalian ventricular cardiomyocytes. J. Mol. Cell. Cardiol. **27**: 1977–1987.

36. KOBAYASHI, R., Y. CHEN, T. D. LEE, M. T. DAVIS, O. ITO & J. H. WALSH. 1994. Degradation of vasoactive intestinal polypeptide by rabbit gastric smooth muscle membranes. Peptides **15**: 323–332.

37. INUKAI, T., C. L. CHIK & A. K. HO. 1994. Vasoactive intestinal polypeptide stimulates cyclic AMP production in mouse NIE-115 neuroblastoma cells: modulation by a protein kinase C activator and ionomycin. Peptides **15**: 1361–1365.

38. SAITO, K., K. YAMATANI, H. MANAKA, K. TAKAHASHI, M. TOMINAGA & H. SASAKI. 1992. Role of Ca^{2+} on vasoactive intestinal peptide-induced glucose and adenosine 3′,5′-monophosphate production in the isolated perfused rat liver. Endocrinology **130**: 2267–2273.

39. WARHURST, G., K. E. FOGG, N. B. HIGGS, A. TONGE & J. GRUNDY. 1994. Ca^{2+}-mobilizing agonists potentiate forskolin-and VIP-stimulated cAMP production in human colonic cell line, HT29-cl.19A: role of $[Ca^{2+}]$i and protein kinase C. Cell Calcium **15**: 162–174.

40. ADLER, E. M. & J. S. FINK. 1993. Calcium regulation of vasoactive intestinal peptide mRNA abundance in SH-SY5Y human neuroblastoma cell. J. Neurochem. **61**: 727–737.

41. WANG, W. Z., A. MATSUMORI, T. YAMADA, T. SHIOI, I. OKADA, S. MATSUI, Y. SATO, H. SUZUKI, K. SHIOTA & S. SASAYAMA. 1997. Beneficial effects of amlodipine in a murine model of congestive heart failure induced by viral myocarditis. A possible mechanism through inhibition of nitric oxide production. Circulation **95**: 245–251.

42. DIJKHORST-OEI, L. T., T. J. RABELINK, P. BOER & H. A. KOOMANS. 1997. Nifedipine attenuated systemic and renal vasoconstriction during nitric oxide inhibition in humans. Hypertension **29**: 1192–1198.

43. QUIGNARD, J. F., J. M. FRAPIER, M. C. HARRICANE, B. ALBAT, J. NARGEOT & S. RICHARD. 1997. Voltage-gated calcium channel currents in human coronary myocytes. Regulation by cyclic GMP and nitric oxide. J. Clin Invest. **99**: 185–193.

44. KIRSTEIN, M., M. RIVET-BASTIDE, S. HATEM, A. BENARDEAU, J. J. MERCADIER & R. J. FISCHMEISTER. 1995. Nitric oxide regulates the calcium current in isolated human atrial myocytes. Clin. Invest. **95**: 794–802.

45. JEN, P. Y., J. S. DIXON & J. A. GOSLING. 1997. Co-localization of nitric oxide synthase, neuropeptides and tyrosine hydroxylase in nerves supplying the human post-natal vas deferens and seminal vesicle. Br. J. Urol. **80**: 291–299.

46. LI, Z. S., H. M. YOUNG & J. B. FURNESS. 1995. Do VIP and nitric oxide synthase-immunoreactive terminals synapse exclusively with VIP cell bodies in the submucous plexus of the guinea-pig ileum? Cell Tissue Res. **281**: 485–491.

47. KERANEN, U., S. VANHATALO, T. KIVILUOTO, E. KIVILAAKSO & S. SOINILA. 1995. Co-localization of NADPH diaphorase reactivity and vasoactive intestinal polypeptide in human colon. J. Auton. Nerv. Syst. **54**: 177–183.

The Protective Effect of Vasoactive Intestinal Peptide (VIP) on Stress-Induced Gastric Ulceration in Rats

NEŞ'E TUNÇEL,[a] NİLÜFER ERKASAP,[a] VAROL ŞAHİNTÜRK,[b]
DİLEK DOĞRUKOL AK,[c] AND MUZAFFER TUNÇEL[c,d]

[a]Department of Physiology, Faculty of Medicine, University of
Osmangazi, Eskişehir, Turkey

[b]Department of Histology-Embryology, Faculty of Medicine, University
of Osmangazi, Eskişehir, Turkey

[c]Department of Analytical Chemistry, Faculty of Pharmacy, University
of Anadolu, Eskişehir, Turkey

ABSTRACT: The pathogenesis of cold-restraint stress ulcer involves various factors and is not completely understood. Mast cell degranulation, increased gastric muscular contractility, diminished mucosal blood flow, release of several biogenic amines, activated polymorphonuclear leukocytes, and lipid peroxidation which results from toxic oxygen molecules were suggested to be related to the production of gastric damage by cold-restraint stress. Recent evidence strongly indicates that VIP has a modulatory effect on tissue injury.

Sprague-Dawley rats were used in two series of experiments. One set of rats was exposed to cold-restraint stress with some of the rats pretreated with VIP. The second set of rats was exposed to cold-restraint stress and then was administered VIP for different durations. Cold-restraint stress induced gastric lesions and mast cell degranulation and also increased lipid peroxidation in gastric tissue. VIP prevented stress-induced ulcers and mast cell degranulation and protected gastric tissue from lipid peroxidation. When VIP was used after induction of stress ulcer it was therapeutically beneficial.

Thanks to its antioxidant and anti-inflammatory activity, VIP can be valuable in the prevention of gastric mucosal damage induced by cold-restraint stress.

Cold-restraint administration is a commonly used experimental model to induce acute gastric damage in rats.[1–6] Although the pathogenesis of gastric lesions in this model is not completely understood, mast cell degranulation,[7–10] increase in gastric muscular contractility,[1,2,11] diminished mucosal blood flow, release of several biogenic amines,[1–3,9–14] activated leukocytes,[2] and lipid peroxidation resulting from free-radical generation have been suggested to explain the acute gastric damage associated with cold-restraint stress. Recent studies suggest that oxygen free radicals might be one of the important factors in inducing gastric mucosal injury during stress.[1,2,15]

[d]Corresponding author: Dr. Neş'e Tunçel, University of Osmangazi, Faculty of Medicine, Dept. of Physiology, 26480, Eskişehir, Turkey. Tel.: 222-239-8082; Fax: 222-229-1179; E-mail: ntuncel@mail.ogu.edu.tr

Vasoactive intestinal peptide (VIP) presents a number of features that offer promise for its potential usefulness in the management of cold-restraint induced gastric ulcers. VIP relaxes vascular and non vascular smooth muscle,[16,17] inhibits mast cell degranulation,[18–20] protects the tissues against oxygen free radical injury, and counteracts the response of inflammatory mediators.[21-23]

In our previous reports, it was stated that VIP administration in experimental hemorrhagic shock models inhibits mast cell degranulation, increases the survival rate, and protects the renal tissue from reperfusion injury.[19,20,24] In our subsequent studies, it was also shown that VIP protects the renal, retinal, and skeletal muscle tissue from ischemia reperfusion injury by acting as an antioxidant.[25–27] In addition to experimental hemorrhagic shock model, VIP also inhibits mast cell degranulation and proliferation and decreases histamine level in testicular tissue of rats exposed to cold-restraint stress.[18,28]

In our present work, we report that cold-restraint stress-induced gastric mucosal damage was inhibited by VIP. VIP prevented stress-induced ulcers and mast cell degranulation and it protected gastric tissue from lipid peroxidation. It was also observed that VIP can be used to cure stress-induced ulcers.

MATERIALS AND METHODS

Experimental Procedures

A group of 52 (Sprague-Dawley) rats (200–220 g) of either sex were used in two series of experiments. In the first experiment, 34 rats were divided into three groups. Group 1 was the controls ($N = 5 + 6$); Group 2 was exposed to cold-restraint stress ($N = 5 + 6$); and in Group 3, prior to stress, 25 ng kg^{-1} VIP was given intraperitoneally ($N = 6 + 6$).

The animals were fasted for 24 h with water *ad libitum* and kept at room temperature. The rats were exposed to restraint-stress in the cold for 3 h (from 08.30 to 11.30 h). Restraint stress was employed by enclosure of rats in a flexible wire mesh (3×3 mm) initially formed into a cone and then bent to conform to the size of the individual animals; the rats was then placed into a refrigerator (+4°C). VIP was administered just before the 3 h exposure to stress. The control rats received intraperitoneal injections of saline. The rats were decapitated at the end of this period and the stomach was removed. The stomachs of 5 rats from groups 1 and 2 and of 6 rats from group 3 were examined morphologically and histologically. Six rats were used from all three groups for the determination of tissue lipid peroxidation [malondialdehyde (MDA)] and antioxidant enzyme (superoxide dismutase and catalase) activity.

In the second set of experiments, 18 rats were divided into three groups. All rats in these groups were exposed to stress for 3 h, then VIP was administered. VIP was administered for one, two, and three days for the groups A, B and C, respectively. The rats were decapitated 24 h after the last administration of VIP. The stomach of each rat was removed and examined macroscopically.

The stomach was opened along the lesser curvature, washed with saline, and pinned out flat in a standard position for macroscopic examination and scoring of ulcer with the help of a magnifying glass. Lesion size (mm) was determined by measuring each lesion along its greatest diameter; in the case of petechial lesions, five such lesions were considered the equivalent of a 1 mm ulcer. The sum of the total severity scores in each group of rats divided by the number of animals was expressed as the mean ulcer index (UI). The stomach tissue was then fixed in 10% neutral formalin and embedded in paraffin. Sections from tissue blocks taken from ulcerated areas were stained with hematoxylin-eosin and alcian blue-safranin for routine histologic and mast cell examination, respectively.

Light Microscopy and Histochemistry

From paraffin blocks, 6μm-thick sections were cut every 30μm. Mast cells were assigned to the one of the following categories: (1) alcian blue-stained cells, showing only blue granules; (2) alcian blue and safranin–stained cells, showing a mixture of blue and red granules or granules of intermediate color; and (3) safranin–stained cells, showing only red granules. Mast cells are counted by using an eyepiece micrometer (OC-M) ($\times 40$, 3,960 μm^2) in selected sets of six serial sections of each animal. Polymorphonuclear leukocytes (PMNL) infiltration was scored on sets of six serial sections stained with hematoxylin-eosin for each rat. If the PMNL infiltration was mild it was graded "+" and if it was severe it was graded "++."

Preparations of Tissue Samples for Determination of Antioxidant Enzyme Activities and Malondialdehyde (MDA) Levels

Stomach tissues were homogenized in 10 volumes ice-cold 50 mM potassium phosphate buffer, pH 7.4. These tissue extracts were then subjected to ultrasonication (15 sn) followed by centrifugation (20,000 $\times$ g, 20 min). Catalase was measured polarographically. Superoxide dismutase, malondialdehyde, and protein were measured spectrophometrically in the supernatant.

TABLE 1. Histochemical heterogeneity of submucosal mast cell

Groups	Red-Stained Mast Cells		Blue-Stained Mast Cells	
	Degranulated	Granulated	Degranulated	Granulated
Control ($N = 5$)	0.4 ± 0.2	4.4 ± 0.9**	–	3.8 ± 0.2****
Stressed ($N = 5$)	2.6 ± 0.5*	2.2 ± 0.3	1.8 ± 0.8***	1.2 ± 0.5
VIP treated ($N = 6$)	1.1 ± 0.4	3.8 ± 0.5	1.5 ± 0.4	2.1 ± 0.7

* $p < 0.05$, significantly different from control and VIP-treated groups; ** $p < 0.05$, significantly different from stressed group; *** $p < 0.05$, significantly different from control group; and **** $p < 0.05$, significantly different from stressed group.

TABLE 2. Total submucosal mast cell numbers

Groups	Granulated	Degranulated
Control ($N = 5$)	8.2 ± 1.0***	0.4 ± 0.2*
Stress ($N = 5$)	3.4 ± 0.9	4.4 ± 0.9
VIP treated ($N = 6$)	5.6 ± 1.2	2.1 ± 0.4**

* $p < 0.05$, significantly different from stressed and VIP-treated groups; ** $p < 0.05$, significantly different from stressed group; and *** $p < 0.05$, significantly different from stressed group.

TABLE 3. The effect of VIP on gastric lesions induced by cold-restraint stress

Animals of G_A	Animals of G_B	Animals of G_C
1. 1 mm lesion	1. No lesion	1. No lesion
2. No lesion	2. No lesion	2. No lesion
3. No lesion	3. No lesion	3. No lesion
4. No lesion	4. No lesion	4. No lesion
5. No lesion	5. No lesion	5. No lesion
6. No lesion	6. 1 mm lesion	6. No lesion

G_A: After stress exposure one day 25 ng kg^{-1} VIP (i.p.) given; G_B: After stress exposure two consecutive days VIP (i.p.) given; G_C: After stress exposure three consecutive days VIP (i.p.) given.

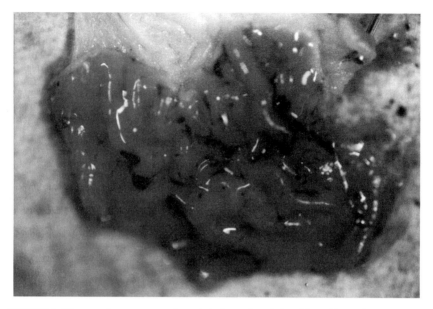

FIGURE 1. Macroscopic appearance of gastric lesions of rats induced by cold-restraint stress.

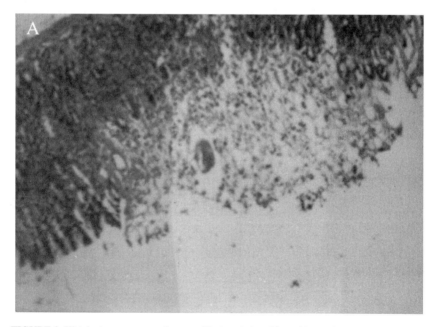

FIGURE 2. Histologic appearance of mucosal lesions induced by cold-restraint stress in rats. (**A**) mucosal damage, HE × 33. (*continued*)

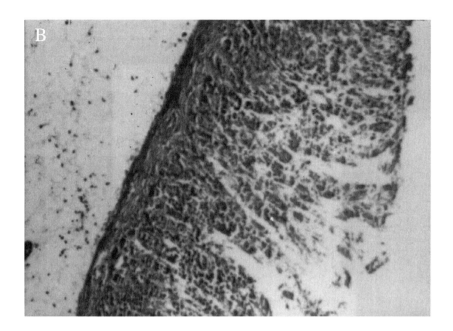

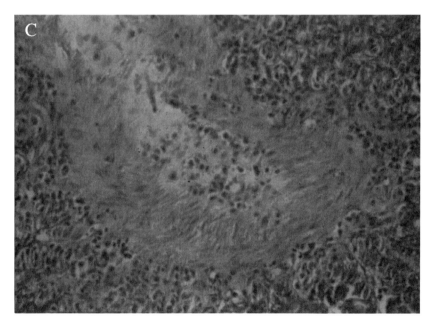

FIGURE 2. (*continued*) (**B**) mucosal damage with hemorrhage, HE × 33; (**C**) increased PMNL infiltration, HE × 66.

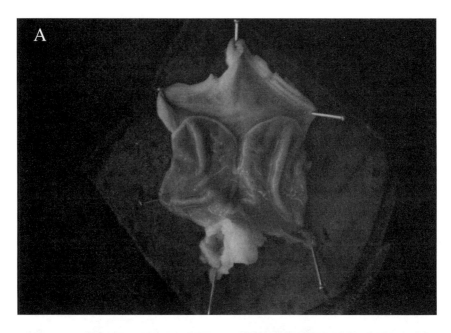

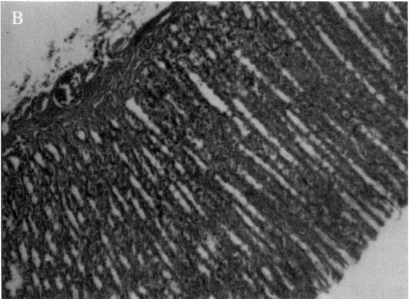

FIGURE 3. The effect of VIP on gastric mucosal ulcers induced by cold-restraint stress in rats. (**A**) Macroscopic appearance; (**B**) Light microscope appearance, HE × 33.

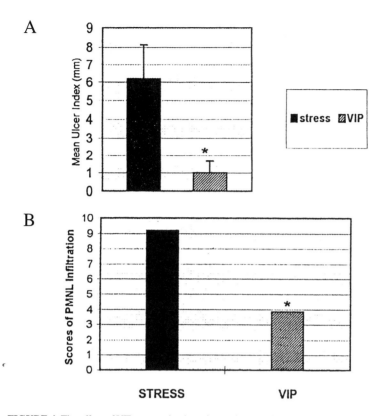

FIGURE 4. The effect of VIP on gastric ulceration and PMNL infiltration in stressed rats. (**A**) mean ulcer indices. * $p < 0.05$, significantly different from stressed group. (**B**) Scored PMNL infiltration, U = 2.0 *$p = 0.0087$, significantly different from stressed group.

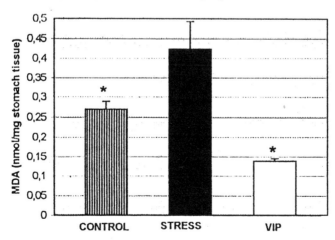

FIGURE 5. The effect of VIP on MDA level of gastric tissue in stressed rats. * $p < 0.05$, stressed group significantly different from VIP-treated and control groups.

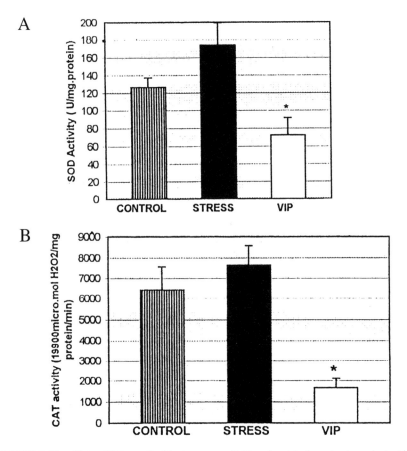

FIGURE 6. The effect of VIP on antioxidant enzyme activities of gastric tissue in stressed rats. (**A**) SOD activity. * $p < 0.05$, significantly different from stressed group. (**B**) Catalase activity. * $p < 0.05$, significantly different from control and stressed groups.

Superoxide Dismutase Determination

Superoxide dismutase was determined by the modified spectrophotometric method.[29] This method depends on quercetin oxidation at pH 9.2. Inhibition of quercetin oxidation was monitored at 406 nm on a Shimadzu 160 A Spectrophotometer. Superoxide dismutase activity was calculated per mg of protein.

Catalase Determination

Catalase activity was determined by a polarographic method.[30] Briefly, a carbon-paste electrode prepared with samples was used as the working electrode of the polarograph (Tacussel model PRG-S), which monitored oxygen produced by the enzymatic disappearance of hydrogen peroxide.[31] Catalase activity was calculated in terms of micromole of hydrogen peroxide consumed/mg of protein/min.

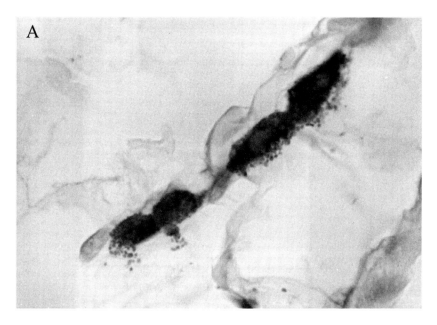

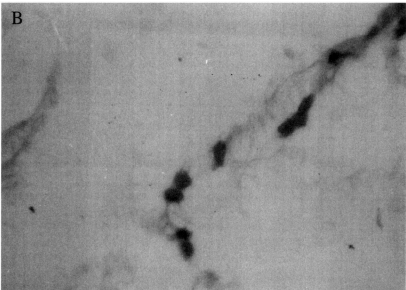

FIGURE 7. The effect of VIP on gastric submucosal mast cell degranulation stressed in rats. (**A**) Stress-induced mast cell degranulation. (red-stained cell) Alcian blue-safranin, ×330. (**B**) Inhibition of mast cell degranulation by (red and blue stained cells) Alcian blue-safranin, ×132.

MDA Determination

Malondialdehyde was determined spectrophotometrically using phenobarbituric acid test. MDA levels were expressed as nmoles/mg tissue.[32] Protein was determined by the Lowry method.[33]

Statistics

The data are expressed as mean ±SE. Statistical analyses were performed using one-way ANOVA and Duncan's multiple range test. For ulcer index data, Student's t-test and for non-parametrical data, Mann-Whitney U-test were used.

RESULTS

Exposure of rats to cold-restraint stress at 4°C for three hours induced hemorrhagic ulcers in the glandular mucosa. Macroscopic and histologic appearance of gastric ulcers are illustrated in FIGURES 1 and 2, respectively. Pretreatment of rats with VIP prevented gastric mucosal ulcers induced by cold-restraint stress (FIG. 3). Ulcer indices and PMNL infiltration are shown in FIGURE 4. In stressed rats, PMNL infiltration and ulcer index were significantly different from that of control and VIP-treated rats. In gastric tissue, MDA level as a metabolic product of lipid peroxide was significantly higher in stressed group than control and VIP-administered group (FIG. 5). Antioxidant enzyme activity was not changed significantly in stressed rats compared to controls. On the other hand, VIP treatment significantly decreased antioxidant enzyme activity of gastric tissue (FIG. 6); VIP decreases the formation of superoxide anion and hydrogen peroxide. Histochemical heterogeneity, degranulation pattern, and the numbers of mast cells are reported in TABLE 1. TABLE 2 shows total numbers of blue and red stained mast cells. TABLES 1 and 2 and FIGURE 7 illustrate that while stress causes degranulation of mast cells, VIP acts against it. The inhibitory effect of VIP is more potent on red-stained mast cells than on blue-stained cells. This indicates that VIP potentially inhibits degranulation of connective tissue mast cells (CTMC). The results obtained from groups A, B, and C are given in TABLE 3. When VIP was used for one and two days, only one of six animals had gastric lesions. On the other hand, after administration of VIP for three days none of the animals had lesions.

DISCUSSION

VIP treatment of rats significantly inhibited development of gastric mucosal ulcers induced by cold-restraint stress. Lipid peroxidation and PMNL infiltration were significantly decreased by VIP. VIP effectively inhibited stress-induced mast cell degranulation, its inhibitory effect is prominent on red-stained cells.

Recent studies suggest that toxic oxygen molecules might be one of the important factors in inducing gastric mucosal injury during cold-restraint stress and the increase of xanthine oxidase activity might be responsible for the production of oxygen free radicals.[1,2,15] Lipid peroxidation, which results from the generation of reactive oxygen species, was suggested to be closely related to production of the gastric damage by cold-restraint stress.

In this study, it was shown that MDA levels increased in gastric tissue of the stressed rat, indicating that cold-restraint stress induces lipid peroxidation. Decreases in MDA and antioxidant enzyme levels (superoxide dismutase and catalase) in gastric tissues of VIP-treated group can be explained by the antioxidant activity of VIP. VIP has a potent protective activity against injury triggered by xanthine/xanthine oxidase and effectively scavenges singlet oxygen and the hydroxyl radical and also inhibits superoxide radical formation by inflammatory cells.[22,34–37] Therefore VIP, by either scavenging or inhibiting the production of reactive oxygen species, decreased the MDA level and antioxidant enzyme activity of gastric tissue. Because it is generally known that superoxide dismutase and catalase activity are induced by the level of their individual reactive species, i.e., superoxide anion and hydrogen peroxide, respectively, the reduced level of superoxide anion and hydrogen peroxide may lead to decreases in the activity of superoxide dismutase and catalase.[38] Hence, our results suggest that the formation of superoxide anion and hydrogen peroxide are decreased by VIP as a result of a decrease in antioxidant enzyme activities and MDA level.

It was recently reported that PMNL infiltration is one of the important mediators of cold-restraint induced gastric damage.[2] In addition to xanthine/xanthine oxidase system, activated PMNL induce tissue injury through the production of reactive oxygen species and release of cytotoxic molecules.[2] These products may derange the structure and function of surrounding cells and act as mediators to exacerbate the subsequent inflammatory reactions.

Thus, VIP possibly prevents PMNL infiltration and thus protects gastric mucosa from the deleterious effect of activated PMNL. Our results are in agreement with reports that explain the antiinflammatory activity of VIP.[22]

There are some reports stating that mast cell degranulation has been shown to be important in the pathogenesis of stress-induced gastric ulcer formation.[7–10,13,14] Different mast cell membrane stabilizers have been used to prevent gastric mucosal damage by stress in animals.[13,14,39] Studies suggested that mast cell stabilizers can be effective against gastric lesions. Most researchers generally use toluidine blue staining technique for the examination of mast cells in their studies. In the present study, we used alcian blue–safranin staining technique to examine the histochemical heterogeneity of mast cells in the stomach of rats exposed to cold-restraint stress. As far as we know, this is the first report that provides information about histochemical heterogeneity of mast cells and stress ulcer development.

Mast cells from different locations are shown to vary in their histochemical, ultrastructural, cytochemical, and functional properties. Depending on their heterogeneity, they are categorized as mucosal (MMC) and connective tissue (CTMC), which are also called atypical and typical mast cells, respectively.[40,41] Differences in the charge distribution of the proteoglycan matrix of mast cells may also be revealed by staining with combinations of dyes such as alcian-blue and safranin. Connective tissue mast cells can be stained with alcian blue-safranin blue to red according to their heparin content.[40,41] When heparin synthesis increases, cells stain only with safranin (red-stained) and in case of a decrease in heparin synthesis cells stain with only alcian blue (blue-stained). Mucosal mast cells, containing little or no heparin, are stained with only alcian blue (blue-stained). In the present study, stress induced degranulation of both red and blue stained mast cells and VIP effectively inhibited red-stained mast cell degranulation. Our results indicate that gastric lesion formation correlates with connective tissue mast cell activity. Our results also indicate that in addition to histamine, connective tissue mast cell–derived heparin participates in the pathogenesis of stress-induced gastric ulcers. It can be suggested that connective tissue mast cell–derived heparin can induce the formation of acute hemorrhagic gastric lesions. Our study also suggests that some of the blue-stained mast cells may be mucosal mast cells and VIP cannot inhibit their degranulation as effectively as it does for CTMC. It has been reported that VIP localizes in connective tissue mast cells but not in mucosal mast cells and has a modulatory effect on degranulation and mediator release for this type of mast cell.[42–44] Our data are somewhat in agreement with the suggestion of a modulatory effect of VIP in the release of mediators from connective tissue mast cells.

Both types of mast cells are involved in tissue repair and wound healing.[45] However, depending on the intensity of degranulation and the amount of released histamine/heparin, mast cells can have different effects on wound healing.[45] A study related to the role of mast cell mediators in the pathogenesis of anaphylactic gastric ulcers reported that mucosal anaphylaxis causes ulceration by greatly increasing the rate of mucosal cell turnover.[46] Mucosal cell turnover was stimulated by low histamine and high heparin concentrations. At higher concentrations of histamine there is a significant inhibitory effect on cell turnover. Thus, we can conclude that VIP, by modulating the activity of mast cells, may regulate the effects of mast cell mediators for the benefit of gastric mucosa.

Another important finding of this study is related to the effect of VIP on gastric lesions when it was used after lesion formation. Our results strongly indicate that VIP is very effective in curing the gastric lesions induced by cold-restraint stress.

In addition to the beneficial effects of VIP mentioned above, it has further favorable actions that counteract gastric ulcer development induced by cold-restraint stress. VIP inhibits the release of gastrin hormone and the action of histamine on acid-secreting parietal cells, relaxes the fundic muscle, and increases blood flow.[17] Thus, VIP has combined effects and it would be more effective in treating stress-induced gastric ulcers than agents that block only one factor contributing to ulcer formation.

REFERENCES

1. DAS, D. & R. K. BANERJEE. 1993. Effect of stress on the antioxidant enzymes and gastric ulceration. Molec Cell. Biochem. **125:** 115–125.
2. ÇOŞKUN, T., I. ALICAN, B. Ç. YEGEN, T. SAN, S. CETNEL & H. KURTEL. 1994. Cyclosporin A reduced the severity of cold-restraint induced gastric lesions: Role of leukocytes. Digestion **336:** 1–10.
3. AL-MASHHADANI, W. M., K. H. KARIM, R. I. AL-TAIE & H. AL-ZAHAWI. 1991. Nifedipine versus cimetidine in prevention of stress-induced gastric ulcers in rats. Eur. J. Pharmacol. **192:** 117–121.
4. CHO, C. H. & C. W. OGLE. 1989. Paracetamol potentiates stress-induced gastric ulceration in rats. J. Pharm. Pharmacol. **42:** 505–507.
5. RAY, A., P. G. HENKE & R. M. SULLIVAN. 1990. Noradrenergic mechanisms in the central amygdalar nucleus and gastric stress ulcer formation in rats. Neurosci. Lett. **110:** 331–336.
6. RAY, A., S. PURI, A. K. CHAKRAVARTY & P. SEN. 1992. Central histaminergic involvement during stress in rats. Indian J. Exp. Biol. **30:** 724–728.
7. GARG, G. P., C. H. CHO & C. W. OGLE. 1992. Ethacrynic acid and sulphasalazine inhibit the generation of leukotriene C_4 in rat stomachs: A possible gastric anti-ulcer mechanism in cold-restraint stressed rats. Pharmacology **44:** 177–189.
8. OGLE, C. W. & C. H. CHO. 1989. The protective mechanism of FPL55712 against stress-induced gastric ulceration in rats. Agents & Actions **26:** 3–4.
9. CHO, C. H. & C. W. OGLE. 1979. Cholinergic-mediated gastric mast cell degranulation with subsequent histamine H_1- and H_2-receptor activation in stress ulceration in rats. Eur. J. Pharmacol. **55:** 23–33.
10. OGLE, C. W & S. C. G. HUI. 1995. The influence of peripheral or central administration of ondansetron on stress-induced gastric ulceration in rats. Experienta **51:** 786–789.
11. FORDTRAN, S. 1989. Pathogenesis of stress ulcer. In Gastrointestinal Disease. 4th edit. p. 785. W. B. Saunders Company. Philadelphia.
12. BROWN, P. A., T. H. BROWN & J. VERNIKOS-DANELLIS. 1976. Histamine H_2 receptor: Involvement in gastric ulceration. Life Sci. **18:** 339–344.
13. KARMELI, F., R. ELIAKIM, E. OKON & D. RACHMILEWITZ. 1991. Gastric mucosal damage by ethanol is mediated by substance P and prevented by ketotifen, a mast cell stabilizer. Gastroenterology **100:** 1206–1216.
14. TABUCHI, Y. & Y. KUREBAYASHI. 1992. Effect of DS-4574, a novel peptidoleukotriene antagonist with mast cell stabilizing action, on gastric lesions and gastric secretion in rats. Jpn. J. Pharmacol. **60:** 335–340.
15. LI, T. & X. J. ZHANG. 1993. Role of oxygen-derived free radicals in stress-induced gastric ulceration. Acta Physiol. Sin. **45:** 286–291.

16. SAID, S. I. 1995. Vasoactive intestinal peptide. Airway smooth muscle, peptide receptors, ion channels and signal transduction. Roseburn & M. A. Giembyez, Eds. Birkhjuser Verlag, Basel.

17. SAID, S. I. 1986.Vasoactive intestinal peptide. J. Endocrinol Invest. **9**: 191–200.

18. TUNÇEL, N., F. GÜRER, E. ARAL, K. UZUNER, Y. AYDIN & C. BAYÇU. 1995. The effect of vasoactive intestinal peptide (VIP) on mast cell invasion/degranulation in testicular interstitium of immobilized + cold stressed and β-endorphin-treated rats. Peptides **17**: (5): 817–824.

19. TIKIZ, H., N. TUNÇEL, F. GÜRER & C. BAYÇU. 1991. Mast cell degranulation in hemorrhagic shock in rats and the effects of vasoactive intestinal peptide, aprotinin and H_1 and H_2 receptor blockers on degranulation. Pharmacology **43**: 47–52.

20. TIKIZ, H., N. TUNÇEL, M. Z. AKIN & F. GÜRER. 1992. The effect of vasoactive intestinal peptide (VIP) and naloxone combination on survival rates exposed to severe hemorrhage. Peptides **13**: (1): 83–89.

21. SAID, S. I. 1994. VIP and nitric oxide: Physiological co-transmitters with antagonistic roles in inflammation. Biomed. Res. **15** (Suppl. 2): 79–84.

22. PAKBAZ, H., H. BERISHA, H. D. FODA, A. ABSOOD & S. I. SAID. 1994. Vasoactive intestinal peptide (VIP) and related peptides: A new class of anti-inflammatory agents? *In* International Symposium on Vasoactive Intestinal Peptide Pituitary Adenylate Cyclase Activating Polypeptide and Related Regulatory Peptide. G. Rosselin, Ed.: 597–605. World Scientific. Singapore.

23. SAID, S. I. 1996. Vasoactive intestinal peptide and nitric oxide divergent roles in relation to tissue injury. Ann. N.Y. Acad. Sci. **805**: 379–388.

24. AKIN, M. Z., N. TUNÇEL, F. GÜRER, N. KURAL & S. USLU. 1993. Effect of vasoactive intestinal peptide and naloxone combination on urinary N-acetyl-β-D-glucosaminidase level and kidney histology of rats exposed to severe hemorrhage. Pharmacology **47**: 194–199.

25. UZUNER, K., N. TUNÇEL, Y. AYDIN, M. TUNÇEL, F. GÜRER, P. BENLI & D. AK. 1995. The effect of vasoactive intestinal peptide (VIP) on superoxide dismutase and catalase activities in renal tissues of rats exposed to hemorrhagic ischemia-reperfusion. Peptides **16** (5): 911–915.

26. TUNÇEL, N., H. BAŞMAK, K. UZUNER, M. TUNÇEL, G. ALTIKOKKA, V. ZAIMOĞLU, A. ÖZER & F. GÜRER. 1996. Protection of rat retina from ischemia-reperfusion injury by vasoactive intestinal peptide (VIP): The effect of VIP on lipid peroxidation and antioxidant enzyme activity of retina and choroid. Ann. N.Y. Acad. Sci. **805**: 489–497.

27. TUNÇEL, N., S. ERDEN, K. UZUNER, G. ALTIOKKA & M. TUNÇEL. 1997. Ischemic-reperfused rat skeletal muscle: The effect of vasoactive intestinal peptide (VIP) on contractile force, oxygenation and antioxidant enzyme systems. Peptides **18** (2): 269–275.

28. TUNÇEL, N., Y. AYDIN, M. KOŞAR & M. TUNÇEL. 1997. The effect of vasoactive intestinal peptide (VIP) on the testicular tissue histamine level of immobilized + cold stressed rats. Peptides **18** (6): 913–915.

29. KOSTYUK, A. V. & A. I. POTAPOVICH. 1989. Superoxide-driven oxidation of quercetine and a simple sensitive assay for determination of superoxide dismutase. Biochem. Int. **19**: 1117–1134.

30. GLICK, D. 1982. Methods of biochemical analysis. *In* Adaptation of Polarographic Oxygen Sensors for Biochemical Assays, Vol 28. M. A. Lessler, Ed.: 175–199. New York. Wiley-Interscience.

31. WANG, J. & S. M. LIN. 1988. Mixed plant tissue-carbon paste bioelectrode. Anal. Chem. **60**: 1545–1548.

32. RAO, N. A. 1990. Role of oxygen free radicals in retinal damage associated with experimental uveitis. Trans. Am. Ophthalmol. Soc. **88**: 797–850.

33. LOWRY, O. H., N. J. ROSEBROUGHT, A. L. FARR & R. J. RANDAL. 1951. Protein measurement with the folin phenol reagent. J. Biol. Chem. **193**: 265–275.

34. BERISHA, H., H. D. FODA, H. SAKAKIBARA, M. TROTZ, H. PAKBAZ & S. I. SAID. 1990. Vasoactive intestinal peptide (VIP) prevents lung injury due to xanthine/xanthine oxidase. Am. J. Physiol. **259**: L151–L155.

35. MISRA, R. B. & P. H. MISRA. 1990. Vasoactive intestinal peptide (VIP), a singlet oxygen quencher. J. Biol. Chem. **265**: 15371–15374

36. KUROSAWA, M. & T. ISHIZUKA. 1993. Inhibitory effects of vasoactive intestinal peptide on superoxide anion formation by N-formyl-methionyl-leucyl-phenylalaline-activated inflammatory cells in vitro. Int. Arch. Allergy Immunol. **100**: 28–34.

37. SAKAKIBARA, H., J. TAKAMATSU & S. I. SAID. 1990. Vasoactive intestinal peptide (VIP) inhibits superoxide anion release from rat alveolar macrophages. Am. Rev. Respir. Dis. **141:** A645.
38. WHITESIDE, C. & M. H. HASSAN. 1987. Induction and inactivation of catalase and superoxide dismutase of Eschericha coli by ozone. Arch. Biochem. Biophys. **257:** 464–471.
39. GOOSSENS, J., J. V. REEMPTS & J. P. VAN WAUME. 1987. Cytoprotective effects of disodium cromoglycate on rat stomach mucosa. Br. J. Pharmac. **91:** 165–169.
40. GALLI, S. 1990. Biology of disease. Lab. Invest. **62:** 5–25.
41. PEARCE, F. L. 1986. On the heterogeneity of mast cells. Pharmacology **32:** 61–71.
42. DEREK, M., M. MCKAY & J. BIENENSTOCK. 1994. The interaction between mast cells and nerves in the gastrointestinal tract. Immunol. Today. **15:** 11.
43. AIUTI, F., C. CARINI & R. PAGANELLI. 1989. Report on the XIII ICACI Montreux, Switzerland on 17–23 October 1988. Immunol. Today **10:** 71–73.
44. CUTZ, E., W. CHAN, N. S. TRACK, A. GOTH. & S. I. SAID. 1978. Release of vasoactive intestinal polypeptide in mast cells by histamine liberator. Nature **275:** 661–662.
45. TRABUCCHI, E., E. RADAELLI, M. MARAZZI, D. FOSCHI, M. MUSAZZI & A. M. VERONESI. 1988. The role of mast cells in wound healing. Int. J. Tiss. Reac. **X (6):** 367–372.
46. ANDRE, F., J. GILLON, C. ANDRE & S. FOURNIER. 1983. Role of mast cell mediators in pathogenesis of anaphylactic gastric ulcer. Digestion **28:** 108–113.

Is There Appetite after GLP-1 and PACAP ?

JEAN CHRISTOPHE[a]

*Department of General and Human Biochemistry, Institute of
Pharmacy, Université Libre de Bruxelles, CP 205/3, Boulevard du
Triomphe, B-1050 Brussels, Belgium*

ABSTRACT: Anitobesity drugs must increase the sensitivity of the hypothalamic satiety center towards leptin and antagonize the synthesis and action of NPY. The array
of pharmacologic tools available is vast and presently ineffective. Among peptide
analogs considered for evaluation [NPY-5 antagonists and CCK-A, bombesin, amylin
and melanocyte-stimulating hormone-4 (or melanin-concentrating hormone?) ago-
nists], is there a place for GLP-1 and PACAP? GLP-1 receptors present in ARC,
PVN, VMN, and SON are the target for both central and blood-borne GLP-1 in those
hypothalamic neurons endowed with GLUT-2 and glucokinase. GLP-1, hypersecreted by L-cells after a meal, is a potent insulinotropic agent and, together with
glucose, reduces food intake and induces c-fos in the ARC. PACAP is present in the
ARC, PVN, and SCH, and its hypothalamic type I receptor elevates cAMP and inositol triphosphate in the PVN, where it may perhaps antagonize NPY-induced food
intake and hyperinsulinemia. However, irrelevant neuroendocrine, autonomic, and
circadian functions are also activated by this peptide, making it a less than ideal base
on which to build an obesity treatment.

CENTRAL ROLE OF HYPOTHALAMUS IN EATING PATTERNS

The "set point" hypothesis of food intake and energy expenditure (FIG. 1) considers that
two vegetative centers in the hypothalamus are reciprocally regulated: the PVN-DMN
(paraventricular-dorsomedial nucleus of the hypothalamus) in the mediodorsal area initi-
ates and maintains feeding and has an anabolic role, while the satiety VMN (ventromedial
hypothalamic nucleus) center in the laterobasal area stops feeding and plays a catabolic
role. Accordingly, goldthioglucose (GTG) or electrolytic lesions of the VMN lead to
hyperphagia in mice and rats, respectively. The resulting dynamic obesity is accompanied
by cholinergic vagal hyperactivity with hyperinsulinemia,[1] and decreased sympathetic
activity with diminished energy expenditure through body heat.

The subfornical organ, vascular organ of the lamina terminalis, and area postrema (AP)
lack a blood-brain barrier so that the hypothalamus can respond to blood-borne peptides
and proteins. Systemic insulin crosses the saturable blood barrier by transcytosis through
fenestrated endothelia in the median eminence (ME) and choroid plexus. Together with
glucose it acts as glucostat at the VMN level. Another afferent signal, the protein hormone
leptin secreted by white adipose tissue (WAT), acts as a lipostat at the arcuate nucleus
(ARC) level, the important center producing neuropeptide tyrosine(1-36) (NPY). These
signals constantly indicate the quantity and composition of energy stores to the hypothalamus. Efferent hypothalamic mechanisms control appetite through neuropeptides (such as
corticotropin-releasing factor) and other neurotransmitters, as well as energy expenditure
(through the autonomic nervous system).

Corresponding author: Tel.: 322-650.52.99; Fax: 322-650.53.24; E-mail: jdehaye@ulb.ac.be

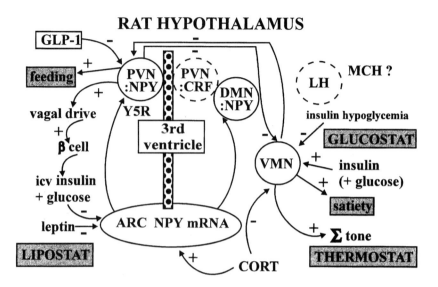

FIGURE 1. The hypothalamic control of food intake and energy expenditure (Σ: sympathetic).

The hypothalamic disorders of obesity often include, beside hyperphagia, vagal hyper-insulinism, a low central orthosympathetic tone (with reduced thermogenesis), low serotonin efficacy, a hyperactive hypothalamo-hypophyso-adrenal axis, a hypoactive GHRH-GH-IGF (growth hormone-releasing factor–growth hormone–insulin-like growth factor) axis, and hypogonadism of central origin. Hyperlipogenesis, glucose intolerance, and excessive gluconeogenesis are secondary features.

HYPOTHALAMIC GLUCOREGULATORY NEURONS, INSULIN, AND NPY

Glucose transporter 4 (GLUT 4) in the hypothalamus is sensitive to circulating insulin.[2] In rats, intravenous insulin when given alone brings about glucoprivic feeding while the same doses of insulin are anorexigenic if combined with doses of glucose that prevent hypoglycemia (FIG. 1). In the rat there is a delay of 30 min between a single injection of insulin and the stimulation of food intake due to hypoglycemia. Glucopenia increases c-fos in the VMN, then elevates corticotropin-releasing factor(1-41) (CRF) and somatostatin (SS) levels in the PVN.[3] This lack of glucose decreases the firing rate in the VMN, increases firing in lateral hypothalamus area (LH) and nucleus of the tractus solitarius (NTS) glucosensors, and increases sympathetic outflow through the splanchnic nerves, provoking increased hepatic glucose production. Intracerebroventricular 2-deoxy-D-glucose (2-DG) also produces central glucoprivation, but without hypoglycemia. The resulting local hypothalamic glucopenia induces c-fos gene expression in 20% of NPY neurons in the ARC, and 3–6% of SS neurons in the ARC and PVN.[4] A 3-h direct perfusion of the forebrain with a moderate glucose concentration via the carotid leads to a doubling in c-fos neurons in the VMN, parvocellular PVN (rich in CRF neurons), and DMN.[5] An increase in portal vein glucose decreases the rate of vagal visceral signals to the NTS (FIG. 2). The NTS is a glucose-sensitive nucleus adjacent to the fourth ventricle in the medulla oblongata which is a relay center to the PVN, ARC and DMN. PVN axons project

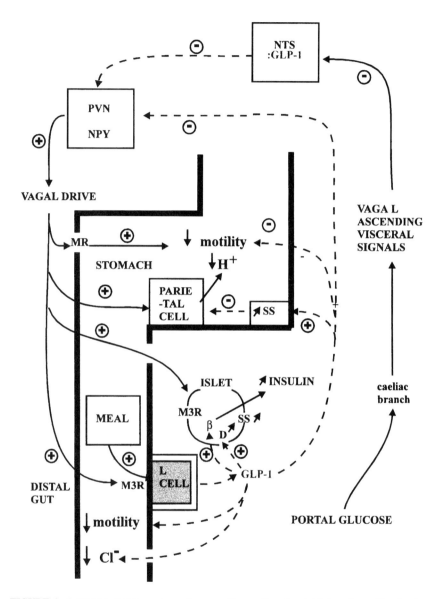

FIGURE 2. A scheme showing autonomic connections with gut and islets and resulting *in vivo* effects (MR: muscarinic receptors; β and D cells secrete insulin and SS, respectively).

dorsocaudally to the locus ceruleus, central gray and lateral parabrachial nucleus (LPBN). The locus ceruleus and LPBN project to the NTS and the DMN. The central gray and LPBN areas project to sympathetic neurons in the intermediolateral cell column of the spinal cord. In addition, these complex autonomic connections interplay with several neuropeptides. For instance, the ARC contains NPY, GHRH, proopiomelanocortin (POMC), galanin (GAL), and norepinephrine (NE) cell groups. The dorsal parvocellular division of the PVN contains CRF and SS neurons. CRF neurons project to parasympathetic and sympathetic preganglionic cell groups in the brainstem and spinal cord.

NPY is the most potent and long-lasting orexigenic agent yet identified in rats. This abundant brain peptide is notably produced in the ARC, whose nerve fibers project into the PVN, DMN, VMN, PFH (perifornical hypothalamus), and MPOA (medial preoptic area). When administered centrally NPY releases insulin acutely via the efferent parasympathetic nervous system (from the LH to the vagus of the medulla then the pancreatic vagus), even in the absence of food.[6,7] NPY-induced feeding is attenuated by peripheral glucose infusion.[8] The injection of insulin into the third ventricle or hypothalamus prevents the rise in NPY mRNA and NPY levels observed in fasting rats (FIG. 3). Thus, circulating insulin acts as a negative feedback loop to inhibit hypothalamic NPY mRNA levels and peptide secretion as it readily enters the ARC, which contains many insulin receptors.[9–17] On the other hand, dexamethasone treatment increases hypothalamic NPY mRNA and the NPY content in ARC and PVN through CORT receptors (while it suppresses CRF mRNA levels within the PVN).[18,19] NPY synthesis also depends on cAMP[20]: CREB (cAMP-regulated response element–binding protein), when phosphorylated by PKA (protein kinase A), dimerizes showing afterwards a tenfold higher affinity for CRE. Besides, activators of Ca^{2+}- or PL (phospholipase)-dependent protein kinases, such as TPA and NGF (nerve growth factor), also regulate NPY expression.[21]

NPY infusion into the PVN elicits feeding within 30 min, which is associated with hypothermia[22] and c-fos expression, mostly in the parvocellular (SS-CRF–rich) region of the PVN but also in the magnocellular (arginine vasopressin–rich) PVN region, DMN, SON (supraoptic nucleus), and BNST (bed nucleus of the stria terminalis). At low con-

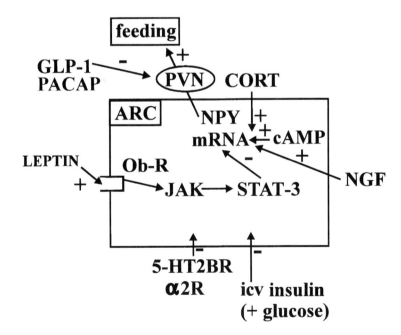

FIGURE 3. The control of NPY synthesis in the ARC and that of GLP-1 (and PACAP) on the PVN influence feeding (5-HT2BR and α2R: serotonin and α2 adrenergic-receptors).

centration (0.15–15 nM) NPY and [Pro[34]]-NPY increase discharge rates in single PVN neurons.[23] Among extrahypothalamic sites, NPY-induced feeding augments c-fos in the lateral subdivisions of the central amygdaloid nucleus and solitary tract.[24,25]

NPY may also inhibit the satiety center in VMN after food deprivation. The electrophysiological unit activity in VMN neurons is intensively suppressed by NPY applied *in vitro.*[26] Besides, the orexigenic effect of intracerebroventricularly administered NPY is lacking in fasted rats with bilateral electrolytic lesions in the VMN (and too much NPY already produced in the ARC),[27,28] while administration of NPY antibodies abolishes hyperphagia in these VMN-lesioned rats.[29] GTG-sensitive VMN neurons play little role in the induction of NPY mRNA by fasting in the ARC, as fasting increases NPY levels mRNA in the ARC of GTG-treated mice.[30]

NPY-induced feeding is mediated by Y5 receptors. The rat hypothalamic Y5 receptor mRNA encodes a 456–467 amino acid protein with less than 33% overall identity to the other NPY receptors and is found primarily in the PVN and lateral hypothalamus. The Y5 receptor occupancy stimulates food intake, being therefore the main "feeding" receptor subtype involved in obesity.[31,32] The mouse NPY Y5 receptor shares 60% amino acid identity to the Y1 receptor and its pharmacology resembles that of the anxiolytic Y1 receptor. It is abundantly expressed in the PVN, but also within the VMN, SCH (suprachiasmatic nucleus), BNST, and anterior hypothalamus.[33] Hypothalamic NPY receptors are coupled to cAMP-dependent PKA and Ca^{2+}/calmodulin-dependent protein kinase II that activate CREB, allowing increased CRE binding. Indeed, NPY injected in the rat PFH provokes an intense phosphoCREB signal in the PVN and VMN of rats, similar to that observed after food deprivation.[34]

GLP-1 IN THE ENTERO-HYPOTHALAMIC AXIS

Most frequently in obese rodents and obese humans the ARC reacts poorly to the leptin hypersecreted by adipose tissue, so that the local synthesis of NPY is unchecked (FIG. 3). Feeding is induced neither in ob/ob mice by intracerebroventricular injection of 2-DG nor in fa/fa rats by intracerebroventricular injection of insulin. This may reflect a secondary reduced insulin receptor concentration and/or impaired glucose utilization in the hypothalamus of these genetically obese rodents.[15,35] Anti-obesity drugs must therefore increase the sensitivity of the ARC towards leptin and insulin, thereby antagonizing the synthesis of NPY. They could as well inhibit the action of NPY on the PVN. Is GLP-1 worth evaluation in this respect?

In rat, pro-GLP-1–glucagon mRNA and GLP-1 are localized in both open-type L cells of the distal intestine and in NTS neurons, whose efferent fibers go to the hypothalamus[36–38] (FIG. 2). Rat brain GLP-1 receptors may therefore act as the target for blood borne as well as for central GLP-1 in those hypothalamic nuclei involved in appetite.[39–41] The mRNAs for GLP-1 receptors, GLUT-2, and GK are indeed colocalized in the PVN, ARC, VMN, and SON[42–44] and there is only one GLP-1 receptor isoform known.[45,46]

Intracerebroventricular administration of GLP-1 powerfully inhibits feeding in fasted rats as well as angiotensin II-induced drinking behavior and induces c-fos accumulation in the PVN and central amygdaloid nucleus. Intracerebroventricular administration of exendin-4(9-39)amide, the N-shortened derivative of exendin-4 (a peptide from heloderma suspectum venom), acts as a specific GLP-1 antagonist. It blocks the inhibitory effect of GLP-1 on food intake, doubles food intake in satiated rats, and augments the feeding response to NPY.[43,47–49] GLP-1 produces an immediate secretion of aspartic acid and glutamine (from astrocytes?) in the VMN.[50]

Are peripheral effects of GLP-1 contributing to satiety: perhaps but indirectly (FIG. 2). There is a pulsatile secretion of GLP-1 from intestinal L cells in fasted man with a frequency of 5–7/h. GLP-1 is hypersecreted after a carbohydrate or fat-rich meal. More specifically, the amplitude of plasma GLP-1 pulses increases after a 100 g oral glucose load but atropine infusion delays and reduces this increased amplitude,[51] suggesting a direct muscarinic (M3?)[52] stimulatory control of L cells. Hence, an oral glucose load may provoke a higher GLP-1 response when the vagal celiac branch allows a cholinergic-sensitizing effect.

GLP-1 is the most potent insulinotropic hormone known and is also able to induce proinsulin biosynthesis. Accordingly, exendin-4(9-39) reduces insulin secretion in vivo in rats after a standard meal.[53] In vitro, in the presence of glucose, the immediate effects of GLP-1 include an elevation of cAMP and a potentiation of membrane depolarization. This β cell depolarization is due to (1) the closure of ATP-sensitive K^+ channels by ATP derived from glycolysis that provokes Ca^{2+} influx through the opening of voltage-dependent Ca^{2+} channels (VDCCs) and (2) the PKA phosphorylation of VDCCs increasing opening probability and the activation of non selective cation channels (NSCCs) promoting an inward Na^+ (Ca^{2+})-dependent current.[54,55] This glucose-dependent insulin secretion is enhanced by carbachol releasing Ca^{2+} from intracellular pools through phospholipase C.[56] In fa/fa pancreas the glucose threshold for the insulinotropic action of GLP-1 is reduced and, in the presence of 4.4 mM glucose, GLP-1 increases insulin secretion 25-fold, whereas there is only a fivefold stimulation in Fa/? pancreas, showing increased GLP-1 sensitivity in the β cells from obese animals.[57] GLP-1 inhibits glucagon secretion in vivo in rat and pig, another way to lower blood glucose. This inhibitory action of GLP-I on intact islets is mediated by a paracrine release of SS from D cells.[58,59]

GLP-1 also inhibits gastric acid secretion, delays gastric emptying and secondary exocrine pancreatic secretion, and inhibits upper gastrointestinal secretion and motility in humans (the "ileal-brake" effect).[60-62] The inhibitory effects of GLP-1 on the stomach are observed not only after pentagastrin stimulation but also under vagal induction through sham feeding in man. A centrally mediated mechanism is implied, considering the absence of effects in vagotomized subjects.[63,64] In isolated rat parietal cells GLP-1 stimulates H^+ production (aminopyrine accumulation) via cAMP, a finding not duplicated in vivo in rat,[65] suggesting that the direct stimulatory effect on parietal cells is overcome by SS release that mediates Gi-adenylyl cyclase inhibitory transduction.[66,67]

PACAP AS A POSSIBLE LINK IN OBESITY

The neuropeptide pituitary adenylyl cyclase activating peptide (PACAP), first isolated from ovine hypothalamus,[68] is widely distributed in the brain. It is notably present in the rat hypothalamus. PACAP perikarya are found in the PVN, ARC, SCH, VMN, SON, preoptic periventricular zone, retrochiasmatic area, tuber cinereum, and mammillary nuclei. PACAP axons are present in an ascending pathway from the SCH (the brain's biological clock) to the PVN and in the BNST (a relay between limbic and autonomic structures).

PACAP type 1 receptors (PACAP-R) are distributed even more homogeneously in brain. Such receptors are present in the PVN, SON, SCH, retrochiasmatic area, ME, and posterior periventricular nucleus as well as the BNST, dorsal vagal nucleus, and amygdaloid complex. The PACAP type I receptor gene is expressed in both the magnocellular (arginine vasopressin) and parvocellular (SS/CRF) parts of the PVN and in the SON in rats.[69-72] Alternative splicing of two exons for the region encoding the third intracellular loop of rat PACAP type 1 receptors generates five isoforms with differential signal transduction properties. A sixth splice variant produces a 21 amino acid deletion in the N-terminal extracellular domain showing altered receptor selectivity with respect to PACAP-27 and -38 binding.[73] PACAP present in nerve terminals around and within rat islets activates L-type Ca^{2+} channels through a seventh short PACAP receptor isoform with discrete sequences in TM-II and TM-IV.[74]

The type-1 PACAP-R and heavier PACAP-R-hop splice variant mRNAs operating in the rat PVN[75] stimulate cAMP and inositol triphosphate formation. PACAP as well as feeding inhibits strongly the binding of rat hypothalamic nuclear extracts to CRE while NPY administration or food deprivation exerts the opposite effect, suggesting that CREB dephosphorylation is the satiety signal that negatively modulates genes involved in feeding.[34] In food-deprived mice, intracerebroventricularly administered PACAP reduces food intake and increases motor activity (rearing and grooming through CRF release?). It also antagonizes NPY-induced food intake: an intrahypothalamic injection of PACAP 10 min prior to the injection of NPY reduces feeding during the 4-h measurement period.[76-78]

PACAP is the most potent glucose-dependent insulinotropic neuropeptide known.[79] It stimulates insulin and glucagon secretion *in vivo* and *in vitro* in rat and mouse.[80,81] The mode of action of PACAP on β cells is similar to that exerted by GLP-1 acting via its own receptor (it augments insulin secretion glucose-dependently by increasing cAMP and Ca^{2+}; see above).

PACAP present in the gut "little brain" (in submucous more frequently than in myenteric ganglia)[82] inhibits pentagastrin- and histamine-stimulated gastric acid secretion by interacting with histamine effects on parietal cells.[83] In the rat jejunal mucosa PACAP promotes anion secretion via submucous neurons.[84,85] PACAP also relaxes smooth muscle in all portions of the rat gastrointestinal tract by activating apamin-sensitive K^+ channels through PACAP type I receptors and inhibits the contraction mediated by VIP (acting on type II receptors in cholinergic neurons).[85,86] The involvement of PACAP in several more autonomic and neuroendocrine functions, as well as regulation of the circadian pacemaker, is demonstrated by several observations (this volume and Christophe[87]).

GLP-1 AGONISTS AS THERAPEUTIC ANOREXIGENIC AGENTS?

The United States population (with 33% obese adult subjects) spends 33 billion dollars each year on weight loss products and services while drug companies are spending hundreds of millions of dollars on developing obesity treatments. Alongside other peptides considered for evaluation (FIG. 4) as anorexigenic agents [NPY-5 antagonists and cholecystokinin A (CCK-A), bombesin, amylin, and melanocyte-simulating hormone-4 (or melanin-concentrating hormone?) agonists], is there any clinical usefulness conceivable for GLP-I or PACAP? The half-life of GLP-I in human blood is only 5 min, due to its rapid degradation by enzymes derived from plasma cell membranes, such as

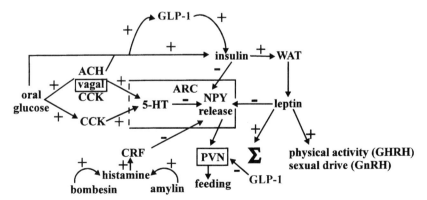

FIGURE 4. The positions of GLP-1 as a candidate among other anorexigenic agents (Σ: sympathetic tone).

dipeptidyl-peptidase IV, producing the desHis-Ala derivative.[88] Besides endopeptidase 24.11 could cleave GLP-1 at Phe6-Thr7, Tyr13-Leu14, and Glu21-Phe22. Therefore, drug design for longer acting, peptidase-resistant GLP-I/exendin-4 analogs with protracted efficacy after subcutaneous, sublingual, or oral administration may be required before considering efficient clinical trials.

The multiple irrelevant neuroendocrine, autonomic of PACAP, and circadian functions, including those exerted through PACAP-VIP type 2 receptors of PACAP, prevent the development of obesity treatment based on this anorexigenic neuropeptide.

SUMMARY

Antiobesity drugs must inter alia increase the sensitivity of the hypothalamic satiety center towards leptin and antagonize the synthesis and action of NPY. Unfortunately, the array of pharmacologic tools available is as vast as, presently, ineffective. Among peptide analogs considered for evaluation [NPY-5 antagonists and CCK-A, bombesin, amylin and melanocyte-simulating hormone-4 (or melanin-concentrating hormone?) agonists], is there a place for GLP-1 and PACAP? GLP-1 receptors present in ARC, PVN, VMN, and SON are the target for both central and blood-borne GLP-1 in those hypothalamic neurons endowed with GLUT-2 and glucokinase. GLP-1, hypersecreted by L-cells after a meal, is a potent insulinotropic agent and, together with glucose, reduces food intake and induces c-fos in the ARC. Concerning PACAP present in the ARC, PVN, and SCH, its hypothalamic type I receptor elevates cAMP and inositol triphosphate in the PVN where it may perhaps antagonize NPY-induced food intake and hyperinsulinemia. However, irrelevant neuroendocrine, autonomic and circadian functions are also activated by this peptide.

REFERENCES

1. KIBA, T., K. TANAKA, K. NUMATA, M. HOSHINO, K. MISUGI & S. INOUE. 1996. Ventromedial hypothalamic lesion-induced vagal hyperactivity stimulates rat pancreatic cell proliferation. Gastroenterology **110**: 885–893.

2. LIVINGSTONE, C., H. LYALL & G. W. GOULD. 1995. Hypothalamic GLUT 4 expression: a glucose- and insulin-sensing mechanism? Mol. Cell. Endocrinol. **107**: 67–70.

3. BAHJAOUI BOUHADDI, M., D. FELLMANN & C. BUGNON. 1994. Induction of Fos-immunoreactivity in prolactin-like containing neurons of the rat lateral hypothalamus after insulin treatment. Neurosci. Lett. **28**: 11–15.

4. MINAMI, S., J. KAMEGAI, H. SUGIHARA, N. SUZUKI, H. HIGUCHI & I. WAKABAYASHI. 1995. Central glucoprivation evoked by administration of 2-deoxy-D-glucose induces expression of the c-fos gene in a subpopulation of neuropeptide Y neurons in the rat hypothalamus. Mol. Brain Res. **33**: 305–310.

5. DUNN MEYNELL, A. A., E. GOVEK & B. E. LEVIN. 1997. Intracarotid glucose selectively increases Fos-like immunoreactivity in paraventricular, ventromedial and dorsomedial nuclei neurons. Brain Res. **748**: 100–106.

6. ABE, M., M. SAITO & T. SHIMAZU. 1989. Neuropeptide Y and norepinephrine injected into the paraventricular nucleus of the hypothalamus activate the endocrine pancreas. Biomed. Res. **10**: 431–436.

7. MARKS, J. L., K. WAITE & L. DAVIES. 1996. Intracerebroventricular neuropeptide Y produces hyperinsulinemia in the presence and absence of food. Physiol. Behav. **60**: 685–692.

8. ROWLAND, N. E. 1988. Peripheral and central satiety factors in neuropeptide Y-induced feeding in rats. Peptides **9**: 989–992.

9. BRADY, L. S., M. A. SMITH, P. W. GOLD & M. HERKENHAM. 1990. Altered expression of hypothalamic neuropeptide mRNAs in food restricted and food-deprived rats. Neuroendocrinology **52**: 441–447.

10. CALZA, L., L. GIARDINO, N. BATTISTINI, M. ZANNI, S. GALETTI, F. PROTOPAPA & A. VELARDO. 1989. Increase of neuropeptide Y-like immunoreactivity in the paraventricular nucleus of fasting rats. Neurosci. Lett. **104**: 99–104.

11. CUSIN, I., S. DRYDEN, Q. WANG, F. ROHNER JEANRENAUD, B. JEANRENAUD & G.WILLIAMS. 1995. Effect of sustained physiological hyperinsulinaemia on hypothalamic neuropeptide Y and NPY mRNA levels in the rat. J. Neuroendocrinol. **7**: 193–197.

12. MARKS LI, J. L. M., M. SCHWARTZ, JR. D. PORTE & D. G. BASKIN. 1992. Effect of fasting on regional levels of neuropeptide Y mRNA and insulin receptors in the rat hypothalamus: an autoradiobiographic study. Mol. Cell. Neurosci. **3**: 199–205.

13. MARKS, J. L., K. WAITE, D. CAMERON SMITH, S. C. BLAIR & G. J. COONEY. 1996. Effects of gold thioglucose on neuropeptide Y messenger RNA levels in the mouse hypothalamus. Am. J. Physiol. **270**: 1208–1214.

14. SAHU, A., C. A. SNINSKY, P. S. KALRA & S. P. KALRA. 1990. Neuropeptide Y concentrations in microdissected hypothalamic regions and in vitro from the medial basal hypothalamic-preoptic area of streptozotocin-diabetic rats with and without insulin substitution therapy. Endocrinology **126**: 192–198.

15. SCHWARTZ, M. W., J. L. MARKS, A. J. SIPOLS, D. J. BASKIN, S. C. WOODS, S. E. KAHN & JR. D. PORTE. 1991. Central insulin administration reduces neuropeptide Y mRNA expression in the arcuate nucleus of food- deprived lean (Fa/Fa) but not obese (fa/fa) Zucker rats. Endocrinology **128**: 2645–2647.

16. SCHWARTZ, M. W., A. J. SIPOLS, J. L. MARKS, G. SANACORA, J. D. WHITE, A. SCHEURINK, S. E. KAHN, D. G. BASKIN, S. C. WOODS, D. P. FIGLEWICZ & JR. D. PORTE. 1992. Inhibition of hypothalamic neuropeptide Y gene expression by insulin. Endocrinology **130**: 3608–3616.

17. WILLIAMS, G., J. S. GILL, Y. C. LEE, H. M. CARDOSO, B. E. OKPERE & S. R. BLOOM. 1989. Increased neuropeptide Y concentrations in specific hypothalamic regions of streptozotocin-induced diabetic rats. Diabetes **38**: 321–327.

18. CORDER, R., F. PRALONG, D. TURNILL, P. SAUDAN, A. F. MULLER & R. C. GAILLARD. 1988. Dexamethasone treatment increases neuropeptide Y levels in rat hypothalamic neurons. Life Sci. **43**: 1879–1886.

19. WILDING GILBEY, J. P. H. S. G., N. ASLAM, M. A. GHATEI & S. R. BLOOM. 1991. Dexamethasone induced increases in hypothalamic NPY-reversal with insulin administration. Diabetologia **34**: 115–115.

20. AKABAYASHI, A., C. T. ZAIA, S. M. GABRIEL, I. SILVA, W. K. CHEUNG & S. F. LEIBOWITZ. 1994. Intracerebroventricular injection of dibutyryl cyclic adenosine 3′,5′-monophosphate increases hypothalamic levels of neuropeptide Y. Brain Res. **17**: 323–328.

21. SABOL, S. L. & H. HIGUCHI. 1990. Transcriptional regulation of the neuropeptide Y gene by nerve growth factor. Mol. Endocrinol. **4**: 384–392.

22. CURRIE, P. J. & D. V. COSCINA. 1995. Dissociated feeding and hypothermic effects of neuropeptide Y in the paraventricular and perifornical hypothalamus. Peptides **16**: 599–604.

23. ARAMAKIS, V. B., B. G. STANLEY & J. H. ASHE. 1996. Neuropeptide Y receptor agonists: multiple effects on spontaneous activity in the paraventricular hypothalamus. Peptides **17**: 1349–1357.

24. LAMBERT, P. D., P. J. PHILLIPS, J. P. WILDING, S. R. BLOOM & J. HERBERT. 1995. c-fos expression in the paraventricular nucleus of the hypothalamus following intracerebroventricular infusions of neuropeptide Y. Brain Res. **23**: 59–65.

25. LI, B. H., B. XU, N. E. ROWLAND & S. P. KALRA. 1994. c-fos expression in the rat brain following central administration of neuropeptide Y and effects of food consumption. Brain Res. **5**: 277–284.

26. NISHIMURA, F., M. NISHIHARA, K. TORII & M. TAKAHASHI. 1996. Changes in responsiveness to serotonin on rat ventromedial hypothalamic neurons after food deprivation. Physiol. Behav. **60**: 7–12.

27. TANAKA, Y., M. EGAWA, S. INOUE & Y. TAKAMURA. 1994. Effects of intraventricular administration of neuropeptide Y on feeding behavior in fasted female rats with ventromedial hypothalamic lesions. Regul. Pept. **52**: 47–52.

28. YOUNG, J. K., J. C. MCKENZIE, L. S. BRADY & M. HERKENHAM. 1994. Hypothalamic lesions increase levels of neuropeptide Y mRNA in the arcuate nucleus of mice. Neurosci. Lett. **3**: 13–17.

29. DUBE, M. G., P. S. KALRA, W. R. CROWLEY & S. P. KALRA. 1995. Evidence of a physiological role for neuropeptide Y in ventromedial hypothalamic lesion-induced hyperphagia. Brain Res. 4: 275–278.

30. BERGEN, H. T. & C. V. MOBBS. 1996. Ventromedial hypothalamic lesions produced by gold thioglucose do not impair induction of NPY mRNA in the arcuate nucleus by fasting. Brain. Res. 29: 266–271.

31. GERALD, C., M. W. WALKER, L. CRISCIONE, E. L. GUSTAFSON, C. BATZL HARTMANN, K. E. SMITH, P. VAYSSE, M. M. DURKIN, T. M. LAZ, D. L. LINEMEYER, A. O. SCHAFFHAUSER, S. WHITEBREAD, K. G. HOFBAUER, R. I. TABER, T. A. BRANCHEK & R. L. WEINSHANK. 1996. A receptor subtype involved in neuropeptide-Y-induced food intake. Nature 382: 168–171.

32. HU, Y., B. T. BLOOMQUIST, L. J. CORNFIELD, L. B. DECARR, J. R. FLORES RIVEROS, L. FRIEDMAN, P. JIANG, L. LEWIS HIGGINS, Y. SADLOWSKI, J. SCHAEFERJ, N. VELAZQUEZ & M. L. MCCALEB. 1996. Identification of a novel hypothalamic neuropeptide Y receptor associated with feeding behavior. J. Biol. Chem. 271: 26315–26319.

33. WEINBERG, D. H., D. J. SIRINATHSINGHJI, C. P. TAN, L. L. SHIAO, N. MORIN, M. R. RIGBY, R. H. HEAVENS, D. R. RAPOPORT, M. L. BAYNE, M. A. CASCIERI, C. D. STRADER, D. L. LINEMEYER & D. J. MACNEIL. 1996. Cloning and expression of a novel neuropeptide Y receptor. J. Biol. Chem. 12: 16435–16438.

34. SHERIFF, S., W. T. CHANCE, J. E. FISCHER & A. BALASUBRAMANIAM. 1997. Neuropeptide Y treatment and food deprivation increase cyclic AMP response element-binding in rat hypothalamus. Mol. Pharmacol. 51: 597–604.

35. ANGEL, I., R. L. HAUGER, B. A. GIBLIN & S. M. PAUL. 1992. Regulation of the anorectic drug recognition site during glucoprivic feeding. Brain Res. Bull. 28: 201–207.

36. JIN, S. L. C., V. K. M. HAN, J. G. SIMMONS, A. C. TOWLE, J. M. LAUDER & P. K. LUND. 1988. Distribution of glucagon-like peptide 1 (GLP-1), glucagon, and glicentin in the rat brain: an immunocytochemical study. J. Comp. Neurol. 271: 519–532.

37. KREYMANN, B., M. A. GHATEI, P. BURNET, G. WILLIAMS, S. KANSE, A. R. DIANI & S. R. BLOOM. 1989. Characterization of glucagon-like peptide-1-(7-36) amide in the hypothalamus. Brain Res. 502: 325–331.

38. STOBIE HAYES, K. M. & P. L. BRUBAKER. 1992. Control of proglucagon-derived peptide synthesis and secretion in fetal rat hypothalamus. Neuroendocrinology 56: 340–347.

39. BULLOCK, B. P., R. S. HELLER & J. F. HABENER. 1996. Tissue distribution of messenger ribonucleic acid encoding the rat glucagon-like peptide-1 receptor. Endocrinology 137: 2968–2978.

40. GOKE, R., P. J. LARSEN, J. D. MIKKELSEN & S. P. SHEIKH. 1995. Distribution of GLP-1 binding sites in the rat brain: evidence that exendin-4 is a ligand of brain GLP-1 binding sites. Eur. J. Neurosci. 1: 2294–2300.

41. ØRSKOV, C., S. S. POULSEN, M. MOLLER & J. J. HOLST. 1996. Glucagon-like peptide I receptors in the subfornical organ and the area postrema are accessible to circulating glucagon-like peptide I. Diabetes 45: 832–835.

42. ALVAREZ, E., I. RONCERO, J. A. CHOWEN, B. THORENS & E. BLAZQUEZ. 1996. Expression of the glucagon-like peptide-1 receptor gene in rat brain. J. Neurochem. 66: 920–927.

43. NAVARRO, M., F. RODRIQUEZ DE FONSECA, E. ALVAREZ, J. A. CHOWEN, J. A. ZUECO, R. GOMEZ, J. ENG & E. BLAZQUEZ. 1996. Colocalization of glucagon-like peptide-1 (GLP-1) receptors, glucose transporter GLUT-2, and glucokinase mRNAs in rat hypothalamic cells: evidence for a role of GLP-1 receptor agonists as an inhibitory signal for food and water intake. J. Neurochem. 67: 1982–1991.

44. SHUGHRUE, P. J., M. V. LANE & I. MERCHENTHALER. 1996. Glucagon-like peptide-1 receptor (GLP1-R) mRNA in the rat hypothalamus. Endocrinology 137: 5159–5162.

45. THORENS, B. 1992. Expression cloning of the pancreatic β cell receptor for the glucoincretin hormone Glucagon-like peptide 1. Proc. Natl. Acad. Sci. USA 89: 8641–8645.

46. WEI, Y. & S. MOJSOV. 1995. Tissue-specific expression of the human receptor for glucagon-like peptide-1: brain, heart and pancreatic forms have the same deduced amino acid sequences. FEBS Lett. 358: 219–224.

47. TANG CHRISTENSEN, M., P. J. LARSEN, R. GOKE, A. FINK JENSEN, D. S. JESSOP, M. MOLLER & S. P. SHEIKH. 1996. Central administration of GLP-1-(7-36) amide inhibits food and water intake in rats. Am. J. Physiol. 271: R848–R856.

48. Turton, M. D., D. O'Shea, I. Gunn, S. A. Beak, C. M. Edwards, K. Meeran, S. J. Choi, G. M. Taylor, M. M. Heath, P. D. Lambert, J. P. Wilding, D. M. Smith, M. A. Ghatei, J. Herbert & S. R. Bloom. 1996. A role for glucagon-like peptide-1 in the central regulation of feeding. Nature 379: 69–72.

49. Van Dijk, G., T. E. Thiele, J. C. Donahey, L. A. Campfield, F. J. Smith, P. Burn, I. L. Bernstein, S. C. Woods & R. J. Seeley. 1996. Central infusions of leptin and GLP-1-(7-36) amide differentially stimulate c-FL1 in the rat brain. Am. J. Physiol. 271: 1096–1100.

50. Calvo, J. C., C. V. Gisolfi, E. Blazquez & F. Mora. 1995. Glucagon-like peptide-1(7-36)amide induces the release of aspartic acid and glutamine by the ventromedial hypothalamus of the conscious rat. Brain Res. Bull. 38: 435–439.

51. Balks, H. J., J. J. Holst, A. Von Zur Muhlen & G. Brabant. 1997. Rapid oscillations in plasma glucagon-like peptide-1 (GLP-1) in humans: cholinergic control of GLP-1 secretion via muscarinic receptors. J. Clin. Endocrinol. Metab. 82: 786–790.

52. Abello, J., F. Ye, A. Bosshard, C. Bernard, J. C. Cuber & J. A. Chayvialle. 1994. Stimulation of glucagon-like peptide-1 secretion by muscarinic agonist in a murine intestinal endocrine cell lines. Endocrinology 134: 2011–2017.

53. Wang, Z., R. M. Wang, A. A. Owji, D. M. Smith, M. A. Ghatei & S. R. Bloom. 1995. Glucagon-like peptide-1 is a physiological incretin in rat. J. Clin. Invest. 95: 417–421.

54. Leech, C. A., G. G. Holz & J. F. Habener. 1996. Signal transduction of PACAP and GLP-1 in pancreatic β cells. Ann. N.Y. Acad. Sci. 805: 81–93.

55. Lu, M., M. B. Wheeler, X. H. Leng & A. E. Boyd III. 1993. The role of the free cytosolic calcium level in beta-cell signal transduction by gastric inhibitory polypeptide I(7-37). Endocrinology 132: 94–100.

56. Zawalich, W. S., K. C. Zawalich & H. Rasmussen. 1993. Influence of glucagon-like peptide-1 on β cell responsiveness. Regul. Pept. 44: 277–283.

57. Jia, X., R. Elliott, Y. N. Kwok, R. A. Pederson & C. H. McIntosh. 1995. Altered glucose dependence of glucagon-like peptide I(7-36)-induced insulin secretion from the Zucker (fa/fa) rat pancreas. Diabetes 44: 495–500.

58. Ding, W. G., E. Renstrom, P. Rorsman, K. Buschard & J. Gromada. 1997. Glucagon-like peptide I and glucose-dependent insulinotropic polypeptide stimulate Ca^{2+}-induced secretion in rat alpha-cells by a protein kinase A-mediated mechanism. Diabetes 46: 792–800.

59. Ørskov, C., J. J. Holst & O. V. Nielsen. 1988. Effect of truncated glucagon-like peptide-1 (proglucagon 78-107 amide) on endocrine secretion from pig pancreas, antrum and stomach. Endocrinology 123: 2009–2013.

60. Schirra, J., M. Katschinski, C. Weidmann, T. Schäfer, U. Wank, R. Arnold & B. Göke. 1996. Gastric emptying and release of incretin hormones after glucose ingestion. J. Clin. Invest. 97: 92–103.

61. O'Halloran, D. J., G. C. Nicon, B. Kreymann, M. A. Ghatei & S. R. Bloom. 1990. Glucagon-like peptide-1 (7-36)-NH2: a physiological inhibitor of the gastric acid secretion in man. J. Endocrinol. 126: 169–173.

62. Schjoldager, B. T. G., P. E. Mortensen, J. Myhre, J. Christiansen & J. J. Holst. 1989. Oxyntomodulin from the distal gut: role in the regulation of gastric and pancreatic functions. Dig. Dis. Sci. 34: 1411–1419.

63. Holst, J. J. 1997. Enteroglucagon. Annu. Rev. Physiol. 59: 257–271.

64. Wettergren, A., H. Petersen, C. Ørskov, J. Christiansen, S. P. Sheikh & J. J. Holst. 1994. Glucagon-like peptide-1 (GLP-1) 7-36 amide and peptide YY from the L-cell in the ileal mucosa are potent inhibitors of vagally induced gastric acid in man. Scand. J. Gastroenterol. 29: 501–505.

65. Schmidtler, J., W. Schepp, I. Janczewska, C. Weigert Fürlinger, V. Schusdziarra & M. Classen. 1991. GLP-1(7-36) amide, -(1-37), and -(1-36)amide: potent cAMP-dependent stimuli of rat parietal cell function. Am. J. Physiol. 260: G940–G950.

66. Eissele, R., H. Koop & R. Arnold. 1990. Effect of glucagon-like peptide-1 on gastric somatostatin and gastrin secretion in the rat. Scand. J. Gastroenterol. 25: 449–454.

67. Schepp, W., J. Schmidtler, K. Dehne, V. Schusdziarra & M. Classen. 1992. Pertussis toxin-sensitive and pertussis toxin-insensitive inhibition of parietal cell response to GLP-1 and histamine. Am. J. Physiol. 262: G660–G668.

68. ARIMURA, A. 1992. Pituitary adenylate cyclase activating polypeptide (PACAP): discovery and current status of research. Regul. Pept. **37**: 287–303.
69. MASUO, Y., T. OHTAKI, Y. MASUDA, M. TSUDA & M. FUJINO. 1992. Binding sites for pituitary adenylate cyclase activating polypeptide (PACAP): comparison with vasoactive intestinal polypeptide (VIP) binding sites localization in rat brain sections. Brain Res. **575**: 113–123.
70. MASUO, Y., N. SUZUKI, H. MATSUMOTO, F. TOKITO, Y. MATSUMOTO, M. TSUDA & M. FUJINO. 1993. Regional distribution of pituitary adenylate cyclase activating polypeptide (PACAP) in the rat central nervous system as determined by sandwich-enzyme immunoassay. Brain Res. **602**: 57–63.
71. NOMURA, M., Y. UETA, R. SERINO, N. KABASHIMA, I. SHIBUYA & H. YAMASHITA. 1996. PACAP type I receptor gene expression in the paraventricular and supraoptic nuclei of rats. Neuroreport **20**: 67–70.
72. PIGGINS, H. D., J. A. STAMP, J. BURNS, B. RUSAK & K. SEMBA. 1996. Distribution of pituitary adenylate cyclase activating polypeptide (PACAP) immunoreactivity in the hypothalamus and extended amygdala of the rat. J. Comp. Neurol. **376**: 278–294.
73. PANTALONI, C., P. BRABET, B. BILANGES, A. DUMUIS, S. HOUSSAMI, D. SPENGLER, J. BOCKAERT & L. JOURNOT. 1996. Alternative splicing in the N-terminal extracellular domain of the pituitary adenylate cyclase-activating polypeptide (PACAP) receptor modulates receptor selectivity and relative potencies of PACAP-27 and PACAP-38 in phospholipase C activation. J. Biol. Chem. **6**: 22146–22151.
74. CHATTERJEE, T. K., R. V. SHARMA & R. A. FISHER. 1996. Molecular cloning of a novel variant of the pituitary adenylate cyclase-activating polypeptide (PACAP) receptor that stimulates calcium influx by activation of L-type calcium channels. J. Biol. Chem. **271**: 32226–32232.
75. D'AGATA, D. V., S. CAVALLARO, F. STIVALA & P. L. CANONICO. 1996. Tissue-specific and developmental expression of pituitary adenylate cyclase-activating polypeptide (PACAP) receptors in rat brain. Eur. J. Neurosci. **8**: 310–318
76. CHANCE, W. T., H. THOMPSON, I. THOMAS & J. E. FISCHER. 1995. Anorectic and neurochemical effects of pituitary adenylate cyclase activating polypeptide in rats. Peptides **16**: 1511–1516.
77. MASUO, Y., J. NOGUCHI, S. MORITA & Y. MATSUMOTO. 1995. Effects of intracerebroventricular administration of pituitary adenylate cyclase-activating polypeptide (PACAP) on the motor activity and reserpine-induced hypothermia in murines. Brain Res. **27**: 219–226.
78. MORLEY, J. E., M. HOROWITZ, P. M. K. MORLEY & J. F. FLOOD. 1992. Pituitary adenylate cyclase activating polypeptide (PACAP) reduces food intake in mice. Peptides **13**: 1133–1135.
79. YADA, T., M. SAKURADA, K. IKADA, M. NAKATA, F. MURATA, A. ARIMURA & M. KIKUCHI. 1994. Pituitary adenylate cyclase activating polypeptide is an extraordinarily potent intra-pancreatic regulator of insulin secretion from islet (-cells. J. Biol. Chem. **269**: 1290–1293.
80. FRIDOLF, T., F. SUNDLER & B. AHREN. 1992. Pituitary adenylate cyclase-activating polypeptide (PACAP): occurrence in rodent pancreas and effects on insulin and glucagon secretion in the mouse. Cell Tissue Res. **269**: 275–279.
81. KAWAI, K., C. OHSE, Y. WATANABE, S. SUZUKI, K. YAMASHITA & S. OHASHI. 1992. Pituitary adenylate cyclase polypeptide stimulates insulin release from the isolated perfused pancreas. Life Sci. **50**: 257–261.
82. SUNDLER, F., E. EKBLAD, A. ABSOOD, R. HAKANSON, K. KÖVES & A. ARIMURA. 1992. Pituitary adenylate cyclase activating peptide: a novel vasoactive intestinal peptide-like neuropeptide in the gut. Neuroscience **46**: 439–545.
83. MUNGAN, Z., R. A. HAMMER, U. S. AKARCA, G. KOMAKI, A. ERTAN & A. ARIMURA. 1995. Effect of PACAP on gastric acid secretion in rats. Peptides **16**: 1051–1056.
84. COX, H. M. 1992. Pituitary adenylate cyclase activating polypeptides, PACAP-27 and PACAP-38: stimulators of electrogenic ion secretion in the rat small intestine. Br. J. Pharmacol. **106**: 498–502.
85. MUNGAN, Z., A. ARIMURA, A. ERTAN, W. J. ROSSOWSKI & D. H. COY. 1992. Pituitary adenylate cyclase-activating polypeptide relaxes rat gastrointestinal smooth muscle. Scand. J. Gastroenterol. **27**: 375–380.
86. SCHWÖRER, H., S. KATSOULIS, W. CREUTZFELDT & W. E. SCHMIDT. 1992. Pituitary adenylate cyclase activating peptide, a novel VIP-like gut-brain peptide, relaxes the guinea-pig taenia caeci via apamin-sensitive potassium channels. Naunyn Schmiedeberg's Arch. Pharmacol. **346**: 511–514.

87. CHRISTOPHE, J. 1993. Type 1 receptors for PACAP (a neuropeptide even more important than VIP?). Biochim. Biophys. Acta **1154**: 183–199.
88. MENTLEIN, R., B. GALLWITZ & W. E. SCHMIDT. 1993. Dipeptidyl-peptidase hydrolyses gastric inhibitory polypeptide, glucagon-like peptide-1(7-36) amide, peptide histidine methionine and is responsible for their degradation in human serum. Eur. J. Biochem. **214**: 829–835.

On the Treatment of Diabetes Mellitus with Glucagon-like Peptide-1

JENS JUUL HOLST,[a,d] CAROLYN DEACON,[a] MAJ-BRIT TOFT-NIELSEN,[b] AND LOTTE BJERRE-KNUDSEN[c]

[a]Department of Medical Physiology, The Panum Institute, University of Copenhagen, DK-2200 Copenhagen N, Denmark

[b]Department of Endocrinology, Hvidovre Hospital, DK-2650 Hvidovre, Denmark

[c]NovoNordisk A/S, 2730 Måløv, Denmark

ABSTRACT: As a therapeutic principle, the insulinotropic peptide, GLP-1, of the secretin-glucagon family of peptides, has turned out to possess some remarkably attractive properties, including the capability of normalizing blood glucose concentrations in patients with non-insulin-dependant diabetes mellitus and promoting satiety and reducing food intake in healthy volunteers. Because of rapid and extensive metabolization, the peptide is not immediately clinically applicable and, as a therapeutic principle, GLP-1 is still in its infancy. Some possible avenues for circumventing these difficulties are the development of DPP-IV-resistant analogs, the inhibition of DPP-IV, enhancement of GLP-1 secretion, GLP delivery systems using continuous subcutaneous infusion or buccal tablets, GLP-1 absorption, and orally active, stable analogs. It seems likely that one or more of these approaches could result in a clinically useful development program.

In a previous publication,[1] we introduced the new incretin hormone, glucagon-like peptide-1 and presented its effects on blood glucose regulation in normal and diabetic subjects. Its role as an incretin hormone, being the most potent insulinotropic hormone known, was emphasized, and it was described how intravenous infusions of GLP-1 can completely normalize the hyperglycemia of patients with non–insulin-dependent diabetes mellitus (NIDDM). It was concluded that GLP-1 had great potential as a therapeutic agent. The present communication deals with the attempts that have been made to transform this typical peptide hormone of the glucagon-secretin-VIP-PACAP family into a clinically useful therapeutic agent.

RECENT RESULTS CONCERNING THE POTENTIALLY THERAPEUTIC EFFECTS OF GLP-1

The therapeutic interest in GLP-1 was aroused when it was shown that an intravenous infusion of GLP-1 was capable of reducing to nearly zero the amount of insulin required to maintain normoglycemia in NIDDM patients studied during ingestion of a meal while attached to an artificial pancreas.[2] A significant insulin-sparing effect was noted also in

[d]Corresponding author: J. J. Holst, M.D., Professor of Medical Physiology, Department of Medical Physiology, The Panum Institute, Blegdamsvej3, DK-2200 Copenhagen N, Denmark; Tel.: (45) 3532 7518; Fax: (45) 3532 7537; E-mail: holst@mfi.ku.dk

patients with IDDM. Subsequent studies showed that blood glucose could be completely nor-malized by GLP-1 infusions in fasting subjects with long-standing NIDDM and secondary failure of their sulfonylurea treatment, admitted to hospital for insulin therapy.[3] Very recent studies have shown that near normal blood glucose levels can be achieved in NIDDM patients receiving a continuous infusion of a small dose of GLP-1 followed for 7 consecu-tive days and eating normal meals.[4] This extremely important observation provides proof of the concept that continuous activation of the GLP-1 receptors is a viable approach to the treatment of NIDDM. In addition, because of the powerful reduction of hyperglycemia, GLP-1 treatment potentially surpasses all other strategies currently under evaluation.[5]

Moreover, recent studies have shown that GLP-1 has additional, physiological effects rendering it even more attractive as a therapeutic agent. It has been shown that GLP-1, in addition to its incretin effects, acts as one of the enterogastrone hormones of the so-called ileal brake mechanism, i. e., the endocrine mechanism, activated by the presence of unab-sorbed nutrients in the ileum, that inhibits secretion and motility in the upper gastroin-testinal tract.[6] Thus, physiological infusions of GLP-1 strongly inhibit meal-induced gastric acid secretion, pancreatic secretion, and gastric emptying.[7] In fact, it seems that its enterogastrone effects may outweigh its insulinotropic effects, when evaluated in the form of a dose-response study in relation to meal ingestion.[8] It should be noted that this finding cannot be interpreted to indicate that its incretin effect is insignificant; its importance in this respect was illustrated in a recent study involving a targeted disruption of the GLP-1 receptor gene. The mutated animals had a significantly impaired glucose tolerance.[9] The powerful and extensive effects of GLP-1 on gastrointestinal motility and secretion raise the question of whether GLP-1 also plays a role in the regulation of food intake. Indeed, in a recent study of normal volunteers an infusion of physiological amounts of GLP-1 was demonstrated to significantly promote satiety and suppress energy intake.[10] Whether the satiating effect of GLP-1 is related to its enterogastrone effects is not known, but it is of considerable interest that intracerebroventricular administration of GLP-1 also profoundly and specifically suppresses food intake in rats.[11,12] This effect seems to be due to interac-tion of GLP-1 with GLP-1 receptors in the hypothalamus, normally activated by GLP-1 released from nerve fibers projecting from the nucleus of the solitary tract in the brain stem.[14] In addition, it has been shown that peripheral GLP-1 may reach the subfornical organ and the area postrema, two circumventricular organs of the brain involved in the reg-ulation of food intake—the circumventricular organs are accessible to many circulating substances because of the leaks in the blood brain-barrier in that area.[14] It may be added that the brain GLP-1 receptor is identical to the pancreatic receptor, and that only a single receptor has been identified so far—in spite of brave attempts to identify isoforms.[15] There is no doubt that part of the remarkable effect of GLP-1 on meal-induced blood glucose excursions in NIDDM patients is due to its enterogastrone effects;[16] for a while it was sus-pected that this might represent an untoward side effect. However, the patients experience no discomfort during GLP-1 infusion, neither during short-term infusions nor in the course of prolonged administration as illustrated in the abovementioned 7-day infusion study.[4] On the contrary, it appears that GLP-1 administration may result in a combined normalization of blood glucose and a reduction in energy intake, both of which would be regarded as extremely beneficial.

METABOLISM OF GLP-1

Obviously, intravenous infusions do not represent a realistic approach. Therefore, the effect of subcutaneous injections of GLP-1 has been evaluated in a number of studies. GLP-1 is readily absorbed from the subcutaneous tissue and the plasma concentrations of

GLP-1 after subcutaneous injection vary proportionally with the injected dose.[17] However, the effects on insulin, glucagon, and, hence, blood glucose are surprisingly short-lived. Measurements of plasma concentrations of GLP-1 after injection provided the explanation: the peptide is being metabolized very rapidly. Upon intravenous infusion, the peptide is eliminated by the kidneys with a steady state half-life of 4–5 min.[18,19] After subcutaneous injection, peak concentrations are reached after 30–60 min and little remains in the circulation after 1–2 hours. But the effects on insulin and glucagon secretion are even more short-lived. It turns out that GLP-1 is metabolized by several mechanisms, the earliest and the most extensive being an N-terminal truncation by two amino acid residues catalyzed by the ubiquitous enzyme dipeptidyl peptidase IV (DPP IV).[20] Furthermore, the metabolite has been demonstrated to act as a competitive antagonist at the GLP-1 receptor.[21] Although the physiological importance of this antagonism is not yet clear, it is of sufficient magnitude to deserve further investigation, especially when it comes to the pharmacological use of GLP-1. The conversion of intact, biologically active GLP-1 to its metabolite occurs with an apparent half-life of 1–1.5 min.[19] This means that the majority of exogenous GLP-1, whether administered intravenously or subcutaneously is present in the circulation as the truncated metabolite.[22] Endogenous GLP-1 is also rapidly metabolized; more than half of GLP-1 released in relation to meal ingestion circulates as the truncated metabolite,[20] and there is evidence that the gut actually secretes the metabolite.[23] A limited degradation of GLP-1 may also occur from the C-terminus. As noted above GLP-1, but also the metabolite, are removed by the kidneys with a half-life of about 4–5 min.[19]

STRATEGIES FOR THE DEVELOPMENT OF CLINICALLY USEFUL THERAPIES BASED ON ACTIVATION OF THE GLP-1-RECEPTOR

A number of approaches currently under investigation in various laboratories with the purpose of transforming GLP-1 into a clinically useful drug are presented below. As already mentioned, single subcutaneous injections have an insufficient duration of action to be clinically relevant, but it has also been shown that two subcutaneous injections (of 1.5 nmol/kg, the highest subcutaneous dose without side effects) have an effect on blood glucose in fasting NIDDM patients similar to that of a 4-hour continuous intravenous infusion, i. e., complete normalization of blood glucose.[24] Thus it can be predicted that modifications resulting in even moderate prolongation of survival would be clinically interesting.

DPP-IV Resistant Analogs

The remarkable sensitivity of GLP-1 to the enzymatic activity of DPP IV is due to the presence of Ala in position two.[25] Thus, analogs modified at this position might prove resistant, and such analogs would be predicted to have a half-life in the circulation corresponding to the renal rate of elimination of GLP-1, i. e., 4–5 min, compared to the natural peptide, a prolongation by a factor three to five. We studied four synthetic analogues substituted with alpha-aminoisobutyric acid, glycine, threonine, and serine in position two.[26] All were much more resistant than GLP-1 itself with plasma half-lives in vitro ranging from 159 min to > 6 h, as compared to 28 min for GLP-1. The in vivo half-lives in pigs were similar to due to the renal elimination, i.e, about 4 min. In studies using a receptor binding assay based on transfected human GLP-1 receptors, the analogs bound to the receptor with affinities ranging from 0.45 to 9 nmol/l (GLP-1: 0.78 nmol/l). Thus, each of these analogues fullfil the requirement of prolonged survival and retained biological activity, and are—at least in principle—applicable for subcutaneous administration in NIDDM patients.

Inhibition of DPP-IV

A prolonged survival of GLP-1 could theoretically be achieved through inhibition of DPP IV, in analogy to the inhibition of converting enzyme in order to reduce formation of angiotensin II in the treatment of hypertension. A particular virtue of this approach would be a possible effect on endogenous GLP-1. The endogenous hormone would escape degradation and has the additional advantage of being released at a time when a glucose lowering substance is most needed, i.e., in relation to meal ingestion. DPP IV is thought to be involved in the metabolism of a number of other regulatory peptides and also functions in the immune system, but GLP-1 seems to be particularly sensitive to its degrading activity.[25] Surprisingly, however, rats with a genetic DPP IV deficiency do not exhibit abnormalities of glucose metabolism.[27] Diprotin A, a tripeptide, is a specific inhibitor of DPP IV. It has been shown that diprotin A can completely prevent degradation of GLP-1 in plasma *in vitro* and that infusions of diprotin A significantly reduced the clearance of exogenous GLP-1 in pigs.[20,28] However, diprotin A is itself a substrate for DPP IV and has, therefore, limited effectiveness. We subsequently studied the effect of a stable and selective inhibitor of DPP IV in anesthetized pigs.[29] With adequate doses of the inhibitor, plasma DPP IV activity was decreased by more than 90%, resulting in a complete prevention of the N-terminal truncation of GLP-1. In control experiments the latter amounted to 80%. In order to investigate the effect of the treatment on glucose metabolism we performed an intravenous glucose tolerance test with or without DPP-IV inhibition; inhibitor administration both curtailed blood glucose excursions and more than tripled insulin secretion. Thus, there is little doubt that DPP-IV inhibition with a suitable inhibitor represents a promising approach, clearly as an adjunct to GLP-1 therapy, but perhaps also as monotherapy by enhancing the activity of endogenous GLP-1.

Enhanced Secretion of GLP-1

Little is presently known about the regulation of GLP-1 secretion, except that luminal administration of lipids and carbohydrates is effective. Muscarinic agonists and tachykinins and bombesin may also release some GLP-1,[6] but their effects are unlikely to be exploited because the systemic effects of such agents clearly outweigh the effects on GLP-1 secretion. The GLP-1–producing L cells are characteristically positioned distally in the small intestine and in the colon and agents capable of causing a transfer of a larger proportion of ingested nutrients from the proximal to the distal gut, might be effective. Thus, oral administration of the alpha-glucosidase inhibitor, acarbose, which inhibits the enzymatic cleavage of sucrose to glucose and fructose, whereby a proportion of the ingested sucrose is transferred to the distal gut, is associated with a delayed and reduced glucose reponse to the ingested sucrose as well as a greatly enlarged GLP-1 response.[30,31] Acarbose has been applied with some success in the treatment of NIDDM, and we believe that part of the beneficial effect of this drug is its enhancing effect on GLP-1 secretion.[33]

Continuous Subcutaneous Infusion

One way of providing a constant supply of GLP-1 via the subcutaneous route is to employ constant infusion pumps similar to those used for continuous insulin therapy. In our department such experiments have been carried out in a limited number of patients with two concentrations of GLP-1 in the pumps. GLP-1 infusion resulted in clear-cut reductions in both basal and meal-related blood glucose levels compared to infusion of vehicle, but it is our impression that with this method it is difficult to obtain quite as dramatic a reduction of hyperglycemia as it is with continuous intravenous infusion. Further studies are, however, required, in order to do this approach justice.

Buccal GLP-1 Tablets

The surprising absorptive capacity of the oral mucosa for peptides is the basis for attempts to circumvent the degradation in the stomach and the intestines by utilizing a biodegradable buccal tablet. The tablets, which contain 400 µg GLP-1, also have an adhesive layer for fixation in the oral cavity and contain an absorption enhancer. GLP-1 is absorbed remarkably well from such tablets, and GLP-1 profiles in plasma of both diabetic patients and healthy volunteers can be obtained that are similar to those obtained by subcutaneous injections.[32,33] As a result, insulin secretion is markedly stimulated, glucagon secretion inhibited, and blood glucose declines. As for subcutaneous injections, however, the decrease is relatively small and short-lived. Analyses for intact GLP-1 and its primary metabolite, the N-terminally truncated form, show that the tablet GLP-1 is being degraded to a similar extent. This approach will require protection of the peptide to be useful.

It is noteworthy that other routes of application; e.g., nasal, rectal, and transdermal may also turn out to be useful, but are likely to require protected analogs as well.

Delayed Absorption

The subcutaneous absorption of GLP-1 can be considerably protracted by complexing the GLP-1 solution with zinc or protamine. With the Zn addition, a suspension of crystals is formed, and it appears that a zero-order rate of absorption from the subcutaneous depot results in a plateau-like elevation (and a flattened-out post-injection peak) of plasma GLP-1.[34] However, in this case it also seems that a considerable degradation is occurring, again requiring modifications of the peptide to stabilize it against DPP-IV.

Orally Active, Stable Analogs

Clearly, the most desirable preparation would be orally active and have an appropriate metabolic stability. A number of structure-activity studies have been carried out[35,36] with the purpose, among others, of providing information for the design of "peptoids" or other peptide-based analogs that might be orally active or show a favorable stability, as this is known for somatostatin, for example. The conclusion of these studies, however, has been that there are no domains of the peptide that are specifically reponsible for its biological activity or, in other words, that the entire molecule (except for the ultimate C-terminal amino acids) is required for activation of the receptor. However, the details of the interaction of GLP-1 with its receptor have not yet been worked out, and it cannot be excluded that receptor activation could be obtained with a smaller molecule capable of, for example, bridging a part of the N-terminal extracellular tail of the receptor and one or more of the extracellular loops in a way that could lock the receptor in an active conformation. Efforts have also been invested into screening of chemical libraries for non-peptide substances capable of interacting with cloned GLP-1 receptors transfected in cell preparations, that, one way or the other, could be employed in high volume screening programs. Such screenings are said to have been conducted at very large scales in various medical companies, but so far there is no public information about "hits." Given the unsurpassed capacity of these methods, supplemented by "combinatorial chemistry," "virtual libraries," and "virtual screening,"[37] it seems inevitable that a useful substance will eventually be isolated.

CONCLUSION

As a therapeutic principle, the insulinotropic peptide, GLP-1, of the secretin-glucagon family of peptides, has turned out to possess some remarkably attractive properties, including the capability of normalizing blood glucose concentrations in patients with NIDDM and promoting satiety and reducing food intake in healthy volunteers. Because of rapid and extensive metabolization, the peptide is not immediately clinically applicable and, as a therapeutic principle, GLP-1 is still in its infancy. However, there are numerous ways by which these difficulties can be circumvented, as discussed above, and it seems likely that one or more of these approaches could result in a clinically useful development program.

REFERENCES

1. HOLST, J. J., M. B. TOFT-NIELSEN, C. ØRSKOV, M. NAUCK & B. WILLMS. 1996. On the effects of glucagonlike peptide1 on blood glucose regulation in normal and diabetic subjects. Ann. N. Y. Acad. Sci. **805**: 729–736.

2. GUTNIAK, M., C. ØRSKOV, J. J. HOLST, B. AHREN & S. EFENDIC. 1992. Antidiabetogenic effect of glucagonlike peptide-1(7-36)amide in normal subjects and patients with diabetes mellitus [see comments]. N. Engl. J. Med. **326**: 1316–1322.

3. NAUCK, M. A., N. KLEINE, C. ØRSKOV, J. J. HOLST, B. WILLMS & W. CREUTZFELDT. 1993. Normalization of fasting hyperglycaemia by exogenous glucagonlike peptide 1 (7-36 amide) in type 2 (non-insulin-dependent) diabetic patients. Diabetologia **36**: 741–744.

4. LARSEN, J., N. JALLAD & P. DAMSBO. 1996. One-week continuous infusion of GLP-1(7-37) improves glycemic control in NIDDM. Diabetes **45**(suppl 2): 233A.

5. RACHMAN, J. & R. C. TURNER. 1995. Drugs on the horizon for treatment of type 2 diabetes. Diabetic Med. **12**: 467–478.

6. HOLST, J. J. 1997. Enteroglucagon. Annu. Rev. Physiol. **59**: 257–271.

7. WETTERGREN, A., B. SCHJOLDAGER, P. E. MORTENSEN, J. MYHRE, J. CHRISTIANSEN & J. J. HOLST. 1993. Truncated GLP-1 (proglucagon 72-107amide) inhibits gastric and pancreatic functions in man. Dig. Dis. Sci. **38**: 665–673.

8. NAUCK, M. A., U. NIEDEREICHHOLZ, R. ETTLER, J. J. HOLST, C. ØRSKOV, R. RITZELL & W. H. SCHMIEGEL. 1997. The glucagon-like peptide-1 inhibition of gastric emptying outweighs its insulinotropic effects in healthy humans. Am. J. Physiol. **273**: E981–E988.

9. SCROCCHI, L. A., T. J. BROWN, N. MACLUSKY, P. L. BRUBAKER, A. B. AUERBACH, A. L. JOYNER & D. J. DRUCKER. 1996. Glucose intolerance but normal satiety in mice with a null mutation in the glucagon-like peptide 1 receptor gene. Nat. Med. **2**: 1254–1258.

10. FLINT, A., A. RABEN, A. ASTRUP & J. J. HOLST. 1998. Glucagon-like peptide-1 promotes satiety and suppresses energy intake in humans. J. Clin. Invest. **101**: 515–520.

11. TURTON, M. D., D. O'SHEA, I. GUNN, S. A. BEAK, C. M. EDWARDS, K. MEERAN, S. J. CHOI, G. M. TAYLOR, M. M. HEATH, P. D. LAMBERT, J. P. WILDING, D. M. SMITH, M. A. GHATEI, J. HERBERT & S. R. BLOOM. 1996. A role for glucagon-like peptide-1 in the central regulation of feeding. Nature **379**: 69–72.

12. TANG CHRISTENSEN, M., P. J. LARSEN, R. GOKE, A. FINK JENSEN, D. S. JESSOP, M. MOLLER & S. P. SHEIKH. 1996. Central administration of GLP-1-(7-36) amide inhibits food and water intake in rats. Am. J. Physiol. **271**: R848–56.

13. LARSEN, P. J., M. TANG CHRISTENSEN, J. J. HOLST & C. ØRSKOV. 1997. Distribution of glucagon-like peptide-1 and other preproglucagon-derived peptides in the rat hypothalamus and brainstem. Neuroscience **77**: 257–270.

14. ØRSKOV, C., S. S. POULSEN, M. MOLLER & J. J. HOLST. 1996. Glucagon-like peptide I receptors in the subfornical organ and the area postrema are accessible to circulating glucagon-like peptide I. Diabetes **45**: 832–835.

15. WEI, Y. & S. MOJSOV. 1995. Tissue-specific expression of the human receptor for glucagon-like peptide-I: brain, heart and pancreatic forms have the same deduced amino acid sequences. FEBS Lett. **358**: 219–224.

16. WILLMS, B., J. WERNER, J. J. HOLST, C. ØRSKOV, W. CREUTZFELDT & M. A. NAUCK. 1996. Gastric emptying, glucose responses, and insulin secretion after a liquid test meal: effects of exogenous glucagon-like peptide-1 (GLP-1)-(7-36) amide in type 2 (noninsulin-dependent) diabetic patients. J. Clin. Endocrinol. Metab. **81**: 327–332.

17. RITZEL, R., C. ØRSKOV, J. J. HOLST & M. A. NAUCK. 1995. Pharmacokinetic, insulinotropic, and glucagonostatic properties of GLP-1 [7-36 amide] after subcutaneous injection in healthy volunteers. Dose-response-relationships. Diabetologia **38**: 720–725.

18. ØRSKOV, C., A. WETTERGREN & J. J. HOLST. 1993. The metabolic rate and the biological effects of GLP-1 7-36amide and GLP-1 7-37 in healthy volunteers are identical. Diabetes **42**: 658–661.

19. DEACON, C. F., L. PRIDAL, L. KLARSKOV, M. OLESEN & J. J. HOLST. 1996. Glucagon-like peptide 1 undergoes differential tissue-specific metabolism in the anesthetized pig. Am. J. Physiol. **271**: E458–64.

20. DEACON, C. F., A. H. JOHNSEN & J. J. HOLST. 1995. Degradation of glucagon-like peptide-1 by human plasma in vitro yields an N-terminally truncated peptide that is a major endogenous metabolite in vivo. J. Clin. Endocrinol. Metab. **80**: 952–957.

21. KNUDSEN, L. B. & L. PRIDAL. 1996. Glucagon-like peptide-1-(9-36) amide is a major metabolite of glucagon-like peptide-1-(7-36) amide after in vivo administration to dogs, and it acts as an antagonist on the pancreatic receptor. Eur. J. Pharmacol. **318**: 429–435.

22. DEACON, C. F., M. A. NAUCK, M. TOFT NIELSEN, L. PRIDAL, B. WILLMS & J. J. HOLST. 1995. Both subcutaneously and intravenously administered glucagon-like peptide I are rapidly degraded from the NH2-terminus in type II diabetic patients and in healthy subjects. Diabetes **44**: 1126–1131.

23. HOLST, J. J., L. HANSEN & C. F. DEACON. 1997. The GLP-1 receptor antagonist, GLP-1 (9-36amide), is a primary product of the intestinal L-cell. Diabetologia **40** (suppl.1): A27.

24. NAUCK, M. A., D. WOLLSCHLAGER, J. WERNER, J. J. HOLST, C. ØRSKOV, W. CREUTZFELDT & B. WILLMS. 1996. Effects of subcutaneous glucagon-like peptide 1 (GLP-1 [7-36 amide]) in patients with NIDDM. Diabetologia **39**: 1546–1553.

25. MENTLEIN, R., B. GALLWITZ & W. E. SCHMIDT. 1993. Dipeptidyl-peptidase IV hydrolyses gastric inhibitory polypeptide, glucagon-like peptide-1(7-36)amide, peptide histidine methionine and is responsible for their degradation in human serum. Eur. J. Biochem. **214**: 829–835.

26. DEACON, C. F., L. BJERRE KNUDSEN, N. L. JOHANSEN, K. MADSEN & J. J. HOLST. 1997. Dipeptidyl peptidase IV resistant analogues of glucagon-like peptide-1: in vitro and in vivo studies. Diabetologia **40** (suppl 1): A127.

27. PEDERSON, R. A., T. J. KIEFFER, R. PAULY, H. KOFOD, J. KWONG & C. H. S. MCINTOSH. 1996. The enteroinsular axis in dipeptidyl peptidase IV-negative rats. Metabolism **45**: 1335–1341.

28. DEACON, C. F., L. PRIDAL, M. OLESEN, L. KLARSKOV & J. J. HOLST. 1996. Dipeptidyl peptidase IV inhibition influences GLP-1 metabolism in vivo. Regul. Pept. **64**: 30.

29. DEACON, C. F. , T. E. HUGHES & J. J. HOLST. 1998. Dipeptidyl peptidase IV inhibition potentiates the insulinotropic effect of glucagon-like peptide-1 in the anaesthetized pig. Diabetes **47**: 764–769.

30. QUALMANN, C., M. A. NAUCK, J. J. HOLST, C. ØRSKOV & W. CREUTZFELDT. 1995. Glucagon-like peptide 1 (7-36 amide) secretion in response to luminal sucrose from the upper and lower gut. A study using alpha-glucosidase inhibition (acarbose). Scand. J. Gastroenterol. **30**: 892–896.

31. SEIFARTH, C., J. BERGMANN, J. J. HOLST, R. RITZELT, W. SCHMIEGEL & M. A. NAUCK. 1997. Influence of acarbose on GLP-1 release after a sucrose load in NIDDM patients. Diabetologia **40**(suppl. 1): A131.

32. GUTNIAK, M. K., H. LARSSON, S. J. HEIBER, O. T. JUNESKANS, J. J. HOLST & B. AHREN. 1996. Potential therapeutic levels of glucagon-like peptide I achieved in humans by a buccal tablet. Diabetes Care **19**: 843–848.

33. GUTNIAK, M. K., H. LARSSON, S. W. SANDERS, O. JUNESKANS, J. J. HOLST & B. AHRÉN. 1997. GLP-1 tablet in NIDDM in fasting and postprandial conditions. Diabetes Care **20**: 1874–1879.

34. PRIDAL, L., H. AGERBÆK, L. N. CHRISTENSEN, K. THOMSEN & O. KIRK. 1996. Absorption of glucagon-like peptide-1 can be protracted by zinc or protamine. Int. J. Pharm. **136**: 53–59.

35. ADELHORST, K., B. B. HEDEGAARD, L. B. KNUDSEN & O. KIRK. 1994. Structure-activity studies of glucagon-like peptide-1. J. Biol. Chem. **269**: 6275–6278.
36. HJORTH, S. A., K. ADELHORST, B. B. PEDERSEN, O. KIRK & T. W. SCHWARTZ. 1994. Glucagon and glucagon-like peptide 1: selective receptor recognition via distinct peptide epitopes. J. Biol. Chem. **269**: 30121–30124.
37. FINN, P. W. 1996. Computer-based screening of compound databases for the identification of novel leads. Drug Discovery Today **1**: 363–370.

Binding Sites for VIP in the Reorganizing Mucosa of the Irradiated Bowel[a]

U. HÖCKERFELT,[b] M. HANSSON,[b] S. GULBENKIAN,[d]
L. FRANZÉN,[c] R. HENRIKSSON,[c] AND S. FORSGREN[b,e]

[b]Departments of Anatomy and [c]Oncology, Umeå University, Umeå, Sweden

[d]Unit of Cell Morphology and Image Processing, Gulbenkian Institute of
Science, Oeiras, Portugal

ABSTRACT: Rats were given radiotherapy (total dose 30 Gy) over the abdomen. Seven
days later specimens of the duodenum were prepared for *in vitro* receptor autoradiography using the radioligand [^{125}I]VIP. The autoradiograms were quantitatively
analyzed using a computer system. Histological examination revealed that a very
marked reorganization of the mucosa had occurred in response to irradiation. Using
receptor autoradiography, we found [^{125}I]VIP-specific binding sites in the reorganizing mucosa, except where denudation had occurred. Such binding sites also occurred
in the smooth muscle layer of the duodenal wall. The observations suggest that VIP
has profound effects in radiation-induced enteropathy.

The small intestine is a major organ at risk when radiotherapy treatment is given. The sequelae of gastrointestinal exposure to irradiation range from acute dysfunction to a subacute or chronic irradiation enteropathy months to years later.[1] Thus, external abdominal irradiation is associated with marked alterations in intestinal morphology and function. The severe post-irradiation histologic changes are dose- and/or time-dependent and are related to the high turnover of mucosal cells.[2] The mitotic activity in the crypts is significantly increased in the damaged parts of the intestine, leading to a prolonged but sometimes insufficient compensatory response to maintain mucosal integrity[3]. Repair and recovery, when present, occur within a few days after the end of the irradiation exposure, and the epithelium may regain normal morphology.[4] The detailed mechanisms behind the process of recovery are unclear.

Neuropeptides are widely distributed in the innervation and in endocrine/paracrine cells in the gut.[5] A neuropeptide frequently expressed in the intrinsic innervation of the intestine is vasoactive intestinal polypeptide (VIP). This peptide is a major regulator of motility and fluid transport in the gastrointestinal tract.[6-8] VIP is often implicated in various gastrointestinal pathologies.[9,10] Not least, VIP is considered a diarrhetic agent,[11] this point being relevant when discussing the reasons for the frequently encountered diarrhea after irradiation of the abdomen.[12]

A high concentration of VIP receptors is found in the mucosa of the gut, above all in the small intestine, in both humans and animals.[13-18] Interestingly, exposure of pigs to irradiation leads to modifications in VIP receptor characteristics in isolated membranes from

[a]The study was supported by grants from the Faculty of Medicine, Umeå University, Lions Cancer Research Foundation, Umeå, and the Swedish Medical Research Council (10364).

[e]Corresponding author: Dr. S. Forsgren, Department of Anatomy, Umeå University, S-901 87 Umeå, Sweden; Tel.: 46 90 786-5147; Fax: 46 90 786-5480; E-mail: Sture.Forsgren@anatomy.umu.se

the jejunum of these animals.[11] These effects were observed at doses lower than those normally associated with the manifestation of marked gastrointestinal dysfunction.

To the best of our knowledge, it has never been examined as to whether mucosal VIP receptors are markedly affected by radiotherapy, i.e., when morphologic changes prevail and when possible epithelial denudation occurs. Therefore, we have studied the possible occurrence of [^{125}I]VIP receptor binding sites in the moderately and severely damaged/reorganizing duodenal mucosa of rats given external radiotherapy over the abdomen.

MATERIALS AND METHODS

Animals

Five albino rats of the Sprague-Dawley strain were subjected to irradiation over the upper part of the abdomen, corresponding to the area of the duodenum. Six animals served as controls. All the rats were eight-week-old females and weighed approximately 200 g at the start of the experiments. The rats were fed with water and chow *ad libitum* and kept on a diurnal light schedule. The experiments have been approved by the Local Ethical Committee on Animal Experiments in Northern Sweden (A48/95).

Irradiation

The irradiation was carried out on a medical linear accelerator, 4 MV (dose rate 3.8 Gy min⁻¹), and was administered on one occasion with a total dose of 30 Gy. The rats were anesthetized with an injection of a mixture (1:1) of Hypnorm((fluanisonum 10 mg/ml, fentanylum 0.2 mg/ml) and Dormicum® (midazolam 5 mg/ml), 2.7–3.3 ml/kg body weight, in the peritoneum. They were then secured in a special mold to be firmly positioned during irradiation. The control rats were anesthetized and treated similarly, except for not being irradiated.

Tissue Sampling and Sectioning

Seven days after irradiation, the rats, both controls and irradiated, were anesthetized with an injection of Mebumal (sodium pentobarbital, 40 mg/kg intraperitoneally). The duodenum was dissected out and then the animals were killed by exsanguination. The tissue samples were mounted on thin cardboard, embedded in Tissue-Tek (O.C.T. Miles Inc., Elkhart, IN, USA), frozen in liquid nitrogen, and stored at –70°C.

Receptor Autoradiography

The unfixed samples were cut in a cryostat in 12-μm-thick sections. Some sections were stained with hematoxylin-eosin to show tissue morphology. Most sections were mounted on poly-L-lysine–coated slides and air dried for 2 h at 4°C. Then the sections were preincubated for 30 min at 23°C, in a humid environment, in either a solution of 0.0005% polyethylenamine (vol/vol) in 50 mM Tris-HCl buffer (pH 7.4) or in the same solution also containing 1.250 μM VIP (Sigma), 0.1% bacitracin (wt/vol) and 1% bovine serum albumin (wt/vol), in order to uncouple bound endogenous VIP or to saturate the VIP binding sites, respectively. The sections were then incubated in a humid environment for 60 min at 23°C in 50 mM Tris-HCl containing 0.125 nM [^{125}I]VIP (Amersham, specific activity 2,000 Ci/mmol), 0.1% bacitracin (wt/vol) and 1% bovine serum albumin (wt/vol). By incubating the sections with 0.125 nM [^{125}I]VIP in the presence of 1.250 µM unlabeled VIP, non-specific binding was assessed. After the incubation, the sections were washed 3×10 min in a 4°C solution of 50 mM Tris-HCl, followed by a washing in 4°C distilled water. The sections were then fixed for 30 min at 23°C with 2.5% glutaraldehyde in phosphate-buffered saline (PBS), washed 3×10 min in PBS and air dried. After that, they were covered, by dipping, with LM-1 nuclear emulsion (Amersham) and exposed at 4°C for 24, 48, and 72 h. Following development in Kodak D 19 and fixation in 30% sodium thiosulfate (wt/vol), the sections were stained with Mayer's hemalum solution for 30 sec. Then they were dehydrated in alcohol and xylene and mounted with DPX.

Quantitative Analysis

The autoradiograms were analyzed with a VIDAS computer system combined with a Leitz Aristoplan dark field microscope equipped with a CCD video camera (CCD-72EX, Dage MTI, USA). The optical system was installed so that 512×512 pixels on the monitor screen was equivalent to an analyzed area of 342×342 µm of the tissue specimen, i.e., 1.5 $\times$ 1.5 pixels per 1.00 µm tissue length. The gray scaled image was converted into a binary image, displaying the silver grains against a neutral background, by selecting an appropriate discrimination level, the same level being used in every computation. The presence of artefacts, due to the surface of the glass and crystals on the tissue, was interactively erased from the binary image. By subtracting the surface grain density measured in the non specific binding conditions from the ones in the total binding situation, the specific binding of [^{125}I]VIP was obtained. The total areas analyzed were 0.52×10^6 µm^2 (control muscosa), 1.90×10^6 µm^2 (irradiated muscosa), 0.49×10^6 µm^2 (control circular musculature), 0.30×10^6 µm^2 (control longitudinal musculature), and 0.56×10^6 µm^2 (irradiated longitudinal muscle).

Data Analysis

The data were analyzed statistically using paired Student's t-test and p values lower than 0.05 were considered significant. The values are expressed as mean ± S.E.M.

RESULTS

Morphology

In the duodenum from the irradiated animals the morphology was drastically changed. The degree of morphologic change varied between the different specimens, as well as within the individual specimens. The circular muscular layer and the submucosa were often thickened. In the mucosa, the villi as well as the crypts were seriously affected, the crypt area usually being much decreased. At some places the mucosa was ulcerated and denuded. Where denudation had occurred, no or only occasional epithelial cells were observed.

Autoradiography

The occurrence of specific [^{125}I]VIP binding sites in the different parts of the duodenum of control and irradiated animals was assessed via comparisons between the total and non-specific binding autoradiograms.

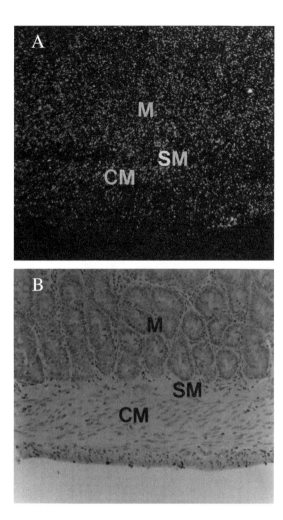

FIGURE 1. (A) Sections of the outer part of the duodenal wall of a control animal. Darkfield image of autoradiograph demonstrating binding sites for [^{125}I]VIP in the absence of 1.250 μM unlabeled VIP (total binding). Specific binding sites, as revealed via comparisons with non-specific binding section, for [^{125}I]VIP are observed in the mucosa (M) and the circular smooth muscle layer (CM). **(B)**Section stained with Mayer's hemalum solution and eosin. SM = submucosa.

Non-Irradiated Animals

Intense [^{125}I]VIP binding was observed in the mucosa. The silver grain density in the mucosa was clearly lower in the non-specific binding sections than in the total binding ones. The autoradiographs also revealed that there was a specific [^{125}I]VIP binding in the circular smooth muscle layer (FIG. 1.). In the other layers, no clear difference in the degree of silver grain density could be seen.

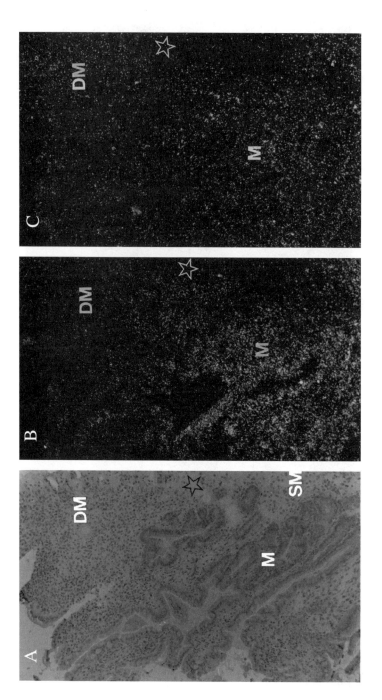

FIGURE 2. Three sections of the mucosa (M) and submucosa (SM) of a specimen of irradiated duodenum. (**A**) Bright field image showing a section stained with Mayer's hemalum solution and eosin. Darkfield images of autoradiograms demonstrating binding sites for [^{125}I]VIP in the (**B**) absence (total binding) or in the (**C**) presence (non-specific binding) of 1.250 μM unlabeled VIP. Specific binding sites for [^{125}I]VIP are observed in large parts of the mucosa (M). The exception is where there is total denudation (DM).

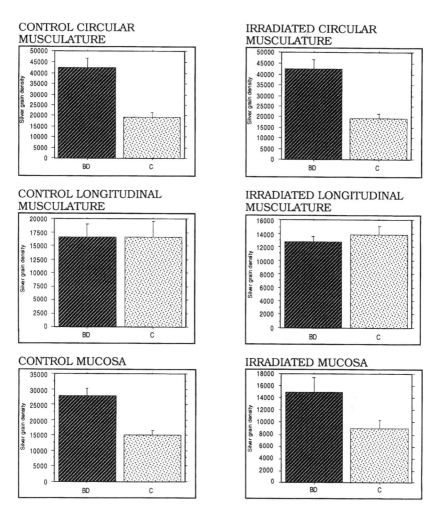

FIGURE 3. Total binding (BD) ([^{125}I]VIP) versus unspecific binding (C) ([^{125}I]VIP+VIP) in the mucosa and in the circular and longitudinal muscle layers, respectively, in controls and after irradiation. All the values are expressed as mean ± S.E.M. and were obtained from three binding experiments. $p < 0.0001$ (control mucosa, control and irradiated circular musculature) and $p = 0.005$ (irradiated mucosa).

Irradiated Animals

Silver grain densities were greater in the circular smooth muscle layer in total than in non-specific autoradiographs also in the duodenum of irradiated animals. In most parts of the mucosa, specific binding sites for [^{125}I]VIP also occurred (FIG. 2). However, in the regions where the mucosa was totally denuded, as well as in the longitudinal smooth muscle layer and the submucosa, no obvious difference in the levels of silver grain density between total and non-specific binding was observed.

Quantitative Analysis

In order to more exactly define the occurrence of specific binding sites for [^{125}I]VIP, computations comparing the different layers were made. These revealed a difference between total and non-specific [^{125}I]VIP binding in the mucosa and in the circular smooth muscle layer of both irradiated and non irradiated animals (FIG. 3).

DISCUSSION

In the present study, specific binding sites for [^{125}I]VIP were observed in parts of the duodenum in normal as well as irradiated animals. In accordance with previous studies on the normal rat duodenum,[16,17] there was a high degree of [^{125}I]VIP binding in the mucosa. We also found a high level of binding in the circular smooth muscle layer, in contrast to Sayadi and colleagues,[16] who observed a low degree of VIP binding in this layer. As judged from estimation of relative density of silver grains, as determined by microscopic examination and image analysis, it was also evident that there were specific VIP receptors in most parts of the mucosa in the duodenum in irradiated animals. In this context, the present study extends the observations described by Griffiths and colleagues.[11] These authors found that irradiation lead to a modification of VIP receptor characteristics in membranes isolated from pig jejunum, the changes included an increase in VIP receptor numbers. In this study, a modest dose was given and the morphological analysis revealed no major structural changes with no apparent loss of mucosal integrity.[11] In the present study, the mucosa was greatly affected, therefore, the results give an indication of what the function of VIP might be in the presence of severe damage. At this point it should be stressed that the existence of specific [^{125}I]VIP binding sites could not be verified in the regions where complete denudation had occurred.

The presence of VIP receptor binding sites in the reorganizing mucosa and the circular smooth muscle layer can be compared with the presence of the VIP peptide. In all parts of the mucosa as well as in the circular smooth muscle layer of the irradiated rat duodenum there is a pronounced VIPergic innervation, as seen immunohistochemically (unpublished observations). Thus, there is a correlation between presence of VIP and its receptors. It is therefore likely that the VIP receptors are functionally related to a release from VIPergic nerves in this situation. With respect to the interpretation of the data presented in this study, it should be recalled that a high degree of binding may be due to the presence of small numbers of high-affinity receptors or large numbers of low-affinity binding sites. Nevertheless, if the binding observed represents functional receptors it is plausible to suggest that VIP has marked effects in the reorganizing mucosa in the situation examined. The cell types to which binding is confined remain to be defined in further studies. In the normal small intestine, [^{125}I]VIP binding occurs in both crypt and villous epithelium.[13]

In the mucosa showing denudation and non-existence of [^{125}I]VIP binding sites there were rare or no epithelial cells. In these areas there is no correlation between presence of the peptide and its receptors. One explanation may be that VIP does not act directly in a mucosa that is deprived of epithelial cells. Alternatively, the affinity and/or number of receptors per cell is inadequate to detect binding. Thirdly, as there is a profound VIPergic innervation in the denuded mucosa, it can not be excluded that high amounts of slowly dissociating, endogenous VIP may mask receptors. An occurrence of some degree of mismatch between VIP and its receptor localization has also previously been reported for the normal rat and rabbit gastrointestinal tract.[16] Such a mismatch is also observed in the human intestine.[18]

The observations of VIP binding sites in the reorganizing mucosa are interesting for different reasons. Firstly, VIP has marked actions on intestinal secretion and smooth muscle contractility.[8,19] Secondly, changes in VIP innervation are suggested to have significant effects on gut function in inflammatory conditions like Crohn´s disease.[20] VIP may also cause diarrhea.[21] At this point it should be recalled that there are varying descriptions as to whether VIP levels in the gut indeed are changed in the various abdominal diseases investigated.[9,22–24] In any case, it is well-known that irradiation of the abdomen leads to adverse symptoms from the gastrointestinal tract, such as diarrhea. These undesired effects lead to limitations in the use of radiotherapy in the treatment of malignant tumors. VIP may be involved in this troublesome radiation-induced enteropathy. Apart from being related to water and electrolyte secretion, motility, and possibly adverse affects of the radiotherapy treatment given, VIP is also likely to be involved in the restoration process after irradiation of the duodenum. Thus, VIP has marked trophic actions and growth-regulatory effects. For example, a population of VIP receptors accumulates during enterocytic differentiation of a human colon adenocarcinoma cell line in culture.[25]

In conclusion, our study shows that [^{125}I]VIP binding sites are present in the reorganizing duodenal mucosa, except where complete denudation had occurred, and in the circular smooth muscle layer in response to irradiation. These observations suggest that VIP has profound effects not only in the normal duodenum but also in radiation-induced enteropathy.

ACKNOWLEDGMENTS

The authors are grateful to Mrs. L. Crenshaw, U. Hedlund and, M. Wranding for excellent technical services. We wish to thank Prof. Jan-Olaf Kellerth for letting us use the VIDAS computer system.

REFERENCES

1. BERTHRONG, M. & L. F. FAJARDO. 1981. Radiation injury in surgical pathology: Part II. Am. J. Surg. Pathol. **5:** 153–178.
2. RUBIO, C. A. & M. JALNAS. 1996. Dose-time-dependent histological changes following irradiation of the small intestine of rats. Dig. Dis. Sci. **41:** 392–401.
3. DEWIT, L. & Y. OUSSOREN. 1987. Late effects in mouse small intestine after a clinically relevant multifractionated radiation treatment. Radiat. Res. **110:** 372–384.
4. BECCIOLINI, A. *et al.* 1986. Cell proliferation and differentiation in the small intestine after irradiation with multiple fractions. Acta Radiol. Oncol. **25:** 51–56.
5. SUNDLER, F. *et al.* 1991. The neuroendocrine system of the gut—An update. Acta Oncol. **30:** 419–427.
6. SAID, S. I. 1986. Vasoactive intestinal peptide. J. Endocrinol. Invest. **9:** 191–200.
7. DONOWITZ, M. & M. J. WELSH. 1987. Regulation of mammalian small intestinal electrolyte secretion. *In* Physiology of the Gastrointestinal Tract. Vol 2, 2nd edit. L. R. Johnson, Ed.:1351–1388. Raven Press. New York.
8. GRINDER, J. R. & G. M. MAKHLOUF. 1988. Vasoactive intestinal peptide: Transmitter of inhibitory motor neurons of the gut. Ann. N. Y. Acad. Sci. **527:** 369–377.
9. MILNER, P. *et al.* 1990. Vasoactive intestinal polypeptide levels in sigmoid colon in idiopathic constipation and diverticular disease. Gastroenterology **99:** 666–675.
10. KOCH, T. R. *et al.*.1991. Plasma vasoactive intestinal polypeptide concentration determination in patients with diarrhoea. Gastroenterology **100:** 99–106.
11. GRIFFITHS, N. M. *et al.* 1996. Exposure to either gamma or a mixed neutron/gamma field irradiation modifies vasoactive intestinal peptide receptor characteristics in membranes isolated from pig jejunum. Int. J. Radiat. Biol. **70:** 361–370.
12. WIDMARK, A. *et al.* Daily-diary evaluated side-effects of conformal versus conventional prostatic cancer technique. 1997. Acta Oncol. **36:** 499–507.

13. LABURTHE, M. *et al.* 1979. VIP receptors in intestinal epithelial cells: Distribution throughout the intestinal tract. *In* Hormone Receptors in Digestion and Nutrition. G. Rosselin, P. Fromageot & S. Bonflis, Eds.: 241–254. Elsevier/North Holland Biomedical Press. Amsterdam.
14. ZIMMERMAN, R. P. *et al.* 1988. Vasoactive intestinal peptide (VIP) receptors in the canine gastrointestinal tract. Peptides **9:** 1241–1253.
15. ZIMMERMAN, R. P. *et al.* 1989. Vasoactive intestinal polypeptide receptor binding sites in the human gastrointestinal tract: Localisation by autoradiography. Neuroscience **31:** 771–783.
16. SAYADI, H. *et al.* 1988. Autoradiographic distribution of vasoactive intestinal polypeptide receptors in rabbit and rat small intestine. Peptides **9:** 23–30.
17. KING, S. C. *et al.* 1989. Autoradiographic localization of binding sites for galanin and VIP in small intestine. Peptides **10:** 313–317.
18. KORMAN, L. Y. *et al.* 1989. Distribution of vasoactive intestinal polypeptide and substance P in human colon and small intestine. Dig. Dis. Sci. **34:** 1100–1108.
19. SCHWARTS, C. J. *et al.* 1974. Vasoactive intestinal peptide stimulation of adenylate cyclase and active electrolyte section in intestinal mucosa. J. Clin. Invest. **54:** 536–554.
20. BISHOP, A. E. *et al.* 1980. Abnormalities of vasoactive intestinal polypeptide-containing nerves in Crohn's disease. Gastroenterology **79:** 853–860.
21. MAKHLOUF, G. M. & S. I. SAID. 1975. The effect of vasoactive intestinal polypeptide (VIP) on digestive and hormonal function. *In* Gastrointestinal Hormones. J. C. Thompson, Ed.: 599–610. University of Texas. Austin, Texas.
22. KOCH, T. R. *et al.* 1987. Distribution and quantification of gut neuropeptides in normal intestine and inflammatory bowel diseases. Dig. Dis. Sci. **32:** 369-376.
23. SJÖLUND, K. *et al.* 1983. Peptide-containing nerve fibres in the gut wall in Crohn's disease. Gut **24:** 724–733.
24. O'MORIAN, C. *et al.* 1984. Vasoactive intestinal peptide concentrations and immunocytochemical studies in rectal biopsies from patients with inflammatory bowel disease. Gut **25:** 57–61.
25. LABURTHE, M. *et al.* 1987. Development of vasoactive intestinal peptide-responsive adenylate-cyclase during enterocytic differentation of Caco-2 cells in culture: evidence for an increased receptor level. J. Biol. Chem. **262:** 10180–10184.

Main Sensory Neuropeptides, but not VIP and NPY, Are Involved in Bone Remodeling during Orthodontic Tooth Movement in the Rat[a]

L. I. NOREVALL,[b,e] L. MATSSON,[c] AND S. FORSGREN[d]

Departments of [b]Orthodontics and [d]Anatomy, Umeå University, S-901 87 Umeå, Sweden

Department of [c]Pedodontics, Faculty of Odontology, Lund University, Sweden

ABSTRACT: During orthodontic tooth movement (OTM) a remodeling of the periodontal ligament (PDL) and the alveolar bone occurs. We have recently observed that the expression of CGRP and substance P (SP) increases in the PDL and dental pulp in response to buccally directed OTM of the upper first molar in the rat. We have now examined whether there is also an involvement of VIP and NPY in this type of OTM. A sectional arch wire cemented to the upper incisors exerted an orthodontic force, mediated by a coil on the lingual side of the tooth, for 24 hours. It was observed that the blood vessels in the periodontal ligament were supplied with VIP- and NPY-immunoreactive (IR) nerve fibers, whereas VIP-IR nerve fibers in contrast to NPY-IR fibers were only occasionally observed in the dental pulp. No significant changes were observed in response to OTM. The observations suggest that VIP and NPY, in contrast to the main sensory neuropeptides CGRP and SP, are not involved in the tissue processes that occur in the remodeling of PDL and alveolar bone during orthodontic tooth movement.

Orthodontic tooth movement (OTM) is a complex process involving the application of a well designed force on a tooth and the subsequent transformation of that force into a resulting movement of the tooth through the alveolar bone into the desired position. This process includes reorganization of the periodontal ligament (PDL), leading to compression sites with an initial hyaline zone, and tension sites, characteristics previously described by Reitan[1] and Rygh.[2] Furthermore, in the surrounding alveolar bone, a gradual remodeling occurs allowing the tooth to take the new position. During this reorganization major changes in the innervation and vascularization of the PDL occur. Accordingly, OTM in the rat has been shown to increase the peripheral expression of the sensory neuropeptides calcitonin gene–related peptide (CGRP) and substance P (SP).[3–5] In our laboratory, the occurrence of a proliferation of CGRP- and SP-nerve fibers in response to OTM, in this case buccally directed OTM of the upper first molar in the rat, has been verified, and furthermore, we noted that there was also an increase in CGRP- and SP-immunoreactivity in the contralateral tooth. [6] It is likely that CGRP and SP not only convey nociceptive stimuli

[a]This work was supported by grants from the J. C. Kempes Minnes Stipendiefond and the Swedish Medical Research Council (10364).

[b]Corresponding author: Lars-Inge Norevall, Department of Orthodontics, Umeå University, S-901 87 Umeå, Sweden; Fax: 46 90 786 54 80; E-mail: Lars-Inge.Norevall@ortodont.umu.se

from the PDL and pulp in this situation but that they also are of importance in the tissue remodeling that occurs in response to OTM.

Apart from what is known of the sensory neuropeptides SP and CGRP in OTM, there is very little information as to whether other types of neuropeptides are affected during this process. This is a drawback since other neuropeptides are also involved in plasticity and inflammatory reactions. One such peptide, which is also important in growth and differentiation, is vasoactive intestinal peptide (VIP).[7] Another neuropeptide likely to be of interest during OTM, is neuropeptide Y (NPY), as it recently has been observed that many NPY-containing nerve fibers coming from the trigeminal ganglion appear in the pulp and PDL in response to injury to the inferior alveolar nerve in the rat.[8,9] These results suggest that PDL mechanoreceptors are a main target of injury-evoked NPY and that NPY might be released from pulpal and PDL nerve endings in situations of damage.

The aim of the present investigation was to evaluate the effect of buccal orthodontic tooth movement of the upper first molar on the occurrence of VIP-ergic and NPY-ergic innervation in the PDL and dental pulp in the rat. The expression of still another neuropeptide, bombesin (BN), was examined as well.

MATERIALS AND METHODS

Orthodontic Procedure

Twenty female Sprague-Dawley rats with a mean age of 3 months and a mean body weight of 280 g (range 220–300 g) were studied. A fixed orthodontic appliance consisting of a band fitting to the maxillary incisors and a supporting palatal wire with an active spring for the exertion of a well designed orthodontic force towards the upper first molar (in the range of 30–50 g) were utilized. Additional details on the set-up of the experimental procedure have been presented previously.[6] After 24 hours of buccal orthodontic tooth movement (OTM) of the upper left first molar, the experimental animals were selected in a randomized fashion for immediate sacrifice with a carbon dioxide overdose (7 animals) or assigned to a 2-week-long restitution period (7 animals). In these latter animals the exerted orthodontic force was measured, whereafter they were allowed a 14-day-long restitution period. Three of the control animals were sacrificed at the same occasion as the experimental animals that were sacrificed after 24 h of OTM (cf. above) and the remaining three were handled together with the experimental animals from the healing group. During the active treatment period with the fixed appliances all animals received a standard pellet diet, mixed with water. The animal experiments had been approved by the Local Ethical Committee at Umeå University (No. A77/91).

Tissue Sampling, Sectioning, and Immunohistochemical Procedures

After sacrifice, the maxilla was dissected free and placed in a solution of 4% formaldehyde in 0.1 M phosphate buffer, pH 7.0, for 24 hours at 4°C. Demineralization in EDTA (the EDTA was changed every second day), as described by Bjurholm and colleagues[10] was accomplished for 3–5 weeks. The progress of the demineralization was checked by the use of a dental explorer and by radiographic evaluation once a week. After demineralization, the specimens were thoroughly washed in Tyrode's solution, containing 10% sucrose, at 4°C overnight. Thereafter, the maxilla was cut just in front of the first molars and just behind the second molars. The specimens were mounted on thin cardboard in OCT embedding medium (Miles Laboratories, Naperville, IL, USA), and frozen in propane chilled with liquid nitrogen. In order to compare specimens of experimental animals with those of controls, some of the maxillae of the former animals were divided in the midline and frozen together with half the maxilla of a control animal, allowing simultaneous sectioning, immunostaining, and examinations of these maxillae. Serial sections, 10 μm thick, of the entire mesio-distal width of the upper first molar were cut using a cryostat. The sections were mounted on slides precoated with chrome-alum gelatin, dried, and processed for indirect immunofluorescence.[6,11] As a positive control of one of the antisera used (BN-antiserum), formaldehyde-fixed specimens of rat larynx were used.

Rabbit antibodies against VIP, NPY, and BN were used (Amersham Int., Buckinghamshire, U.K.). The characteristics of all these antibodies have been described in previous studies.[12,13] As the BN-antiserum to some extent crossreacts with SP,[13] this antiserum was always preabsorbed with SP before incubation on the sections. Specific immunoreaction with BN-antiserum preabsorbed in this way was obtained in the control tissue used (rat larynx) (not shown).

RESULTS

Orthodontic Tooth Movement

The animals subjected to buccal orthodontic tooth movement of the upper first molar for 24 hours were checked for verification of tooth movement. By measuring the width of the PDL in the pressure and tension sides, these studies revealed that an OTM actually was at hand, as measured in all the experimental animals.[6] There was no significant difference in the width of the PDL in the experimental teeth after 14 days of recovery from OTM as compared with the contralateral teeth from the same animals and the teeth from the control animals.

VIP Immunoreaction

VIP-immunoreactivity (VIP-IR) was detected in the walls of the small blood vessels in the periodontal ligament (PDL) of the upper first molar, and especially in the apical third of the PDL, but not in other parts of the PDL (FIG. 1). Also in association with blood vessels in the bone lacunas communicating with the PDL, nerve fibers showing VIP-IR were observed. In the apical part of the dental pulp, no or only a few VIP-IR nerve fibers occurred. The degree of VIP-IR in the PDL and pulp of the experimental (left) side was similar to that seen in the contralateral tooth (right side) after 24 h of OTM. Furthermore, in the samples containing the left half of the maxilla of an experimental animal frozen together with the left half of the maxilla from a control animal at this stage, no differences in VIP-IR in the PDL were seen. The pattern of VIP-IR in the animals subjected to OTM followed by 14 days recovery was similar.

NPY Immunoreaction

NPY-IR was localized to nerve fibers present in the walls of the blood vessels in the apical third of the PDL, in the adjacent bone lacunas (FIG. 2), and in the dental pulp in the controls as well as in animals subjected to OTM and OTM followed by 14 days recovery, respectively. The NPY-IR occurred to the same extent on the OTM side as in association with the contralateral tooth and the tooth of control animals frozen together with the experimental tooth. Apart from the blood vessel–related NPY-IR, no specific immunoreaction was detected in other parts of the PDL (cf. FIG. 2). In some of the sections, NPY-IR

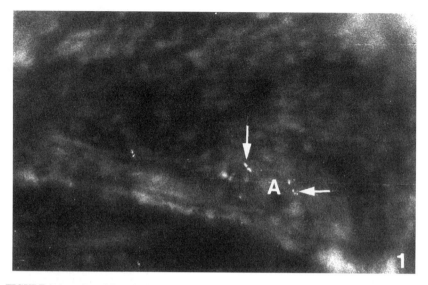

FIGURE 1. A section of the apical third of the PDL of an animal subjected to 24 h of OTM processed for demonstration of VIP. A few VIP-positive varicose nerve fibers (*arrows*) occur in association with an arteriole (**A**). (×400)

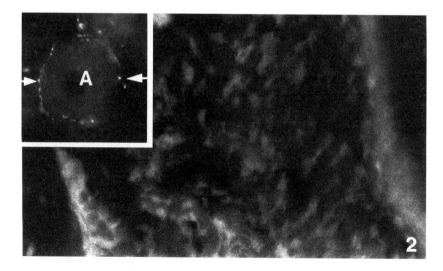

FIGURE 2. The apical third of the PDL of an experimental animal subjected to OTM followed by a recovery period of 14 days. This section was processed for NPY. There is no NPY-IR in the PDL. In the wall of an arteriole in a bone lacuna associated with the PDL there is a marked NPY-IR (*arrows*) (*inset*). (×400)

occurred in long varicose threadlike structures extending from the apices of the dental pulp up to the pulp chamber. In the roof of the pulp chamber, small dot-like NPY-IR varicose structures sometimes occurred. On the whole, there was a higher degree of NPY-IR than VIP-IR in both the PDL and pulp.

BN Immunoreaction

We were not able to detect any BN-positive immunoreaction in the PDL (FIG. 3) nor in the pulp in the control samples or the animals sacrificed immediately after the termination of 24 hours with OTM or the animals subjected to OTM followed by a 14-day recovery period.

DISCUSSION

In the present study, the immunohistochemical expression of VIP, NPY, and BN in the PDL and dental pulp of rats exposed to buccally directed orthodontic tooth movement has been studied.

In accordance with previous studies on various species on maxillary/mandibular teeth,[14,15] VIP-containing nerve fibers were regularly found in close relation to blood vessels in the apical third of the PDL and only sometimes in the pulp. Apart from having vasodilative properties, VIP functions as a growth factor and increases the survival of neurons in culture.[16] Furthermore, VIP has been reported to participate in wound healing processes[17] and to stimulate bone resorption *in vitro*.[18] With this information as a basis, the effect of OTM on VIP expression was examined. We did not observe any differences in the amount and distribution of VIP-positive fibers in the orthodontically treated rats in comparison with the

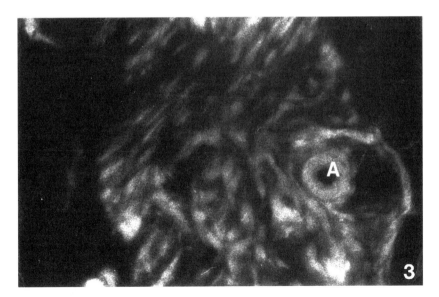

FIGURE 3. The apical PDL in a section processed for demonstration of BN. This animal had been subjected to OTM followed by a 14 day recovery. The reactions that are seen are unspecific. There is no specific BN immunoreaction in the PDL nor in the wall of the small arteriole (**A**). (×400)

non-treated animals. This is in line with previous observations that the distribution of VIP is unaffected in periodontitis-affected human gingiva.[19] On the other hand, during OTM in cats, intense VIP labeling was observed in the compressed PDL near sites of bone resorption and in the pulp of moving teeth.[3, 20] One possible explanation for the diverging results between the present study and the studies on cats may be the fact that different species were used. Thus, differences in VIP expression between species have been reported by Uddman and colleagues[21] who found that VIP-containing nerve fibers occur in close relationship with blood vessels in the dental pulp of various species including cat, but not in the dental pulp of the rat. Furthermore, Kerezoudis and colleagues[22] did not find any VIP-IR nerve fibers in the rat incisor pulp. In the present study, only a few VIP-IR fibers occurred in the pulp, preferably in the apical part, in the upper first molar.

As was previously observed in studies on the rat mandible,[23,24] NPY-IR was detectable in the innervation of blood vessels in the pulp and PDL. To the best of our knowledge, the pattern of NPY innervation has previously never been studied during OTM. The present study shows that the NPY innervation was not influenced during the process of OTM. Thus, it is noteworthy that neither VIP nor NPY, which are messengers of the nervous system to immune cell transactions, appears to have a prominent role in regulating the local inflammatory response in OTM in the rat.

Immunoreaction for BN, the peptide originally isolated from the skin of *Bombina bombina*[25] and having a mammalian homologue termed gastrin-releasing peptide, was not detected in the PDL and pulp of the control animals nor in the orthodontically treated animals. Similarly, only faint immunoreactions for methionine enkephalin have been observed in the PDL of moving teeth as well as control teeth.[26] These observations, and those on NPY/VIP on one hand and those on SP/CGRP[6] on the other, show that the major sensory peptides are the peptides that are affected in force-related tooth movement. In this remodeling process, these peptides are likely to have interactions with cytokines and hormones.[26]

Our results are in line with results obtained in a recent study on cultivated dental pulp cells from rat molars by Bongenhielm and colleagues[27] in which evidence was presented that SP and CGRP, but not VIP and NPY, are involved during pulpal development and in wound healing after pulpal injury. In conclusion, sensory peptides, rather than peptides like VIP and NPY, are of importance in the healing after injuries to the PDL and pulp.

ACKNOWLEDGMENTS

We are grateful to Mrs. Lena Crenshaw, Mai Wranding, Ulla Hedlund, Inga Johansson, and Lisa Arvidsson for skillful technical assistance.

REFERENCES

1. REITAN, K. 1951. The initial tissue reaction incident to orthodontic tooth movement as related to the influence of function. Thesis. Acta Odont. Scand. Suppl. 6.
2. RYGH, P. 1990. The periodontal ligament under stress. *In* The Biology of Tooth Movement, L.A Norton & C. J. Burstone, Eds.: 10–27. CRC Press Inc. Boca Raton, Florida.
3. DAVIDOVITCH, Z. *et al.* 1988. Neurotransmitters, cytokines and the control of alveolar bone remodeling in orthodontics. Dent. Clin. North Am. **32:** 411–435.
4. KVINNSLAND, I. & S. KVINNSLAND. 1990. Changes in CGRP-immunoreactive nerve fibers during experimental tooth movements in rats. Eur. J. Orthod. **12:** 320–329.
5. SAITOH, I. *et al.* 1991. Responses of calcitonin gene-related peptide-immunopositive nerve fibers in the periodontal ligament of rat molars to experimental tooth movement. Arch. Oral Biol. **36:** 689–692.
6. NOREVALL, L.I. *et al.* 1995. Expression of neuropeptides (CGRP, Substance P) during and after orthodontic tooth movement in the rat. Eur. J. Orthod. **17:** 311–325.

7. HÖKFELT, T. *et al.* 1994. Messenger plasticity in primary sensory neurons following axotomy and its functional implications. Trends Neurosci. **17:** 22–30.
8. ITOGAWA, T. *et al.* 1993. Appearance of neuropeptide Y-like immunoreactive cells in the rat trigeminal ganglion following dental injuries. Arch. Oral Biol. **38:** 725–728.
9. FRISTAD, I. *et al.* 1996. Neuropeptide Y expression in the trigeminal ganglion and mandibular division of the trigeminal nerve after inferior alveolar nerve axotomy in young rats. Exp. Neurol. **142:** 276–286.
10. BJURHOLM, A. *et al.* 1989. Fixation and demineralization of bone tissue for immunohistochemical staining of neuropeptides. Calcified Tissue Int. **45:** 227–231.
11. FORSGREN, S. & L. SÖDERBERG. 1987. Immunohistochemical procedures for the demonstration of peptide- and tyrosine hydroxylase-containing nerve fibers in cryostat sections of unfixed rapidly frozen tissue stored for long period of time. Histochemistry **87:** 561–568.
12. FORSGREN, S. *et al.* 1994. Effects of irradiation on neuropeptide expression in rat salivary gland and spinal cord. Histochem. J. **26:** 630–640.
13. LIDEGRAN, M. *et al.* 1995. Bombesin-like immunoreactivity in the rat larynx: increase in response to irradiation. Regul. Pept. **55:** 321–330.
14. KATO, J. *et al.* 1990. The distribution of vasoactive intestinal polypeptides and calcitonin gene-related peptide in the periodontal ligament of mouse molar teeth. Arch. Oral Biol. **35:** 63–66.
15. AKAI, M. & S. WAKISAKA. 1990. Distribution of peptidergic nerves. *In* Dynamic Aspects of Dental Pulp. R. Inoki, T. Kudo & L.M. Olgart, Eds.: 337–348. Chapman and Hall. London.
16. PINCUS, D.W. *et al.* 1990. Vasoactive intestinal peptide regulates mitosis, differentiation and survival of cultured sympathetic neuroblasts. Nature **343:** 564–567.
17. DALSGÅRD, C.J. *et al.* 1989. Neuropeptides as growth factors. Possible roles in human diseases. Regul. Pept. **25:** 1–9.
18. HOHMANN, E. L. *et al.* 1983. Vasoactive intestinal peptide stimulates bone resorption via a cyclic adenosine 3', 5'-monophosphate dependent mechanism. Endocrinology **112:** 1233–1239.
19. LUTHMAN, J. *et al.* 1989. Immunohistochemical study of neurochemical markers in gingiva obtained from periodontitis- affected sites. J. Periodont. Res. **24:** 267–278.
20. MOTAKEF, M. *et al.* 1990. Localization of VIP at bone resorption sites in vivo. J. Dent. Res. **69:** 253.
21. UDDMAN, R. *et al.* 1980. Occurrence of VIP nerves in mammalian dental pulps. Acta Odontol. Scand. **38:** 325–328.
22. KEREZOUDIS, N. P. *et al.* 1995. Haemodynamic and immunohistochemical studies of rat incisor pulp after denervation and subsequent re-innervation. Arch. Oral Biol. **40:** 815–823.
23. UDDMAN, R.T. *et al.* 1984. Neuropeptide Y: occurrence and distribution in dental pulps. Acta Odont. Scand. **42:** 361–365.
24. FRISTAD, I. *et al.* 1994. Nerve fibers and cells immunoreactive to neurochemical markers in developing rat molars and supporting tissues. Arch. Oral Biol. **39:** 633–646.
25. ERSPAMER, V. *et al.* 1970. Some pharmacological actions of alytesin and bombesin. J. Pharm. Pharmac. **22:** 875–876.
26. DAVIDOVITCH, Z. 1991. Tooth movement. Crit. Rev. Oral Biol. Med. **2:** 411–450.
27. BONGENHIELM, U. *et al.* 1995. Effects of neuropeptides on growth of cultivated rat molar pulp fibroblasts. Regul. Pept. **60:** 91–98.

PACAP$_{27}$ and Other Neuropeptides in the Inferior Mesenteric Ganglion[a]

LEONID G. ERMILOV AND JOSEPH H. SZURSZEWSKI[b]

Department of Physiology and Biophysics, Mayo Clinic and Mayo Foundation, Rochester, Minnesota 55905 USA

ABSTRACT: The presence and location of PACAP$_{27}$-like immunoreactivity (PACAP$_{27}$-LI) in the colon-inferior mesenteric ganglion (IMG) reflex pathway and the effect of exogenously administered PACAP$_{27}$ on the excitability of IMG are reported. The results provide morphological and electrophysiological support for the hypothesis that PACAP modulates reflex activity between the large intestine and IMG. The intense excitatory effect would be expected to increase the rate of action potential discharge in IMG neurons, increasing sympathetic drive to the colon thereby decreasing of colonic activity.

Mammalian prevertebral ganglia have the capacity to integrate and support peripheral reflexes with the gastrointestinal tract without any influence of the spinal cord and higher levels of the central nervous system.[1] Most of the work examining this reflex pathway has been *in vitro* using the guinea pig inferior mesenteric ganglion (IMG) with an attached segment of colon. The afferent limb of the colon-IMG reflex consists of mechanosensory, enteric cholinergic neurons that discharge action potentials during increases in colonic intraluminal pressure and during propulsive colonic motor activity.[2,3] The inhibitory IMG sympathetic neurons constitute the efferent limb of the reflex. Thus, the colon-IMG reflex functions as a negative feedback control system. A similar peripheral reflex loop between colonic enteric and sympathetic IMG neurons is present in the cat[3–5] and dog,[6] and there is reason to believe one is present in humans as well.[7–9] Mechanosensory colonic enteric neurons (colonofugal neurons) project also to sympathetic neurons in the celiac and superior mesenteric ganglia.[3,10] Furthermore, the superior mesenteric ganglion neurons and especially the celiac ganglion neurons receive synaptic input from mechanosensory afferents located in the small intestine and distal stomach.[11,12] Thus, there is evidence supporting the notion that neurons of all three major prevertebral ganglia receive mechanosensory afferent synaptic input from a subset of enteric neurons (viscerofugal neurons) distributed throughout the gastrointestinal tract.[11,12]

Increases in colonic intraluminal pressure and ongoing contractile activity release neurotransmitters and neuromodulators in the IMG. Low pressure (< 10 cm H$_2$O) colonic distension releases acetylcholine (ACh),[13] GABA,[14-15] and VIP.[13] The release of SP and calcitonin gene–related peptide (CGRP) from en passant synapses of capsaicin-sensitive dorsal root nerve fibers requires high pressure (> 20 cm H$_2$O) colonic distension.[16,17] This can be taken as direct evidence that high colonic pressures release SP from putative visceral nociceptors. An important finding was the demonstration that neurotensin potentiated colonic distension-evoked

[a]This work supported by National Institute of Diabetes and Digestive and Kidney Diseases Grant DK 17632.

[b]Corresponding author: J. H. Szurszewski, Ph.D., Department of Physiology and Biophysics, Mayo Clinic and Mayo Foundation, 200 First Street SW, Rochester, MN 55905; Tel.: 507 284-3927; Fax: 507 284-0266; E-mail: gijoe@mayo.edu

release of SP and CGRP whereas the enkephalins inhibited their release.[16,17] In contrast, neither neurotensin nor the enkephalins had any effect on distension-evoked release of VIP.[16,17] Thus, prevertebral ganglia form an extended neural network that connects the lower intestinal tract to the upper gastrointestinal tract. ACh, GABA,[14,15] and a large number of peptide neuromodulators are utilized to finely tune this peripheral reflex pathway.

The majority of the mechanosensory viscerofugal neurons that project to prevertebral ganglia selectively target -/NA (i.e., neurons containing neither NPY or SOM) and SOM/NA sympathetic neurons.[1] The -/NA neurons control motility because -/NA terminals surround myenteric excitatory motor neurons that innervate smooth muscle. SOM/NA sympathetic neurons control secretion because their terminals surround inhibitory submucous secretomotor neurons (SOM/NA).[18] Sympathetic neurons (NPY/NA), which are vasomotor, do not receive inputs from viscerofugal neurons. Thus, from a functional point of view, activation of viscerofugal neurons inhibits intestinal motility and fluid secretion from the intestine. Unlike the NPY/NA neurons, the -/NA and SOM/NA neurons receive few if any inputs from central preganglionic nerves. Thus, ongoing sympathetic control of motility and fluid secretion is peripheral rather than of central nervous system origin. It can be suggested that the entero-prevertebral ganglion reflex limits or reduces fluid loading of the large intestine and therefore reduces the incidence of watery diarrhea.[1] Interruption of the entero-prevertebral ganglion reflex could be the basis for paralytic secretion, first noted by Wright and colleagues in 1940.[19]

Although the postsynaptic neuron has a number of intrinsic mechanisms to alter its excitability and firing properties and patterns, it is the nerve terminal region that is the key to the overall performance of the colon-IMG reflex. Most, if not all, synapses in the IMG are plurichemical. While acetylcholine acting on nicotinic receptors is the primary transmitter mediating fast transmission, a growing number of neuropeptides, many co-localized in cholinergic nerve terminals, are found to mediate slow synaptic transmission. Neuropeptides such as SP, VIP, and CCK mediate slow excitatory postsynaptic potentials increasing neuronal excitability.[1] Others such as the enkephalins and neurotensin modulate the release of other peptides such as SP.[1]

Recently, pituitary adenylate cyclase activating peptide (PACAP) has been found in a number of autonomic ganglia, including rat sympathetic preganglion neurons of the thoracolumbar spinal cord,[20] superior cervical ganglion,[21] celiac ganglion,[21] myenteric ganglion neurons,[20] and several parasympathetic ganglia.[20] There have been no studies to date on whether PACAP is present in the guinea pig IMG, and if so the effect of PACAP on IMG neurons. The results that follow report on the presence and location of PACAP$_{27}$-like immunoreactivity (PACAP$_{27}$-LI) in the colon-IMG reflex pathway and the effect of exogenously administered PACAP$_{27}$ on the excitability of IMG neurons.

PACAP IMMUNOREACTIVITY IN THE IMG

Serial cryosections (15 μm) of the guinea pig IMG were processed for immunofluorescence using a polyclonal rabbit antiserum against PACAP$_{27}$Amide (diluted 1:200). The reaction product was visualized with anti-rabbit IgG conjugated to rhodamine and observed with an Axiophot upright light microscope (Carl Zeiss, Inc.) using plan-neofluor oil-immersion lens with a high numerical apperture.

The vast majority of nerve cell bodies in the IMG were surrounded by PACAP immunoreactive varicosities (FIG. 1A). Although there were nerve cell bodies with PACAP$_{27}$-LI, they were uncommon.

Intense staining for PACAP$_{27}$-LI was present in nerve fibers of the lumbar colonic nerve trunk (FIG. 1B) and the hypogastric nerve trunk. PACAP$_{27}$-LI was also detected in a few inferior splanchnic nerves dorsal root ganglion (T_{13}-L_2) neurons.

PACAP$_{27}$-LI was found in myenteric ganglion neurons (FIG. 1C). Occasionally, a ganglion cell body with PACAP$_{27}$-LI was observed in the submucosal ganglion plexus.

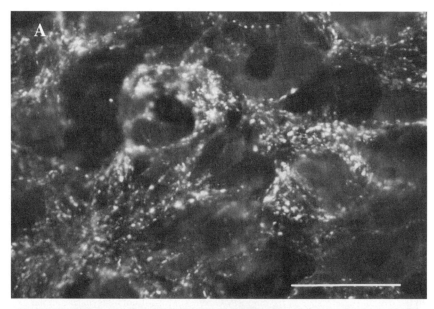

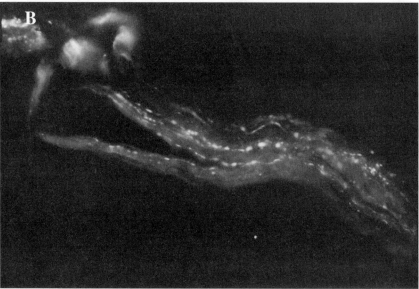

FIGURE 1. Immunofluorescence for $PACAP_{27}$ in (**A**) the guinea pig inferior mesenteric ganglion, (**B**) lumbar colonic nerve trunk, and (**C**) myenteric plexus of the distal colon. In (**A**), $PACAP_{27}$-positive varicosities surround principal ganglion cells. In (**B**), $PACAP_{27}$-positive nerve fibers can be seen in the lumbar colonic nerve.

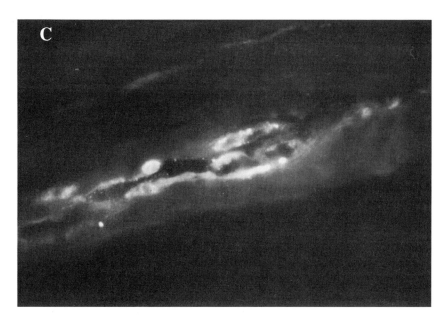

FIGURE 1. (*continued*) In (**C**), PACAP$_{27}$-positive nerve cell bodies can be seen in the myenteric plexus. In (**C**), the longitudinal muscle layer is to the top and the circular muscle layer is to the bottom. No immunofluorescence staining was observed when anti-PACAP$_{27}$ antiserum was omitted. Calibration bar, for both panels 50 µm. Objective 63×, oil, NA1.4

EFFECT OF PACAP ON IMG NEURONS

To study the effect of exogenous administration of PACAP$_{27}$ on the excitability of IMG neurons, separate experiments were done on the IMG with and without a segment of attached colon. Intracellular recordings of transmembrane potential were obtained using conventional intracellular recording methods. PACAP$_{27}$ was applied by N$_2$ pressure microejection (5 µM, 50 to 150 msec "puffs") from a 10-µm diameter glass micropipette brought close to the impaled neuron.

When the IMG was not attached to the colon, the resting membrane potential was stable at 50.5 ± 0.4 mV and there were no spontaneous fast excitatory postsynaptic potentials (F-EPSPs). All ($N = 12$) neurons tested responded to PACAP$_{27}$. A single "puff" of PACAP$_{27}$ caused an intense and long lasting (6 to 11 min) spike discharge with or without a slow depolarization of the membrane potential. When there was a PACAP$_{27}$-induced depolarization, it ranged from 4 to 20 mV in the different neurons tested. The duration of the PACAP$_{27}$-induced spike discharge did not seem to depend on a PACAP$_{27}$-induced membrane depolarization. Repeated application of PACAP$_{27}$ led to desensitization. Typically, after two or three applications, spike intensity was greatly reduced and the duration of response considerably shorter. An example of the response to a 50 msec "puff" of PACAP$_{27}$ is shown in FIGURE 2.

When a segment of colon was left attached to the IMG, all neurons tested exhibited ongoing F-EPSPs, which occasionally reached threshold for firing an action potential (FIG. 3, upper trace, left of arrow). Previous studies have shown that ongoing F-EPSPs in IMG neurons are due to cholinergic nicotinic synaptic input from projections of colonic myenteric ganglion neurons, which are slow-adapting and mechanosensory.[1] A single "puff" of

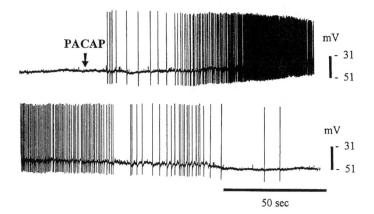

FIGURE 2. Response of a guinea pig IMG neuron to PACAP$_{27}$. Application of PACAP by pressure microinjection evoked a long-lasting membrane depolarization and an intense spike discharge. The beginning of the bottom trace is 6 minutes after the end of the top trace. A 50 msec pressure pulse of PACAP$_{27}$ (5 µM) was applied at the arrow.

PACAP$_{27}$ caused an intense spike discharge (FIG. 3, upper panel). When ongoing nicotinic F-EPSP activity was blocked by hexamethonium (100 µM), PACAP$_{27}$ evoked a depolarization of the membrane potential (FIG. 3, lower panel). The amplitude of the depolarization ranged from 8 to 20 mV in the different neurons tested.

DISCUSSION

The results of the present study show that PACAP$_{27}$ evoked a potent and long lasting excitatory effect in IMG neurons. Similar responses to exogenously administered PACAP have been observed in rat sympathetic preganglionic neurons[22] and guinea pig myenteric ganglion neurons.[23] The intense discharge and the immunohistochemical localization of PACAP$_{27}$ in varicosities surrounding neurons in the IMG support the notion that PACAP is

FIGURE 3. Effect of PACAP$_{27}$ on an IMG neuron receiving excitatory synaptic input from the colon. In normal Krebs solution, a 50 msec pressure application of PACAP$_{27}$ (5 µM) evoked a long-lasting excitatory response. After 30 minutes of superfusion with hexamethonium (100 µM), ongoing spike activity was abolished (left of arrow). The spike discharge caused by PACAP$_{27}$ seen in control was abolished, but PACAP$_{27}$ still caused membrane depolarization.

a messenger molecule that functions as an excitatory neuromodulator of ganglion cell excitability. PACAP$_{27}$ appears to act presynaptically and postsynaptically to facilitate release of neurotransmitter (most likely acetylcholine) and cause membrane depolarization, respectively. There also is the possibility that PACAP may have released one of several neuromodulatory peptides that mediate the slow-excitatory postsynaptic potential.[1]

The present study does not identify whether PACAP acted on type I, II, or III receptors. The PACAP type I receptor is a strong candidate as this receptor type has been found in other sympathetic ganglia.[20,24,25] The present study also does not address the mechanism of action of PACAP$_{27}$. In previous studies, SP and VIP have been shown to act postsynaptically to evoke spike discharge and membrane depolarization by closing K channels.[1] Whether the postsynaptic effect of PACAP$_{27}$ was through a similar mechanism is at present unknown.

The morphological data obtained in the present study by immunostaining for PACAP$_{27}$ provide support for the hypothesis that PACAP is a neuromodulatory peptide released from colonofugal neurons projecting to the IMG. PACAP$_{27}$-LI was found in nerves connecting the IMG to the colon, and PACAP$_{27}$-LI was found in cell bodies of myenteric ganglion neurons. Based on these observations, it seems reasonable to suggest that axons of some of the PACAP-positive myenteric ganglion neurons reached the IMG via the lumbar colonic and hypogastric nerves.

SUMMARY

The results provide morphological and electrophysiological support for the hypothesis that PACAP modulates reflex activity between the large intestine and IMG. The intense excitatory effect would be expected to increase the rate of action potential discharge in IMG neurons increasing sympathetic drive to the colon thereby decreasing of colonic motility.

ACKNOWLEDGMENTS

The authors are grateful to Jan Applequist and LeAnn Lake for assistance in typing this paper.

REFERENCES

1. SZURSZEWSKI, J. H. & S. M. MILLER. 1994. Physiology of prevertebral ganglia. *In* Physiology of the Gastrointestinal Tract. 3rd edit. L. R. Johnson, Ed.: 795–877. Raven Press. New York.
2. SZURSZEWSKI, J. H. & W. A. WEEMS. 1976. A study of peripheral input to and its control by postganglionic neurones of the inferior mesenteric ganglion. J. Physiol. Lond. **256**: 541–556.
3. WEEMS, W. A. & J. H. SZURSZEWSKI. 1977. Modulation of the colonic motility by peripheral neural inputs to neurons of the inferior mesenteric ganglion. Gastroenterology **73**: 273–278.
4. JULÉ, Y., J. KRIER & J. H. SZURSZEWSKI. 1983. Patterns of innervation of neurones in the inferior mesenteric ganglion of the cat. J. Physiol. Lond. **344**: 293–304.
5. JULÉ, Y. & J. H. SZURSZEWSKI. 1983. Electrophysiology of neurones of the inferior mesenteric ganglion of the cat. J. Physiol. Lond. **344**: 277–292.
6. KING, B. F. & J. H. SZURSZEWSKI. 1984. An electrophysiological study of the inferior mesenteric ganglion of the dog. J. Neurophysiol. **51**: 607–615.
7. DELFIACCO, M., M. C. LEVANTI, G. BROTZU & R. MONTISU. 1984. Substance P-like immunoreactivity in human sympathetic ganglia. Brain Res. **321**: 143–146.
8. LEVANTI, M. C., R. MONTISCI, M. D. ROSA, P. SEDDA & M. DELFIAICO. 1988. Substance P-like immunoreactive innervation of the human colon and coeliac superior mesenteric and inferior mesenteric ganglia. Basic Appl. Histochem. **32**: 145–152.

9. QUARTU, M. J., J. M. POLAK & M. DELFIACCO. 1993. Morphology and distribution of efferent vagal innervation of rat pancreas as revealed with anterograde transport of DiI. J. Chem. Neuroanat. **6**: 79–99.
10. KREULEN, D. L. & J. H. SZURSZEWSKI. 1979. Reflex pathways in the abdominal prevertebral ganglia: evidence for a colo-colonic inhibitory reflex. J Physiol. Lond. **295**: 21–32.
11. KING, B. F. & J. H. SZURSZEWSKI. 1984. Mechanoreceptor pathways from the distal colon to the autonomic nervous system in the guinea-pig. J. Physiol. Lond. **350**: 93–107.
12. KREULEN, D. L., T. C. MUIR & J. H. SZURSZEWSKI. Peripheral sympathetic pathways to gastroduodenal region of the guinea pig. Am. J. Physiol. **245**: G369–G375.
13. PARKMAN, H. P., R. C. MA, W. H. STAPELFELDT & J. H. SZURSZEWSKI. 1993. Direct and indirect mechanosensory pathways from the colon to the inferior mesenteric ganglion. Am. J. Physiol. **265**: G499–G505.
14. PARKMAN, H. P., W. H. STAPELFELDT, C. L. WILLIAMS, V. A. LENNON & J. H. SZURSZEWSKI. 1993. Enteric GABA-containing nerves projecting to the guinea-pig inferior mesenteric ganglion modulate acetylcholine release. J. Physiol. Lond. **471**: 191–207.
15. STAPELFELDT, W. H., H. P. PARKMAN & J. H. SZURSZEWSKI. 1993. The electrophysiological effects of endogenous GABA in the guinea-pig inferior mesenteric ganglion. J. Physiol. Lond. **471**: 175–189.
16. MA, R. C. & J. H. SZURSZEWSKI. 1996. Release of calcitonin gene–related peptide in guinea pig inferior mesenteric ganglion Peptides **17**: 161–170.
17. MA, R. C. & J. H. SZURSZEWSKI. 1996. Modulation by opioid peptides of mechanosensory pathways supplying the guinea-pig inferior mesenteric ganglion. J. Physiol. Lond. **491**: 435–445.
18. KEAST, J. R., E. M. MCLACHLAN & R. L. MECKLER. 1993. Relation between electrophysiological class and neuropeptide content of guinea pig sympathetic prevertebral neurons. J. Neurophysiol. **69**: 384–394.
19. WRIGHT, R. D., M. A. JENNINGS, H. W. FLOREY & R. LIUM. 1940. The influence of nerves and drugs on secretion by the small intestine and an investigation of the enzymes in intestinal juice. Q. J. Exp. Physiol. **30**: 73–120.
20. SUNDLER, F., E. EKBLAD, J. HANNIBAL, K. MOLLER, Y-Z ZHANG, H. MULDER, T. ELSAS, T. GRUNDITZ, N. DANIELSEN, J. FAHRENKRUG & R. UDDMAN. 1996. Pituitary adenylate cyclase–activating peptide in sensory and autonomic ganglia: Localization and regulation. Ann. N.Y. Acad. Sci. **805**: 410–428.
21. KLIMASCHEWSKI, L., C. HAUSER & C. HEYM. 1996. PACAP immunoreactivity in the rat superior cervical ganglion in comparison to VIP. Autonomic Nervous System Vegetative Control **7**: 2797–2801.
22. LAI, C. C., S. Y. WU, H. H. LIN & N. J. DUN. 1997. Excitatory action of pituitary adenylate cyclase activating peptide on rat sympathetic preganglionic neurons in vivo and in vitro. Brain Res. **748**: 189–194.
23. CHRISTOFI, F. L. & J. D. WOOD. 1993. Effects of PACAP on morphologically identified myenteric neurons in guinea pig small bowel. Am. J. Physiol. **27**: G414–G421.
24. SPENGLER, D., C. WAEBER, C. PANTALONI, F. HOLSBOER, J. BOCKAERT, P. H. SEEBURG & L. JOURNOT. 1993. Differential signal transduction by five splice variants of the PACAP-receptor. Nature **365**: 170–175.
25. MOLLER, K. & F. SUNDLER. 1996. Expression of pituitary adenylate cyclase activating peptide (PACAP) and PACAP type I receptors in the rat medulla. Regul. Pept. **63**: 129–139.

Induction of Multiple Pituitary Adenylate Cyclase Activating Polypeptide (PACAP) Transcripts through Alternative Cleavage and Polyadenylation of proPACAP Precursor mRNA[a]

SUSAN A. HARAKALL, CYNTHIA A. BRANDENBURG,
GREGORY A. GILMARTIN,[b] VICTOR MAY, AND
KAREN M. BRAAS

*Department of Anatomy and Neurobiology and [b]Department of
Microbiology and Molecular Genetics, University of Vermont College of
Medicine, Given Health Science Complex, Burlington, Vermont 05405
USA*

The vasoactive intestinal peptide (VIP)/pituitary adenylate cyclase activating polypeptide (PACAP)/secretin/glucagon family of bioactive peptides has important physiological roles in autonomic nervous system development, function, and target tissue regulation.[1-6] PACAP cDNA has been cloned from several species and the transcript has been identified in many tissues. The human proPACAP gene is composed of five exons and four introns; the mature PACAP peptides and the entire 3′ untranslated region (UTR) are encoded by one exon.[7] Two proPACAP mRNA species have been observed by Northern analysis. In nervous tissues, the predominant form of proPACAP mRNA is approximately 2.2 kb, whereas a 0.9 kb message was reported in testes.[1,8,9] Cloning of proPACAP cDNA from these tissues demonstrated that the two forms of the message contain the same coding region, but differ in the 5′ and 3′ UTR. Specifically, the testicular transcript has a truncated 3′ noncoding region and a unique 5′ end not observed in the neuronal transcript.

Many autonomic nervous system target tissues respond to PACAP and a number of ganglia express PACAP peptides.[1,2,10-13] As part of our ongoing studies examining the neuromodulatory and neurotrophic roles of PACAP peptides in the sympathetic nervous system, we have investigated the regulatory mechanisms that guide PACAP expression in the sympathetic neurons of the superior cervical ganglion (SCG).[1] Of the many possible transcriptional and translational sites for regulation, the PACAP system surprisingly appears to have adopted an alternative precursor mRNA (pre-mRNA) processing mechanism, which may have significant consequences in peptide production during altered neuronal states.

[a]This work was supported by grants HD-27468 (K.M.B. and V.M.), NS-01636 (V.M.), and GM-46624 (G.A.M.) from the National Institutes of Health and Grant 94015540 (K.M.B.) from the American Heart Association.

DEPOLARIZATION OF SYMPATHETIC NEURONS INCREASES PROPACAP mRNA EXPRESSION THROUGH INDUCTION OF MULTIPLE TRANSCRIPTS

Previous studies in this laboratory and others investigated the expression of PACAP by SCG neurons. Endogenous sympathetic neuron expression of PACAP peptides and proPACAP mRNA was established and PACAP release from primary cultured sympathetic neurons was demonstrated.[1,2,14-16] To further examine potential regulated changes in neuronal PACAP expression, a well established depolarization paradigm was employed: treatment of cultured sympathetic neurons with 40 mM KCl augmented cellular PACAP content with a concomitant increase in PACAP peptide release.[1] To assess whether these changes in secretion and cellular PACAP content reflected an increase in cellular proPACAP mRNA expression, Northern analysis was performed using poly(A$^+$) mRNA from control and depolarized cultured SCG neurons. Control sympathetic neurons expressed predominantly a 2.2 kb form of proPACAP mRNA, the primary form observed in the hypothalamus and other nervous system regions. Depolarization of sympathetic neurons not only elicited an increase in the 2.2 kb proPACAP transcript over the control neurons, but also induced unexpectedly high levels of an approximately 0.9 kb proPACAP mRNA transcript not observed previously in neuronal tissues, which appeared similar in size to the form of PACAP mRNA reported in testes (FIG. 1). Accordingly, the challenges were to begin to establish the identity of the shortened neuronal PACAP mRNA, the mechanisms underlying the production of the smaller transcript, and the implications of shortened transcript expression with respect to peptide production.

THE SHORTER PROPACAP mRNA TRANSCRIPT INDUCED IN SCG SYMPATHETIC NEURONS DOES NOT REPRESENT THE TESTICULAR FORM

To examine whether the shorter proPACAP mRNA induced in depolarized sympathetic neurons represented the testicular form of the transcript, reverse transcription PCR and sequence-specific hybridization were performed (FIG. 2).[1] A distinguishing feature of the testicular message, a unique 5′ end, was utilized in the design of primers that would dif-

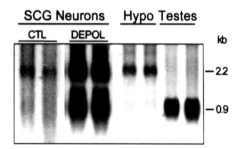

FIGURE 1. Depolarization of sympathetic neurons induces multiple proPACAP mRNA transcripts. The regulatory effects of depolarization on SCG neuronal proPACAP mRNA expression were evaluated by Northern blot analysis. Primary cultured rat SCG neurons were incubated in medium containing 40 mM NaCl (CTL) or 40 mM KCl (DEPOL) for 96 h beginning on day 9 of culture. Poly(A$^+$) RNA was extracted from the neurons, fractionated on denaturing agarose gels, and transferred to Nytran-Plus membrane for hybridization with a radiolabeled proPACAP specific riboprobe. Two micrograms of poly(A$^+$) mRNA from hypothalamus (Hypo) and testes were analyzed for proPACAP mRNA in parallel.

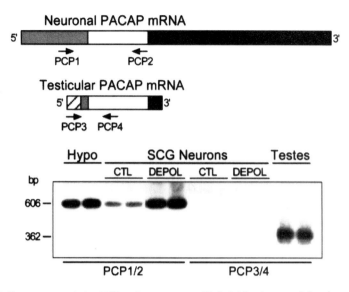

FIGURE 2. Reverse-transcription PCR and sequence-specific hybridization reveal that the smaller proPACAP transcript induced in SCG neurons is not the testicular form of PACAP mRNA. (*Top*) Schematic representation of the neuronal and testicular proPACAP transcripts demonstrating the positions of the neuron-specific (PCP1/PCP2) and testes-specific (PCP3/PCP4) oligonucleotide primer pairs used for the reverse transcription PCR amplification. *White*, coding region; *grey*, 5′ UTR; *black*, 3′ UTR; *hatched*, testes-specific 5′ UTR. (*Bottom*) Total RNA was isolated from individual SCG neuronal cultures treated for 96 h with 40 mM NaCl (CTL) or 40 mM KCl (DEPOL), hypothalamus, and testes. First strand cDNA was synthesized from total RNA using the SuperScript II Preamplification System (GIBCO-BRL) and amplified with primer pairs PCP1/PCP2 or PCP3/PCP4. The amplified products were separated on 1.6% agarose gels, transferred to Nytran-Plus membrane, and hybridized to a radiolabeled internal oligonucleotide probe corresponding to a segment within the proPACAP coding region. The expected amplified products are 606 base pairs for the neuronal transcript-specific primers PCP1/PCP2 and 362 base pairs for the testicular transcript-specific primers PCP3/PCP4. (Modified from Brandenburg *et al.*[1] and reproduced with permission.)

ferentiate between the neuronal PACAP transcript and the testicular form (FIG. 2). Amplification of control and depolarized cultured sympathetic neuronal cDNA using primers PCP1 and PCP2, specific for the neuronal transcript, resulted in the expected 606 bp product also observed in the hypothalamus (FIG. 2). Testicular cDNA, amplified using primers PCP3 and PCP4, specific to the testicular proPACAP transcript, produced the predicted product (FIG. 2). However, in key experiments using the heterologous pairing of primers and tissues, the testicular transcript-specific primers failed to amplify products from control or depolarized SCG neuronal cDNA (FIG. 2). Thus, the short proPACAP transcript induced in the depolarized sympathetic neurons was not the result of an induced testicular form of the message, but instead was related to changes in the neuronal PACAP message. The underlying mechanisms generating the short transcripts, however, were unclear but encompassed several possibilities, including alternative splicing of the proPACAP gene, transcription initiation at different start sites, and alternative usage of pre-mRNA cleavage and polyadenylation signals.

3′ RACE IDENTIFIES MULTIPLE PROPACAP TRANSCRIPTS IN SYMPATHETIC NEURONS, HYPOTHALAMUS, AND TESTES

We hypothesized that the shortened transcripts identified in the depolarized sympathetic neurons arose from alternative polyadenylation of the primary proPACAP transcript. Cleavage and polyadenylation occur post-transcriptionally downstream of a consensus polyadenylation hexamer signal.[17–19] While the rat proPACAP 3′ UTR does not contain the canonical AAUAAA polyadenylation signal, multiple potential variant signal sequences are present at sites that could produce the short and long transcripts.[20] Accordingly, shortened proPACAP transcript polyadenylation was investigated using rapid amplification of 3′ cDNA ends (3′ RACE) followed by cloning and sequencing of the 3′ RACE products. Polyadenylated RNA was prepared from control and depolarized sympathetic neurons, hypothalamus, and testicular tissues and reverse-transcribed using a lock-docking oligo dT primer (FIG. 3). Following second strand cDNA synthesis and creation of an oligonucleotide adaptor ligated double stranded (ds) cDNA library, 3′ RACE was performed using an upstream PACAP-specific primer and the downstream AP1 primer complementary to a segment of the ligated adaptor piece (FIG. 3). DNA sequencing confirmed alternative polyadenylation of the neuronal proPACAP transcript. In control and depolarized sympathetic neurons and brain hypothalamus, both long and short 3′ RACE products were identified (TABLE 1), corresponding to the two proPACAP transcript forms observed by

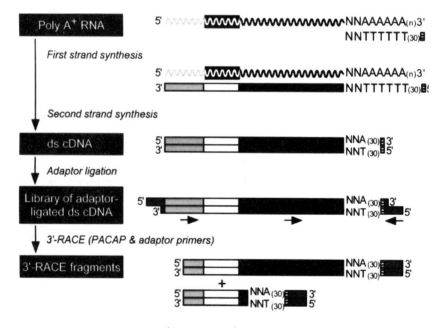

FIGURE 3. Rapid amplification of 3′ cDNA ends (3′ RACE) identifies alternative polyadenylation of proPACAP pre-mRNA leading to short and long transcripts. Poly(A⁺) RNA purified from control and depolarized SCG neurons, hypothalamus, and testes was used to generate first and second strand cDNA for ligation using the Marathon cDNA Amplification Kit (Clontech). 3′ RACE using the resulting double-stranded cDNA library was performed using a neuronal PACAP-specific primer together with the adaptor primer AP1, and the Expand High Fidelity PCR System (Boehringer Mannheim). The diagram illustrates PACAP cDNA: *light grey*, 5′ UTR; *white*, coding region; *black*, 3′ UTR; *dark grey*, oligonucleotide adaptor.

TABLE 1. Long and short proPACAP mRNA transcripts exhibit tissue-specific expression and differential stability.

ProPACAP Transcript	Tissue			Stability
	SCG	Hypo	Testis	
Long 3' UTR Neuronal PACAP mRNA 5'▇▇▇▇▇▇▇▇▇▇▇aaaaaa(n) 3'	+	+	–	<
Short 3' UTR Neuronal PACAP mRNA 5'▇▇▇▇▇▇▇aaaaaa(n) 3'	+	+	+	>
Testicular PACAP mRNA 5'◩▇▇▇▇aaaaaa(n) 3'	–	–	+	?

Northern analysis. In contrast to the neuronal tissues, amplification of testicular cDNA identified only short transcripts and failed to generate detectable levels of the long 3' RACE product, which was consistent with the absence of the larger PACAP mRNA transcripts in testes (FIG. 1 and TABLE 1).

A number of genes use alternative polyadenylation in the 3' UTR to generate different forms of mRNA.[17–19,21] Although the same product may be encoded, the resulting differences in the structure of the 3' UTR may impact on the stability, translation, or transport of mRNAs. Destabilizing elements in the 3' UTR are one of the primary determinants of mRNA stability. Accordingly, elimination of these elements through upstream alternative polyadenylation in the production of shortened transcripts should have significant impact on mRNA stability and define the ultimate amount of protein produced. Differences in the stability between short and long 3' UTR proPACAP transcripts were evaluated in depolarized sympathetic neurons. To examine proPACAP mRNA half-lives, transcription was inhibited with actinomycin D, RNA was extracted from the neurons at specific times following treatment, and the short and long proPACAP mRNA levels were examined by Northern analysis and quantitated by storage phosphor imaging. The half-life of the short proPACAP transcripts was extended, exhibiting greater stability than the corresponding longer mRNA (TABLE 1). Induction of short alternative polyadenylated 3' UTR proPACAP mRNA thus results in expression of more stable transcripts, which may be essential to increased PACAP synthesis required under different physiological conditions. In the testes for example, which express only short proPACAP transcripts, transcription initiation is limited in testicular germ cells, and increased mRNA stability with ensuing protracted translational processes has decided advantages.

CHANGES IN POLYADENYLATION FACTOR EXPRESSION ARE CORRELATED WITH PROPACAP PRE-mRNA POLYADENYLATION SITE USAGE

One mechanism that determines alternative polyadenylation site usage relies on the relative expression of polyadenylation factors.[21] Two components, a cleavage and polyadenylation specificity factor (CPSF) and a cleavage stimulatory factor (CstF), are essential to the processing reaction.[17–19] CPSF is the only factor capable of binding to the pre-mRNA upon recognition of the hexamer polyadenylation signal, while CstF binds to downstream GU-rich elements. The interactions between these factors form part of a multi component complex that also includes cleavage factors and poly(A) polymerase. Assembly of the complex facilitates mRNA cleavage downstream of the consensus hexamer and addition of the poly(A) tract at the 3′ end (FIG. 4). Processing events in the 3′ UTR appear to depend in part on levels of polyadenylation factors and their relative binding efficiencies at competing alternative polyadenylation sites. Since the primary polyadenylation site recognition proteins are CPSF and CstF, we have begun to explore whether changes in expression of these factors may play a role in proPACAP mRNA polyadenylation site usage. Polyadenylation factor protein and mRNA expression were examined in control and depolarized sympathetic SCG neurons using Western analysis and reverse-transcription PCR. Both the protein and mRNA levels of the 64 kD subunit of CstF (CstF-64) were decreased in depolarized sympathetic neurons compared to control cells (FIG. 4). In contrast, protein levels of the 160 kD subunit of CPSF (CPSF-160) were augmented. These results suggest that specific ratios of polyadenylation factor levels could influence alternative polyadenylation site usage; changes in the expression and binding of these factors to the short and long 3′ UTR

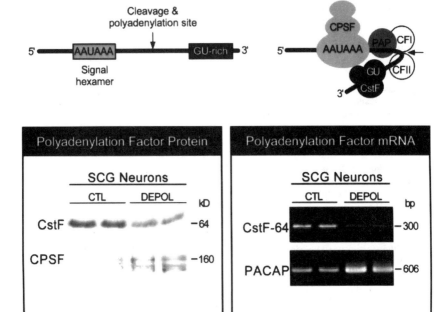

FIGURE 4. Changes in polyadenylation factor expression are correlated with proPACAP pre-mRNA polyadenylation site choice. (*Top*) Several elements within a precursor mRNA 3′ UTR are involved in cleavage and polyadenylation, including the signal sequence hexamer and downstream GU-rich regions. The cleavage and polyadenylation specificity factor (CPSF) recognizes the polyadenylation signal, whereas the cleavage stimulation factor (CstF) binds the GU-rich element. *Arrows*, sites of cleavage; PAP, poly(A) polymerase; CF, cleavage factor. (*Bottom, left*) Protein levels of the 160 kD subunit of CPSF and the 64 kD subunit of CstF were analyzed in control (CTL) and depolarized (DEPOL) SCG neurons by SDS-PAGE and Western blot. (*Bottom, right*) Messenger RNA from control (CTL) and depolarized (DEPOL) neurons was reverse-transcribed and amplified by PCR using primers for the 64 kD subunit of CstF (CstF-64) or PACAP.

proPACAP polyadenylation sites may contribute to polyadenylation site usage with important downstream ramifications in mRNA stability and peptide biosynthesis.

SUMMARY

Many regulated events guide neuropeptide biosynthesis, processing, and secretion. For PACAP peptides, these events have not been well examined. In our studies of PACAP expression in sympathetic neurons, we discovered that neuronal depolarization not only increased the levels of the 2.2 kb form of proPACAP mRNA identified in neuronal tissues, but also induced a novel 0.9 kb PACAP transcript, which appeared similar in size to a form present in testes. Using reverse-transcription PCR and 3′ RACE studies, we demonstrated that the 0.9 kb PACAP mRNA in depolarized SCG neurons was not identical to the testicular PACAP mRNA, but represented shortened, more stable, forms of the 2.2 kb transcript resulting from alternative upstream polyadenylation site usage. These results demonstrate that post-transcriptional mechanisms play important roles in determining cellular PACAP levels and provide several important insights. For example, alternative upstream polyadenylation can elicit a major influence on the amount of bioactive peptide that can be synthesized, since short 3′ UTR transcripts are usually more stable due to elimination of destabilizing elements present in the longer messages. In cells such as testicular germ cells, which have restricted transcriptional periods, stable mRNAs allow longer translational events and extended periods of peptide production. The neuronal PACAP system adopts a similar post-transcriptional strategy following neuronal depolarization, and although the roles of PACAP remain unclear, this suggests important roles for PACAP peptides during increased neuronal activity. Additionally, unlike alternative polyadenylation described for many genes, alternative site usage in the proPACAP transcript does not result from alternative splicing. The mechanism of alternative site usage may be related to changes in the expression and binding of polyadenylation factors to the short and long 3′ UTR proPACAP sites leading to production of more stable transcripts and increased PACAP precursor biosynthesis. The implications of increased PACAP production following altered neurophysiological states and the mechanisms underlying alternative polyadenylation site choice are important considerations for future inquiries.

REFERENCES

1. BRANDENBURG, C. A., V. MAY & K. M. BRAAS. 1997. Identification of endogenous sympathetic neuron pituitary adenylate cyclase activating polypeptide (PACAP): depolarization regulates production and secretion through induction of multiple propeptide transcripts. J. Neurosci. **17**: 4045–4055.
2. SUNDLER, F., E. EKBLAD, J. HANNIBAL, K. MOLLER, Y-Z. ZHANG, H. MULDER, T. ELSAS, T. GRUNDITZ, N. DANIELSEN, J. FAHRENKRUG & R. UDDMAN. 1996. Pituitary adenylate cyclase activating polypeptide in sensory and autonomic ganglia: localization and regulation. Ann. N. Y. Acad. Sci. **805**: 410–428.
3. ZHANG, Y-Z., J. HANNIBAL, Q. ZHAO, K. MOLLER, N. DANIELSEN, J. FAHRENKRUG & F. SUNDLER. 1996. Pituitary adenylate cyclase activating peptide (PACAP) expression in the rat dorsal root ganglia. Upregulation after peripheral nerve injury. Neuroscience **74**: 1099–1110.
4. MOLLER, K., M. REIMER, J. HANNIBAL, J. FAHRENKRUG, F. SUNDLER & M. KANJE. 1997. Pituitary adenylate cyclase activating peptide (PACAP) and PACAP type I receptor expression in regenerating adult mouse and rat superior cervical ganglia in vitro. Brain Res. **75**: 156–165.
5. MAY, V., C. A. BRANDENBURG & K. M. BRAAS. 1996. Axotomy and decentralization regulate pituitary adenylate cyclase activating polypeptide (PACAP) expression in rat superior cervical ganglion. Abstr. Soc. Neurosci. **22**: 1764.
6. DICICCO-BLOOM, E. 1996. Region-specific regulation of neurogenesis by VIP and PACAP: direct and indirect modes of action. Ann. N. Y. Acad. Sci. **805**: 244–258.

7. OHKUBO, S., C. KIMURA, K. OGI, K. OKAZAKI, M. HOSOYA, H. ONDA, A. MIYATA, A. ARIMURA & M. FUJINO.1992. Primary structure and characterization of the precursor to human pituitary adenylate cyclase activating polypeptide. DNA Cell Biol. **11**: 21–30.

8. GHATEI, M. A., K. TAKAHASI, Y. SUZUKI, J. GARDINER, P. M. JONES & S. M. BLOOM. 1993. Distribution, molecular characterization of pituitary adenylate cyclase activating polypeptide and its precursor endocine mRNA in human and rat tissues. J. Endocrinol. **136**: 159–166.

9. HURLEY, J. D., J. V. GARDINER, P. M. JONES & S. M. BLOOM. 1995. Cloning and molecular characterization of complementary deoxyribonucleic acid corresponding to a novel form of pituitary adenylate cyclase activating polypeptide messenger ribonucleic acid in the rat testis. Endocrinology **136**: 550–557.

10. TOBIN G., A. ASZTELY, V. EDWARDS, J. EKSTROM, R. HAKANSON & F. SUNDLER. 1995. Presence and effects of pituitary adenylate cyclase activating peptide in the submandibular gland of the ferret. Neuroscience **66**: 227–235.

11. ELSAS, T., R. UDDMAN, H. MULDER & F. SUNDLER. 1997. Pituitary adenylate cyclase activating polypeptide and nitric oxide synthase are expressed in the rat ciliary ganglion. Brit. J. Ophthalmol. **81**: 223–227.

12. MULDER, H., R. UDDMAN, K. MOLLER, T. ELSAS, E. EKBLAD, J. ALUMETS & F. SUNDLER. 1995. Pituitary adenylate cyclase activating polypeptide is expressed in autonomic neurons. Regul. Pept. **59**: 121–128.

13. HARDWICK, J. C., K. M. BRAAS, S. A. HARAKALL, V. MAY & R. L. PARSONS. 1997. Endogenous pituitary adenylate cyclase activating polypeptide (PACAP) regulates guinea pig cardiac neuron excitability through PACAP$_1$-selective receptors. Abstr. Soc. Neurosci. **23**: 428.

14. KLIMASCHEWSKI, L., C. HAUSER & C. HEYM. 1996. PACAP immunoreactivity in the rat superior cervical ganglion in comparison to VIP. Neuroreport **7**: 2797–2801.

15. MOLLER, K., M. REIMER, J. HANNIBAL, J. FAHRENKRUG, F. SUNDLER & M. KANJE. 1997. Pituitary adenylate cyclase activating peptide (PACAP) and PACAP type I receptor expression in regenerating adult mouse and rat superior cervical ganglia in vitro. Brain Res. **75**: 156–165.

16. MOLLER, K., M. REIMER, E. EKBLAD, J. HANNIBAL, J. FAHRENKRUG, M. KANJE & F. SUNDLER. 1997. The effects of axotomy and preganglionic denervation on the expression of pituitary adenylate cyclase activating peptide (PACAP), galanin and PACAP type I receptors in the rat superior cervical ganglion. Brain Res. **75**: 166–182.

17. PROUDFOOT, N. 1996. Ending the message is not so simple. Cell **87**: 779–781.

18. WAHLE, E. 1995. 3' end cleavage and polyadenylation of mRNA precursors. Biochim. Biophys. Acta **1261**: 183–194.

19. WAHLE, E. & W. KELLER. 1996. The biochemistry of polyadenylation. TIBS **21**: 247–251.

20. OGI, K., C. KIMURA, H. ONDA, A. ARIMURA & M. FUJINO. 1990. Molecular cloning and characterization of cDNA for the precursor of rat pituitary adenylate cyclase activating polypeptide (PACAP). Biochem. Biophys. Res. Commun. **173**: 1271–1279.

21. TAKAGAKI, Y., R. L. SEIPELT, M. L. PETERSON & J. L. MANLEY. 1996. The polyadenylation factor CstF-64 regulates alternative processing of IgM heavy chain pre-mRNA during B cell differentiation. Cell **87**: 941–952.

A Model of the Receptors in the VIP Receptor Family

J. W. TAMS,[a] S. M. KNUDSEN, AND J. FAHRENKRUG

Department of Clinical Biochemistry, Bispebjerg Hospital, University of Copenhagen, DK-2400 Copenhagen NV, Denmark

A three-dimensional (3D) model of the VIP receptor family (VRF) is important for the understanding of the molecular mechanisms of ligand binding and receptor activation. It will probably take many years before a resolute 3D structure of a member of the VRF will be revealed by electron microscopy or X-ray diffraction. However, alignment of homologous sequences can give hints to the 3D structure of the helical arrangement of transmembrane helices.

RESULTS AND DISCUSSION

Transmembrane α-Helices

The free energy of transferring a side chain from a random coil conformation in water to an α-helix conformation in a lipophilic phase,[1] using a window size of 18, was used to set the middle position of a transmembrane α-helix (TMH). In order to have a comparable numbering scheme, each TMH is set to have a length of 26 residues (FIG. 1). The most external positions of each TMH are not necessarily aligned properly or arranged in an α-helical conformation. At all positions in the alignment, the following parameters were determined: (1) mean free energy for transferring the side chain from a random coil conformation in water to an α-helix conformation in a lipophilic phase, (2) maximum and minimum volume of the side chain,[2] and (3) number of strong hydrophilic residues that would be expected not to be in contact with lipid (Asp, Asn, Glu, Gln, His, Arg, and Lys).[3]

Orientation of the Seven TMH of the VIP Receptor Family

The features of each TMH are summarized by plotting them around helical wheels (FIG. 2). We have initially used the topography of the proposed 3D structure of bovine rhodopsin, which has been determined by electron microscopy,[4] as a template for the seven TMH. The final arrangement of the helices is then fitted so the extent of exposure to the lipid of a helix is determined by the size of its lipophilic face. If TMH are put in order of decreasing lipid face the order is: IV > I > V > VI > VII > II > III. Positions numbered 8–19 (FIG. 2) have been used to orient the seven TMH, so conserved, hydrophilic residues and positions with restricted volume changes face the other TMH and positions with low identity and large volume change are at the lipid face or, to a lesser extent, at the inter-helical cavities.

[a]Corresponding author: Tel.: 45-35312645; Fax: 45-35313955; E-mail: jwtams@biobase.dk

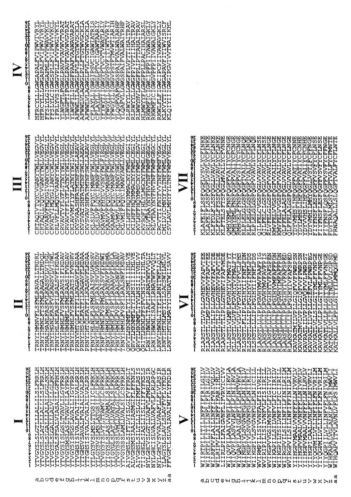

FIGURE 1. Alignment of sequences of the transmembrane helices (TMH) of the VIP receptor family (VRF). Sequences with more than 90% identity to other members of the family were removed from the alignment to avoid bias from subfamilies with many sequences. Gaps were removed near the TMH to give an unbroken block of sequences with 26 residues. Sequences used in the alignment are hamster gastric inhibitory peptide receptor (a), rat gastric inhibitory peptide receptor (b), human glucagon receptor (c), rat glucagon receptor (d), human glucagon-like peptide 1 receptor (e), opossum parathyroid hormone receptor (f), rat parathyroid hormone receptor (g), human PTH2 receptor (h), human growth hormone-releasing hormone receptor (i), pig growth hormone-releasing hormone receptor (j), rat growth hormone-releasing hormone receptor (k), rat vasoactive intestinal peptide receptor 2 (l), human vasoactive intestinal peptide receptor 2 (m), human pituitary adenylate cyclase activating polypeptide receptor (n), rat vasoactive intestinal peptide receptor 1 (o), human vasoactive intestinal peptide receptor 1 (p), human secretin receptor (q), rat secretin receptor (r), human calcitonin receptor (s), rat calcitonin receptor (t), pig calcitonin receptor (u), rat calcitonin-like receptor (v), gallus gallus corticotrophin releasing factor receptor (w), human corticotrophin releasing factor receptor (x), rat corticotrophin releasing factor receptor subtype 2 (y), manduca sexta diuretic hormone receptor (z), acheta domesticus diuretic hormone receptor (aa).

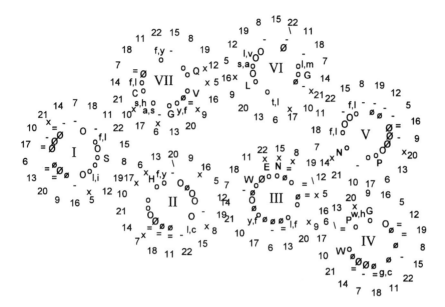

FIGURE 2. Arrangement of the seven transmembrane helices (TMH) of the VIP receptor family (VRF). The arrangement of the TMH, with the middle 18 residues, is viewed from the extracellular side of the TMH. The orientation is accomplished by turning the individual TMH to an optimal fit of the helix- and lipid-facing positions in the middle span of 12 residues. Helix-facing positions are positions that have (1) small variation in side chain volume, below 15 Å³ (o) or below 26 Å³ (O), (2) conserved residues, or (3) can accommodate strong hydrophilic residues, less that 10% (\) or more than 10% (x). Lipid-facing positions are positions that can accommodate large volume variation, above 65 Å³ (ø) or above 86 Å³ (Ø), and have less than 30% (-) or less than 20% (=) identity. Positions at cavities in the interior of the seven TMH have to some extent the same characteristics as lipid-facing positions, with respect to the ability to accommodate large volume change and low identity, but these positions are also able to accommodate strong hydrophilic residues.

REFERENCES

1. VON HEIJNE, G. & BLOMBERG, C. 1979. Trans-membrane translocation of proteins. The direct transfer model. Eur. J. Biochem. **97**: 175–181.
2. RICHARDS, F. M. 1974. The interpretation of protein structures: total volume, group volume distributions and packing density. J. Mol. Biol. **82**: 1–14.
3. BALDWIN, J. M. 1993. The probable arrangement of the helices in G protein-coupled receptors. EMBO J. **12**: 1693–1703.
4. SCHERTLER, G. F. X. & P. A. HARGRAVE. 1995. Projection structure of frog rhodopsin in two crystal forms. Proc. Natl. Acad. Sci. USA **92**: 11578–11582.

Site-Directed Mutagenesis of Human VIP1 versus VIP2 Receptors

PASCAL NICOLE, KAI DU, ALAIN COUVINEAU, AND
MARC LABURTHE[a]

*Laboratoire de Neuroendocrinologie et Biologie Cellulaire Digestives,
Institut National de la Santé et de la Recherche Médicale, INSERM
U410, Faculté de Médecine Xavier Bichat, B.P. 416, 75870 Paris Cedex
18, France*

VIP receptors belong to the class II subfamily of G protein–coupled receptors.[1] Two VIP receptor subtypes have been identified by molecular cloning and are now referred to as VIP1 and VIP2 receptors.[1] In humans, these subtypes have 49% amino acid identity and mediate activation of adenylyl cyclase in their target tissues.[1] Both subtypes have a high affinity for VIP and PACAP and there is not yet evidence for major differences in their pharmacological profiles with respect to recognition of natural agonists of the VIP family of peptides.[1–5] However, anatomical mapping supported that VIP1 and VIP2 receptors may have complementary tissue distribution.[6]

Previous studies of the human VIP1 receptor by site-directed mutagenesis and/or construction of receptor chimeras provided evidence for an important role of the N-terminal extracellular domain for VIP binding with several crucial residues probably positioned in a tertiary structure maintained by multiple disulfide bonds.[1,7,8] This domain also contains two N-glycosylation sites necessary for correct delivery of the receptor to the plasma membrane.[9] Finally, the first and second extracellular loops are also involved in VIP binding[8,10] or pharmacological selectivity of the receptor.[11]

The rationale of this study is based on the idea that residues conserved in VIP1 and VIP2 receptors, but poorly or not at all conserved in other members of the class II family of G protein–coupled receptors, may be essential for VIP binding. Since the N-terminal extracellular domain and extracellular loops appear to be crucial in VIP binding to the VIP1 receptor, we selected and mutated such amino acid residues in extracellular domains. This approach allowed us to characterize a new crucial residue (glutamate) in the N-terminal domain of both receptors. Moreover, quite unexpectedly, we also characterized three conserved residues in the VIP1 or VIP2 receptor, the mutation of which decreased the affinity of VIP for a VIP receptor subtype but not for the other subtype.

RESULTS AND CONCLUSION

Residues conserved in extracellular domains of VIP1 and VIP2 receptors but poorly or not at all conserved in other members of the class II G protein–coupled receptor family were found in the N-terminal extracellular domain (E36, I43, S64, D132, and F138 in the VIP1 receptor corresponding, respectively, to E24, I31, S53, D116, and F122 in the VIP2

[a]Corresponding author: Tel.: 33 01 44 85 61 35; Fax: 33 01 44 85 61 24; E-mail: laburthe@bichat.inserm.fr

receptor) and the second extracellular loop (T288 and S292 in the VIP1 receptor corresponding respectively to T274 and S278 in the VIP2 receptor). All these residues have been mutated into alanine by site-directed mutagenesis and the mutated cDNAs transfected into Cos cells for subsequent functional studies. Scatchard analysis of competitive inhibition of ^{125}I-VIP binding to transfected cells by unlabeled VIP gave straight lines for the wild-type VIP1 receptor and all VIP1 receptor mutants (I43A, S64A, D132A, F138A, T288A, S292A) but one (E36A) for which no specific ^{125}I-VIP binding could be detected (TABLE 1). Dissociation constants were similar for the wild-type receptor and all mutants that bound ^{125}I-VIP but one (S64A), for which a significant threefold increase of the dissociation constant was observed (TABLE 1). Binding studies were then performed for VIP2 receptor mutants (TABLE 1). Scatchard analysis gave straight lines for the wild-type receptor and VIP2 receptor mutants S53A, D116A, F122A, and S278A which had K_d similar to that of the wild-type VIP2 receptor (TABLE 1). In contrast, the K_d of I31A and T274A mutants were 11 and 5 times higher than the K_d of the wild-type VIP2 receptor. No specific binding could be detected with the mutant E24A. Confocal laser microscopy immunofluorescence studies of the tagged mutants showed that all mutants of VIP1 and VIP2 receptors were expressed by Cos cells in the same way as the wild-type VIP1 or VIP2 receptors (not shown).

Further experiments (not shown) were designed to study VIP-stimulated cAMP production in cells transfected with wild-type and mutated receptors. VIP (1 μM) promoted cAMP production with similar efficacies in wild-type VIP1 receptor and all mutants but one (E36A). There was therefore a good correlation between binding data and cAMP production. With regard to the E36A VIP1 receptor mutant, VIP displayed a very low efficacy, if any, for stimulating cAMP production. We further investigated the S64A VIP1 receptor mutant that exhibited an increased dissociation constant for VIP (TABLE 1) by performing dose-reponse experiments of VIP in stimulating cAMP production. It appeared that half-maximal stimulation was obtained for 0.3 ± 0.1 and 0.9 ± 0.1 nM VIP for the wild-type receptor and the S64A receptor mutant, respectively. The shift of potency of VIP in stimulating cAMP production through the S64A mutant is therefore identical to the shift of affinity of VIP for binding to this mutant (TABLE 1). Further studies showed that VIP (1 μM) promoted cAMP production with similar efficacies in wild-type VIP2 receptor and all mutants but one (E24A). VIP displayed a very low efficacy, if any, for stimulating cAMP through the E24A mutant. The dose-responses of VIP for stimulating cAMP production through the I31A and T274A mutants were shifted to the right as compared to the wild-type receptor. Half-maximal stimulation was obtained for 1.9 ± 0.6, 66 ± 16, and 16 ± 6 nM VIP for the wild-type receptor, and the I31A and T274A receptor mutants, respectively. This important shift of potency was similar to the shift of affinity measured in binding studies (TABLE 1).

The present site-directed mutagenesis study provides new information regarding the structure-function relationship of recombinant human VIP1 and VIP2 receptor subtypes. Although mutagenesis studies, such as the present one, can not determine whether the effects of mutations on receptor phenotypes are direct or indirect via allosteric mechanisms,[12] this work supports for the first time that human VIP receptor subtypes that had been identified owing to molecular cloning and have a very similar pharmacological profile do exhibit differences in the structure-function relationship with respect to VIP binding. The differences involve specific residues present in the N-terminal extracellular domain and also the second extracellular loop.

Our study also points out a new crucial amino acid residue in the N-terminal extracellular domain of the human VIP1 receptor for VIP binding and subsequent stimulation of cAMP production, i.e., glutamate 36. This further emphasizes the importance of the N-terminal extracellular domain of the human VIP1 receptor for VIP binding with four crucial residues [glutamate 36 (this paper), aspartate 68, tryptophan 73, and glycine 109] probably

TABLE 1. Binding parameters of wild-type and mutant human VIP1 and VIP2 receptors expressed in Cos cells

Constructs		Dissociation Constant (nM)	Binding Capacity (pmoles/mg of protein)
VIP1	wt	1.4 ± 0.7	4.14 ± 0.33
	E36A	ND	ND
	I43A	1.3 ± 0.3	7.07 ± 0.71
	S64A	$4.4 \pm 0.5^*$	7.52 ± 0.76
	D132A	1.7 ± 0.2	1.56 ± 0.31
	F138A	1.3 ± 0.9	1.36 ± 0.38
	T288A	1.3 ± 0.5	1.22 ± 0.15
	S292A	3.1 ± 1.1	5.19 ± 0.84
VIP2	wt	1.8 ± 0.9	1.3 ± 0.2
	E24A	ND	ND
	I31A	$20.2 \pm 2.9^*$	4.1 ± 1.1
	S53A	1.3 ± 0.4	1.9 ± 0.2
	D116A	1.7 ± 0.3	1.7 ± 0.3
	F122A	1.4 ± 0.3	0.7 ± 0.4
	T274A	$8.5 \pm 0.5^*$	3.9 ± 0.8
	S278A	1.0 ± 0.5	2.5 ± 0.2

Note: Binding parameters were determined by Scatchard analysis of competitive inhibition of specific ^{125}I-VIP binding by increasing concentrations of native VIP. The concentration of ^{125}I-VIP was 0.05 nM except for the mutants E36A (VIP1) and E24A, I31A, and T274A (VIP2) for which 0.4 nM was used. ND, not detectable, wt, wild-type. Results are mean $\pm$ SE of three experiments.
* Values significantly different, $p < 0.01$, from that of corresponding wt.

positioned in a tertiary functional structure maintained by multiple disulfide bridges formed by six cysteine residues.[1] The presence of two important acidic residues (glutamate 36 and aspartate 68) in the N-terminal VIP binding domain is consistent with the unusual isoelectric point (> 11) of VIP and the importance of many basic residues of VIP for biological activity (author's unpublished results). The importance of tryptophan 73 is also consistent with previous work suggesting the role of hydrophobic interactions in VIP binding to receptors.[13]

Finally, this paper provides the first site-directed mutagenesis study of VIP2 receptors. Our study indicates that both the N-terminal extracellular domain (glutamate 24 and isoleucine 31) and second extracellular loop (threonine 274) of the human VIP2 receptor may be involved in VIP binding. It will be interesting to determine whether residues previously reported to be important for VIP binding to the human VIP1 receptor in the N-terminal extracellular domain[1,7–11] are also important for VIP binding to the VIP2 receptor subtype.

In conclusion, current knowledge indicates that the VIP2 receptor is distinct in sequence,[2–5] distribution,[6] and structure-function relationship (this paper) from the VIP1 receptor. However, no natural or synthetic ligand selective for VIP receptor subtypes has yet been described. In this context, the respective physiological roles of VIP1 and VIP2 receptors are still conjectural. The availability of comparative studies of the structure-function relationship of VIP receptor subtypes such as this one will facilitate the development of selective agonists and antagonists and ultimately will contribute to a better knowledge of their physiological roles.

REFERENCES

1. LABURTHE, M., A. COUVINEAU, P. GAUDIN, J. J. MAORET, C. ROUYER-FESSARD & P. NICOLE. 1996. Ann. N. Y. Acad. Sci. **805**: 94–111.

2. COUVINEAU, A., C. ROUYER-FESSARD, D. DARMOUL, J. J. MAORET, I. CARRERO, E. OGIER-DENIS & M. LABURTHE. 1994. Biochem. Biophys. Res. Commun. **200**: 769–776.
3. LUTZ, E. M., W. J. SHEWARD, K. M. WEST, J. A. MORROW, G. FINK & A. J. HARMAR. 1993. FEBS Lett. **334**: 3–8.
4. SVOBODA, M., M. TASTENOY, J. VAN RAMPELBERGH, J. F. GOOSENS, P. DE NEEF, M. WAELBROECK & P. ROBBERECHT. 1994. Biochem. Biophys. Res. Commun. **205**: 1617–1624.
5. ADAMOU, J. J., N. AIYAR, S. VAN HORN & N. A. ELSHOURBAGY. 1995. Biochem. Biophys. Res. Commun. **209**: 385–392.
6. USDIN, T. B., T. I. BONNER & E. MEZEY. 1994. Endocrinology **135**: 2662–2680.
7. COUVINEAU, A., P. GAUDIN, J. J. MAORET, C. ROUYER-FESSARD, P. NICOLE & M. LABURTHE. 1995. Biochem. Biophys. Res. Commun. **206**: 246–252.
8. GAUDIN, P., A. COUVINEAU, J. J. MAORET, C. ROUYER-FESSARD & M. LABURTHE. 1995. Biochem. Biophys. Res. Commun. **211**: 901–908.
9. COUVINEAU, A., C. FABRE, P. GAUDIN, J. J. MAORET & M. LABURTHE. 1996. Biochemistry **35**: 1745–1752.
10. DU, K., P. NICOLE, A. COUVINEAU & M. LABURTHE. 1997. Biochem. Biophys. Res. Commun. **230**: 289–292.
11. COUVINEAU, A., C. ROUYER-FESSARD, J. J. MAORET, P. GAUDIN, P. NICOLE & M. LABURTHE. 1996. J. Biol. Chem. **271**: 12795–12800.
12. FONG, T. M., M. R. CANDELORE & C. D. STRADER. 1995. *In* Receptor Molecular Biology. S. C. Sealfon, Ed.: 263–277. Academic Press. San Diego.
13. BODANSZKY, M., A. BODANSZKY, Y. S. KLAUSNER, & S. I. SAID. 1974. Bioorg. Chem. **3**: 1185–1193.

Constitutive Activation of the Human VIP1 Receptor

PASCALE GAUDIN, CHRISTIANE ROUYER-FESSARD, ALAIN
COUVINEAU, JEAN-JOSE MAORET, AND MARC LABURTHE[a]

*Laboratoire de Neuroendocrinologie et Biologie Cellulaire Digestives,
Institut National de la Santé et de la Recherche Médicale, INSERM
U410, Faculté de Médecine Xavier Bichat, B.P. 416, 75870 Paris Cedex
18, France*

During the past few years a subfamily of the superfamily of G protein–coupled receptors has emerged that shares the seven membrane-spanning domain topography but has a low overall amino acid sequence homology (<20%) with other members of the superfamily.[1,2] This subfamily, now referred to as the class II G protein–coupled receptor family, comprises receptors for a family of structurally related peptides that includes vasoactive intestinal peptide (VIP), pituitary adenylate cyclase-activating polypeptide (PACAP), glucagon, secretin, glucagon-like peptide (GLP-1), gastric inhibitory polypeptide (GIP), growth hormone-releasing peptide (GRF), and, more unexpectedly, also comprises receptors for parathyroid hormone (PTH) and calcitonin.[1,2] Recent studies have extended this subfamily[1] with the discovery of subtypes of the abovementioned receptors as well as two new members having an extraordinarily long N-terminal domain: the putative EGF module-containing, mucin-like hormone receptor EMR1 and the leukocyte activation antigen CD97.

Constitutively active mutants of several G protein–coupled receptors have been characterized experimentally by site-directed mutagenesis[3] and have been also described as pathogenic in humans.[4,5] A constitutively active mutant of a class II G protein receptor has been reported for the PTH-PTH-related peptide receptor in Jansen-type metaphyseal chondrodysplasia that affects a strictly conserved histidine residue in the first intracellular loop of this class of receptor.[6]

In this context, we investigated the constitutive activation of the human VIP1 receptor, a class II G protein–coupled receptor whose structure-function relationship has been previously documented.[1] In this paper, we demonstrate that specific mutation of histidine 178 in the VIP receptor causes its constitutive activation and show that constitutive activation requires the integrity of the natural ligand binding site in the N-terminal extracellular domain.

RESULTS AND CONCLUSION

Histidine (H) 178 has been mutated into arginine (mutant H178R) by site-directed mutagenesis and the mutated cDNA transfected into Cos cells for subsequent intracellular cAMP measurement. The basal cAMP level was 3.5-fold higher in cells transfected with H178R than the basal cAMP level measured after transfection of the wild type human VIP1 receptor (TABLE 1) or the vector alone (not shown). This latter observation suggests

[a]Corresponding author: Tel.: 33 01 44 85 61 35; Fax: 33 01 44 85 61 24; E-mail: laburthe@
bichat.inserm.fr

TABLE 1. Basal cyclic AMP level and VIP binding parameters of wild type and mutated human VIP1 receptors expressed in Cos cells

Constructs	cAMP Basal Level (pmoles/10⁶ cells)	Binding Parameters	
		Dissociation Constant (nM)	Binding Capacity (pmoles/mg of protein)
wt	18.8 ± 3.7	0.6 ± 0.1	6.1 ± 1.5
H178R	67.4 ± 9.9*	0.2 ± 0.1	2.0 ± 0.9
H178A	25.8 ± 9.5	ND	ND
H178D	18.3 ± 5.3	ND	ND
H178K	13.4 ± 4.6	ND	ND
H178R-D68A	20.1 ± 5.5	ND	ND
D68A	16.4 ± 5.3	ND	ND
H178R-E36A	22.2 ± 2.6	ND	ND
E36A	18.8 ± 2.6	ND	ND
H178R-D132A	52.6 ± 4.4*	1.1 ± 0.6	3.0 ± 0.6
D132A	22.2 ± 4.8	0.7 ± 0.1	1.6 ± 0.3

Note: Cyclic AMP levels were measured by radioimmunoassay as described.[7] Binding parameters were determined by Scatchard analysis of competitive inhibition of specific ^{125}I-VIP binding by increasing concentrations of native VIP.[7] ND, not detectable; wt, wild type. Results are mean ± SE of at least three experiments. * Values significantly different, $p < 0.01$, from that of wt.

that the wild type receptor was not itself constitutively activated. The cAMP response in Cos cells transfected by H178R mutant and wild type receptors was tested upon stimulation by VIP. Maximal cAMP responses were identical and half-maximal stimulations above basal level were obtained for similar concentrations of VIP in cells transfected with the H178R receptor or the wild type receptor, i.e., $1.5\pm0.5 \times 10^{-10}$ M and $0.4\pm0.1 \times 10^{-10}$ M, respectively (not shown). Scatchard analysis of competitive inhibition of ^{125}I-VIP binding to Cos cell membranes by native VIP indicated that the H178R receptor mutant bound VIP with a similar dissociation constant as compared to the wild type receptor (TABLE 1). In these experiments the concentration of VIP binding sites was higher in cells transfected with the wild type receptor than in cells transfected with the H178R receptor mutant (TABLE 1). This observation did not favor the hypothesis that the higher basal cAMP level in cells transfected with H178R could be merely related to a higher expression of the receptor as compared to cells transfected with the wild type receptor inasmuch as the wild type receptor is not constitutively activated by itself (see above). However, in order to document this issue, we transfected Cos cells with increasing concentrations of cDNA encoding the H178R mutant or the wild type receptor. For the receptor mutant, we observed that the more cDNA transfected the more cAMP level in Cos cells (not shown). In sharp contrast, the basal cAMP level in Cos cells transfected with the wild type receptor was constant regardless of the amount of cDNA transfected. This latter observation further argued against constitutive activity of the wild type human VIP1 receptor itself.

Further experiments were carried out to determine whether mutation of H178 into residues other than arginine also resulted in constitutive activation of the human VIP1 receptor. We mutated H178 into a neutral residue (H178A), an acidic one (H178D), or a basic one (H178K). Among these mutants, none was constitutively activated after transfection in Cos cells (TABLE 1). Moreover, these mutants were unable to mediate VIP-stimulated cAMP production (not shown) likely because they no longer bound VIP (TABLE 1). In order to determine the pattern of expression of these inactive mutants in transfected Cos cells, immunofluorescence studies were performed after insertion of a Flag at the intracellular C-terminus of the mutated and wild type receptors as previously described.[7] Data indicated that the mutants H178A, H178D, and H178K, which no longer bound VIP, exhibited the same pattern of expression as the wild type receptor and the mutant H178R (not shown). As expected for proteins in an active phase of synthesis in transfected cells,[7,8] immunofluorescence could be detected at the plasma membrane and also in intracellular compartments.

That constitutive activation of the human VIP1 receptor by mutation of H178 into arginine did not happen with mutations into alanine, aspartate, and even lysine and that such mutations also abolished VIP binding suggested that such constitutive activation might be dependent on the integrity of the VIP binding site. In order to investigate this issue we constructed double mutants in which the mutation resulting in the constitutive activation of the receptor (H178R) was associated with point mutations in the N-terminal extracellular domain of the receptor, which abolished VIP binding as shown, i.e., D68A[9] or E36[10]. As expected the H178R-D68A mutant no longer bound VIP like the single mutant D68A (TABLE 1). Nor did it mediate the stimulation of cAMP production by VIP (not shown). TABLE 1 further shows that this double mutant is no longer constitutively activated when expressed in Cos cells. Like the single mutant E36A, the H178R-E36A mutant no longer bound VIP (TABLE 1) and did not mediate the stimulation of cAMP production by VIP (not shown). The H178R-E36A double mutant was not constitutively activated when expressed in Cos cells (TABLE 1). As a control, we developed another double mutant in which H178R was associated with a point mutation (D132A) in the N-terminal extracellular domain which was previously shown not to alter VIP binding.[10] As expected, the H178R-D132A mutant, like the single mutant D132A, bound VIP with a dissociation constant similar to that of the wild type receptor (TABLE 1). This mutant also mediated VIP-stimulated cAMP production (not shown). Interestingly the double mutant H178R-D132A did exhibit constitutive activation upon transfection in Cos cells (TABLE 1).

Constitutive activation evoked by replacement of histidine 178 by arginine in the human VIP1 receptor is much less efficient than activation induced by the natural agonist VIP in the H178R mutant receptor or the wild type receptor. Nevertheless, the construction of double mutants supports a close relationship between the ligand-dependent activation triggered by VIP and the ligand-independent one evoked by mutation of histidine 178 into arginine. Indeed, our original approach consisting of constructing double mutants suggests that the conformational change triggering constitutive activation in the H178R receptor mutant may transit over the ligand binding site in the N-terminal extracellular domain and requires the structural and functional integrity of this domain. This double mutation approach, which invalidates the agonist binding domain, was feasible for the VIP receptor because the binding site in the N-terminal extracellular domain[1] is well separated from the site of the mutation evoking constitutive activation.

Since histidine 178 in the human VIP1 receptor is strictly conserved in all members of the class II family of receptors, it may be of general functional importance in these G protein-coupled receptors. Whether mutation of the equivalent histidine in other peptide receptor of this family results in ligand-independent activation has been only reported for the PTH receptor.[6] In this context, it is worth pointing out that two members of this family are receptors having an extraordinary long N-terminal extracellular domain with unique features: the putative EGF module-containing, mucin-like hormone receptor EMR1[11] and the leukocyte activation antigen CD97.[12] Since the signaling pathway(s) are not known, experimental mutation of the equivalent histidine in these two receptors could be instrumental in demonstrating their possible coupling with adenylyl cyclase.

REFERENCES

1. LABURTHE, M., A. COUVINEAU, P. GAUDIN, J. J. MAORET, C. ROUYER-FESSARD & P. NICOLE. 1996. Ann. N. Y. Acad. Sci. **805**: 94–111.
2. SEGRE, G. V. & S. R. GOLDRING. 1993 Trends Endocrinol. Metab. **4**: 309–314.
3. LEFKOWITZ, R. J., S. COTECCHIA, P. SAMAMA & T. COSTA. 1993. Trends Pharmacol. Sci. **14**: 303–307.

4. PARMA, J., L. DUPREZ, J. VAN SANDE, R. PASCHKE, M. TONACHERRA, J. DUMONT & G. VASSART. 1994. Mol. Cell. Endocrinol. **100**: 159–162.

5. SPIEGEL, A. M. 1995. Annu. Rev. Physiol. **58**: 143–170.

6. SCHIPANI, E., K. KRUSE & H. JUPPNER. 1995. Science **268**: 98–100.

7. COUVINEAU, A., C. ROUYER-FESSARD, D. DARMOUL, J. J. MAORET, I. CARRERO, E. OGIER-DENIS & M. LABURTHE. 1994. Biochem. Biophys. Res. Commun. **200**: 769–776.

8. COUVINEAU, A., C. FABRE, P. GAUDIN, J. J. MAORET & M. LABURTHE. 1996. Biochemistry **35**: 1745–1752.

9. COUVINEAU, A., P. GAUDIN, J. J. MAORET, C. ROUYER-FESSARD, P. NICOLE & M. LABURTHE. 1995. Biochem. Biophys. Res. Commun. **206**: 246–252.

10. NICOLE, P., K. DU, A. COUVINEAU & M. LABURTHE. 1997. Regul. Pept. **64**: 139.

11. BAUD, V., S. L. CHISSOE, E. VIEGAS-PEQUIGNOT, S. DIRIONG, V. C. N'GUYEN, B. A. ROE & M. LIPINSKI. 1995. Genomics **6**: 334–344.

12. HAMMAN, J., W. EICHLER, D. HAMMAN, H. M. J. KERSTEN, P. J. PODDHIGE, J. M. N. HOOVERS, E. HARTMANN, M. STRAUSS & R. A. W. VAN VLIER. 1995. J. Immunol. **155**:1942–1950.

Construction of Chimeras between Human VIP1 and Secretin Receptors: Identification of Receptor Domains Involved in Selectivity towards VIP, Secretin, and PACAP

KAI DU, PASCAL NICOLE, ALAIN COUVINEAU, AND MARC LABURTHE[a]

Laboratoire de Neuroendocrinologie et Biologie Cellulaire Digestives, Institut National de la Santé et de la Recherche Médicale, INSERM U410, Faculté de Médecine Xavier Bichat, B.P. 416, 75870 Paris Cedex 18 France

Human VIP1 and secretin receptors belong to the new class II subfamily of G protein–coupled receptors and exhibit 48% amino acid identity in their sequences.[1–3] Both receptors bind VIP, secretin, and PACAP but with different selectivity. Structure-function relationship studies on this subfamily of G protein–coupled receptors have revealed that the large N-terminal extracellular domain and the first extracellular loop play an important role in ligand binding.[1] In particular, site-directed mutagenesis studies on VIP1 receptor have shown that (1) several highly conserved amino acid residues probably positioned in a tertiary functional structure maintained by disulfide bonds were crucial for VIP binding to the N-terminal domain of the receptor[4–6] and that (2) two N-glycosylation sites were involved in the receptor protein delivery to the plasma membrane.[7] Recent chimeric receptor studies within this receptor subfamily have shown that three nonadjacent amino acid residues in the first extracellular loop and third transmembrane domain of the VIP1 receptor constitute a structural determinant for the selectivity towards the VIP agonist PHI.[8] Such studies also indicated that the first and second extracellular loops as well as the N-terminal domain of secretin receptor contribute to secretin recognition.[9–11]

In this study, we have constructed chimeras between the human VIP1 and secretin receptors to further identify the critical determinants in the N-terminal domain of VIP1 and secretin receptors for their selectivity towards VIP, secretin, and PACAP.

RESULTS AND CONCLUSION

Several chimeras between the human VIP1 receptor[2] and the human secretin receptor[3] were constructed by exchanging the whole N-terminal extracellular domain or parts of this domain as shown in FIGURE 1. The cDNA encoding the wild-type receptors as well as

[a]Corresponding author: Tel.: 33 01 44 85 61 35; Fax: 33 01 44 85 61 24; E-mail: laburthe@bichat.inserm.fr

TABLE 1. Wild type (wt) and chimeric receptors transfected in COS-7 cells: cAMP response and ^{125}I- VIP competition binding experiments.

Receptor Constructs	V-wt	S-wt	V144S	S45V	S123V	S144V
cAMP Response (EC$_{50}$, nM)						
VIP	0.2 ± 0.1	60 ± 40	3.0 ± 1.5	0.3 ± 0.1	2.0 ± 0.5	2.5 ± 0.8
Secretin	>200	0.15 ± 0.1	80 ± 20	100 ±50	300 ± 200	2.0 ± 0.5
PACAP27	0.15 ± 0.1	50 ± 30	2.5 ± 1.0	0.35 ± 0.1	60 ± 20	100 ± 50
^{125}I-VIP Competition Binding						
K_d (VIP, nM)	0.51 ± 0.14	ND	8.9 ± 2.4	1.12 ± 0.29	5.76 ± 3.77	2.67 ± 1.03
K_i (Secretin, nM)	>500		>500	>500	>500	1.54 ± 0.33
K_i (PACAP27, nM)	3.36 ± 0.28		7.8 ± 4.6	0.77 ± 0.14	26.6 ± 4.5	14.2 ± 0.04
B_{max} (pmole/mg protein)	3.12 ± 0.60		0.36 ± 0.22	2.40 ± 0.36	0.25 ± 0.12	0.14 ± 0.04

Values are expressed as mean ± S.E. of at least three experiments. K_d, K_i, and B_{max} values were determined by Scatchard analysis using RADLIG version 4.0 computer software (BIOSOFT, UK). ND: binding not detected. See text for details.

chimeras were transfected into Cos cells for subsequent studies consisting of stimulation of cAMP production by VIP, secretin, or PACAP and of inhibition of ^{125}I-VIP binding by unlabeled VIP, secretin, or PACAP.

FIGURE 1 shows that the order of potency of peptides for stimulating cAMP production in cells expressing the wild-type human VIP receptor was VIP = PACAP27 >> secretin as previously reported.[2] As expected, the order of potency is quite different for the wild-type human secretin receptor, i.e., secretin >> VIP ≅ PACAP27. The chimera V144S constructed with the N-terminal extracellular domain of the VIP receptor and the core of the secretin receptor retains a peptide selectivity similar to that of the wild-type VIP receptor, i.e., VIP = PACAP-27 >> secretin (FIG. 1). This observation supported that the N-terminal extracellular domain of the human VIP1 receptor contains the structural determinant(s) necessary and sufficient for keeping the selective profile of the wild-type receptor. It may be noticed however that the potencies of VIP and PACAP27 for stimulating cAMP were slightly decreased in this chimera as compared to the wild-type receptor. The mirror chimera S144V in which the N-terminal extracellular domain of the secretin receptor was associated with the core of the VIP receptor exhibited an original peptide selectivity, i.e., secretin ≅ VIP >> PACAP27. This interesting observation supported that the N-terminal extracellular domain of the secretin receptor cannot discriminate between VIP and secretin and does not contain the structural determinant(s) necessary for recognizing PACAP with high affinity. It also suggested that VIP and PACAP bind to different domains in the VIP receptor. Further chimeras were then constructed by replacing only small parts of the N-terminal domain of the VIP receptor by the corresponding parts of the secretin receptor, i.e., S45V and S123V (FIG. 1). The chimera S45V exhibited a pattern of selectivity almost identical to that of the wild-type VIP receptor, supporting that no structural determinant for peptide selectivity is present in the first 45 amino acid residues at the N-terminal end of the VIP receptor. In sharp contrast, the order of potency of peptides for stimulating cAMP production via the chimera S123V was VIP >> PACAP27 > secretin (FIG.1), a pattern that was different from that of wild-type VIP1 or secretin receptors. This observation supported that the 1-123 sequence of the human VIP1 receptor contains a domain crucial for PACAP binding but not for VIP binding and that this domain cannot be provided by the corresponding secretin receptor sequence. In that respect the S123V chimera provides the first example of a VIP-specific receptor that discriminates between VIP and PACAP. Indeed, the native VIP1 receptors and also VIP2 receptors characterized in humans and rodents have a similar affinity for VIP and PACAP.[1] Another interesting property of chimera S123V was that the potency of secretin was very low, like in the wild-type VIP receptor. Since the S144V chimera had a high affinity for secretin, these data suggested that a structural determinant present in the 124-144 part of the N-terminal extracellular of the human VIP1 receptor is crucial for discriminating between VIP and secretin.

We further explored the wild-type receptors and all chimeras by binding experiments using ^{125}I-VIP as a tracer. The data are shown in FIGURE 1 and can be summarized as follows. Competitive inhibition of ^{125}I-VIP binding by unlabeled VIP, PACAP27, or secretin gave patterns that were very similar to those observed in cAMP experiments. This suggested that the construction of chimeras did not alter the transduction of the signal between peptide binding and the adenylyl cyclase system and further validated the conclusions drawn from cAMP experiments. As expected, the wild-type secretin receptor did not bind specifically ^{125}I-VIP (0.05 nM) due to its very low affinity for VIP (FIG. 1).

CONCLUSION

In conclusion, the N-terminal extracellular domain of the human VIP1 receptor is crucial for the high affinity of this receptor for PACAP since its exchange with the corresponding domain of the secretin receptor results in an important decrease of PACAP affinity. In contrast, the N-terminal extracellular domain of the secretin receptor, when associated with the core of the human VIP1 receptor, does not cause a dramatic change in the affinity for VIP. This suggests that amino acid residues conserved in the N-terminal domains of secretin and VIP receptors may be crucial for high affinity VIP binding. This is consistent with previous site-directed mutagenesis studies of the human VIP1 receptor.[1,4,5] A small box (between residues 124 and 144) located at the distal portion of the N-terminal extracellular domain of the secretin receptor contains amino acids that play a pivotal role for selective high affinity recognition of secretin. It is worth pointing out that this domain of the secretin receptor contains four basic residues instead of only one in the corresponding domain of the VIP1 receptor. Indeed, previous studies suggested that basic residues made a contribution to secretin binding.[10,11] Site-directed mutagenesis studies are currently performed in our laboratory to determine which specific residues within this box are involved in the selective recognition of secretin.

REFERENCES

1. LABURTHE, M., A. COUVINEAU, P. GAUDIN, J. J. MAORET, C. ROUYER-FESSARD & P. NICOLE. 1996. Ann. N. Y. Acad. Sci. 805: 94–111.
2. COUVINEAU, A., C. ROUYER-FESSARD, D. DARMOUL, J. J. MAORET, I. CARRERO, E. OGIER-DENIS & M. LABURTHE. 1994. Biochem. Biophys. Res. Commun. 200: 769–776.
3. JIANG, S. & C. ULRICH. 1995. Biochem. Biophys. Res. Commun. 207: 883–890.
4. COUVINEAU, A., P. GAUDIN, J. J. MAORET, C. ROUYER-FESSARD, P. NICOLE & M. LABURTHE. 1995. Biochem. Biophys. Res. Commun. 206: 246–252.
5. GAUDIN, P., A., COUVINEAU, J. J. MAORET, C. ROUYER-FESSARD & M. LABURTHE. 1995. Biochem. Biophys. Res. Commun. 211: 901–908.
6. DU, K., P. NICOLE, A. COUVINEAU & M. LABURTHE. 1997. Biochem. Biophys. Res. Commun. 230: 289–292.
7. COUVINEAU, A., C. FABRE, P. GAUDIN, J. J. MAORET & M. LABURTHE. 1996. Biochemistry 35: 1745–1752.

8. COUVINEAU, A., C. ROUYER-FESSARD, J. J. MAORET, P. GAUDIN, P. NICOLE & M. LABURTHE. 1996. J. Biol. Chem. **271**: 12795–12800.
9. HOLTMANN, M. H., E. HADAC & L. MILLER. 1995. J. Biol. Chem. **270**: 14394–14398.
10. VILARDAGA, J. P., E. DI PAOLO, P. DE NEEF, M. WAELBROECK, A. BOLLEN & P. ROBBERECHT. 1996. Biochem. Biophys. Res. Commun. **218**: 842–846.
11. HOLTMANN, M. H., S. GANGULI, E. HADAC, V. DOLU & L. MILLER. 1996. J. Biol. Chem. **271**: 14944–14949.

Involvement of a Pit-1 Binding Site in the Regulation of the Rat Somatostatin Receptor 1 Gene Expression

H. BAUMEISTER[a,c] AND W. MEYERHOF[a,b]

[a]Department of Molecular Genetics, German Institute of Human Nutrition, [b]University of Potsdam, D-14558 Potsdam-Rehbrücke, Germany

The pituitary-specific transcription factor Pit-1, a member of the POU-domain protein family, is essential for the expression of the growth hormone (GH), prolactin, and thyroid stimulating hormone b-subunit genes in the adenohypophysis.[1] Pit-1 also activates the transcription of the GH releasing hormone (GHRH) receptor gene.[2] Thus, Pit-1 may have a general role as an important regulator of GH secretion at the transcriptional level. The peptide hormone somatostatin inhibits release of various hormones, including GH.[3] The actions of somatostatin are mediated by at least five transmembrane receptors, sst1–sst5, which have been cloned recently.[4] All five receptors are present in many tissues including the pituitary.[5] Sst1 transcripts are fairly abundant in pituitary GH3[6] and pancreatic RIN1046-38 cells,[7] which express the GH and the insulin gene, respectively. In order to elucidate the molecular mechanisms underlying sst1 gene expression, we have recently cloned the rat sst1 gene, including the 2.2 kb upstream sequence.[6] Computer analysis of the nucleotide sequence revealed the presence of a consensus sequence for Pit-1 binding sites. In the present study we examined whether Pit-1 is involved in the transcriptional control of the sst1 gene.

IDENTIFICATION OF A 48 BP REGION IMPORTANT FOR THE ACTIVITY OF THE SST1 GENE PROMOTER

A 2,175 bp DNA fragment located 5′ to the translational start site of the sst1 gene was subcloned upstream of the luciferase coding region (sst1Luc). Transfection of GH3 and RIN 1046-38 cells, which express the sst1 gene endogenously with sst1Luc plasmid DNA, resulted in an about 14-fold induction of luciferase activity (TABLE 1). In contrast, only a twofold induction was observed when this construct was used for transfection of Chinese hamster ovary (CHO) cells, which do not express the sst1 gene endogenously. To localize the regulatory elements responsible for the cell type-specific activity of the sst1 gene promoter, various 5′-deletion mutants of the sst1Luc construct were analyzed (TABLE 1). Interestingly, the promoter activity increased about sixfold in GH3 cells but not in RIN and CHO cells when 1,661 bp have been eliminated (-324sst1Luc). This observation suggests the presence of GH3 cell-specific negative regulatory elements between -1985 and -324. Elements that stimulate transcription in GH3 cells are located between -324 and -165

[c]Corresponding author: Tel.: 49-3320088374; Fax: 49-3320088384; E-mail: baumeist@www.dife.de

TABLE 1. Promoter activities of 5'-deletion mutants of the sst1Luc construct

Plasmid	Relative Luciferase Activity[d] (fold induction) after Transfection of		
	GH3 cells	RIN1046-38 cells	CHO cells
-Luc[a]	1	1	1
sst1Luc[b]	14 ± 1	13 ± 2	1.6 ± 0.5
-324sst1Luc[c]	89 ± 16	13 ± 3	2.4 ± 0.4
-165sst1Luc[c]	39 ± 6	18 ± 2	n.d.
-117sst1Luc[c]	11 ± 1	8.4 ± 0.9	2 ± 0.2
-48sst1Luc[c]	7.5 ± 3	8 ± 3	n.d.
+52sst1Luc[c]	3.3 ± 0.5	2.5 ± 0.8	1 ± 0.2

[a]The -Luc plasmid lacks any regulatory element.

[b]The sst1 gene fragment covers the 5' untranslated region of 190 bp and 1985 bp of 5' upstream sequences.

[c]The numbers refer to the positions of the 5' deletions with respect to the transcriptional start site.

[d]For determination of the relative luciferase activity the transfection efficiency was taken into account.

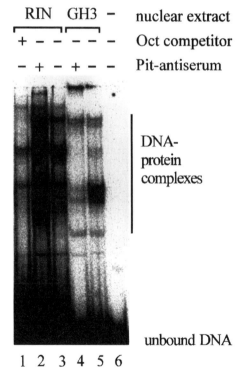

FIGURE 1. Binding of Pit-1 to the 48 bp region between -165 to -117. Nuclear extracts obtained from RIN1046-38 and GH3 cells were incubated with radiolabeled DNA from the 48 bp region that contains the putative Pit-1 binding site. The DNA-protein complexes were analyzed in a native 4% acrylamide gel. The positions of the DNA-protein complexes and the unbound DNA are indicated. The Pit-1 antiserum and the Oct competitor DNA were added to the reactions before the radiolabeled DNA was included.

(twofold), -165 and -117 (fourfold), and between -48 and +52 (twofold). Moreover, the 48 bp region between -165 and -117 harbors most of the promoter activity in RIN cells. This region contains the putative Pit-1 binding site.

BINDING OF PIT-1 TO THE 48 BP REGION

In order to characterize the proteins that bind to the 48 bp region, electrophoretic mobility shift assays were performed in the absence and presence of an antiserum directed against Pit-1. FIGURE 1 reveals the formation of several DNA-protein complexes. Most of them were specific for only one of the nuclear extracts from GH3 and RIN cells. A single complex seems to be common to both of them (lanes 5 and 3). Addition of the Pit-1 antiserum eliminated the two most abundant GH3 cell-specific DNA-protein complexes (lane 4). It is a characteristic property of Pit-1 to form two comigrating DNA-protein complexes. No such an effect was seen with nuclear proteins of RIN cells (lane 2). This indicates that the 48 bp region represents a binding site for nuclear proteins and that Pit-1 is a major binding protein at this site in pituitary but not pancreatic cells. The identity of the other binding proteins in GH3 and RIN cell extracts is still unknown and needs to be elucidated. However, the binding proteins in RIN cell extracts probably belong to the family of POU-domain proteins. This is suggested by the ability of a DNA fragment containing a consensus sequence for binding sites of POU-domain proteins (Oct) to compete with the radiolabeled probe (lane 1).

SUMMARY

It was shown that at least four regions in the 2.2 kb upstream DNA of the sst1 gene are important for the cell type-specific promoter activity in GH3 and RIN cells. Moreover, the 48 bp region located between -165 to -117 harbors positive regulatory elements that are active in RIN and GH3 cells. This region is recognized by the pituitary-specific transcription factor Pit-1. It is therefore concluded that Pit-1 represents a major regulator of GH secretion at the genetic level by regulating transcriptional activity not only of the GH gene itself but also of the genes for the receptors that mediate stimulation and inhibition of GH secretion.

REFERENCES

1. ANDERSEN, B. & M. G. ROSENFELD. 1994. Pit-1 determines cell types during development of the anterior pituitary gland. J. Biol. Chem. **269:** 29335–29338.
2. LIN, C., S.-C. LIN, C.-P. CHANG & M. G. ROSENFELD. 1992. Pit-1- dependent expression of the receptor for growth hormone releasing factor mediates pituitary cell growth. Nature **360:** 765–768.
3. GILLIES, G. 1997. Somatostatin: The neuroendocrine story. Trends Pharmacol. Sci. **18:** 87–95.
4. PATEL, Y. C., M. GREENWOOD, R. PANETTA, N. HUKOVIC, S. GRIGORAKIS, L.-A. ROBERTSON & C. B. SRIKANT. 1996. Molecular biology of somatostatin receptor subtypes. Metabolism **45:** 31–38.
5. MEYERHOF, W. 1997. The elucidation of somatostatin receptor functions: a current view. Rev. Physiol. Biochem. Pharmacol. **133:** 55–108.
6. HAUSER, F., W. MEYERHOF, S. FEHR & D. RICHTER. 1994. Sequence analysis of the promoter region of the rat somatostatin receptor subtype 1 gene. FEBS Lett. **345:** 225–228.
7. ROOSTERMAN, D., G. GLASSMEIER, H. BAUMEISTER, H. SCHERÜBL & W. MEYERHOF. 1998. A somatostatin receptor 1 selective ligand inhibits Ca^{2+} currents in rat insulinoma 1046 - 38 cells. FEBS Lett. **425:** 137–140.

Relaxant Responses of VIP and PACAP in Rat Ileum: Receptors and Adaptive Supersensitivity

E. EKBLAD,[a] M. EKELUND, AND F. SUNDLER

Department of Physiology and Neuroscience, Section Neuroendocrine Cell Biology, University Hospital, Experimental Research Center, E-blocket vån 5, S-221 85 Lund, Sweden

VIP-containing nerve fibers are numerous throughout the intestinal wall, while PACAP-containing are few in number. In several species, subpopulations of VIP-containing enteric neurons also harbor NPY, PACAP, and/or NO synthase.[1–4] PACAP and VIP are, together with NO, important relaxant messengers of intestinal smooth muscle.[5–8] Our knowledge of the VIP/PACAP receptors involved in regulation of motor activity of the gastrointestinal tract and their signal transduction is still incomplete.

In this study our aims were to characterize different types of PACAP/VIP receptors in the rat ileal longitudinal muscle *in vitro* and to examine the motor effects of PACAP and VIP in an experimental model of inactive (bypassed) rat ileum.

METHODS

Motor Activity

Ileal longitudinal muscle strips with adherent myenteric ganglia were mounted *in vitro* and registered for isometric tension. The strips were precontracted with prostaglandin $F_{2\alpha}$ and concentration-response curves for PACAP-38, PACAP-27, VIP, and the NO donor sodium-nitroso-N-acetylpenicillamine (SNAP) were constructed. Pretreatment was by addition of tetrodotoxin (10^{-6} M), apamin (10^{-6} M), atropine (10^{-6} M), forskolin (3×10^{-7} M), NPY (10^{-7} M), VIP (10^{-7} M), PACAP-27 (10^{-7} M), or PACAP-38 (10^{-7} M) 5 min prior to precontraction. L-NAME (10^{-4} M) was added 30 min prior to precontraction.

Surgery

The distal ileum was bypassed by anastomosis of proximal ileum and colon. Sham-operated rats served as control (FIG. 2). Rats were killed 1 week postoperatively and muscle strips from distal ileum were tested *in vitro* as described above.

RESULTS

VIP, PACAP-38, and PACAP-27 all caused concentration-dependent relaxations of the rat ileal longitudinal smooth muscle (FIG. 1, A). PACAP-27 was, however, much more

[a]Corresponding author: Tel.: 46-46-17 77 11; Fax: 46-46-17 77 20; E-mail: eva.ekblad@mphy.lu.se

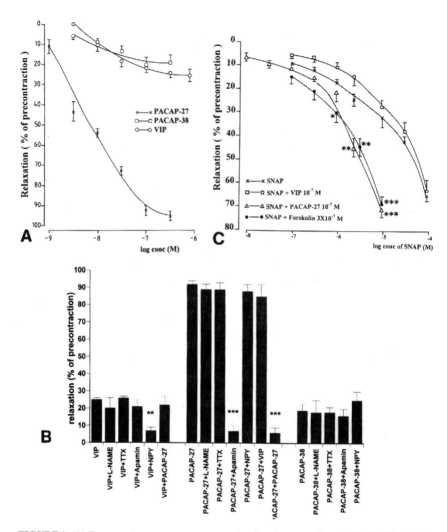

FIGURE 1. (**A**) Concentration-response curves showing the relaxatory effects of PACAP-27, PACAP-38, and VIP on rat ileal longitudinal muscle. (**B**) Relaxatory effects of VIP (10^{-6} M), PACAP-27 (10^{-7} M), and PACAP-38 (10^{-7} M) in the presence of L-NAME (10^{-4} M), TTX (10^{-6} M), apamin (10^{-6} M), or NPY (10^{-7} M). Cross-desensitization was tested by the addition of VIP (10^{-7} M) or PACAP-27 (10^{-7} M). (**C**) Concentration-response curves showing the relaxatory effects of SNAP on rat ileal longitudinal muscle. Presence of PACAP-27 or forskolin, but not VIP, caused a marked potentiation of the relaxations. Each value is the mean of 12–24 experiments. Vertical bars give SEM. * $p < 0.05$, ** $p < 0.01$, *** $p < 0.001$.

potent than PACAP-38 and VIP. The relaxations were unaffected by pretreatment with L-NAME, TTX, or atropine (FIG. 1, B). Pretreatment with apamin abolished the PACAP-27–induced relaxations, while relaxations induced by VIP or PACAP-38 were unaffected (FIG. 1, B). Pretreatment with NPY almost totally abolished the VIP-induced relaxations

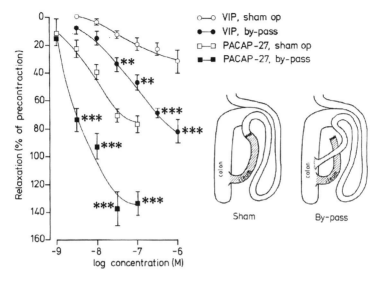

FIGURE 2. Concentration-response curves showing the relaxatory effects of PACAP-27 and VIP on rat ileal longitudinal muscle after sham operation or by-pass. Each value is the mean of 14–16 experiments. Vertical bars give S.E.M. ** $p < 0.01$, *** $p < 0.001$ as compared to the effects of VIP and PACAP-27, respectively, on sham-operated ileum. Right panel is a schematic illustration of the surgical procedures used.

but did not influence relaxations induced by PACAP-38 or PACAP-27 (FIG. 1, B). No cross-desensitization between PACAP-27 and VIP could be revealed.

SNAP induced a relaxatory response. Pretreatment with PACAP-27 (10^{-7} M), PACAP-38 (10^{-7} M), or forskolin (3×10^{-7} M) significantly potentiated the SNAP-induced relaxation causing a leftward shift of the concentration-response curve (FIG. 1, C).

One week after bypassing the distal ileum, the relaxant responses to both VIP and PACAP-27 were significantly increased (FIG. 2).

CONCLUSIONS

At least three different types of PACAP/VIP receptors mediate relaxation of rat ileal longitudinal muscle as studied *in vitro*. (1) A PACAP-27 preferring receptor coupled to apamin-sensitive Ca^{2+}-dependent K^+ channels. A strong relaxant response is elicited upon activation of this receptor. (2) A PACAP receptor with equal affinity to both PACAP-27 and PACAP-38, but with no affinity to VIP. Activation of this receptor gives *per se* a weak relaxation, but the receptor may be involved in the potentiation of the cGMP-mediated relaxation (induced by SNAP), since pretreatment with PACAP-27 or PACAP-38 causes a significant leftward shift of the concentration-response curve of the SNAP-evoked relaxation. This PACAP receptor probably operates through activation of cAMP production, since forskolin also potentiates the SNAP-induced relaxation. (3) A VIP-specific receptor. Activation of this receptor evokes a weak relaxation that can be blocked by pretreatment with NPY. A marked supersensitivity to both VIP and PACAP is seen as an adaptive change to intestinal inactivity.

REFERENCES

1. SUNDLER, F., E. EKBLAD, A. ABSOOD, R. HÅKANSON, K. KÖVES & A. ARIMURA. 1991. Pituitary adenylate cyclase activating peptide: A novel vasoactive intestinal peptide-like neuropeptide in the gut. Neuroscience **46**: 439–454.
2. EKBLAD, E., P. ALM & F. SUNDLER. 1994. Distribution, origin and projections of nitric oxide synthase-containing neurons in gut and pancreas. Neuroscience **63**: 233–248.
3. FURNESS, J. B., H. M. YOUNG, S. POMPOLO, J. C. BORNSTEIN, W. A. A. KUNZE & K. MCCONALOGUE. 1995. Plurichemical transmission and chemical coding of neurons in the digestive tract. Gastroenterology **108**: 554–563.
4. HANNIBAL, J., E. EKBLAD, H. MULDER, F. SUNDLER & J. FAHRENKRUG. 1998. Pituitary adenylate cyclase activating polypeptide (PACAP) in the gastrointestinal tract of the rat: Distribution and effects of capsaicin or denervation. Cell Tissue Res. **291**: 65–79.
5. SHUTTLEWORTH, C. W. R. & K. D. KEEF. 1995. Roles of peptides in enteric neuromuscular transmission. Regul. Pept. **56**: 101–120.
6. KATSOULIS, S., A. CLEMENS, H. SCHWÖRER, W. CREUTZFELDT & W. SCHMIDT. 1993. Pituitary adenylate cyclase activating peptide (PACAP) is a potent relaxant of the rat ileum. Peptides **14**: 587–592.
7. EKBLAD, E. & F. SUNDLER. 1997. Distinct receptors mediate pituitary adenylate cyclase-activating peptide- and vasoactive intestinal peptide-induced relaxation of rat ileal longitudinal muscle. Eur. J. Pharmacol. **33**: 61–66.
8. EKBLAD, E. & F. SUNDLER. 1997. Motor responses in rat ileum evoked by nitric oxide donors versus field stimulation: Modulation by pituitary adenylate cyclase-activating peptide, forskolin and guanylate cyclase inhibitors. J. Pharm. Exp. Ther. **283**: 23–28.

VIP$_1$ and VIP$_2$ Receptors but not PVR1 Mediate the Effect of VIP/PACAP on Cytokine Production in T Lymphocytes

XIAOMING JIANG, HONG-YING WANG, JIAN YU, AND DOINA GANEA[a]

Department of Biological Sciences, Rutgers University, 101 Warren Street, Newark, New Jersey 07102 USA

Neuroimmune interactions are mediated primarily through a network of soluble molecules and specific receptors shared to a large degree by neuronal/glial cells and immune cells. The neuropeptide vasoactive intestinal peptide (VIP) present in both primary and secondary immune organs[1] affects a variety of immune parameters including lymphocyte migration,[2-5] proliferation of thymocytes and peripheral T cells,[6-8] immunoglobulin production,[9,10] and macrophage activity.[11] Some of the immune functions of VIP are mediated through the modulation of cytokines. Previous reports from our laboratory indicated that VIP and the related neuropeptide pituitary adenylate cyclase-activating polypeptide (PACAP) inhibit IL-2, IL-4, and IL-10 production by T cell receptor (TCR)-engaged CD4+ T cells and thymocytes.[12-18] Three types of PACAP/VIP receptors (PVR) have been cloned recently.[19-24] PVR1 functions as a PACAP receptor, expressing a much higher affinity for PACAP than for VIP, whereas PVR2 (VIP$_1$) and PVR3 (VIP$_2$) do not distinguish between the two peptides.[20] The distribution, modulation, and role of these receptors in T lymphocytes remain to be determined. In the present study we examine the involvement of the three PVR in the inhibitory effect of VIP/PACAP on IL-2, IL-4, and IL-10 production, and the expression of these receptors at both mRNA and protein levels in purified CD4+ and CD8+ T lymphocytes.

MATERIALS AND METHODS

Peptides

VIP, PACAP-38, and secretin were purchased from American Peptide Co. (Sunnyvale, CA). The PVR2 (VIP$_1$) agonist [R^{16}]-chicken secretin was a generous gift from Dr. Patrick Robberecht (Université Libre de Bruxelles, Brussels, Belgium), and the PVR3 (VIP$_2$) agonists RO 25-1553 and RO 25-1392 were generous gifts from Drs. Ann Welton and D.R. Bolin (Hoffmann-La Roche Inc., Nutley, NJ). The PVR1 agonist maxadilan was a generous gift from Dr. Ethan A. Lerner (Harvard Medical School, Cambridge, MA). Anti-murine CD3-ε monoclonal antibody (clone 145-2C11) was purchased from Pharmingen

[a]Corresponding author: Tel.: 973-353-1162; Fax: 973-353-1007; E-mail: dganea@andromeda.rutgers.edu

(San Diego, CA). The polyclonal antibodies VIPR1-A and VIPR2-6, and the control absorbed sera were generous gifts from Dr. Edward J. Goetzl (UCSF, San Francisco, CA).

Cells

Spleen cell suspensions were prepared from Balb/c mice in RPMI containing 2% fetal bovine serum. Purified CD4+ and CD8+ T cells were prepared by positive selection using anti-CD4 or anti-CD8 coupled magnetic microbeads (Miltenyi Biotec, Auburn, CA) according to the manufacturer's instructions. The purity of the preparations was routinely >95% (FACS). The spleen cells or the purified CD4+ and CD8+ T cells ($2-5 \times 10^6$ cells/ml) were stimulated with anti-CD3 (1 µg/ml) or anti-CD3 plus PMA (10 ng/ml), respectively. Supernatants and cells were collected at 24 or 48 h for ELISA and RT-PCR.

Cytokine Determination

The amounts of IL-2, IL-4, and IL-10 were determined by ELISA assays (Pharmingen) according to the manufacturer's recommended protocols. The ELISA assays were specific for murine IL-2, IL-4, and IL-10 (they did not cross-react with each other, with human IL-2, IL-4, and IL-10, or with other murine cytokines, such as GM-CSF, IFNγ, IL-3, IL-5, IL-6, and TNFα). The sensitivity limits were 5 pg/ml for IL-2, 10 pg/ml for IL-4, and 0.8 ng/ml for IL-10.

Measurement of Cyclic AMP

Splenocytes (5×10^6 cells/ml) were treated with VIP, PACAP, or agonists in the presence of the cAMP phosphodiesterase inhibitor, 3-isobutyl-1-methyl-xanthine (IBMX, Sigma, St. Louis, MO). At different times (as indicated in text) the reaction was stopped and the intracellular cAMP content was determined by using a cAMP radioimmunoassay kit purchased from Amersham (Arlington Heights, IL). The assay system has a sensitivity of 0.05 pmoles cAMP.

RT-PCR: Design of Oligonucleotide Primers for VIP₁, VIP₂, and PVR1

The primers were designed by using Macvector software. The VIP-R1 sense and antisense primers are: sense 5'-CCTTCTTCTCTGAGCGGAAGTACTT, antisense 5'-CTGCTTCCTCAATGGTGAGGTGCAGG, which generate a 450 bp fragment. The VIP-R2 primers are: sense 5'-GTCAAGGACAGCGTGCTCTACTCC, antisense 5'-CCTTACAAT-GCTGATGAAGAGGGC, which generate a 572 bp fragment. The PVR1 primers are: sense 5'-CAAGAAGGAGCAAGCCATGTGC, antisense: 5'-CATCGAAGTAATGGGGGAAG-GG, which generate a 317 bp fragment. The primers used for β-actin are: sense 5'-TGC-TAAGGAGGCCCCTCATGACGT, antisense: 5'-GCTCATTGCCGATAGTGATG ACCG, which generate a 660 bp fragment. We showed previously that the VIP₁ and VIP₂ primers are specific through sequencing and/or Southern blots.[18,26] The PVR1 fragment amplified from brain was also sequenced and found identical to the published sequence. Two micrograms of RNA was reverse-transcribed to cDNA for 1 h at 42°C in a 30 µl reaction mixture containing 50 mM Tris-HCl (pH 8.3), 60 mM KCl, 3 mM $MgCl_2$, 10 mM dithiothreitol, 0.1 mg/ml BSA, 40 U RNasin (Promega), 3.3 µM random hexamer primers (Promega), 0.5 mM of each deoxynucleotide triphosphate, and 10 U/ml MMLV reverse transcriptase (GIBCO BRL). Then 3 µl of cDNA was amplified in a 30 µl PCR reaction mixture containing 25 µM of each deoxynucleotide triphosphate, 50 mM Tris-HCl (pH 9.0), 1.5 mM $MgCl_2$, 20 mM $(NH_4)_2SO_4$, 0.005% BSA, 1 µM each oligonucleotide primer, and 1 U of pyrostase polymerase (Molecular Genetic Resources, Tampa, FL). The PCR reaction was processed in a GeneAmp PCR system 2400 (Perkin Elmer) programmed to the following conditions: 94°C 40 sec, 56°C 30 sec, and 72°C 50 sec for 35–40 cycles, and a final extension at 72°C for 7 min. Brain RNA used as positive control for the RT-PCR and the no cDNA containing PCR reactions were used as negative controls. Either 10 µl of the PCR products (for actin amplification of all samples and VIP_1, VIP_2, and PVR1 amplification of brain cDNA) or the entire PCR mixture (for all other PCR reactions) was run on 1.5% agarose gels for size analysis and visualization with EtBr.

Immunofluorescence

Purified CD4+ and CD8+ splenic T cells were washed with PBS, resuspended in 5% albumin at a concentration of 5×10^7 cells/ml and smears were prepared on Superfrost plus slides (Fisher, Springfield, NJ). The dried slides were washed in PBS and fixed in 4% paraformaldehyde. The fixed cells were incubated with the primary Abs VIPR1-A and VIPR2-6 at a dilution of 1/2000 to 1/4000 for 60 min at room temperature, followed by FITC-goat anti-rabbit IgG (Gibco/BRL) for 30 min before being placed in mounting fluid and viewed in a fluorescent microscope.

RESULTS AND DISCUSSION

Both VIP-R1 (PVR2) and VIP-R2 (PVR3) Mediate the Inhibition of Cytokine Production

Previous studies indicated that VIP inhibits IL-2, IL-4, and IL-10 production in murine T cells stimulated through the TCR *in vitro,*[12–18] although through different molecular mechanisms. VIP inhibits IL-2 and IL-10 at the level of transcription,[13,15–17] whereas IL-4 is affected post-transcriptionally and indirectly through the reduction in available IL-2.[25] Since naive T cells express both VIP_1 and VIP_2 mRNA, and the expression of the two receptors appears to be differentially regulated during antigen stimulation,[26,27] the two receptors may play different roles in modulating T cell activity. Since IL-2 and IL-10 are produced sequentially during T cell activation and are independently downregulated by VIP/PACAP, the question arose whether both or only one of the two VIP receptors mediate the inhibitory effects of VIP/PACAP. Due to the lack of strong and highly specific antagonists for the lymphocytic VIP receptors, the role of the two receptors in the inhibition of IL-2 and IL-10 was investigated through the use of specific agonists. Spleen cells were stimulated with anti-CD3 Abs in the presence or absence of different concentrations of VIP, PACAP-38, R^{16}-chicken secretin (a strong agonist for rat, and to a lesser degree for human VIP_1; Dr. P. Robberecht, personal communication), porcine secretin (a weak VIP_1 agonist[28]), RO 25-1392 or RO 25-1553 (VIP_2 agonists[29,30]). The IL-2 and IL-10 production was quantitated in supernatants harvested at 24 and 48 h, respectively. Both VIP_1 and VIP_2 agonists inhibited IL-2 and IL-10 production (FIG. 1). VIP, PACAP-38, and R^{16}-secretin appear to share similar potencies, whereas porcine secretin and the two RO compounds are approximately ten times less efficient (FIG. 1). These results suggest that both VIP_1 and VIP_2 can mediate the inhibition of IL-2 and IL-10 in murine T cells.

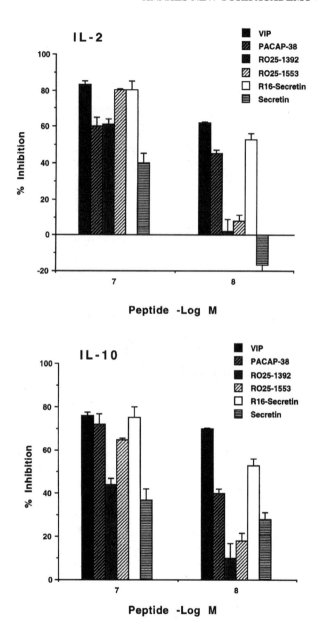

FIGURE 1. VIP$_1$ and VIP$_2$ agonists inhibit IL-2 and IL-10 production. Spleen cells (2×10^6 cells/ml) were stimulated with anti-CD3 (1 µg/ml) in the presence or absence of VIP, PACAP-38, the VIP$_1$ agonists secretin and R[16]-chicken secretin, or the VIP$_2$ agonists RO 25-1553 and RO 25-1392. Supernatants were collected 24 or 48 h later and assayed for IL-2 and IL-10, respectively, by ELISA. Percent inhibition was calculated in reference to the cultures without peptides. Results from one representative experiment out of three are shown.

The Specific PACAP Receptor PVR1 Does Not Contribute to the Inhibition of Cytokine Production

Since PACAP-38 and VIP can also bind to the PACAP-specific receptor, PVR1 may contribute to the inhibitory effect of the two neuropeptides on cytokine production. However the participation of PVR1 is doubtful, since PACAP-38, instead of being 100–1,000-fold more efficient, is actually slightly less efficient than VIP in inhibiting IL-2 and IL-10 production.[16,17] To investigate the possible participation of PVR1, we used the PVR1 specific agonist, maxadilan.[31] Spleen cells were stimulated with anti-CD3 and treated with different concentrations of VIP, PACAP-38, or maxadilan. IL-2, IL-4, and IL-10 were quantitated in supernatants harvested at 24 h (for IL-2) and 48 h (for IL-4 and IL-10). Both VIP and PACAP-38 were inhibitory, whereas maxadilan (in the concentration range from 10^{-6} to 10^{-9} M) was inactive (FIG. 2). These results argue against the participation of PVR1 in the inhibition of cytokine production by VIP/PACAP.

VIP$_1$ and VIP$_2$ Agonists, but not the PVR1 Agonist Maxadilan, Induce cAMP in T Cells

Previous studies showed that VIP, PACAP-38, and the VIP$_2$ agonist RO 25-1553 induce cAMP in murine T cells.[13,15,16] We investigated whether R^{16}-secretin, the strong VIP$_1$ agonist, and maxadilan, the PVR1 agonist, induce cAMP in murine T cells. As expected from the cytokine inhibition data, R^{16}-secretin induces cAMP at levels similar to VIP, whereas maxadilan does not affect cAMP above medium control levels (FIG. 3). The effect of the VIP$_1$ and VIP$_2$ agonists and the lack of effect of maxadilan on intracellular cAMP levels suggest that VIP$_1$ and VIP$_2$, but not PVR1, are expressed in T cells.

Expression of VIP$_1$ and VIP$_2$, but not PVR1, mRNA in CD4+ and CD8+ T Cells

To further determine which type of VIP/PACAP receptors are expressed in murine T cells, we immunomagnetically purified splenic CD4+ and CD8+ T cells, and performed RT-PCR on unstimulated and anti-CD3 stimulated cultures. Both VIP$_1$- and VIP$_2$-specific fragments were amplified from CD4+ and CD8+ T cells, whereas PVR1 fragments were amplified from brain mRNA but not from T cells (FIG. 4). These results indicate that CD4+ and CD8+ T cells express VIP$_1$ and VIP$_2$, but not PVR1 mRNA. Several variants of PVR1 resulting from differential splicing have been identified.[32,33] Five of the six variants differ in the third intracellular loop,[32] and the sixth variant differs in the N-terminal end.[33] Our choice of PCR primers amplifies a fragment coding for part of the N-terminus, the first transmembrane domain, and part of the first intracellular loop. These primers amplify all six splicing variants, resulting in a 317 bp fragment for the variants spliced within the third intracellular loop, and a 254 bp fragment for the N-terminal spliced variant. Therefore, the RT-PCR results, together with the lack of effect of maxadilan, and the equal potency of PACAP and VIP in inhibiting cytokine production, strongly suggest that murine T cells do not express PVR1.

CD4+ and CD8+ Murine T Cells Express Surface VIP$_1$ and VIP$_2$ Receptors

Previous binding studies indicated the presence of functional VIP receptors on lymphocytes.[34] Based on the expression of VIP$_1$ and VIP$_2$ mRNA in CD4+ and CD8+ T cells, we investigated the expression of the two receptors on the cell surface. Two polyclonal antibodies[35] prepared against the extracellular amino terminus of the human VIP$_1$ and VIP$_2$

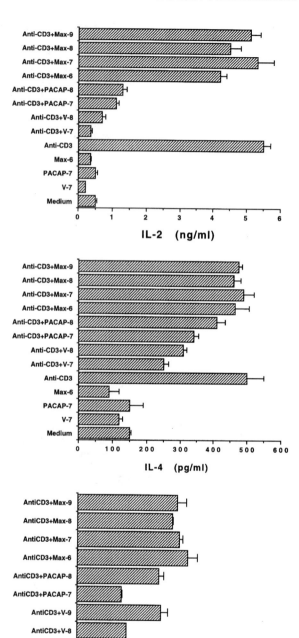

IL-2 (ng/ml)

IL-4 (pg/ml)

IL-10 (ng/ml)

FIGURE 2. Maxadilan, a PVR1 agonist, does not inhibit IL-2, IL-4, or IL-10 production. Spleen cells (2×10^6 cells/ml) were stimulated with anti-CD3 (1 µg/ml) in the presence or absence of VIP, PACAP-38, or maxadilan. Supernatants were collected 24 h later for IL-2 and 48 h later for IL-4 and IL-10 determinations. Percent inhibition was calculated in reference to the cultures without peptides. Results from one representative experiment out of three are shown.

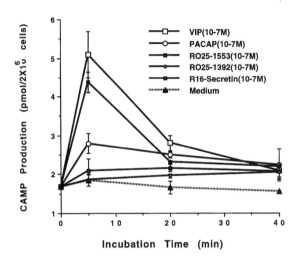

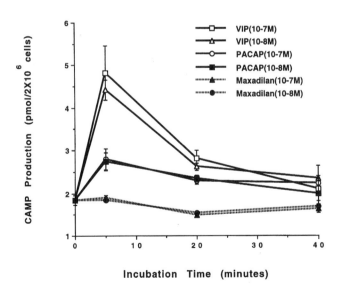

FIGURE 3. VIP$_1$ and VIP$_2$ agonists, but not maxadilan, induce cAMP in spleen cells. Spleen cells (5×10^6 cells/ml) were treated with VIP, PACAP-38, secretin, R[16]-secretin, RO 25-1553, RO 25-1392, or maxadilan, and intracellular cAMP levels were determined at different times as described in *Methods*. Results from one representative experiment out of three are shown.

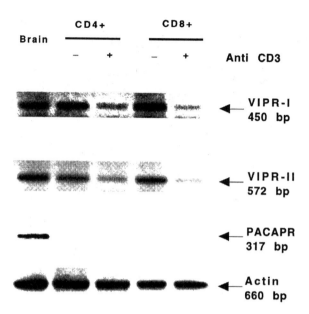

FIGURE 4. RT-PCR for VIP_1, VIP_2, and $PVR1$. Purified CD4+ and CD8+ splenic T cells (5×10^6 cells/ml) were cultured for 48 h either with medium (unstimulated) or with anti-CD3 (1 μg/ml) plus PMA (10 ng/ml) (stimulated). Total RNA was prepared and subjected to RT-PCR as described in *Methods*. Brain RNA was used as a positive control.

and which crossreact with murine and rat lung tissues were used in immunofluorescence studies. Immunomagnetically purified CD4+ and CD8+ splenic T cells were fixed and stained separately with the two antisera, or with control absorbed serum.[35] Both CD4+ and CD8+ T cells stained positively with the VIP_1 and VIP_2 specific antibodies (FIG. 5). These results indicate that both subsets of murine peripheral T cells translate the VIP_1 and VIP_2 mRNA into cell surface expressed proteins.

SUMMARY AND CONCLUSIONS

Neuropeptides such as VIP and PACAP produced or released within the lymphoid microenvironment modulate the immune response through their effect on immune cells bearing specific receptors. In response to antigenic stimulation, CD4+ T cells, and to a lesser degree CD8+ T cells, produce cytokines that play essential roles in the initiation and amplification of various immune responses. VIP/PACAP downregulate the expression of a variety of cytokines such as IL-2, IL-4, and IL-10, by directly affecting the cytokine-producing T cells. Since three types of receptors, PVR1 (the PACAP-preferring receptor), PVR2 (VIP_1), and PVR3 (VIP_2) bind PACAP/VIP, this study investigated the expression of these receptors in murine T lymphocytes and their role in mediating the inhibition of cytokines. VIP_1 and VIP_2 agonists, but not PVR1 agonists, inhibit IL-2, IL-4, and IL-10 production, and VIP_1 and VIP_2, but not PVR1 mRNA, were identified in purified CD4+ and CD8+ splenic T cells. In addition, immunofluorescence studies confirmed the presence of VIP_1 and VIP_2 on CD4+ and CD8+ T cells. These results indicate that both subsets of peripheral T lymphocytes express VIP_1 and VIP_2, but not PVR1 receptors, and that the

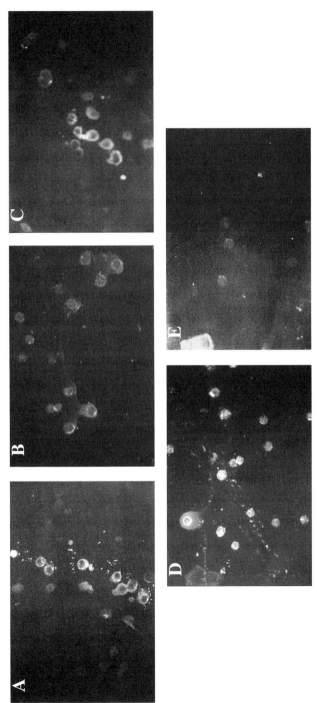

FIGURE 5. Murine T cells express both VIP$_1$ and VIP$_2$. Suspensions of fixed CD4+ (**A** and **B**) and CD8+ (**C** and **D**) T cells were incubated with antiserum VIPR1-A and antiserum VIPR2-6, respectively, or with the control absorbed serum (**E**), followed by FITC-labeled secondary antibody.

inhibitory effect of VIP/PACAP on IL-2 and IL-10 production is mediated by both VIP$_1$ and VIP$_2$ receptors.

ACKNOWLEDGMENTS

We are grateful to Dr. Patrick Robberecht (Université Libre de Bruxelles) for providing us with the R^{16}-chicken secretin, Dr. Ethan A. Lerner (Harvard Medical School) for maxadilan, Drs. Ann Welton and D. R. Bolin (Hoffmann-La Roche, Inc.) for RO 25-1553 and RO 25-1392, and Dr. Edward J. Goetzl (University of California, San Francisco) for providing us with the VIP$_1$ and VIP$_2$ antibodies.

REFERENCES

1. BELLINGER, D. L., D. LORTON, S. BROUXHON, S. FELTEN & D. L. FELTEN. 1996. Adv. Neuroimmunol. **6**: 5–27.
2. MOORE, T. C., C. H. SPRUCK & S. I. SAID. 1988. Immunology **64**: 475–478.
3. OTTAWAY, C. A. 1985. Adv. Exp. Med. Biol. **186**: 637–645.
4. DE LA FUENTE, M., M. DEL RIO, E. GARRIDO, J. LECETA, A. HERNANZ & R. P. GOMARIZ. 1994. Peptides **15**: 1157–1163.
5. DELGADO, M., M. DE LA FUENTE, C. MARTINEZ & R. P. GOMARIZ. 1995. J. Neuroimmunol. **62**: 137–146.
6. OTTAWAY, C. A & G. R. GREENBERG. 1984. J. Immunol. **132**: 417–423.
7. BOUDARD, F. & M. BASTIDE. 1991. J. Neurosci. Res. **29**: 29–41.
8. METWALI, A., A. BLUM, R. MATHEW, M. SANDOR, R. G. LYNCH & J. V. WEINSTOCK. 1993. Cell. Immunol. **149**: 11–23.
9. STANITZ, A. M., D. BEFUS & J. BIENENSTOCK. 1986. J. Immunol. **136**: 152–156.
10. KIMATA, H. 1996. Adv. Neuroimmunol. **6**: 107–115.
11. DE LA FUENTE, M., M. DELGADO & R. P. GOMARIZ. 1996. Adv. Neuroimmunol. **6**: 75–91.
12. GANEA, D. & L. SUN. 1993. J. Neuroimmunol. **47**: 147–158.
13. SUN, L. & D. GANEA. 1993. J. Neuroimmunol. **48**: 59–70.
14. XIN, Z., H. TANG & D. GANEA. 1994. J. Neuroimmunol. **54**: 59–68.
15. TANG, H., A. WELTON & D. GANEA. 1995. J. Interferon Cytokine Res. **15**: 993–1003.
16. TANG, H., L. SUN, Z. XIN & D. GANEA. 1996. Ann. N.Y. Acad. Sci. **805**: 768–778.
17. MARTINEZ, C., M. DELGADO, R. P. GOMARIZ & D. GANEA. 1996. J. Immunol. **156**: 4128–4136.
18. XIN, Z., X. JIANG, H-Y. WANG, T. N. DENNY, B. N. DITTEL & D. GANEA. 1997. Regul. Pept. **72**: 41–54.
19. CHRISTOPHE, J. 1993. Biochim. Biophys. Acta **1154**: 183–199.
20. RAWLINGS, S. R. & M. HEZAREH. 1996. Endocrine Rev. **17**: 4–29.
21. ISHIHARA, T., R. SHIGEMOTO, K. MORI, K. TAKAHASHI & S. NAGATA. 1992. Neuron **8**: 811–819.
22. SREEDHARAN, S. P., D. R. PATEL, J. X. HUANG & E. J. GOETZL. 1993. Biochem. Biophys. Res. Commun. **193**: 546–553.
23. LUTZ, E. M., W. J. SHEWARD, K. M. WEST, J. A. MORROW, G. FINK & A. J. HARMAR. 1993. FEBS Lett. **334**: 3–8.
24. INAGAKI, N., H. YOSHIDA, M. MIZUTA, N. MIZUNO, Y. FUJII, T. GONOI & S. SEINO. 1994. Proc. Natl. Acad. Sci USA **91**: 2679–2683.
25. WANG, H-Y., Z. XIN, H. TANG & D. GANEA. 1996. J. Immunol. **156**: 3243–3253.
26. JOHNSON, M., R. MCCORMACK, M. DELGADO, C. MARTINEZ & D. GANEA. 1996. J. Neuroimmunol. **68**: 109–119.
27. DELGADO, M., C. MARTINEZ, M. JOHNSON, R. P. GOMARIZ & D. GANEA. 1996. J. Neuroimmunol. **68**: 27–38.
28. USDIN, T. R., T. I. BONNER & E. MRZEY. 1994. Endocrinology **135**: 2662–2680.
29. GOURLET, P., P. VERTONGEN, A. VANDERMEERS, M. C. VANDERMEERS-PIRET, J. RATHE, P. DE NEEF, M. WAELBROECK & P. ROBBERECHT. 1997. Peptides **18**: 403–408.
30. XIA, M., S. P. SREEDHARAN, D. R. BOLIN, G. O. GAUFO & E. J. GOETZL. 1997. J. Pharmacol. Exp. Ther. **281**: 629–633.

31. Moro, O. & E. A. Lerner. 1997. J. Biol. Chem. **272**: 966–970.
32. Spengler, D., C. Waeber, C. Pantaloni, F. Holsboer, J. Bockaert, P. H. Seeburg & L. Journot. 1993. Nature **365**: 170–175.
33. Pantaloni, C., P. Brabet, B. Bilanges, A. Dumuis, S. Houssami, D. Spengler, J. Bockaert & L. Journot. 1996. J. Biol. Chem. **271**: 22146–22151.
34. Calvo, J. R., D. Pozo & J. M. Guerrero. 1996. Adv. Neuroimmunol. **6**: 39–47.
35. Goetzl, E. J., D. R. Patel, J. L. Kishiyama, A. C. Smoll, C. W. Turck, N. M. Law, S. A. Rosenzweig & S. P. Sreedharan. 1994. Mol. Cell. Neurosci. **5**: 145–152.

Distribution and Ultrastractural Localization of Pituitary Adenylate Cyclase–Activating Polypeptide (PACAP) and Its Receptor in the Rat Retina[a]

TAMOTSU SEKI,[b,e] SEIJI SHIODA,[c,d] YASUMITSU NAKAI,[c] AKIRA ARIMURA,[d] AND RYOHEI KOIDE[b]

[b]Departments of Ophthalmology and [c]Anatomy, Showa University School of Medicine, Tokyo 142, Japan

[d]U.S.-Japan Biomedical Research Laboratories, Tulane University Hebert Center, Belle Chasse, Louisiana 70037, USA

Pituitary adenylate cyclase-activating polypeptide (PACAP) is a member of the secretin/glucagon/vasoactive intestinal polypeptide (VIP) family.[1,2] It was recently reported that PACAP stimulates accumulation of cAMP in the mammalian retina and increases the blood flow in the rabbit choroid plexus.[3] VIP is reported to be localized in the inner nuclear layer (INL), the inner plexiform layer (IPL), and ganglion cell layer (GCL).[4] A recent immunohistochemical study has reported that PACAP receptor–like immunoreactivity (PACAPR-LI) is localized in the retina and ciliary body.[5] Cavallero and colleagues[5] found that the strongest immunostaining was detected in the epithelium of ciliary processes. However the exact localization of PACAP and PACAPR within the retina remains to be determined. Thus, the present study was designed to identify the distribution and ultrastructural localization of PACAP and its receptor in the rat retina by immunocytochemistry.

MATERIALS AND METHODS

Adult male Sprague-Dawley rats (250–300 g) were housed in a temperature- and light-controlled room and supplied with standard Oriental Chow and water *ad libitum*. The animals were sacrificed by overdose of ketamine (Aveco, Fort Dodge, IA, USA) and the retinas were rapidly dissected. For immunocytochemistry, the tissues were immediately fixed by immersion in 4% paraformaldehyde and 0.2% glutaraldehyde in 0.1 M phosphate buffer (pH 7.2). Frozen sections cut at 8 µm were used for light microscopic study. Vibratome 40 µm sections were used for electron microscopic study. Both sections were incubated with the PACAP and PACAPR antisera (1:1000–1:2000) and then immunostained by the ABC method. After immunostaining, the vibratome sections were osmicated

[a]This study supported in part by grants from the Ministry of Education, Science, Sports, and Culture of Japan.

[e]Corresponding author: Tamotsu Seki M.D. Ph.D., Department of Ophthalmology, Showa University School of Medicine, 1-5-8 Hatanodai, Shinagawa-ku, Tokyo 142, Japan; Tel.: 81-3-3784-8553; Fax: 81-3-3784-5048; E-mail:tamseki@med.showa-u.ac.jp

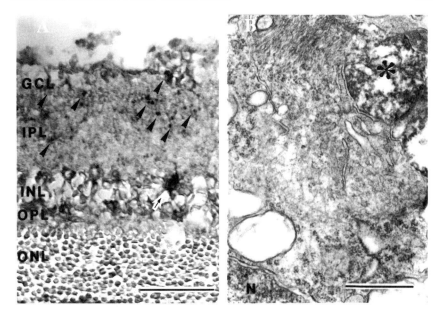

FIGURE 1. (**A**)PACAP-like immunoreactive structures (*arrowheads*) in the rat retina PACAP-like immunoreactivity is seen in the GCL, IPL, and INL. PACAP-like immunoreactive cell body (*arrow*). Scale bar = 50 μm. (**B**)PACAP-like immunoreactive axon terminal (*asterisk*) making synaptoid contact with ganglion cell. N: Nucleus of ganglion cell. Scale bar = 1 μm.

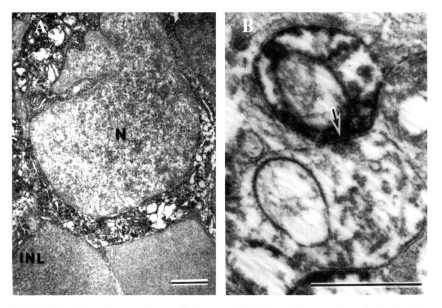

FIGURE 2. (**A**)An electron micrograph showing PACAPR-positive amacrine cell in the INL. PACAPR-like immunoreactivity is visible in cytoplasmic matrix. (**B**)PACAPR-positive cell processes in the IPL. These axon terminals make synaptic contact (*arrow*) with unlabeled nerve terminals. N: nucleus. Scale bar = 1 μm.

and dehydrated through ethanols and embedded in an Epon-Araldite mixture. Ultrathin, 70–80 nm, sections were stained with uranyl acetate and observed with Hitachi H-7000 or JEOL JEM1200 EXII electron microscopes. The PACAPR antisera (nos. 93093-2 and 93094-2) were raised in rabbits against a synthetic peptide that corresponds to the sequence from Lys[411] to Ala[435] of PACAPR, which is presumed to be its carboxy terminal intracellular domain. Immunoblot analysis indicated that the antisera reacted specifically with PACAPR.[2] To verify the specificity of immunostaining, the sections were incubated with non-immune serum instead of the primary antiserum, incubated with the primary antiserum without the second and third antisera, and incubated with the primary antiserum that had been preabsorbed with the antigen (50 μg of peptide in 1 ml of the antiserum at the working dilution).

RESULTS AND DISCUSSION

We have observed high level PACAPR mRNA expression in the ganglion cell layer and inner nuclear layer (INL), but weak expression in the inner plexiform layer (IPL), outer plexiform layer (OPL), outer nuclear layer (ONL), and outer segment by *in situ* hybridization histochemistry.[6,7] PACAP-positive fibers and terminals were visible in the ganglion cell layer (GCL), nerve fiber layer (NFL), and IPL (FIG. 1). Distribution of PACAP-positive fibers and terminals in the rat retina was different from that of VIP; VIP-like immunoreactivity (VIP-LI) is found in the cells of GCL, inner part of the IPL, and INL.[4] We found PACAPR-LI in the GCL and amacrine cells in the INL. Intense PACAPR-LI appeared to be concentrated predominantly in the ganglion cells, amacrine cells, and the IPL. PACAPR-LI appeared to be concentrated predominantly in the neuronal perikarya and processes. No PACAPR-LI was found in photoreceptors. At the ultrastructural level, PACAPR-LI was detected in the plasma membrane rough endoplasmic reticulum, and the cytoplasmic matrix of ganglion cells and neurons in the INL (FIG. 2). PACAPR-LI was also seen in the cell processes in the NFL and IPL. Morphologically, PACAP-like immunoreactivity (PACAP-LI) and PACAP mRNA were found in a population of sensory neurons in the rat uvea [8,9] and in the rabbit trigeminal ganglion.[9] Nerve fibers with PACAP-LI were also seen in the uvea of the rat eye.[6] Our immunocytochemical study with an antiserum against PACAP27 showed immunostaining for PACAP-positive fibers and terminals in the GCL, NFL, and IPL.[6] PACAP appears to be involved in the inflammatory response since PACAP-induced responses mimic the symptoms of inflammation and the concentration of PACAP-LI in the aqueous humor was increased greatly in inflamed eyes.[8] The survival of rat retinal ganglion cells *in vitro* is not promoted by peptide trophic factors unless their intracellular cAMP is increased or they are depolarized by K[+] or glutamate agonists.[10] Moreover, it has been shown that cAMP plays an important role in the differentiation and growth of ganglion cells in the postnatal rat retina.[10] Taken together, these findings strongly suggest that PACAP has a different function from VIP and that PACAP may function as a trophic factor in addition to its role as a transmitter and/or modulator through PACAPR.

REFERENCES

1. ARIMURA, A. 1992. Pituitary adenylate cyclase activating polypeptide (PACAP): discovery and current status of research. Regul. Pept. **37:** 287–303.
2. ARIMURA, A. & S. SHIODA. 1995. Pituitary adenylate cyclase activating polypeptide (PACAP) and its receptors: neuroendocrine and endocrine interaction. Front. Endocrinol. **16:** 53–88.
3. NILSSON, S. F. E., M. AMÉR & A. VANDERMEERS. 1996. Adenylate cyclase activation by VIP and PACAP in the retina and choroid: effects of antagonists. Ann. N.Y. Acad. Sci. **805:** 749–752.
4. CASINI, G. & N. C. BRECHA. 1991. Vasoactive intestinal polypeptide-containing cells in the rabbit retina: immunohistochemical localization and quantitative analysis. J. Comp. Neurol. **305:**

313–327.

5. CAVALLARO, S., V. D'AGATA, F. DRAGO, S. MUSCO, G. NUCIFORO, F. RICCIARDOLO, S. TRACALI, F. STICALA, A. ARIMURA & P. L. CANONICO. 1996. Ocular expression of type-1 pituitary adenylate cyclase-activating polypeptide (PACAP) receptors, Ann. N.Y. Acad. Sci. **805:** 555–557.

6. SEKI, T., D. OGINO, S. SHIODA, Y. NAKAI, A. ARIMURA & R. KOIDE. 1996. Localization of PACAP receptor and its transcript in the rat retina determined by immunochemistry and *in situ* hybridization. Acta Histochem. Cytochem. **29:** 882–883.

7. SEKI, T., D. OGINO, S. SHIODA, Y. NAKAI, A. ARIMURA & R. KOIDE. 1997. Distribution and ultrastructural localization of a receptor for pituitary adenylate cyclase activating polypeptide and its mRNA in the rat retina. Neurosci. Lett. **238:** 127–130.

8. MULDER, H., R. UDDMAN, K. MOLLER, Y.-Z. ZHANG, E. EKBLAD, J. ALUMETS & F. SUNDLER. 1994. Pituitary adenylate cyclase activating polypeptide expression in sensory neurons. Neuroscience **63:** 307–312.

9. WANG, Z.-Y., P. ALM & R. HÅKANSON. 1995. Distribution and effects of pituitary adenylate cyclase-activating peptide in the rabbit eye. Neuroscience **69:** 297–308.

10. MAYER-FRANKE, A., M. R. KAPLAN, F. W. PFRIEGER & B. BARRES. 1995. Characterization of the signaling interactions that promote the survival and growth of developing retinal ganglion cells in culture. Neuron **15:** 805–819.

Autoradiographic Visualization of the Receptor Subclasses for Vasoactive Intestinal Polypeptide (VIP) in Rat Brain

PASCALE VERTONGEN,[a,c] SERGE N. SCHIFFMANN,[b] PHILIPPE GOURLET,[a] AND PATRICK ROBBERECHT[a]

[a]Department of Biochemistry and Nutrition, and [b]Laboratory of Neuropathology and Neuropeptides Research, Faculty of Medicine, Université Libre de Bruxelles, Brussels, Belgium

Vasoactive intestinal polypeptide (VIP) exerts its biological effects through interaction with monomeric G protein–coupled receptors named the PACAP type II receptors.[1] Two receptor subtypes have been cloned and named the VIP_1 and VIP_2 receptors.[2,3] They recognized with a high affinity pituitary adenylate cyclase activating polypeptide (PACAP) and were distinct from the selective PACAP type I receptor that recognized VIP with a low affinity.[1] PACAP I and PACAP II receptors were both expressed in brain. PACAP II receptors were tenfold less abundant than the PACAP I receptors.[1] Distribution and mapping of the PACAP II receptors (the "VIP receptors") were performed[4] before it was discovered that these receptors were indeed heterogeneous. The location of the mRNA coding for the two VIP receptors was performed by *in situ* hybridization: VIP_1 and VIP_2 receptor mRNA were present in different brain areas.[5,6] However, as selective ligands for each receptor subtype were not available their direct visualization was not possible. We recently discovered[7] that a long-acting VIP analogue, RO 25-1553, developed as a bronchorelaxant and an antiinflammatory agent,[8] was selective for the VIP_2 receptor subtype and that [Arg^{16}]chicken secretin ([R^{16}]Ch.Sn) was selective for the rat VIP_1 receptor subclass.[9] These two iodinated molecules were used for direct visualization of the VIP_1 and the VIP_2 receptors in rat brain.

MATERIALS AND METHODS

Slide-mounted brain sections (25-μm thick) were preincubated at 20°C, twice for 10 min in a 50 mM Tris-HCl buffer pH 7.4 containing 0.5% bovine serum albumin. The incubation was performed in the same buffer but also containing 5 mM $MgCl_2$, 0.05% bacitracin, 2 mM EGTA, and 100 pM of tracer for 90 min at 20°C. The slides were then washed twice for 10 min in ice-cold preincubation buffer, then dipped in cold water. The dried slide-mounted brain sections were exposed to ^{3}H-hyperfilm (Amersham) for one week at -20°C. The non-specific binding was defined as the binding that remains in the presence of 1 μM VIP for ^{125}I-VIP, ^{125}I-[R^{16}]Ch.Sn and ^{125}I-RO 25-1553 (Fig.1).

[c]Corresponding author: Pascale Vertongen, Laboratoire de Chimie Biologique et de la Nutrition, Faculté de Médecine de l'Université Libre de Bruxelles, Bât. G/E, CP 611, 808 Route de Lennik, B-1070 Brussels, Belgium; Tel.: 32.2.555.62.15; Fax: 32.2.555.62.30; E-mail: probbe@ulb.ac.be

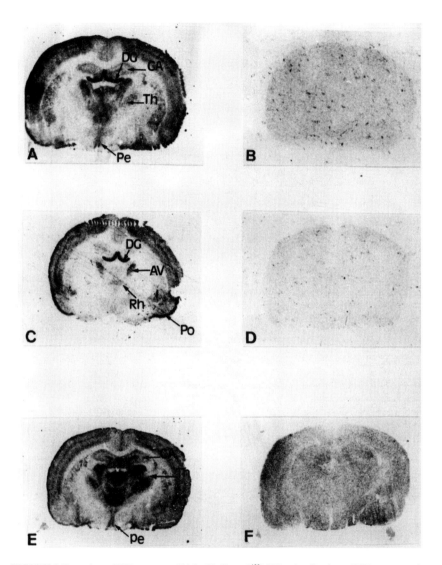

FIGURE 1. Detection of VIP receptors (**A**) by binding of ^{125}I-VIP; visualization of VIP$_1$ receptors by binding ^{125}I-[R^{16}]Ch.Sn (**C**) and visualization of VIP$_2$ receptors by binding of ^{125}I-RO 25-1553 (**E**). **B, D,** and **F** represent the non-specific binding of ^{125}I-VIP, ^{125}I-[R^{16}]Ch.Sn, and ^{125} RO 25-1553, respectively, obtained in the presence of 1 μM VIP. *Abbreviations*: AV, anteroventral thalamic nucleus; BL, basolateral amygdaloid nucleus; CA, Ammon's horn; CE, central amygdaloid nucleus; DG, dentate gyrus; LA, lateral amygdaloid nucleus; Pe, hypothalamic periventricular nucleus; Po, pyriform cortex; Rh, rhomboid thalamic nucleus; Th, thalamus.

RESULTS

Serial coronal sections were labeled with ^{125}I-[R^{16}]Ch.Sn or ^{125}I-RO 25-1553 for identification of the VIP$_1$ and VIP$_2$ receptors, respectively. VIP$_1$ receptors were identified in the cerebral and the pyriform cortex, the dentate gyrus, and the lateral amygdaloid nucleus. In the thalamus, VIP_1 receptors were identified only in the anteroventral and the rhomboid nucleus (FIG. 1, TABLE 1). The VIP_1 receptors were also expressed in the claustrum, caudate putamen, supraoptic nucleus, choroid plexus, and pineal gland. The VIP_2 receptors were abundant in the cerebral cortex, lateral septal nuclei, bed nucleus of the stria terminalis, basolateral amygdaloid nucleus, most of the thalamic nuclei except some centromedial nuclei, mammillary nuclei, and superficial grey layer of the superior colliculus (FIG.1, TABLE 1). In the hypothalamus, VIP_2 receptors were identified in the suprachiasmatic nucleus, periventricular nucleus, and dorsomedial compact nucleus. They were detected at lower levels in the claustrum, nucleus accumbens, caudate putamen, choroid plexus, medial habenula nucleus, and Ammon's horn.

DISCUSSION

We recently discovered by a systematic screening of VIP derivatives on recombinant VIP_1, VIP_2, PACAP I and secretin receptors expressed in CHO cells that the cyclic derivative of VIP, RO 25-1553, recognized the VIP_1 and VIP_2 receptors with K_d values of 100 and 0.8 nM, respectively.[7] RO 25-1553 did not interact with the secretin and the PACAP I receptors. The selective VIP_1 receptor agonist [R[16]]Ch.Sn resulted from two original findings: first, that the introduction of an arginine residue in position 16 increased the affinity of the ligand for the VIP receptor and decreased the affinity for the secretin receptor,[10] second, that chicken secretin was equipotent on rat VIP_1 and secretin receptors but did not interact with the VIP_2 receptor. In the present study, we used ^{125}I-[R[16]]Ch.Sn and ^{125}I-RO 25-1553 to determine the distribution on rat brain sections of VIP_1 and the VIP_2 receptors, respectively. The mapping of the VIP_1 and VIP_2 receptors in rat brain revealed a clear difference in receptor localization: some areas expressed only one receptor type (TABLE 1), other areas expressed both receptors but at this stage of mapping a colocalization cannot be proposed. In TABLE 1, we compare the distribution of VIP_1 and VIP_2 receptors (present work) to the previously described mRNA distribution.[5,6] In most brain areas, the distribution of each VIP receptor

TABLE 1. Comparison of VIP_1 and VIP_2 receptors localization to their mRNA distribution[5,6] and to the peptide (VIP and PACAP) distribution

	VIP_1 Receptor	VIP_1 Receptor mRNA	VIP_2 Receptor	VIP_2 Receptor mRNA	VIP (peptide)	PACAP (peptide)
Cortex	+++	+++	++	+++	++++	++
Accumbens	+	N.D.	++	++	+	N.D.
Bed stria terminalis	-	-	+++	+++	++	++
Lateral septum	-	N.D.	++	N.D.	+	++
Caudate putamen	+	N.D.	+	N.D.	+	-
Amygdaloid	LA +++	LA++	BL +++	CE ++	CE ++	LA ++
Suprachiasmatic	-	-	++++	++++	++++	-
Paraventricular	-	-	-	++	-	+++
Supraoptic	+	-	-	-	++	+++
Periventricular	-	-	++	++	+	++
Dentate gyrus	+++++	+++++	-	-	+++	-
CA hippocampus	+	+	+	++	+	+
Thalamus nuclei	- AV,Rh ++	- AV ++	++++++	+++++	-	++++
Habenula	-	N.D.	++	N.D.	N.D.	++
Median geniculate	-	-	+++	+++	-	N.D
Lateral geniculate	-	-	+++	+++	++	N.D
Mammillary N.	-	-	+++	+++	+	+
Superior colliculus	-	-	+++	+++	++	N.D.
Choroid plexus	++	++	++	++	+	N.D.

subtype was superimposable to the mRNA localization except in the supraoptic nucleus for the VIP_1 receptor and in the paraventricular nucleus for the VIP_2 receptor. Indeed, in the supraoptic nucleus, the VIP_1 receptor was present but not the VIP_1 receptor mRNA. It could be possible that the VIP_1 receptor was present only on the fibers; therefore the mRNA should be detected on the A1 dopaminergic cells. No data on this region were available. In the paraventricular nucleus, the mRNA of the VIP_2 receptor was present,[5,6] but we did not detect the VIP_2 receptor that could be located in the posterior pituitary gland. However, binding studies performed on total pituitary gland revealed PACAP I receptors only.[11] In TABLE 1, we also compared the VIP receptor distribution to the VIP and PACAP localization. In all the regions tested, at least one of the two peptides was expressed. It was remarkable that in the thalamus, where a high density of VIP_2 receptors was detected, only PACAP was expressed, and yet on the other hand that in the dentate gyrus, where only VIP_1 receptor was expressed, only VIP was present, suggesting that in these regions, VIP was the ligand for the VIP_1 receptor and PACAP was the ligand for the VIP_2 receptor.

REFERENCES

1. CHRISTOPHE, J. 1993. Type I receptors for PACAP (a neuropeptide even more important than VIP?). Biochim. Biophys. Acta **1154**: 183–199.
2. ISHIHARA, T. *et al.* 1992. Functional expression and tissue distribution of a novel receptor for vasoactive intestinal polypeptide. Neuron **8**: 811–819.
3. LUTZ, E. M. *et al.* 1993. The VIP_2 receptor: molecular characterisation of a cDNA encoding a novel receptor for vasoactive intestinal peptide. FEBS Lett. **334**: 3–8.
4. BESSON, J. *et al.* 1986. Characterization and autoradiographic distribution of vasoactive intestinal peptide binding sites in the rat central nervous system. Brain Res. **398**: 329–336.
5. SHEWARD, W. J., E. M. LUTZ & A. J. HARMAR. 1995. The distribution of vasoactive intestinal peptide 2 receptor messenger RNA in the rat brain and pituitary gland as assessed by *in situ* hybridization. Neurosci. **67**: 409–418.
6. USDIN, T. B., T. I. BONNER & E. MEZEY. 1994. Two receptors for vasoactive intestinal polypeptide with similar specificity and complementary distributions. Endocrinology **135**: 2662–2680.
7. GOURLET, P. *et al.* 1997. The long-acting vasoactive intestinal polypeptide agonist RO 25-1553 is highly selective of the VIP2 receptor subclass. Peptides **8**: 403–408.
8. O'DONNELL, M. *et al.* 1994. RO 25-1553: a novel long-acting vasoactive intestinal peptide agonist. Part II: Effect on in vitro and in vivo models of pulmonary anaphylaxis. J. Pharmacol. Exp. Ther. **270**: 1289–1294.
9. GOURLET, P. *et al.* 1997. Development of high affinity selective VIP_1 receptor agonists. Peptides **18**: 1539–1545.
10. GOURLET, P. *et al.* 1996. Effect of introduction of an arginine[16] in VIP, PACAP and secretin on ligand affinity for the receptors. Biochim. Biophys. Acta **1314**: 267–273.
11. VERTONGEN, P. *et al.* 1996. Pituitary adenylate cyclase-activating polypeptide/vasoactive intestinal polypeptide receptor subtypes are differently expressed in rat transplanted pituitary tumours (SMtTW) and in the normal gland. J. Mol. Endocrinol. **16**: 239–248.

Pituitary Adenylate Cyclase-Activating Polypeptide Receptors in the Fetal Human Adrenal Gland[a]

L. YON,[b] L. BREAULT,[c] V. CONTESSE,[b] G. BELLANCOURT,[b]
C. DELARUE,[b] A. FOURNIER,[d] J.G. LEHOUX,[c] N. GALLO-PAYET,[c]
AND H. VAUDRY[b,e]

[b]European Institute for Peptide Research (IFRMP N°23), Laboratory of
Cellular and Molecular Neuroendocrinology, INSERM U413, UA
CNRS, University of Rouen, 76821 Mont-Saint-Aignan, France

[c]Service of Endocrinology, Department of Medicine, Faculty of Medicine,
University of Sherbrooke, Sherbrooke, Quebec, Canada J1H 5N4

[d]Institut National de la Recherche Scientifique-Santé, University of
Quebec, Pointe-Claire, Quebec, Canada H9R 1G6

The occurrence of PACAP and PACAP binding sites has been demonstrated in the adrenal gland of several species.[1,2] Moreover, PACAP has been shown to modulate proliferation and differentiation of adrenal chromaffin cells,[3,4] suggesting a role of the peptide in the ontogeny of the adrenal gland. The fetal human adrenal gland is characterized by a large fetal zone, representing 85% of the gland, and a thin neocortex. During fetal life, cells derived from the ectoderm migrate inside the gland to form the medulla. After birth, the fetal zone involutes and the cortex develops and differentiates into the three characteristic zones present in the adult adrenal gland. These profound changes in the organization of the gland suggest that trophic factors, such as PACAP, may control the modeling of the tissue during ontogenesis. However, the distribution and function of PACAP binding sites have not yet been studied in the adrenal gland during development. The aim of the present study was to localize PACAP receptors in the fetal human adrenal by autoradiography and to investigate the effect of PACAP on second messenger systems in suspended fetal adrenal cells.

MATERIALS AND METHODS

Fetal human adrenal glands were obtained from fetuses aged between 14 and 20 weeks (post-conceptional) at the time of therapeutic abortion. Autoradiographic experiments

[a]This work was supported by grants from INSERM (U413), the Medical Research Council of Canada (MRC MT13679), a France-Québec exchange programme (Coopération Scientifique et Technologique Franco-Québécoise, n° PVP-33-9) and the Conseil Régional de Haute-Normandie. L.B. was the recipient of a studentship from the Fonds de la Recherche en Santé du Québec and the Fonds pour la Formation de Chercheurs et l'Aide à la Recherche (FRSQ/FCAR). G.B. was the recipient of a studentship from the Conseil Régional de Haute-Normandie.

[e]Corresponding author: Tel.: (33) 235-14-6624; Fax: (33) 235-14-6946; E-mail: hubert.vaudry@univ-rouen.fr

416

were performed using HPLC-purified monoiodinated PACAP27 as a tracer. Fetal adrenal cells were dissociated and used after a 12-h resting period (suspended cells) as previously described.[5] Cyclic AMP production was determined by measuring the conversion of [^{3}H]ATP to [^{3}H]cAMP. The effect of PACAP on the formation of total inositol phosphates (InsPs) was investigated after incubation of fetal adrenal cells in the presence of 2 µCi of *myo*-[^{3}H]inositol.

RESULTS AND DISCUSSION

Localization of PACAP Binding Sites

Autoradiographic labeling using [^{125}I]PACAP27 showed that in the adrenal gland of 14- to 20-week-old fetuses PACAP binding sites are exclusively borne by medullary cells (FIG. 1). Neither the neocortex nor the fetal zone displayed any labeling. In agreement with this observation, it has previously been reported that in the neonatal and adult rat adrenal gland PACAP receptors are exclusively located on chromaffin cells.[6]

Characterization of PACAP Binding Sites

Scatchard plot analysis of displacement curves, using PACAP27 as a competitor, indicated the existence of a single class of high affinity PACAP receptors with K_d values ranging from 0.32 to 0.74 nM and B_{max} values of 0.30 to 0.81 pmol/mg wet tissue. PACAP27 and PACAP38 were equipotent in displacing the binding of the tracer (IC$_{50}$ = 0.28 to 0.64 nM and 0.15 to 0.81 nM, respectively) while VIP was a poor competitor. These data demonstrate that the binding sites evidenced with [^{125}I]PACAP27 correspond predominantly to type I PACAP receptors. Thus, the situation in the fetal human adrenal resembles that of the rat adrenal medulla, which expresses mRNA for type I PACAP receptors.[6,7]

Effect of PACAP on Second Messenger Systems

PACAP38 stimulated, in a dose-dependent manner, cAMP production from suspended fetal adrenal cells (ED$_{50}$ = 0.07 ± 0.02 nM; FIG. 2). These data indicate that the binding

A   **B**

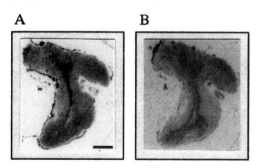

FIGURE 1. Typical autoradiograms showing the distribution of [^{125}I]PACAP27 binding in the fetal human adrenal gland (**A**) and the displacement of radioligand binding by synthetic PACAP27 (10^{-8} M) (**B**). Autoradiograms were obtained from 10-µm consecutive sections of an adrenal gland from a 19-week old human fetus. Bar: 2 mm.

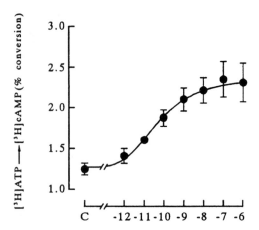

FIGURE 2. Effect of PACAP38 on cAMP production by adrenal cells from 18- to 20-week-old human fetuses. Suspended cells used after a 12-h resting period were labeled with [³H]adenine and then incubated for 15 min at 37°C in the absence (control, C) or in the presence of graded concentrations of PACAP38. Results are the mean (± S.E.) of triplicate determinations from one representative experiment out of three.

sites visualized by autoradiography actually correspond to functional receptors positively coupled to adenylyl cyclase. While PACAP has been shown to activate both adenylyl cyclase and phospholipase C in porcine adrenochromaffin cells,[8] we did not observe any effect of PACAP on InsPs formation in suspended human adrenal cells and only a slight increase of InsPs in cells cultured for 5 days.

Functional Implications

The role of PACAP in the human adrenal during ontogenesis is currently unknown. However, other studies conducted in animal models suggest that in the fetal human adrenal gland PACAP may stimulate catecholamine release and/or modulate the migration or differentiation of immature chromaffin cells.

CONCLUSION

The present results show that type I PACAP receptors are expressed in the fetal human adrenal medullary zone during the second trimester of gestation and that these binding sites are functionally coupled to adenylyl cyclase. These data suggest that PACAP may play a role in the ontogenesis of the human adrenal medulla and/or in the regulation of catecholamine secretion.

REFERENCES

1. YON, L. *et al.* 1993. Localization, characterization and activity of pituitary adenylate cyclase-activating polypeptide in the frog adrenal gland. J. Endocrinol. **139**: 183–194.
2. TABARIN, A. *et al.* 1994. Pituitary adenylate cyclase-activating peptide in the adrenal gland of mammals: distribution, characterization and responses to drugs. Neuroendocrinology **59**: 113–119.

3. TISCHLER, A. S., J. C. RISEBERG & R. GRAY. 1995. Mitogenic and antimitogenic effects of pituitary adenylate cyclase-activating polypeptide (PACAP) in adult rat chromaffin cell cultures. Neurosci. Lett. **189**: 135–138.
4. WOLF, N. & K. KRIEGLSTEIN. 1995. Phenotypic development of neonatal rat chromaffin cells in response to adrenal growth factors and glucocorticoids: focus on pituitary adenylate cyclase-activating polypeptide. Neurosci. Lett. **200**: 207–210.
5. BREAULT, L., J. G. LEHOUX & N. GALLO-PAYET. 1996. The angiotensin AT2 receptor is present in the fetal human adrenal gland throughout the second trimester of gestation. J. Clin. Endocrinol. Metab. **81**: 3914–3922.
6. MOLLER, K. & F. SUNDLER. 1996. Expression of pituitary adenylate cyclase-activating polypeptide (PACAP) and PACAP type I receptors in the rat adrenal medulla. Regul. Pept. **63**: 129–139.
7. SPENGLER, D. *et al.* 1993. Differential signal transduction by five splice variants of the PACAP-receptor. Nature **365**: 170–175.
8. ISOBE, K., T. NAKAI & Y. TAKUWA. 1993. Ca^{2+}-dependent stimulatory effect of pituitary adenylate cyclase-activating polypeptide on catecholamine secretion from cultured porcine adrenal medullary chromaffin cells. Endocrinology **132**: 1757–1765.

Neuropeptides in Developmental Tumors of the Central and Peripheral Nervous System

MICHAEL C. FRÜHWALD,[a,b,f] M. SUE O'DORISIO,[a,e]
SANDRA L. COTTINGHAM,[c,g] STEPHEN J. QUALMAN,[c]
THOMAS M. O'DORISIO[d]

[a]*Department of Pediatrics, Hematology and Oncology, The Ohio State University, Columbus, Ohio 43205, USA*

[b]*Department of Pediatrics, University of Würzburg, 97080 Würzburg, Germany*

[c]*Department of Pathology and Laboratory Medicine, The Ohio State University, Columbus, Ohio 43205, USA*

[d]*Department of Internal Medicine, The Ohio State University, Columbus, Ohio 43210, USA*

Neuropeptides such as VIP, somatostatin, NPY, neurotensin, GRP (bombesin), and substance P fulfill an important role in the regulation of normal neuronal development.[1] Moreover they have been implicated in the progression and differentiation of malignant tumors of the central and peripheral nervous system.[2] While there are extensive data available on the expression of these regulatory peptides and their respective receptors in adult nervous system tumors and in neuroblastoma of childhood, little is known about their occurrence and significance in other neuronal malignancies of childhood.

Central primitive neuroectodermal tumors (cPNET), especially medulloblastoma, and the Ewing's sarcoma family of tumors (ESF tumors) make up a large group of solid neoplasms in childhood and contribute significantly to the mortality and morbidity of childhood cancer. There is mounting evidence that the Ewing's sarcoma family of tumors is derived, as is neuroblastoma, from cells of the neural crest. The cell of origin for medulloblastoma and other PNET of the CNS has not been determined to date, but is generally thought to be neuroepithelial.[3]

We have previously demonstrated the correlation of increased VIP and somatostatin levels in neuroblastomas with the grade of differentiation and also with favorable disease stage in this particular type of tumor.[4] Since neuroblastoma is the prototype of developmental childhood malignancies, we speculated that there could be a similar relationship in other tumors of the developing nervous system.

The detection of neuropeptides and their receptors in childhood malignancies of the nervous system may help to elucidate the origin of these tumors. This would be a big step

[e]Corresponding author: Department of Pediatrics, Children's Hospital, 700 Children's Drive, Columbus, Ohio 43205; e-mail: modorisi@chi.osu.edu
[f]M. C. F. is supported by a fellowship of the Dr. Mildred Scheel Stiftung für Krebsforschung.
[g]Current address: Blodgett Memorial Medical Center, Dept. of Laboratory Medicine, Grand Rapids, Michigan 49506.

in our comprehension of the molecular pathology of these tumors. With the rapid development of pharmacological agonists and antagonists for these peptides, a window of diagnostic and therapeutic opportunities may open up.

STUDY DESIGN

Tumor tissue specimens of human tumors were obtained through the Cooperative Human Tissue Network (CHTN). Samples had been collected fresh at the time of surgery and immediately frozen to -80°C. A total of 39 nervous system tumors were assayed (central primitive neuroectodermal tumors including medulloblastomas, $N = 9$; pilocytic astrocytomas, $N = 3$; high grade gliomas, $N = 6$; ependymoma, $N = 4$; Ewing's sarcoma family of tumors, $N = 15$; neuroblastomas, $N = 17$). Whenever sufficient tumor material was available tissues were assayed for NPY, somatostatin (SRIF), neurotensin, VIP, substance P (SP), and GRP.

Radioimmunoassay (RIA) was performed as described earlier.[4,5] Briefly, tissue from tumors was extracted in 1 M acetic acid:ethanol (1:1). The resulting supernatant was lyophilized and the residue was dissolved in 0.05 M acetic acid for the assay. DNA concentration was determined by a standard diphenylamine analysis.[6] Peptide contents are expressed in pg/µg DNA. Synthetic porcine antibodies against VIP, SRIF, NPY, GRP, neurotensin, and SP were used. Radioactive peptides were purchased from Amersham.

In selected cases, immunocytochemical analysis was performed as described in our previous paper.[4] Staining for neuropeptides was performed on formaldehyde-fixed deparaffinized tumor sections and the avidin-biotin-peroxidase reaction was used for visualization (FIG. 1).

RESULTS

Thirty-nine pediatric tumors were examined including 9 central and 15 peripheral PNET, 3 low grade and 5 high grade brain tumors, 4 ependymomas, and 17 neuroblastomas. A compilation of the results is given in TABLE 1. In each single tumor type one or, at maximum, two neuropeptides predominated.

Neuropeptide Y was, as expected, positive in all of 17 neuroblastomas; the levels ranged between 13.1 and 90.2 pg/µg DNA. Eight of 14 tumors of the Ewing's sarcoma family of tumors (57.1%) also expressed NPY (0.5–8.5 pg/µg DNA). None of three pilocytic astrocytomas and only one of four ependymomas expressed measurable NPY. Two of seven cPNET had levels for NPY at the limit of detection (0.2 and 0.5 pg/µg DNA). Four of six high grade gliomas exhibited detectable levels for this peptide (2.0–16.3 pg/µg DNA).

Somatostatin was detected in 100% of the central nervous system tumors. Levels for cPNET ranged from 0.2–7.9 pg/µg DNA; pilocytic astrocytoma levels ranged from 0.8–34.0 pg/µg DNA. In high grade gliomas the lowest level was 1.1; the highest 8.1 pg/µg DNA. Ependymomas showed levels from 0.5 to 8.4 pg/µg DNA; 11 of 16 neuroblastomas (1.0–10.0 pg/µg DNA) and 9 of 15 ESF tumors (0.2–9.5 pg/µg DNA) also exhibited detectable somatostatin.

VIP was expressed in a high number of neuroblastomas (13/17, 3.7–34.2 pg/µg DNA). Only three of nine cPNET (0.1–0.2 pg/µg DNA), three of six high grade gliomas (0.4–3.3 pg/µg DNA), one of 4 ependymomas (3.7 pg/µg DNA), and four of 15 ESF tumors (0.1–1.5 pg/µg DNA) presented VIP immunoreactivity. None of the pilocytic astrocytomas showed immunoreactivity for VIP.

Like somatostatin, substance P was positive in all central PNET (0.1–2.8 pg/µg DNA) and all but one ESF tumor (0.1–10.2 pg/µg DNA). Two of the three pilocytic astrocytomas

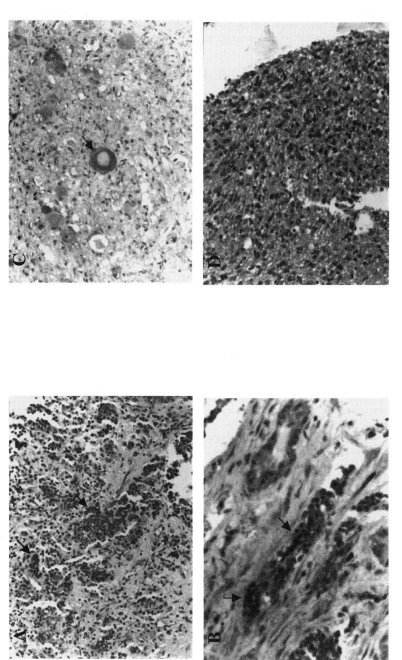

FIGURE 1. Immunocytochemical studies on neuronal tumors of childhood. Formaldehyde-fixed deparaffinized tumor sections were marked with the appropriate first antibody and the avidin-biotin-peroxidase reaction was applied followed by H&E stain. Brown pigment represents the immunoreactivity. (**A**) Peripheral PNET stained for substance P (SP). (**B**) Magnification of a section from the same tumor. Significant immunoreactivity around small, blue, round tumor cells. (**C**) Presence of neuropeptide Y (NPY) in a classic neuroblastoma. (**D**) Demonstration of gastrin releasing peptide (GRP) in the case of a high grade glioma.

(12.3 and 83.3 pg/μg DNA), three of the six high grade gliomas (0.1–265.0 pg/μg DNA) and three of four ependymomas (0.5–10.1 pg/μg DNA) were also positive. All neuroblastomas were negative.

As anticipated, each of the six high grade gliomas showed immunoreactivity for GRP/bombesin (1.0–8.2 pg/μg DNA). GRP was also positive in four of nine cPNET (0.6–33.0 pg/μg DNA), two of three pilocytic astrocytomas (1.8 and 54.3 pg/μg DNA), three of four ependymomas (0.7–8.1 pg/μg DNA), and six of 15 ESF tumors (0.9–45.9 pg/μg DNA). Only two neuroblastomas were assayed of which one was positive with 19.3 pg/μg DNA.

Interestingly none of the tumors measured showed immunoreactivity for neurotensin.

At least one regulatory peptide was expressed in all but one of the 39 tumors. This neoplasm (a Ewing's sarcoma) was largely necrotic. When examined with the more sensitive method of RT-PCR the sample showed expression of mRNA for VIP (SRIF was negative, other peptides were not tested).

DISCUSSION

Our results clearly identify a neuropeptide pattern for each tumor subtype (TABLE 2), suggesting a specific origin or developmental stage for the individual tumor types. The origin of cPNET has not yet been clearly established. Ever since the description by Harvey and Cushing in 1925, the cell of origin for medulloblastomas and the other cPNET has been somewhat elusive.[7] By light microscopy alone, central PNET are indistinguishable from their peripheral counterpart, which is grouped with the ESF tumors. In our study we found substance P as a unifying peptide for these two tumor types. Gould and colleagues had found a similar pattern when examining the neuropeptide status of 10 cPNET; somatostatin, substance P, and VIP were the most frequently found neuropeptides, suggesting a neuroendocrine origin for these neoplasms.[8]

Little data are available on substance P and its receptors in the ESF tumors and neuroblastoma. Hennig and colleagues[9] found no receptors on neuroblastoma tumor cells in eight specimens when tested by autoradiography, while all of them showed positive signals in the tumor vessels. In our study, substance P could not be detected at all in neuroblastoma.

Somatostatin peptide expression is characteristic of the CNS tumors. All central nervous system tumors in this study had measurable levels of somatostatin. This finding is clearly supported by the literature. Allen and colleagues[10] and Bateman[11] identified somatostatin in central nervous system tumors. While Allen had speculated a differentiation dependence of somatostatin expression, Bateman found high levels for somatostatin in one medulloblastoma, which is by definition a WHO grade-four tumor. Developmental studies have shown that receptors for somatostatin are abundant in the immature human cerebellum, which is the primary location for the most important cPNET, the medulloblastoma. The concentration of these receptors reaches a maximum around birth and decreases in parallel with maturation of the cerebellum.[12] As we have previously shown, somatostatin receptors are found in high densities in medulloblastomas,[13] suggesting a developmental arrest of certain cerebellar cells that produce somatostatin or bear its receptors. Our findings in this provide further support for this concept, since all cPNET expressed somatostatin to some extent.

Another interesting finding is that all of the neuroblastomas expressed NPY, while not a single one of them showed substance P immunoreactivity. Even though neuroblastomas and ESF tumors are possibly both derived from the neural crest, our findings clearly distinguish these tumors from each other. While a role for NPY has been established in neuroblastoma,[14] the role of substance P in the ESF tumors remains undefined. Even though we found little NPY expression by RIA in the ESF tumors, several studies on cell lines of these

TABLE 1. Neuropeptides in neuronal tumors of childhood as measured by radioimmunoassay

Peptide	cPNET[a]	ESF Tumors	High Grade Glioma	Pilocytic Astrocytoma	Ependymoma	NB
NPY	2/7	8/14	4/6	0/3	1/4	17/17
SRIF	9/9	9/15	6/6	3/3	4/4	11/16
Neurotensin	0/7	0/12	0/4	0/3	0/2	0/17
VIP	3/9	4/15	3/6	0/3	1/4	13/17
Substance P	9/9	14/15	3/6	2/3	3/4	0/17
GRP	4/9	6/15	6/6	2/3	3/4	1/2

[a]Number of tumors expressing peptide/number of tumors examined.

tumors employing molecular biology techniques provided data for the existence of NPY- and related peptide receptors on these lineages.[15,16] Therefore our RIA may not be sensitive enough to detect very low levels of NPY that may be downregulated by a negative feedback inhibition mechanism when high concentrations of NPY receptors are present.

The high levels for somatostatin and GRP in high grade gliomas are somewhat surprising and the origin of the peptides remains to be determined. Generally the content of ligands and the corresponding receptors for neuropeptides is lower in less differentiated, more primitive neuronal cells. Accordingly, Reubi and colleagues found no receptors for somatostatin on glioblastomas.[17] In contrast receptors for GRP have been identified in cell lines of glioblastomas by several groups.[18,19] The role of GRP in normal neurophysiology and in the pathogenesis of brain tumors needs further investigation.

The physiological function of neurotensin has not yet been fully elucidated. There is, however, evidence for its role in nociception and regulation of blood flow.[20] Therefore our finding of no detectable levels for this regulatory peptide in any of the assayed neoplasms is unexpected. The validity of our assay was tested by running a VIPoma sample in which neurotensin was cosecreted with VIP as a positive control.[21]

We discovered VIP in a high percentage of neuroblastomas and we have previously shown the significance of this peptide in neuroblastoma.[22] Even if our data do not show extraordinarily high levels for VIP in any of the CNS tumors, evidence suggests the importance of this peptide in the growth regulation of these malignancies.[23] Recently, two groups independently demonstrated an antiproliferative effect of VIP on human glioblastoma cell lines.[24,25] Therefore VIP's involvement in developmental tumors certainly deserves further investigations.

Neuropeptides and their receptors have a significant potential as molecular targets in the diagnosis and treatment of various tumor types. Somatostatin analogs (especially octreotide) have been successfully used in the treatment of acromegaly, chemotherapy-induced pancreatitis and diarrhea states in children. The visualization of neuroblastomas and lymph node metastases with somatostatin receptor scintigraphy has become standard care in some institutions.[26] Radiolabeled VIP has recently been shown to be a promising candidate in the scintigraphic visualization of gastroentero-pancreatic tumors.[27] Preliminary data employing radiolabeled substance P as an imaging agent suggest a similar role for this neuropeptide.[28,29] We propose to use this radioligand in the diagnosis of

TABLE 2. Neuropeptide pattern in developmental tumors of the nervous system

Tumor Type	Neuropeptides
cPNET	Somatostatin and substance P (100%)
High grade glioma	Somatostatin and GRP (100%)
Neuroblastoma	NPY (100%); no substance P
ESF tumors	Substance P (93.3%)

central and peripheral PNET. It may prove a suitable agent for the diagnosis of occult lymph nodes and or help distinguish between therapy-related tissue changes and biologically active tumor tissue. Receptors of the specific neuropeptides may also be used as therapeutic targets for radiotherapy. Krenning and colleagues report on the successful use of ^{111}In octreotide as therapy in a patient with a disseminated GEP tumor.[30]

Taken together our findings indicate characteristic neuropeptide profiles in developmental tumors of the nervous system. These results suggest that neuropeptides and neuropeptide receptors should be considered as diagnostic and therapeutic targets in pediatric nervous system tumors.

REFERENCES

1. O'DORISIO, M. & T. O'DORISIO. 1993. Neural crest tumors: Rationale for somatostatin and its analogs in diagnosis and therapy. *In* Endocrine Tumors. E. Mazzaferri & N. Jamaan, Eds.: 531–542. Blackwell Scientific Publications. Cambridge.
2. O'DORISIO, M. S., F. CHEN, T. M. O'DORISIO, D. WRAY & S. QUALMAN. 1994. Characterization of somatostatin receptors on human neuroblastoma tumors. Cell Growth Diff. **5**: 1–8.
3. RORKE, L. B, J. Q. TROJANOWSKI, V. M. LEE *et al.* 1997. Primitive neuroectodermal tumors of the central nervous system. Brain Pathol. **7**: 765–784.
4. QUALMAN, S. J., M. S. O'DORISIO, D. J. FLESHMAN, J. LABANOWSKI, H. SHIMADA & T. M. O'DORISIO. 1992. Neuroblastoma: Correlation of neuropeptide expression in tumor tissue with other prognostic factors. Cancer **70**: 2005–2012.
5. O'DORISIO, T. M., H. S. MEKHJIAN, E. C. ELLISON, M. S. O'DORISIO, T. S. GAGINELLA & E. A. WOLTERING. 1987. Role of peptide radioimmunoassay in understanding peptide-peptide interactions and clinical expression of gastroenteropancreatic endocrine tumors. Am. J. Med. **82**: 60–67.
6. BURTON, K. 1972. Determination of DNA concentration with diphenylamine. Methods Enzymol. **7**: 163–167.
7. BAILEY, P. & H. CUSHING. 1925. Medulloblastoma cerebelli: A common type of midcerebellar glioma of childhood. Arch. Neurol. Psychiatry **14**: 192–224.
8. GOULD, V. E., D. S. JANSSON, W. M. MOLENAAR *et al.* 1990. Primitive neuroectodermal tumors of the central nervous system. Patterns of expression of neuroendocrine markers, and all classes of intermediate filament proteins. Lab. Invest. **62**: 498.
9. HENNIG, I. M., J. A. LAISSUE, U. HORISBERGER & J-C. REUBI. 1995. Substance-P receptors in human primary neoplasms: tumoral and vascular localization. Int. J. Cancer **61**:786–792.
10. ALLEN, J. M., N. R. HOYLE, J. C. YEATS, M. A. GHATEI, D. G. THOMAS & S. R. BLOOM. 1985. Neuropeptides in neurological tumours. J. Neurooncol. **3**: 197–202.
11. BATEMAN, D. E., J. R. MCDERMOTT, R. H. PERRY, R. DIMALINE, J. A. BIGGINS & J. A. EDWARDSON. 1986. Neuropeptides in gliomas: identification of somatostatin 14 in a medulloblastoma. J. Neurol. Neurosurg. Psych. **49**: 1074–1076.
12. LAQUERRIERE, A., B. LEROUX, B. GONZALEZ, C. BODENANT, J. TAYOT & H. VAUDRY. 1992. Somatostatin receptors in the human cerebellum during development. Brain Res. **573**: 251–259.
13. MÜLLER, H. L., M. C. FRÜHWALD, M. SCHEUBECK *et al.* 1997. A possible role for somatostatin receptor (sstr) scintigraphy in the follow-up of children with medulloblastoma (MB). J. Neurooncol. (In press.)
14. KOGNER, P., O. BJORK & E. THEODORSSON. 1993. Neuropeptide Y in neuroblastoma: increased concentration in metastasis, release during surgery, and characterization of plasma and tumor extracts. Med. Pediatr. Oncol. **21**: 317–322.
15. VAN VALEN, F., W. WINKELMANN & H. JURGENS. 1992. Expression of functional Y1 receptors for neuropeptide Y in human Ewing's sarcoma cell lines. J. Cancer Res. Clin. Oncol. **118**: 529–536.
16. INUI, A., K. SANO, M. MIURA *et al.* 1992. Evidence for further heterogeneity of the receptors for neuropeptide-Y and peptide- W in tumor cell lines derived from neural crest. Endocrinology **131**: 2090–2096.

17. REUBI, J., W. LANG, R. MAURER, J. KOPER & S. LAMBERTS. 1987. Distribution and biochemical characterization of somatostatin receptors in tumors of the human central nervous system. Cancer Res. **47**: 5758–5764.

18. MOODY, T. W., S. MAHMOUD, J. STALEY et al. 1989. Human glioblastoma cell lines have neuropeptide receptors for bombesin/gastrin-releasing peptide. J. Mol. Neurosci. **1**: 235–242.

19. SHARIF, T. R., W. LUO & M. SHARIF. 1997. Functional expression of bombesin receptor in most adult and pediatric human glioblastoma cell lines; role in mitogenesis and in stimulating the mitogen-activated protein kinase pathway. Mol. Cell Endocrinol. **13**: 119–30.

20. REINECKE, M. 1985. Neurotensin: Immunohistochemical localization in central and peripheral nervous system and in endocrine cells and its functional role. Prog. Histochem. Cytochem. **16**: 1–172.

21. O'DORISIO, T. M., B. A. HOWE, D. S. HILL, J. OBERMEYER & S. CATALAND. 1987. Vasoactive intestinal peptide (VIP) and neurotensin radioimmunoassays. J. Clin. Immunoassay **10**: 85–90.

22. O'DORISIO, M. S., D. J. FLESHMAN, S. J. QUALMAN & T. M. O'DORISIO. 1992. Vasoactive intestinal peptide: autocrine growth factor in neuroblastoma. Regul. Pept. **37**: 213–226.

23. GIANGASPERO, F., P. C. BURGER, D. A. BUDWIT, L. USELLINI & A. M. MANCINI. 1985. Regulatory peptides in neuronal neoplasms of the central nervous system. Clin. Neuropathol. **4**: 111–115.

24. VERTONGEN, P., I. CAMBY, F. DARRO, R. KISS & P. ROBBERECHT. 1996. VIP and pituitary adenylate cyclase activating polypeptide (PACAP) have an antiproliferative effect on the T98G human glioblastoma cell line through interaction with VIP2 receptor. Neuropeptides **30**: 491–496.

25. MOODY, T. W., J. LEYTON, I. GOZES et al. 1997. VIP receptor antagonists inhibit the growth of glioblastoma cells. In Proceedings of the Society for Neuroscience (27th Annual Mtg.): 2239(Abstract).

26. ALBERS, A. R. & M. S. O'DORISIO. 1996. Clinical use of somatostatin analogues in paediatric oncology. Digestion **57**(Suppl. 1): 38–41.

27. VIRGOLINI, I., M. RADERER, A. KURTARAN et al. 1994. Vasoactive intestinal peptide-receptor imaging for the localization of intestinal adenocarcinomas and endocrine tumors. N. Engl. J. Med. **331**: 1116–1121.

28. BREEMAN, W. A., M. P. VAN HAGEN, H. A. VISSER-WISSELAAR et al. 1996. In vitro and in vivo studies of substance P receptor expression in rats with the new analog [indium-111-DTPA-Arg1] substance P. J. Nucl. Med. **37**: 108–117.

29. VAN HAGEN, P. M., W. A. P. BREEMAN, J. C. REUBI et al. 1996. Visualization of the thymus by substance P receptor scintigraphy in man. Eur. J. Nucl. Med. **23**: 1508–1513.

30. KRENNING, E.P., P. P. M. KOOIJ, W. H. BAKKER et al. 1994. Radiotherapy with a radiolabeled somatostatin analogue, [^{111}In-DTPA-D- Phe^1]octreotide: A case history. Ann. N.Y. Acad. Sci. **733**: 496–506.

PACAP Increases Cytosolic Calcium in Vasopressin Neurons: Synergism with Noradrenaline[a]

SEIJI SHIODA,[b,c,g] TOSHIHIKO YADA,[d,e] SHIGEO NAKAJO,[f]
YASUMITSU NAKAI,[b] AND AKIRA ARIMURA[c]

[b]Department of Anatomy, Showa University School of Medicine, Tokyo 142, Japan

[c]U.S.-Japan Biomedical Research Laboratories, Tulane University Hebert Center, Belle Chasse, Louisiana 70037, USA

[d]Department of Physiology, School of Medicine, Kagoshima University, Kagoshima 890, Japan

[e]Laboratory Intracellular Metabolism, National Institute of Physiological Science, Okazaki 444, Japan

[f]Laboratory of Biological Chemistry, Showa University School of Pharmaceutical Sciences, Tokyo 142, Japan

PACAP-containing axon terminals have been shown to innervate arginine-vasopressin (AVP)-containing neurons in the rat hypothalamic supraoptic nucleus (SON).[1,2] PACAP receptor (PACAPR) mRNA was expressed at high levels in the central nervous system including the hypothalamic SON[3] and it was expressed in AVP-containing neurons, but at very low levels in oxytocin-containing neurons.[2] Doses of PACAP in the nanomolar range increased cytoplasmic Ca^{2+} concentration ($[Ca^{2+}]_i$) in AVP-containing neurons; the increase in $[Ca^{2+}]_i$ is inhibited by a protein kinase A (PKA) blocker (H89).[2] These findings suggest that PACAP serves as a transmitter and/or modulator and the activation of PACAPR stimulates a cAMP-PKA pathway that in turn evokes the Ca^{2+} signaling system. We have shown noradrenergic innervation of AVP-containing neurons in the SON by immunocytochemistry at the ultrastructural level.[4] Noradrenaline (NA), dibutyryl-cAMP, and forskolin increased $[Ca^{2+}]_i$ in isolated AVP-containing neurons in the SON.[5] The NA-induced increase in $[Ca^{2+}]_i$ in AVP-containing neurons was abolished by the α_1-receptor antagonist prazosin and was markedly reduced when treated with H89.[5] NA activates AVP-containing neurons via the α_1-receptor, which is linked to the stimulation of the cAMP-PKA–regulated Ca^{2+} signaling pathway. PACAP and catecholamine are found to coexist in the same axon terminals in the SON (data not shown) and they are suggested to originate from the caudal ventrolateral medulla (CVLM).[6] Although the pathway arising from PACAP/NA-containing neurons in the medulla is involved in the regulation of AVP, the precise mechanism of action and physiological significance of the coexistence of PACAP and NA still must be investigated.

[a]This study supported in part by grants from the Ministry of Education, Science, and Culture of Japan.

[g]Corresponding author: Department of Anatomy, Showa University School of Medicine, 1-5-8 Hatanodai, Shinagawa-ku, Tokyo 142, Japan; Tel.: 81-3-3784-8104; Fax: 81-3-3784-6815; E-mail: shioda@med.showa-u.ac.jp

427

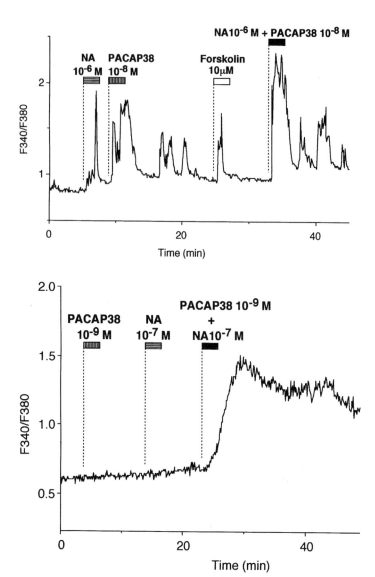

FIGURE 1. Synergistic effect of PACAP and NA on AVP neurons. (**A**) When PACAP (10 nM) and NA (1 μM) were added together, they evoked a large increase in $[Ca^{2+}]_i$. (**B**) PACAP (1.0 nM) and NA (0.1 μM) evoked a larger increase in $[Ca^{2+}]_i$ in AVP neurons when added together.

Unilateral electrical stimulation of CVLM induced bilateral increase of Fos-like immunoreactivity (Fos-LI) in hypothalamic neurons.[7] Strong Fos-LI was visible in many AVP-containing neurons but rarely in oxytocin neurons.[7] *In situ* hybridization revealed high levels of AVP mRNA in the hypothalamus at 3 h, which peaked at 6 h after stimulation.[7] Both PACAP (≥ 1 nM) and NA ($1 \geq$ μM) induced large increases in the $[Ca^{2+}]_i$ (peak:

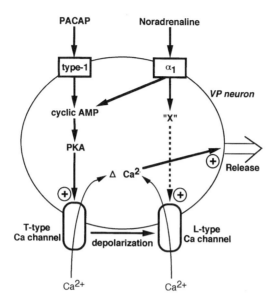

FIGURE 2. Proposed status of PACAP and NA as regulators of AVP secretion. Signaling transduction underlying synergistic action of PACAP and nonadrenaline in SON vasopressin neurons. PACAP and NA, co-released from the same axon terminals, act synergistically to stimulate calcium signaling in AVP neurons, which is mediated by both the cAMP-PKA pathway and L-type Ca^{2+} channels through an unidentified factor "X".

300–800 nM) in isolated AVP-containing neurons. When 1 nM PACAP and 1 µM NA were added together, they evoked larger increases in $[Ca^{2+}]_i$ in AVP neurons (FIG. 1,A). PACAP at 0.1 nM and NA at 0.1 µM had much lesser effects, if any, on $[Ca^{2+}]_i$. However, when 0.1 nM PACAP and 0.1 µM NA were added together, they evoked a large increase in $[Ca^{2+}]_i$ in AVP neurons (FIG. 1,B). An inhibitor of PKA, H89, completely inhibited the PACAP-induced increase in $[Ca^{2+}]_i$, and partly inhibited the NA-induced increase in AVP-containing neurons. Characterization of elicited Ca^{2+} currents in neurosecretory cells in the SON has demonstrated that certain types are defined pharmacologically in somata or axon terminals.[8] We found that T-type channels might contribute to Ca^{2+} influx during action potentials through the cAMP-PKA signaling pathway and L-type currents might contribute to the generation of bursting activity (data not shown). These findings suggest that PACAP and NA, co-released from the same axon terminals, act synergistically to stimulate calcium signaling in AVP neurons, which is mediated by both the cAMP-PKA pathway and the L-type Ca^{2+} channel through an unidentified factor "X" (FIG. 2). The synergism between PACAP and NA may constitute the regulation of secretion and gene transcription of AVP under physiological conditions. It is hypothesized that PACAP/NA regulate the functions of AVP-containing neurons, which participate in the control of plasma osmolarity and blood pressure.

REFERENCES

1. SHIODA, S. *et al.* 1996. Pituitary adenylate cyclase-activating polypeptide and its type I receptors in the rat hypothalamus: Neuroendocrine interactions. Ann. N.Y. Acad. Sci. **805**: 670–676.

2. SHIODA, S. *et al.* 1997. Pituitary adenylate cyclase-activating polypeptide (PACAP): a novel regulator of vasopressin-containing neurons. Brain Res. **765**: 81–90.
3. SHIODA, S. *et al.* 1997. Localization and gene expression of the receptor for pituitary adenylate cyclase-activating polypeptide in the rat brain. Neurosci. Res. **28**: 345–354.
4. SHIODA, S. & Y. NAKAI. 1992. Noradrenergic innervation of vasopressin-containing neurons in the rat hypothalamic supraoptic nucleus. Neurosci. Lett. **140**: 215–218.
5. SHIODA, S. *et al.* 1997. Noradrenaline activates vasopressin neurons via α_1-receptor-mediated Ca^{2+} signaling pathway. Neurosci. Lett. **226**: 210–212.
6. SHIODA, S. & Y. NAKAI. 1996. Direct projections from catecholaminergic neurons in the caudal ventrolateral medulla to vasopressin-containing neurons in the supraoptic nucleus: a triple-labeling electron microscope study in the rat. Neurosci. Lett. **221**: 45–48.
7. SHIODA, S. *et al.* 1998. Vasopressin neuron activation and Fos expression of the caudal ventrolateral medulla. Brain Res. Bull. **45**: 443–450.
8. FISHER, T. E. & BOURQUE, C. W. 1996. Calcium-channel subtypes in the soma and axon terminals of magnocellular neurosecretory cells. TINS **19**: 440–444.

Effect of PACAP-27 on Adenylate Cyclase in Ductal and Acinar Cells of Rat Submandibular Gland

E. KABRÉ,[a,c] N. CHAIB,[a] H. AMSALLEM,[b] A. MORAN,[b] M. C VANDERMEERS,[a] AND J. P DEHAYE[a]

[a]Université Libre de Bruxelles, Bruxelles, Belgium

[b]Ben-Gurion University of the Negev, Beer-Sheva, Israel

P ACAP-38 and PACAP-27 have been first isolated from ovine hypothalamus.[1] Structural analysis of these peptides revealed that the first 28 residues of PACAP-38 share a 68% homology with the vasoactive intestinal polypeptide (VIP). Hence the two PACAP join the family of peptides including VIP, secretin, glucagon, and GH releasing hormone.[2] Radioimmunoassay and immunochemical studies have shown that these regulatory peptides are abundant in the central nervous system, and also in peripheral tissues. PACAP has diverse biological effects that are tissue specific.[3]

Transduction of peptidergic signals is frequently mediated by intracellular pathways involving adenylate cyclase and/or phospholipase C. Since the salivary secretion is triggered by increases of the intracellular concentration of calcium and/or cyclic AMP[4] and since PACAP is widely distributed through the digestive tract, we have studied its possible role in salivary gland function.

MATERIALS AND METHODS

Fura 2/AM and BCECF/AM were purchased from Molecular Probes (Eugene, OR). Thapsigargin, Dowex, alumina, [^{3}H]adenine, epinephrine, forskolin, isoproterenol, and IBMX were from Sigma Chemical Co. (St. Louis, MO). PACAP-27 was a gift of Dr. M. C. Vandermeers (Université Libre de Bruxelles). Collagenase P was supplied by Boehringer Mannheim (Darsmstadt, Germany).

Male Wistar rats were from the Proefdierencentrum ofthe Katholieke Universiteit (Heverlee, Belgium). They had free access to water and food.

Preparation of the Crude Suspension of Rat Submandibular Glands

The cells were obtained as previously described.[5] Briefly, the rat was killed by exsanguination and the submandibular glands were excised and dissected. The glands were finely minced in 0.5 ml of HEPES-buffered saline (HBS) containing (mM): NaCl 96, KCl 16, MgCl$_2$ 1, NaH$_2$PO$_4$ 2.5, glucose 11, Na-pyruvate 5, Na-glutamate 5, Na-fumarate 5,

[c]Corresponding author: E. Kabré, Laboratoire de Biochimie générale et humaine, Institut de pharmacie C.P. 205/3, Université libre de Bruxelles, Boulevard du Triomphe, B-1050, Bruxelles, Belgium; Tel.: 32 2 6505294; Fax: 32 2 6505305; E-mail: ekabre@ulb.ac.be

HEPES 24.5 (pH 7.4), amino acid mixture 0.1% (w/vol). The minced tissue was then incubated for 20 min under constant shaking (160 cycles/min) at 37°C in 10 ml of the same medium but containing 1.4 mg collagenase P (2.8 U/mg). The tissue was finally aspirated through glass pipets, and filtered through four layers of gauze. The suspension was then washed three times by centrifugation and the supernatant removed. The cells were finally resuspended in 6 ml medium and kept at 4°C until use.

Preparation of Pure Suspensions of Acinar and Ductal Cells

Six ml of 40% isotonic Percoll were placed in two 15-ml tubes. Three ml of the crude cellular suspension were layered on the top of the Percoll and the tubes were centrifuged at 4°C for 8 min at 1500 g. At the end of the centrifugation two populations of cells could be observed: the ductal cells had remained on the top of the Percoll while the acinar cells were in the pellet. The two cellular fractions were aspirated and washed four times with isotonic saline solution.

Measurement of cAMP Production

Cells were prelabeled for 40 min with [^{3}H]adenine (3 μCi/ml) in HBS solution containing 0.5 mM CaCl$_2$. They were then washed three times with isotonic saline and resuspended in 10 ml of the same solution containing 0.5 mM IBMX. After stimulation with agonists the reaction was stopped by adding 50 μl of ice-cold 10% TCA to the medium. The [^{3}H]cAMP was isolated by the modified method of Salomon.[6]

Measurement of the Intracellular Calcium Concentration

The intracellular calcium concentration was assayed using the Ca^{2+}-sensitive fluorescent probe fura-2.[7]

Measurement of the Na$^+$-K$^+$-2Cl$^-$ Cotransport Activity

The cells were loaded with the pH-sensitive probe BCECF and their intracellular pH measured as previously described.[8] The ability of NH$_4^+$ to substitute for K$^+$ on the Na$^+$-K$^+$-2Cl$^-$ was used to study NH$_4^+$ uptake. The cells were resuspended in HBS medium containing 1 mM ouabain (to block the Na$^+$/K$^+$-ATPase) and 10 μM N-ethyl isopropyl amiloride (EIPA) (to block the Na$^+$/H$^+$ exchanger). They were exposed to agonists, and then to 1 mM NH$_4$Cl. At this concentration, no alkalinization is observed, but a rapid acidification follows the exposure to NH$_4$Cl due to ammonium uptake. The NH$_4^+$ acidification was inhibited by bumetanide, suggesting an involvement of the cotransporter in the transport of NH$_4^+$. The results were plotted as the intracellular pH versus time. The rate of ammonium influx was estimated by linear regression of the decay of the pH observed between 20 and 40 sec after the addition of NH$_4$Cl to the medium.

RESULTS AND DISCUSSION

The variation of the free intracellular calcium concentration [Ca^{2+}]$_i$ in isolated submandibular acinar cells was measured in the presence of 1 mM extracellular calcium (FIG. 1).

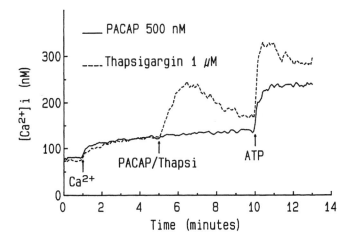

FIGURE 1. Effect of PACAP-27 on the intracellular concentration of calcium in rat submandibular acinar cells. Rat submandibular acinar cells were loaded with Fura-2. The cells were washed and resuspended in HBS medium. The measurement of the $[Ca^{2+}]_i$ was performed at 25°C. One minute after the beginning of the measurement, 1 mM $CaCl_2$ was added to the cuvette. The cells were then exposed at 5 min to 500 nM PACAP (*solid line*) or 1 μM thapsigargin (*dashed line*). ATP (1 mM) was added at 10 min in both conditions.

PACAP-27 at a 500 nM concentration had no effect on the $[Ca^{2+}]_i$ while both the intracellular Ca^{2+}-ATPases inhibitor thapsigargin (1 μM) and the purinergic agonist ATP (1 mM) induced significant increases of the $[Ca^{2+}]_i$.

We next examined the effects of PACAP-27 on cAMP generation in either a crude cellular suspension which comprises both acinar and ductal cells or in pure acinar or ductal cell preparations. We also compared these effects to those of adrenergic agonists. In the

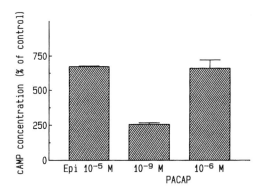

FIGURE 2. Response of a crude rat submandibular cellular suspension to epinephrine and PACAP. Submandibular cells were loaded with [³H]adenine and stimulated with 10 μM epinephrine and 1 nM or 1 μM PACAP. The cAMP produced was determined as described in *Materials and Methods*. Results are expressed as percentage of cAMP produced when compared to the control and are the means ± S.E.M. of three experiments performed in triplicate.

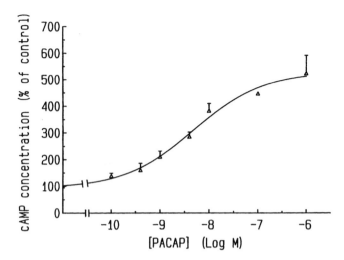

FIGURE 3. Dose-response curve for PACAP on cAMP production by acinar cells. Acinar cells were loaded with [³H]adenine and stimulated with different concentrations of PACAP. The cAMP produced was determined as described in *Materials and Methods*. Results are expressed as percentage of cAMP produced when compared to the control and are the means ± S.E.M. of three experiments performed in triplicate.

crude cellular suspension, 1 μM PACAP-27 and 10 μM epinephrine increased cAMP production six- and sevenfold, respectively (FIG. 2). In pure acinar cells, both PACAP-27 and the β-adrenergic agonist isoproterenol dose-dependently increased cAMP production. The

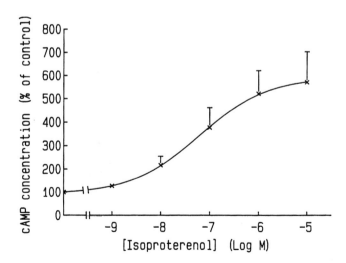

FIGURE 4. Dose-response curve for isoproterenol on cAMP production by acinar cells. Acinar cells were loaded with [³H]adenine and stimulated with different concentrations of isoproterenol. The cAMP produced was determined as described in *Materials and Methods*. Results are expressed as percentage of cAMP produced when compared to the control and are the means ± S.E.M. of three experiments performed in triplicate.

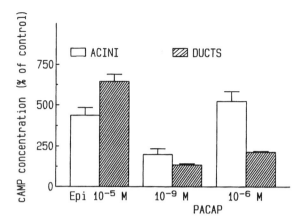

FIGURE 5. Response of rat submandibular acini and ducts to epinephrine and PACAP. Comparative effect of 10 μM epinephrine and 1 nM and 1 μM PACAP on cAMP production in acinar cells (*open bars*) and ductal cells (*hatched bars*). The results are expressed as percentage of cAMP produced when compared to the control. They are the means ± S.E.M. of at least three experiments performed in triplicate.

half-maximal and maximal concentrations were 10 nM and 1 μM for PACAP-27 (FIG. 3) and 100 nM and 10 μM for isoproterenol (Fig. 4). The maximal stimulation averaged 5.2-fold and 5.8-fold for PACAP-27 and isoproterenol, respectively.

FIGURE 5 shows a comparative effect of epinephrine and PACAP-27 on both acinar and ductal cells. In response to epinephrine, the cAMP production was lower in acinar cells than in ductal cells. The inverse effect was observed with PACAP-27 (1 nM or 1 μM).

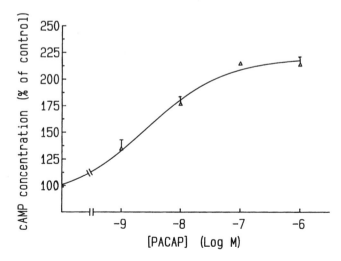

FIGURE 6. Dose-response curve for PACAP on cAMP production by ducts. Ductal cells were loaded with [³H]adenine and stimulated with different concentrations of PACAP. The cAMP produced was determined as described in *Materials and Methods*. Results are expressed as percentage of cAMP produced when compared to the control and are the means ± S.E.M. of two experiments performed in triplicate.

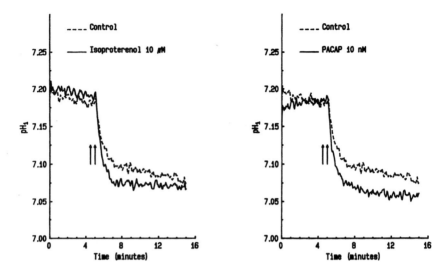

FIGURE 7. Effect of isoproterenol and PACAP on the uptake of ammonium by rat submandibular acini. The acinar cells were loaded with the pH-sensitive probe BCECF and their intracellular pH measured as described in *Materials and Methods*. The cells were resuspended in HBS medium containing 1 mM ouabain (to block the Na^+/K^+-ATPase) and 10 μM 5-(*N*-ethyl-*N*-isopropyl)-amiloride (EIPA) (to block the Na^+/H^+ exchanger). The agonist or the vehicle was added to the cuvette 5 min after the beginning of the measurement (*first arrow*). Thirty seconds later the cells were exposed to 1 mM NH_4Cl (*second arrow*). The results are representative of three experiments. (*Left*) Control (*dashed line*) and 10 μM isoproterenol (*solid line*). (*Right*) Control (*dashed line*) and 10 nM PACAP (*solid line*).

Indeed, as shown in FIGURE 6, PACAP-27, in contrast to epinephrine, only weakly affected the intracellular cAMP production in pure ductal cells.

Like isoproterenol, 10 nM PACAP-27 increased the activity of the Na^+-K^+-$2Cl^-$ cotransporter in acinar cells (FIG. 7).

Taken together, our results provide the evidence for the existence of PACAP receptors on the plasma membrane of acinar and ductal cells of rat submandibular gland. Considering that PACAP-27 has a different efficacy on the production of cAMP by these two cellular populations, our results suggest either that acini and ducts have different types of PACAP receptors or that the coupling between these receptors and the effector is different in these two types of cells. The existence of two distinct PACAP receptors (types I and II) in rat salivary glands has been recently reported,[9] but the cellular distribution of these receptors has not been investigated. Since acini and ducts have distinct functions during the two phases of exocrine secretion, the presence of different PACAP receptors on acini and ducts might increase the selectivity of the response to this peptide.

REFERENCES

1. MIYATA, A., A. ARIMURA, R. R. DAHL, N. MINAMINO, A. UEHARA, L. JIANG, M. CULLER & D. H. COY. 1989. Isolation of a novel 38 residue-hypothalamic polypeptide which stimulates adenylate cyclase in pituitary cells. Biochem. Biophys. Res. Commun. **164**: 567–574.
2. KIMURA, C., S. OHKUBO, K. OGI, M. HOSOYA, Y. ITOH, H. ONDA, A. MIYATA, L. JIANG, R. R. DAHL, H. H. STIBBS, A. ARIMURA & M. FUJIMO. 1990. A novel peptide which stimulates adeny-

late cyclase: molecular cloning and characterization of the ovine and human cDNAs. Biochem. Biophys. Res. Commun. **166**: 88–89.

3. BRENDA, D. S., J. G. TAMAS, E. G. PAUL & A. AKIRA. 1991. Two high affinity binding sites for pituitary adenylate cyclase-activating polypeptide have different tissue distribution. Endocrinology **128**: 3055–3065.

4. MCKINNEY, J. S., M. S. DESOLE & R. P. RUBIN. 1989. Convergence of cAMP and phosphoinositide pathways during rat parotid secretion. Am. J. Physiol. **26**: C651–C657.

5. AMSALLEM, H., M. METIOUI, A. VANDENABEELE, A. ELYAMANI, A. MORAN & J. P. DEHAYE. 1996. Presence of a metabotrophic and an ionotropic receptor on rat submandibular ductal cells. Am. J. Physiol. **271**: C1546–C1555.

6. GRYNKIEWICZ, G., M. POENIE & R. Y. TSIEN. 1985. A new generation of Ca^{2+} indicators with greatly improved fluorescence properties. J. Biol. Chem. **260**: 3440–3450.

7. SALOMON, Y., 1979. Adenylate cyclases. Adv. Cycl. Nucl. Res. **10**: 36–55.

8. LACHISH, M., E. ALZOLA, N. CHAIB, M. METIOUI, K. GROSFILS, E. KABRE, A. MORAN, A. MARINO & J. P. DEHAYE. 1996. Study of the non-specific cation channel coupled to P_{2z} purinergic receptors using an acid load technique. Am. J. Physiol. **271**: C1920–C1926.

9. MIRFENDERESKI, S., G. TOBIN, R. HAKANSON & J. EKSTROM. 1997. Pituitary adenylate cyclase activating peptide in salivary gland of the rat: origin and secretory and vascular effect. Acta Physiol. Scand. **160**: 15–22.

Distribution and Ultrastructural Localization of PACAP Receptors in the Rat Pancreatic Islets[a]

MASAYO MUROI,[b] SEIJI SHIODA,[b,c,g] TOSHIHIKO YADA,[d,e]
CHENG JI ZHOU,[b] YASUMITSU NAKAI,[b] SHIGEO NAKAJO,[f] AND
AKIRA ARIMURA[c]

[b]Department of Anatomy, Showa University School of Medicine, Tokyo
142, Japan

[c]U.S.-Japan Biomedical Research Laboratories, Tulane University
Hebert Center, Belle Chasse, Louisiana 70037, USA

[d]Department of Physiology, School of Medicine, Kagoshima University,
Kagoshima 890, Japan

[e]Laboratory Intracellular Metabolism, National Institute of
Physiological Science, Okazaki 444, Japan

[f]Laboratory of Biological Chemistry, Showa University School of
Pharmaceutical Sciences, Tokyo 142, Japan

R adioimmunoassays revealed the presence of PACAP38 in pancreatic tissues.[1] Two binding sites, type I receptors (PACAPRs) specific for PACAP and type II receptors (VIP$_1$/PACAPR, VIP$_2$/PACAPR) shared with VIP, have been demonstrated,[2] and cDNAs

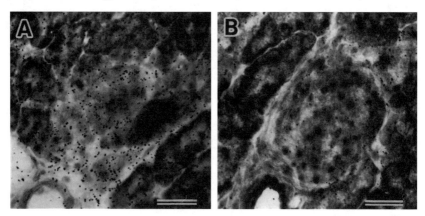

FIGURE 1. *In situ* hybridization in the rat pancreas Sections were hybridized with [^{35}S]-labeled antisense and sense synthetic oligonucleotide probes for PACAPR. (A) Strong PACAPR mRNA expression is visible in the entire pancreatic islet and moderate labeling is detected in exocrine pancreatic tissues. (B) Control section. Section was incubated with a sense probe for PACAPR mRNA. No specific labeling is visible in the pancreas. Bars represent 30 μm.

[a]This study supported by grants (08458249 and 09558098) from the Ministry of Education, Science and Culture of Japan to S. S.

[g]Corresponding author: Dr. Seiji Shioda, Department of Anatomy, Showa University School of Medicine, 1-5-8 Hatanodai, Shinagawa-ku, Tokyo 142, Japan, Tel.: 81-3-3784-8104; Fax: 81-3-3784-6815; E-mail: shioda@med.showa-u.ac.jp

encoding the PACAPR[3] and VIP/PACAPR[4] have been cloned. It was recently reported that islet β-cells expressed novel PACAPR variant with different transmembrane domain IV, designated PACAPR TM4, that stimulates Ca^{2+} influx through L-type Ca^{2+} channels.[5] The distribution and localization of PACAPR remain unclear, as does expression of its gene in the pancreas. Therefore, we studied the distribution of PACAPR and expression of the gene for PACAPR in the rat pancreas by immunocytochemistry and *in situ* hybridization.

Adult male Sprauge-Dawley rats (250–300 g) were anesthetized with sodium pentobarbital and perfusion fixed with a fixative of 2% paraformaldehyde with or without 0.1% glutaraldehyde in phosphate buffer. Immunohistochemical staining of PACAPR was performed using rabbit antiserum against a synthetic peptide that corresponded to the C-terminal intracellular cytoplasmic domain of PACAPR.[6] For electron microscopic observation, pre-embedding immunogold method was used.[7] For *in situ* hybridization of PACAPR mRNA, the animals were anesthetized and killed by decapitation and fresh frozen sections were hybridized with [35S]-labeled sense and antisense synthetic oligonucleotide probes for PACAPR.[8]

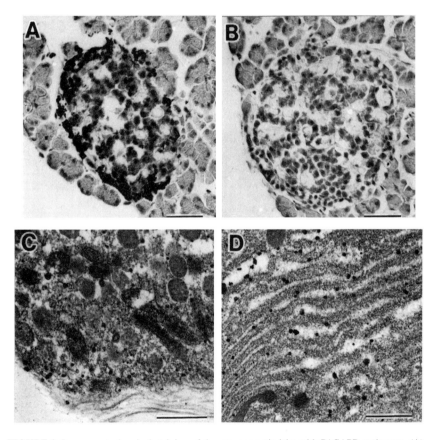

FIGURE 2. Immunocytochemical staining of the rat pancreatic islet with PACAPR antiserum. (**A**) PACAPR immunoreactivity is visible in the islet. (**B**) Control section. Sections was incubated with rabbit antiserum preabsorbed with the antigen. (**C** and **D**) Many gold particles with PACAPR immunoreactivity are visible in the cytoplasm underneath the cell membrane (**C**) and rough endoplasmic reticulum (**D**). Bars represent 30 μm in (**A**) and (**B**) and 0.5 μm in (**C**) and (**D**).

In situ hybridization with an oligonucleotide probe for PACAPR, mRNA yielded intense signals in the entire region of islets (FIG. 1,A). Moderately strong signals were also seen in the exocrine pancreas (FIG. 1,A). No specific labeling was seen in the pancreas when tissue sections were incubated with sense probe (FIG. 1,B). In the immunohisto-chemical study with the PACAPR-specific antiserum, PACAPR immunoreactivity was demonstrated in the entire region of rat islets (FIG. 2,A), but not when the antiserum was preabsorbed with the antigen (FIG. 2,B). At the ultrastructural level, gold particles with PACAPR immunoreactivity were densely localized to the basal face of cell membranes (FIG. 2,C) and of rough endoplasmic reticulum (FIG. 2,D) in the islet cells.

Islet endocrine cells are composed of β-, α-, δ-, and pp-cells. It is currently unknown from which cells PACAP could be released. However, PACAP is localized in rat islets including the central portion exclusively occupied by β-cells.[9] We have shown PACAP gene transcripts and PACAP-38 biosynthesis in MIN6 cells, as well as in islets.[10] Thus, it is likely that PACAP-38 is produced and released at least by β-cells. Since PACAP-38 also directly targets β-cells, PACAP-38 appears to operate an autocrine function in β-cells. However, the immunoreactivity for PACAP and that for PACAPR found in this study are observed not only in the center but also in the periphery of the islets where α-cells are abundant. PACAP is reported to stimulate glucagon release in mice.[11] Thus, PACAP may also be released from α-cells and participate in the paracrine interaction between α- and β-cells and/or in the autocrine regulation of α-cells.

REFERENCES

1. ARIMURA, A. *et al.* 1991. Tissue distribution of PACAP as determined by RIA: high abundant in the rat brain and testes. Endocrinology **129**: 2787–2789.
2. AREN, B. *et al.* 1986. Neuropeptidergic versus cholinergic and adrenergic regulation of islet hormone secretion. Diabetologia **30**: 827–836.
3. ARIMURA, A. & S. SHIODA. 1995. Pituitary adenylate cyclase activating polypeptide (PACAP) and its receptors: Neuroendocrine and endocrine interaction. Front. Neuroendocrinol. **16**: 53–88.
4. ISHIHARA, T. *et al.* 1992. Functional expression and tissue distribution of a novel receptor for vasoactive intestinal polypeptide. Neuron **8**: 811–819.
5. CHATTERRJEE, T. K. *et al.* 1996. Molecular cloning of a novel variant of the pituitary adenylate cyclase-activating polypeptide (PACAP) receptor that stimulate calcium influx by activation of L-type calcium channels. J. Biol. Chem. **271**: 32226–32232.
6. LI, M. *et al.* 1997. Specific antibody recognition of rat pituitary adenylate cyclase activating polypeptide receptors. Endocrine **7**: 183–190.
7. BERMARD, V. 1997. Cellular, subcellular, and subsynaptic distribution of AMPA-type glutamate receptor subunits in the neostriatum of the rat. J. Neurosci. **15**: 819–833.
8. SHIODA, S. *et al.* 1997. Localization and gene expression of the receptor for pituitary adenylate cyclase activating polypeptide in the rat brain. Neurosci. Res. **28**: 345–354.
9. YADA, T. *et al.* 1997. PACAP is an islet substance serving as an intra-islet amplifier of glucose-induced insulin secretion in rats. J. Physiol. (London) **505**(2): 319–328.
10. YADA, T. *et al.* 1996. Current status of PACAP as a regulator of insulin secretion in pancreatic islets. Ann. N.Y. Acad. Sci. **805**: 329–342.
11. FRIDOLF, T. *et al.* 1992. Pituitary adenylate cyclase activating polypeptide (PACAP): Occurrence in rodent pancreas and effects on insulin and glucagon secretion in the mouse. Cell Tissue Res. **269**: 275–279.

Protein Kinase A Inhibition and PACAP–Induced Insulin Secretion in HIT-T15 Cells[a]

KARIN FILIPSSON[b] AND BO AHRÉN

Department of Medicine, Lund University, Malmo University Hospital, SE-205 02 Malmo, Sweden

Pituitary adenylate cyclase-activating polypeptide (PACAP) occurs in pancreatic nerves, both in the exocrine and the endocrine tissues.[1,2] In the endocrine pancreas, PACAP potently stimulates insulin secretion.[2,3] In a previous study, we showed that PACAP38-stimulated insulin secretion in HIT-T15 cells is accompanied by activation of several intracellular signaling pathways, *viz.*, formation of cAMP and increase in both cytoplasmic calcium and inositol phosphates.[3] The importance of the increase in cellular cAMP content for the stimulation of exocytosis by PACAP38 may be studied by inhibiting protein kinase A (PKA), as PKA is activated by cAMP and thought to mediate its actions. For this purpose it is necessary to use a reliable PKA inhibitor. In this study, we have examined the effects of three PKA inhibitors: Rp-cAMPS, Rp-8-Br-cAMPS, and H89 (N-[2-(p-bromo-cinnamylamino)ethyl]-5-isoquinoline sulfonamide), on PACAP38 and forskolin-stimulated insulin secretion in HIT-T15 cells. Rp-cAMPS and Rp-8-Br-cAMPS are competitive antagonistic isomers of cAMP, which previously have been used to examine the contribution of PKA to the insulinotropic effect of glucagon-like peptide-l.[4,5] Of the two, Rp-8-Br- cAMPS is more lipophilic, and thus more cell membrane permeable. In contrast, the structure of H89 is different from Rp-isomers, and H89 exerts its inhibitory effect by directly binding to PKA.[6]

MATERIALS AND METHODS

Cell Culture and Experiments

HIT-T15 cells, a clonal hamster β-cell line, were cultured as previously described.[3] Cells were seeded in 24-well plates at a concentration of 0.5 million cells per well and cultured for 48 h (about 80% confluency). The cells were washed twice in a HEPES buffer[3] and preincubated in a volume of 200 µl with 1 mM glucose with or without 20 µM of each of the PKA inhibitors (Rp-cAMPS was from Calbiochem, La Jolla, CA, USA, Rp-8-Br-cAMPS from Biolog, Life Science Institute, Bremen, Germany; and H89 was from Seikagaku Corporation, Tokyo, Japan) for 30 min at 37°C. The wells were then washed once in HEPES buffer and then incubated with 10 mM glucose and 100 nM PACAP38

[a]This study was supported by the Swedish Medical Research Council (grant no. 14X-6834), Ernhold Lundström, Albert Påhlsson, and Novo Nordic Foundations, Swedish Diabetes Association, Malmö University Hospital, and the Faculty of Medicine, Lund University.

[b]Corresponding author: Karin Filipsson, Department of Medicine, Wallenberg Lab, 2nd floor, Malmo University Hospital, SE-205 02 Malmo, Sweden; Tel.: 46 40 337212; Fax: 46 40 337041; E-mail: Karin.Filipsson@medforsk.mas.lu.se

(Peninsula Europe Labs, Merseyside, England) or 1 μM forskolin (Sigma Chemical Co., St. Louis, MO, USA) at 37°C for 15 or 60 min. After incubation, medium was collected and analyzed for insulin by radioimmunoassay (Linco Research, St. Charles, MO, USA).

Statistics

The results are reported as mean ± S.E.M. Statistical evaluations of differences between groups were performed by one-way non-parametric ANOVA followed by Student-Newman-Keul *post hoc* analysis. A probability level of $p < 0.05$ was considered significant.

RESULTS AND DISCUSSION

In the first series of experiments, we verified our previous results that after 60-min incubation, PACAP38 or forskolin at 10 mM glucose potentiates insulin secretion. When incubating the cells for 60 min in the presence of either of the three PKA inhibitors, we found that none of them inhibited insulin secretion after stimulation with PACAP38, and

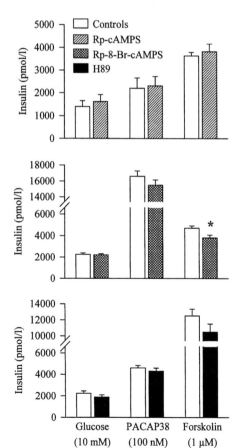

FIGURE 1. The effect of different PKA inhibitors on insulin secretion from HIT-T15 cells with or without PACAP38 (100 nM) or forskolin (1 μM) at 10 mM glucose during a 60-min incubation. (*Top*) Rp-cAMPS (20 μM); (*Middle*) Rp-8-Br-cAMPS (20 μM); and (*Bottom*) H89 (20 μM). $N = 8–20$ for the glucose and PACAP38 groups, and $N = 4–8$ for forskolin groups. Asterisk indicates a probability level of $p < 0.05$ for random difference of observations with Rp-8-Br-cAMPS versus controls without inhibitor.

that only Rp-8-Brc-AMPS could slightly inhibit the response to forskolin (FIG. 1). This inhibition was, however, not reliable, since it was not seen in all individual experiments. The effect of glucose on insulin secretion was not affected by any of the three PKA inhibitors (FIG. 1). The failure of the PKA inhibitors to inhibit insulin secretion might be explained by a too long (60 min) incubation time, since the increase of cAMP induced by PACAP38 peaks at 2 min after stimulation, suggesting involvement of the nucleotide only in the early events after stimulation.[3] Furthermore, a decreased negative feedback by PKA on adenylate cyclase might result in an exaggerated formation of cAMP during a long-term incubation with the inhibitors, which could override the direct inhibition of PKA. Therefore, we shortened the incubation time to 15 min. We then found that Rp-cAMPS and Rp-8-Br-cAMPS still had no effect on glucose-, PACAP38-, or forskolin-induced insulin secretion (FIG. 2). However, H89 decreased insulin levels at 10 mM glucose from 1,170 ± 90 pmol/l in controls to 1,010 ± 110 pmol/l ($p < 0.05$), and PACAP38-induced insulin secretion was inhibited from 2,230 ± 340 pmol/l in controls to 1,410 ± 230 pmol/l ($p < 0.05$). Similarly, the forskolin-induced insulin secretion was inhibited from 2690 ± 210 pmol/l in the absence of H89 to 1,920 ± 160 pmol/l ($p < 0.05$) with the inhibitor (FIG. 2).

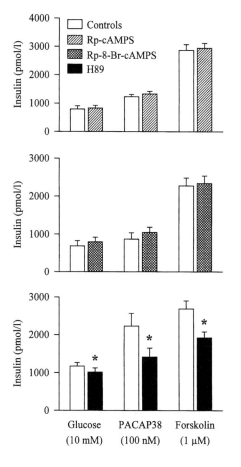

FIGURE 2. The effect of different PKA inhibitors on insulin secretion from HIT-T15 cells with or without PACAP38 (100 nM) or forskolin (1 μM) at 10 mM glucose during a 15-min incubation. (*Top*) Rp-cAMPS (20 μM); (*Middle*) Rp-8-Br-cAMPS (20 μM); and (*Bottom*) H89 (20 μM). $N = 12$–24. Asterisks indicate a probability level of $p < 0.05$ for random difference of observations with H89 versus controls without inhibitor.

In conclusion, both the PACAP38- and the forskolin-induced insulin secretion were inhibited by approximately 35% by H89 after 15-min incubation, whereas the Rp-isomers of cAMP were ineffective. The failure of H89 to abolish PACAP38- and forskolin-induced insulin secretion might be due to cAMP-independent effects of the two agents,[7,8] or, alternatively, PKA might not mediate all effects of cAMP.[9] Hence, the PKA inhibitor most suitable for our cell system is H89, and the inhibition of PKA is best studied during shorter time periods. With H89, therefore, the contribution of cAMP for PACAP-induced insulin secretion in HIT-T15 cells might be examined in further studies.

ACKNOWLEDGMENTS

We thank Lilian Bengtsson and Kerstin Knutsson for expert technical assistance.

REFERENCES

1. FRIDOLF, T., F. SUNDLER & B. AHRÉN. 1992. Pituitary adenylate cyclase activating polypeptide (PACAP): occurrence in rodent pancreas and effects on insulin and glucagon secretion in the mouse. Cell Tissue Res. **269**: 275–279.
2. YADA, T., M. SAKURADA, K. IHIDA, M. NAKATA, F. MURATA, A. ARIMURA & M. KIKUCHE. 1994. Pituitary adenylate cyclase-activating polypeptide is an extraordinarily potent intrapancreatic regulator of insulin secretion from islet β-cells. J. Biol. Chem. **269**: 1290–1293.
3. AF KLINTEBERG, K., S. KARLSSON & B. AHRÉN. 1996. Signaling mechanisms underlying the insulinotropic effect of pituitary adenylate cyclase-activating polypeptide in HIT-T15 cells. Endocrinology **137**: 2791–2798.
4. HOLZ, G. G., C. A. LEECH & J. F. HABENER. 1995. Activation of a cAMP-regulated Ca^{2+}–signaling pathway in pancreatic beta-cells by the insulinotropic hormone glucagon-like peptide-l. J. Biol. Chem. **270**: 17749–17757.
5. DING, W. G., E. RENSTROM, P. RORSMAN, K. BUSCHARD & J. GROMADA. 1997. Glucagon–like peptide I and glucose-dependent insulinotropic polypeptide stimulate Ca^{2+}-induced secretion in rat alpha-cells by a protein kinase A-mediated mechanism. Diabetes **45**: 792–800.
6. HIDAKA, H. & R. KOBAYASHI. 1994. Protein kinase inhibitors. Essays Biochem. **28**: 73–97.
7. STRAUB, S. G. & G. W. SHARP. 1996. A wortmannin-sensitive signal transduction pathway is involved in the stimulation of insulin release by vasoactive intestinal polypeptide and pituitary adenylate cyclase-activating polypeptide. J. Biol. Chem. **271**: 1660–1668.
8. ZUNKLER, B. J., G. TRUBE & T. OHNO-SHOSAKU. 1988. Forskolin-induced block of delayed rectifying K^+ channels in pancreatic beta-cells is not mediated by cAMP. Pflugers Arch. **411**: 613–619.
9. ÄMMÄLÄ, C., F. M. ASHCROFT & P. RORSMAN. 1993. Calcium-independent potentiation of insulin release by cyclic AMP in single beta cells. Nature **363**: 356–358.

PACAP and GLP-1 Protect Islet β-Cells against Ca²⁺ Toxicity Induced by High K⁺ᵃ

KAZURO YAEKURA,[b,c] KAZUHIRO YANAGIDA,[b,d] AND
TOSHIHIKO YADA[b,e]

[b]Department of Physiology, Kagoshima University School of Medicine,
8-35-1 Sakuragaoka, Kagoshima 890-8520, Japan

[c]First Department of Internal Medicine, Kagoshima University School of
Medicine, Kagoshima 890-8520, Japan

[d]Second Department of Internal Medicine, Kagoshima University School
of Medicine, Kagoshima 890-8520, Japan

Impairment of islet β-cell responsiveness to glucose (glucose insensitivity) induced by chronic hyperglycemia, a phenomenon defined as glucose toxicity, is thought to play an important role in pathogenesis of non–insulin-dependent diabetes mellitus (NIDDM).[1,2] The mechanisms by which glucose exerts toxic effects on β-cells are yet to be elucidated.[1,2] Under physiological conditions, glucose is the primary stimulus for insulin release, and an increase in cytosolic Ca²⁺ concentration ($[Ca^{2+}]_i$) in β-cells plays a central role in coupling the glucose stimulation to insulin release.[3,4] Therefore, it is possible that a sustained increase in $[Ca^{2+}]_i$ is produced by chronic hyperglycemia and acts as a factor that develops glucose toxicity. However, whether a sustained $[Ca^{2+}]_i$ increase causes glucose insensitivity in β-cells (Ca²⁺ toxicity) is unknown. Furthermore, it is of particular importance to find substances or conditions that protect β-cells against glucose and/or Ca²⁺ toxicity.

In the present study, to study the mechanisms underlying the induction of glucose insensitivity and to search for a condition that prevents glucose insensitivity, we developed a simple *in vitro* system in which glucose insensitivity is produced in single β-cells in primary culture within 2–3 days. Using this system, it was examined whether a sustained $[Ca^{2+}]_i$ increase induced by high KCl causes glucose insensitivity and whether the presence of pituitary adenylate cyclase–activating polypeptide (PACAP), a pancreatic substance of islet and neural origin,[5–7] and glucagon-like peptide-1(7-36)amide (GLP-1), an incretin hormone,[8] could prevent the development of glucose insensitivity in β-cells. The glucose sensitivity of β-cells was assessed by their $[Ca^{2+}]_i$ responses to glucose.

MATERIALS AND METHODS

Preparation and Culture of Single Islet Cells

Islets of Langerhans were isolated from Wistar rats aged 8–12 weeks by collagenase digestion. Islets were collected and immediately dispersed into single cells in Krebs Ringer bicarbonate buffer (KRB) without Ca²⁺.[5,7] The single cells were plated on coverslips and maintained in culture for 2 days in control conditions (Eagle's minimal essential medium

ᵃThis work was supported by grants from the Ministry of Education, Science and Culture of Japan (T.Y.).

ᵉCorresponding author: Dr. Toshihiko Yada, Department of Physiology, Kagoshima University School of Medicine, 8-35-1 Sakuragaoka, Kagoshima 890-8520, Japan. Tel.: 81-99-275-5225; Fax: 81-99-275-5231; E-mail: yada@med2.kufm.kagoshima-u.ac.jp containing 5.6 mM glucose and 10% fetal bovine serum), in conditions added with 25 mM KCl, and in conditions added with 25 mM KCl plus nifedipine, PACAP-38, or GLP-1.

Solutions and Chemicals

Measurements were carried out in KRB composed of 129 mM NaCl, 5.0 mM NaHCO$_3$, 4.7 mM KCl, 1.2 mM KH$_2$PO$_4$, 1.0 mM CaCl$_2$, 1.2 mM MgSO$_4$, and 10 mM HEPES at pH 7.4 supplemented with 0.1% bovine serum albumin. Fura-2 and Fura2/AM was obtained from Dojin Chemical (Kumamoto, Japan) and PACAP-38 from Peptide Institute (Osaka, Japan). All other chemicals were from Sigma (St. Louis, MO, USA).

Measurements of [Ca²⁺]ᵢ in β-Cells and Criteria for the Glucose Response

After culture for 2 days, the cells were first incubated in KRB containing 2.8 mM glucose and 5.9 mM K$^+$ for 1 h for stabilization, and then subjected to [Ca^{2+}]$_i$ measurements. [Ca^{2+}]$_i$ was measured by fura-2 microfluorometry as previously reported.[5,8] Briefly, cells on coverslips were incubated with 1 μM fura-2/acetoxymethylester in KRB for 30 min. Cells were then mounted in a chamber and superfused at a rate of 1 ml/min at 37°C in KRB. Cells were excited at 340 and 380 nm alternately every 2.5 sec, emission signals at 510 nm (F340 and F380, respectively) were detected with an intensified charge-coupled device (ICCD) camera, and ratio (F340/F380) images were produced by an Argus-50 system (Hamamatsu Photonics, Hamamatsu, Japan). Ratio values were converted to [Ca^{2+}]$_i$ according to calibration curves.

β-Cells were selected according to the previously reported procedure.[5,8] Briefly, single islet cells on coverslips that had a diameter of 12.5–17.5 μm and responded to tolbutamide (300 μM) with increases in [Ca^{2+}]$_i$ were found to be immunocytochemically positive for insulin as reported previously.[5] Data were taken from the cells that fulfilled these morphological and physiological criteria for β-cells.

Only a [Ca^{2+}]$_i$ increase that took place within 10 min of stimulation with 8.3 mM glucose and whose amplitude was greater than 0.3 ratio (F340/F380) unit was considered as a response.

RESULTS AND DISCUSSION

It was first confirmed that a depolarizing concentration of KCl increases [Ca^{2+}]$_i$ in β-cells. At a basal glucose concentration of 2.8 mM, addition of 25 mM KCl induced a sharp initial peak followed by a sustained elevation of [Ca^{2+}]$_i$ in single β-cells (FIG. 1,A). The KCl-induced sustained elevation of [Ca^{2+}]$_i$ was almost completely inhibited by 1 μM nifedipine, a L-type Ca^{2+} channel blocker (FIG. 1,B).

Isolated single β-cells were cultured in control conditions (5.6 mM glucose and 5.9 mM KCl) and those added with 25 mM KCl. After culture for 2 days, the cells were first stabilized in KRB with 2.8 mM glucose and normal K$^+$, and then subjected to [Ca^{2+}]$_i$ measurements. Upon elevation of glucose concentration from a basal (2.8 mM) to a stimulatory level (8.3 mM), a large increase in [Ca^{2+}]$_i$ was induced (FIG. 1) and the [Ca^{2+}]$_i$ response occurred in 75% of β-cells cultured in control conditions (FIG. 3). In the high K$^+$ culture group, in contrast, [Ca^{2+}]$_i$ responses to glucose were impaired in many β-cells (FIG. 2,A) and only 43% of β-cells exhibited [Ca^{2+}]$_i$ responses to glucose (FIG. 3). On the

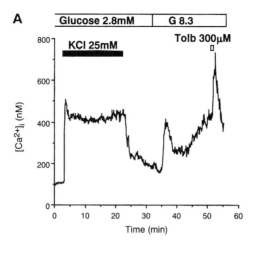

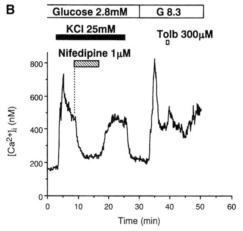

FIGURE 1. Sustained elevation of $[Ca^{2+}]_i$ in single β-cells induced by high KCl and its inhibition by a Ca^{2+} channel blocker. (**A**) Addition of 25 mM KCl, under superfusion conditions, induced sustained elevation of $[Ca^{2+}]_i$ in a single β-cell. (**B**) Nifedipine (1 μM) inhibited the KCl-induced sustained $[Ca^{2+}]_i$ elevation in a single β-cell. The cells subsequently responded to 8.3 mM glucose (G8.3) and 300 μM tolbutamide (Tolb).

other hand, the responses to 300 μM tolbutamide and 10 mM arginine were little altered by high K^+ culture (FIG. 2,A). When β-cells were cultured with high K^+ plus 1 μM nifedipine, the condition in which the $[Ca^{2+}]_i$-increasing ability of high K^+ was inhibited (FIG. 1,B), the glucose response was preserved (FIG. 2,B) and the percentage of β-cells responding to glucose (78%) was indistinguishable from that in the control (FIG. 3). These results indicate that the sustained elevation of $[Ca^{2+}]_i$ plays a critical role in the induction of glucose insensitivity in β-cells by high K^+ culture.

It has recently been shown that PACAP-38 is a novel islet hormone serving as an autocrine/paracrine amplifier of glucose-induced insulin release.[7] As an additional possible function, PACAP could act as an endogenous trophic factor for β-cells. We therefore

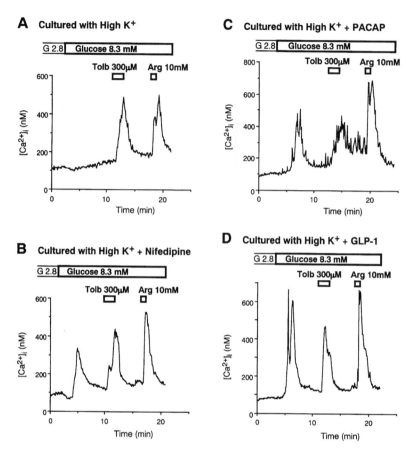

FIGURE 2. Impairment of $[Ca^{2+}]_i$ responses to 8.3 mM glucose, but not those to 300 μM tolbutamide (Tolb) or 10 mM arginine (Arg), in β-cells cultured with high KCl for 2 days (**A**). Impairment of $[Ca^{2+}]_i$ responses to glucose induced by high K^+ culture was prevented by the presence of 1 μM nifedipine (**B**), 10^{-9} M PACAP (**C**), and 10^{-9} M GLP-1 (**D**).

examined whether PACAP, when present in high K^+ culture, could prevent the occurrence of glucose insensitivity. The effect of the presence of GLP-1, another potent insulinotropic peptide, was also examined. When PACAP-38 at 10^{-9} M or GLP-1 at 10^{-9} M was present in high K^+ culture, induction of glucose insensitivity was substantially attenuated (FIG. 2, C and D versus A). The percentage of β-cells responding to glucose with $[Ca^{2+}]_i$ increases was significantly higher (68% with PACAP and 76% with GLP-1) than that in high K^+ culture (FIG. 3). The amplitude of the $[Ca^{2+}]_i$ increase in response to glucose, when averaged for the responding cells, was decreased in high K^+ culture, and it was partially restored by the presence of 10^{-9} M PACAP or 10^{-9} M GLP-1 in the culture (data not shown). Thus, physiological insulinotropins PACAP-38 and GLP-1, when added in the single β-cell culture, prevented the development of high K^+–induced glucose insensitivity.

Since high K^+–induced glucose insensitivity depended on sustained elevation of $[Ca^{2+}]_i$ and it was prevented by GLP-1 and PACAP, a possibility was raised that these peptides

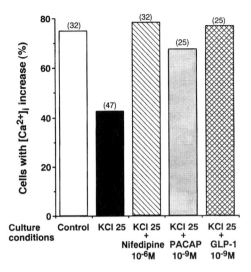

FIGURE 3. Percentage of β-cells exhibiting $[Ca^{2+}]_i$ responses to 8.3 mM glucose following culture for 2 days under control conditions and those with high K^+, high K^+ + 10^{-6} M nifedipine, high K^+ + 10^{-9} M PACAP, and high K^+ + 10^{-9} M GLP-1. The number above each bar indicates the number of cells examined.

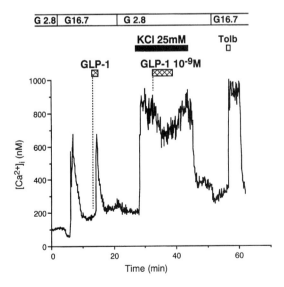

FIGURE 4. Attenuation of high K^+–induced sustained $[Ca^{2+}]_i$ elevation by 10^{-9} M GLP-1 in a single β-cell. At a stimulatory glucose concentration of 16.7 mM, 10^{-9} M GLP-1 also induced an increase in $[Ca^{2+}]_i$ in the same cell. $[Ca^{2+}]_i$ responses to 16.7 mM glucose and Tolb were also obtained.

could attenuate the sustained elevation of $[Ca^{2+}]_i$. We found that 10^{-9} M GLP-1 (FIG. 4) and 10^{-9} M PACAP-38 (data not shown) significantly attenuated the KCl-induced sustained elevation of $[Ca^{2+}]_i$, in addition to the well-known action to increase $[Ca^{2+}]_i$ in a glucose-dependent manner (FIG. 4).[5,8] The $[Ca^{2+}]_i$-attenuating effects of GLP-1 and PACAP-38 could be mediated by cAMP, since cAMP has recently been shown to reduce KCl-induced $[Ca^{2+}]_i$ elevation.[9] The amplitude of $[Ca^{2+}]_i$ reduction by GLP-1 and PACAP-38 was in a range of 50–200 nM, and was much smaller than that by nifedipine (FIG. 1,B). Thus, the relatively smaller effect of GLP-1 and PACAP-38 to attenuate sustained $[Ca^{2+}]_i$ elevation is an apparent discrepancy with their potent β-cell protective action against Ca^{2+} toxicity. A possible explanation for this is that there may be a threshold $[Ca^{2+}]_i$ level beyond which a toxic action of Ca^{2+} operates, and that these peptides are capable of reducing $[Ca^{2+}]_i$ to the level below this threshold. Alternatively or additionally, GLP-1 and PACAP-38 may stimulate yet unknown mechanisms in β-cells that counteract the toxic action of the elevated $[Ca^{2+}]_i$.

REFERENCES

1. LEAHY, J. L., S. BONNER-WEIR & G. C. WEIR. 1992. Beta-cell dysfunction induced by chronic hyperglycemia. Current ideas on mechanism of impaired glucose-induced insulin secretion. Diabetes Care **15**: 442–455.
2. EIZIRIK, D. L., G. S. KORBUTT & C. HELLERSTROM. 1992. Prolonged exposure of human pancreatic islets to high glucose concentrations in vitro impairs the beta-cell function. J. Clin. Invest. **90**: 1263–1268.
3. GILON, P., R. M. SHEPHERD & J. C. HENQUIN. 1993. Oscillations of secretion driven by oscillations of cytoplasmic Ca^{2+} as evidenced in single pancreatic islets. J. Biol. Chem. **268**: 22265–22268.
4. WOLLHEIM, C. B. & G. W. G. SHARP. 1981. Regulation of insulin release by calcium. Physiol. Rev. **61**: 914–973.
5. YADA, T., M. SAKURADA, K. IHIDA, M. NAKATA, F. MURATA, A. ARIMURA & M. KIKUCHI. 1994. Pituitary adenylate cyclase activating polypeptide is an extraordinarily potent intrapancreatic regulator of insulin secretion from islet β-cells. J. Biol. Chem. **269**: 1290–1293.
6. YADA, T., M. SAKURADA, M. NAKATA, K. YAEKURA & M. KIKUCHI. 1996. Current status of PACAP as a regulator of insulin secretion in pancreatic islets. Ann. N.Y. Acad. Sci. **805**: 329–342.
7. YADA, T., M. SAKURADA, H. ISHIHARA, M. NAKATA, S. SHIODA, K. YAEKURA, N. HAMAKAWA, K. YANAGIDA, M. KIKUCHI & Y. OKA. 1997. Pituitary adenylate cyclase-activating polypeptide (PACAP) is an islet substance serving as an intra-islet amplifier of glucoseinduced insulin secretion in rats. J. Physiol. (Lond.) **505(2)**: 319–328.
8. YADA, T., K. ITOH & M. NAKATA. 1993. Glucagon-like peptide-1-(7-36)amide and a rise in cyclic adenosine 3',5'-monophosphate increase cytosolic free Ca^{2+} in rat pancreatic β-cells by enhancing Ca^{2+} channel activity. Endocrinology **133**: 1685–1692.
9. YAEKURA, K. & T. YADA. 1998. Cytosolic Ca^{2+}-reducing action of cAMP in rat pancreatic β-cells: involvement of thapsigargin-sensitive stores. Am. J. Physiol. **274**: C513–C521.

Autocrine Action of PACAP in Islets Augments Glucose-Induced Insulin Secretion[a]

T. YADA,[b,g,h] M. SAKURADA,[e] M. NAKATA,[b,c] S. SHIODA,[f]
K. YAEKURA,[b,d] AND M. KIKUCHI[e]

Departments of [b]Physiology, [c]Laboratory Medicine and [d]First Internal Medicine, Kagoshima University School of Medicine, Kagoshima 890-8520, Japan

[e]The Institute for Adult Diseases, Asahi Life Foundation, Tokyo 160, Japan

[f]Department of Anatomy, Showa University School of Medicine, Tokyo 142, Japan

[g]Laboratory of Intracellular Metabolism, National Institute for Physiological Sciences, Okazaki 444, Japan

Glucose-induced insulin release from pancreatic β-cells takes place with greater efficiency in islets or cell clusters than in single cells.[1,2] This assembly-dependent potentiation is thought to be mediated partly by a humoral factor released from islet cells.[1,2] However, a physiologically relevant potentiator of islet origin has yet to be identified.

Pituitary adenylate cyclase–activating polypeptide with 38 or 27 residues (PACAP-38 or PACAP-27) is a new member of secretin/glucagon/VIP family.[3,4] PACAP potentiates glucose-induced insulin release *in vivo* in mouse,[5] in isolated rat pancreas and islets.[6,7] PACAP-38 and -27 also increase cytosolic Ca^{2+} concentration ($[Ca^{2+}]_i$) in rat β-cells, and the $[Ca^{2+}]_i$ increase triggers the insulin release.[8] The concentrations at which PACAP exerts these effects are as low as 10^{-14}–10^{-13} M,[7] exhibiting an outstanding potency. Furthermore, the PACAP immunoreactivity is localized in pancreatic islets and nerve fibers.[5,7]

The aim of the present study was to examine whether PACAP functions as an intra-islet regulator of glucose-induced insulin secretion. We examined (1) the effect of a specific PACAP antiserum on glucose-induced insulin release from rat islets, (2) the release of PACAP from glucose-stimulated islets, and (3) the biosynthesis and localization of PACAP in islets. For the release study, since the concentrations of PACAP that can stimulate insulin release (10^{-14}–10^{-13} M) are below the detection limit by conventional assay methods, we adopted a bioassay system using the $[Ca^{2+}]_i$ response of single β-cells, a system sensitive enough to detect sub-picomolar levels of PACAP.[7]

[a]This work was supported by grants from the Ministry of Education, Science, Sports and Culture of Japan (T.Y.).

[h]Corresponding author: Dr. Toshihiko Yada, Department of Physiology, Kagoshima University School of Medicine, 8-35-1 Sakuragaoka, Kagoshima 890-8520, Japan, Tel.: 81-99-275-5225; Fax: 81-99-275-5231; E-mail: yada@med2.kufm.kagoshima-u.ac.jp

METHODS

Preparation of Islets and Measurement of Insulin Release

Islets of Langerhans were isolated from Wistar rats aged 8–12 weeks by collagenase digestion. Measurement of insulin release was carried out as previously described.[7] Briefly, groups of 30 islets were first incubated for 1 h in Krebs Ringer bicarbonate buffer (KRB) with 2.8 mM glucose. Islets were then incubated at 37°C for 30 min in 1 ml of KRB with test agents. Insulin concentration was determined by radioimmunoassay.

Solutions, Antibodies, and Chemicals

Insulin release experiments were carried out in KRB composed of (in mM): NaCl 121.7, KCl 4.4, KH_2PO_4 1.2, $CaCl_2$ 2.0, $MgSO_4$ 1.2, $NaHCO_3$ 23.0 at pH 7.4 supplemented with 0.1% bovine serum albumin (BSA) and equilibrated against a mixture of 95% O_2–5% CO_2 at 37°C. For other experiments, KRB containing 10 mM 4-(2-hydroxyethyl)-1-piperazineethanesulfonic acid (HEPES) and a reduced HCO_3^- (5.0 mM) was used to stably maintain pH at 7.4. Rabbit polyclonal PACAP antisera (No. 88121-6 and No. 92112-6) were gifts from Dr. A. Arimura. Synthetic PACAP-38 was from American Peptide Co. (Sunnyvale, CA, USA) and Peptide Institute (Osaka, Japan).

Preparation of Single β-Cells and Measurement of $[Ca^{2+}]_i$

Isolated islets were dispersed into single cells in Ca^{2+}-free KRB. The single cells were plated on coverslips and maintained in short-term culture for up to 3 days in Eagle's minimal essential medium (MEM) with 10% fetal bovine serum (FBS). Measurement of $[Ca^{2+}]_i$ were carried out by previously described procedures.[7,9] Briefly, single cells on coverslips were mounted in an open chamber and superfused in KRB, and $[Ca^{2+}]_i$ was measured by dual-wavelength fura-2 microfluorometry using an intensified charge-coupled device (ICCD) camera, and the ratio image produced by Argus-50 system (Hamamatsu Photonics, Hamamatsu, Japan). Data were taken only from the β-cells that were confirmed either immunocytochemically or by morphological and physiological criteria.[7,9]

Bioassay of PACAP Release

Fifty to 400 islets were incubated with 8.3–16.7 mM glucose in 100–200 μl of KRB. At 60 min, following cooling and centrifugation at 100 g for 1 min, the medium supernatant was collected and stored on ice for up to 2 h until use. It was diluted at 1:100 in KRB (Med) without or with PACAP antiserum (1:10,000), in which the final concentration of glucose was adjusted to 8.3 mM, and applied to single islet cells subjected to $[Ca^{2+}]_i$ measurements under superfusion conditions.

de Novo Protein Synthesis

Six hundred islets, after culture for 1 day in Eagle's MEM with 10% FBS, were incubated in Met/Cys-free medium for 30 min, followed by incubation with 100 μCi [^{35}S] Met/Cys (Amersham Co., Arlington Heights, IL) for 4 hours. After washing, the cells were treated with 1 ml NP40 lysis buffer: 100 mM NaCl, 0.5% NP40, 1 mM EDTA, 10 mM Tris-HCl (pH 7.5), 0.5 U aprotinin, and 1 mM PMSF. The lysate was centrifuged (12,000 g, 30 min, 4°C) and precleared with protein A–sepharose for 1 hour. PACAP antiserum was then added, followed by incubation at 4°C for 1 hour. Following addition of protein A–sepharose, total immunoprecipitates were eluted in Laemmli's SDS-PAGE sample buffer and subjected to 4–20% SDS-PAGE slab gel. The gel was autoradiographed on Fuji film.

Immunohistochemistry of PACAP

Rat pancreatic tissues were removed after the animals were perfusion-fixed with a mixture solution of 2% paraformaldehyde and 2% acrolein in 0.1 M phosphate buffer. Ten-μm frozen sections were cut and picked up on gelatin-coated slides. The sections were incubated first with PACAP antiserum (No. 92112-6) at a dilution of 1:2,000, and then with an avidin-biotin peroxidase complex. Subsequently, they were developed with True Blue (Vector).

RESULTS AND DISCUSSION

Attenuation of Insulin Release from Islets by PACAP Antiserum and Bioassay of PACAP Release from Islets

Stimulation of insulin release by 8.3 mM glucose from isolated rat islets, under static incubation conditions, was attenuated by 20–30% by a specific PACAP antiserum (anti-P)

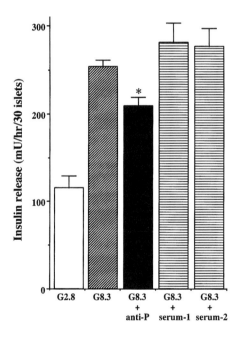

FIGURE 1. Attenuation of glucose-induced insulin release from rat islets by PACAP antiserum. Stimulation of insulin release from 30 rat islets by 8.3 mM glucose (G8.3), under conditions of static incubation, was attenuated by PACAP antiserum (anti-P) (No. 88121-6) at a 1:100 dilution (G8.3+anti-P), whereas it was unaltered by control rabbit sera (G8.3+serum-1 and -2). Values are means ± SEM of 11 experiments for G2.8, G8.3, and G8.3+anti-P, and experiments of 7 experiments for G8.3+serum-1, and -2 groups. *$p < 0.005$ versus G8.3, G8.3+serum-1, and G8.3+serum-2, by Student's t-test.

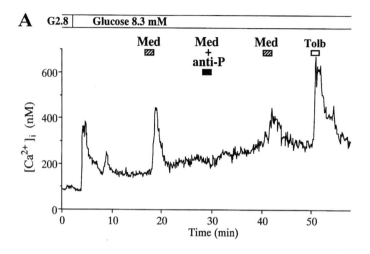

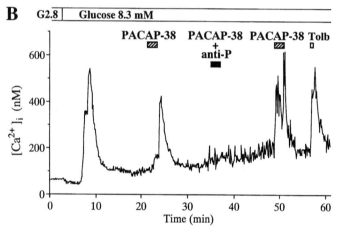

FIGURE 2. PACAP release from glucose-stimulated islets bioassayed by $[Ca^{2+}]_i$ responses of single β-cells. (**A**) A 1:100 dilution of the medium in which 200 islets were incubated with 16.7 mM glucose (Med) evoked increases in $[Ca^{2+}]_i$ in single rat β-cells in the presence of 8.3 mM glucose, whereas Med mixed with PACAP antiserum (anti-P) (No. 88121-6) (1:10,000) was without effect. (**B**) Anti-P (1:10,000) inhibited $[Ca^{2+}]_i$ responses to 10^{-9} M synthetic PACAP-38 in β-cells. Anti-P had no effect on subsequent $[Ca^{2+}]_i$ responses to 300 μM tolbutamide (Tolb). Representative results from 18 and 7 similar experiments are shown in **A** and **B**, respectively. (Reproduced from Yada *et al.*[12] with permission.)

(No. 88121-6),[10,11] whereas control rabbit sera were without effect (FIG. 1). Thus, a fraction of the glucose-induced insulin release from islets was neutralized by the PACAP antiserum. It was previously shown that this PACAP antiserum reacts with PACAP-38 and PACAP-27 but not with VIP, glucagon, secretin, and other structurally related peptides, or with somatostatin, thus behaving as a highly selective PACAP antiserum.[10,11]

Release of PACAP from glucose-stimulated islets was studied by a bioassay system using the $[Ca^{2+}]_i$ response of single β-cells. A 1:100 dilution of the medium in which isolated rat islets were incubated with high glucose (Med) evoked an increase in $[Ca^{2+}]_i$ in single β-cells under superfusion conditions in the presence of 8.3 mM glucose (FIG. 2,A). When Med was mixed with the PACAP antiserum, its $[Ca^{2+}]_i$-increasing capacity was either abolished or markedly reduced (FIG. 2,A). Synthetic PACAP-38 increased $[Ca^{2+}]_i$ in a glucose-dependent manner, and this effect was also antagonized by the PACAP antiserum (FIG. 2,B). The results indicate that the glucose-stimulated islets release a β-cell–stimulating factor, which is neutralized by the specific PACAP antiserum and therefore is very likely to be PACAP-38 and/or -27.

Biosynthesis and Localization of PACAP in Islet β-cells

Biosynthesis of PACAP was studied by metabolic labeling with ^{35}S-Met/Cys in rat islets after 1 day of culture in which proteins of neural origin are excluded (FIG. 3,A). A protein showing a clear metabolic labeling was immunoprecipitated specifically with the PACAP antiserum (lane 2), but not with control rabbit serum (lane 1), and its apparent molecular weight, around 5 kD, corresponded to that of synthetic PACAP-38. Essentially the same results were obtained in a β-cell line, MIN-6 cells (data not shown). Next, storage of PACAP in islets was studied immunochemically using a specific PACAP antiserum (No. 92112-6).[4] Immunostaining for PACAP was observed in the rat islet, including its central portion exclusively occupied by β-cells (FIG. 3,B). Thus, the localization of PACAP in islets, the finding originally demonstrated using another specific PACAP antiserum (No. 88121-6),[7] has been further confirmed. These results reveal *de novo* synthesis and storage of PACAP-38 in islet β-cells.

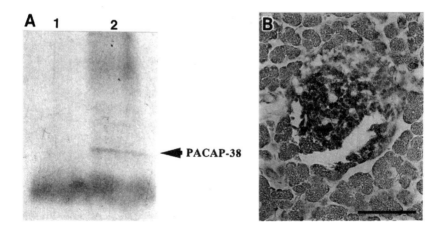

FIGURE 3. Biosynthesis and localization of PACAP in islets. (**A**) Metabolic labeling with ^{35}S-Met/Cys in rat islets after culture for one day, indicating biosynthesis of PACAP-38 by islet cells. Islet cell lysate was immunoprecipitated with PACAP antiserum (No. 88121-6) (lane 2) and control rabbit serum (lane 1). Bar at right shows the position of synthetic PACAP-38. (**B**) Immunohistochemical staining of rat pancreas with PACAP antiserum (No. 92112-6) (1:2,000). PACAP immunoreactivity was observed in islets including the central portion exclusively occupied by β-cells, but not in the exocrine portion of the rat pancreas. Bar indicates 100 μm.

CONCLUSION

In the present study, we have shown that PACAP-38 is synthesized and stored by islet β-cells, that PACAP-like activity is released from glucose-stimulated islets, and that the neutralization of this activity by the specific antiserum attenuates glucose-induced insulin release. PACAP-38 directly targets β-cells.[7,12] It has been shown that the messages for a PACAP receptor (PACAP-R) variant, PACAP-R TM4,[13] and VIP2/PACAP3-R[14,15] are expressed in islet β-cells and that PACAP-R and/or PACAP-R-TM4 are localized in islet β-cells.[12] Based on these findings, we conclude that PACAP-38 is a novel islet hormone serving as an autocrine regulator, wherein this peptide is released by glucose-stimulated islet β-cells and acts on islet β-cells, resulting in augmentation of glucose-induced insulin secretion. It is proposed that glucose-induced insulin secretion in islets is auto-amplified by the autocrine action of PACAP.

ACKNOWLEDGMENTS

We thank Dr. A. Arimura for kindly providing PACAP antisera (No. 88121-6 and No. 92112-6).

REFERENCES

1. PIPELEERS, D. 1987. The biosociology of pancreatic B cells. Diabetologia **30**: 277–291.
2. HENQUIN, J. C. 1994. Cell biology of insulin secretion. *In* Joslin's Diabetes Mellitus. C. R. Kahn & G. C. Weir. Eds.: 56–80. Lea & Febiger. Philadelphia, PA.
3. MIYATA, A., A. ARIMURA, R. R. DAHL, N. MINAMINO, A. UEHARA, L. JIANG, M. D. CULLER & D. H. COY. 1989. Isolation of a novel 38-residue hypothalamic polypeptide which stimulates adenylate cyclase in pituitary cells. Biochem. Biophys. Res. Commun. **164**: 567–574.
4. ARIMURA, A. & S. SHIODA. 1995. Pituitary adenylate cyclase activating polypeptide (PACAP) and its receptors: neuroendocrine and endocrine interactions. Front. Neuroendocrinol. **16**: 53–88.
5. FRIDOLF, T., F. SUNDLER & B. AHREN. 1992. Pituitary adenylate cyclase-activating polypeptide (PACAP): occurrence in rodent pancreas and effects on insulin and glucagon secretion in the mouse. Cell Tissue Res. **269**: 275–279.
6. KAWAI, K., C. OHSE, Y. WANTABE, S. SUZUKI, K. YAMASHITA & S. OHASI. 1992. Pituitary adenylate cyclase activating polypeptide stimulates insulin release from isolated perfused rat pancreas. Life Sci. **50**: 257–261.
7. YADA, T., M. SAKURADA, K. IHIDA, M. NAKATA, F. MURATA, A. ARIMURA & M. KIKUCHI. 1994. Pituitary adenylate cyclase activating polypeptide is an extraordinarily potent intra-pancreatic regulator of insulin secretion from islet β-cells. J. Biol. Chem. **269**: 1290–1293.
8. YADA, T., M. SAKURADA, M. NAKATA, K. YAEKURA & M. KIKUCHI. 1996. Current status of PACAP as a regulator of insulin secretion in pancreatic islets. Ann. N. Y. Acad. Sci. **805**: 329–342.
9. YADA, T., K. ITOH & M. NAKATA. 1993. Glucagon-like peptide-1-(7-36)amide and a rise in cyclic adenosine 3′,5′-monophosphate increase cytosolic free Ca^{2+} in rat pancreatic β-cells by enhancing Ca^{2+} channel activity. Endocrinology **133**: 1685–1692.
10. KOVES, K., A. ARIMURA, A. SOMOGYVARI-VIGH, S. VIGH & J. MILLER. 1990. Immunohistochemical demonstration of a novel hypothalamic peptide, pituitary adenylate cyclase-activating polypeptide, in the ovine hypothalamus. Endocrinology **127**: 264–271.
11. KOVES, K., A. ARIMURA, T. G. GORCS & A. SOMOGYVARI-VIGH. 1991. Comparative distribution of immunoreactive pituitary adenylate cyclase activating polypeptide and vasoactive intestinal peptide in rat forebrain. Neuroendocrinology **54**: 159–169.
12. YADA, T., M. SAKURADA, H. ISHIHARA, M. NAKATA, S. SHIODA, K. YAEKURA, N. HAMAKAWA, K. YANAGIDA, M. KIKUCHI & Y. OKA. 1997. Pituitary adenylate cyclase-activating polypeptide (PACAP) is an islet substance serving as an intra-islet amplifier of glucose-induced insulin secretion in rats. J. Physiol. (Lond.) **505**(2): 319–328.

13. CHATTERJEE, T. K., R. V. SHARHAJ & R. A. FISHER. 1996. Molecular cloning of a novel variant of the pituitary adenylate cyclase-activating polypeptide (PACAP) receptor that stimulates calcium influx by activation of L-type calcium channels. J. Biol. Chem. **271**: 32226–32232.

14. USDIN, T. B., T. I. BONNER & E. MEZEY. 1994. Two receptors for vasoactive intestinal polypeptide with similar specificity and complementary distributions. Endocrinology **135**: 2662–2680.

15. INAGAKI, N., H. YOSHIDA, M. MIZUTA, N. MIZUNO, Y. FUJII, T. GONOI, J. MIYAZAKI & S. SEINO. 1994. Cloning and functional characterization of a third pituitary adenylate cyclase-activating polypeptide receptor subtype expressed in insulin-secreting cells. Proc. Natl. Acad. Sci. USA **91**: 2679–2683.

A Target Cell to Oxyntomodulin and Glicentin: The Antral Smooth Muscle Cell

G. RODIER,[a,d] R. MAGOUS,[b] T. MOCHIZUKI,[c] J. P. BALI,[b]
D. BATAILLE,[a] AND C. JARROUSSE[a]

[a]Laboratory Endocrinologie des Peptides et Régulation Génique
INSERM U 376, 34,295 Montpellier, France

[b]Laboratory Biologie des Membranes, CNRS EP 612, 34060
Montpellier, France

[c]College of Pharmacy, Laboratory of Bioorganic Chemistry, 422
Shizuoka, Japan

Oxyntomodulin (OXM) and glicentin (GLIC) are two digestive hormones that belong to the secretin-VIP-glucagon family. They are composed of the glucagon sequence and of an additional C-terminal octapeptide KRNRNNIA.[1,2] A N-terminus extension called glucagon-related pancreatic peptide (GRPP) is present in the GLIC molecule.[2]

OXM crossreacts with the glucagon receptor in liver or other tissues[3] and with the glucagon-like peptide 1 (tGLP1) receptor on a cell line secreting somatostatin.[4] A direct consequence is that studies where OXM acts through the adenylyl cyclase system cannot be easily interpretated. A specific target cell to OXM has been examined for many years on epithelial and muscular digestive tissues. Inhibition of acid secretion in stomach is not mediated by an OXM effect on the parietal cell[5,6]; inhibition of caerulein-induced proteases secretion in rat exocrine pancreas is not mediated by an OXM effect on acinus (personal observation); inhibition of chloride secretion in rat jejunum is partly mediated by the intrinsic nervous system,[7] suggesting that enterocytes and intrinsic neurons may be target cells. A slowing of human gastric emptying by OXM[8] and a reduction of the canine post-prandial antral contractions by GLIC[9] demonstrate smooth muscle control by OXM and GLIC. On isolated smooth muscle cells from rabbit antrum (free of neurons and composed of 95% muscle cells), human GLIC, OXM and their common C-terminal fragment OXM(19-37) have a contractile activity. Their respective EC_{50} are: 5 pM, 80 pM, and 72 pM.[10] Thus the smooth muscle cell from rabbit antrum is a target cell to GLIC and OXM. If the C-terminal moiety is required for getting the biological effect, the GRPP only present in GLIC increases the affinity of this hormone to its binding site. Glucagon or tGLP1 are devoided of contractile activity on these cells.

A binding site in visceral smooth muscle can be a G protein–coupled receptor, a tyrosine kinase–related receptor, or a ligand–gated ion channel. This work addressed the possibility of a GLIC/OXM G protein–coupled receptor. Contraction was studied on permeabilized smooth muscle cells to discard the ligand-gated ion channel hypothesis and to validate the G protein–coupled receptor hypothesis. The effect of a G protein inactivator was tested on contraction. At last, the eventuality of a link between the GLIC/OXM G protein–coupled receptor and the

[d]Corresponding author: Dr. Geneviève Rodier, Laboratoire d'Endocrinologie des Peptides et de Régulation Génique, CHU Arnaud de Villeneuve, 371 rue du doyen Giraud, 34 295 Montpellier cedex, France; Tel.: 33 4 67 41 52 22; Fax: 33 4 67 41 52 26; E-mail: jarrou@u376.montp.inserm.fr

inositol phosphates, intracellular messengers resulting from activation of phosphoinositide-specific phospholipase C β by G proteins was investigated.

MATERIALS AND METHODS

Materials

Guanosine 5'[γ-thio]-triphosphate (GTPγS) and guanosine 5'-O(2-thiodiphosphate) (GDPβS), heparin (MW 4,000), inositol (1,4,5)-triphosphates, and saponin were purchased from Sigma Chemical Co. (St. Louis, MO, USA). Synthetic human glicentin was the generous gift of N. Yanaihara and synthetic oxyntomodulin and OXM (19-37) were made by J. Martinez and our laboratory.[10]

Preparation of Smooth Muscle Cells and Measurement of Contraction

Smooth muscle cells (SMC) were isolated by enzymatic digestion from the muscle layers of the rabbit antrum as previously described.[10]

Muscle cells were permeabilized with saponin, a nonionic detergent that retains receptor functionality. Fifty μg/ml saponin was added for 7–8 min in a cytosol-like medium containing 180 nM Ca^{2+} to get 70% permeabilized cells. The medium consisted of 20 mM NaCl, 100 mM KCl, 5 mM NaH_2PO_4, 1 mM NaH_2PO_4, 0.5 mM $CaCl_2$, 1 mM EGTA, 10 mM HEPES, 2% BSA, pH 7.4. The cells were washed free of saponin by centrifugation at 150 g and resuspended in the presence of ATP (1.5 mM) and of an ATP-regenerating system consisting of creatine phosphate (5 mM) and creatine kinase (10 U/ml).

Cells (125,000 cells/450 μl) were incubated for 30 sec at 30°C in the presence of the effectors. Cells were preincubated for 5 min with GDPβS (10^{-4} M) or 2 min with heparin (10 μg/ml). After addition of glutaraldehyde (final concentration, 2.5%), the length of 100 cells was measured in successive microscopic fields.

Cyclic AMP Determination

Intracellular cAMP content of muscle cells was determined as followed: cells (250,000/500 μl) were incubated in a saline medium: 116 mM NaCl, 5.4 mM KCl, 0.81 mM $MgSO_4$, 1.8 mM $CaCl_2$, 1 mM NaH_2PO_4, 10 mM HEPES, 5.5 mM glucose, 0.5% BSA, pH 7.4. After a 10-min preincubation with 100 μM isobutylmethylxanthine, a phosphodiesterase inhibitor, peptides were incubated 30 sec at 30°C. Vasoactive intestinal polypeptide was used as control. Reaction was stopped by addition of ice-cold 60% $HClO_4$. After centrifugation, the supernatant was adjusted to pH 9 with KOH 9N, centrifuged again, and the resulting supernatant radioimmunoassayed.[4]

RESULTS AND DISCUSSION

Contractile Activity of GLIC and OXM on Permeabilized Cells

A maximal contraction on isolated smooth muscle cells was achieved at 10^{-10} M GLIC, 10^{-9} M OXM or OXM(19-37)[10] and was equal at 15% of decrease in cell length for the three molecules. A maximal contraction on the saponin-permeabilized cells was also observed with 10^{-8} M GLIC or OXM and 10^{-9} M OXM(19-37). As shown in FIGURE 1, the amplitude

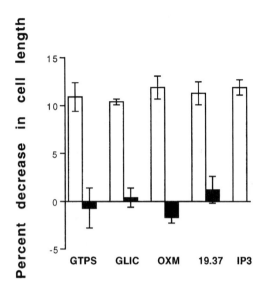

FIGURE 1. Contraction of permeabilized smooth muscle cells with (*black bars*) or without (*white bars*) GDPβS (10^{-4} M). Cells were incubated for 30 sec at 30°C. Contraction obtained with GTPγS (10^{-4} M) or Ins(1,4,5)P$_3$ (IP3) 10^{-6} M were compared to that obtained with GLIC (10^{-8} M), OXM (10^{-8} M), or OXM(19-37) (10^{-9} M). Results were expressed as the mean decrease in cell length ± SEM of 4–5 experiments.

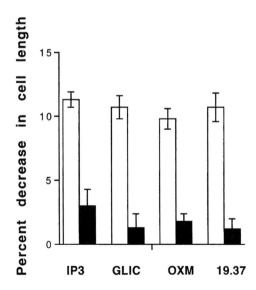

FIGURE 2. Contraction of permeabilized smooth muscle cells in function of the presence (*black bars*) or absence (*white bars*) of heparin (10 µg/ml). Contraction obtained with Ins(1,4,5)P$_3$ (IP3) 10^{-6} M, GLIC (10^{-9} M), OXM (10^{-8} M), or OXM(19-37) (10^{-9} M). Results were expressed as the mean decrease in cell length ± SEM of 10 experiments for Ins(1,4,5)P$_3$ (IP3) and 4–7 experiments for peptides.

of contraction reached 11%. At 10^{-8} M OXM(19-37), a reversion of contraction was observed due to an expected intracellular diffusion of the smallest peptide. To evaluate the maximal capacity of contraction of permeabilized cells, a direct activation of G protein α subunits by GTPγS 10^{-4} M, a poorly hydrolyzable analogue of GTP, was achieved and the same amplitude of contraction (10.9 ± 1.5%) was obtained. A maximal release of sarcoplasmic reticulum calcium is another way to evaluate the contractile capacity of permeabilized cells. Inositol triphosphates(1,4,5) at a maximal concentration (10^{-6} M) induced a 11.9 ± 0.8 % decrease in cell length. Thus we conclude that GLIC and OXM kept the ability to contract the smooth muscle cells in the absence of ionic gradients between the two sides of the plasma membrane. In such conditions, the contractile activity of GLIC and OXM cannot be mediated by a ligand gated ion channel. Thus the G protein–coupled receptor hypothesis has been considered and validated using a G protein inactivator GDPβS. At 10^{-4} M, this inactivator did not induce contraction by itself (2.9±1%, N=7). On the contrary, it counteracted the contractile activity of GTPγS, GLIC, OXM, and OXM(19-37) (Fig. 1). These results confirmed the involvement of a G protein-coupled receptor.

G-proteins mediate signaling between receptors and effectors such as adenylyl cyclases, phospholipases, phosphodiesterases, and ions channels. In smooth muscle cells, stimulation of adenylyl cyclases is related to relaxation and stimulation of phospholipases to contraction. We looked for both possibilities.

Activity of OXM, Glucagon, tGLP1, and VIP on cAMP Content

The cAMP content was estimated at 30 sec on intact cells with the same 10^{-7} M concentration for each peptide. Under a basal state, the cAMP content was equal at 47±6 pmol/10^6 cells. VIP induced a marked cytosolic increment of cAMP (+33±5 pmol/10^6 cells), while OXM did not (-2±2 pmol/10^6 cells). Thus, if OXM belongs to the glucagon family, the bound receptor was not activating adenylyl cyclases as the glucagon or tGLP1 receptor. On antral smooth muscle cells, glucagon or tGLP1 exhibited no significant variations of the basal cAMP content (+9±9 and +3±2 pmol/10^6 cells, respectively).

Activity of GLIC and OXM on the Inositol Triphosphate Pathway

In circular smooth muscle cells, the pathway initiating contraction is generally an activation of a G protein–coupled phospholipase C β.[11] This enzyme produces inositol triphosphates and diacylglycerol. Inositol(1,4,5) triphosphate [Ins $(1,4,5)P_3$] is recognized by a specific calcium channel on sarcoplasmic reticulum and the opening of this channel induces a cytosolic calcium increase and contraction of the cells. Heparin prevents the binding of Ins(1,4,5)P_3 and, as a consequence, contraction.

On the permeabilized smooth muscle cells, the contractile activity of GLIC, OXM, OXM(19-37), or exogenous Ins(1,4,5)P_3 was abolished by preincubating cells with heparin (Fig. 2). Thus the GLIC/OXM contraction was linked to the production of Ins(1,4,5)P_3. In smooth muscle cells from intestine, PLCβ activated by G proteins generates Ins(1,4,5)P_3. Such a signaling pathway is also coupled to the glucagon receptor[12] but not to the described tGLP1 receptor.[13]

REFERENCES

1. BATAILLE, D., K. TATEMOTO, C. GESPACH, H. JORNVALL, G. ROSSELIN & V. MUTT. 1982. Isolation of glucagon-37 (bioactive enteroglucagon/oxyntomodulin) from porcine jejuno-ileum. Characterisation of the peptide. FEBS Lett. **146**: 79–86.

2. THIM, L. & A. J. MOODY. 1982. Porcine glicentin related pancreatic peptide (GRPP): Purification and chemical characterization of a glicentin-related pancreatic polypeptide (proglucagon fragment). Biochim. Biophys. Acta **703**: 134–141.
3. BATAILLE, D. 1996. Oxyntomodulin and its related peptides. *In* Glucagon III. P.J. Lefèbvre, Ed.: 327–340. Handb. Exp. Pharmacol. **123**: Springer. Heidelberg.
4. GROS, L., B. THORENS, D. BATAILLE & A. KERVRAN. 1993. Glucagon-like peptide 1(7-36) amide, oxyntomodulin and glucagon interact with a common receptor in a somatostatin-secreting cell line. Endocrinology **133**: 631–638.
5. GROS, L., F. HOLLANDE, B. THORENS, A. KERVRAN & B. BATAILLE. 1995. Comparative effects of GLP-1-(7-36) amide, oxyntomodulin and glucagon on rabbit gastric parietal cell function. Eur. J. Pharmacol. **288**: 319–327.
6. GEHL, J., J. L. JEPPESEN, S. S. POULSEN & J. J. HOLST. 1988. The gastric acid secretagogue gastrin-releasing peptide and the inhibitor oxyntomodulin do not exert their effect directly on the parietal cell in the rat. Digestion **40**: 144–151.
7. BEAUCLAIR, F., D. PANSU, G. RODIER, T. MOCHIZUKI, J. MARTINEZ, D. BATAILLE & C. JARROUSSE. 1998. Oxyntomodulin reduces the hydromincral transport through rat small intestine. Dig. Dis. Sci. **43**: 1814–1823.
8. SCHJOLDAGER B, P. E. MORTENSEN, J. MYHRE, J. CHRISTIANSEN & J. J. HOLST. 1989. Oyntomodulin from distal gut. Role in regulation of gastric and pancreatic functions. Dig. Dis. Sci. **34**: 1411–1419.
9. OHTANI, N., I. SASAKI, H. NAITO, M. TAKAHASHI, C. SHIBATA, T. DOI, S. MATSUNO, A. OHNEDA, T. NAGASAKI & K. SASAKI. 1994. Inhibitory effect of peptide YY, neurotensin and glicentin on upper gastrointestinal motility in dogs. Biomed. Res. **15**: 299–302.
10. RODIER, G., R. MAGOUS, T. MOCHIZUKI, J. MARTINEZ, D. LE NGUYEN, J. P. BALI, D. BATAILLE & C. JARROUSSE. 1997. Effect of glicentin, oxyntomodulin and related peptides on isolated gastric smooth muscle cells. Pflüg. Arch. Eur. J. Physiol. 1997. **434**: 729–734.
11. MAKHLOUF, G. & K. MURTHY. 1997. Signal transduction in gastrointestinal smooth muscle. Cell Signal. **9**: 269–276.
12. PECKER, F. & C. PAVOINE. 1996. Mode of action of glucagon revisited. Handb. Exp. Pharmacol. **123**: 75–104.
13. GÖKE, B., R. GÖKE, H. C. FEHMANN & H. P. BODE. 1996. Physiology and pathophysiology of GLP-1. Handb. Exp. Pharmacol. **123**: 275–309.

Effects of PACAP/VIP/Secretin on Pancreatic and Gastrointestinal Blood Flow in Conscious Dogs[a]

S. NARUSE,[b] O. ITO, M. KITAGAWA, H. ISHIGURO,
M. NAKAJIMA, AND T. HAYAKAWA

*Department of Internal Medicine II, Nagoya University School of
Medicine, 65 Tsurumaicho, Showa-ku, Nagoya 466, Japan*

Pituitary adenylate cyclase activating polypeptide (PACAP) and vasoactive intestinal polypeptide (VIP) are members of the secretin-glucagon family of neuropeptides.[1] PACAP-immunoreactive neurons are present in the myenteric and submucous ganglia of the stomach, small intestine, and pancreas.[1-3] The small arteries and arterioles in the gastrointestinal tract and pancreas are also innervated by VIP- or PACAP-positive fibers.[2,3] Both peptides are very potent vasodilators of the gastrointestinal blood vessels in conscious dogs.[4,5] These findings suggest that PACAP may participate in the regulation of the gastrointestinal circulation but its effect on pancreatic blood flow is not known in dogs. In the present investigation we have compared the effects of PACAP38 and PACAP27 on pancreatic and gastrointestinal blood flow with those of VIP and secretin in conscious beagle dogs.

METHODS

Five beagle dogs (8–14 kg) of either sex were used. Following induction by thiamylal (20 mg/kg) and atropine (0.5 mg), anesthesia was maintained with N_2O–O_2–ether. Flow probes of an ultrasound transit-time blood flowmeter (Transonic Systems, New York) were placed around a pancreatic branch of the splenic artery and the left gastric and superior mesenteric arteries (FIG. 1). The connectors of the probes were pulled out of the abdominal cavity through a subcutaneous tunnel and fixed at the chest. After a recovery period of 4 weeks, the animals were restrained in Pavlov stands and the experiments were conducted in the conscious state. PACAP38, PACAP27, VIP, and secretin (Peptide Institute, Inc., Osaka; 1, 2.5, 5, 10, 25, 50, and 100 pmol/kg in 1 min) were infused intravenously and blood pressure, heart rate, and blood flows were measured. Blood flow responses to oral ingestion of 300 ml milk served as control. In another study PACAP(6-38), a PACAP receptor antagonist,[6] was infused intravenously (1 nmol/kg) before the administration of each peptide.

RESULTS

Basal blood flows of the pancreatic, left gastric, and superior mesenteric arteries were 3.6 ± 0.9, 22.6 ± 5.5, 103 ± 13 (mean $\pm$ SE, $N = 5$) ml/min, respectively. Oral ingestion of

[a]This work was supported by a grant from the Ministry of Education, Science and Culture of Japan.
[b]Corresponding author: Satoru Naruse, M.D., Internal Medicine II, Nagoya University School of Medicine, 65 Tsurumaicho, Showa-ku, Nagoya 466, Japan; Tel.: 81-52-744-2170; Fax: 81-52-744-2179; E-mail: snaruse@tsuru.med.nagoya-u.ac.jp

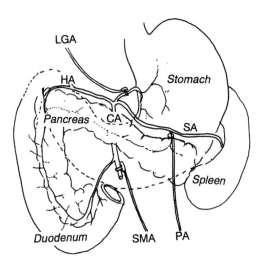

FIGURE 1. Measurement of pancreatic (PA), left gastric (LGA), and superior mesenteric arterial (SMA) blood flows of dog. CA: celiac artery, SA: splenic artery, HA: hepatic artery.

300 ml milk increased pancreatic, gastric and intestinal blood flows to 830 ± 334%, 308 ± 56%, 386 ± 19% of the basal, respectively.

PACAP38, PACAP27, and VIP, but not secretin, increased heart rate in a dose-related manner but systemic blood pressure remained unchanged. PACAP38, PACAP27, and VIP increased pancreatic and gastric arterial blood flows in a dose-related manner (FIG. 2). Their maximal effects were comparable to or greater than those observed after milk ingestion. Secretin was a weak but significant pancreatic and intestinal vasodilator. The intesti-

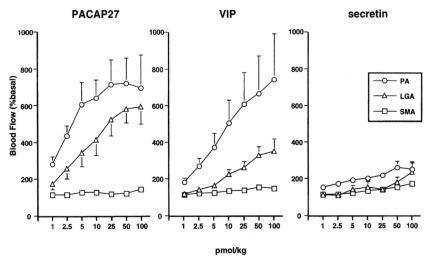

FIGURE 2. The effects of PACAP27, VIP, and secretin on pancreatic (PA), left gastric (LGA), and superior mesenteric arterial (SMA) blood flows in conscious dogs. Mean responses ± SE ($N = 5$) expressed as percent of the basal flow are shown.

nal vascular effect of VIP was comparable to that of secretin. PACAP38 decreased intestinal blood flow transiently. Vasodilator effects of PACAP38 and PACAP27, but not of VIP or secretin, on pancreatic and gastric blood vessels were inhibited by PACAP(6-38).

DISCUSSION

The present study has clearly shown that blood vessels of the pancreas are very sensitive to PACAP38 and PACAP27. In the intestinal vasculature, in contrast to calcitonin gene-related peptide,[7] PACAPs and VIP were very weak agonists. Pancreatic blood flow was more than doubled by as low as 1 pmol/kg of PACAPs, which caused no pancreatic secretory response in dogs.[8] At doses of 10–100 pmol/kg, pancreatic blood flow reached a plateau that was comparable to postprandial blood flow, whereas fluid and bicarbonate secretion was only 10% that of secretin. Though secretin increased pancreatic blood flow, its maximal effect was only 20% that of PACAP 38 or PACAP27. In most vascular systems PACAPs act via VIP/PACAP common (type 2) receptors.[1,9] In agreement with previous studies on gastric circulation,[4,5] the maximal effects of VIP on pancreatic and gastric blood flows were observed at 100 pmol/kg. Below this dose, PACAPs were much more effective than VIP, suggesting the presence of PACAP specific receptors (type 1) in pancreatic and gastric vasculatures. The inhibition of pancreatic and gastric blood flow response to PACAPs, but not to VIP, by PACAP(6-38) further supports this view. Thus, it appears that both PACAP and VIP/PACAP specific receptors are present in blood vessels of the pancreas and stomach in dogs.

REFERENCES

1. ARIMURA, A. 1992. Pituitary adenylate cyclase activating polypeptide (PACAP): discovery and current status of research. Regul. Pept. **37**: 287–303.
2. SUNDLER, F., E. EKBLAD, A. ABSOOD, R. HÅKANSON, K. KÖVES & A. ARIMURA. 1992. Pituitary adenylate cyclase activating peptide: a novel vasoactive intestinal peptide-like neuropeptide in the gut. Neuroscience **46**: 439–454.
3. TORNØE, K., J HANNIBAL, M. GIEZEMANN, P. SCHMIDT & J. J. HOLST. 1996. PACAP 1-27 and PACAP 1-38 in the porcine pancreas: Occurrence, localization, and effects. Ann. N.Y. Acad. Sci. **805**: 521–535.
4. NARUSE, S., T. NAKAMUR, K. NOKIHARA, E. ANDO & V. WRAY. 1994. The effect of PACAP on gastrointestinal blood flow in conscious dogs. *In* Vasoactive Intestinal Peptide, Pituitary Adenylate Cyclase Activating Polypeptide & Related Regulatory Peptides. G. Rosselin, Ed.: 470–473. World Scientific. Singapore.
5. NARUSE, S., T. NAKAMURA, M. WEI *et al.* 1996. Effects of PACAP-VIP hybrid peptides on gastric blood flow in conscious dogs. Ann. N.Y. Acad. Sci. **805**: 511–515.
6. ROBBERECHT, P., P. GOURLET, P. DE NEEF *et al.* 1992. Structural requirements for the occupancy of pituitary adenylate cyclase activating polypeptide (PACAP) receptors and adenylate cyclase activation in human neuroblastoma NB-OK-1 cell membranes. Discovery of PACAP(6-38) as a potent antagonist. Eur. J. Biochem. **207**: 239–246.
7. NAKAMURA, T., S. NARUSE, T. OZAKI & K. KUMADA. 1996. Calcitonin gene-related peptide is a potent intestinal, but not gastric, vasodilator in conscious dogs. Regul. Pept. **65**: 211–217.
8. NARUSE, S., T. SUZUKI & T. OZAKI. 1992. The effect of pituitary adenylate cyclase activating polypeptide (PACAP) on exocrine pancreatic secretion in dogs. Pancreas **7**: 543–547
9. NARUSE, S., T. SUZUKI, T. OZAKI & K. NOKIHARA. 1993. Vasodilator effect of pituitary adenylate cyclase activating polypeptide (PACAP) on femoral blood flow in dogs. Peptides **14**: 505–510.

Pituitary Adenylate Cyclase Activating Polypeptide Stimulates Insulin Secretion in a Glucose-Dependent Manner *In Vivo*[a]

BO AHRÉN AND KARIN FILIPSSON

Department of Medicine, Lund University, Malmö, Sweden

Nerves harboring pituitary adenylate cyclase–activating polypeptide (PACAP) have been localized to intrapancreatic nerve ganglia, to single nerves in the exocrine pancreatic parenchyma, around blood vessels, and in conjunction with pancreatic islets.[1,2] Furthermore, PACAP has been demonstrated to stimulate insulin secretion both *in vivo* and *in vitro*,[2-4] mainly through activating adenylate cyclase.[5] Hence, PACAP is a pancreatic neuropeptide with potential involvement in the regulation of insulin secretion.

Under *in vitro* conditions, PACAP stimulates insulin secretion in a glucose-dependent manner. For example, we have shown in the insulin-producing cell line HIT-T15 that the insulinotropic action of PACAP is exaggerated by 50% by increasing the glucose level in the extracellular medium from zero to 10 mmol/l.[5] However, the glucose dependency of the insulinotropic action of PACAP under *in vivo* conditions has not been studied before. Therefore, we have compared the increase in circulating insulin induced by an intravenous administration of PACAP27 at different glucose concentrations in mice.

METHODS

Non-fasted anesthetized NMRI mice (Bomholtgaard Breeding and Research Center, Ry, Denmark, 20–25 g) were injected intravenously with saline or D-glucose (British Drug Houses, Poole, U.K.; 0.3, 1.0, or 3.0 g/kg) either alone or together with synthetic ovine PACAP27 (Peninsula Europe Laboratories, Merseyside, U.K.; 1.3 nmol/kg). Blood was sampled and plasma separated and analyzed for insulin with a radioimmunoassay (Linco, St. Charles, MO, USA) or glucose with a glucose oxidase method. Means+SEM are shown. Statistical comparisons were performed with Student's *t*-test. The acute insulin secretion (AIR) was calculated as the mean of suprabasal 1 and 5 min insulin levels and the glucose elimination rate after the glucose injection (K_G) was calculated using the $t_{1/2}$ for the 1–20 min after glucose injection after logarithmic transformation of the individual plasma glucose values.

RESULTS AND DISCUSSION

Under baseline conditions, PACAP27(1.3 nmol/kg) slightly increased circulating insulin after 20 min (237 ± 23 pmol/l versus 167 ± 18 pmol/l in controls; $p = 0.040$) but

[a]This study was supported by the Swedish Medical Research Council (grant no. 14X-6834), Ernhold Lundström, Albert Påhlsson, and Novo Nordic Foundations, Swedish Diabetes Association, Malmö University Hospital, and the Faculty of Medicine, Lund University.

[b]Corresponding author: Dr. Bo Ahrén, Department of Medicine, Malmö University Hospital, S-205 02 Malmö, Sweden; Tel.: 4640336454, Fax: 4640337041; E-mail bo.ahren@medforsk.mas.lu.se

not after 1 or 5 min (FIG. 1). Plasma glucose was also slightly increased after 20 min (9.2 ± 0.3 mmol/l versus 7.7 ± 0.3 mmol/l in controls; $p = 0.046$). When PACAP27(1.3 nmol/kg) was injected intravenously together with glucose (0.3, 1.0, or 3.0 g/kg), the insulin response to glucose was potentiated because both 1 and 5 min values of plasma insulin were higher after PACAP27 than in controls at all three glucose doses ($p < 0.05$ or less). In contrast, the glucose disposal after glucose administration was not affected by PACAP27, in spite of its marked insulinotropic action (FIG. 1). TABLE 1 shows AIR and K_G in the mice given glucose without or with addition of PACAP27. FIGURE 2 shows the relation between the AIR and the peak glucose at 1 min after glucose injection in animals given glucose alone and glucose plus PACAP27. The small insert shows the potentiating effect of PACAP27 on glucose-stimulated insulin secretion as a function of the amount of glucose injected. Thus, PACAP27 glucose-dependently stimulates insulin secretion *in vivo* in mice, as previously was demonstrated for its *in vitro* action in insulin-producing cells.[5] In a previous study, we showed that PACAP38 did not potentiate glucose-stimulated insulin secretion in unanesthetized mice,[1] which contradicts to our present study. This may be explained by the higher degree of stress in unanesthetized than in anesthetized animals.

The insulin response to glucose was maximal at glucose levels exceeding approximately 25 mmol/l, which is similar to previous studies in humans where a maximal insulin secretory capacity of the islet B cell is seen at such a high glucose level.[6] PACAP27 could clearly potentiate insulin secretion also at these high glucose levels, showing that the PACAP27 in conjunction with glucose also exceeds the action of a maximal dose of glucose. This is similar to the effect of glucagon-like peptide 1 (GLP-1).[7] Both these peptides stimulate insulin secretion mainly by augmenting the formation of cyclic AMP.[5,8] This illustrates the power of amplifying the formation of cyclic AMP for the B cell secretory capacity.

In spite of stimulating insulin secretion, PACAP27 did not enhance the glucose disposal. This might be due to indirect compensatory actions induced by PACAP27 under these conditions, such as inducing hepatic glucose delivery either directly or indirectly through glucagon and/or catecholamines, or inhibiting the peripheral uptake of glucose. More studies are required to solve this surprising action of PACAP. The present study shows, however, that in spite of a marked insulinotropic action of PACAP, the peptide does not seem to offer a possibility for use as a treatment of diabetes, unless B cell-specific PACAP receptor agonists are developed.

TABLE 1. The acute insulin response and the K_G value for glucose elimination

Experimental Series	Parameter	Glucose Injection	p	PACAP27 + Glucose Injection
Glucose 0.3 g/kg ($N = 13$ in each group)	AIR (nmol/l)	0.58 ± 0.09	0.045	1.02 ± 0.19
	K_G (%/min)	1.9 ± 0.1	n.s.	1.6 ± 0.2
Glucose 1.0 g/kg ($N = 14$ in each group)	AIR (nmol/l)	1.08 ± 0.16	0.002	2.37 ± 0.32
	K_G (%/min)	2.8 ± 0.1	n.s.	2.9 ± 0.2
Glucose 3.0 g/kg ($N = 11$ in each group)	AIR (nmol/l)	1.45 ± 0.32	0.039	3.26 ± 0.73
	K_G (%/min)	2.3 ± 0.1	n.s.	2.2 ± 0.1

Note: The acute insulin response (mean suprabasal 1 and 5 min plasma insulin after injection) and the K_G value for the glucose elimination during 1–20 min after glucose administration in the three experimental series given glucose at 0.3, 1.0, or 3.0 g/kg without or with addition of PACAP27 (1.3 nmol/kg). Means±SEM are shown. p values indicate probability level of random difference between the groups. n.s. indicates not significantly different.

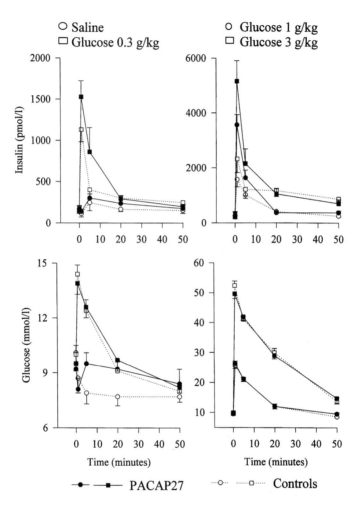

FIGURE 1. Plasma insulin and glucose immediately before and at 1, 5, 20, and 50 min after an intravenous injection of saline or glucose (0.3, 1.0, or 3.0 g/kg) alone or together with PACAP27 (1.3 nmol/kg) in mice anesthetized with an intraperitoneal injection of midazolam (Dormicum[R], Hoffman-La-Roche, Basel, Switzerland, 0.4 mg/mouse) and a combination of fluanison (0.9 mg/mouse) and fentanyl (0.02 mg/mouse; Hypnorm[R], Janssen, Beerse, Belgium). Means ± SEM are shown. There were 11–14 animals in each group.

ACKNOWLEDGMENTS

The authors are grateful to Lilian Bengtsson, Ulrika Gustavsson, and Lena Kvist for expert technical assistance.

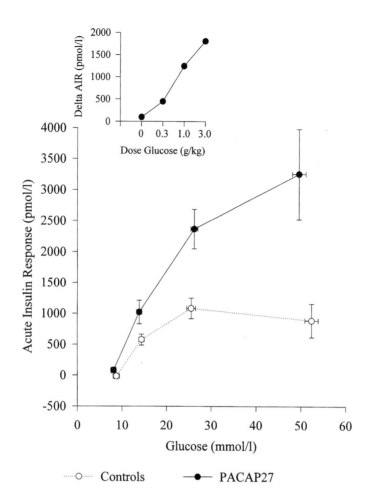

FIGURE 2. Relation between acute insulin response (=suprabasal mean 1 and 5 min value after injection) to saline or glucose (0.3, 1.0, or 3.0 g/kg) alone or together with PACAP27 (1.3 nmol/kg) and the plasma glucose level at 1 min after injection. Means ± SEM are shown. There were 11–14 animals in each group. (*Insert*) Difference in mean AIR between groups with or without injection of PACAP27 plotted against the amount of glucose injected.

REFERENCES

1. FRIDOLF, T., F. SUNDLER & B. AHRÉN. 1992. Pituitary adenylate cyclase activating polypeptide (PACAP): occurrence in rodent pancreas and effects on insulin and glucagon secretion in the mouse. Cell Tissue Res. **269**: 275–279.
2. YADA, T., M. SAKURADA, K. IHADA, M. NAKATA, F. MURATA, A. ARIMURA & M. KIKUCHI. 1994. Pituitary adenylate cyclase-activating polypeptide is an extraordinarily potent intra-pancreatic regulator of insulin secretion from islet β-cells. J. Biol. Chem. **269**: 1290–1293.

3. FILIPSSON, K., G. PACINI, A. J. W. SCHEURINIC & B. AHRÉN. 1998. PACAP stimulates insulin secretion but inhibits insulin sensitivity in mice. Am. J. Physiol. **274**: E834–E842.
4. FILIPSSON, K., K. TORNØE, J. HOLST & B. AHREN. 1997. Pituitary adenylate cyclase activating polypeptide (PACAP) stimulates insulin and glucagon secretion in humans. J. Clin. Endocrinol. Metab. **82**: 3093–3098.
5. AF KLINTEBERG, K., S. KARLSSON & B. AHRÉN. 1996. Signaling mechanisms underlying the insulinotropic effect of pituitary adenylate cyclase-activating polypeptide in HIT-T15 cells. Endocrinology **137**: 2791–2798.
6. WARD, W. K., D. C. BOLGIANO, B. MCKNIGHT, J. B. HALTER & D. PORTE, JR. 1984. Diminished B cell secretory capacity in patients with noninsulin-dependent diabetes mellitus. J. Clin. Endocrinol. Metab. **74**: 1318–1328.
7. AHRÉN, B. 1995. Insulinotropic action of truncated glucagon-like peptide-1 in mice. Acta Physiol. Scand. **153**: 205–206.
8. AHRÉN, B. 1996. Glucagon-like peptide $1_{(7-36)}$ amide increases cyclic AMP acumulation in normal islets. Pancreas **12**: 211–213.
9. SCHUIT, F. C. 1996. Factors determining the glucose sensitivity and glucose responsiveness of pancreatic beta cells. Horm. Res. **46**: 99–106.

Effects of Pituitary Adenylate Cyclase-Activating Polypeptide (PACAP) on cAMP Formation and Growth Hormone Release from Chicken Anterior Pituitary Cells[a]

K. PEETERS,[b] L. LANGOUCHE, F. VANDESANDE, V.M. DARRAS, AND L.R. BERGHMAN

Laboratory of Neuroendocrinology and Immunological Biotechnology, Katholieke Universiteit Leuven, Zoological Institute, Naamsestraat 59, B-3000 Leuven, Belgium

Several morphological and physiological observations in mammals are indicative that pituitary adenylate cyclase-activating polypeptide (PACAP) may function as a releasing factor for pituitary hormones. A dense PACAP immunoreactive network in the median eminence as well as the presence of PACAP in hypophyseal portal blood and the presence of PACAP-specific receptors on anterior pituitary cells were demonstrated. The effect of PACAP on the pituitary hormone secretion was demonstrated in *in vitro* and *in vivo* systems.[4] However the role of PACAP within the hypothalamo-hypophyseal system of birds is unknown. Therefore, in the present study we have determined the effect of chicken PACAP38 and PACAP27 on intracellular cAMP accumulation in a monolayer cell culture system and we have compared the effects of hGHRH and PACAP on GH release from individual anterior pituitary cells, using a reverse hemolytic plaque assay (RHPA).

MATERIALS AND METHODS

Anterior pituitary glands were removed from 4-week-old male broiler chickens (Hybro, Euribrid, Aarschot, Belgium) and dissociated enzymatically into single cells as described previously by Denef and coworkers.[1] The dispersed pituitary cells were resuspended in DMEM supplemented with 10% horse serum and 2.5% fetal calf serum (FCS) (Gibco) at a density of 2.5×10^5 cells/ml. The cells were transferred to 24-well culture plates (1 ml/well) and were incubated at 37°C in an atmosphere of 95% air and 5% CO_2 for 3 days prior to further experimentation. Cell viability was greater than 95% at the time of experiments as indicated by the trypan blue dye–exclusion test. The cells were washed with serum-free DMEM and incubated in the presence of increasing concentrations of synthetic chicken PACAP38, PACAP27 (10^{-12} M–10^{-6} M), or vehicle for 10 min. Chicken PACAP27 and PACAP38[3] were synthesized by the continuous flow technique and were purified by

[a]This work was supported by grants from the Fonds voor Wetenschappelijk Onderzoek—Vlaanderen (to K.P. and L.B.).

[b]Corresponding author: Kristel Peeters, Zoological Institute, Laboratory of Neuroendocrinology and Immunological Biotechnology, Naamsestraat 59, B-3000 Leuven, Belgium; Tel.: (32) 16.32.39.26; Fax: (32) 16.32.39.02; E-mail: Kristel.Peeters@bio.kuleuven.ac.be

reverse-phase high performance liquid chromatography in our laboratory. The sequence was verified utilizing Pico-Tag™ amino acid analysis and Maldi mass spectrometry. Intracellular cAMP was extracted with 20% ice-cold trifluoroacetic acid. The samples were centrifuged and the supernatant was removed, frozen, and lyophilized. Each lyophilized sample was washed twice by adding distilled, deionized water and relyophilized to remove trifluoroacetic acid completely. cAMP was measured by RIA (Biotrak™ cAMP worksystem, Amersham).

The RHPA was developed using a polyclonal antiserum specific for chGH (a gift of Dr. V. Darras) as previously described by Smith and colleagues.[5] Omission of any of the assay reagents, i.e. anti-GH antiserum, complement, protein-A, or replacement of the antiserum with preimmune rabbit serum resulted in failure to form hemolytic plaques. Plaque formation was attenuated and eventually abolished by further dilution of the antiserum. Cells were incubated for 2 h at 37°C with the antiserum (1:50) and 100 nM synthetic hGHRH (Peninsula Laboratories, Belmont, CA, USA), 100 nM chPACAP38, 100 nM chPACAP27, or control vehicle.

RESULTS AND DISCUSSION

The neuropeptide PACAP has been shown to stimulate the cAMP/PKA pathway in mammalian and amphibian anterior pituitary cells.[4,7] In the present study, we showed, for the first time, that PACAP stimulates intracellular cAMP accumulation in a dose-dependent manner in avian and more specifically in chicken anterior pituitary cells (FIG. 1).

It is generally accepted that GH secretion from somatotropes is stimulated by GHRH and inhibited by SRIF. In addition, PACAP has been shown to stimulate GH release in mammals[4] and frog.[7] Wei and colleagues[6] demonstrated that PACAP was more effective at stimulating GH secretion in a static culture of male rat anterior pituitary cells at shorter (15 min) rather than longer (4 h) incubation times. These results contradict those of Hart and colleagues,[2] who found only a statistically significant increase of GH release after 24 h of incubation. This delayed effect of PACAP on GH release may possibly be explained by the stimulation of GH synthesis or by paracrine actions on other pituitary cell types, such as

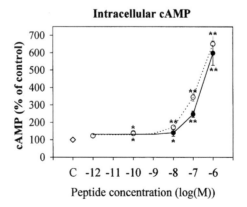

FIGURE 1. Effect of synthetic chicken PACAP27 (O) and PACAP38 (●) on the intracellular cAMP accumulation in a monolayer cell culture of chicken pituitary cells (basal secretion (◇)). The data points show the mean ± SE ($N = 4$). The results are expressed as a percentage of the control (untreated) value. $^*p < 0.05$; $^{**}p < 0.01$; $^{***}p < 0.001$ versus control (basal secretion).

the non-hormone secreting folliculo-stellate cells. It has been demonstrated that PACAP38 stimulates the interleukin-6 production of these cells and that IL-6 in turn can stimulate GH release.[4]

RHPA has been widely used for the study of pituitary hormone secretion. Because in RHPA hormone release is measured from single cells, paracrine effects of neighboring cells are avoided. In the present study we have demonstrated that 10^{-7} M chPACAP38 and 10^{-7} M chPACAP27 increased GH release from individual chicken anterior pituitary cells in a RHPA (FIG. 2). After 2 h chPACAP increased both the number of cells secreting GH (plaque-forming cells, %PFC) and the mean amount of GH released per cell (mean plaque area, MPA), resulting in an increase of the total GH secretion (TSI). In comparison with the effect of hGHRH, chPACAP38 and chPACAP27 were less potent in stimulating the GH release from somatotropes. Analysis of the frequency distribution of plaque area showed that hGHRH stimulation caused an important shift toward the large plaque classes (IV–VI, 76%) in comparison with control conditions (II–IV, 69%). Most of the plaques formed after chPACAP27 and chPACAP38 stimulation were distributed one or two classes higher (III–V) than under basal conditions (FIG. 2).

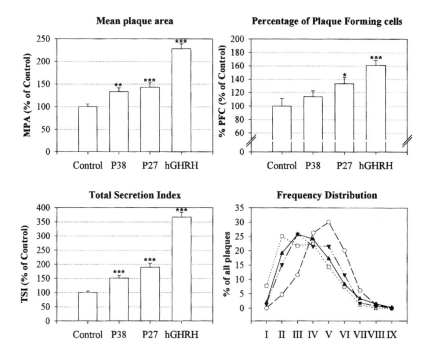

FIGURE 2. (A–C) Effects of hGHRH, PACAP27, and PACAP38 on GH plaque formation. The data represent the mean ± SE (N = 3,900 cells/treatment for GH %PFC and N = 3,420 plaques/treatment for GH MPA). The MPA was determined using a computer-assisted image system (BioQuant, VR&M Biometrics, Nashville, TN). Plaque areas were measured by tracing the circumference of individual plaques. Statistical analysis was performed by unpaired Student's *t*-test. *$p < 0.01$; **$p < 0.001$ versus control (basal secretion). (D) Frequency distribution of hemolytic plaque areas under basal conditions (□) and following stimulation with 100 nM hGHRH (○) and 100 nM PACAP27 (▼) or 100 nM PACAP38 (▲). Plaque areas were calculated from their circumference and distributed in nine classes (I < 1,000 μm^2 < II < 2,000 μm^2 < III < 4,000 μm^2 < IV < 8,000 μm^2 < V < 16,000 μm^2 < VI < 32,000 μm^2 < VII < 64,000 μm^2 < VIII < 128,000 μm^2 < IX).

In summary, we have shown that chPACAP markedly stimulates intracellular cAMP production and increases GH secretion from chicken anterior pituitary cells. Taken together, these results provide the first evidence for a role of PACAP as a hypophysiotropic hormone in birds.

REFERENCES

1. DENEF, C., P. MAERTENS, W. ALLAERTS, A. MIGNON, W. ROBBERECHT, L. SWENNEN & P. CARMELIET. 1989. Cell-to-cell communication in peptide target cells of anterior pituitary. Methods Enzymol. **168**: 47–71.
2. HART, G. R., H. GOWING & J. M. BURRIN. 1991. Effects of a novel hypothalamic peptide, pituitary adenylate cyclase-activating polypeptide, on pituitary hormone release in rats. J. Endocrinol. **134**: 33–41.
3. MCRORY, J. E., R. L. PARKER & N. M. SHERWOOD. 1997. Expression and alternative processing of a chicken gene encoding both growth hormone-releasing hormone and pituitary adenylate cyclase-activating polypeptide. DNA Cell Biol. **16**: 95–102.
4. RAWLINGS, S. R. & M. HEZAREH. 1996. Pituitary adenylate cyclase-activating polypeptide (PACAP) and PACAP/vasoactive intestinal polypeptide receptors: actions on the anterior pituitary gland. Endocr. Rev. **17**: 4–29.
5. SMITH, P. F., E. H. LUQUE & J. D. NEILL. 1986. Detection and measurement of secretion from individual neuroendocrine cells using a reverse hemolytic plaque assay. Methods Enzymol. **124**: 443–465.
6. WEI, L., W. W-S. CHAN, B. BUTLER & K. CHENG. 1993. Pituitary adenylate cyclase-activating polypeptide–induced desensitization on growth hormone release from rat primary pituitary cells. Biochem. Biophys. Res. Commun. **197**: 1396–1401.
7. YON, L., L. JEANDEL, N. CHARTREL, M. FEUILLOLEY, J. M. CONLON, A. ARIMURA, A. FOURNIER & H. VAUDRY. 1993. Neuroanatomical and physiological evidence for the involvement of pituitary adenylate cyclase-activating polypeptide in the regulation of the distal lobe of the frog pituitary. J. Neuroendocrinol. **5**: 289–296.

Localization of Pituitary Adenylate Cyclase-Activating Polypeptide in the Central Nervous System of the European Eel *Anguilla anguilla*: Stimulatory Effect of PACAP on GH Secretion[a]

M. MONTERO,[b] L. YON,[b] K. ROUSSEAU,[c] A. ARIMURA,[d]
A. FOURNIER,[e] S. DUFOUR,[c] AND H. VAUDRY[b,f]

[b]*European Institute for Peptide Research (IFRMP 23), Laboratory of Cellular and Molecular Neuroendocrinology, INSERM U413, UA CNRS, University of Rouen, 76821 Mont-Saint-Aignan, France*

[c]*Laboratory of General and Comparative Physiology, CNRS UA 90, National Museum of Natural History, 75005 Paris, France*

[d]*US- Japan Biomedical Research Laboratories, Tulane University, Hebert Center, Belle Chasse, Louisiana 70037, USA*

[e]*Institut National de Recherche Scientifique Santé, University of Québec, Pointe-Claire, Québec, Canada H9R 1G6.*

PACAP was first isolated from the ovine hypothalamus[1] and the primary structure of PACAP38 appears to be identical in all mammalian species studied so far.[2] The sequence of PACAP has also been determined in tunicate,[3] fish,[4] amphibian,[5] and avian species;[6] in all these groups, PACAP27 exhibits more than 95% sequence similarity with its mammalian counterpart.

The distribution of PACAP-immunoreactive structures has been reported in the brain of mammals[7,8] and amphibians[9] but has never been investigated in fish. The aim of the present study was to determine the localization and the biochemical characteristics of PACAP-like immunoreactivity in the central nervous system of a primitive teleost fish, the European eel *Anguilla anguilla*. Immunofluorescence studies revealed that PACAP-positive perikarya are exclusively located in the diencephalon, i.e., in the preoptic nucleus of the hypothalamus (FIG. 1A) and in the dorsal and ventral nuclei of the thalamus. Immunoreactive processes were detected in most regions of the eel brain, notably in the ventral telencephalon, the diencephalon, the mesencephalon, and the medulla oblongata. A dense accumulation of PACAP-containing nerve terminals was also found in the pars distalis of the pituitary (FIG.1B) indicating that PACAP may act as a hypophysiotropic factor in the eel. The molecular forms of PACAP present in the eel brain were characterized by HPLC analysis combined with RIA detection of PACAP-like immunoreactivity. The

[a]This work was supported by INSERM (U 413), CNRS (UA 90), the Conseil Supérieur de la Pêche, and the Conseil Régional de Haute-Normandie.

[f]Corresponding author: Tel.: (33) 235.14.66.24; Fax: (33) 235.14.69.46; E-mail: hubert.vaudry@univ-rouen.fr

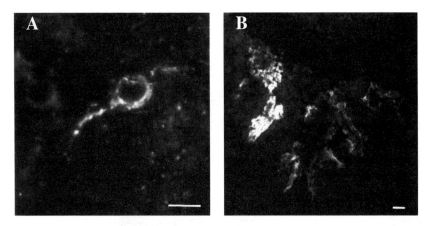

FIGURE 1. Confocal laser scanning microscope microphotographs illustrating the localization of PACAP-immunoreactive structures in the eel brain. (**A**) Bipolar PACAP-positive neuron located in the posterior parvocellular preoptic nucleus of the hypothalamus. (Scale bar = 10 μm.) (**B**) Transverse section through the distal lobe of the pituitary showing a dense accumulation of nerve terminals in the neurohypophysial digitations. (Scale bar = 30 μm.)

major form of PACAP contained in the diencephalon and mesencephalon co-eluted with mammalian PACAP38.

It has been recently demonstrated that, in teleost fish, a single gene encodes for both PACAP and GHRH[4] while, in mammals, PACAP and GHRH are encoded by two distinct genes.[10] In addition, PACAP was reported to stimulate the activity of somatotrope cells in

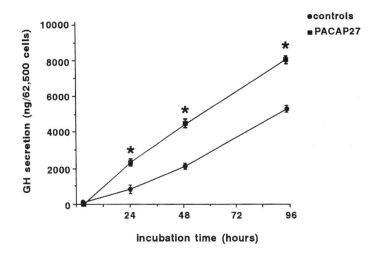

FIGURE 2. Time-course effect of PACAP27 (10^{-6} M) on GH secretion from cultured eel pituitary cells. GH release into the culture medium was measured by using a homologous RIA. The results represent the mean (±S.E.) of 6 independent determinations from a single experiment (*$p<0.001$ vs. respective controls).

various vertebrate species. We have thus investigated the effect of PACAP on growth hormone (GH) secretion by eel pituitary cells in primary culture. PACAP27 (10^{-6} M) significantly increased GH release after 24 h of incubation (2.4 times over controls) (FIG. 2). After 48 and 96 h of incubation, PACAP still stimulated GH secretion, although to a lesser extent, suggesting the occurrence of a desensitization phenomenon. Dose-response experiments performed at 24 h of incubation revealed that PACAP27 and PACAP38 possess the same efficacy. However, PACAP38 appeared to be 12 times more potent than PACAP27 in stimulating GH release. VIP only stimulated GH secretion at micromolar concentrations. These data indicate that eel pituitary cells express PACAP receptors that exhibit pharmacological characteristics similar to mammalian type I receptors. Interestingly, human GHRH (10^{-6} M) had no effect on GH secretion from cultured eel pituitary cells.

In conclusion, in the primitive teleost fish *Anguilla anguilla*, PACAP-containing cell bodies are exclusively located in the diencephalon. PACAP-immunoreactive processes are widely distributed in the brain and, in particular, a dense network of PACAP-containing fibers innervates the pars distalis of the pituitary. These observations suggest that, in the eel, PACAP may act both as a neuromodulator and a hypophysiotropic neurohormone. As a matter of fact, PACAP appears to be a potent stimulator of GH secretion from cultured eel adenohypophysial cells. Taken together, our data indicate that PACAP may play a major role in the neuroendocrine regulation of somatotrope cell activity and thus may be involved in the control of growth, osmoregulation, and reproduction.

REFERENCES

1. MIYATA, A. *et al*. 1989. Isolation of a novel 38 residue hypothalamic peptide which stimulates adenylate cyclase in pituitary cells. Biochem. Biophys. Res. Commun. **164**: 567–574.
2. ARIMURA, A. 1992. Pituitary adenylate cyclase-activating polypeptide (PACAP): discovery and current status of research. Regul. Pept. **37**: 287–303.
3. MCRORY, J. E. & N. M. SHERWOOD. 1997. Two protochordate genes encode pituitary adenylate cyclase-activating polypeptide and related family members. Endocrinology **138**: 2380–2390.
4. PARKER, D. B. *et al*. 1993. Two salmon neuropeptides encoded by one brain cDNA are structurally related to members of the glucagon superfamily. Eur. J. Biochem. **215**: 439–448.
5. CHARTREL, N. *et al*. 1991. Primary structure of frog pituitary adenylate cyclase-activating polypeptide (PACAP) and effect of ovine PACAP on frog pituitary. Endocrinology **129**: 3367–3371.
6. MCRORY, J. E., R. L. PARKER & N. M. SHERWOOD. 1997. Expression and alternative processing of a chicken gene encoding both growth hormone-releasing hormone and pituitary adenylate cyclase-activating polypeptide. DNA Cell Biol. **16**: 95–102.
7. KOVES, K. *et al*. 1990. Immunohistochemical demonstration of a novel hypothalamic peptide, pituitary adenylate cyclase activating polypeptide, in the ovine hypothalamus. Endocrinology **127**: 264–271.
8. VIGH, S. A. *et al*. 1991. Immunohistochemical localization of the neuropeptide, pituitary adenylate cyclase-activating polypeptide (PACAP), in human and primate hypothalamus. Peptides **12**: 313–318.
9. YON, L. *et al*. 1992. Immunohistochemical distribution and biological activity of pituitary adenylate cyclase-activating polypeptide (PACAP) in the central nervous system of the frog *Rana ridibunda*. J. Comp. Neurol. **324**: 485–499.
10. HOSOYA, M. *et al*. 1992. Structure of the human pituitary adenylate cyclase activating polypeptide (PACAP) gene. Biochim. Biophys. Acta **1129**: 199–206.

Effect of Pituitary Adenylate Cyclase-Activating Polypeptide (PACAP) on Tyrosine Hydroxylase Gene Expression in the Rat Adrenal Medulla

M. HONG,[a] L. YON,[b] A. FOURNIER,[c] H. VAUDRY,[b] AND
G. PELLETIER[a,d]

[a]MRC Group in Molecular Endocrinology, CHUL Research Center,
Quebec, Canada G1V 4G2

[b]European Institute for Peptide Research (IFRMP No. 23), Laboratory
of Cellular and Molecular Neuroendocrinology, INSERM U413, UA
CNRS, University of Rouen, 76821 Mont-Saint-Aignan, France

[c]Institut National de la Recherche Scientifique-Santé, University of
Quebec, Pointe-Claire, Quebec, Canada H9R 1G6

It is well known that the secretion of catecholamine by adrenal medullary cells is under the regulation of neurotransmitters, mainly acetylcholine. During the last years, the localization of peptides and peptidergic receptors in the adrenal medulla has raised the possibility of a peptidergic regulation of catecholamine secretion by adrenal chromaffin cells. For example, we have recently shown that neuropeptide Y can increase mRNA levels of tyrosine hydroxylase (TH), the enzyme that catalyzes the first step in the biosynthesis of catecholamines in the rat adrenal medulla.[1] Adenylate cyclase–activating polypeptide (PACAP), a 38 amino acid polypeptide isolated from the ovine hypothalamus on the basis of its stimulatory effect on cAMP production by pituitary cells,[2] has been found in the brain and peripheral organs, particularly the gastrointestinal tract, testis, and adrenal medulla.[3–5] The occurrence of PACAP receptors (type 1) in the rat adrenal medulla[6] and the stimulating influence of PACAP on the release of catecholamines from cultured porcine adrenal chromaffin cells[7] strongly suggest that this peptide might play a role in the control of catecholamine secretion by the adrenal medulla. In order to study the possible involvement of PACAP in the regulation of catecholamine biosynthesis by the adrenal gland, we have studied the effects of intravenous injections of PACAP and its potent antagonist PACAP(6-38)[8] on the genetic expression of TH in the adrenal medulla of the adult intact male rat. To determine the site of action of PACAP, we performed bilateral adrenalectomy and homologous implantation of the adrenals under the kidney capsules. Those animals with implanted adrenals were injected with the same peptides.

In all the experiments, eight animals per group were used. The levels of TH mRNA levels were measured by quantitative *in situ* hybridization. The hybridization technique was performed as previously described.[1] The hybridized sections were exposed to Kodak X-Omat

[d]Corresponding author: Dr. Georges Pelletier, MRC Group in Molecular Endocrinology, 2705 Laurier Boulevard, Québec, G1V 4G2, Canada; Tel.: (418) 654-2296, Fax: (418) 654-2761; E-mail: LREM@crchul.ulaval.ca

films for 2 days before being coated with liquid photographic emulsion (Kodak NTB-2). Densitometric measurements of autoradiographs (X-ray films) were obtained using a digitized Amersham RAS image analysis system.

The hybridization signal obtained on X-ray films was rather uniform through the adrenal medulla, the cortex being devoid of any reaction (FIG. 1). At the light microscopic level, the silver grains appeared to be localized exclusively over the chromaffin cells of the adrenal medulla.

Administration of graded doses of PACAP (from 25 to 100 µg/kg body weight injected 4 h before sacrifice) indicated that the maximal effect was obtained with the dose of 50 µg/kg body weight (FIG. 1). Time-course studies revealed that the maximum effect of PACAP (50 µg/kg body weight) was achieved between 2 and 4 h after intravenous injection. In the experiments aimed at elucidating the mechanism(s) of action of PACAP, both the peptide and its potent antagonist PACAP(6-38) were injected at the dose of 50 µg/kg body weight 4 h before

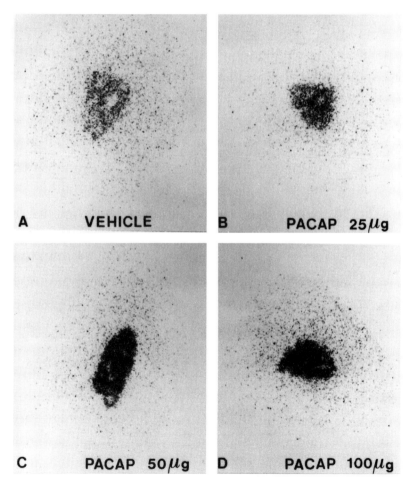

FIGURE 1. X-ray autoradiographs illustrating the effect of increasing doses (/kg body weight) of PACAP injected intravenously 4 h before sacrifice.

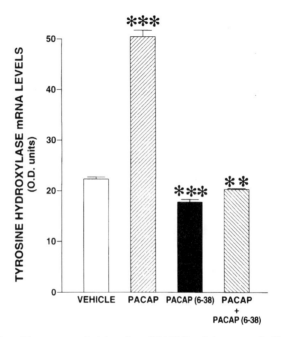

FIGURE 2. Effect of intravenous administration of PACAP and the antagonist PACAP(6-38) (50 µg/kg, body weight) 4 h before sacrifice. *** $p<0.001$, ** $p<0.01$ control versus all of the other groups. The group treated with PACAP and PACAP(6-38) is significantly ($p<0.001$) different from the group that received only PACAP.

sacrifice. The injection of PACAP induced a 125% increase in the hybridization signal detected in the adrenal medulla (FIG. 2). On the other hand, the administration of the antagonist PACAP(6-38) produced a 20% decrease in the optical density. When PACAP and its antagonist were administered simultaneously, the effect of PACAP was completely prevented, the mRNA levels being slightly lower than those observed in the vehicle-treated animals. In animals with adrenals implanted under the kidney capsule, a stimulatory influence of PACAP on TH mRNA levels (70% over control) was also observed.

It then appears that, on the basis of the present and previous reports indicating the occurrence of PACAP receptors in the rat adrenal medulla, PACAP can directly interact with adrenal medullary chromaffin cells to stimulate the biosynthesis of catecholamines via specific receptors. Moreover, the inhibitory effect of the PACAP antagonist strongly suggests that PACAP exerts a tonic stimulatory influence on TH gene expression.

REFERENCES

1. HONG, M. *et al.* 1995. Role of neuropeptide Y in the regulation of tyrosine hydroxylase gene expression in rat adrenal glands. Mol. Neuroendocrinol. **61**: 85–88.
2. MIYATA, A. *et al.* 1981. Isolation of a novel 38 residue-hypothalamic polypeptide which stimulates adenylate cyclase in pituitary cells. Biochem Biophys. Res. Commun. **164**: 567–574.
3. YON, L. *et al.* 1994. Pituitary adenylate cyclase-activating polypeptide (PACAP) stimulates both adrenocortical cells and chromaffin cells in the frog adrenal gland. Endocrinology **135**: 2749–2758.
4. ARIMURA, A. *et al.* 1991. Tissue distribution of PACAP as determined by IRA: highly abundant in the rat brain and testes. Endocrinology **129**: 2787–2789.

5. SHIOTANI, Y. *et al.* 1995. Immunohistochemical localization of pituitary adenylate cyclase-activating polypeptide (PACAP) in the adrenal medulla of the rat. Peptides **16**: 1045–1050.
6. ARIMURA, A. 1992. Pituitary adenylate cyclase-activating polypeptide (PACAP): discovery and current status of research. Regul. Pept. **37**: 287–303.
7. ISOBE, K., T. NAKAI & T. TAKUWA. 1993. Ca^{2+}-dependent stimulatory effect of pituitary adenylate cyclase-activating polypeptide on catecholamines secretion from cultured porcine adrenal medullary chromaffin cells. Endocrinology **132**: 1757–1765.
8. ROBBERECHT, P. *et al.* 1992. Structural requirements for the occupancy of pituitary adenylate cyclase-activating polypeptide (PACAP) receptors and adenylate cyclase activation in human neuroblastoma NB-OK-1 cell membranes. Eur. J. Biochem. **207**: 239–246.

The Stimulatory Effect of VIP and PACAP on Adrenal Aldosterone Release[a]

M. RADZIKOWSKA, E. WASILEWSKA-DZIUBIŃSKA, AND
B. BARANOWSKA

Neuroendocrinology Department, Medical Centre of Postgraduate Education, Fieldorfa 40, 04-158 Warsaw, Poland

Vasoactive intestinal peptide (VIP) and pituitary adenylate cyclase–activating polypeptide (PACAP) were demonstrated in the central and peripheral nervous system.[1,2] VIP has been identified in the nerve cells that connect the adrenocortical cells to the medulla.[3,4] It has been demonstrated that this peptide is released in response to the stimulation of the splanchnic nerve.[4] PACAP-containing nerve fibers have also been demonstrated in the adrenal gland of several species.[5] Both peptides stimulate adrenal catecholamines and aldosterone (ALDO) release.[6–8] VIP, PACAP, NPY, noradrenaline, and adrenaline stimulated ALDO secretion by zone glomerulosa cells *in vitro* through interaction with β1-adrenergic receptors.[9] It was postulated that the adrenal cortex can be stimulated independently of the hypothalamus-pituitary-adrenal axis via a neuroadrenocortical local axis.[10]

In our earlier experiments we observed an increased serum ALDO level after intravenous VIP administration on ovariectomized and hysterectomized (OVX+HTX) rats.

The aim of this study was to evaluate the effect of VIP and PACAP38 as well as the effect of blockers of the receptors, that is, β-adrenergic-alprenolol (ALP) and opioid-naltrexone (NAL) on aldosterone release in adult OVX+HTX female rats.

MATERIAL AND METHODS

Experiments were performed on OVX+HTX rats weighing 220–250 g. The indoor temperature was 20–22°C and lighting conditions were 14:10 hours of light to dark. Food and tap water were *ad libitum*. Animals were treated in accordance with principles and procedures outlined in the National Institutes of Health Guidelines for Care and Use of Experimental Animals.

Three weeks after ovariectomy and hysterectomy, VIP and PACAP38 were administered intravenously into the tail vein to subject rats (groups 1–9):

Group 1, control (placebo), 300 μl 0.9% NaCl, $N = 8$; Group 2, VIP 10 μg VIP in 300 μl 0.9% NaCl, $N = 8$; Group 3, PACAP38, 10 μg PACAP38 in 300 μl 0.9% NaCl, $N = 6$; Group 4, alprenolol (ALP), β-adrenergic blocker, 250 μg in 300 μl 0.9% NaCl, $N = 8$; Group 5, naltrexone (NAL), opioid blocker, 10 μg in 300 μl 0.9% NaCl, $N = 8$; Group 6, VIP+ALP, $N = 8$; Group 7, VIP+NAL, $N = 8$; Group 8, PACAP38 + ALP, $N = 7$; and Group 9, PACAP38+NAL, $N = 7$.

The blood samples were collected to determine ALDO by RIA method (CIS, France) at 15 min after intravenous injection of VIP and PACAP, as well as after injection of β-adrenergic and opioid receptor blockers.

[a]This work was supported by Program No. 501-2-2-25-48/97.

The statistical analysis was carried out by means of Student's *t*-test for unpaired samples.

RESULTS

VIP injected intravenously in OVX+HTX rats induces a significant increase in ALDO serum level ($p < 0.001$). A significant decrease of ALDO serum level after VIP+ALP administration was stated ($p < 0.01$) as compared with the response of ALDO after VIP administration (FIG. 1). Serum ALDO level after VIP+NAL was significantly lower in comparison with the response of ALDO with VIP injection alone (FIG. 1).

After PACAP38 administration, a significant increase of ALDO level was observed ($p < 0.001$). Serum ALDO level in response to PACAP38+ALP did not differ from serum ALDO level in response to PACAP38 injected alone. The response of ALDO to PACAP38+NAL was significantly lower ($p < 0.01$) as compared with the response to PACAP38 injection alone (FIG. 2).

DISCUSSION

Aldosterone secretion is known to be regulated in a paracrine manner by several neuropeptides released by the nerve plexuses distributed in the outer zone of the adrenal cortex.[4] The increase of ALDO secretion after VIP administration into OVX+HTX rats confirmed our earlier results[2] and other data obtained in different preparations of isolated adrenal glands of the rat.[6] A significant decrease of ALDO serum level after VIP+ALP administration in our experiments confirms the inhibitory effect of alprenolol, a β-adrenergic antagonist, on ALDO secretion in response to VIP stimulation in adrenal capsular preparation.[6,9]

The release of ALDO secretion after administration of VIP+NAL was significantly lower in comparison with the response of ALDO after VIP injection alone. It was observed that in the isolated perfused adrenal glands naloxone caused a concentration-dependent

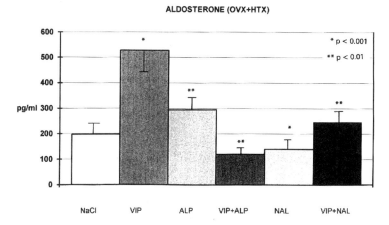

FIGURE 1. Serum aldosterone level after VIP, ALP, VIP+ALP, NAL, and VIP+NAL administration in OVX+HTX rats. Mean ± SEM.

ALDOSTERONE (OVX+HTX)

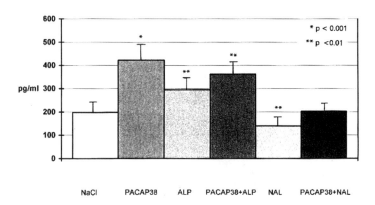

FIGURE 2. Serum aldosterone level after PACAP38, ALP, PACAP38+ALP, NAL, and PACAP38+NAL administration in OVX+HTX rats. Mean ± SEM.

inhibition of catecholamine secretion evoked by stimulation of splanchnic nerves.[11] We suppose that the stimulatory mechanism of VIP on ALDO secretion is dependent on β-adrenergic and opioid activity, probably connected with catecholamine release. Both VIP and PACAP stimulated ALDO secretion in rat adrenal slices and PACAP stimulated ALDO in human adrenal slices, but not by dispersed adrenocortical cells, suggesting a rather indirect role of VIP and PACAP mediated by the chromaffin cells of the adrenal medulla in rats and humans.[7,8]

The treatment of bovine adrenal cortical cells with PACAP demonstrated that this neuropeptide could directly stimulate ALDO production.[12] In our experiments the increase of ALDO secretion after PACAP38 administration in our experiments confirmed results obtained with PACAP on human adrenal slice preparations. However, alprenolol did not suppress the ALDO response to PACAP as has been seen in dispersed human adrenocortical cells.[8] The response of ALDO to PACAP38+NAL was similar to VIP+NAL response. We supposed the stimulatory mechanism of PACAP on ALDO release is independent of β-adrenergic receptors, but it was dependent on opioid activity.

In our experiments both VIP and PACAP stimulated ALDO release. The differences observed by us after β-receptor blockade may be connected with the direct effect of released catecholamines on the VIP and PACAP receptors.

CONCLUSIONS

(1) The stimulatory mechanism of VIP on ALDO secretion is dependent on β-adrenergic and opioid activity in OVX+HTX rats.

(2) The stimulatory mechanism of PACAP on ALDO secretion is independent of β-adrenergic activity but it was dependent on opioid activity in OVX+HTX rats.

REFERENCES

1. ARIMURA, A. 1992. Pituitary adenylate cyclase activating polypeptide (PACAP): Discovery and current status of research. Regul. Pept. **37**: 287–303.
2. GOZES, I. & D. E. BRENNEMAN. 1989. VIP molecular biology and neurobiological function. Mol. Neurobiol. **3**: 201–236.

3. HÖKFELT, T., J. M. LUNDBERG, M. SCHULTZBERG & J. FARENKRUG. 1981. Immunohistochemical evidence for local VIPergic neuron system in the adrenal gland of the rat. Acta Physiol. Scand. **116**: 575–576.
4. HOLZWARTH, A., L. A. GUNNINGHAM & N. KLEITMAN. 1987. The role of adrenal nerves in the regulation of adrenocortical functions. Ann. N.Y. Acad. Sci. **512**: 449–463.
5. TABARIN, A., D. CHEN, R. HAKANSON & F. SUNDLER. 1994. Pituitary adenylate cyclase activating polypeptide in the adrenal gland of mammals: distribution, characterization and responses to drugs. Neuroendocrinol. **59**: 113–119.
6. HINSON, J. P., S. KAPAS, C. D. ORFORD & G. P. VINSON. 1992. Vasoactive intestinal peptide stimulation of aldosterone secretion by the rat adrenal cortex may be mediated by the local release of catecholamines. J. Endocrinol. **133**: 253–258.
7. ANDREIS, P. G., L. K. MALENDOWICZ, A. S. BELLONI & G. G. NUSSDORFER. 1995. Effects of pituitary adenylate cyclase activating peptide (PACAP) on the rat adrenal secretory activity: preliminary in vitro studies. Life Sci. **56**: 135–142.
8. NERI, G., P. G. ANDEIS, T. PRAYER-GALETTI, G. P. ROSSI, L. K. MALENDOWICZ & G. G. NUSSDORFER. 1996. Pituitary adenylate cyclase activating peptide enhances aldosterone secretion of human adrenal gland: Evidence for an indirect mechanism, probably involving the local release of catecholamines. J. Clin. Endocrinol. Metab. **81**: 169–173.
9. BERNET, F., J. BERNARD, C. LABORIE, V. MONTEL, E. MALBERT & J. P. DUPOUY. 1994. Neuropeptide Y (NPY) and vasoactive intestinal peptide (VIP) induced aldosterone secretion by rat capsule/glomerular zone could be mediated by catecholamines via $\beta 1$ adrenergic receptors. Neurosci. Lett. **166**: 109–112.
10. ERHART-BORNSTEIN, M., S. R. BORNSTEIN, W. A. SCHEURBAUM, E. E. PFEIFFER & J. J. HOLST. 1991. Role of the vasoactive intestinal peptide in a neuroendocrine regulation of the adrenal cortex. Neuroendocrinol. **54**: 623–628.
11. MALHOTRA, R. & T. D. WAKADE. 1987. Non-cholinergic component of rat splanchnic nerves predominated at low neuronal activity and is eliminated by naloxone. Neuroscience **10**: 973–978.
12. BODART, V., K. BABINSKI, H. ONG & A. DE LEAN. 1997. Comparative effect of pituitary adenylate cyclase activating polypeptide on aldosterone secretion in normal bovine and human tumorous adrenal cells. Endocrinology **138**: 566–573.

Evidence for PACAP to Be an Autocrine Factor on Gonadotrope Cells[a]

ANNEMONE RADLEFF-SCHLIMME,[b] SABINE LEONHARDT,
WOLFGANG WUTTKE, AND HUBERTUS JARRY

*Division of Clinical and Experimental Endocrinology, Department of
Obstetrics and Gynecology, University of Göttingen, D-37075
Göttingen, Germany*

Pituitary adenylate cyclase activating polypeptide (PACAP) is expressed in the hypothalamus and the pituitary and functions as a neurotransmitter and neuromodulator. PACAP was first isolated from ovine hypothalamus showing potent activity in stimulating cAMP formation in rat anterior pituitary cells.[1] PACAP belongs to the VIP/secretin/glucagon family and exerts pleiotropic physiological effects. It stimulates pituitary hormone secretion *in vivo* and *in vitro*.[2] Two forms are known, PACAP-38 and PACAP-27, which are derived from a common precursor protein by posttranslational cleavage. At the mRNA level PACAP has been shown to be expressed in the lung, gut, testis, prostate, pancreas, adrenal gland, eye, as well as in lymphocytes from the peripheral blood.[3] Among the cell population in the pituitary, it is suggested that PACAP may be produced by the gonadotrope cells.[4] Recently the expression of the type I PACAP receptor (PVR1) in the gonadotrope-derived cell line alphaT3-1 has been demonstrated.[5]

In the present study we wanted to assess whether PACAP is secreted by gonadotropes in the pituitary and modulates their functions in an autocrine manner. Therefore we determined PACAP release by alphaT3-1 cells as a model system for normal gonadotropes and we studied the ability of the peptide to modulate its own expression.

MATERIALS AND METHODS

In Vitro *Experiments: Cell Culture*

AlphaT3-1 cells were kindly provided by Prof. O. Orthmann (University of Lübeck). The cells were grown to confluence in Dulbecco's Modified Eagle's Medium (DMEM) with 10% fetal calf serum (FCS) at 37°C in a humidified atmosphere containing 10% CO_2 and 90% air. For static culture experiments, cells (2.5×10^5/ml in 24-well culture plates) were cultivated for 24 h in DMEM with 1% BSA. To stimulate PACAP release, cells were incubated for 1 h with PMA (100 ng/ml) or cAMP (1 mM). To perform perfusion culture experiments, cells (2.5×10^5/ml in 24-well culture plates) were grown in DMEM with 10% FCS for 24 to 48 h on cytodex microcarrier beads in 10-mm tissue culture inserts (8 μm,

[a]This work supported by DFG grant JA 398/5-2.
[b]Corresponding author: Dr. Annemone Radleff-Schlimme, Division of Clinical and Experimental Endocrinology, Department of Obstetrics and Gynecology, University of Göttingen, Robert-Koch-Str. 40, D-37075 Göttingen; Tel.: 049/551/392144; Fax: 049/551/396518; E-mail: aradschl@med.uni-goettingen.de

486

polycarbonate membrane, NUNC). The cells were transferred to the perfusion culture apparatus. After 90 min of perfusion with DMEM, stimulation with cAMP (1 mM) was started. Fractions were collected every 15 min. PACAP concentrations were measured in the supernatants and perfusates by radioimmunoassay (RIA).[6] Total RNA was isolated using the RNeasy Kit (Qiagen).

In Vivo *Experiments*

Intact male rats (Sprague-Dawley, 3 months old, 12 animals per group) were implanted with a jugular vein catheter. PACAP (10 µg per bolus) was injected intravenously (i.v.) every 40 min over a period of 6 hours. Blood samples were taken at 10-min intervals to analyze the LH release by RIA[7] starting 3 h after the first bolus. At the end of the experiment, the animals were decapitated and the total RNA of the anterior pituitary was isolated using the RNeasy Kit (Qiagen).

Oligonucleotide Primer

The PACAP forward primer used was 5'-GATGTCGCCCACGAAATCCT-3'. The reverse primer used was 5'-AGTGATGACTGGTCAGTCAA-3'.[6]

Reverse Transcription

For reverse transcription (RT), 300 ng total RNA (alphaT3-1 cells, anterior pituitary, control tissues) was preincubated with 1 µl oligodT primer (0.5 mg/ml) and H_2O in a total volume of 10 µl for 10 min at 70°C and then rapidly cooled on ice. To this mixture 0.1 µl RNasin (40 U/µl), 2 µl H_2O, 4 µl 5 × RT buffer, 2 µl DTT (100 mM), and 1 µl deoxyribonucleotides (dNTP, 10 mM of each) were added. Finally, 1 µl Moloney murine leukemia virus reverse transcriptase (M-MLV RT, 200 U/µl) was added to give a total volume of 10 µl. The reaction was incubated for 10 min at 22°C, for 50 min at 42°C, for 10 min at 95°C, and then stopped by putting the samples on ice.

Polymerase Chain Reaction

The cDNA produced in the RT reaction was amplified using the oligonucleotide primers described above and the expected size of the PCR product was 375 base-pairs (bp). For the polymerase chain reaction (PCR) 3 µl cDNA, 0.5 µl of both oligonucleotide primers (50 pmol), 0.5 µl *Taq* DNA polymerase (5 U/µl), 50 µl 2 × PCR buffer, and 45.5 µl H_2O were added to each reaction tube to give a final volume of 100 µl. The PCR reaction was run at 95°C for 3 min, followed by 38 cycles at 95°C for 1 min, 50°C for 1 min, 72°C for 2 min, a final cycle at 72°C for 10 min, and then finished at 4°C. A 10-µl sample of the reaction mixture was analyzed by 1.5% agarose gel electrophoresis, stained with ethidium bromide, visualized under UV illumination, and photographed. PACAP type 1 receptor PCR was conducted according to Rawlings and colleagues.[5]

RESULTS AND DISCUSSION

We were able to demonstrate PACAP gene expression not only in rat anterior pituitary cells but also in the gonadotrope-derived clonal cell line alphaT3-1 by RT PCR. cDNA was amplified in a PCR reaction with PACAP-specific pairs of oligonucleotide primers. As shown in FIGURE 1, PCR products derived from rat anterior pituitary, rat liver, and alphaT3-1 cells show the predicted fragment length of 375 bp for PACAP. In addition, as described previously by Rawlings and colleagues,[5] alphaT3-1 cells express the PACAP type I receptor (PVR1) (FIG. 1).

AlphaT3-1 cells secrete PACAP *in vitro* as measured by RIA under static as well as under perfusion culture conditions. Unexpectedly, it could be observed that the amount of PACAP secreted by the cells under static culture conditions was lower after 24 h of cultivation than after 1 h (data not shown). Therefore we stimulated alphaT3-1 cells for 1 h with either PMA (100 nM) or cAMP (1 mM) and could measure a significant increase of PACAP release from the cells (FIG. 2). PMA induced 1.7-fold higher PACAP levels (19.6 pg/100 µl) compared to control cultures (11.6 pg/100 µl) whereas cAMP caused a 1.4-fold increase (18.8 pg/100 µl; controls: 13.9 pg/100 µl). However, to achieve similar effects on the PACAP release from alphaT3-1 cells, the concentration of cAMP (a protein kinase A–activating substance) was 10^4-fold higher than PMA.

Under perfusion culture conditions, PACAP secretion from alphaT3-1 cells was rapidly detectable in sampling intervals of only 15 min. After approximately 90 min of perfusion with DMEM, the system reached a steady-state level (data not shown) and the stimulation with 1 mM cAMP was started. A rapid and sustained increase of PACAP release to 14.4

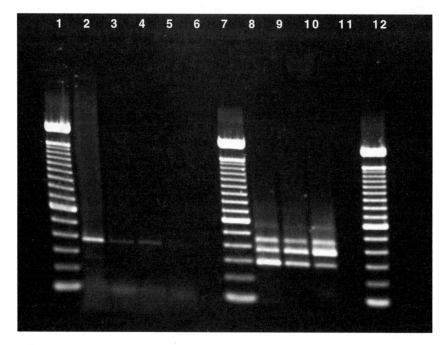

FIGURE 1. PACAP gene expression detected by RT PCR in rat anterior pituitary cells and the clonal cell line alphaT3-1. The figure shows the ethidium-bromide stained products of the PCR experiments resolved using 1.5% agarose gel electrophoresis. cDNA produced from a RT reaction using total RNA from alphaT3-1 cells, rat anterior pituitary, and rat liver tissue was amplified using PCR with pairs of oligonucleotide primers specific for PACAP and PVR1. In the controls cDNA has been omitted. The predicted fragment length for PACAP is 375 bp. For the PVR1 the expected PCR product sizes are 280 bp for the basic receptor, 364 bp for a single cassette insert (hiphop1 or hiphop2), and 448 bp for a double insert (hip, hop1, or hop2) Lanes 1, 7, 12: LMW marker; Lanes 2, 8: rat anterior pituitary; Lanes 3, 9: rat liver; Lanes 4, 5, 10: alphaT3-1 cells; and Lanes 6, 11: PCR negative control.

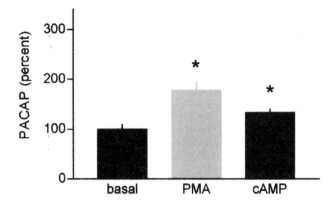

FIGURE 2. PACAP secretion from alphaT3-1 cells under static culture conditions after 1 h of stimulation with either PMA (100 nM) or cAMP (1 mM). Results are expressed as a percentage of the basal PACAP release. $N = 12$ wells per group. Mean ± SEM; *$p < 0.05$ versus basal.

pg/100 µl could be induced while the controls without stimulus remained on a basal level of 7.3 pg/100 µl.

To evaluate whether PACAP is able to modulate its own expression, we performed an *in vivo* experiment using 3-month-old, intact male rats. Animals were treated with 10 intravenous boluses of 10 µg PACAP each at 40-min intervals. The control group received saline. LH levels were measured in the blood by RIA. In the anterior pituitary the PACAP gene expression was detected by RT PCR. PACAP application caused a pronounced increase of LH (7.2 ± 0.75 ng/ml; control group: 0.16 ± 0.024 ng/ml) (FIG. 4). These high levels of LH secretion are associated with a significantly elevated PACAP gene expression (1.6 ± 0.29; control group: 0.24 ± 0.08) in the anterior pituitary of the animals (FIG. 4).

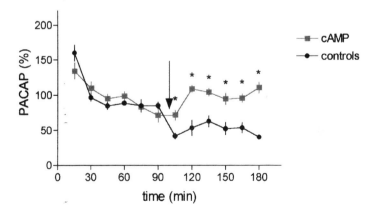

FIGURE 3. PACAP secretion from alphaT3-1 cells under perfusion culture conditions stimulated with 1 mM cAMP. Fractions were collected in 15-min intervals. Results are expressed as a percentage of basal PACAP release. $N = 12$ wells per group. Mean ± SEM; *$p < 0.05$ versus time point of 95 min.

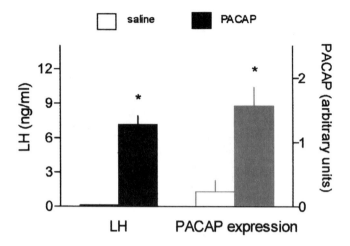

FIGURE 4. LH levels in the blood and PACAP gene expression in the anterior pituitary of intact male rats after 10 intravenous applied boluses of 10 μg PACAP each or saline in 40-min intervals. $N = 12$ animals per group. Mean ± SEM; *$p < 0.05$ versus saline.

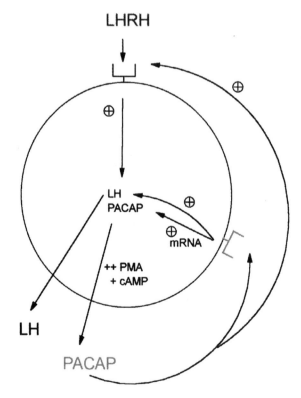

FIGURE 5. Effects of PACAP as an autocrine factor on gonadotrope cells.

SUMMARY

In summary, PACAP actions on gonadotrope cells are schematically outlined in FIGURE 5. PACAP is produced by and secreted from gonadotropes. The peptide increases LH release in an autocrine manner by a direct stimulatory action via the PACAP receptor and, as suggested by Culler and Paschall,[9] by amplification of the LHRH stimulus. Hence, our data support the hypothesis that PACAP acts as an autocrine regulator on gonadotrope cell function.

REFERENCES

1. MIYATA, A., A. ARIMURA, D. H. DAHL, N. MINAMINO, A. UEHARA, L. JIANG, M. D. CULLER & D. H. COY. 1989. Isolation of a novel 38 residue hypothalamic peptide which stimulates adenylate cyclase in pituitary cells. Biochem. Biophys. Res. Commun. **164**: 567–574.
2. RAWLINGS, S. R. & M. HEZAREH. 1996. Pituitary adenylate cyclase-activating polypeptide (PACAP) and PACAP/vasoactive intestinal polypeptide receptors: Actions on the anterior pituitary gland. Endocr. Rev. **17**(1): 4–29.
3. ARIMURA, A. 1992. Pituitary adenylate cyclase-activating polypeptide (PACAP): discovery and current status of research. Regul. Pept. **37**: 287–303.
4. JARRY, H., M. BÖTTNER, S. BENTER & W. WUTTKE. 1995. Rat gonadotropes express PACAP: In vivo upregulation of mRNA levels by a GnRH agonist. Presented at the Annual Meeting of American Endocrine Society. Abstr. P2-51.
5. RAWLINGS, S. R., I. PIUZ, W. SCHLEGEL, J. BOCKAERT & L. JOURNOT. 1995. Differential expression of pituitary adenylate cyclase-activating polypeptide/vasoactive intestinal polypeptide receptor subtypes in clonal pituitary somatotrophs and gonadotrophs. Endocrinology **136**(5): 2088–2098.
6. STEINHOFF, M., G. P. MCGREGOR, A. RADLEFF-SCHLIMME, A. STEINHOFF, H. JARRY & W. E. SCHMIDT. 1997. Evidence and distribution of PACAP and PACAP type 1 receptor in human skin: PACAP-38 is increased in patients with psoriasis. J. Invest. Dermatol. In press.
7. LEONHARDT, S., H. JARRY, G. FALKENSTEIN, J. R. PALMER & W. WUTTKE. 1991. LH release in ovariectomized rats is maintained without noradrenergic neurotransmission in the preoptic/anterior hypothalamic area: Extreme functional plasticity of the GnRH pulse generator. Brain Res. **562**: 105–110.
8. OGI, K., C. KIMURA, H. ONDA, A. ARIMURA & M. FUJINO. 1990. Molecular cloning and characterization of the cDNA for the precursor of rat pituitary adenylate cyclase-activating polypeptide (PACAP). Biochem. Biophys. Res. Commun. **173**: 1271–1279.
9. CULLER, M. D. & C. S. PASCHALL. 1991. Pituitary adenylate cyclase-activating polypeptide (PACAP) potentiates the gonadotropin-releasing activity of luteinizing hormone-releasing hormone. Endocrinology **129**: 2260–2262.

Peptidergic Component of Non-Adrenergic Non-Cholinergic Relaxation of the Rat Gastric Fundus[a]

DIEGO CURRÒ[b] AND PAOLO PREZIOSI

Institute of Pharmacology, Catholic University School of Medicine, L. go F. Vito, 1, I-00168, Rome, Italy

Findings from a number of studies suggest that non-adrenergic non-cholinergic (NANC) relaxation of the rat gastric fundus includes a peptidergic component, which seems to be mediated by vasoactive intestinal polypeptide (VIP).[1,2] Release of VIP is produced only by high-frequency (> 4 Hz) electrical field stimulation (EFS) of rat gastric fundus strips,[3] while NANC relaxation provoked by low-frequency stimulation is greatly reduced by nitric oxide synthase inhibitors.[2] Involvement of VIP thus seems to be more considerable in the former response. The aim of the present study was to define the role of this peptide in the high-frequency, electrically induced response of rat fundus strips.

Longitudinal muscle strips (3 × 20 mm) were prepared from gastric fundi removed from Wistar rats (180–320 g) subjected to an overnight fast. The strips were mounted under a 1-g load between parallel platinum electrodes inside 5-ml organ baths containing Krebs solution maintained at 37°C and bubbled with a 95:5 mixture of O_2/CO_2. The Krebs solution contained 1 μM atropine, 5 μM guanethidine, and 0.1 μM U46619 (this latter to increase strip tone). Smooth-muscle activity was recorded with isotonic transducers. EFS parameters were 1 msec, 120 mA, 13 Hz, 2-min pulse train duration. All experiments began with a 60-min equilibration period, during and after which the bath medium was renewed every 10 min. In a first series of experiments, the strips were exposed to consecutive 2-min incubations with increasing concentrations of VIP (0.3 nM–0.3 μM). The strips were allowed to recover basal tone before subsequent concentration. Studies were then conducted to identify the effects of α-chymotrypsin (1 U/ml) and anti-VIP serum (1:100) on the inhibitory response evoked by 13 Hz EFS. In these experiments, each strip was subjected to control and post-treatment stimulations. The effect of anti-VIP serum was compared with that of normal rabbit serum (NRS, 1:100), while the relaxations observed in the presence of α-chymotrypsin were compared with those observed in untreated time controls. All relaxant responses were expressed as both amplitudes and areas under the curves (AUC) and the results were evaluated by means of Student's paired and unpaired t test.

VIP (0.3 nM–0.3 μM) induced concentration-dependent relaxations that showed a latency of a few seconds after the addition of this substance to the bath medium and proceeded slowly throughout the 2-min incubation. Similarly, the recovery of the basal tone after VIP removal from the bath was slow and the recovery time appeared to be concentration-dependent. For this reason, two different concentration-response curves were plotted, one based on amplitude and the second on the AUC of each relaxation (FIG. 1). The potency of VIP revealed by the former curve was approximately 15 times greater than that

[a]This study was financially supported by CNR Grant No 94.02848.CT04.

[b]Corresponding author: Dr. Diego Currò, Institute of Pharmacology, Catholic University School of Medicine, L.go F. Vito, 1, I-00168, Rome, Italy; Tel.: 39-0630154367; Fax: 39-06-3050159.

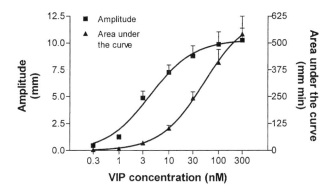

FIGURE 1. Mean concentration-response curves for VIP (0.3 nM–0.3 μM)-induced relaxations of U46619 (0.1 μM)-precontracted longitudinal muscle strips of the rat gastric fundus under NANC conditions. Each point represents the mean ± S.E.M. of responses observed in 8 strips. Relaxations were measured as peak amplitude (■) and as area under the curve (▲).

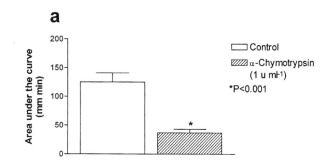

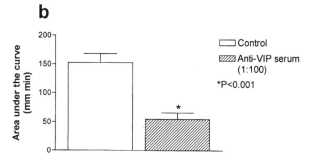

FIGURE 2. Effects of α-chymotrypsin (1 U/ml) (**a**) and anti-VIP serum (1:100) (**b**) on NANC relaxation (expressed as area under the curve) induced by EFS (1 msec, 120 mA, 13 Hz, pulse trains of 2 min) of rat longitudinal gastric fundus strips. Each column represents the mean ± S.E.M. of responses observed in 6 (**a**) and 8 (**b**) strips. *$p < 0.001$ vs. control relaxations.

seen in the latter (EC_{50}s of 3.59 ± 0.43 nM and 55.2 ± 7.8 nM, respectively, N = 8). The response produced by the maximal VIP concentration used (0.3 μM) displayed an amplitude of 10.25 ± 1.12 mm and an AUC of 540.5 ± 84.0 mm min. Also, 13-Hz EFS induced a long-lasting response, but the relaxation, unlike that evoked by VIP, had an immediate beginning after the initiation of the stimulation and a rapid development to a near-maximal level. α-Chymotrypsin (1 U/ml) and anti-VIP serum (1:100) significantly reduced the AUCs with respect to pre-treatment controls (FIG. 2), but neither had any effect on the amplitude (percentage of pre-treatment value: 100.0 ± 5.7%, N = 6 and 106.0 ± 5.8%, N = 8, respectively) or the rapidity of the NANC inhibitory effect. The AUC of the relaxant response was reduced slightly, but significantly ($p < 0.05$), by NRS (1:100), but the reduction achieved with anti-VIP serum was significantly greater (64.7 ± 5.5% of the control AUC vs. 17.6 ± 5.5% with NRS, $p < 0.001$). In the strips that served as time controls, the AUCs of the first and second NANC inhibitory responses were not significantly different (data not shown). Neither NRS (1:100) nor vehicle for α-chymotrypsin (distilled water, time control strips) affected the amplitude of NANC relaxation (data not shown).

These data clearly indicate that a peptidergic component is present in the NANC inhibitory response of the rat gastric fundus to high-frequency EFS and is mainly involved in maintenance of the response. Although VIP appears to be the major peptide mediator of this response, the attenuating effects of α-chymotrypsin are slightly greater than those that can be achieved with anti-VIP serum, indicating the involvement of at least one other peptide mediator. PHI, a peptide co-synthesized with VIP, that has been shown to be released by the rat gastric fundus in response to EFS,[4] might thus play a neurotransmitter role in this physiological response. The fact that neither the amplitude nor the initiation of the NANC relaxation was affected by α-chymotrypsin or anti-VIP serum demonstrates that other neurotransmitter(s) are capable of compensating for the abolished peptidergic component. The most likely candidate is obviously nitric oxide, which is thought to be primarily responsible for the rapidity of this relaxant response.

REFERENCES

1. DE BEURME, F. A. & R. A. LEFEBVRE. 1988. Vasoactive intestinal polypeptide as possible mediator of relaxation in the rat gastric fundus. J. Pharm. Pharmacol. **40**: 711–715.
2. LI, C. G. & M. J. RAND. 1990. Nitric oxide and vasoactive intestinal polypeptide mediate non-adrenergic, non-cholinergic inhibitory transmission to smooth muscle of the rat gastric fundus. Eur. J. Pharmacol. **191**: 303–309.
3. D'AMATO, M., D. CURRÒ, P. MONTUSCHI, G. CIABATTONI, E. RAGAZZONI & R. A. LEFEBVRE. 1992. Release of vasoactive intestinal polypeptide from the rat gastric fundus. Br. J. Pharmacol. **105**: 691–695.
4. CURRÒ, D., P. PREZIOSI, E. RAGAZZONI & G. CIABATTONI. 1994. Peptide histidine isoleucine-like immunoreactivity release from the rat gastric fundus. Br. J. Pharmacol. **113**: 541–549.

Effect of VIP and PACAP on Vascular and Luminal Release of Serotonin from Isolated Perfused Rat Duodenum[a]

M. FUJIMIYA,[b,d] H. YAMAMOTO,[b] AND A. KUWAHARA[c]

[b]Department of Anatomy, Shiga University of Medical Science, Otsu, Shiga 520-21, Japan

[c]Laboratory of Environmental Physiology, Graduate School of Nutritional and Environmental Sciences, University of Shizuoka, Shizuoka 422, Japan

Although a number of previous studies have shown the cholinergic and adrenergic neuronal involvement in the release of serotonin (5HT), a relatively small number of studies have shown nonadrenergic, noncholinergic (NANC) mechanisms to regulate the release of 5HT. In isolated vascularly perfused guinea pig intestine, direct inhibitory action of vasoactive intestinal polypeptide (VIP) on the release of 5HT into the portal circulation has been shown, therefore this peptide was a candidate for the non-cholinergic, non-adrenergic inhibitory neurotransmitter regulating 5HT release.[1,2] Other candidates for NANC inhibitory neurotransmitter are adenosine triphosphate (ATP) studied in isolated sheets of rabbit intestine[1,3] and GABA studied in isolated vascularly perfused guinea pig intestine.[4] In previous studies that investigated NANC mechanisms affecting the release of 5HT, only vascular release of 5HT has been examined. Since vascular and luminal release of 5HT from enterochromaffin cells (EC) are known to be mediated by independent mechanisms,[5,6] the effects of a NANC inhibitory mechanism on the luminal release of 5HT should be investigated in comparison to the vascular release of 5HT.

In the present study we aimed to examine the mechanism of action of VIP, pituitary adenylate cyclase activating peptide (PACAP) 38, and PACAP27 on the release of 5HT into the intestinal lumen as well as into the portal circulation. PACAP38 and its shorter form, PACAP27, are newly isolated polypeptides and show structural homology with VIP.[7,8] Immunohistochemical localization of PACAP 27[9] resembles that of VIP[10] in the gut, however the role of this peptide in the release of 5HT has not yet been examined.

METHODS

Male Wistar rats weighing 200–300 g were used. Animals were housed in a light-controlled room with free access to laboratory food and water, but were fasted overnight (16–18 h) before the operation. The animal was anesthetized with an intraperitoneal injection

[a]This work supported by a grant from the Joint Research Program of the National Institute for Physiological Sciences (M. F. and A. K.).

[d]Corresponding author: Mineko Fujimiya, Department of Anatomy, Shiga University of Medical Science, Seta, Otsu, Shiga 520-21, Japan; Tel.: 81-775-48-2136; Fax: 83-775-48-2139; E-mail: fujimiya@belle.shiga-med.ac.jp

of sodium pentobarbital (60 mg/kg; Nembutal®, Abbott Labs., USA). The duodenum, between the pylorus and Treitz' ligament, was prepared for both vascular and intraluminal perfusion as described previously.[11] Arterial perfusion was achieved through an aortic cannula with the tip lying adjacent to the celiac and superior mesenteric arteries and effluent perfusate was collected through a portal vein cannula. All other vasculature apart from that leading into the duodenal segment was cut between double ligatures. Stomach, jejunum, ileum, colon, pancreas, and spleen were removed. Luminal perfusion was performed through a cannula inserted into the pylorus; effluent perfusate was collected through a cannula placed into the duodenal lumen at the level of Treiz' ligament.

The vascular perfusate consisted of Krebs' solution containing 3% dextran, 0.2% bovine serum albumin (RIA grade; Sigma Chemical Co., St. Louis, MO, USA), and 5 mM glucose. The perfusate was saturated with 95% O_2/5% CO_2 gas to maintain a pH of 7.4. Phosphate buffered saline (PBS, 0.1 M, pH 7.4) was used as a luminal perfusate. Both perfusates and the preparation were kept at 37°C throughout the experiment by thermostatically controlled heating apparatus. The flow rates for vascular and luminal perfusion were maintained at 3 ml/min and 1 ml/min, respectively. After a 25-min equilibration period, both vascular and luminal effluents were collected at 3-min intervals for 33 min into ice-cold vials. Each vial contained 10 µl of 57 mmol/l ascorbic acid, 10 µl of 10 mmol/l EDTA-2Na, 10 µl of 1 mol/l perchloric acid, and 10 µl of 51 mmol/l pargyline hydrochloride in 1 ml of samples. VIP (human and porcine, Peptide Institute, Inc. Osaka, Japan), PACAP38 (human, Peptide Institute Inc.), or PACAP27 (human, Peptide Institute, Inc.) was introduced into the vasculature via a side-arm infusion at a final concentration of 0.1 µM during the fifth to seventh periods, as indicated in the figures. In some experiments, 1 µM atropine sulfate (Sigma Chemical Co.), 100 µM hexamethonium bromide (Sigma Chemical Co.), 1 µM tetrodotoxin (TTX, Sankyo Co. Ltd., Tokyo, Japan), or 100 µM N^G-nitro-L-arginine (L-NA) was introduced into the vasculature during the fourth to seventh periods in addition to the infusion of each peptide as described above.

The determination of 5HT was performed by HPLC. Vascular effluents were filtrated with Ultrafree-MC® (30,000 NMWL, Nihon Millipore Ltd., Yonezawa, Japan) by centrifuging for 30 min at 10,000 rpm at 4°C. Luminal effluents were filtrated manually with 0.22 µm pore disk filter (Millex-GV, Nihon Millipore Ltd.). Next, 100 µl aliquots of filtrates were injected into HPLC and 5HT content was measured.[12] Results were expressed as mean ± SEM in ng/min in each fraction.

Statistical analysis of the data shown in figures was performed using the single-factor ANOVA for repeated measures followed by the Scheffe's F test. A paired t-test (two-tail) was used to compare the values of mean basal release (during periods 1–3, or 1–4) and mean 5HT release during periods as indicated in the results. In both cases, a value of $p < 0.05$ was considered statistically significant.

RESULTS

VIP at the concentration of 0.1 µM was introduced into the vasculature during perfusion periods 5–7. The basal release of 5HT into the lumen (4.87 + 0.61 ng/min, $N = 9$) was significantly decreased during perfusion periods 7 and 8 (2.21 ± 0.32 ng/min, $N = 9$) and then gradually returned to the basal levels (FIG. 1,A). The mean levels of 5HT release into the vasculature (periods 5–7; 1.46 ± 0.17 ng/min, $N = 9$) during infusion of VIP, on the other hand, did not change from the mean basal release of 5HT into the vasculature (1.79 ± 0.18 ng/min, $N = 9$). The effect of atropine (1 µM) as well as hexamethonium (100 µM) on VIP-induced decrease of 5HT was tested. Each was introduced 3 min before an administration of VIP, as shown in FIGURE 1(B and C). The basal release of 5HT into the lumen before atropine infusion (4.74 ± 0.69 ng/min, $N = 3$) became significantly lower during

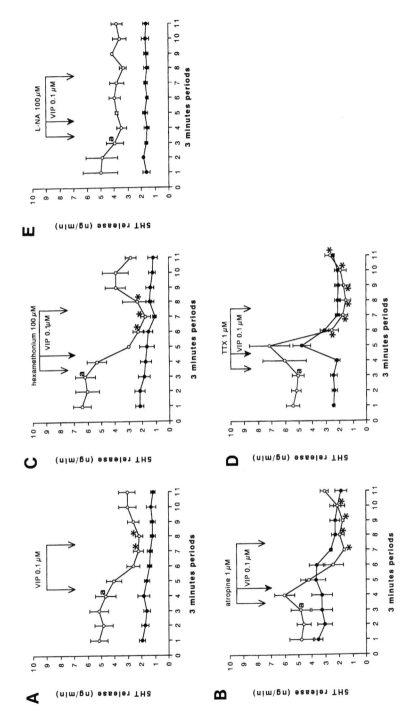

FIGURE 1. Effect of VIP ($N = 9$, **A**), VIP + atropine ($N = 3$, **B**), VIP + hexamethonium ($N = 3$, **C**), VIP + TTX ($N = 3$, **D**), and VIP + L-NA ($N = 3$, **E**) on luminal (O) and vascular (●) release of 5HT from isolated perfused rat duodenum. Each value represents mean ± SEM of 3-min samples. The luminal release of 5HT (*stars*) is significantly ($p < 0.05$) decreased when compared with the period prior to drug infusion (**a**).

periods 7–10 (1.86 ± 0.12 ng/min, N = 3) as shown in FIGURE 1(B), whereas the basal release of 5HT into the lumen before hexamethonium infusion (6.20 ± 0.72 ng/min, N = 3) became significantly lower during periods 6–8 (2.11 ± 0.60 ng/min, N = 3), as shown in FIGURE 1(C). There was no further reduction in the VIP response when atropine or hexamethonium was present at the same time. A similar phenomenon was observed in the treatment of TTX combined with VIP, where the basal release of 5HT into the lumen (5.24 ± 0.41 mg/min, N = 3) became significantly lower during periods 6–11 (2.03 ± 0.41 ng/min, N = 3) as shown in FIGURE 1(D). On the other hand, when nitric oxide synthase inhibitor L-NA (100 µM) was introduced prior to VIP (FIG. 1,E), the basal release of 5HT into the lumen (4.63 ± 1.02 ng/min, N = 3) was not changed from the mean levels of luminal 5HT release during periods 5–7 (3.86 ± 0.34 ng/min, N = 3).

PACAP38 at the concentration of 0.1 µM was introduced into the vasculature during perfusion periods 5–7. The basal release of 5HT into the lumen (4.53 ± 0.28 ng/min, N = 4) was significantly decreased during perfusion periods 7–9 (2.20 ± 0.23 ng/min, N = 4) (FIG. 2,A). However, the basal release of 5HT into the vasculature (2.37 ± 0.16 ng/min, N = 4) was not changed during infusion of PACAP38 (periods 5–7; 2.91 ± 0.37 ng/min, N = 4). The effect of atropine (1 µM), hexamethonium (100 µM), TTX (1 µM), or L-NA (100 µM) on the inhibitory effect of PACAP38 on the luminal 5HT release was examined. The basal release of 5HT into the lumen before atropine infusion (4.25 ± 0.47 ng/min, N = 3) became significantly lower during periods 6–11 (0.97 ± 0.15 ng/min, N = 3) (FIG. 2,B). Whereas the basal release of 5HT into the lumen before hexamethonium infusion (6.78 ± 1.33 ng/min, N = 4) became significantly lower during periods 7–9 (1.52 ± 0.16 ng/min, N = 4) (FIG. 2,C). The basal release of 5HT into the lumen before TTX infusion (4.74 ± 0.36 ng/min, N = 3) became significantly lower during periods 7–11 (1.55 ± 0.26 ng/min, N = 3) (FIG. 2,D). Yet, when L-NA was introduced prior to PACAP38 (FIG. 2,E), the basal release of 5HT into the lumen (5.48 ± 0.42 ng/min, N = 3) was not changed from the mean levels of luminal 5HT release during periods 5–7 (5.77 ± 0.75 ng/min, N = 3).

PACAP27 at the concentration of 0.1 µM was introduced into the vasculature during perfusion periods 5–7. The basal release of 5HT into the lumen (4.43 ± 0.80 ng/min, N = 5) was significantly decreased during periods 7–9 (1.39 ± 0.27 ng/min, N = 5, FIG. 3,A). However, the basal release of 5HT into the vasculature (2.07 ± 0.28 ng/min, N = 5) was not changed during infusion of PACAP27 (periods 5–7; 2.04 ± 0.28 ng/min, N = 5, FIG. 3,A). The effect of atropine (1 µM, FIG. 3,B), hexamethonium (100 µM, Fig. 3, C), TTX (1 µM, FIG. 3,D) or L-NA (100 µM, FIG. 3,E) on the inhibitory effect of PACAP27 on luminal 5HT release was examined. Again the inhibitory response caused by PACAP27 was not altered by the administration of atropine, hexamethonium, or TTX, however completely antagonized with L-NA.

DISCUSSION

The present results showed that a considerable amount of 5HT was released from isolated perfused rat duodenum into both the duodenal lumen and portal circulation and that the basal release of 5HT into the lumen was always higher than that into the vasculature in all experiments. 5HT detected in the present study seems to originate in the epithelial EC cells, because 5HT levels contained in the EC cells are much higher than those in 5HT-containing neurons in the gut.

VIP inhibited the release of 5HT into the lumen but did not affect the vascular release of 5HT. This inhibitory effect of VIP was not affected by the presence of atropine, hexamethonium, or TTX. These results suggest that VIP exerts a direct inhibitory effect on the luminal release of 5HT from the EC cells and that other neuronal mechanisms, including

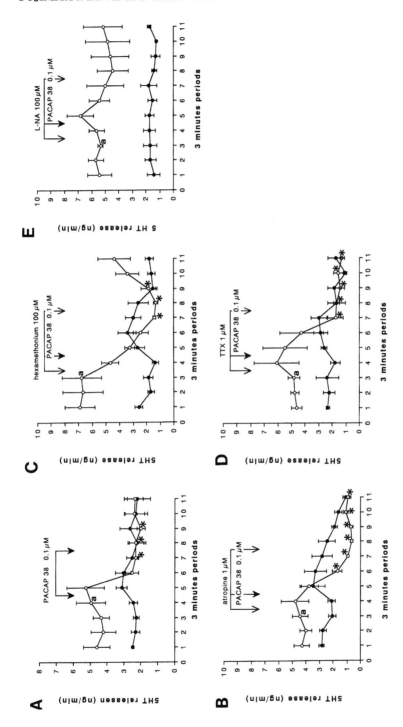

FIGURE 2. Effect of PACAP 38 ($N = 4$, **A**), PACAP38 + atropine ($N = 3$, **B**), PACAP38 + hexamethonium ($N = 4$, **C**), PACAP38 + TTX ($N = 3$, **D**), and PACAP38 + L-NA ($N = 3$, **E**) on luminal ($\bigcirc$) and vascular ($\bullet$) release of 5HT from isolated perfused rat duodenum. Each value represents mean ± SEM of 3-min samples. The luminal release of 5HT (*stars*) is significantly ($p < 0.05$) decreased when compared with the period prior to drug infusion (**a**).

cholinergic neurons, do not mediate this mechanism. The effect of VIP on the release of 5HT has previously been examined in the isolated guinea pig small intestine, however only vascular release of 5HT was examined,[2] in which VIP inhibited the vascular release of 5HT and this effect was not affected by TTX. Although the direction of the release of 5HT from the EC cells caused by VIP was different between the previous and present studies, it might be concluded that VIP exerts a direct inhibitory action on the luminal release of 5HT from the EC cells. Endogenous VIP is involved in the neuronal cell bodies in both myenteric and submucous plexuses and the number of VIP-positive neurons was much higher in the submucous plexus than the myenteric plexus.[10] Furthermore, it has been reported that VIP-containing nerve fibers are located close to the EC cells in the rat and guinea pig intestine.[13] These morphological findings may suggest the direct action of VIP neurons on the EC cells.

The inhibitory effect of VIP on the luminal release of 5HT was antagonized by NO synthase inhibitor L-NA. This result suggests that NO involves the effect of VIP on the release of 5HT from the EC cells. It is known that NO and VIP are functionally linked cotransmitters, because VIP and NOS colocalize in the myenteric neurons[14-16] and VIP release from myenteric neurons is regulated by the production of NO.[17] The present results imply that the functional linkage between VIP and NO previously shown in muscle relaxation[17,18] can also be seen in the regulation of 5HT release from the EC cells.

Both PACAP38 and PACAP27 inhibited the release of 5HT into the lumen but did not affect the vascular release of 5HT. These inhibitory effects exerted by PACAP38 or PACAP27 on the luminal 5HT release were not changed by the presence of atropine, hexamethonium, or TTX. These results suggest that PACAP38 as well as PACAP27 exert a direct inhibitory effect on the luminal release of 5HT from the EC cells, but no other neuronal mechanisms, including cholinergic neurons, involve this mechanism. The effects observed in PACAP38 and PACAP27 were quite similar to those observed in VIP. PACAP27-containing neurons have been shown to be distributed in the myenteric and submucous plexus and the number of cell bodies was numerous in the submucous plexus.[9] Such distribution was quite similar to that of VIP-containing neurons in the intestine.[10] Furthermore some PACAP27 neurons are known to colocalize with VIP neurons.[9] The inhibitory effects seen in PACAP38 and PACAP27 were antagonized by L-NA. This result suggests that properties such as the functional interaction of VIP with NO are shared by PACAPs.

In conclusion, the present results provide the evidence that VIP and PACAPs directly inhibit the luminal release of 5HT from EC cells of the rat duodenum and NO seems to be essential to exert this function.

SUMMARY

Effects of CCK, VIP, PACAP38, and PACAP27 on the release of 5HT into the intestinal lumen and into the portal circulation were examined in *in vivo* experiments of isolated rat duodenum perfused vascularly and luminally. VIP, PACAP 38 and 27 reduced the release of 5HT into the lumen but did not affect the vascular release of 5HT. These effects were not affected by the presence of atropine, hexamethonium, or TTX, suggesting that VIP, PACAP 38 and 27 exert a direct inhibitory effect on the luminal release of 5HT from the EC cells. Nitric oxide synthase inhibitor, N^G-nitro-L-arginine, antagonized the inhibitory effects of VIP, PACAP 38 and 27, suggesting that nitric oxide seems to be essential to exert the inhibitory action of VIP and PACAPs on the release of 5HT into the intestinal lumen from the EC cells.

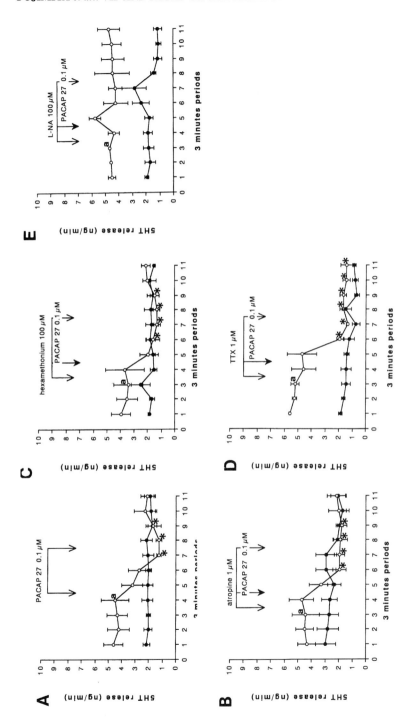

FIGURE 3. Effect of PACAP27 ($N = 5$, **A**), PACAP27 + atropine ($N = 3$, **B**), PACAP27 + hexamethonium ($N = 4$, **C**), PACAP27 + TTX ($N = 3$, **D**), and PACAP27 + L-NA ($N = 3$, **E**) on luminal (○) and vascular (●) release of 5HT from isolated perfused rat duodenum. Each value represents mean ± SEM of 3-min samples. The luminal release of 5HT (*stars*) is significantly ($p < 0.05$) decreased when compared with the period prior to drug infusion (**a**).

REFERENCES

1. RACHÉ, K., H. SCHWÖRER, D. V. AGOSTON & H. KILBINGER. 1991. Evidence that neuronally released vasoactive intestinal polypeptide inhibits the release of serotonin from enterochromaffin cells of the guinea pig small intestine. Acta Endocrinol. (Copenh) **124**: 203–207.
2. SCHWÖRER, H., K. RACKÉ & H. KILBINGER. 1989. Effect of vasoactive intestinal polypeptide on the release of serotonin from the in vitro vascularly perfused small intestine of guinea pig. Naunyn-Schmied. Arch. Pharmacol. **339**: 540–545.
3. KELLUM, J. M., J. WU & M. DONOWITZ. 1984. Enteric neural pathways inhibitory to rabbit duodenal serotonin release. Surgery **96**: 139–145.
4. SCHWÖRER, H., K. RACKÉ & H. KILBINGER. 1989. GABA receptors are involved in the modulation of the release of 5-hydroxytryptamine from the vascularly perfused small intestine of the guinea-pig. Eur. J. Pharmacol. **165**: 29–37.
5. GRÖNSTAD, K. O., M. J. ZINNER, O. NILSSON, A. DAHLSTRÖM, B. M. JAFFE & H. AHLMAN. 1988. Vagal release of serotonin into gut lumen and portal circulation. J. Surg. Res. **44**: 146–151.
6. MONEY, S. R., K. KIMURA, A. PETROIANU & B. M. JAFFE. 1988. Effects of intravenous calcium on release of serotonin into jejunal lumen and portal circulation. Dig. Dis. Sci. **33**: 977–981.
7. MIYATA, A., A. ARIMURA, R. R. DAHL, N. MINAMINO, A. UEHARA, L. JIANG, M. D. CULLER & D. H. COY. 1989. Isolation of a novel 38 residue-hypothalamic polypeptide which stimulates adenyl cyclase in pituitary cell. Biochem. Biophy. Res. Commun. **164**: 567–574.
8. MIYATA, A., L. JIANG, R. D. DAHL, C. KITADA, K. KUBO, M. FUJINO, N. MIMAMINO & A. ARIMURA. 1990. Isolation of neuropeptide corresponding to the n-terminal 27 residues of the pituitary adenylate cyclase activating polypeptide with 38 residues (PACAP 38). Biochem. Biophy. Res. Commun. **170**: 643–648.
9. SUNDLER, F., E. EKBLAD, A. ABSOOD, R. HAKANSON, K. KÖVES & A. ARIMURA. 1992. Pituitary adenylate cyclase activating peptide: a novel vasoactive intestinal peptide-like neuropeptide in the gut. Neurosci. **46**: 439–454.
10. COSTA, M., J. B. FURNESS, R. BUFFA & S. I. SAID. 1980. Distribution of enteric nerve cell bodies and axons showing immunoreactivity for vasoactive intestinal polypeptide in the guinea-pig intestine. Neurosci. **5**: 587–596.
11. FUJIMIYA, M., C. H. S. MCINTOSH, H. KIMURA & Y. N. KWOK. 1992. Effect of carbachol on luminal release of serotonin from isolated perfused rat duodenum. Neurosci. Lett. **145**: 229–233.
12. FUJIMIYA, M., T. MAEDA & H. KIMURA. 1991. Serotonin-containing epithelial cell in the rat duodenum. II. Quantitative study of the effect of 5HTP administration. Histochemistry **95**: 225–229.
13. IWANAGA, T., H. HAN, O. HOSHI, H. KANAZAWA, I. ADACHI & T. FUJITA. 1994. Topographic relation between serotonin-containing paraneurons and peptidergic neurons in the intestine and urethra. Biol. Signals **3**: 259–270.
14. BARBIERS, M., J-P. TIMMERMANS, D. ADRIAENSEN, M. H. A. DE GROODT-LASSEEL & D. W. SCHEUERMANN. 1995. Projections of neurochemically specified neurons in the porcine colon. Histochemistry **103**: 115–126.
15. SANG, Q., S. WILLIAMSON & H. M. YOUNG. 1997. Projections of chemically identified myenteric neurons of the small and large intestine of the mouse. J. Anat. **190**: 209–222.
16. SCHEMANN, M., C. SCHAAF & M. MÄDER. 1995. Neurochemical coding of enteric neurons in the guinea pig stomach. J. Comp. Neurol. **353**: 161–178.
17. MAKHLOUF, G. M. 1994. Neuromuscular function of the small intestine. *In* Physiology of the Gastrointestinal Tract. L. R. Johnson, Ed.: 977–990. Raven Press. New York.
18. BOECKXSTRAENS, G. E., P. A. PELCKMANS, J. G. DE MAN, H. BULT, A. G. HERMAN & Y. M. VAN MAERCKE. 1992. Evidence for a differential release of nitric oxide and vasoactive intestinal polypeptide by nonadrenergic noncholinergic nerves in the rat gastric fundus. Arch. Int. Pharmacodyn. **318**: 107–115.

Sites of Actions of Contractile and Relaxant Effects of Pituitary Adenylate Cyclase Activating Peptide (PACAP) in the Internal Anal Sphincter Smooth Muscle[a]

SATISH RATTAN[b] AND SUSHANTA CHAKDER

*Department of Medicine, Division of Gastroenterology and Hepatology,
Jefferson Medical College, Thomas Jefferson University, Philadelphia,
Pennsylvania, 19107, USA*

The role of vasoactive intestinal polypeptide (VIP) as an inhibitory neurotransmitter in the internal anal sphincter (IAS) is well known.[1,12] Recently, PACAP-38, closely related to VIP, has been identified in the central as well as the peripheral nervous systems.[9,13,15,18–21] Endogenously, PACAP may also occur in the form of PACAP-27 (27 amino acid form).[9] Furthermore, PACAP has been shown to be distributed in the myenteric neurons throughout the gastrointestinal tract from the esophagus[13,21] to the small and large intestines in animals[15] as well as humans.[19] PACAP is released from the myenteric neurons of the gastrointestinal tract[4,8] and is a potent and direct relaxant of different smooth muscles of the gut.[4,6,8,13] Furthermore, it has been suggested that PACAP plays a significant role in the inhibitory neurotransmission of the gut.[4,5,8] The inhibitory actions of PACAP and VIP may share a common receptor.

Actions of PACAP in the gut may vary from no effect[16] to either relaxation[4,6,8,13] or contraction.[7,10,14] The sites and mechanism of actions of PACAP may also vary widely in different preparations of the gastrointestinal tract: the inhibitory effects may be exerted directly at the smooth muscle,[4,6,8,13] and the excitatory effects of PACAP may be explained on the bases of acetylcholine (ACh) and substance P release either by the direct activation of myenteric neurons or by the negative coupling of adenosine receptors to adenylate cyclase in the myenteric neurons of the gut.[3]

VIP is well known to produce frank relaxation of the IAS smooth muscle.[1,2,12] The actions of PACAP in the IAS however have not been investigated. Interestingly, in the IAS, PACAP, especially in the higher concentration range, caused a biphasic response, an initial contraction followed by a relaxation. The purpose of the present investigation was to examine the sites of divergent actions of PACAP in the IAS smooth muscle.

MATERIALS AND METHODS

Preparation of IAS Smooth Muscle Strips and Measurement of Isometric Tension

Studies were performed on the IAS circular smooth muscle strips (~2 mm wide and 8 mm long) obtained from adult opossums (*Didelphis virginiana*) of either sex. The muscle strips were tied at both ends with silk sutures for measurement of isometric tension.

[a]This work was supported by U.S. Public Health Service Grant DK-35385 from the National Institutes of Health and an institutional grant from Thomas Jefferson University.

[b]Corresponding author: Dr. Satish Rattan, Professor of Medicine and Physiology, 901 College, Department of Medicine, Division of Gastroenterology & Hepatology, 1025 Walnut Street, Philadelphia, PA 19107; Tel.: 215 955-6944; Fax: 215 923-7697.

The IAS smooth muscle strips were transferred to thermostatically controlled 2-ml muscle baths containing oxygenated Kreb's solution that was bubbled constantly with a mixture of 95% oxygen and 5% carbon dioxide. The composition of the Kreb's solution was as follows (in mM): NaCl, 118.07; KCl, 4.69; $CaCl_2$, 2.52; $MgSO_4$, 1.16; NaH_2PO_4, 1.01; $NaHCO_3$, 25; and glucose, 11.10. The isometric tension was recorded using a force transducer (model FT03, Grass Instruments Co., Quincy MA) on a Dynograph recorder (Model R411, Beckman Instruments, Schiller Park, IL). After an equilibration period of 1 h, with intermittent washings, the optimal length and the baseline of each smooth muscle strip were determined as described previously.[11]

Electrical Field Stimulation and Drug Responses

Electrical field stimulation (EFS) was carried out using a pair of platinum electrodes placed at both sides of the smooth muscle strips via a Grass stimulator (Model S88; Grass Instruments Co.) at various frequencies (0.5–20 Hz; 4 sec train, 0.5 msec pulse duration; 20–30 V).

To examine the relaxant effects of VIP, PACAP-38, and PACAP-27 on the basal tone of the IAS, the agonists were added cumulatively to the muscle bath. For the examination of the contractile effect of PACAP-38 on the IAS smooth muscle, it was added in single-bolus doses and after the maximal response to an added dose was achieved, the smooth muscle strips were washed for at least 1 h before the next dose was added. The smooth muscle strips were washed appropriately to ensure the removal of any of the residual effects of an agent and to avoid any tachyphylaxis.

The effects of PACAP and the related peptides were examined before and after application of different neurohumoral antagonists, neurotoxin, or an ion-channel blocking agent, in the concentrations that have previously been documented to be maximally effective and selective. Tachyphylaxis with PACAP or VIP was achieved by repeated administration (usually four to five) of single doses of 1×10^{-6} M PACAP or VIP.

Drugs and Chemicals

VIP (porcine), VIP(10-28) (porcine), PACAP(1-38) amide (PACAP or PACAP-38; human, ovine, rat), PACAP(1-27) amide (PACAP-27; human, ovine, rat), PACAP(6-38) amide (PACAP 6-38; human, ovine, rat), PACAP(6-27), and ω-conotoxin GVIA were from Bachem (Bachem California Inc., Torrance, CA). Tetrodotoxin (TTX), apamin, N_ω-nitro-L-arginine (L-NNA), and atropine methyl bromide were from Sigma Chemical Co. (St. Louis, MO). Guanethidine monosulfate was from Ciba Pharmaceuticals (Summit, NJ).

Data Analysis

The relaxation and contraction of the smooth muscle strips in response to different stimuli was expressed as percentage of maximal relaxation and contraction (100%) caused by 5 mM EDTA and bethanechol, respectively. All the values are expressed as mean ± S.E. The significance of differences was determined by Student's t test or analysis of variance (ANOVA).

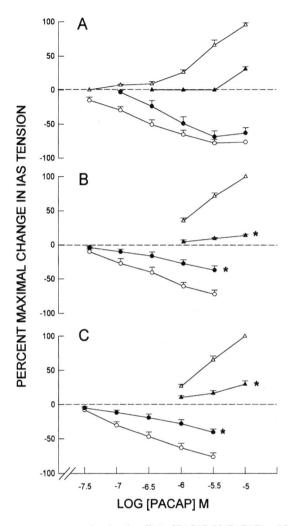

FIGURE 1. Dose-response curves showing the effect of PACAP-38 (PACAP) and PACAP-27 on the basal IAS tension. The changes (mean ± SE) in IAS tension were expressed as percent of maximal changes. (**A**) PACAP-38 produced biphasic responses (an initial rise followed by a fall). The rise in the basal tension of the IAS was observed only with higher concentrations of PACAP. PACAP-27 on the other hand produced only a fall in the IAS tension up to a concentration of 3×10^{-6} M (○, PACAP-38, Relaxation; ●, PACAP-27, Relaxation; △, PACAP-38, Contraction; ▲, PACAP-27, Contraction). (**B**) Influence of the PACAP antagonist PACAP 6-38 on both the inhibitory and the excitatory responses of PACAP-38 in the IAS. Note, significant rightward shifts in the dose-response curves showing that both types of effects of PACAP in the IAS smooth muscle were sensitive to PACAP 6-38 (*) (○, Control Relaxation; ●, PACAP 6-38 3×10^{-5} M, Relaxation; △, Control, Contraction ▲, PACAP6-38 3×10^{-5} M, Contraction). (**C**) VIP 10-28 caused a significant antagonism of the changes in the basal IAS tension by PACAP (*) (○, Control, Relaxation; ●, VIP 10-28 3×10^{-5} M, Relaxation; △, Control, Contraction; ▲, VIP 10-28 3×10^{-5} M, Contraction).

RESULTS

Effect of PACAP-38 and PACAP-27 on the Basal Tension of the IAS Smooth Muscle

PACAP-38 and PACAP-27 caused a concentration-dependent fall of the basal tension of the smooth muscle strips (FIG. 1,A). However, at concentrations higher than 10^{-6} M, PACAP-38 caused a biphasic response, an initial contraction followed by a relaxation. In order to determine the sites of excitatory and inhibitory actions of PACAP in the IAS, we examined the effects of the different concentrations of PACAP as single doses. Unless otherwise stated, the term PACAP has been used as synonymous to PACAP-38 throughout the manuscript.

Typical tracings to show the effects of PACAP-38, VIP, and PACAP-27 are given in FIGURE 2.

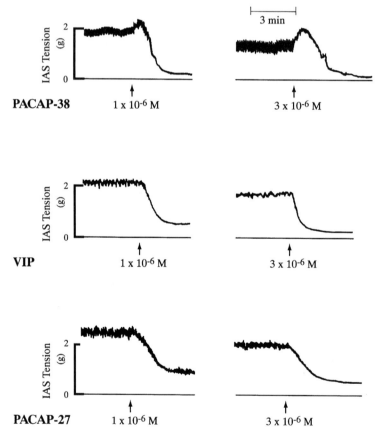

FIGURE 2. An actual tracing to show the biphasic nature of PACAP responses on the basal tone of the IAS smooth muscle. Note an initial rise followed by a fall in the basal tension of the IAS. Shown are responses to 1×10^{-6} and 3×10^{-6} M concentrations of PACAP.

Effect of PACAP 6-38 and VIP 10-28 on PACAP-38–Induced Fall and Rise in the Basal Tension of the IAS Smooth Muscle

PACAP 6-38 (3×10^{-5} M), a well known antagonist of PACAP, caused significant and selective antagonism of both the inhibitory and excitatory actions of PACAP-38 ($N = 5$ to 6; $p < 0.05$; Fig. 1,B).

The VIP antagonist VIP 10-28 (3×10^{-5} M) caused significant antagonism of the fall and the rise in the smooth muscle tension caused by different concentrations of PACAP-38 ($N = 5$; $p < 0.05$; Fig. 1,C) suggesting that VIP and PACAP-38 share a common receptor.

Effect of PACAP Tachyphylaxis, Tetrodotoxin, and Nitric Oxide Synthase Inhibitor L-NNA on PACAP-38–Induced Fall and Rise in the Basal IAS Tension

The fall or rise in basal tension of the IAS smooth muscle strips by PACAP-38 was significantly blocked by PACAP tachyphylaxis. In the control experiments, 1×10^{-6} and 3×10^{-6} M PACAP-38 caused 49.0 ± 6.5 and $58.7 \pm 6.4\%$ fall of the basal tension of the smooth muscle strips. After tachyphylaxis with PACAP, these values were 5.8 ± 1.9 and $7.5 \pm 2.2\%$, respectively ($N = 4$; $p < 0.05$).

For the contractile actions, in the controls, 1×10^{-6} and 3×10^{-6} M PACAP-38 produced 33.4 ± 8.4 and $84.9 \pm 8.6\%$ rise of the basal tensions of the IAS smooth muscles and after PACAP tachyphylaxis, these values were 8.7 ± 4.6 and $12.0 \pm 1.1\%$, respectively ($N = 4$; $p < 0.05$).

The neurotoxin TTX (1×10^{-6} M) had no significant effect on PACAP-38–induced fall or rise in the basal tension of the IAS smooth muscle strips ($N = 5$ to 7; $p > 0.05$).

L-NNA (3×10^{-5} M) caused a decrease in the PACAP-38–induced fall of the basal IAS tension suggesting that a part of the inhibitory action of PACAP-38 is NO-dependent ($N = 7$; $p < 0.05$; Fig. 3,A). In contrast to its effect on relaxation, L-NNA had no overall significant effect on the IAS smooth muscle contractions caused by different concentrations of PACAP-38 ($N = 4$; $p > 0.05$; Fig. 3,A). One exception is 1×10^{-5} M PACAP-38, a concentration that was significantly augmented by L-NNA ($N = 4$; $p > 0.05$).

Effect of Atropine, Guanethidine, and Substance P Antagonist Spantide on PACAP-38–Induced Fall and Rise in the Basal Tension of the IAS Smooth Muscle Strips

The muscarinic receptor antagonist atropine (1×10^{-6} M) and the adrenergic blocking agent guanethidine (3×10^{-6} M) had no significant effect on PACAP-38–induced fall and rise in the IAS smooth muscle tension. PACAP-38 at concentrations of 1×10^{-6}, 3×10^{-6}, and 1×10^{-5} M, caused 51.6 ± 5.1, 58.8 ± 5.1, and $63.8 \pm 5.4\%$ fall in the basal IAS tension. After atropine, the same concentrations of PACAP caused 47.9 ± 7.5, 62.8 ± 7.9, and $75.5 \pm 7.8\%$ fall of the smooth muscle tension respectively ($N = 6$; $p > 0.05$). For the contractile actions, in the control experiments 1×10^{-6}, 3×10^{-6}, and 1×10^{-5} M PACAP-38 produced 30.3 ± 6.1, 63.1 ± 8.8, and $93.5 \pm 6.5\%$ rise in the basal IAS tension, respectively. After atropine, these values were 37.2 ± 7.0, 76.5 ± 6.0, and $92.2 \pm 4.5\%$, respectively ($N = 6$; $p > 0.05$).

After guanethidine pretreatment, the same concentrations of PACAP-38 produced 37.4 ± 8.0, 50.5 ± 7.9, and $62.2 \pm 8.7\%$ fall and 26.6 ± 6.8, 59.9 ± 6.4, and $100.0 \pm 0\%$ rise of the IAS tensions, respectively ($N = 6$; $p < 0.05$).

PACAP-38–induced fall of the basal tension of the IAS smooth muscle strips was not affected significantly by the substance P antagonist spantide (3×10^{-5} M) ($N = 4$, $p > 0.05$; Fig. 3,B).

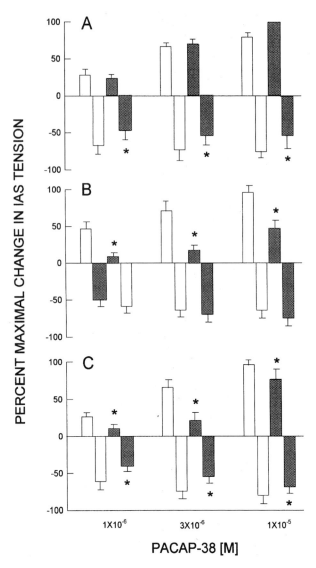

FIGURE 3. Percent maximal fall and rise in the IAS tension by PACAP before and after the NOS inhibitor L-NNA (**A**), spantide (3×10^{-5} M) (**B**), and ω-conotoxin (1×10^{-6} M) (**C**). (**A**) Note that the fall but not the rise in the IAS tension caused by PACAP was significantly attenuated by L-NNA (*) (*White bars*, Control; *Gray bars*, L-NNA 3×10^{-5} M). (**B**) Spantide in the concentration that blocked the effect of substance P in the IAS had no significant effect on the fall of the IAS tension by PACAP. However, it caused a significant and concentration-dependent antagonism of the rise in the IAS tension by PACAP (*) (*White bars*, Control; *Gray bars*, Spantide 3×10^{-5} M). (**C**) ω-Conotoxin caused significant antagonism of the fall as well as the rise in the IAS tension caused by PACAP (*) (*White bars*, Control; *Gray bars*, ω-Conotoxin GVIA 1×10^{-6} M).

Conversely, spantide caused a significant inhibition of PACAP-38 induced contractions of the smooth muscle strips ($N = 4$; $p < 0.05$; FIG. 3,B). Spantide also caused a shortening of the latencies of onsets of IAS relaxations by PACAP-38. The fall in IAS tension by 3×10^{-6} and 1×10^{-5} M PACAP in control began within 74.2 ± 6.1 and 80.0 ± 4.1 sec, respectively, and following spantide, these values decreased significantly to 48.3 ± 4.2 and 52.5 ± 3.2 sec, respectively ($p < 0.05$).

Effect of the Ganglionic Blocker Hexamethonium, Apamin, and ω-Conotoxin GVIA on the Fall and Rise in the Basal IAS Tension Caused by PACAP-38

Hexamethonium (1×10^{-4} M) had no effect on either the fall or the rise in the basal tension of the IAS smooth muscle strips caused by PACAP-38. In the control experiments, 1×10^{-6} and 1×10^{-5} M PACAP-38 caused 34.0 ± 4.4 and $95.0 \pm 5.0\%$ rise of the smooth muscle tensions, respectively, while after treatment with hexamethonium these values were 20.9 ± 5.7 and $81.1 \pm 9.5\%$, respectively ($N = 7$; $p > 0.05$)).

The Ca^{2+}-activated K^+ channel blocker apamin (1×10^{-5} M) had no significant effect on either PACAP-38–induced fall or rise in the basal IAS tension ($N = 5$; $p > 0.05$).

The N-type calcium channel blocker ω-conotoxin GVIA (1×10^{-6} M) caused a significant suppression of PACAP-38–induced rise and fall in the basal IAS tension ($N = 6$; $p < 0.05$; FIG. 3,C).

DISCUSSION

The studies show that PACAP-38 (PACAP) in the IAS smooth muscle not only produces relaxation but also causes contraction of the sphincteric smooth muscle. Both the inhibitory and the excitatory actions of PACAP may share the common receptor with VIP. The inhibitory action of PACAP in the IAS appears to be exerted primarily via its direct action at the smooth muscle cells. The excitatory action of PACAP appears to be via the activation of a distinct PACAP receptor on substance P–containing postganglionic nerve terminals.

The inhibitory effect of PACAP and the receptor responsible for this action in the gastrointestinal smooth muscle is well known. Conversely, the contractile action of PACAP in the smooth muscle is relatively unknown. Before the present study, the contractile action of PACAP had only been demonstrated in the guinea-pig ileum[7] and gall bladder[14] *in vitro* and on gall bladder motility in the conscious dog.[10] In these tissues, only a frank contraction with PACAP was seen. The data emphasize the importance of species differences in the action of PACAP.

The biphasic effect (an initial contraction followed by a relaxation) of PACAP-38 on the smooth muscle has not been shown before. These effects of PACAP in the IAS were selectively antagonized by the PACAP antagonist PACAP 6-38 and VIP antagonist VIP 10-28. A major part of the inhibitory action of PACAP in the IAS may be via direct action at the smooth muscle cells since the TTX and other neurohumoral antagonists had no significant influence on the fall in IAS tension by PACAP. However, a part of the inhibitory action of PACAP in the IAS appears to be exerted via the activation of PACAP receptor at the myenteric inhibitory nerve terminals since the N-type Ca^{2+}-channel blocker ω-conotoxin partly attenuated the inhibitory action of PACAP.

The excitatory action of PACAP-38, on the other hand, appears to be mediated via the activation of nonadrenergic, noncholinergic neurons since it was not modified by the cholinergic and adrenergic blockade. The excitatory action of PACAP on the basal tone of the IAS may occur via the activation of a specific PACAP receptor at the substance P–containing nerve terminals. The contractile actions of PACAP in the IAS were not significantly modified by the ganglionic blocking agent hexamethonium and TTX. These observations suggest the site of the excitatory action of PACAP to be below the level of the post-ganglionic cell body and axon. ω-Conotoxin, an agent well known to specifically block the N-type Ca^{2+} channel at the postganglionic nerve terminal, caused significant attenuation of the contractile effect of PACAP. Moreover, the specific antagonist of substance P, spantide, caused almost complete obliteration of the rise in the IAS tone by PACAP-38. The involvement of substance P in the mediation of PACAP excitatory response in the IAS in part is similar to that in the guinea-pig ileum. However, unlike the actions of PACAP in the IAS, a significant portion of the contractile action of PACAP in the guinea-pig ileum was also via the release of ACh.

The fall, but not the rise, in the IAS tension by PACAP-38 was significantly antagonized by the NOS inhibitor L-NNA, suggesting the involvement of NOS pathway in the mediation of PACAP-induced IAS relaxation similar to that of VIP.[17]

The present study in IAS smooth muscle shows that PACAP-27 primarily produced relaxation with potency almost similar to that of PACAP-38. Actions of other PACAP fragments and the mechanism of the inhibitory action of PACAP in the IAS smooth muscle remain to be investigated. Two main types of PACAP receptors have been described in the literature. For the PACAP I receptor, PACAP-38 and PACAP-27 each act preferentially with higher potency than that of VIP. PACAP II receptor on the other hand is characterized with equally high potency for PACAP-38, PACAP-27, and VIP. PACAP I receptor may be further subclassified into PACAP IA, where PACAP-38 and PACAP-27 show equally high potency, and PACAP IB, where PACAP-38 is 100 to 1,000 times more potent than PACAP-27. It is speculated that the PACAP receptor for the inhibitory actions of PACAP in the IAS is PACAP II and for the excitatory action, PACAP I. The activation of PACAP receptors for the IAS contraction, on the other hand, may involve either PACAP IB or a yet another PACAP I receptor subtype, since PACAP-27 either causes no or only a small contraction of the IAS smooth muscle. The insensitivity of the inhibitory effect of PACAP to apamin (unlike that of VIP) may further suggest that the involvement of the PACAP receptor in the IAS smooth muscle is distinct from the VIP receptor. It is also possible that the comparison of the potencies of PACAP and VIP by examining their effects on the basal IAS tension may not provide exact characterization of PACAP and VIP receptors since the biphasic effect of PACAP in the IAS smooth muscle may undermine the actual potency of its inhibitory effect.

SUMMARY

In summary, PACAP exerts a biphasic effect (an initial contraction followed by a relaxation) in the IAS. The initial contractile effect with higher concentrations of PACAP was found to be mediated by the activation of PACAP receptor at the substance P–containing nerve terminals. The PACAP receptor(s) responsible for the inhibitory action of the neuropeptide is(are) hypothesized to be present in the IAS smooth muscle cells and on the myenteric nerve terminals. The exact nature and the role of PACAP and the PACAP receptors in the inhibitory neurotransmission, the relationship of PACAP receptors with substance P–containing neurons and IAS smooth muscle cells, and interactions with the NOS pathway and VIP remain to be determined.

REFERENCES

1. BIANCANI, P., J. H. WALSH & J. BEHAR. 1985. Vasoactive intestinal polypeptide: a neurotransmitter for relaxation of the rabbit internal anal sphincter. Gastroenterology **89**: 867–874.

2. CHAKDER, S. & S. RATTAN. 1993. The entire vasoactive intestinal polypeptide molecule is required for the activation of the vasoactive intestinal polypeptide receptor: Functional and binding studies on opossum internal anal sphincter smooth muscle. J. Pharmacol. Exp. Ther. **266**: 392–399.

3. CHRISTOFI, F. L. & J. D. WOOD. 1993. Effects of PACAP on morphologically identified myenteric neurons in guinea pig small bowel. Am. J. Physiol. Gastrointest. Liver Physiol. **264**: G414–G421.

4. GRIDER, J. R., S. KATSOULIS, W. E. SCHMIDT & J.-G. JIN. 1994. Regulation of the descending relaxation phase of intestinal peristalsis by PACAP. J. Auton. Nerv. Syst. **50**: 151–159.

5. JIN, J. -G., S. KATSOULIS, W. E. SCHMIDT & J. R. GRIDER. 1994. Inhibitory transmission in tenia coli mediated by distinct vasoactive intestinal peptide and apamin-sensitive pituitary adenylate cyclase activating peptide receptors. J. Pharmacol. Exp. Ther. **270**: 433–439.

6. KATSOULIS, S., A. CLEMENS, H. SCHWÖRER, W. CREUTZFELDT & W. E. SCHMIDT. 1993. Pituitary adenylate cyclase activating polypeptide (PACAP) is a potent relaxant of the rat ileum. Peptides **14**: 587–592.

7. KATSOULIS, S., A. CLEMENS, H. SCHWÖRER, W. CREUTZFELDT & W. E. SCHMIDT. 1993. PACAP is a stimulator of neurogenic contraction in guinea pig ileum. Am. J. Physiol. Gastrointest. Liver Physiol. **265**: G295–G302.

8. KATSOULIS, S., W. E. SCHMIDT, R. SCHWARZHOFF, U. R. FOLSCH, J.-G. JIN, J. R. GRIDER & G. M. MAKHLOUF. 1996. Inhibitory transmission in guinea pig stomach mediated by distinct receptors for pituitary adenylate cyclase-activating peptide. J. Pharmacol. Exp. Ther. **278**: 199–204.

9. MIYATA, A., L. JIANG, R. R. DAHL, C. KITADA, K. KUBO, M. FUJINO, N. MINAMINO & A. ARIMURA. 1990. Isolation of a neuropeptide corresponding to the N-terminal 27 residues of the pituitary adenylate cyclase activating polypeptide with 38 residues (PACAP 38). Biochem. Biophys. Res. Commun. **170**: 643–648.

10. MIZUMOTO, A., M. FUJIMURA, M. OHTAWA, S. UEKI, N. HAYASHI, Z. ITOH, M. FUJINO & A. ARIMURA. 1992. Pituitary adenylate cyclase activating polypeptide stimulates gallbladder motility in conscious dogs. Regul. Pept. **42**: 39–50.

11. MOUMMI, C. & S. RATTAN. 1988. Effect of methylene blue and N-ethylmaleimide on internal anal sphincter relaxation. Am. J. Physiol. **255**: G571–G578.

12. NURKO, S. & S. RATTAN. 1988. Role of vasoactive intestinal polypeptide in the internal anal sphincter relaxation of the opossum. J. Clin. Invest. **81**: 1146–1153.

13. NY, L., B. Larsson, P. ALM, P. EKSTRÖM, J. FAHRENKRUG, J. HANNIBAL & K. E. ANDERSSON. 1995. Distribution and effects of pituitary adenylate cyclase activating peptide in cat and human lower oesophageal sphincter. Br. J. Pharmacol. **116**: 2873–2880.

14. PARKMAN, H. P., A. P. PAGANO & J. P. RYAN. 1997. Dual effects of PACAP on guinea pig gallbladder muscle via PACAP-preferring and VIP/PACAP-preferring receptors. Am. J. Physiol. Gastrointest. Liver Physiol. **272**: G1433–G1438.

15. PORTBURY, A. L., K. MCCONALOGUE, J. B. FURNESS & H. M. YOUNG. 1995. Distribution of pituitary adenylyl cyclase activating peptide (PACAP) immunoreactivity in neurons of the guinea-pig digestive tract and their projections in the ileum and colon. Cell Tissue Res. **279**: 385–392.

16. PRADHAN, T., Z. F. GU, R. T. JENSEN & P. N. MATON. 1991. Pituitary adenylate cyclase activating peptide (PACAP) increases cyclic AMP, but does not cause relaxation of gastric smooth muscle cells. Gastroenterology **100**: A483. (Abstract)

17. RATTAN, S. & S. CHAKDER. 1992. Role of nitric oxide as a mediator of internal anal sphincter relaxation. Am. J. Physiol. Gastrointest. Liver Physiol. **262**: G107–G112.

18. RAWLINGS, S. R. 1994. PACAP, PACAP receptors, and intracellular signalling. Mol. Cell. Endocrinol. **101**: C5–C9.

19. SHEN, Z., L. T. LARSSON, G. MALMFORS, A. ABSOOD, R. HÅKANSON & F. SUNDLER. 1992. A novel neuropeptide, pituitary adenylate cyclase-activating polypeptide (PACAP), in human intestine: Evidence for reduced content in Hirschsprung's disease. Cell Tissue Res. **269**: 369–374.

20. SUDA, K., D. M. SMITH, M. A. GHATEI & S. R. BLOOM. 1992. Investigation of the interaction of VIP binding sites with VIP and PACAP in human brain. Neurosci. Lett. **137**: 19–23.

21. UDDMAN, R., A. LUTS, A. ABSOOD, A. ARIMURA, M. EKELUND, H. DESAI, R. HÅKANSON, G. HAMBREAUS & F. SUNDLER. 1991. PACAP, a VIP-like peptide, in neurons of the esophagus. Regul. Pept. **36**: 415–422.

PACAP Inhibits Spontaneous Contractions in the Intestine of the Atlantic Cod, *Gadus morhua*

CATHARINA OLSSON AND SUSANNE HOLMGREN

Department of Zoophysiology, University of Göteborg, Medicinare gatan 18, S-413 90 Göteborg, Sweden

Immunohistochemical studies have revealed a high density of VIP and PACAP immunoreactive nerve fibers, in addition to intrinsic nerve cell bodies, throughout the gastrointestinal tract of the Atlantic cod, *Gadus morhua*. Double staining has shown that VIP and PACAP coexist to 100% in these nerve cells.[1] A subpopulation of the VIP/PACAP immunoreactive nerves is also NADPH-diaphorase reactive or NOS immunoreactive, indicating that they have the ability to synthesize nitric oxide.[2] In mammals, VIP, PACAP, and nitric oxide inhibit gastrointestinal motility and interact with each other in different ways depending on species and tissue examined.[3–5] In the cod, the effect of VIP varies between tissues. In perfusion experiments, VIP inhibits the spontaneous contractions of the stomach,[6] but has no effect on the motility of the intestine.[7] In these studies mammalian VIP, which differs in five amino acids from the cod sequence,[8] was used. No studies on motility have previously been performed with the endogenous cod VIP.

The L-arginine analogue L-NAME increases the mean tension in cod intestine by raising the resting tone and/or increasing the frequency of spontaneous contractions. L-NAME likewise abolishes the anal relaxation in response to electrical stimulation and augments oral contractions. This indicates that nitric oxide is involved in the descending inhibitory pathway in the cod intestine.[9]

The aim of this study was to examine the effect of PACAP and to compare it with the effects of mammalian and cod VIP on the intestine of the Atlantic cod, *Gadus morhua*. In addition, we wanted to see how PACAP/VIP interact with nitric oxide.

MATERIAL AND METHODS

Atlantic cod of either sex were captured off the Swedish west coast and kept in aerated circulating sea water at 10°C. The fish were killed by a sharp blow to the head and preparations were made from smooth muscles of the proximal part of the intestine. Ring preparations (3–5 mm wide) of circular muscles or strips (approx. 2 × 10 mm) of longitudinal muscles were mounted in organ baths containing cod Ringer's solution at 10°C bubbled with 0.3% CO_2 in air. The tension was recorded via a FT03 force transducer on a Grass polygraph. An initial tension of 10 mN was applied. During recovery the preparations developed spontaneous contractions, usually within 2 h.

The drugs were added as single doses or in cumulative concentrations. After wash-out the preparations were allowed to re-equilibrate until the spontaneous activity returned or for at least 1 h.

The results are presented as mean tension (± SEM) after treatment subtracted by the resting tension during the control period. Wilcoxon signed-ranks test was used for statistical evaluation of the results. Differences where $p < 0.05$ were regarded as statistically significant.

RESULTS AND DISCUSSION

In the ring preparations the rhythmic activity was reduced by both PACAP 27 and PACAP 38 while neither mammalian nor the endogenous VIP (10^{-10} – 10^{-6} M) had any significant effect on the contractions (FIG. 1). PACAP 27 seemed to be more potent than PACAP 38, causing a reduction in mean tension at 10^{-8} M and abolishing the contractions at 10^{-7} M, whereas some activity still remained after addition of 10^{-7} M of PACAP 38. Usually PACAP did not relax the preparations below the resting tone. Preincubation with the NOS inhibitors L-NAME or L-NOARG (3×10^{-4} M) for 15–25 min did not alter the response to PACAP (FIG. 2). However, after addition of L-NAME ($N = 11$) the mean tension increased to $183 \pm 25\%$, compared to the control, and to $162 \pm 20\%$ after addition of L-NOARG ($N = 5$). This indicates a nitrergic inhibitory tonus in the preparations.

The longitudinal preparations showed a similar response to PACAP as the ring preparations. PACAP 27 (10^{-7} M) reduced the mean tension from 1.58 ± 0.22 mN to 0.44 ± 0.25 mN ($N = 8$) and PACAP 38 (10^{-7} M) similarly reduced the tension from 1.61 ± 0.47 mN to 0.75 ± 0.28 mN ($N = 5$). L-NAME (3×10^{-4} M) increased the activity ($146 \pm 18\%$ compared to control; $N = 6$) but considering this, the decrease in mean tension in response to PACAP was not affected. Usually the longitudinal preparations had a higher resting tension than the ring preparations.

The results suggest that PACAP is an important contributor to the inhibitory control of the cod intestine while the role of VIP is still unclear. Since the NOS inhibitors did not affect the response to PACAP it is not likely that the peptide acts via stimulating the synthesis of nitric oxide but is probably acting directly on the smooth muscles.

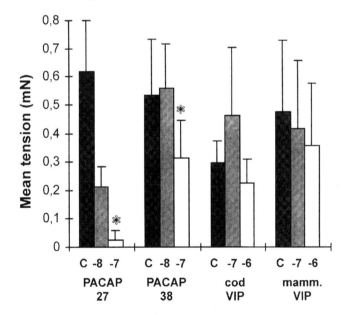

FIGURE 1. The effect of PACAP 27 (10^{-8} – 10^{-7} M; $N = 10$), PACAP 38 (10^{-8} – 10^{-7} M; $N = 7$), cod VIP (10^{-7} – 10^{-6} M; $N = 6$) and mammalian VIP (10^{-7} – 10^{-6} M; $N = 5$) on the circular smooth muscles of the proximal intestine of the Atlantic cod, *Gadus morhua*. PACAP (10^{-7} M) decreased the mean tension by reducing the spontaneous contractions while no significant effect was seen with either cod or mammalin VIP compared to the control periods (C) ($p \leq 0.05$).

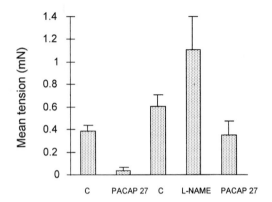

FIGURE 2. Preincubation of the circular muscle preparations with the NOS inhibitor L-NAME $(3 \times 10^{-4}$ M) caused an increase in the mean tension but did not affect the response to PACAP 27 $(10^{-7}$ M) compared to a prior exposure of the same preparations to PACAP $(p \leq 0.05)$.

REFERENCES

1. OLSSON, C. & S. HOLMGREN. 1994 . Distribution of PACAP (pituitary adenylate cyclase-activating polypeptide)-like and helospectin-like peptides in the teleost gut. Cell Tissue Res. **277**: 539–547.
2. OLSSON, C. & P. KARILA. 1995. Coexistence of NADPH-diaphorase and vasoactive intestinal polypeptide in the enteric nervous system of the Atlantic cod (*Gadus morhua*) and the spiny dogfish (*Squalus acanthias*) Cell Tissue Res. **280**: 297–305.
3. DANIEL, E. E., J. A. E. T. FOX-THRELKELD, Y. K. MAO, Y. F. WANG, F. CAYABYAB, M. JIMINEZ, P.VERGARA & C. MEMEH. 1994. Interactions of VIP (vasoactive intestinal polypeptide) and nitric oxide (NO) in mediating intestinal inhibition. Biomed. Res. **15**: 69–77.
4. MURTHY, K. S., J. R. GRIDER, J. G. JIN & M. MAKHLOUF. 1995. Interplay of VIP and nitric oxide in the regulation of neuromuscular activity in the gut Arch. I17
nt. Pharmacodyn. **329**: 27–38.
5. CHAKDER, S. & S. RATTAN. 1996. Evidence for VIP-Induced increase in NO production in myenteric neurons of opossum internal anal sphincter. Am. J. Physiol. **270**: G492–G497.
6. JENSEN, J., M. AXELSSON & S. HOLMGREN. 1991. Effects of substance P and vasoactive intestinal polypeptide on gastrointestinal blood flow in the Atlantic cod, *Gadus morhua*. J. Exp. Biol. **156**: 361–373.
7. JENSEN, J. & S. HOLMGREN. 1985. Neurotransmitters in the intestine of the Atlantic cod, *Gadus morhua*. Comp. Biochem. Physiol. **82C**: 81–89.
8. THWAITES, D. T., J. YOUNG, M. C. THORNDYKE & R. DIMALINE. 1988. The isolation and chemical characterization of a novel vasoactive intestinal peptide-related peptide from a teleost fish, the cod, *Gadus morhua*. Biochim. Biophys. Acta **999**: 217–220.
9. KARILA, P. & S. HOLMGREN. 1995. Enteric reflexes and nitric oxide in the fish intestine J. Exp. Biol. **198**: 2405–2411.

Effect of Sodium Depletion by Frusemide on Tissue Concentrations and Metabolism of VIP[a]

K. A. DUGGAN[b] AND V. Z. C. YE

Hypertension Laboratory, Liverpool Hospital, Sydney, Australia

Vasoactive intestinal peptide (VIP) is a 28 amino acid peptide of the secretin-glucagon group first isolated by Said and Mutt in 1970.[1] In addition to its role as a neurotransmitter, VIP appears to participate in sodium homeostasis as a humoral agent. It is natriuretic when infused in physiological concentrations by both the intrarenal and intravenous routes.[2] Further, acute and chronic sodium loading by the orogastric route affect both the secretion of VIP and the rate at which it is removed from the circulation. VIP is released from the upper gastrointestinal tract by acute gastric sodium ingestion.[3] In addition, this stimulus affects the degradation of VIP as both the hepatic and extrahepatic metabolic clearance rates have been demonstrated to decrease.[4,5] Chronic increases in sodium ingestion are also associated with changes in the rate of metabolism of VIP.[6] The metabolic clearance rate of VIP was lower after animals were maintained on a high sodium diet compared with the clearance rate after a similar period on a diet with a normal sodium content.[6]

Thus far studies that have suggested a role for VIP as a humoral agent in sodium homeostasis have investigated the effects of sodium loading either acutely or chronically. The effects of sodium depletion on the plasma concentration, the metabolic clearance rate, and the rate at which VIP is secreted into the circulation have not been documented. Such studies may provide additional evidence in support of a role for VIP in sodium homeostasis. Thus we sought to determine the effects of sodium depletion achieved by enhanced urinary sodium excretion and dietary sodium restriction on plasma concentration, secretion and metabolism of VIP. The effects of sodium depletion on the concentration of VIP in three tissues, the heart, lung, and kidney were also investigated. The heart was chosen as it secretes VIP, being responsible for diurnal variations in plasma concentration of VIP.[7] Concentrations of VIP in the lung and kidney were also measured as these are the major sites for removal of VIP from the circulation.[8]

METHODS

Male Sprague-Dawley rats, each weighing 250–300 g ($N = 8$, each group), were equilibrated on a normal (2.2%) sodium chloride diet and distilled drinking water or on a low (0.008%) sodium diet and frusemide 1 mg kg^{-1} day^{-1} in the drinking water. The low and normal sodium diets contained all essential nutrients and trace elements for adequate nutrition and were identical except for their sodium chloride content (Janos Chemicals, Forbes,

[a]This work supported by the National Health and Medical Research Council of Australia.
[b]Corresponding author: Professor K. A. Duggan, Renal Unit, Liverpool Hospital, P.O. Box 103 Liverpool, NSW 2170, Australia.

NSW). The rats were housed four to a box, in a room with a 12 hour light-dark cycle and allowed access to water and diet *ad libitum* for seven days. Urinary sodium excretions were determined from metabolic cage collections on days 6 and 7. These experiments complied with the Australian Code of Practice for the Care and Use of Animals for Scientific Purposes (1990) and were approved by the Animal Ethics Committee of the University of New South Wales.

The Effect of Sodium Depletion on Plasma and Tissue Concentrations of VIP

On the day of experiment the rats were anesthetized using gaseous anesthesia (halothane 2.5%, nitrous oxide 70%, and oxygen 30%) and truncal blood was sampled to determine plasma VIP concentrations. The blood was collected into pre-cooled syringes containing 100 KIU of trasylol (Bayer, Leverkusen) and 50 units of lithium heparin per ml. The heart, lungs, and kidneys were then removed, weighed, and immediately frozen by immersion in liquid nitrogen. Tissues were homogenized and VIP measured by radioimmunoassay as previously described.[10]

The Effect of Sodium Depletion on the Metabolism of VIP

On the day of experiment the rats were anesthetized as above and polyethylene cannulae inserted into the carotid artery and jugular vein. An infusion of hemaccell (Hoechst, Australia) was started at 0.017 ml min^{-1} through the jugular venous cannula. One hour later an infusion of VIP at 10 pmol kg^{-1} min^{-1} began. After 60 more minutes, arterial blood was sampled to determine the plasma VIP concentration.

The concentration of VIP was measured in plasma and infusate by radioimmunoassay. The metabolic clearance rate was calculated by the method of Tait and colleagues.[9]

RESULTS

Plasma and Tissue Concentrations of VIP

Sodium depletion by oral frusemide and a low sodium diet caused a significant decrease in the circulating concentration of VIP (6.0 ± 0.4 pmol l^{-1}) compared with the sodium-replete animals (10.1 ± 1.2 pmol l^{-1}, $p < 0.025$). A similar decrease in the concentration of VIP in the kidney and lung was observed in the group that received frusemide and the low sodium diet. Renal VIP concentration in this group was 64.7 ± 7.6 fmole/g tissue compared with 10.6 ± 12.7 fmole/g tissue in the rats maintained on the normal diet ($p < 0.01$). Pulmonary VIP decreased from 434.4 ± 64.1 fmol/g tissue to 78.4 ± 5.0 fmol/g tissue ($p < 0.005$). In contrast the concentration of VIP in the heart increased nonsignificantly in the sodium-depleted rats.

Metabolic Clearance Studies

The metabolic clearance rate was decreased in response to sodium depletion. Following the seven-day treatment protocol with frusemide and a low sodium diet, the clearance rate was 7.4 ± 0.8 ml min^{-1} 100g^{-1} compared with 11.2 ± 1.2 ml min^{-1} 100g^{-1} in the rats maintained on the normal diet ($p < 0.05$). The decrease in metabolic clearance rate was reflected in the plasma concentrations of VIP at the end of the 60-min infusion period. Plasma VIP was significantly higher in the sodium depleted (152.7 ± 15.6 pmol l⁻¹) than in the sodium-replete rats (65.5 ± 8.5 pmol l⁻¹; $p < 0.005$).

The theoretical or calculated secretion rate was also decreased in response to sodium depletion. In the sodium-replete rats secretion rate was 89.9 ± 11.9 pmol min⁻¹ 100g⁻¹, while in the rats receiving frusemide and a low sodium diet it was 38.3 ± 1.9 pmol min⁻¹ 100g⁻¹ ($p < 0.01$).

DISCUSSION

This study provides further support for a role for VIP as a humoral mediator in sodium homeostasis. As might be predicted for a humoral natriuretic agent, the plasma concentration of VIP was decreased after sodium depletion as a result of treatment with frusemide and a low sodium diet, compared with those rats fed a normal sodium diet.

This decrease in the plasma concentration appears to be due to a decrease in the rate at which VIP is secreted into the plasma. The theoretical or calculated secretion rate was decreased in the rats receiving frusemide and the low sodium diet when compared with those rats maintained on the diet with a normal sodium content. The non-significant increase in the myocardial concentration of VIP would also support this hypothesis. VIP is synthesized and secreted by the heart.[7] However, experiments in which injection of radioiodinated VIP (¹²⁵I-VIP) and ¹²⁵I were compared showed the same pattern of accumulation of radioactivity for both.[8] This suggests that the heart is not a significant site of VIP metabolism. Thus myocardial VIP concentration probably reflects the difference between the rate at which VIP is synthesized and at which it is secreted from the heart. An increase in the myocardial concentration of VIP would be consistent with decreased secretion and some continued synthesis.

The reduction in the rate of VIP metabolism as well as the decrease in the concentrations of VIP in metabolizing tissues, such as lung and kidney, would be in keeping with a downregulated system. A natriuretic system would be predicted to downregulate in the presence of sodium depletion, the decreased secretion of VIP and circulating concentration

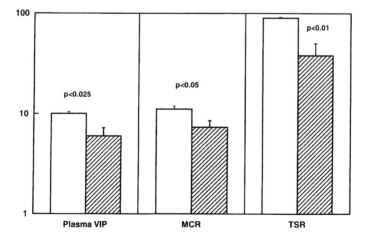

FIGURE 1. (*Left*) Plasma VIP concentration (pmol/l), (*middle*) metabolic clearance rate (MCR ml/min/100g), (*right*) theoretical secretion rate (TSR pmol/min/100g). Open bars represent rats treated with a normal sodium diet. Hatched bars represent rats treated with a low sodium diet on frusemide.

requiring a lower rate of clearance. The reductions in metabolic clearance rate and in lung and kidney concentrations support this hypothesis.

Thus we conclude that sodium depletion reduces the secretion and circulating concentration of VIP. This lends further support for a role for VIP in the maintenance of sodium balance.

REFERENCES

1. SAID, S. & V. MUTT. 1970. Polypeptide with a broad biological activity: isolation from the small intestine. Science **169**: 1217–1218.
2. DUGGAN, K. A. & G. J. MACDONALD. 1987. Vasoactive intestinal peptide: a direct renal natriuretic substance. Clin. Sci. **72**: 195–200.
3. DUGGAN, K. A., C. M. HAWLEY, G. J. MACDONALD & S. SHELLEY. 1989. Sodium depletion decreases hepatic clearance of vasoactive intestinal peptide in the rabbit. J. Physiol. **418**: 251–259.
4. HAWLEY, C. M., K. A., DUGGAN, G. J. MACDONALD & S. SHELLEY. 1991. Oral sodium regulates extrahepatic metabolism of vasoactive intestinal peptide. Clin. Sci. **81**: 79–83.
5. HAWLEY, C. M., K. A. DUGGAN, G. J. MACDONALD & S. SHELLEY. 1991. Acute but not chronic gastric sodium administration regulates vasoactive intestinal peptide metabolism by the liver. Acta Physiol. Scand. **143**: 295–298.
6. DAVIS, R. E., S. SHELLEY, G. J. MACDONALD & K. A. DUGGAN. 1992. The effects of a high sodium diet on the metabolism and secretion of vasoactive intestinal peptide in the rabbit. J. Physiol. **451**: 17–23.
7. CUGINI, P., P. LUCIA, G. SCIBILIA, L. DI PALMA, A. R. CIOLI, A. CIANETTI, L. GASBARRONE, R. CANOVA & B. MARINO. 1993. Lack of circadian rhythm of plasma concentrations of vasoactive intestinal peptide in patients with orthotopic heart transplants. Br. Heart J. **70**: 363–365.
8. HUMPHREY, C. S., P. MURRAY, A. M. EBEID & J. E. FISCHET. 1979. Hepatic and pulmonary clearance of exogenous vasoactive intestinal peptide in the rat. Gastroenterology **77**: 55–60.
9. TAIT, J. F., B. LITTLE, S. A. S TAIT & C. FLOOD. 1962. The metabolic clearance rate of aldosterone in pregnant and non-pregnant subjects estimated by both single injection and constant infusion methods. J. Clin. Invest. **41**: 2093–2100.
10. DUGGAN, K. A., D. M. JONES, V. Z. C. YE, R. E. DAVIS & G. J. MACDONALD. 1995. Effects of endopeptidase 24.11 inhibition on plasma and tissue concentrations of vasoactive intestinal peptide. Clin. Sci. **89**: 267–271.

Neuromodulator Role of VIP in Recovery of Rat Behavior and Brain Neurotransmitters Level after Frontal Lobectomy

M. KARGANOV,[a] G. ROMANOVA, W. BRASLAWSKY,
D. TARSHITZ, AND G. TELEGDY[b]

*Institute of General Pathology and Pathophysiology, Russian Academy
of Medical Sciences, Moscow, Russia*

[b]*Department of Pathophysiology, Albert Szent-Gyorgyi Medical
University, Szeged, Hungary*

The presence of vasoactive intestinal polypeptide (VIP) was demonstrated in the brain, with high concentrations found in cortical regions, and low to moderate concentrations found in the hypothalamus and hypophysis.[1,2] VIP plays a possible role as neurotransmitter or neuromodulator in the central nervous system (CNS) since it was detected in nerve terminals in subcortical regions and specific receptors for VIP were identified in cortical synaptosomes.[2] VIP may also affect behavioral activities after brain damage.[3] In order to prove this possible effect, the exploratory and conditional reflex behavior of normal and neofrontal decorticated rats was investigated after daily (from the next day after operation) intracerebroventricular (i.c.v.) injections of picomole quantities of VIP or VIP antiserum. The aims of this study were to examine such issues as: (1) the frontal lobectomy effects on VIP level in rat brain; (2) the effect of VIP and VIP antiserum effects on highly integrative activity and on neurotransmitters level in brain structures in normal and neofrontal decorticated rats; and (3) evaluation of the possible use of VIP and VIP antiserum intracerebroventricular injections to promote the restoration of disturbed functions of the CNS.

METHODS

Bilateral frontal lobectomy was performed in 35 male Wistar rats weighing 180–200 g, narcotized with pentobarbital (50 mg/kg, i.p.). After retraction of the dura mater, the frontal cortex, including precentral medial, precentral lateral, and prelimbic areas, was carefully removed bilaterally. Sham-operated animals were narcotized too; craniotomy without the brain damage took place. Implantation of stainless-steel cannula (d = 0.8 mm) into the lateral brain ventricle was carried out simultaneously with decortication. The cannula was placed 1.5 mm laterally and 1.0 mm caudally from the bregma and 4 mm deep into the brain. Animals with frontal lobectomy were divided into groups, which were given 2 µl intracerebroventricular injection by means of a microinjector of

[a]Corresponding author: Dr. M. Karganov, Institute of General Pathology and Pathophysiology, RAMS, 8, Baltiyskaya st. 125315 Moscow, Russia; Tel./Fax 095-155-4783

physiological saline, VIP, or VIP antiserum (Amersham, England) for each rat. The VIP injection dose was of 80 pg for each animal. VIP antiserum dose was sufficient for 5 picomoles peptide neutralization.

"Open Field" Test

Animals' behavior was studied under the "open field" conditions, to estimating additional parameters: number of crossed squares and vertical rearings. Each animal was tested for these parameters for 5 min in a rectangular "open field." For statistical analyses of the results, ANOVA was used, followed by Tukey's test.

Active Avoidance Behavior

The rats were trained in a shuttle box to study active avoidance reflex. The acquisition session was completed when criterion of learning (10 consecutive successful conditioned avoidance responses) was achieved. Saving coefficient was calculated according to the formula $(N1-N2) / N1 \times 100\%$, where N1 is the number of the trials necessary for the achievement of learning criterion during the acquisition, N2 is the same during the testing (the seventh preoperative day) or retesting (the ninth postoperative day).[4]

Radioimmunoassay

VIP content in cortex and hypothalamus of sham-operated rats and animals with frontal lobectomy on the ninth day after operation was determined by radioimmunoassay (RIA). RIA was performed with standard kits (Instar, USA). VIP content was calculated as pg/mg tissue weight. For statistical analyses Student's *t*-test was used.

Fluorimetry

The fluorimetric method for brain noradrenaline (NA), dopamine (DA), and serotonin (5-HT) simultaneous estimation was used. Standard procedure of measurement with some modifications was used. For statistical analyses, Student's *t*-test was used.

CONCLUSIONS

(1) RIA of VIP content in brain structures of rats revealed significant decrease of cortex peptide level (from 28.3 to 5.7 pg/mg) and its increase in hypothalamus (from 29.8 to 48.2 pg/mg) after frontal lobectomy with respect to sham-operated animals on the ninth day after operation.

(2) Daily intracerebroventricular VIP injection after frontal lobectomy caused a normalizing effect on the behavior of rats in the "open field" test: hyperactivity decreased in the decorticated animals. VIP antiserum injection resulted in a considerable decrease in the number of vertical rearings (FIG. 1).

(3) Increase of VIP brain content promoted faster and more complete recovery of active avoidance reflex, disturbed by frontal lobectomy (FIG. 1).

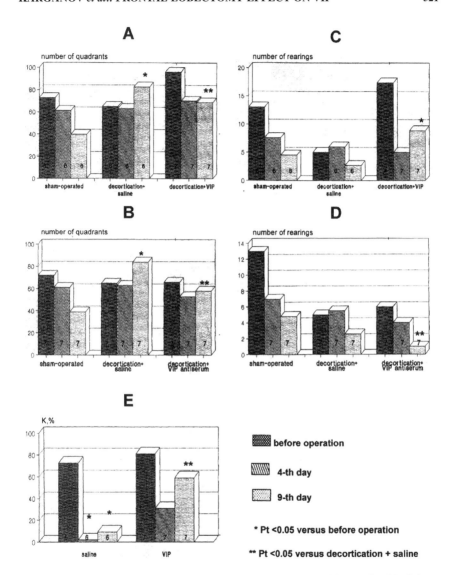

FIGURE 1. Changes of behavior in "open field" test (**A–D**) and active avoidance reflex (**E**) of sham-operated and decorticated (frontal lobectomy) rats before operation and on the fourth and ninth post-operative days. Saline, VIP, and VIP antiserum were injected intracerebroventricularly daily during the nine post-operative days.

(4) A modulatory VIP effect was more remarkable on the subcortical level and was expressed by normalizing of neurotransmitters balance in brain structures. VIP antiserum caused a normalizing effect in the cortex (FIG.2).

(5) The submitted experimental data of the VIP modulatory effect on neurotransmitters processes in brain and behavior during the early period after frontal lobectomy indicated a possible use for VIP in complex pathogenetic therapeutics of disturbed CNS functions.

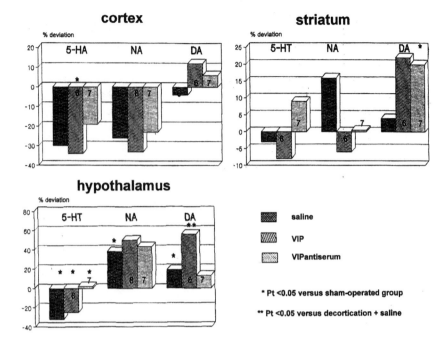

FIGURE 2. Changes of neurotransmitters (5-HT, NA, and DA) levels in rat brain structures after intracerebroventricular injections of saline, VIP, or VIP antiserum on the ninth day after sham-operation or frontal lobectomy.

REFERENCES

1. ASHMARIN, I. P. 1987. Hypothesis of the existence of a new higher category in the hierarchy of regulatory peptides. Neurochemistry **6**: 23–27.
2. BESSON, J. et al. 1979. Vasoactive intestinal peptide (VIP): brain distribution, subcellular localization and effect of deafferentation of the hypothalamus in male rats. Brain Res. **165**: 79–85.
3. KATSUURA, G. & S. ITOH. 1985. Behavioral effects of cholecystokinin and vasoactive intestinal peptide in neofrontal-decorticated rats. Ann. N.Y. Acad. Sci. **446**: 616–620.
4. FLEXNER, J. B. et al. 1963. Memory in mice as affected by intracerebral puramycin. Science **141**: 57–59.

Immunohistochemical Evidence for PACAP and VIP Interaction with Met-Enkephalin and CRF Containing Neurons in the Bed Nucleus of the Stria Terminalis

TAMÁS KOZICZ,[a] SANDOR VIGH, AND AKIRA ARIMURA

U.S.-Japan Biomedical Research Laboratories, Tulane University
Hebert Center, Belle Chasse, Louisiana 70037, USA

The bed nucleus of the stria terminalis (BST) appears to be an important relay nucleus in the modulatory control of the hypothalamic and brainstem centers responsible for the behavioral, endocrine, and physiological aspects of stress responses.[1-3] The BST exhibits immunoreactivity for numerous neuropeptides. Dense PACAP[9] and VIP[12,17] immunoreactive fiber terminals have been described in the lateral division of the bed nucleus of the stria terminalis (BSTL). The many Met-enkephalinergic and CRF immunoreactive perikarya found in the BST[8] appear to play an important role in determining the output of the BSTL. Since the concentrations of PACAP and VIP in the BSTL are the second highest in the rat brain, these axon terminals are likely to be involved in the modulation of the physiological responses of BSTL neurons. The origin of PACAP-ir fibers to the BST is not known. Many PACAP-ir neurons have been found in the paraventricular hypothalamic nucleus,[9] dorsal vagal nucleus,[11] and amygdaloid nuclei. Since all of these nuclei have anatomical connections with the BSTL, they are all possible sources of the PACAP-ir fiber terminals seen in the BSTL. The literature is inconsistent regarding the origin of the VIP-ir fiber terminals in the BSTL. Some authors have postulated that they originate in the amygdala,[14] but others have indicated that the VIP-ir perikarya in the dorsal and linear raphe nuclei are their source.[15] The aims of the present study were to investigate the possible interactions between the CRF-ir and Met-enkephalin-ir nervous structures and the VIP- and PACAP-ir axon terminals, as well as to find the origin of the PACAP-ir and VIP-ir fiber terminals in the BSTL.

MATERIALS AND METHODS

Injection of Cholera Toxin B Subunit

In order to trace the source of origin of the PACAP- and VIP-ir fiber terminals in the BSTL, cholera toxin B subunit, a retrograde tracer, was injected into the BSTL of rats. The survival time was 10–14 days. Two days before sacrificing the animals colchicine was injected as described below.

[a]Corresponding author: Tel.: (504) 394-7199; Fax: (504) 394-7169; E-mail: tkozicz@-tmcpop.tmc.tulane.edu

Colchicine Treatment

A stereotaxic instrument was used for injection of colchicine into the lateral cerebral ventricle. The animals, under Nembutal anesthesia (50 mg/kg, intraperitoneally), received 60–80 μg colchicine per animal. The survival time was 24–36 hours.

Fixation

The anesthetized animals were perfused transcardially with 50 ml of 0.1 M phosphate-buffered saline (PBS, pH 7.4), followed by 250 ml ice-cold 4% paraformaldehyde. The brains were kept in 4% paraformaldehyde solution for 4–24 h after fixation. For cryostat sectioning, following postfixation the brains were kept in a cryoprotective solution of 30% sucrose until they completely submerged.

Immunohistochemistry

Coronal sections of the forebrain were cut using a Vibratome at 30 μm. After four 15-min washes in PBS, the sections were immersed in a 0.1% Triton X-100 solution for 10 min to enhance antibody penetration. After an additional set of four 15-min washes in PBS, the sections were placed into a solution of 3% normal goat serum in PBS for 1 h. After a brief wash in PBS, the sections were transferred into vials containing the primary antisera. The primary antisera dilutions were as follows: aCRF 1:7,500 (the antibody was generated in our laboratory) and aMet-enkephalin 1:4,000 (Yanaihara Institute, Fujinomiya, Japan). The sections were incubated for 48 h in the primary antisera at 4°C, followed by four 15-min washes in PBS. The Vector ABC Elite Kit was used for visualization of the immunoreaction. To develop immunostaining, 10 mg 3-3'-diaminobenzidine (DAB, Sigma Chemical Co., St. Louis, MO) in 50 ml Tris buffer (pH 7.6) was used for approximately 10 min. After several washes in PBS, the sections were subjected to silver-gold intensification. The sections were washed in 10% thioglycolic acid (Sigma) for 1 h. After four 15-min washes in 2% sodium acetate solution the sections were placed in a physical developer. The reaction was stopped by placing the sections in a 1% acetic acid bath for 1 min. The sections were then washed in 2% sodium acetate for 10 min, followed by gold toning in 0.05% gold chloride in 2% acetic acid for 10 min at 4°C. After a 10-min wash in 3% Na-thiosulfate (Sigma) and rinsing in PBS, the sections were incubated in the second antibody for 48 h at 4°C. The dilutions for the second antibody were: aPACAP 1:8,000 and aVIP 1:50,000 (both antibodies were generated in our laboratory). The immunoreactivity of the second antibody was detected using the ABC technique with DAB chromogen as described above.

Double Immunofluorescence Labeling

Eight to 12 days after the injection of the cholera toxin B subunit, and 24 h after colchicine treatment, the animals were sacrificed and fixed as described above. After cryoprotection, 10-μm thick sections were cut using a cryostat. After four 15-min washes in PBS and 0.1% Triton X-100 solution, the sections were placed into a solution of 4–5% normal donkey serum in PBS for 1 h. The sections were incubated for 48 h at 4°C in either a mixture of anti–cholera toxin B subunit (List Laboratories-goat) at 1:600 and anti-PACAP (rabbit) at 1:500, or anti–cholera toxin B subunit at 1:600 and anti-VIP (rabbit) at 1:2,000. After incubation, the sections were washed for 10 min in 4–5% normal donkey serum followed by several washes in PBS. A secondary antiserum cocktail [fluorescein conjugated anti-goat IgG (1:80) and lissamine rhodamide conjugated anti-rabbit IgG (1:80)] was used for 2–4 h. Following several washes in PBS, the sections were mounted on glass slides and coverslipped with an antifade mounting media (Aqua Mount, Lerner Labs). An Olympus BH2 fluorescent microscope was used to study the sections. Color negative pictures (Kodak Color Gold, ASA 1000) were taken immediately. The negatives were digitally processed using Adobe Photoshop 4.0.

RESULTS AND DISCUSSION

The regions of the brain responsible for the central regulation of stress responses form the limbic-hypothalamic-pituitary-adrenal system. Some nuclei of the telencephalon (the septal area, BST, and amygdala), as well as some diencephalic nuclei (the hypothalamic paraventricular nucleus), and brain stem regions (the parabrachial nucleus, dorsal vagal nucleus, and monoaminergic cell groups) belong to this system.[2]

In the present study, numerous Met-enkephalin-ir neurons were labeled in the BSTL (FIG. 1,A). Several CRF-ir perikarya were also labeled in the BSTL, which contains a substantial proportion of the forebrain population of CRF-ir neurons.[7] Synaptic connections between PACAP-ir or VIP-ir axon terminals and CRF-ir dendrites and perikarya in the BSTL were demonstrated. At light microscopic level, juxtapositions were seen between Met-enkephalin-ir profiles and VIP-ir axon terminals (FIG. 1, A–E). This suggests the existence of synaptic connections between VIP-ir fiber terminals and Met-enkephalin-ir perikarya and dendrites. Using double-labeling fluorescense immunohistochemistry, the PACAP-ir fiber terminals in the BSTL were found to originate in the paraventricular hypothalamic nucleus (FIG. 2,A and B). However, neither the VIP-ir nor the PACAP-ir axons were found to originate in the amygdaloid nuclei. We have begun further studies to investigate various brainstem centers as possible sources of origin of PACAP-ir and VIP-ir fiber terminals in the BSTL. The CRF neurons of the BNSTL are known to project to the dorsal vagal nucleus,[7] parabrachial nucleus,[13] and midbrain central gray.[6] These CRF-ir afferents could play an important role in modifying cardiovascular changes and behavioral reflexes during stress. In addition, the PACAP-ir and VIP-ir afferents synapsing onto CRF-ir neurons could influence their control over autonomic changes during stress. Numerous studies have indicated that the hippocampus exercises inhibitory control over the HPA axis.[4,5,16] The numerous GABA-ergic neurons in the BSTL can turn the excitatory subicular projections into inhibitors in the hypothalamic nuclei.[1,3] Enkephalin and GABA very often colocalize in neurons,[18,19] and enkephalins inhibit GABA release.[10,18,20] Thus VIP-ir terminals synapsing onto Met-enkephalin-ir cells could modulate the activity of these neurons, influencing their inhibitory effect on GABA release.

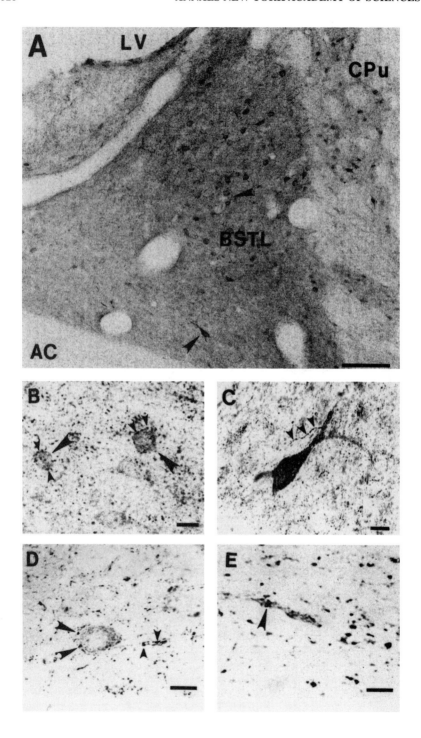

FIGURE 1. (**A**) Met-enkephalin-ir neurons and VIP-ir fiber terminals in the BSTL of the rat. Note the several VIP-ir terminals surrounding the Met-enkephalin-ir cells. The arrowheads point to cells enlarged in **B** and **C**. BSTL=bed nucleus of the stria terminals laterodorsal division; Cpu=caudate putamen; AC=anterior commissure; LV=lateral ventricle. Scale bar: 100 μm. (**B**) A high power picture from the BSTL. The Met-enkephalin-ir perikarya (*large arrowheads*) are surrounded by several VIP-ir axon terminals. The possible synaptic sites are marked by *small arrowheads*. Scale bar: 100 μm. (**C**) A high power micrograph of Met-enkephalin-ir neuron. A VIP-ir axon terminal runs toward the cell (*arrowheads*), and gives a synapse onto its main dendrite. Scale bar: 20 μm. (**D**) A semithin section showing a Met-enkephalin-ir neuron and VIP-ir fiber terminals. The *small arrowheads* point to possible axodendritic synaptic sites, whereas the *large arrowheads* point to possible axosomatic synaptic sites. Scale bar: 20 μm. (**E**) Met-enkephalin-ir dendritic process surrounded by VIP-ir axon terminals. The *arrowhead* points to a possible synaptic contact site. Scale bar: 20 μm.

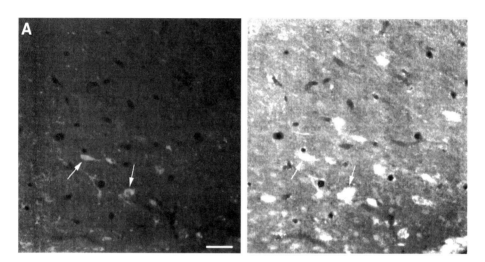

FIGURE 2. Double fluorescence staining in the PVN. (**A**) The FITC fluorescence labeling represents retrograde labeled, cholera toxin B subunit-ir neurons. (**B**) The rhodamide fluorescence indicates PACAP-ir neurons in PVN. *Arrows* point to double-labeled cells in **A** and **B**. Scale bar: 100 μm.

REFERENCES

1. BOUDABA, C., K. SZABO & J. G. TASKER. 1996. Physiological mapping of local inhibitory inputs to the hypothalamic paraventricular nucleus. J. Neurosci. **16**: 7151–7160.
2. CARPENTER, W. T. & P. H. GRUEN. 1978. The Limbic-Hypothalamic-Pituitary-Adrenal System and Human Behavior: 109–145. Plenum Press. New York.
3. CULLINAN, W. E., J. P. HERMAN & S. J. WATSON. 1993. Ventral subicular interaction with the hypothalamic paraventricular nucleus: evidence for a relay in the bed nucleus of the stria terminalis. J. Comp. Neurol. **332**: 1–20.
4. DUNN, J. D. 1987. Plasma corticosterone responses to electrical stimulation of the bed nucleus of the stria terminalis. Brain Res. **407**: 327–331.
5. ENDRÖCZI, E., K. LISSAK, B. BOHUS & S. KOVACS. 1959. The inhibitory influence of archicortical structures on pituitary-adrenal function. Acta Phisiol. Hung. **16**: 17–22.
6. GRAY, T. S. 1986. Autonomic neuropeptide connections of the amygdala. *In* Neuropeptides and Stress. Y. Tache, Ed. Springer-Verlag. New York.
7. GRAY, T. S. & D. J. MAGNUSON. 1987. Neuropeptide neuronal efferents from the bed nucleus of the stria terminalis and central amygdaloid nucleus to the dorsal vagal complex in the rat. J. Comp. Neurol. **262**: 365–374.

8. GRAY, T. S. & D. J. MAGNUSON. 1992. Peptide immunoreactive neurons in the amygdala and the bed nucleus of the stria terminalis project to the midbrain central gray in the rat. Peptides **13**: 451–460.

9. KOVES, K., A. ARIMURA, T. G. GORCS & A. SOMOGYVARI-VIGH. 1991. Comparative distribution of immunoreactive pituitary adenylate cyclase activating polypeptide and vasoactive intestinal polypeptide in rat forebrain. Neuroendocrinology **54**: 159–169.

10. LAM, D. M. K., Y. Y. T. SU & C. B. WATT. 1986. The self-regulating synapse: a functional role for the co-existence of neuroactive substances. Brain Res. Rev. **11**: 249–257.

11. LEGRADI, G., S. SHIODA & A. ARIMURA. 1994. Pituitary adenylate cyclase-activating polypeptide-like immunoreactivity in autonomic regulatory areas of the rat medulla oblongata. Neurosci. Lett. **176**: 193–196.

12. LOREN, I., P. C. EMSON, J. FAHRENKRUG, A. BJORKLUND, J. ALUMETS, R. HAKANSON & F. SUNDLER. 1979. Distribution of vasoactive intestinal polypeptide in the rat and mouse brain. Neuroscience **4**: 1953–1976.

13. MOGA, M. M. & T. S. GRAY. 1985. Evidence for corticotropin-releasing factor, neurotensin, and somatostatin in the neural pathway from the central nucleus of the amygdala to the parabrachial nucleus. J. Comp. Neurol. **241**: 275–284.

14. PALKOVITS, M., J. BESSON & W. ROTSZTEJN. 1981. Distribution of vasoactive intestinal polypeptide in intact, stria terminalis transected and cerebral cortex isolated rats. Brain Res. **213**: 455–459.

15. PETIT, J. M., P. H. LUPPI, C. PEYRON, C. RAMPON & M. JOUVET. 1995. VIP-like immunoreactive projections from the dorsal raphe and caudal linear raphe nuclei to the bed nucleus of the stria terminalis demonstrated by a double immunohistochemical method in the rat. Neurosci. Lett. **193**: 77–80.

16. SAPHIER, D. & S. FELDMAN. 1987. Effects of septal and hippocampal stimuli on paraventricular neurons. Neuroscience **20**: 740–755.

17. SIMS, K. B., D. L. HOFFMAN, S. I. SAID & E. A. ZIMMERMAN. 1980. Vasoactive intestinal polypeptide (VIP) in mouse and rat brain: an immunocytochemical study. Brain Res. **186**: 165–183.

18. WATT, C. B. & V. J. FLORACK. 1994. Interaction between enkephalin and GABA in the chicken retina: further analyses of coexisting relationships. Brain Res. **634**: 317–324.

19. WATT, C. B., P. A. GLAZEBROOK & D. M. K. LAM. 1986. Enkephalins in the teleost retina: localization and coexistence with gamma-aminobutyric acid. Invest. Ophthalmol. Vis. Sci. Suppl. 27: 231.

20. WATT, C. B., Y. Y. T. SU & D. M. K. LAM. 1984. Interaction between enkephalin and GABA in avian retina. Nature **311**: 761–763.

Distribution and Somatotopical Localization of Pituitary Adenylate Cyclase Activating Polypeptide (PACAP) in the Trigeminal Ganglion of Cats and Rats[a]

M. KAUSZ,[b,d] A. ARIMURA,[c] AND K. KÖVES[b]

[b]Department of Human Morphology and Developmental Biology,
Semmelweis University Medical School, Budapest, Hungary 10941

[c]US-Japan Biomedical Research Laboratories, Tulane University Hebert
Center, Belle Chasse; Departments of Anatomy and Physiology, Tulane
University Medical School, New Orleans, Louisiana 70112, USA

Pituitary adenylate cyclase activating polypeptide (PACAP) is a member of the secretin family.[1] The presence of PACAP in the primary sensory neurons of rats has been demonstrated.[2,3] In the present work we have investigated the somatotopical localization of PACAP immunoreactive (ir) neurons of the trigeminal ganglion, which project to various cutaneous and visceral areas of the head in cats and rats. Three days after the injection of tracers (wheat germ agglutinin [WGA], diamidino-yellow [DY], or fast blue [FB]) into various areas of the head, cats were perfused by 4% paraformaldehyde, but the rats received colchicine injection intracerebroventricularly and they were sacrificed two additional days later. Trigeminal ganglia were studied for tracers (WGA, DY, FB) and PACAP immunoreactivity (IR).

It was observed that a subpopulation of small- and medium-sized cells were PACAP ir. They were present throughout the ganglion in both rats and cats (FIG. 1) and represented about 7% of the total cell population in the ganglion. Some of PACAP ir neurons exhibited a partial colocalization with the tracers (WGA, DY, FB) after several tracer injections into various regions of the head.

After WGA or DY injection into the forehead, the PACAP ir cells were present throughout the ganglion, while the double labeled (PACAP and tracer) and only retrogradely labeled perikarya were located along the medial margin of the ganglion in both species (FIG. 2,a). FB injection into the vitreous body of the eyes resulted in retrogradely and double labeled perikarya in the same location.

After WGA injection into the area of whiskers the retrogradely labeled cells were evenly distributed all over the ganglion in rat and the majority of them were PACAP ir as

[a]This work was partially supported by OTKA grant T020403 (K. K. and M. K.) and by a National Institutes of Health grant DK-09094 (A. A.).

[d]Corresponding author: Maria Kausz, M.D., Department of Human Morphology and Developmental Biology, Semmelweis University, Medical School, Tuzoltó u. 58, Budapest, Hungary H-1094; Tel.: 36 1-215 6920; Fax: 36 1-215 3064; E-mail: Kausz@ana2.sote.hu

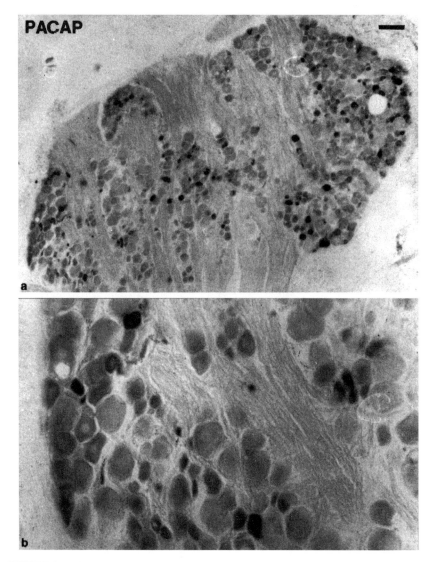

FIGURE 1. PACAP immunoreactive (ir) cells in the trigeminal ganglion of a cat. The microphotographs are derived from the lateral margin of the ganglion. Scale: **a**, 250 μm; **b**, 25 μm.

well. DY or WGA injections into the area of the whiskers in cats resulted in retrogradely
labeled cells only in the middle part of the ganglion and some of them were also PACAP
ir (FIG. 2,b and c).

DY or WGA injections into the skin of the chin resulted in retrogradely labeled cells in
the lateral part of the ganglion. Some of them were PACAP ir (FIG. 2,d and e). After the
WGA injection into the submucosa of the tongue or after the DY injection into the masseter
muscle, retrogradely and the double-labeled cells were found in the rostro-lateral part of the
ganglion both in cats and rats.

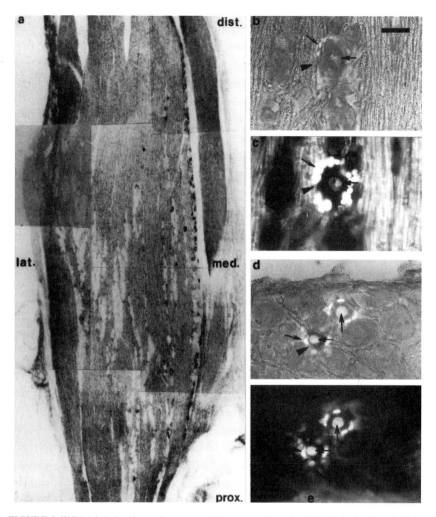

FIGURE 2. WGA-labeled cells are demonstrated in a rat ganglion after WGA injection into the forehead. WGA-labeled cells were located in the medial margin of the ganglion (**a**). Double-labeled cells are demonstrated in rats after DY injection into the area of the whiskers (**b** and **c**) and into the chin (**d** and **e**). The *arrowheads* indicate PACAP immunreactivity (IR) and *arrows* indicate DY in the double-labeled neurons. The nuclei of the DY-labeled neurons glow. The cytoplasm of the double-labeled neurons is dark due to the PACAP reaction product. The small glial cells around the neurons may have been labeled by leakage of DY from the labeled neurons. Scale: **a**, 500 μm; **b–e**: 25 μm.

CONCLUSION

We have concluded from the data presented above that: (1) PACAP ir cells are present in the trigeminal ganglion in cats and rats. Most of the cells are small or middle sized. They are easily stained in intact cats; however, we had to use colchicine treatment in rats for demonstration of PACAP immunoreactive cells.[4] (2) PACAP immunopositive perikarya exhibited a partial colocalization with the retrograde tracers (WGA; DY or FB) in the ganglion revealed by the combination of PACAP immunohistochemistry and retrograde tracing technique. Namely, some of PACAP immunoreactive cells participate in the innervation of various regions of the head provided by the trigeminal nerve. (3) PACAP immunoreactive cells of the medio-dorsal portion of the ganglion innervate the area of the forehead and the eyes. The cells of the middle portion of the ganglion innervate the area of the whiskers and the cells of the lateral portion of the ganglion supply the skin of the chin, the mucosa of the tongue, and the muscles of mastication.

It is known from the literature that the small- and middle-sized sensory neurons contain neuropeptides and these cells were also immunolabeled for PACAP.[5] According to our study, the PACAP ir perikarya represent about 7% of the total cell population not only in the trigeminal but also in the spinal ganglia.

The results available in the literature[6] had shown that PACAP was involved in nociceptive mechanisms modulating neural transmission.

REFERENCES

1. MIYATA, A., A. ARIMURA, R. R. DAHL, N. MINAMINO, A. UEHARA, P. E. GOTTSCHALL, L. JIANG, M. D. CULLER & D. H. COY. 1989. Isolation of a novel 38 residue-hypothalamic polypeptide which stimulates adenylate cyclase in the pituitary cells. Biophys. Biochem. Res. Commun. **164**: 567–574.

2. SUNDLER, F., K. MOLLER, H. MULDER, Y. ZHANG, A. LUTS J. ALUMETS, R. HAKANSON, B. SJÖLUND & R. UDDMAN. 1994. Pituitary adenylate cyclase activating peptide is a sensory neuropeptide and suppresses spinal nociceptive reflexes. *In* International Symposium on Vasoactive Intestinal Peptide, Pituitary Adenylate Cyclase Activating Polypeptide & Related Regulatory Peptides. G. Rosselin, Ed.: 304–309. World Scientific. Singapore.

3. LÉGRADI, G., S. SHIODA & A. ARIMURA. 1994. Pituitary adenylate cyclase activating polypeptide-like immunoreactivity in autonomic regulatory areas of the rat medulla oblongata. Neurosci. Lett. **176**: 193–196.

4. KÖVES, K., T. J. GÖRCS, M. KAUSZ & A. ARIMURA. 1994. Present status of knowledge about the distribution and colocalization of PACAP in the forebrain. Acta Biol. **45**: 297–321.

5. MULDER, H., R. UDDMAN, K. MOLLER, Y. Z. ZHANG, E. EKBLAD, J. ALUMETS & F. SUNDLER. 1994. Pituitary adenylate cyclase activating polypeptide expression in sensory neurons. Neuroscience **63**: 307–312.

6. ZHANG, Y. Z., B. SJOLUND, K. MOLLER, R. HAKANSON & F. SUNDLER. 1996. Pituitary adenylate cyclase activating peptide produces a marked and long-lasting depression of a C-fibre-evoked flexion reflex. Neuroscience **57**: 733–737.

Prenatal Expression of Pituitary Adenylate Cyclase Activating Polypeptide (PACAP) in Autonomic and Sensory Ganglia and Spinal Cord of Rat Embryos[a]

H.S. NIELSEN, J. HANNIBAL, AND J. FAHRENKRUG[b]

Department of Clinical Biochemistry, Bispebjerg Hospital, University of Copenhagen, DK-2400 Copenhagen NV, Denmark

Pituitary adenylate cyclase activating polypeptide (PACAP), a member of the VIP/secretin/glucagon family of regulatory peptides,[1] is expressed in the central nervous system,[2–5] as well as in cell bodies of autonomic and sensory ganglia,[6-8] and in neural elements of several peripheral organs.[9,10]

PACAP may be important in neural development since PACAP stimulates proliferation and differentiation of cultured neuroblasts,[11] promotes survival of cerebellar granule cells *in vivo*,[12] stimulates dendritic outgrowth of PC-12 cells,[13,14] and functions as growth and survival factor during brain development.[15] Although these data indicate a role for PACAP during development, it remains to be established that PACAP is expressed in early embryonic life.

In this study, using immunohistochemistry and *in situ* hybridization histochemistry, PACAP expression was investigated in autonomic and sensory ganglia and in spinal cord of rat embryos.

MATERIAL AND METHODS

Fetuses from pregnant Wistar rats were used. The study period lasted from prenatal day 12 to day 21 (E12–E21).

Immunocytochemistry

PACAP immunostaining was performed by a specific monoclonal PACAP antibody (code Mab JHH1). Immunoreaction was visualized with the DAB-chromogen combined with biotinyl tyramide (DuPont, Boston, MA; TSA-indirect kit, cat no. NEL700) according to the principles described by Berghorn and colleagues.[16]

[a]This study was supported by the Danish Biotechnology Center for Cellular Communication.
[b]Corresponding author: Tel.: 45-35312640; Fax: 45-35312099; E-mail: bbhjanf@inet.uni2.dk.

In Situ *Hybridization Histochemistry*

PACAP mRNA was detected by a ^{33}P-labeled riboprobe containing the whole PACAP cDNA sequence. Cryostat sections were processed for *in situ* hybridization histochemistry according to the protocol described previously.[17]

RESULTS AND CONCLUSION

A general summary of the occurrence and distribution of PACAP-immunoreactive neurons in sensory and autonomic ganglia and spinal cord is given in TABLE 1.

Weakly stained PACAP-immunoreactive nerve fibers were visible in the spinal cord as early as E13. At E14, PACAP-positive nerve fibers occurred in the sympathetic trunk, and a marked increase in the number of these fibers was seen at E15 (FIG. 1, a and c). At the same embryonic age, the PACAP-immunoreactive nerve fibers in the spinal cord could with certainty be identified to an area close to the intermediolateral column, in which PACAP-immunoreactive nerve cell bodies were demonstrated for the first time (FIG. 1, a and d). PACAP immunoreactivity was present in the parasympathetic ganglia of the cranial nerves at E16. At this time, PACAP-positive sympathetic nerve fibers were innervating the adrenal medulla, and PACAP immunoreactivity was demonstrated in cell bodies of sensory ganglia, such as the trigeminal ganglia and dorsal root ganglia (FIG. 1, b). The distribution of PACAP-immunoreactive cell bodies was confirmed by *in situ* hybridization histochemistry.

CONCLUSION

In conclusion, PACAP appears in the autonomic nervous system in embryonic development, and the earliest PACAP gene expression was found to be at E13. No transient expression of PACAP is observed in any of the areas examined and the distribution pattern

TABLE 1. Occurrence of PACAP immunoreactivity in neurons of rat embryos at day E12–E21

	E12	E13	E14	E15	E16	E17	E18	E19	E20	E21
Sensory ganglia										
Dorsal root ganglia	–	–	–	–	+	+	+++	+++	nd	+++
Trigeminal ganglion	–	–	–	–	+	+	++	+++	+++	+++
Parasympathetic ganglia										
Sphenopalatine ganglion	–	–	–	–	+	+++	+++	+++	+++	++++
Otic ganglion	–	–	–	–	–	+	++	+++	+++	+++
Submandibular ganglion	–	–	–	–	+	+	++	++	++	++
Ciliary ganglion	–	–	–	–	–	–	–	–	–	–
Sympathetic ganglia										
Sympathetic trunk	–	–	–	+	+	+	+	+	+	+
Adrenal gland	–	–	–	–	–	–	–	–	–	–
Superior cervical ganglion	nd	nd	nd	nd	–	–	+	+	nd	+
Mixed ganglion										
Nodose ganglion	nd	nd	nd	nd	+	nd	++	++	++	++
Spinal cord	–	–	–	+	+	++	+++	+++	nd	+++

Number of immunohistochemically stained nerve cells is estimated as: – none, + few, ++ some, +++ many, ++++ numerous. (nd = not done)

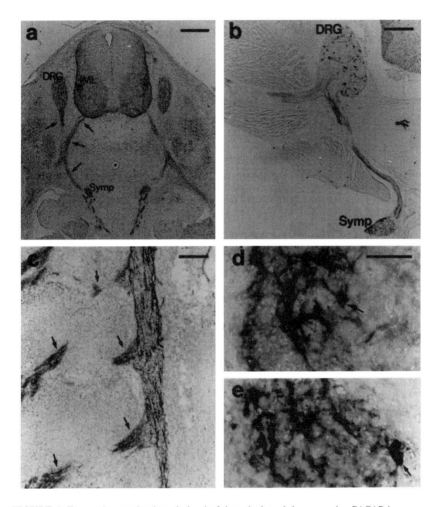

FIGURE 1. Transections at the thoracic level of the spinal cord demonstrating PACAP immunostaining in the intermediolateral column (IML) of the spinal cord at E15 (**a,d**), the dorsal root ganglia (DRG) at E15 (**a**) and E21 (**b**), and in sagittal and transversal sections of the sympathetic trunk (Symp) at E16 (**c**) and E15 (**e**), respectively. (**a**) At E15, few moderately to intensely stained PACAP-immunoreactive cell bodies are detected in the intermediolateral column. (**d**) High magnification of the same section, illustrating the PACAP-positive cell bodies (*arrows*) in the intermediolateral column. (**a**) From the lateral horn of the spinal cord and the dorsal root ganglia, PACAP-immunoreactive nerve fibers (*arrows*) seem to project towards the sympathetic trunk. (**b**) At day E21, these PACAP-positive nerve fibers are markedly increased in number and staining intensity. (**c**) PACAP-immunoreactive nerve fibers innervate the sympathetic trunk either as single or bundles of fibers, some projecting towards the dorsal root ganglia (*arrows*). (**e**) A single PACAP-immunoreactive cell body is occasionally observed in the sympathetic trunk (*arrow*). Scale bars: **a,b** =200 μm, **c** = 100 μm, **d,e** = 25 μm.

is similar to that described in the adult rat. The early expression of PACAP suggests that the peptide may have a role during the prenatal period.

REFERENCES

1. MIYATA, A., A. ARIMURA, R. R. DAHL et al. 1989. Isolation of a novel 38 residue-hypothalamic polypeptide which stimulates adenylate cyclase in pituitary cells. Biochem. Biophys. Res. Commun. **164**: 567–574.
2. ARIMURA, A., A. SOMOGYVÁRI-VIGH, A. MIYATA, K. MIZUNO, D. H. COY & C. KITADA. 1991. Tissue distribution of PACAP as determined by RIA: highly abundant in the rat brain and testes. Endocrinology **129**: 2787–2789.
3. GHATEI, M. A., K. TAKAHASHI, Y. SUZUKI, J. GARDINER, P. M. JONES & S. R. BLOOM. 1993. Distribution, molecular characterization of pituitary adenylate cyclase-activating polypeptide and its precursor encoding messenger RNA in human and rat tissues. J. Endocrinol. **136**: 159–166.
4. HANNIBAL, J., J. D. MIKKELSEN, H. CLAUSEN, J. J. HOLST, B. S. WULFF & J. FAHRENKRUG. 1995. Gene expression of pituitary adenylate cyclase activating polypeptide (PACAP) in the rat hypothalamus. Regul. Pept. **55**: 133–148.
5. HANNIBAL, J., J. M. DING, D. CHEN et al. 1997. Pituitary adenylate cyclase activating peptide (PACAP) in the retinohypothalamic tract. A daytime regulator of the biological clock. J. Neurosci. **17**: 2637–2644.
6. MOLLER, K., Y. Z. ZHANG, R. HÅKANSON et al. 1993. Pituitary adenylate cyclase activating peptide is a sensory neuropeptide—immunocytochemical and immunochemical evidence. Neuroscience **57**: 725–732.
7. MULDER, H., R. UDDMAN, K. MOLLER et al. 1994. Pituitary adenylate cyclase activating polypeptide expression in sensory neurons. Neuroscience **63**: 307–312.
8. MULDER, H., R. UDDMAN, K. MOLLER et al. 1995. Pituitary adenylate cyclase activating polypeptide is expressed in autonomic neurons. Regul. Pept. **59**: 121–128.
9. ARIMURA, A. 1992. Pituitary adenylate cyclase activating polypeptide (PACAP): discovery and current status of research. Regul. Pept. **37**: 287–303.
10. ARIMURA, A. & S. SHIODA. 1995. Pituitary adenylate cyclase activating polypeptide (PACAP) and its receptors: neuroendocrine and endocrine interaction. Front. Neuroendocrinol. **16**: 53–88.
11. DICICCO-BLOOM, E. & P. J DEUTSCH. 1992. Pituitary adenylate cyclase activating polypeptide (PACAP) potently stimulates mitosis, neuritogenesis and survival in cultured rat sympathetic neuroblasts. Regul. Pept. **37**: 319.
12. VILLALBA, M., J. BOCKAERT & L. JOURNOT. 1997. Pituitary adenylate cyclase-activating polypeptide (PACAP-38) protects cerebellar granule neurons from apoptosis by activating the mitogen-activated protein kinase (MAP kinase) pathway. J. Neurosci. **17**: 83–90.
13. DEUTSCH, P. J. & Y. SUN. 1992. The 38-amino acid form of pituitary adenylate cyclase-activating polypeptide stimulates dual signaling cascades in PC12 cells and promotes neurite outgrowth. J. Biol. Chem. **267**: 5108–5113.
14. HERNANDEZ, A., B. KIMBALL, G. ROMANCHUK & M. W. MULHOLLAND. 1995. Pituitary adenylate cyclase-activating peptide stimulates neurite growth in PC12 cells. Peptides **16**: 927–932.
15. ARIMURA, A., A. SOMOGYVÁRI-VIGH, C. WEILL et al. 1994. PACAP functions as a neurotrophic factor. Ann. N.Y. Acad. Sci. **739**: 228–243.
16. BERGHORN, K. A., J. H. BONNETT & G. E. HOFFMAN. 1994. cFos immunoreactivity is enhanced with biotin amplification. J. Histochem. Cytochem. **42**: 1635–1642.
17. LARSEN, J. O., J. HANNIBAL, S. M. KNUDSEN & J. FAHRENKRUG. 1997. Expression of pituitary adenylate cyclase-activating polypeptide (PACAP) in the mesencephalic trigeminal nucleus of the rat after transection of the masseteric nerve. Mol. Brain Res. **46**: 109–117.

VIP and NPY Expression during Differentiation of Cholinergic and Noradrenergic Sympathetic Neurons[a]

B. SCHÜTZ,[b,d] M.K.-H. SCHÄFER,[b] L. E. EIDEN,[c] AND E. WEIHE[b]

[b]Institute of Anatomy & Cell Biology, Philipps University, Robert-Koch-Str. 6, D-35033 Marburg, Germany

[c]Section on Molecular Neuroscience, Laboratory of Cellular and Molecular Regulation, National Institute of Mental Health, National Institutes of Health, Bethesda, Maryland 20892, USA

The neuropeptides vasoactive intestinal peptide (VIP) and neuropeptide Y (NPY) are often co-stored in mature cholinergic or noradrenergic neurons, respectively. We have re-examined the sympathetic ontogeny of VIP and NPY[1] to extend the recent hypothesis that cholinergic chemical coding in the sympathetic nervous system occurs prenatally in a target-independent fashion.[2,3,4] For this purpose the expression of these neuropeptides in neurons of developing sympathetic ganglia and in fibers innervating forepaw sweat glands was compared with new markers for noradrenergic (vesicular monoamine transporter type 2, VMAT2) and cholinergic (vesicular acetylcholine transporter, VAChT) neurons.

MATERIALS AND METHODS

Tissue preparation, fixation, and immunohistochemical analysis were performed as described previously.[2] The antibodies used in this study were obtained from the following sources: anti-VIP (rabbit polyclonal; Dr. Yanaihara, Shizuoka, Japan), anti-NPY (sheep polyclonal; Auspep, Parkville, Australia), and anti-tyrosine hydroxylase (TH) (mouse monoclonal; Boehringer Mannheim, Germany). Antibodies against VMAT2 and VAChT have been described earlier.[5,6]

RESULTS

NPY immunoreactivity (IR) was first detected in neurons of the primary sympathetic chain at E12.5 (FIG. 1,A). Also at this day, noradrenergic VMAT2-IR neurons are present (FIG. 1,B) and innervated by preganglionic cholinergic, VAChT-positive fibers (FIG. 1,C). At E12.5, VIP staining is not visible in neurons or fibers (data not shown). VIP-IR (FIG. 1,D) and VAChT-IR (FIG. 1,F) co-appear at E14.5 in a subpopulation of neurons in the stellate

[a]This work supported in part by Volkswagen-Stiftung, German Research Foundation, Kempkes-Stiftung, and the National Institute of Mental health Intramural Research Program

[d]Corresponding author: Dr. Burkhard Schütz, Institute of Anatomy and Cell Biology, Robert-Koch-Str. 6 D-35033 Marburg; Tel.: 49-6421-28-4030; Fax: 49-6421-28-8965; E-mail: schuetzb@mailer.uni-marburg.de

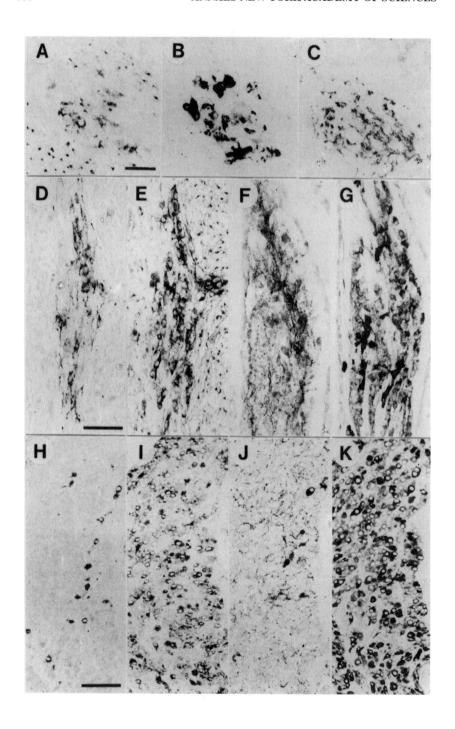

FIGURE 1. (**A–C**) Immunocytochemistry (ICC) for NPY (**A**), VMAT2 (**B**) and VAChT (**C**) on adjacent sections in the primary sympathetic chain at E12.5. (**D–G**) ICC for VIP (**D**), NPY (**E**), VAChT (**F**) and VMAT2 (**G**) in the paravertebral sympathetic chain at E14.5. (**D,E**) and (**F,G**) are pairs of adjacent sections. (**H–K**) ICC for VIP (**H**), NPY (**I**), VAChT (**J**), and VMAT2 (**K**) on adjacent sections in the stellate ganglion at P2. Note that VMAT2 is expressed with different levels in sympathetic neurons at E14.5 (**G**) and P2 (**K**). The NPY antiserum shows dotted nonspecific staining in the surrounding of the ganglion in (**A,E**). The scale bar in **A** = 25 μm and applies to panels **A–C**; scale bar in **D** = 50 μm and applies for panels **D–G**; scale bar in **H** = 100 μm and applies to panels **H–K**.

and paravertebral thoracolumbar ganglia. In contrast, staining for VMAT2 (Fig. 1,G) and NPY (Fig. 1,E) is strong in nearly all cells. During the next several days of gestation and into early postnatal life, VIP-IR (Fig. 1,H) and VAChT-IR (Fig. 1,J) remain restricted to a minority of principal ganglion cells of the stellate and paravertebral ganglia, whereas NPY (Fig. 1,I) and VMAT2 (Fig. 1,K) are expressed in the majority of postganglionic neurons.

In the sweat gland fields of the forepaw, which are a peripheral target of cholinergic sympathetic innervation, VIP-IR (Fig. 2,A) and VAChT-IR (Fig. 2,B) co-appear in sudomotor fibers around sweat gland coils at around P4. NPY-IR (Fig. 2,C) and VMAT2-IR (Fig. 2,D) are co-restricted to varicose fibers of blood vessels and are never detected in sudomotor fibers around sweat gland epithelia at any time of development. During the second postnatal week, the VIP (Fig. 2,E) and VAChT (Fig. 2,F) fiber plexus become more dense and reach the adult pattern of innervation at around P21 (data not shown). The sympathetic cholinergic innervation of the sweat glands (Fig. 2,H) is accompanied by a partial noradrenergic phenotype into the second postnatal week, which is characterized by tyrosine hydroxylase (TH)-IR (Fig. 2,I), but no VMAT2-IR (Fig. 2,J).

CONCLUSIONS

This study demonstrates the parallel onset of coexpression of NPY/VMAT2 and of VIP/VAChT in the developing paravertebral sympathetic chain with a transient partial overlap of these two paired phenotypes during later prenatal stages and the mutual segregation of the two paired phenotypes from early postnatal days into adulthood. This suggests that sympathetic neuropeptide/amine cophenotypes are differentially regulated by developmentally specific intrinsic or extrinsic signals. The detection of VIP and VAChT at E14.5 shows that cholinergic differentiation in the sympathetic nervous system of the rat occurs very early during development, long before fibers appear around cholinergic targets, e.g., sweat glands. Thus, cholinergic and associated peptidergic (VIP) differentiation in the sympathetic nervous system appears to be target independent. This view is supported by the observation that the acquisition of the cholinergic innervation of the sweat glands is not obliterated in TH or DBH (dopamine-ß-hydroxylase) knockout mice.[7] The detection of TH-IR around sweat gland coils between P4 and P11 in conjunction with absence of VMAT2 suggests that sympathetic cholinergic neurons retain a partial noradrenergic phenotype into the second postnatal week, which is, however, not functional in the sense of regulated vesicular noradrenaline transport.

ACKNOWLEDGMENTS

The authors thank Dr. Yanaihara for the generous supply of the anti-VIP antiserum and C. Brett for the supply of paraffin sections of rat embryos.

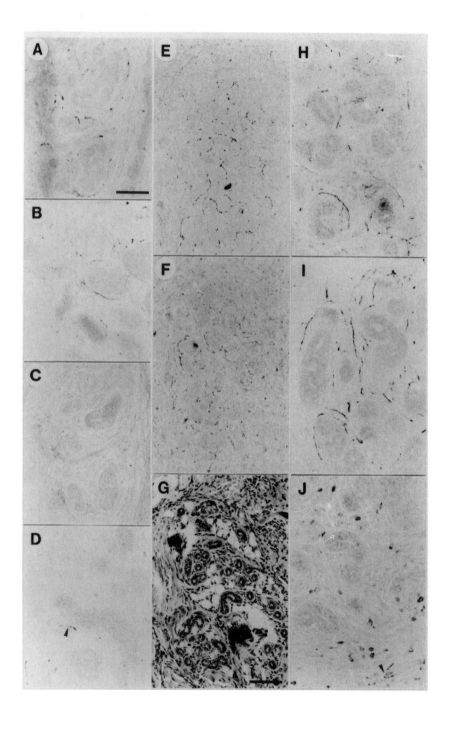

FIGURE 2. (A–D) ICC for VIP (**A**), VAChT (**B**), NPY (**C**), and VMAT2 (**D**) in the sweat gland fields at P4. The *arrowhead* in **D** points toward VMAT2-positive nerve terminals targeting a blood vessel, running between sweat gland coils. (**E,F**) ICC for VIP (**E**) and VAChT (**F**) innervation around sweat gland epithelium at P14. The Giemsa-stained section in **G** is adjacent to panels **E** and **F** and shows the corresponding sweat gland field. (**H–J**) ICC for VAChT (**H**), TH (**I**) and VMAT2 (**J**) innervation around sweat gland epithelium at P8. The *arrowhead* in **J** points toward VMAT2-positive perivascular nerve terminals, adjacent to the sweat gland field. Note also staining of putative mast cells in **J**. The scale bar in **A** = 50 μm and applies for panels **A–D** and **H–J**; the scale bar in **G** = 100 μm and applies for panels **E–G**. (Panels **B**, **H**, **J** reprinted from Schäfer *et al.* with permission.)

REFERENCES

1. TYRRELL, S. & S. C. LANDIS. 1994. The appearance of NPY and VIP in sympathetic neuroblasts and subsequent alterations in their expression. J. Neurosci. **14**: 4529–4547.

2. SCHÄFER, M. K.-H. *et al.* 1997. Target-independent cholinergic differentiation in the rat sympathetic nervous system. Proc. Natl. Acad. Sci. USA **94**: 4149-4154.

3. SCHÜTZ, B. *et al.* 1998. Vesicular amine transporter expression and isoform selection in developing brain, peripheral nervous system and gut. Dev. Brain Res. **106**: 181–204.

4. SCHÜTZ, B. *et al.* 1998. Ontogeny of vesicular amine transporter expression in the rat: new perspectives on aminergic neuronal and neuroendocrine differentiation. Adv. Pharmacology **42**: 903–908.

5. WEIHE, E. *et al.* 1994. Localization of vesicular monoamine transporter isoforms (VMAT1 and VMAT2) to endocrine cells and neurons in rat. J. Mol. Neurosci. **5**: 149–164.

6. WEIHE, E. *et al.* 1996. Visualization of the vesicular acetylcholine transporter in cholinergic nerve terminals and its targeting to a specific population of small synaptic vesicles. Proc. Natl. Acad. Sci. USA **93**: 3547–3552.

7. TAFARI, A. T. *et al.* 1997. Norepinephrine facilitates the development of the murine sweat response but is not essential. J. Neurosci. **17**: 4275–4281.

PACAP in Visceral Afferent Nerves Supplying the Rat Digestive and Urinary Tracts[a]

JAN FAHRENKRUG[b] AND JENS HANNIBAL

Department of Clinical Biochemistry, Bispebjerg Hospital, University of Copenhagen, DK-2400 Copenhagen NV, Denmark

PACAP is a new member of the vasoactive intestinal polypeptide (VIP) family of peptides[1] present in the brain[2–7] as well as in neuronal elements of a number of peripheral organs.[8–12] PACAP is expressed in nerve cell bodies of the dorsal root ganglia,[13–14] trigeminal ganglion,[15] and mesencephalic trigeminal nucleus,[16] suggesting a sensory role for the peptide. In the present study we have determined the contribution of PACAP sensory nerve fibers to the innervation of rat visceral organs by (1) examining the effect of neonatal treatment with the sensory neurotoxin capsaicin on PACAP immunoreactivity, determined by radioimmunoassay and immunohistochemistry, and (2) using double-immunostaining for PACAP and calcitonin gene–related peptide (CGRP), a peptide known to be stored in capsaicin-sensitive sensory nerves.

MATERIALS AND METHODS

Animals and Tissue Specimens

The digestive and urinary tracts of Wistar rats (weighing 180–200 g), housed in 12 h light:12 h dark, were examined. Animals were killed by decapitation and the organs given in FIGURE 1 were quickly removed. The tissues were frozen on dry ice and stored at –80°C until radioimmunoassay for PACAP-38. For immunohistochemistry, animals were perfusion fixed in Stefanini fixative. Subsequently, specimens from the abovementioned regions were removed and postfixed in the same fixative for 24 h. The specimens were then rinsed repeatedly in Tyrode solution, containing 30% sucrose, and frozen on dry ice for cryostat sectioning.

To determine if PACAP immunoreactivity is present in sensory fibers, animals were treated neonatally with the sensory neurotoxin capsaicin. Rats were injected intraperitoneally with capsaicin (50 mg/kg body weight) at day 2 after birth and litter-matched rats were injected with vehicle solution (Tween 80, saline, absolute alcohol, 1:8:1 by volume) at the same time as the experimental animals and were used as controls. After 8–12 weeks, rats were killed and the various regions from the digestive and urinary tracts collected for radioimmunoassay or fixed for immunohistochemistry.

[a]This study was supported by the Danish Biotechnology Center for Cellular Communication.
[b]Corresponding author: Tel.: 45-35 31 26 40; Fax: 45-35 31 20 99; E-mail: bbhjanf@inet.uni2.dk

Radioimmunoassays

Before peptide analysis the frozen tissue specimens were weighed and extracted in boiling water/acetic acid. The extracted samples were reconstituted and analyzed by radioimmunoassay specific for PACAP-38.[4]

Immunohistochemistry

Immunohistochemical visualization of PACAP immunoreactivity was carried out by the methods described previously[4,9] using biotinylated tyramide (Tyramide System amplification, DuPont NEN®, Boston, MA, USA). The PACAP antibody was a mouse anti-PACAP antibody (code Mab JHH1). The anti-CGRP-antiserum (code B47-1) was obtained from Euro diagnostica (Malmö, Sweden).

RESULTS AND CONCLUSION

Treatment with the sensory neurotoxin capsaicin reduced the concentration of PACAP-38 by the following percentages (FIG. 1): the esophagus 25%, the stomach 31%, the duodenum 22%, and the colon 49%, while no changes were observed in the jejunum or ileum. In the pancreas, capsaicin reduced the PACAP-38 concentration by 34%. The

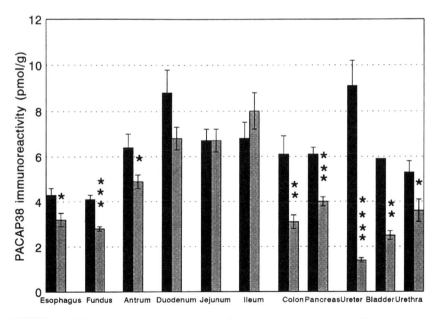

FIGURE 1. PACAP-38 immunoreactivity in the rat digestive tract, the pancreas, and the urinary tract after neonatal capsaicin treatment. Concentration (pmol/wet weight) of immunoreactive PACAP-38 in tissue from capsaicin-treated rats (dark bars) and litter-mate controls (light bars). Each bar represents the mean ± SEM of extracts from seven rats. Student's *t*-test was used to compare the PACAP-38 concentration in treated animals with controls in each region. * $p < 0.05$, ** $p < 0.01$, *** $p < 0.005$, **** $p < 0.0001$.

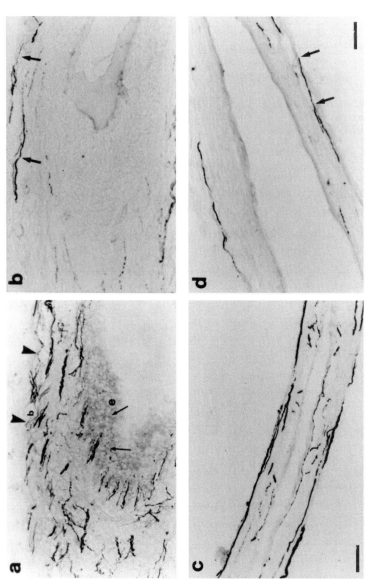

FIGURE 2. PACAP-immunoreactive nerve fibers in the ureter of normal rats and in animals treated with capsaicin. (**a**) Oblique section of the ureter showing PACAP-positive nerve fibers and bundles in the smooth muscle layers and below the epithelium. Delicate varicose nerve fibers (*arrows*) are present between the cells of the epithelium. Immunoreactive nerve fibers (*arrowheads*) are also seen around adventitial blood vessels (**b**). (**b**) Comparable section from a capsaicin-treated rat demonstrating that the number of immunostained fibers was markedly reduced, but some remain along the outer edge of the muscle (*arrows*) and around adventitial blood vessels. (**c**) Longitudinal section of the ureter showing networks and bundles of PACAP-positive nerve fibers running parallel to the long axis in the subepithelial connective tissue and in the smooth muscle layers, mainly along the outer aspect of the muscle. (**d**) Comparable sections from a capsaicin-treated rat illustrating the depletion of PACAP-immunoreactive nerve fibers particularly in the subepithelial layer with only occasional fibers remaining in the muscle layers and adventitia (*arrows*). Scale bars: **a**, **b**, and **d** = 50 μm; **c** = 100 μm.

reduction was, however, most prominent in all regions of the urinary tract, amounting to 85% in the ureter.

By double-staining immunohistochemistry, PACAP was found to be almost completely co-localized with CGRP. Immunohistochemically the number and the intensity of the PACAP-immunoreactive nerve fibers were reduced in the same organs in which the PACAP concentration was lowered, as illustrated in FIGURE 2 for the ureter. The CGRP immunoreactivity was identically affected except for CGRP in endocrine cells of the pancreas, which was not influenced by capsaicin treatment.

Our data indicate that a major fraction of PACAP-immunoreactive nerve fibers of the gastrointestinal tract, pancreas, and urogenital tract is co-storing CGRP immunoreactivity and that most of these nerve fibers are sensitive to neonatal capsaicin treatment, suggesting a sensory role in these organs.

REFERENCES

1. MIYATA, A., A. ARIMURA, R. R. DAHL, N. MINAMINO, A. UEHARA, L. JIANG, M. D. CULLER & D. H. COY. 1989. Isolation of a novel 38 residue-hypothalamic polypeptide which stimulates adenylate cyclase in pituitary cells. Biochem. Biophys. Res. Commun. **164**: 567–574.

2. ARIMURA, A., A. SOMOGYVÁRI-VIGH, A. MIYATA, K. MIZUNO, D. H. COY & C. KITADA. 1991. Tissue distribution of PACAP as determined by RIA: highly abundant in the rat brain and testes. Endocrinology **129**: 2787–2789.

3. GHATEI, M. A., K. TAKAHASHI, Y. SUZUKI, J. GARDINER, P. M. JONES & S. R. BLOOM. 1993. Distribution, molecular characterization of pituitary adenylate cyclase-activating polypeptide and its precursor encoding messenger RNA in human and rat tissues. J. Endocrinol. **136**: 159–166.

4. HANNIBAL, J., J. D. MIKKELSEN, H. CLAUSEN, J. J. HOLST, B. S. WULFF & J. FAHRENKRUG. 1995. Gene expression of pituitary adenylate cyclase activating polypeptide (PACAP) in the rat hypothalamus. Regul. Pept. **55**: 133–148.

5. MASUO, Y., N. SUZUKI, H. MATSUMOTO, F. TOKITO, Y. MATSUMOTO, M. TSUDA & M. FUJINO. 1993. Regional distribution of pituitary adenylate cyclase activating polypeptide (PACAP) in the rat central nervous system as determined by sandwich-enzyme immunoassay. Brain Res. **602**: 57–63.

6. HANNIBAL, J., J. M. DING, D. CHEN, M. U. GILLETTE, J. FAHRENKRUG, P. J. LARSEN & J. D. MIKKELSEN. 1997. Pituitary adenylate cyclase activating peptide (PACAP) in the retinohypothalamic tract. A daytime regulator of the biological clock. J. Neurosci. **17**: 2637–2644.

7. MIKKELSEN, J. D., J. HANNIBAL, P. J. LARSEN & J. FAHRENKRUG. 1994. Pituitary adenylate cyclase activating peptide (PACAP) mRNA in the rat neocortex. Neurosci. Lett. **171**: 121–124.

8. FRÖDIN, M., J. HANNIBAL, B. S. WULFF, S. GAMMELTOFT & J. FAHRENKRUG. 1995. Neuronal localization of pituitary adenylate cyclase activating polypeptide 38 in the adrenal medulla and growth-inhibitory effect on chromaffin cells. Neuroscience **65**: 599–608.

9. FAHRENKRUG, J. & J. HANNIBAL. 1996. PACAP in the rat female genital tract: Effect of capsaicin. Neuroscience **73**: 1049–1060.

10. UDDMAN, R., A. LUTS, A. ARIMURA & F. SUNDLER. 1991. Pituitary adenylate cyclase-activating peptide (PACAP), a new vasoactive intestinal peptide (VIP)-like peptide in the respiratory tract. Cell Tissue Res. **265**: 197–201.

11. HANNIBAL, J., E. EKBLAD, H. MULDER, F. SUNDLER & J. FAHRENKRUG. 1997. Pituitary adenylate cyclase activating polypeptide (PACAP) in the gastrointestinal tract of the rat: Distribution and effects of capsaicin or denervation. Cell Tissue Res. **291**: 65–79.

12. TORNØE, K., J. HANNIBAL, J. FAHRENKRUG & J. J. HOLST. 1997. PACAP 1-38 as a neurotransmitter in the pig pancreas; PACAP receptor activation revealed by the antagonist PACAP 6-38. Am. J. Physiol. **273**: G436–G446.

13. MOLLER, K., Y.-Z. ZHANG, R. HÅKANSON, A. LUTS, B. SJÖLUND, R. UDDMAN & F. SUNDLER. 1993. Pituitary adenylate cyclase activating peptide is a sensory neuropeptide: Immunocytochemical and immunochemical evidence. Neuroscience **57**: 725–732.

14. ZHANG, Y.-Z., J. HANNIBAL, Q. ZHAO, K. MOLLER, N. DANIELSEN, J. FAHRENKRUG & F. SUNDLER. 1996. Pituitary adenylate cyclase activating polypeptide (PACAP) in the rat dorsal root ganglia: Up-regulation after peripheral nerve injuries. Neuroscience **74**: 1099–1110.
15. MULDER, H., R. UDDMAN, K. MOLLER, Y.-Z. ZHANG, E. EKBLAD, J. ALUMETS & F. SUNDLER. 1994. Pituitary adenylate cyclase activating polypeptide expression in sensory neurons. Neuroscience **63**: 307–312.
16. LARSEN, J. O., J. HANNIBAL, S. M. KNUDSEN & J. FAHRENKRUG. 1997. Expression of pituitary adenylate cyclase activating polypeptide (PACAP) in the mesencephalic trigeminal nucleus of the rat after transection of the masseteric nerve. Mol. Brain Res. **46**: 109–117.

Galanin Controls Excitability of the Brain

SERGUEI A. CHEPURNOV,[a] NINA E. CHEPURNOVA, RUSTAM K. BERDIEV

Moscow State University, Human and Animal Physiology Department, Moscow 119899 Russia

Galanin is a 29-amino-acid neuropeptide that coexists with ACh in neurons. Galanin inhibits cholinergic action, including blocking EPSPs of hippocampus pyramidal cells and attenuation of scopolamine-induced release of ACh in ventral hippocampus. The inhibitory actions of galanin forebrain cholinergic function have stimulated interest in this peptide as a possible contributing factor in behavior and cognitive deficits.[1-3] As the synaptic concentration of galanin increased, galanin-induced inhibition of surviving cholinergic neurons would further reduce cholinergic functions, contributing to great cognitive dysfunctions.[1] It was proposed that galanin serves as an inhibitory modulator of cholinergic function under normal conditions, possibly as a rate-dependent, negative feedback mechanism. In the diseased state, the activity of surviving cholinergic neurons could increase their firing rate, possibly inducing greater release of galanin at the higher activities.[4] The biological activity of galanin resides in the N-terminal portion of the peptide. The first galanin receptor antagonist, M15 based on the N-terminal fragment (1-13) of galanin and on SP(5-11), was recently tested. M15 reversed the action of galanin on hippocampal ACh release, spinal flexor reflex, and on hyperpolarization of LC neurons.[5] The intracerebroventricular administration of antagonist M40 (spantide) caused body twisting behavior, termed "barrel rolling."[2] It is known that the barrel-rolling behavior is the part of epileptiform generalized activity. The role of galanin in the epilepsy has not been studied. The inhibitory effect of galanin was demonstrated on the picrotoxin-induced kindling[6] and hyperthermia-induced seizures.[7] The aim of our investigation was to research the anticonvulsant effect of galanin in rat brain. It is reasonable to suppose that galanin is involved in the processes of epileptogenesis, because it was discovered[8] that galanin reduced release of glutamate in rat hippocampus.

METHODS

Eighteen male adult rats (albino), each weighing 250 g, were used. Under deep anesthesia (Phenobarbital, 47 mg/kg, i.p.), rats were equipped with bipolar electrodes in the amygdala and temporal cortex. A stainless-steel cannula was implanted into the right lateral ventricle. As chemoconvulsant penthylenetetrazole (PTZ) was used twice in a 20 mg/kg dose. Galanin(1-29) and galantide (M15) were synthesized by M.P. Smirnova (Saint

[a]Corresponding author: Prof. S. A. Chepurnov, Russia Moscow 119899, Vorobievy Gory, Moscow State University, Biological Faculty, Human & Animal Physiology Department; Tel.: 7 095 939 27 92; Fax: 7 095 939 56 59; E-mail: CHAS@RTM-CS.NPI.MSU. SU

Petersburg, Russia) and were prepared at the appropriate concentration for microinjection of a volume of 4 µl. Galanin and M15 were both prepared at concentrations of 0.16 nmol/µl. The experiment was carried out in unrestrained animals. EEG signals were filtered and recorded by "CONAN" software.[9] Galanin was microinjected 5 min before the intraperitoneal injection of PTZ. Saline or M15 was microinjected 5 min before saline or galanin. Each rat was used four times with 5–7 days between the experiments. The compounds were administered to each rat before the PTZ test in following combinations: saline+saline, galanin+saline, galanin+M15, and M15+saline. The EEG recording was made for 1.5 hours. Latency of the first spike-wave discharge (SWD) and average SWD's duration/min were measured. During the recording of EEG the behavioral control of seizures was performed.

RESULTS AND DISCUSSION

The anticonvulsive effects of galanin were demonstrated in adults rat after intracerebroventricular administration using the model of PTZ-induced seizures. Galanin controls the spike-wave discharges induced by PTZ in electrical activity of temporal cortex and amygdala. The galanin intracerebroventicular administration significantly increased the

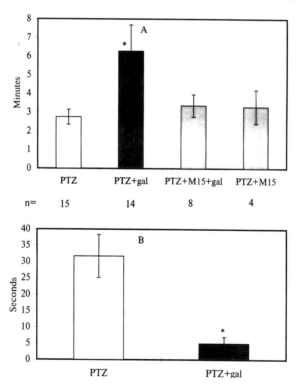

FIGURE 1. (A) Effects of intracerebroventricular galanin1-29 and M15 administration of the first PTZ-induced SWD latent period in rats. *$p < 0.05$. (B) Decrease of the duration of SWD after intracerebroventricular galanin1-29 administration at the maximum of paroxysmal PTZ-induced EEG activity in rats ($N = 8$, $p < 0.05$).

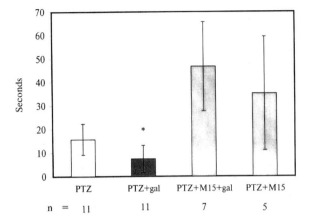

FIGURE 2. Effects of the intracerebroventricular administration of galanin1-29 and M15 on duration of PTZ-induced generalized seizures in rats. *$p<0.05$.

latency of the first SWD compared with control (FIG. 1,A). Intracerebroventricular administration of M15+galanin and only M15 significantly decreased the latency period of the first SWD to the level of the control value. The decreasing influence of galanin on the duration of SWD at the maximum of paroxysmal EEG activity (grand mal) is presented in FIGURE 1B. As compared to control values, intracerebroventricular galanin administration induced the significant decrease of duration of the PTZ-induced generalized seizures. M15 microinjected before galanin significantly inhibits the effect of the last one on the motor convulsions caused by PTZ. A microinjection of M15 alone and before galanin provoked the increase of duration of generalized seizures compared to the control, but these data are not significant (FIG. 2). It is possible that M15 has the tendency to act as proconvulsive agent. The percentage of rats with PTZ-induced grand mal seizure was calculated in each case. After intracerebroventricular administration of galanin only 20% of rats have the generalized convulsions. There were no differences between the percentage of rats with grand mal seizures after intracerebroventricular administration of M15+galanin (75%) and only M15 (71%), and the percentage of rats with grand mal seizures in control (66%). The conclusion is that galanin acts as anticonvulsive agent. The galanin antagonist M15 was found to block the inhibitory effects of endogenous and intracerebroventricularly administered galanin on seizures evoked by chemoconvulsant application. Blockade of postsynaptic galanin receptors by M15 reduces the anticonvulsive galanin effects and points to the specific action of galanin in this model of seizures.

REFERENCES

1. CRAWLEY, J. N. & G. L. WENK. 1989. Coexistence of galanin and acetylcholine: Is galanin involved in memory processes and dementia? Trends Nuerosci. **12**: 278–282.
2. Galanin: A New Multifunctional Peptide in the Neuroendocrine System. 1991. T. Hokfelt, T. Bartfai, D. Jacowitz & D. Otteson, Eds. Macmillan Education. London.
3. CRAWLEY, J. N. 1996. Galanin-acetylcholine interactions: relevance to memory and Alzheimer's disease. Life Sci. **58**: 2185–2199.
4. HOKFELT, T. & X. ZHANG *et al.* 1992. Coexistence and interaction of neuropeptides with SP and other tachykinins, with special reference to galanin. Regul. Pept. (Suppl.) **1**: S 11.

5. BARTFAI T., U. LANGEL & T. HOKFELT. 1994. Galanin: receptor subtypes, agonists and antagonists. Neuropeptides **26**: 21.
6. SHANDRA, A. A. & A. M. MAZARATI. 1994. Neurophysiological effects of neuropeptide galanin. Sechenov Russian Physiol. J. **80**: 8–18. (in Russian)
7. CHEPURNOV, S. A., N. E. CHEPURNOVA., K. R. ABBASOVA & M. P. SMIRNOVA. 1997. Neuropeptide galanin and seizure states of developing brain. Uspekhi Physiol. Nauk. **28**: 3–20. (in Russian)
8. ZINI, S., M. P. ROISIN, U. LANGEL, T. BARTFAI & Y. BEN-ARI. 1993. Galanin reduces release of endogenous excitatory amino acids in the rat hippocampus. Eur. J. Pharmacol. **245**: 1–7.
9. KULAICHEV, A. P. 1997. Computer electrophysiology in clinical practice—CONAN-m, MNMP. "Informatika and computers." Moscow. (in Russian)

Vasoactive Intestinal Peptide Supports Spontaneous and Induced Migration of Human Keratinocytes and the Colonization of an Artificial Polyurethane Matrix[a]

UWE WOLLINA

Department of Dermatology, Friedrich-Schiller-University of Jena, 07740 Jena, Germany

Wound healing is a complex process initialized by trauma and leading to eventual tissue repair. In skin the final wound closure is realized by re-epithelialization. This consists of several steps beginning with the activation and mobilization of keratinocytes along the wound edges, migration onto the wound bed, and adaption to the new microenvironment leading to colonization.[1,2]

Nervous system–skin interactions are well known to be trophic for normal tissue. In adult wounds a temporary hyperinnervation is found and in neonatal wounds a permanent hyperinnervation is found. During traumatization nerves are damaged and neuropeptides are released, which may have an impact on wound healing. In disorders with a reduced skin innervation and/or reduction of cutaneous neuropeptides, like diabetes or leprosy, delay of healing and the development of chronic wounds are not uncommon.[3,4]

In mammalian, including human, skin a large number of neuropeptides have been identified. One of the major neuropeptides found in human skin is vasoactive intestinal peptide (VIP), a peptide of 28 amino acids.[4–8] VIP is synthesized from a primary mRNA translation product prepro-VIP (MW 20,000). There is evidence for the role of VIP as a growth factor in regulation of embryonic growth *in vivo* and for several cell lines *in vitro*, like astroglial cells, B lymphocytes, fibroblasts, and gut epithelial cells.[9–12] VIP binds to specific receptors on the cell surface and there is evidence for heterogeneity of these receptors.[13]

VASOACTIVE INTESTINAL PEPTIDE AS A GROWTH FACTOR FOR KERATINOCYTES

An increasing body of evidence has suggested trophic effects of peripheral nerves on skin. Haegerstrand and colleagues[14] first showed that among several cutaneous neuropeptides only VIP can stimulate the proliferation of human keratinocytes in the presence of 3T3 feeder cells and epidermal growth factor (EGF) in a dose-dependent manner when applied for several days. They found VIP stimulated adenylate cyclase activity, indicating that cAMP is one of the second messengers involved. ^{125}I-labeled VIP was bound to keratinocytes and could be displaced by addition of unlabeled VIP, suggesting the presence of specific receptors.

[a]The study was supported by a grant from the German Ministry of Science and Technology (Klinisch orientierte Neurowissenschaften) (U.W.).
[b]Corresponding author: E-mail: UWOL@derma.uni-jena.de

Pincelli and colleagues[15] employed normal human keratinocytes for the investigation of different VIP fragments. They demonstrated that VIP, [10-12]VIP, and [22-28]VIP can stimulate keratinocyte proliferation in the concentration range between 1 nM to 1 μM, whereas [1-12]VIP, [NAc- Tyr^1,D-Phe^2]VIP, and substance P significantly inhibited these VIP effects. We have used the human non-tumorigenic keratinocyte cell line HaCaT,[16] which has a higher proliferative activity than normal keratinocytes. The HaCaT cells bind ^{125}I-labeled VIP and we found that the binding is only slightly influenced by culture conditions: 4.8×10^{-12} M/10^6 cells for serum-free medium and 4.6×10^{-12} M/10^6 cells in medium supplemented with 2% fetal calf serum. When VIP was added to the culture medium for 48 h, we observed a significant inhibition of ^{3}H-thymidine uptake. This was completely abolished by either the VIP receptor antagonist [4Cl-D-Phe^6,Leu^{17}] at 4×10^{-4} M or the VIP receptor antibody 109.10, and partially abolished by the calmodulin antagonist W-7 (10^{-5} M). In contrast, VIP could stimulate cell proliferation in serum-free cultures reaching a maximum (approx. 150% of the ^{3}H-thymidine uptake of controls at 10^{-11} M VIP). A doubling of the phosphoinositol 4-kinase activity vs. controls has been detected at 10^{-7} M VIP.[17,18]

In another set of experiments, we investigated the effect of VIP on the cell growth fraction. HaCaT keratinocytes seeded on chamber slides with either serum-free or serum-supplemented (2% and 10%) Dulbecco's modified Eagles' medium were exposed to VIP at various concentrations for 48 h. Eventually, the cells were immunostained with monoclonal antibody against the nuclear proliferation–associated Ki-67 antigen.[19] To characterize the cell cycle, we followed the classification of staining patterns by Wersto and colleagues.[20] In cultures with 2% serum, VIP (10^{-11} to 10^{-7} M) almost doubled the number of Ki-67–positive keratinocytes but was ineffective in cultures with 10% serum or under serum-free conditions. On the other hand, we observed a change in the nuclear staining patterns. There was a shift in serum-free cultures from type III to type II and vice versa in 2% serum cultures. This indicates that VIP interferes with a switch in G1-S phase.[21]

These findings suggest that VIP is a growth factor with either stimulatory or inhibitory activity on keratinocytes after binding to specific receptors on the cell surface. VIP seems to activate different intracellular second messengers, including cAMP, the phosphoinositol phosphate cycle, and calmodulin.[17,18]

VASOACTIVE INTESTINAL PEPITDE AS A MOBILITY FACTOR FOR KERATINOCYTES

VIP has been characterized as a growth factor. It is well known that growth factors like EGF under certain circumstances may also modulate the mobility of cells.[22,23] HaCaT cells migrate spontaneously in subconfluent cultures. The migrating keratinocytes can easily be identified since they develop characteristic protusions. In addition, proliferating cells have been stained for Ki-67 antigen. We observed that HaCaT cell mobility was inversely related to the serum content of the culture medium. The concentrations optimal for spontaneous migration were less effective for proliferation (TABLE 1).[24] VIP increased the migratory activity significantly in the range of 10^{-7} to 10^{-9} M, when added to cultures with 10% serum supplementation (maximum: + 121.1%) but decreased migratory activities 40 to 70% in cultures with 2% serum or under serum-free conditions (range: 10^{-13} to 10^{-7} M). These effects were associated with an inverse change in the proportion of Ki-67–labeled keratinocytes.[24] In ongoing experiments, HaCaT keratinocytes have been propagated in Dulbecco's modified Eagle's medium (DMEM) supplemented with 1% antibiotic-antimycotic solution (Gibco), 4 mM L-glutamine (Gibco), and 10% fetal calf serum (Gibco) under a humidified atmosphere with 5% CO_2 at 37°C in 175-ml

TABLE 1. Spontaneous migration of human HaCaT keratinocytes—the influence of serum supplementation

Percentage of Migrating Cells (%)	Fetal Serum Supplement (%)
3.8 ± 2.0	10
18.6 ± 9.6	2
17.4 ± 9.8	0

Falcon tubes. Cells were seeded on chamber slides cultivated for 72 h. We investigated the effect of VIP, [D-Phe$_2$]-VIP, [Lys-Pro-Arg-Tyr]-VIP, and VIP fragment [1-12] on migration. All peptides were added together with the serum-free medium at concentrations of 1 µM to 1 fM. A control without neuropeptides was run in parallel. In this model, confluent keratinocyte cultures were wounded (scratched) with a sterile razor blade. Care was taken to get a sharp wound edge. Peptides were added with serum-free medium immediately after wounding. The migration was monitored by light microscopy. We calculated the area covered by keratinocytes at different time points.[25] In VIP-treated samples keratinocytes of the wound margin migrated more rapidly, i.e., within the first 4 h (1 nM, 1 µM). Elongated cells with limited cell-cell contacts but without mitosis were visible. The wounds were almost closed within 24 h (80% of the wounded area covered by keratinocytes). In contrast, VIP derivatives were not significantly different from controls (10 to 18% of the wounded area covered, $p > 0.05$). After 48 h there was no difference between VIP and the other peptides tested.

In conclusion, there is evidence that VIP can stimulate spontaneous and induced migration of human keratinocytes. The effect is different from the growth factor activities shown above. These findings may be of significance for wound healing *in vivo*, since VIP is released from cutaneous nerve endings during injury and may reach remarkable concentrations *in loco*. Since VIP and other neuropeptides have a short half-life, the physiological role may be limited to the initiation of re-epithelialization.

VASOACTIVE INTESTINAL PEPTIDE AS A FACTOR FOR COLONIZATION

To investigate colonization in more detail we used a different model. Half of the bottom of chamber slides was coated with a sterile self-adhesive polyurethane foil, which served as a synthetic matrix and the medium was changed to serum-free medium supplemented with 5.0 µg recombinant human epidermal growth factor [1-51] and 25 mg bovine pituitary extract. In previous experiments, we have been able to exclude cytotoxicity by the polyurethane foil using a wound cell biological profile.[26] The polyurethane foil served as an artificial matrix for migrating keratinocytes and for colonization. For light microscopy the chamber slides were stained with Pappenheim's stain. Migrating HaCaT cells have been defined by a spindle-shaped morphology. Colonizing cells have been defined as small and large epithelial cells visible on the polyurethane matrix. The matrix has been removed after different time intervals (48 h to 144 h). The percentage of covered area has been calculated.[25] In controls, we were able to observe migration of keratinocytes on the margins of the matrix within the first 24 h. Migrating cells were spindle-shaped. The first morphological event was the formation of protrusions without mitosis. The cells however failed to migrate into the center of the matrix or to survive. The addition of VIP changed the situation remarkably. There was not only an increase in spindle-shaped cells, but during 48 h the matrix area was totally covered by keratinocytes vs. less than 1% in controls ($p < 0.001$). The induction of migrating cells was dose dependent. Changing the N-terminal ending of VIP by introduction of either phenylene or lysine residues decreased

TABLE 2. The influence of VIP and VIP derivatives on HaCaT migration on a polyurethane matrix[a]

Peptide	Concentration	Migration	Colonization
VIP	1 fM	5,568 ± 1,300***	98 ± 15***
	1 nM	1,024 ± 205***	75 ± 23**
	1 µM	487 ± 190*	53 ± 25*
VIP[1-12]	1 fM	97 ± 58	87 ± 31*
	1 nM	99 ± 60	65 ± 15*
	1 µM	89 ± 71	53 ± 17*
[D-Phe₂]-VIP	1 fM	101 ± 67	68 ± 17*
	1 nM	99 ± 80	61 ± 18*
	1 µM	11 ± 40	48 ± 21
[Lys-Pro-Arg-Tyr] -VIP	1 fM	81 ± 38	72 ± 16*
	1 nM	100 ± 43	63 ± 19*
	1 µM	96 ± 41	48 ± 20
Control		100 ± 67	6 ± 17

[a]In percent of controls as measured by the counts of spindle-shaped keratinocytes after 48 h. Colonization is expressed as percentage of area covered by cells.
p value vs. control: ***$p < 0.001$; **$p < 0.01$; *$p < 0.05$.

the ratio of migrating to total cell counts. The absolute number of visible cells on the matrix, however, was in the range of controls. Addition of [1-12]VIP fragment with unchanged N-terminus but loss of C-terminus resulted in reduced colonization and failed to improve migration (TABLE 2).[25] The data indicate that VIP-induced migration is dependent on the integrity of the N-terminus, but loss of the C-terminus abrogates stimulation of migration and reduces colonization. The results favor the assumption that survival and mitotic activity are more critical for colonization than migration.

We could demonstrate that among the cutaneous neuropeptides, VIP is a growth factor for keratinocytes binding to specific receptors and initiating different second messengers in the cell. VIP is also capable of affecting spontaneous and induced migration. Finally, VIP stimulates colonization. Therefore VIP interferes with early steps of epidermal wound healing *in vitro*. Our investigations have implications for our current understanding of wound healing and for the bioengineering of skin.[27,28]

REFERENCES

1. MARTIN, P. 1993. Wound healing—aiming for perfect skin regeneration. Science **276**: 75–78.
2. NAVAMO, X., E. VEDU, G. WENDELSCHAEFERCRABB & W. R. KENNEDY. 1997. Immuno-histochemical study of skin reinnervation by regenerative axons. J. Comp. Neurol. **380**: 164–174.
3. PINCELLI, C., F. FANTINI & A. GIANETTI. 1997. Neuropeptides, nerve growth factor and the skin. Pathol. Biol. **44**: 856–859.
4. SAID, S. I. 1984. Vasoactive intestinal polypeptide (VIP): current status. Peptides **5**: 143–150.
5. HARTSCHUH, E., E. WEIHE & M. REINECKE. 1983. Peptidergic (neurotensin, VIP, substance P) nerve fibres in the skin. Immunohistochemical evidence of an involvement of neuropeptides in nociception, pruritus and inflammation. Br. J. Dermatol. **109** [Suppl. 25]: 14–17.
6. WOLLINA, U. 1992. Zum Einfluß von Neuropeptiden auf die Wundheilung in vitro und in vivo. Hautarzt. **43**: 616–620.
7. WALLENGREN, J. 1997. Vasoactive peptides in the skin. J. Invest. Dermatol. (Symp. Proc.) **2**: 49–55.
8. ANSEL, J. C., C. A. ARMSTRONG, I. S. SONG, K. L. QUINLAN, J. E. OLERUD. S. W. CAUGHNUN & N. W. BUNNETT. 1997. Interactions of the skin and nervous system. J. Invest. Dermatol. (Symp. Proc.) **2**: 23–26.

9. HILL, J. M., S. K. McCUNE, R. J. ALVERO, G. W. GLANZER, K. A. HENINS, S. F. STANZIALE, J. R. KEIMOWITZ & D. E. BRENNEMAN. 1996. Maternal vasoactive intestinal peptide and the regulation of embryonic growth in the rodent. J. Clin. Invest. **97**: 202–208.

10. WALLENGREN, J. 1990. Sensory neuromediators in human skin. Role in inflammation and other disorders. Thesis, University of Malmö. Sweden.

11. BERNSTEIN, J. E. 1991. Neuropeptides and the skin. *In* Physiology, Biochemistry, and Molecular Biology of the Skin. 2nd Edit. L.A. Goldsmith, Ed.: 816–835. Oxford University Press. New York.

12. WENGER, G. D., M. S. O'DORISIO & E. J. GOETZL. 1990. Vasoactive intestinal peptide. Messenger in a neuroimmune axis. Ann. N. Y. Acad. Sci. **594**: 104–119.

13. ROBBERECHT, P., A. CAUVIN, P. GOURLET & J. CHRISTOPHE. 1990. Heterogeneity of VIP receptors. Arch. Int. Pharmacodyn. **303**: 51–66.

14. HAEGERSTRAND, A., B. JONZON, C.-J. DAALSGARD & J. NILSSON. 1989. Vasoactive intestinal peptide stimulates cell proliferation and adenylate cyclase activity of cultured human keratinocytes. Proc. Natl. Acad. Sci. USA **86**: 5993–5996.

15. PINCELLI, C., F. FANTINI, P. ROMUALDI, C. SEVIGNANI, G. LESA, L. BENASSI & A. GIANETTI. 1992. Substance P is diminished and vasoactive intestinal peptide is augmented in psoriatic lesions and these peptides exert disparate effects on the proliferation of cultured human keratinocytes. J. Invest. Dermatol. **98**: 421–427.

16. BOUKAMP, P., R. T. PETRUSSEVSKA, D. BREITKREUTZ, J. HORNUNG, A. MARKHAM & N. E. FUSENIG. 1988. Normal keratinization in a spontaneously immortalized aneuploid human keratinocyte cell line. J. Cell Biol. **106**: 761–771.

17. WOLLINA, U., B. BONNEKOH, R. KLINGER, R. WETZKER & G. MAHRLE. 1992. Vasoactive intestinal peptide (VIP) acting as a growth factor for human keratinocytes. Neuroendocrinol. Lett. **14**: 21–31.

18. CHRISTOPHE, J., M. SVOBODA, M. LAMBERT, M. WAELBROECK, J. WIUNAND, J.-P. DEHAYE, M.-C. VANDERMEERS-PIRET, A. VANDERMEERS & P. ROBBERECHT. 1986. Effector mechanisms of peptides of the VIP family. Peptides **7** [Suppl. 1]: 101–107.

19. GERDES, J. 1990. Ki-76 and other proliferation markers useful for immunohistological diagnostic and prognostic evaluations in human malignancies. Cancer Biol. **1**: 199–206.

20. WERSTO, R. P., F. HERZ, R. E. GALLAGHER & L. G. KOSS. 1988. Cell cycle-dependent reactivity with the monoclonal antibody Ki-67 during myeloid cell differentiation. Exp. Cell Res. **179**: 79–88.

21. WOLLINA, U., B. BONNEKOH & G. MAHRLE. 1992. Vasoactive intestinal peptide (VIP) modulates the growth fraction of epithelial skin cells. Int. J. Oncol. **1**: 17–24.

22. SARRET, Y., D. T. WOODLEY, K. GRIGSBY, K. WYNN & E. J. O'KEEFE. 1992. Human keratinocyte locomotion: the effect of selected cytokines. J. Invest. Dermatol. **98**: 12–16.

23. OGASAWARA, M., J. MURATA, K. AYUKAWA & I. SAIMI. 1997. Differential effect of intestinal neuropeptides on invasion and migration of colon carcinoma cells in vitro. Cancer Lett. **116**: 111–116.

24. WOLLINA, U. & B. KNOPF. 1993. Vasoactive intestinal peptide (VIP) modulates early events of migration in human keratinocytes. Int. J. Oncol. **2**: 229–232.

25. WOLLINA, U., J. HUSCHENBECK, B. KNÖLL, B. STERNBERG & U.-C. HIPLER. 1997. Vasoactive intestinal peptide supports migration of human keratinocytes and their colonization of an artificial matrix. Regul. Pept. **70**: 29–36.

26. WOLLINA, U., B. KNÖLL, K. PRÜFER, A. BARTH, D. MÜLLER & J. HUSCHENBECK. 1996. Synthetic wound dressings—evaluation of interactions with epithelial and dermal cells in vitro. Skin Pharmacol. **9**: 35–42.

27. MERTZ, P. M., S. C. DAVIS, L. FRANZEN, F.-D. UCHIMA, M. P. PICKETT, M. D. PIERSCHBACHER & J. W. POLAREK. 1996. Effects of an arginine-glycine-aspartatic acid peptide-containing artificial matrix on epithelial migration in vitro and experimental second-degree burn wound healing in vivo. J. Burn Care Rehabil. **17**: 199–206.

28. BERNSTAM, L. I., F. L. VAUGHAN & I. A. BERNSTEIN. 1990. Stratified cornified primary cultures of human keratinocytes grown on microporous membranes at the air-liquid interface. J. Dermatol. Sci. **1**: 173–181.

The Stimulatory Effect of VIP on Progesterone Release in Rats after Adrenalectomy, Ovariectomy, and Hysterectomy: Influence of VIP and PACAP38 on Progesterone Release in Rats after Ovariectomy and Hysterectomy[a]

E. WASILEWSKA-DZIUBIŃSKA, M. RADZIKOWSKA, AND B. BARANOWSKA

Neuroendocrinology Department, Medical Centre of Postgraduate Education, Fieldorfa 40, 04-158 Warsaw, Poland

In our earlier experiments performed on ovariectomized and hysterectomized (OVX+HTX) Wistar-Kyoto rats in response to intravenous administration of VIP, serum progesterone (PROG) level was raised[1-3] but it was unaffected by PACAP38 and PACAP27 administration.[3]

VIP and PACAP are structurally related peptides and are members of the secretin/glucagon/VIP family of peptides.[4] PACAP was originally isolated from the ovine hypothalamus in two biologically active amidated forms: with 38 residues (PACAP 38) and 27 residues (PACAP 27).[5,6] Both VIP and PACAP are distributed widely throughout the mammalian body and may function as neurotransmitters and/or modulators in the peripheral and central nervous system.[4,7] VIP and PACAP stimulate ovarian steroidogenesis and release of PROG and estrogens from granulosa cells[8,9] and release of aldosterone from adrenal cortex preparations.[10,11] In 1987, E. Bauliev and P. Robel considered PROG one of the neurosteroids, a class of steroids synthesized and acting in the central and peripheral nervous system,[12] particularly in myelinating glial cells, astrocytes, and Schwann cells.[13,14]

All steroidogenic tissues have a cholesterol side-chain cleavage enzyme cytochrome system (P450scc), which converts cholesterol to pregnenolone, and 3β-hydroxysteroid dehydrogenase/isomerase (3β HSD), which has both 3β-hydroxysteroid dehydrogenase and isomerase activities, thus converting pregnenolone to progesterone.[15] The evidence for the presence of P450scc involved in cholesterol side-chain cleavage was obtained in the white matter[16] in rat brain.[17,18] Incubation of granulosa cells of hen with VIP resulted in an increase of cyclic AMP accumulation, PROG synthesis, and P450scc mRNA levels.[19]

We suppose that VIP may stimulate release of PROG, which is synthesized in the nervous system structure.

The aim of this study was to evaluate the effect of VIP on PROG release in adrenalectomized, ovariectomized, and hysterectomized (ADX+OVX+HTX) rats and the effect of VIP and PACAP38, as well as blockers of the receptors [β-adrenergic-alprenolol (ALP) and opioid naltrexone (NAL)], on PROG release in OVX+HTX rats.

[a]This work was supported by Program No. 501-2-2-25-48/97.

MATERIAL AND METHODS

Experiments were performed on OVX+HTX rats and on ADX+OVX+HTX Wistar-Kyoto rats weighing 220–250 g. The indoor temperature was 20–22°C lighting conditions L:D, 14:10; food and tap water were *ad libitum*. Animals were treated in accordance with principles and procedures outlined in the National Institutes of Health Guidelines for Care and Use of Experimental Animals.

Three weeks after surgery, OVX+HTX animals were bilaterally adrenalectomized. The ADX+OVX+HTX rats were given free access to isotonic saline postoperatively. Two weeks after adrenalectomy, VIP was administered intravenously into the tail vein of ADX+OVX+HTX rats (group 1 and 2). Three weeks after ovariectomy and hysterectomy, VIP and PACAP38 were administered intravenously to OVX+HTX rats (groups 3–11).

The blood samples were collected to determine PROG by EIA method (bioMerieux) at 15 min after intravenous injection of VIP, PACAP, as well as after injection of β-adrenergic and opioid receptor blockers. The eleven experimental groups were: Group 1, control (placebo) 300 μl 0.9% NaCl, $N = 14$; Group 2, VIP, 10 μg VIP in 300 μl 0.9% NaCl, $N = 14$; Group 3, control (placebo), 300 μl 0.9% NaCl, $N = 8$; Group 4, VIP, 10 μg VIP in 300 μl 0.9% NaCl, $N = 8$; Group 5, PACAP 38, 10 μg PACAP 38 in 300 μl 0.9% NaCl, $N = 6$; Group 6, alprenolol (ALP), β-adrenergic blocker, 250 μg in 300 μl 0.9% NaCl, $N = 8$; Group 7, naltrexone (NAL), opioid blocker, 10 μg in 300 μl 0.9% NaCl, $N = 8$; Group 8, VIP+ALP; $N = 8$; Group 9, VIP+NAL, $N = 8$; Group 10, PACAP 38 + ALP, $N = 7$; Group 11, PACAP 38+NAL, $N = 7$.

The statistical analysis was carried out by means of Student *t* test for unpaired samples.

RESULTS

After intravenous VIP administration, a significant increase of PROG serum level was demonstrated in ADX+OVX+HTX rats ($p < 0.001$) (FIG. 1). An increase of PROG serum level was also observed after VIP, ALP, ALP+VIP, NAL, NAL+VIP intravenous administration in OVX+HTX rats ($p<0.001$ in all groups) (FIG. 2). Serum PROG levels in response

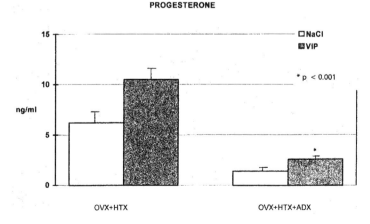

FIGURE 1. Serum progesterone level after VIP administration in OVX+HTX and OVX+HTX+ADX rats. Mean ± SEM.

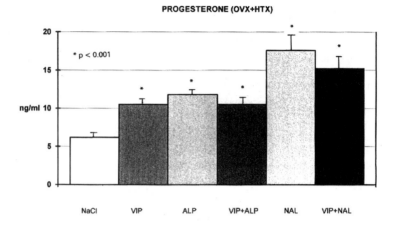

FIGURE 2. Serum progesterone level after VIP, ALP, VIP+ALP, NAL, and VIP+NAL administration in OVX+HTX rats. Mean ± SEM.

to combined therapy of VIP+ALP and VIP+NAL did not differ from the levels after injection of ALP and NAL administered alone (FIG. 2).

After PACAP38 administration we did not observe any differences in PROG serum level as compared with the control group. Serum PROG levels did not change after injection of both PACAP+ALP and PACAP+NAL (FIG. 3).

DISCUSSION

In our experiments, adrenalectomy, ovariectomy, and hysterectomy considerably diminished serum PROG level in comparison with the results of ovariectomy and histerectomy

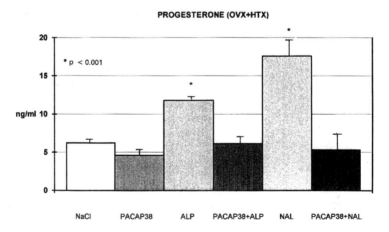

FIGURE 3. Serum progesterone level after PACAP38, ALP, PACAP38+ALP, NAL, and PACAP38+NAL administration. Mean ± SEM.

alone. The persistence of serum PROG after adrenalectomy, ovariectomy, and hysterectomy may confirm a possibility of PROG synthesis independently of gonadal and adrenal steroidogenesis.[20] Moreover in the same conditions we observed an increase in PROG serum level after intravenous VIP administration. This may confirm our hypothesis concerning a stimulatory effect of VIP on PROG release from the nervous system or extrasteroidogenic gland cells sources.

In our experiments performed on OVX+HTX rats, serum PROG level increased after VIP, ALP, NAL, VIP+ALP, and VIP+NAL administration.

It has been reported that VIP stimulated secretion of aldosterone, corticosterone and adrenaline in different preparations of rat adrenocortical tissue.[10,21] It was observed that the addition of alprenolol, a β-adrenergic antagonist to the incubation medium, significantly inhibited the adrenal capsular response to VIP and diminished aldosterone secretion.[21] It was also stated that in the isolated perfused adrenal glands naloxone caused a concentration-dependent inhibition of catecholamine secretion evoked by stimulation of splanchnic nerves.[22] It seems that in our experiments the stimulatory mechanism of VIP on PROG release is not dependent on β-adrenergic and opioid activity.

We confirmed our earlier observation that PACAP38 administered intravenously to OVX+HTX rats does not change PROG serum level[3] and does not affect PROG release by the human adrenal slices.[23] However, PACAP38 abolished a stimulatory influence of ALP and NAL on PROG release in our experiment. It may be connected with direct interaction at receptor level between PACAP specific binding sites present in the adrenal glands and β-adrenergic and opioid receptors.

Differences in PROG response to VIP and PACAP are believed to be dependent on specific modes of action of this peptides. This may be due to the existence of multiple subtypes of PACAP and VIP receptors and multiple target cells for PACAP and VIP in the nervous system and in the adrenal glands.

Our results may confirm our hypothesis concerning a stimulatory effect of VIP on PROG release from the nervous system.

CONCLUSIONS

(1) VIP stimulated PROG secretion in ADX+OVX+HTX rats.

(2) The stimulatory mechanism of VIP on PROG secretion is not dependent on β-adrenergic and opioid activity on OVX+HTX rats.

(3) PACAP38 did not affect PROG release in OVX+HTX rats.

(4) The blockade of β-adrenergic and opioid receptors led to an increase of PROG secretion.

(5) PACAP38 inhibited a stimulatory effect of β-adrenergic and opioid receptors blockers on PROG release.

REFERENCES

1. BARANOWSKA, B., E. WASILEWSKA-DZIUBINSKA & M. RADZIKOWSKA. 1993. Hormonal and cardiovascular effects of vasoactive intestinal peptide. Presented at the International Symposium on Vasoactive Intestinal Peptide Pituitary Adenylate Cyclase Activating Polypeptide & Related Regulatory Peptides, Euroconference. Strasbourg (Bischenberg), September 19–23, 1993.
2. WASILEWSKA-DZIUBINSKA, E., M. RADZIKOWSKA, E. STAFF-ZIELINSKA, A. PLONOWSKI, K. ROGUSKI & B. BARANOWSKA. 1996. Vasoactive intestinal peptide stimulates adrenal progesterone and aldosterone secretion in ovariectomised and histerectomised Wistar-Kyoto rats. Polish J. Endocrinol. **47**: 163–169.

3. BARANOWSKA, B., E. WASILEWSKA-DZIUBINSKA, M. RADZIKOWSKA, A. PLONOWSKI, K. ROGUSKI, E. KRAWCZYK & M. KAWALEC. 1996. Effects of PACAP and VIP on adrenal progesterone release. Ann. N.Y. Acad. Sci. **805**: 628–633.
4. ARIMURA, A. 1992. Pituitary adenylate cyclase activating polypeptide (PACAP): Discovery and current status of research. Regul. Pept. **37**: 287–303.
5. MIYATA, A., A. ARIMURA, R. D. DAHL, N. MINAMINO, A. UEHARA, L. JIANG, M. D. CULLER & D. H. COY. 1989. Isolation of a novel 38-residue hypothalamic polypeptide which stimulates adenylate cyclase in pituitary cells. Biochem. Biophys. Res. Commun. **164**: 567–574.
6. MIYATA, A., L. JIANG, R. D. DAHL, C. KITADA, K. KUBO, M. FUJINO, N. MINAMINO & A. ARIMURA. 1990. Isolation of a neuropeptide corresponding to the N terminal 27 residues of the pituitary adenylate cyclase activating polypeptide with 38 residues. Biochem. Biophys. Res. Commun. **170**: 643–648.
7. GOZES, J. & D. E. BRENNEMAN. 1989. VIP molecular biology and neurobiological function. Mol. Neurobiol. **3**: 201–236.
8. DAVOREN, J. B. & A. J. W. HSUEH. 1985. VIP: a novel stimulator of steroidogenesis by cultured rat granulosa cells. Biol. Reprod. **33**: 33–37.
9. ZONG, Y. & B. G. KASSON. 1994. Pituitary adenylate cyclase activating polypeptide stimulates steroidogenesis and adenosine 3' 5'- monophosphate accumulation in cultured rat granulosa cells. Endocrinology **135**: 207–213.
10. CUNNINGHAM, L. A. & M. A. HOLZWARTH. 1988. Vasoactive intestinal peptide stimulates adrenal aldosterone and corticosterone secretion. Endocrinology **122**: 2090–2097.
11. ANDREIS, P. G., L. K. MALENDOWICZ, A. S. BELLONI & G. G. NUSSDORFER. 1995. Effects of pituitary adenylate cyclase activating peptide (PACAP) on the rat adrenal secretory activity: preliminary in vitro studies. Life Sci. **56**: 135–142.
12. MELLON, S. H. 1994. Neurosteroids: biochemistry, modes of action and clinical relevance. J. Clin. Endocrinol. Metab. **78**: 1003–1008.
13. JUNG-TESTAS, J., Z. Y. HU, E. E. BALIEV & P. ROBEL. 1989. Neurosteroids: biosynthesis of pregnenolone and progesterone in primary cultures of rat glial cells. Endocrinology **125**: 2083–2091.
14. BAULIEV, E. E. 1997 Neurosteroids: Of the nervous system, by the nervous system, for the nervous system. Rec. Prog. Horm. Res. **52**: 1–32.
15. STOCCO, D. M. & B. J. CLARK. 1996. Regulation of the acute production of steroids in steroidogenic cells. Endocrine Rev. **17**: 221–244.
16. LE GOASCOGNE C., P. ROBEL, M. GOVEZOU, N. SANANES, E. E. BAULIEV & M. WATERMAN. 1987. Neurosteroids: cytochrome P450scc in rat brain. Science **237**: 1212–1215.
17. MELLON, S. H., & C. F. DESCHEPPER. 1993. Neurosteroid biosynthesis: genes for adrenal steroidogenic enzymes are expressed in the brain. Brain Res. **629**: 283–270.
18. ZONGH, Y. H., E. BOURREAU, I. JUNG-TESTAS, P. ROBEL & E. E. BAULIEV. 1987. Neurosteroids: Oligodendrocyte mitochondria convert cholesterol to pregnenolone. Proc. Natl. Acad Sci. USA **84**: 8215–8219.
19. JOHNSON, A. L., Z. LI , J. A. GIBNEY & S. MALAMED. 1994. Vasoactive intestinal peptide induces expression of cytochrome P450 cholesterol side-chain cleavage and 17 alpha-hydrolase enzyme activity in hen granulosa cells. Biol. Reprod. **51**: 327–333.
20. CORPECHOT, C., M. SYNGUELAKIS, S. TAHLA, M. AXELSON, J. SJOVALL, R. VIHKO, E. E. BAULIEV & P. ROBEL. 1983. Pregnenolone and its sulfate ester in the rat brain. Brain Res. **270**: 119.
21. HINSON, J. P., S. KAPAS, C. D. ORFORD & G. P. VINSON. 1992. Vasoactive intestinal peptide stimulation of aldosterone secretion by the rat adrenal cortex may be mediated by the local release of catecholamines. J. Endocrinol. **133**: 253–258.
22. MALHOTRA, R. & T. D. WAKADE. 1987. Non-cholinergic component of rat splanchnic nerves predominated at low neuronal activity and is eliminated by naloxone J. Physiol. **383**: 639–652.
23. NERI, G., P. G. ANDERS, T. PRAYER-GALETT, G. P. ROSSI, L. K. MALENDOWICZ & G. G. NUSSDORFER. 1996. Pituitary adenylate cyclase activating peptide enhances aldosterone secretion of human adrenal gland: Evidence for an indirect mechanism, probably involving the local release of catecholamines. J. Clin. Endocrinol. Metab. **81**: 169–173.

Characterization of a PACAP-like Immunoreactive Component in Red Ginseng Root

N. TAKASHIMA,[a,d] Y. ARAKAWA,[a] K. KATAOKA,[a] N.
KUROKAWA,[a] C. YANAIHARA,[b] AND N. YANAIHARA[c]

[a]Laboratory of Pharmaceutical Sciences, Osaka University School of
Medicine, Suita Osaka 565, Japan

[b]Department of Pharmacy, Hyogo College of Medicine, Nishinomiya
Hyogo 663, Japan

[c]Yanaihara Institute Inc., Fujinomiya Shizuoka 418, Japan

Korean ginseng root (*Panax ginseng* C.A. Meyer) is a widely used herb in traditional Chinese and Korean medicine. Ginseng apparently alleviates symptoms of a variety of degenerative diseases, particularly of the endocrine, gastrointestinal, and cardiovascular systems.[1,2] Many recent pharmacological studies indicate that the ginseng saponins explain some of the biological activities of the herb. However, other studies suggest that certain peptide and protein-like substances may also be responsible for the pharmacological activities of ginseng.[3–6] We have previously found that Korean red ginseng root extract contained the immunoreactivities of various regulatory peptides, such as PACAP, galanin, VIP, and PYY at concentrations between 0.2 and 20.7pmol/g. Among them, PACAP-like immunoreactivity (PACAP-IR) was found to be one of the major components. The present study describes purification of the PACAP-IR component in red ginseng root powder, which increased cyclic AMP production in human neuroblastoma NB-OK-1 cells.[7] Furthermore, it was demonstrated that oral administration of the PACAP-IR component in rats resulted in substantial increase in PACAP-IR levels in blood.

MATERIALS AND METHODS

Purification of a PACAP-IR Component in Red Ginseng Powder

Korean red ginseng root powder (Korea T. & Ginseng Corp., Korea), which had been prepared from steamed *Panax ginseng* C.A. Meyer root, was used as starting material. The powder was heated at 60°C in ninefold weight of H_2O for 1 h (red ginseng root powder extract). After centrifugation at 20,000 rpm for 30 min at 4°C, the supernatant was lyophilized and 50% acetonitrile in 0.1% trifluoroacetic acid (TFA) was added. After

[d]Corresponding author: Noriko Takashima, Laboratory of Pharmaceutical Sciences, Osaka University School of Medicine, 2-15 Yamadaoka, Suita, Osaka 565, Japan; Tel.: 81-6-879-3371; Fax: 81-6-879-3379; E-mail: noriko@drug.med.osaka-u.ac.jp

centrifugation at 3,000 rpm for 10 min at 4°C, the supernatant was lyophilized, and then dissolved in 1 M acetic acid. The solution was submitted to gel filtration on a Sephadex G25 superfine column using 1 M acetic acid as eluent, followed by CM-ion exchange chromatography on a Sep-Pak cartridge (Waters, Milford, MA, USA) using as eluent 10 mM Tris-HCl buffer (pH gradient from 7 to 9.9). The fractions containing PACAP-IR component were then subjected to reverse-phase HPLC on a puresil C18 column (Waters, Milford, MA, USA). The column was equilibrated with 10% acetonitrile in 0.1% TFA and eluted by acetonitrile in 0.1% TFA (10%–60%, 30 min). The peak fractions of PACAP-IR were then applied to an anti-PACAP(1-15) antibody-coupled Sepharose 4B affinity column and the column was washed with 10 mM sodium phosphate–buffered saline (pH 7.4) and eluted with 2 M acetic acid.

Radioimmunoassay

Radioimmunoassay (RIA) specific for PACAP was developed using anti-synthetic PACAP(1-15) antibody (Y042, Yanaihara Institute Inc., Fujinomiya, Japan) (final dilution 1:140,000), ^{125}I-PACAP27 as tracer, and PACAP(1-15) as standard. The standard diluent used for the assay was 10 mM phosphate buffer (pH 7.4) containing 0.14 M NaCl, 25 mM EDTA, and 2% BSA. Antibody-bound and free antigens were separated by precipitation with goat anti-rabbit γ-globulin serum and 5% polyethylene glycol. The minimum detection limit was 8 fmol/ml. The assay system recognized PACAP38, PACAP27, and the N-terminal (1-15) fragment identically.

Measurement of Intracellular Cyclic AMP Concentration

NB-OK-1 cells were cultured in RPMI-1640 medium (Nissui Seiyaku Co., Tokyo, Japan) supplemented with 10% fetal calf serum (FCS) (ICN Pharmaceuticals Inc., Costa Mesa, CA, USA) in a 25 cm^2 Falcon flask at 37°C in an atmosphere of 5% CO$_2$ in air at 100% humidity.[7,8] NB-OK-1 cells (about 10^5 cells) were incubated for 30 min in the presence or absence of each of the gel filtration fractions of red ginseng extract. The incubation medium was removed and the cells were extracted with 80% (vol/vol) ethanol containing 1 mM 3-isobutyl-1-methylxanthine. The extract was assayed for cyclic AMP using enzyme immunoassay kits (Amersham International plc, code RPN225, Bucks, UK).

Animal Experiment

Adult male Wistar rats weighing 200 g–250 g were used. After an overnight fast, red ginseng root powder extract or/and synthetic human PACAP(1-15) was orally administered to rats by using a feeding needle (Natume Seisakusho, Tokyo, Japan). After 10 min, the rats were anesthetized with ether and blood samples were taken from the portal vein into heparinized tubes containing aprotinin. After centrifugation at 3,000 rpm for 10 min at 4°C, the plasma was taken and kept at -70°C until assayed.

RESULTS AND DISCUSSION

The present communication describes the purification of a PACAP-related component in the red ginseng root extract, aiming to identify biologically active components in the ginseng root, one of the most important and popular herb medicines. This work also provided additional evidence of the existence of biologically active peptides in plants.

The purification was carried out consecutively by Sephadex G25 gel filtration, CM-ion exchange chromatography, C18 reverse-phase HPLC, and anti-PACAP(1-15) antibody-coupled Sepharose 4B affinity chromatography, using as probing tool a PACAP (1-15) specific RIA that recognized PACAP38 and PACAP27 as well. The dilution curve of PACAP-IR in the red ginseng root powder extract was parallel to the standard curve of the RIA, indicating immunologically indistinguishable property of the PACAP-IR from PACAP38, PACAP27, or PACAP(1-15). PACAP38 has been shown to increase intracellular cyclic AMP production in human neuroblastoma NB-OK-1 cell.[9] The gel filtration profile indicated that the immunoreactivity (FIG. 1a) paralleled the stimulating activity on cyclic AMP production in NB-OK-1 cells (FIG. 1b). HPLC profile of the major peak fractions (No. 30–40) in gel filtration revealed a large peak (Fr. 1) of PACAP-IR eluting at 13.5 min with a small one at 20.5 min (Fr. 2) (FIG. 1c). The two PACAP-IR components were finally purified by affinity chromatography (FIG. 1d). Both large and small peak components behaved identically in the affinity chromatography. We have thus achieved the isolation of a highly purified PACAP-related component in the red ginseng root extract. The elucidation of the component's structure is now underway in our laboratory.

Traditional herb medicines are used exclusively by oral administration. On the other hand, there has been a question about intestinal absorption of peptides and proteins orally applied. In this context, we examined the change of plasma PACAP-IR in rats 10 min after

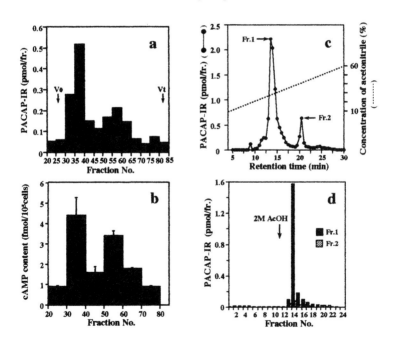

FIGURE 1. Elution profiles of (**a**) PACAP-IR and (**b**) cAMP-producing activity in NB-OK-1 cells in Sephadex G25 gel filtration of the red ginseng root extract, (**c**) PACAP-IR in C18 reverse-phase HPLC of PACAP-IR peak fraction from CM-ion exchange chromatography after gel filtration, and (**d**) PACAP-IR in antibody affinity chromatography of HPLC Fr. 1 and Fr. 2. The vertical bars in (**b**) represent mean ± SEM of three experiments in duplicate.

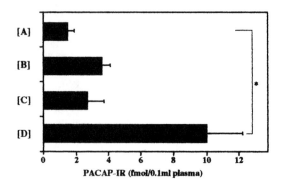

FIGURE 2. Plasma PACAP-IR in rats after oral administration of [A] saline, [B] red ginseng root extract (70 mg powder/rat), [C] synthetic PACAP(1-15) (200 µg/rat) and [D] synthetic PACAP(1-15) (200 µg/rat) with red ginseng root extract (70 mg powder/rat). The vertical bars represent mean ± SEM (*$p < 0.05$).

single oral administration of the red ginseng root powder extract (70 mg powder/rat). The result is shown in FIGURE 2. Plasma PACAP-IR was apparently elevated (240 ± 40%), although it was not statistically significant. On the other hand, the elevation of plasma PACAP-IR after oral synthetic PACAP(1-15) was only 180 ± 73%. However, when synthetic PACAP(1-15) (200 µg/rat) was given concomitantly with the red ginseng root powder extract (70 mg powder/rat), plasma PACAP-IR increased markedly (687 ± 153%) and the increase was significant ($p < 0.05$). The results seem to confirm the difficulty of intestinal absorption of peptides orally given. Interestingly enough it was also found that simultaneous administration of herb medicine could stimulate transport of oral peptide into circulation via digestive tract. This, in turn, supports that the PACAP-related peptide in the red ginseng root powder isolated in this study participates, through intestinal absorption, in the biological events exhibited by this important traditional medicine.

REFERENCES

1. POPOV, I. M. & W. J. GOLDWAG. 1973. A review of the properties and clinical effects of ginseng. Am. J. Chinese Med. **1**: 263–270.
2. LIU, C. X. & P. G. XIAO. 1992. Recent advances on ginseng research in China. J. Ethnopharmacol. **36**: 27–38.
3. ZHANG, T., M. HOSHINO, K. IGUCHI, J. ISHIKAWA, T. MOCHIZUKI, N. TAKATSUKA, C. YANAIHARA, M. YOKOTA, G. H. GREELEY & N. YANAIHARA. 1990. Ginseng root: evidence for numerous regulatory peptides and insulinotropic activity. Biomed. Res. **11**: 49–54.
4. YAGI, A., K. AKITA, T. UEDA, N. OKAMURA & H. ITOH. 1994. Effect of a peptide from *Panax ginseng* on the proliferation of baby hamster kidney-21 cells. Planta Med. **60**: 171–173.
5. KIMURA, M., I. WAKI, O. TANAKA, Y. NAGAI & S. SHIBATA. 1981. Pharmacological sequential trials for the fractionation of components with hypoglycemic activity in alloxan diabetic mice from ginseng radix. J. Pharm. Dyn. **4**: 402–409.
6. KIMURA, M., I. WAKI, T. CHUJO, T. KIKUCHI, C. HIYAMA, K. YAMAZAKI & O. TANAKA. 1981. Effects of hypoglycemic components in ginseng radix on blood insulin level in alloxan diabetic mice and on insulin release from perfused rat pancreas. J. Pharm. Dyn. **4**: 410–417.
7. YANAIHARA, N., T. SUZUKI, H. SATO, M. HOSHINO, Y. OKARU & C. YANAIHARA. 1981. Dibutyryl cAMP stimulation of production and release of VIP-like immunoreactivity in a human neuroblastoma cell line. Biomed. Res. **2**: 728–734.

8. HOSHINO, M., M. LI, L. Q. ZHENG, M. SUZUKI, T. MOCHIZUKI & N. YANAIHARA. 1993. Pituitary adenylate cyclase activating peptide and vasoactive intestinal polypeptide: differentiation effects on human neuroblastoma NB-OK-1 cells. Neurosci. Lett. **159**: 35–38.
9. LI, M., L. Q. ZHENG, M. HOSHINO, M. SUZUKI, A. HABU, K. IGUCHI, C. YANAIHARA, T. MOCHIZUKI & N. YANAIHARA. 1994. Cholera toxin enhanced cAMP production stimulated by PACAP in human neuroblastoma NB-OK-1 cells. Biomed. Res. **15** (Suppl 2): 233–239.

The Effect of Vasoactive Intestinal Polypeptide and Pituitary Adenylate Cyclase Activating Polypeptide on Tolerance to Morphine and Alcohol in Mice[a]

GYULA SZABÓ,[b] MÓNIKA MÁCSAI, ÉVA SCHEK, AND GYULA TELEGDY

Department of Pathophysiology, Albert Szent-Györgyi Medical University, H-6701 Szeged, Hungary

V asoactive intestinal polypeptide (VIP) and pituitary adenylate cyclase activating polypeptide (PACAP) have been detected in the peripheral and central nervous system. The distribution of these peptides throughout the central nervous system suggests the involvement of VIP and PACAP in a number of processes including limbic, autonomic, and neuroendocrine functions.

The relationship between pain sensitivity and peptides belonging to secretin-VIP-glucagon family has been examined before upon intrathecal administration. In spinalized, decerebrated rats VIP facilitated the flexor reflex to thermal stimuli and was implicated in transmission of nociceptive information after peripheral nerve damage.[1] In normal rats a decrease in the reaction latency in tail-flick test was demonstrated.[2] In ovariectomized rats VIP significantly increased tail-flick latency.[3] The effect of VIP proved to be short-lasting in the latest two studies.

The aim of the present study was to investigate the effect of these peptides on analgesia in intact mice upon peripheral and central administration and to extend these studies by investigating the acute interaction of these peptides with morphine, alcohol, and the development of tolerance to morphine and ethanol.

MATERIALS AND METHODS

Animals

Male, albino mice (25±5 g) of an inbred CFLP strain (SZOTE, Animal Husbandry, Szeged, Hungary) were used. The animals were kept under a standard light-dark cycle with food and water available *ad libitum*. At least a week of habituation was allowed before beginning the experiments.

[a]This work supported by Országos Tudományos Kutatási Alap (T 022230 and T 023169) and Egészségügyi Tudományos Tanács (T 02-670/96 and T 661/1996).
[b]Corresponding author: G. Szabó, M.D., Ph.D., Associate Professor of Pathophysiology, Albert Szent-Györgyi Medical University, Semmelweis u. 1, Pf. 531, H-6701 Szeged, Hungary; Tel.: 36-62-455-788; Fax: 36-62-455-695; E-mail: szabog@patph.szote.u-szeged.hu

Surgery

For intracerebroventricular cannulation, mice were anesthetized with sodium pentobarbital (Nembutal, intraperitoneally) and a polyethylene cannula was inserted into the right lateral cerebral ventricle and cemented to the skull with cyanoacrylate-containing instant glue. The experiments were started 5 days after intracerebroventricular cannulation. Animals were lightly anesthetized with ether and morphine pellets were implanted subcutaneously into the sacral area through a small section in the neck area. All procedures were carried out according to the rules of the Ethical Committee for the Protection of the Animals (Albert Szent-Györgyi Medical University, Szeged, Hungary)

Treatments

For subcutaneous treatment, VIP or PACAP was dissolved in isotonic saline and injected in a volume 0.2 ml. For intracerebroventrical treatment, the peptides were dissolved in artificial cerebrospinal fluid and injected in a volume of 2 ml. In the experiments for measuring the antinociceptive effect of morphine, morphine HCl (2.25–5 mg/kg, subcutaneous) was used. In chronic tolerance studies, pellets containing 35 mg morphine were implanted.

Procedures

For testing the analgesic effect, tail-flick tests were performed after single or repeated VIP or PACAP pretreatment. The tail of the animals was placed under a centrally focused radiating heat lamp and the latency of tail avoidance was measured and the degree of analgesic effect was calculated in percentage.

Chronic tolerance to morphine was examined on morphine pellet–implanted animals 72 h after the implantation without removing the tablets. Five mg/kg morphine was given subcutaneously after the determination of the threshold of pain. The analgesic effect of the test dose of morphine was measured at various time intervals. Vasoactive intestinal polypeptide (VIP) or PACAP was administered 30 min before the implantation of the morphine pellets and the treatment was repeated 24, 48, and 72 h later.

For measuring the hypnotic response to ethanol, animals were pretreated with VIP or PACAP (intracerebroventricularly) and 30 min later 4 g/kg ethanol was injected (intraperitoneally). The duration of the loss of the righting reflex was measured.

For determining the development of rapid tolerance to the hypothermic effect of ethanol, mice were pretreated intraperitoneally with 2 g/kg on day 1 in the morning (8 a.m.) and the hypothermic response was detected at various times. Ethanol treatment was repeated at 4 p.m. and the development of tolerance to the hypothermic effect of ethanol was tested the next morning.

Statistical Analysis

Statistical analysis of the data was made by ANOVA. For significant ANOVA values, groups were compared by Tukey's test for multiple comparisons with unequal cell size. A probability level of 0.05 was accepted as indicating significant differences.

TABLE 1. The effect of VIP on analgesic response in tail-flick test

Dose	Subcutaneous Administration	Intracerebroventricular Administration
0	-2.09 ± 1.10[a]	0.34 ± 0.89
0.02 pg	-1.08 ± 5.97	
0.2 pg	5.98 ± 3.60	
2 pg	14.24 ± 3.23*	
20 pg	9.44 ± 1.87*	
200 pg	19.68 ± 2.90*	13.89 ± 1.82*
2 ng	18.99 ± 2.26*	14.98 ± 1.62*
20 ng	17.00 ± 2.11*	11.36 ± 0.76*
200 ng	15.21 ± 1.99*	16.49 ± 1.49*
2 µg	16.56 ± 2.91*	

[a]Mean ± S.E.M. and *$p < 0.05$ compared to control group.

RESULTS AND DISCUSSION

Animals treated with VIP (2 pg–2 µg subcutaneously, 200 pg–200 ng intracerebroventricularly) displayed an increase in tail-flick latency, supporting the observation of Komisaruk and colleagues.[3] However, the effect was not dose dependent and no difference was observed in the effective doses between peripheral and central administration. Peptide administration resulted in a longer lasting effect, compared to earlier studies[2,3] and animals treated with intracerebroventricular VIP displayed longer lasting analgesia (TABLE 1).

VIP pretreatment (subcutaneous or intracerebroventricular) significantly diminished the analgesic effect of a single morphine injection. The analgesic effect of VIP (subcutaneous or intracerebroventricular) can be prevented by naloxone pretreatment or decreased if naloxone was administered after the maximal analgesic effect of VIP (45 min) was reached. Naloxone pretreatment significantly reduced the morphine-induced analgesia in peptide-naive animals. VIP pretreatment reduced the analgesic response of morphine, however VIP treatment could not reduce the analgesic response below the level of the control animals.

Animals injected with PACAP itself did not display analgesia or decrease of the analgesic response to an acute morphine challenge. Subchronic treatment with PACAP for 3 days diminished the acute analgesic effect of morphine.

VIP (intracerebroventricular) blocked the development of chronic tolerance to morphine, but did not affect naloxone-precipitated withdrawal symptoms (loss of body weight, decrease in body temperature, and naloxone-induced withdrawal jumps). PACAP has no effect on the development of tolerance to morphine and on naloxone-induced withdrawal signs (TABLE 2).

Intracerebroventricular administration of PACAP and VIP does not influence the duration of loss of the righting reflex.

TABLE 2. The effect of VIP on development of chronic tolerance to morphine

Time	Morphine Tolerant + Saline	Morphine Tolerant + VIP
30 min	-1.64 ± 1.46[a]	12.58 ± 3.04*
60 min	20.80 ± 3.72	32.58 ± 3.54*
90 min	15.50 ± 2.51	24.89 ± 3.09*
120 min	1.01 ± 2.39	8.17 ± 1.31*

[a]Mean ± S.E.M. and *$p < 0.05$ compared to morphine tolerant + saline group.

Note: The effect of VIP on chronic tolerance to morphine was investigated on animals with morphine pellet implantation. VIP treatment caused a significant increase in tail-flick latency as compared to tolerant control group at all time points.

Acute treatment with PACAP (intracerebroventricular) enhanced, VIP (subcutaneous) decreased the hypothermic response to ethanol on day 1, but did not influence the development of tolerance to the hypothermic effect of ethanol.

Subchronic treatment with PACAP (intracerebroventricular) decreased the hypothermic response of an acute ethanol challenge and a similar effect was observed on the development of tolerance. VIP (subcutaneous) decreased the hypothermic response, but the development of tolerance was not affected. Subchronic subcutaneous treatment with PACAP does not influence the development of tolerance to the hypothermic effect of ethanol.

The results suggest that peptides of the secretin-VIP-glucagon family can modulate the development of tolerance to morphine and ethanol.

REFERENCES

1. WIESENFELD-HALLIN, Z. 1987. Intrathecal vasoactive intestinal polypeptide modulates spinal reflex excitability primarily to cutaneous thermal stimuli in rats. Neurosci. Lett. **80:** 293–297.
2. CRIDLAND, R. A. & L. J. HENRY. 1988. Effects of intrathecal administration of neuropeptides on a spinal nociceptive reflex in the rat: VIP, galanin, CGRP, TRH, somatostatin and angiotensin II. Neuropeptides **11:** 23–32.
3. KOMISARUK, B. R., C. BANAS, S. B. HELLER, B. WHIPPLE, G. F. BARBATO & F. JORDAN. 1988. Analgesia produced by vasoactive intestinal polypeptide administered directly to the spinal cord in rats. Ann. N.Y. Acad. Sci. **527:** 650–654.

The Effect of Vasoactive Intestinal Peptide (VIP) and Inhibition of Nitric Oxide on Renal Tissue Injury of Rats Exposed to Hemorrhagic Ischemia and Retransfusion: A Possible Interaction Mechanism among Mast Cells and Tissue Histamine

S. H. ERDEN,[a] N. TUNÇEL,[a,e] Y. AYDYN,[a] V. ŞAHINTÜRK,[b]
M. KOŞAR,[d] AND M. TUNÇEL[c]

[a]Department of Physiology, [b]Department of Histology and Embryology,
Faculty of Medicine, University of Osmangazi, Eskişehir, Turkey

[c]Department of Analytical Chemistry, Faculty of Pharmacy, University
of Anadolu, Eskişehir, Turkey

[d]Medicinal and Aromatic Plant and Drug Research Center (TBAM),
Eskişehir, Turkey

In hemorrhagic ischemia, which may be viewed as "whole body ischemia," the oxygenation and metabolism of tissues worsen because of inadequate perfusion. Depletion of ATP, release of lysosomal enzymes, and the build up of toxic products of metabolism induce cellular damage during ischemia.[1,2] Reconstitution of blood volume restores tissue perfusion and oxygenation, which must inevitably lead to reactive oxygen products inducing lipid peroxidation, which then disrupts the structural integrity of the cell membrane and thus contributes to tissue injury.[1,3–5]

Kidneys are very susceptible to hemorrhagic ischemia. Reconstitution of blood volume after a period of ischemia appears to exacerbate tissue injury during ischemic period.[6] Renal ischemia and reperfusion are inevitably conditions in many clinical situations like renal transplantation, renal vascular surgery, elective cardioplegia, and sepsis besides hemorrhagic shock.[7,8] For this reason several studies have been focused on protecting and understanding of the fundamental mechanisms of ischemia and reperfusion injury of kidneys.

The interaction of reactive oxygen products, nitric oxide (NO), and the histamine released from activated mast cells that produces the undesirable effects of ischemia-reperfusion is not well known. It has been reported that reactive oxygen products can induce mast cell degranulation and release of histamine, which are shown to inhibit superoxide anion ($O_2^{\bullet}$)

[e]Corresponding author: Dr. Neş'e Tunçel, University of Osmangazi, Faculty of Medicine, Department of Physiology, 26040 Eskişehir, Turkey.

production from neutrophils.[9–11] Thus, there is a bidirectional control mechanism between neutrophils and mast cells. A similar relationship between NO and histamine release from mast cells has been reported as well.[10,12]

Current studies have reported that hypoxia has been identified as a stimulus for NO release and that it enhances the half-life of NO.[13] Interaction of NO with reactive oxygen species generates cytotoxic products that augment tissue injury. In contrast, nitric oxide and superoxide anion interacting rapidly prevents leukocyte and endothelial cell interactions, which protect tissue.[12–17] Nitric oxide has been shown to inhibit histamine release from activated mast cells.[10,18,19] Hence, there has been a complex interaction among NO, histamine released from mast cells, and reactive oxygen species during ischemia-reperfusion.

Increasing evidence indicates that VIP protects rat renal tissue from the undesirable effects of reperfusion injury. The cytoprotective effect of VIP, which inhibits mast cell degranulation induced by hemorrhage, was observed in a dose-dependent manner in rats in an experimental hemorrhagia-reperfusion models.[20,21] In another study, VIP protected the rat renal tissue from ischemia-reperfusion injury without any increase in either superoxide dismutase or catalase activity.[6] An antioxidant activity of VIP is also demonstrated by many other studies in lung, heart, retina, and skeletal muscle tissues.[9,18,22–25]

The purpose of the present study was to investigate the effects of VIP administration and NO inhibition on the activities of mast cells, lipid peroxidation, and histamine levels in the renal tissues of the rat during hemorrhagic ischemia-retransfusion.

MATERIALS AND METHODS

Thirty-seven male Sprague-Dawley rats weighing between 320 and 360 g, were used for the experiment. Rats were fed regular industrial rat chow and allowed free access to water. Anesthesia was induced with an intraperitoneal injection of urethane ($1 g \cdot kg^{-1}$). A heat lamp was used to maintain the body temperature at $36.9 \pm 0.4°C$. The rats had no endotracheal intubation and respiration was spontaneous.

The left femoral vein was cannulated (PE 50 tube) for heparinized saline and drug administration. The left femoral artery was cannulated for withdrawal of blood. The experiments were carried out in six groups: The following procedures were applied for all groups but the control group. After a 20-min stabilization period following surgery, 30% of total blood volume (assuming 80 ml·kg⁻¹ of blood per rat) was withdrawn with a withdrawal pump over a period of 20 min.

(I) Ischemia group ($N = 7$). After 30% hemorrhage, rats were killed following a 35-min hemorrhagic ischemia period.

(IR) Ischemia-retransfusion group ($N = 6$). After 35 min of 30% hemorrhage, shed blood was retransfused and rats were kept alive for 5 h (retransfusion period).

(IR+VIP) Ischemia-retransfusion + VIP group ($N = 6$). After 35 min of 30% hemorrhage and prior to the shed blood retransfusion, 25 ng kg⁻¹ VIP was administered via bolus intravenous injection and rats were kept alive for 5 h.

(IR+L-NAME) Ischemia-retransfusion + L-NAME group ($N = 6$). After 35 min of 30% hemorrhage and prior to the shed blood retransfusion, 50 mg kg⁻¹ L-NAME was administered via bolus intravenous injection and rats were kept alive for 5 h.

(IR+VIP+L-NAME) Ischemia retransfusion +VIP+L-NAME group ($N = 6$). After 35 min of 30% hemorrhage and prior to the shed blood retransfusion, 25 ng kg⁻¹ VIP and 50 mg kg⁻¹ L-NAME was administered via bolus intravenous injection and rats were kept alive for 5 h.

(C) Control (Sham operated) ($N = 6$).

The retransfused blood in appropriate groups was autologous. At the end of each experimental protocol, each rat was killed by cervical dislocation, both kidneys were dissected immediately, blotted on a filter paper, and weighed. One of the kidneys was stored at -80°C for determination of malondialdehyde (MDA). The other kidney was dissected into two halves. One half of the kidney was stored at -80°C to histamine determination and the other half was fixed in neutral formalin for histological evaluation.

Because the experimental protocol of groups 3, 4, 5, and 6 lasted approximately 7 h, the protocol for groups 1 and 2 were adjusted to last 7 h as well for the sake of time matching among the groups. Thus, the effect of time measuring MDA and histamine activity would be eliminated.

Renal tissues were homogenized in 10 volumes of ice-cold 50 mM potassium phosphate buffer, pH 7.4. These tissue extracts were then subjected to ultrasonication (15 sec) followed by centrifugation ($20,000 \times g$ for 20 min). MDA was measured spectrophotometrically and histamine was measured by HPLC in the supernatant.

Malondialdehyde Determination

MDA was determined spectrophotometrically by using thiobarbituric acid test and expressed as nmole/mg kidney tissue.[26]

Histamine Determination

An HPLC combination of a RF 535 model fluorescence detector, CR 4A integrator and an injection part having a 10-μl loop (all Shimadzu, Tokyo, Japan) were used for analysis of histamine. A Phenomenex Hypersil 3μ-C18 column was used for the assay. The mobile phase was 50 volume (0.07 M K_2HPO_4 at pH 9.6) and 50 volume HPLC methanol. Fluorometric detections were performed at $\lambda_{ex} = 345$ nm and $\lambda_{em} = 445$ nm. The flow rate was 1 ml/min. Standard samples of histamine and 1-methylhistamine (π-methylhistamine) were supplied from Sigma Chemical Co. (St. Louis, MO, USA). Other chemicals were analytical grade and they were provided from Merck Co. (Darmstadt, Germany). Double-distilled water and HPLC grade methanol were used for the preparation of mobile phase.

Histamine was extracted from tissue by the *n*-butanol extraction method.[27] Two-ml aliquots containing histamine were transferred to the polyethylene test tubes. The eluates were evaporated to dryness by lyophilization in a Leybold-Heraeus freeze-dryer and stored at -70°C until further use.[28] Lyophilized samples were redissolved in 250 μl double-distilled water, 1×10^{-5} M 250 μl methylhistamine as an internal standard was added and mixed with 200 μl reaction buffer.[29] Then 30 μl *o*-phthalaldehyde (OPA) solution (0.1% in methanol) was added to the mixture. After a 3-min incubation at room temperature, the reaction was stopped by addition of 20 μl 1 N HCl, and 10 μl of the sample were immediately injected to the HPLC column.

During the quantification, internal standard technique was used. π-Methylhistamine was the internal standard as published elsewhere.[29] Recovery rate of histamine from the tissue was calculated with the extraction of 1×10^{-5} M histamine under the same conditions.

Histological Evaluation

Kidney tissue were fixed in a neutral formalin and were embedded in paraffin. Serial sections from each block were cut at a thickness of 5 μm and were stained with hemotoxylin-eosin for general histopathological evaluation. Alcian blue/safranin staining technique was used to demonstrate the histochemical heterogeneity of mast cells and its degree of degranulation.

The following histological evaluation of sections were made under light microscopy in six standardized sections of 18 selected serial sections. Mast cells were assigned to the one of the following categories: Alcian blue–stained cells, showing only blue granules; safranin–stained cells, showing only red granules. Mast cells were counted by using an eyepiece micrometer (OC-M) (X40, 3960 μm^2).

Kidney histology samples obtained from all groups were evaluated using the following scale: "0" there were no tissue damage in renal medullar, corticomedullar and cortical zone; tissue damage was seen in cortical or corticomedullar zone (+); tissue damage was seen both in cortical and corticomedullar zone (++); tissue injury was seen in all three zones (+++).

Statistical Analysis

MDA and histamine levels measured from tissue samples and mast cell count were expressed as the mean ± S.E. Statistical analysis was performed by using one-way ANOVA and Duncan's multiple range test. Data obtained from histological examination of renal tissue stained for evaluation of tissue damage was statistically analyzed by using Kruskal-Wallis one way variance analysis and Mann Whitney U-test.

RESULTS

FIGURE 1 illustrates the effect of hemorrhage, shed blood retransfusion, and blood retransfusion together with 25 ng kg^{-1} VIP, 50 mg kg^{-1} L-NAME, and 25 ng kg^{-1} VIP + 50 mg kg^{-1} L-NAME on the MDA level of renal tissue. Ischemia significantly increased MDA level compared to all other groups. FIGURE 2 illustrates kidney tissue histamine levels in all groups. Histamine levels significantly increased in ischemia retransfusion + VIP group and markedly decreased in ischemia-retransfusion group compared to all groups.

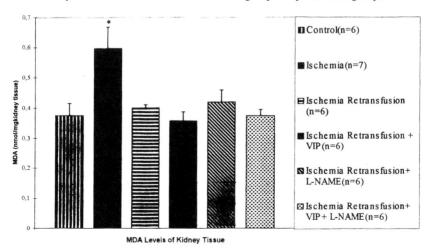

FIGURE 1. MDA levels of kidney tissue. Data are mean ± SE. * $p < 0.05$ Ischemia group compared to the rest of the other groups.

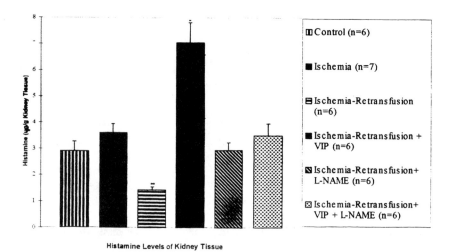

FIGURE 2. Histamine levels of kidney tissue. Data are mean ± SE. * $p < 0.05$ compared to other groups and ** $p < 0.05$ compared to other groups.

Histological Data

TABLE 1 shows the degree of kidney tissue damage obtained from histological examinations of kidney tissue. There was greater damage in animals subjected to hemorrhagic ischemia (FIG. 3) and the tissue damages were gradually decreased in L-NAME, ischemia-retransfusion, VIP + L-NAME, and VIP groups (FIGS. 4–7). The highest tissue protection was observed in the VIP treatment group. Histochemical heterogeneity, degranulation pattern, and the numbers of mast cells are given in TABLE 2. There were few red-stained mast cells in the kidney. For this reason only blue-stained mast cells were counted and evaluated. Ischemia and ischemia-retransfusion induced mast cell degranulation. Administration of VIP either alone or together with L-NAME significantly inhibits mast cell degranulation but L-NAME alone did not inhibit degranulation.

DISCUSSION

Histological evaluations of the present study show that administration of VIP prior to the shed blood retransfusion protected the rat kidney from retransfusion injury following

TABLE 1. The degree of tissue damage obtained from histological examinations of tissue samples

Groups	Degree of Tissue Damage/Number of Rats
C ($N = 6$)	0/6
I ($N = 7$)	++/1, +++/6
IR ($N = 6$)	+/2, ++/3, +++/1
IR+VIP ($N = 6$)	0/1, +/5
IR+L-NAME ($N = 6$)	++/5, +++/1
IR+VIP+L-NAME ($N = 6$)	0/1, +/4, ++/1

Note: See text for details.

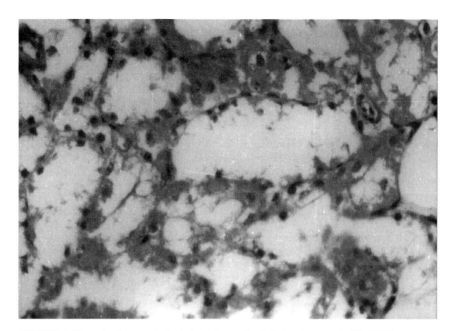

FIGURE 3. Necrosis of the cortical tubules in hemorrhagic ischemia group (×132, H&E).

hemorrhagic ischemia. In the animals subjected to the hemorrhagic ischemia, there was a greater damage, based on the data obtained from histological evaluation and tissue MDA levels as an index of lipid peroxidation. VIP administration together with shed blood totally prevented tissue damage. When VIP was given together with L-NAME, the protection of renal tissue was not as great as in the VIP treatment only groups. Greater damage was observed in the L-NAME group as compared to the hemorrhagic ischemia-retransfusion group. On the basis of these results, we concluded that NO synthase inhibition prior to the retransfusion is not beneficial for the kidney tissue and NO seems to be involved in the protective effect of VIP. When ischemia, ischemia-retransfusion (IR), and IR+L-NAME groups were compared to each other, we observed that tissue damage in the ischemia-retransfusion group was less than in the IR+L-NAME and ischemia groups. This result may indicate that NO increases during the ischemic period and induces tissue damage. In shed blood retransfusion period, NO may be decreased in the tissues by the hemoglobin introduced by retransfused blood.

TABLE 2. Histochemical heterogeneity, degranulation pattern, and the number of mast cells

Groups	Blue-Stained Mast Cells		
	Degranulated	Granulated	Percent Degranulated
C ($N = 6$)	0.41 ± 0.20	6.5 ± 1.46	5.9
I ($N = 7$)	2.14 ± 0.38*	0.64 ± 0.30	76.9
IR ($N = 6$)	1.33 ± 0.60*	5 ± 1.77	21.0
IR +VIP ($N = 6$)	0.16 ± 0.10†	4 ± 0.76	3.8
IR + L-NAME ($N = 6$)	1.41 ± 0.47	2.83 ± 0.82	34.0
IR +VIP+L-NAME ($N = 6$)	0.33 ± 0.21#	5.25 ± 1.31	5.9

Note: *$p < 0.05$, compared to the control; † $p < 0.05$, compared to the IR; and # $p < 0.05$, compared to the IR+L-NAME.

NO, which is an important physiologic regulator of blood flow and perfusion pressure, also has an important role in modulating of tissue injury.[15,30-33] The release and effects of NO in ischemia-retransfusion experimental models are still controversial. Ischemia has been identified as a stimulus for NO release while restoration of blood supply has an inhibiting effect.[13] Excessive amounts of NO lead to various reactions in the presence of reactive oxygen species. NO may react with hydrogen peroxide to form potentially cytotoxic singlet oxygen. NO may increase the release of iron from ferritin, leading to an increase in hydroxyl radical production via superoxide anion and hydrogen peroxide.[34] NO also reacts with superoxide to form peroxynitrite.[16] Ma and colleagues have reported that the rate of reaction between superoxide anion and NO was greater than that of superoxide anion and superoxide dismutase.[34] Beckman and colleagues have proposed that peroxynitrite anion, a product of NO and superoxide anion reaction, oxidize deoxyribose to MDA and the increase in MDA has been suggested as an index of increasing peroxynitrite anion.[35] Additionally, some investigators postulated that large quantities of NO inhibit both aconitase enzyme of the Krebs cycle and complexes I and II of the mitochondrial electron transport chain as well.[16]

In the present study, the increase in MDA and severe tissue injury during ischemia period would suggest NO release might be elevated and NO reacts with superoxide anion to produce peroxynitrite. MDA level did not change significantly during retransfusion followed by hemorrhagic ischemia compared to the control group and shed blood retransfusion decreased the tissue damage compared to the ischemic group. These data indicate that NO generation decreases during restoration of blood supply in kidney tissue. Palmer and colleagues reported that NO was inhibited by hemoglobin.[36] Increasing the amount of hemoglobin concentration during retransfusion followed by ischemia possibly inhibited NO generation, which is increased in ischemia. The decreased MDA level in retransfusion period can be explained by reduced NO levels leading to reduced formation of peroxynitrite anion.

In the present study, we observed that during the period of ischemia and retransfusion mast cell degranulation increases. Consumption of NO by superoxide anion and hemoglobin may in part be responsible for inducing degranulation: some studies suggest that NO has an inhibitory effect on mast cell degranulation.[10,19,37] The effect of L-NAME on mast cell degranulation is somewhat confirmatory that NO could be mast cell stabilizer. On the other hand, administration of VIP alone or together with L-NAME effectively inhibited mast cell degranulation. It indicates that VIP is a potential mast cell stabilizer and that inhibition of mast cell degranulation contributes to the protection of renal tissue against ischemia-retransfusion injury. Histamine is one of the mediators released by mast cells. Unlike many other tissues, the kidney has very few mast cells, suggesting that there are different sources of histamine in kidney. Karlson and Rosengren reported on "nascent histamine," histamine synthesized by L-histidine decarboxylase from L-histidine in non mast cell tissue in situ without being stored in the kidney or, especially, in glomeruli. It is conceivable that histamine may modulate mesengial cell functions, blood flow and glomerular filtration rate. It has been supposed that both mast cell–derived and non mast cell–derived histamine are not accumulated in kidney and are rapidly catabolized by diaminooxidase and histamine methyltransferase.[40] Although histamine has been proposed to have a deleterious effect by increasing epithelial permeability and to cause profound increase in leukocyte adhesion and migration, some in vitro studies indicate that superoxide anion production in leukocytes can be inhibited by histamine via an H_2 receptor.[9,11] Additionally, it has also been reported that histamine inhibits superoxide anion production in renal mesengial cells and also that the inhibitory effect of histamine was potentiated in the presence of phosphodiesterase inhibitor.[41]

In the present study, the results related to mast cell number and tissue histamine level indicate that mast cells are not the only source of histamine in renal tissue, especially for the ischemia period. Decreased tissue histamine level during retransfusion period suggests that histamine may be consumed in order to defend against reactive oxygen species and the harmful effects of leukocytes. VIP increased the tissue histamine level and protected the renal tissue effectively. Thus, it can be suggested that histamine may not be a deleterious agent for ischemic-retransfused kidney. Furthermore, histamine can be a mediator that assists the tissue protective effect of VIP.

Recent studies reported that VIP protects the heart, lung, retina, kidney and skeletal muscle tissue from the injury of ischemia-retransfusion.[6,22–24,42,43] The detailed mechanisms by which VIP may be able to protect tissues against injury are still not known. However, some studies support the view that VIP has a potent protective activity against tissue injury triggered by xanthine/xanthine oxidase, which induces the generation of reactive oxygen species. Additionally, VIP has been shown to inhibit cytokine release by modulating inflammatory cell function, phagocytosis and superoxide radical production in rat alveolar macrophages.[43] VIP also directly inhibits phospholipase A_2 activity.[18] This inhibition could lead to suppressing the liberation of arachidonic acid metabolites. Apart from these special features, VIP is capable of scavenging singlet oxygen and hydroxyl radical by acting as an antioxidant agent.[18,44] In the present study the best tissue protection was observed in the VIP-treated group. When L-NAME was added to VIP, the protective effect of VIP was reduced. Thus, NO may be involved in the protective effect of VIP and NO may not be a deleterious agent when a strong antioxidant agent, like VIP, is present in the environment.

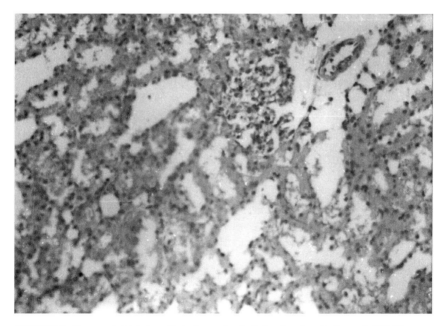

FIGURE 4. Degenerated tubule cells in hemorrhagic ischemia retransfusion + L-NAME group ($\times$ 66, H&E).

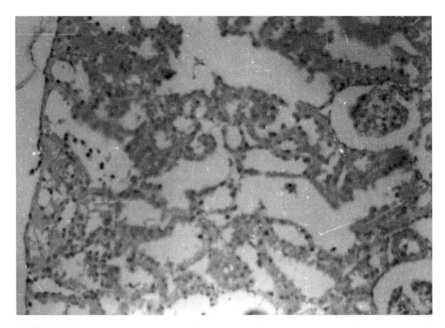

FIGURE 5. Necrosis of the subcapsular tubules in hemorrhagic ischemia retransfusion group (× 66, H&E).

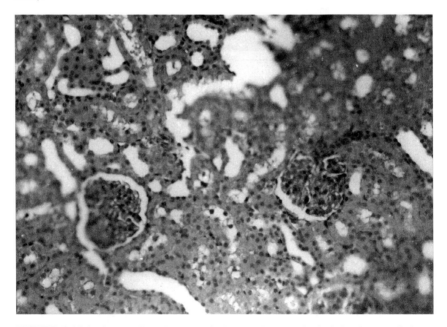

FIGURE 6. Slight degeneration of some cortical tubules in hemorrhagic ischemia retransfusion + VIP + L-NAME group (× 66, H&E).

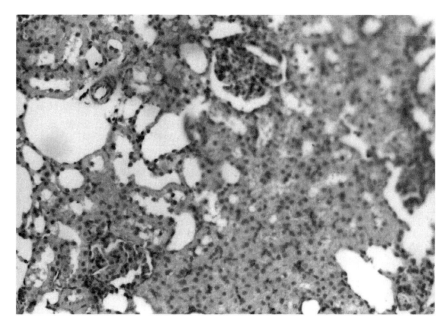

FIGURE 7. Highly preserved, virtually normal histological pattern of kidney in hemorrhagic ischemia retransfusion + VIP group (× 33, H&E).

REFERENCES

1. FARBER, J. L., K. R., CHIEN & S. MITTNACHT. 1981.The pathogenesis of irreversible cell injury in ischemia. Am. J. Pathol. **102**: 271–281.
2. WEISFELDT, M. L. 1987. Reperfusion and reperfusion injury. Clin. Res. **35**(1): 13–20.
3. BAST, A., G. R., M. M. HAENEN & C. J. A. DOELMAN. 1991. Oxidants and antioxidants: State of the art. Am. J. Med. **91**(3): 2–13.
4. CROMPTON, M., A. COSTI & L. HAYAT. 1987. Evidence for the presence of reversible Ca^{2+} dependent pore activated by oxidative stress in heart mitochondria. Biochem. J. **245**: 915–918.
5. WHITESIDE, C. & M. H. HASSAN. 1987. Induction and inactivation of catalase and superoxide dismutase of Escherichia coli by ozone. Arch. Biochem. Biophys. **257**: 464–471.
6. UZUNER, K., N. TUNÇEL, Y. AYDIN, M. TUNÇEL, F. GÜRER, P. BENLI & D. AK. 1995. The effect of vasoactive intestinal peptide (VIP) on superoxide dismutase and catalase activities in renal tissues of rats exposed hemorrhagic ischemia-reperfusion. Peptides. **16**: 911–915.
7. BULKLEY, G. B. 1983.The role of oxygen free radicals in human disease processes. Surgery **94**: 407–411.
8. BULKLEY, G. B. 1987. Free radical mediated reperfusion injury: A selective review. Br. J. Cancer **55**(8):66–73.
9. CHING, T-L., J. G. KOELEMIJ & A. BAST. 1995. The effect of histamine on the oxidative burst of HL60 cells before and after exposure to reactive oxygen species. Inflamm. Res. **44**: 99–104.
10. GABOURY, J., R. C. WOODMAN, D. N. GRANGER, P. REINHARDT & P. KUBES. 1993. Nitric oxide prevents leukocyte adherence: role of superoxide. Am. J. Physiol. **265**: H862–H867.
11. KANWAR, S. & P. KUBES. 1994 Mast cells contribute to ischemia-reperfusion-induced granulocyte infiltration and intestinal dysfunction. Am. J. Physiol. **267**: G316–G321.
12. KUBES, P., S. KANWAR, X-F. NIU & J. P. GABOURY. 1993. Nitric oxide synthesis inhibition induces leukocyte adhesion via superoxide and mast cells. FASEB **7**: 1293–1299.
13. HOSHIDA, S., N. YAMASHITA, J. IGARASHI, M. NISHIDA, M. HORI, T. KAMADA, T. KUZUYA & M. TADA. 1995. Nitric oxide synthase protects the heart against ischemia-reperfusion injury in rabbits. J. Pharmacol. Exp. Ther. **274**: 413–418.

14. COOKE, J. P. & P. S. TSAO. 1993. Cytoprotective effects of nitric oxide. Circulation **88**: 2451–2456.
15. MONCADA, S., R. M. J. PALMER & E. A. HIGGS. 1991. Nitric oxide: Physiology, pathophysiology, and pharmacology. Pharm. Rev. **43**: 109–135,
16. RODEBERG, D. A., M. S. CHAET, R. C. BASS, M. S. ARKOVITZ & V. F. GARCIA. 1995. Nitric oxide: an overview Am. J. Surg. **170**: 292–303.
17. RUBANYI, G. M. & P. M. VANHOUTTE. 1986. Superoxide anions and hyperoxia inactivate endothelium-derived relaxing factor. Am. J. Physiol. **250**: H822–H827.
18. KALFIN, R., N. MAULIK, R. M. ENGELMAN, G. A. CORDIS, K. MILENOV, L. KASAKOV & D. K. DAS. 1986. Protective role of intracoronary vasoactive intestinal peptide in ischemic and reperfused myocardium. J. Pharmacol. Exp. Therap. Immunopath. **39**: 308–318.
19. MASSINI, E., M. G. DI BELLO, A. PISTELLI, S. RASPANTI, F. GAMBASSI, L. MUGNAI, M. LUPINI & P. F. MANNAIONI. 1994. Generation of nitric oxide from nitrovasodilators modulates the release of histamine from mast cells. J. Physiol. Pharmacol. **45**(1):41–52.
20. TIKIZ, H., N. TUNÇEL, F. GÜRER & C. BAYÇU. 1991. Mast cell degranulation in hemorrhagic shock in rats and the effects of vasoactive intestinal peptide (VIP), aprotinin and H1 and H2 receptor blockers on degranulation. Pharmacology **43**: 47–52.
21. TIKIZ, H., N. TUNÇEL, M. Z. AKIN & F. GÜRER. 1992. The effect of VIP and naloxane combination on survival rates in rats exposed to severe hemorrhage. Peptides **13**: 83–90.
22. SAID, S. I. 1994. VIP and nitric oxide: physiological co-transmitters with antagonistic roles in inflammation. Biomed. Res. **15**(Suppl. 2): 79–84.
23. SAID, S. I. 1995. Vasoactive intestinal peptide. In Airway Smooth Muscle, Peptide Receptors, Ion Channels and Signal Transduction. D. Roseburn & M. A. Giembyez, Eds. Birkhjuser Verltag Basel, Switzerland.
24. TUNÇEL, N., H. BASMAK, K. UZUNER, M. TUNÇEL, G. ALTIOKKA, V. ZAIMOĞLU & A. ÖZER. 1996. Protection of rat retina from ischemia-reperfusion injury by vasoactive intestinal peptide (VIP): The effect of VIP on lipid/peroxidation and antioxidant enzyme activity of retina. Ann. N. Y. Acad. Sci. **805**: 489–497.
25. TUNÇEL, N., S. ERDEN, K. UZUNER, G. ALTIOKKA & M. TUNÇEL. 1997. Ischemic-reperfused rat skeletal muscle: The effect of vasoactive intestinal peptide (VIP) on contractile force, oxygenation, and antioxidant enzyme systems. Peptides **18**: 2:269–275.
26. RAO, N. A. 1990. Role of oxygen free radicals in retinal damage associated with experimental uveitis. Trans. Am. Ophthalmol. Soc. **88**: 797–850.
27. SHORE, P. A., A. BURKHALTER & V. H. COHN. 1959. A method for the fluorometric assay of histamine in tissues. J. Pharmacol. Exp. Ther. **127**: 182–184.
28. STEINMANN, N., C. J. ESTLER & O. DANN. 1988. Plasma histamine levels in rats treated with trypanocidal diamidines. Pharmacology **36**: 204–209.
29. TSURATA, Y., K. KOHASHI & Y. OHKURA. 1981. Simultaneous determination of histamine and N-methylhistamine in human urine and rat brain by high-performance liquid chromatography with fluorescence detection. J. Chromatogr. **224**: 105–110.
30. FUKUDA, H., Y. SAWA, K. KADOBA, K. TANIGUCHI, Y. SHIMAZAKI & H. MATSUDA. 1995. Supplement of nitric oxide attenuates neutrophil-mediated reperfusion injury. Circulation **92**(Suppl.II): 413–416.
31. LASZLO, F., B. J. R. WHITTLE & S. MONCADA. 1994. Interactions of constitutive nitric oxide with PAF and thromboxane on rat intestinal vascular integrity in acute endotoxaemia. Br. J. Pharmacol. **113**: 1131–1136.
32. MAY, G. R., P. CROOK, P. K. MOORE & C. P. PAGE. 1991. The role of nitric oxide as an endogenous regulator of platelet and neutrophil activation within the pulmonary circulation of the rabbit. Br. J. Pharmacol. **102**: 759–763.
33. WELCH, G. & J. LOSCALZO. 1994. Nitric oxide and the cardiovascular system. J. Cardiovasc. Surg. **9**: 361–371.
34. MA, T. T., H. ISCHIROPOULOS & C. A. BRASS. 1995. Endotoxin stimulated nitric oxide production increases injury and reduces rat liver chemiluminescence during reperfusion. Gastroenterology **108**: 463–469.
35. BECKMAN, J. S., T. W. BECKMAN, J. CHEN, P. A. MARSHALL & B. A. FREEMAN. 1990. Apparent hydroxyl radical production by peroxynitrite: Implications for endothelial injury from nitric oxide and superoxide. Proc. Natl. Acad. Sci. USA **87**: 1620-1624.

36. PALMER, R. M. J., A. G. FERRIGE & S. MONCADA. 1987. Nitric oxide release accounts for the biological activity of endothelium-derived relaxing factor. Nature **327**: 524–526.
37. MANNAIONI, P. F., E. MASINI, A. PISTELLI, D. SALVEMINI & J. R. VANE. 1991. Rat mast cells inhibit platelet aggregation by releasing a nitric oxide-like factor: Influence of histamine release. New perspectives in histamine research: 423–428. Birkhäuser Verlag. Basel.
38. KARLSON, G. & E. ROSENGREN. 1971. Biogenesis and Physiology of Histamine. Arnold. London.
39. KARLSON, G. & E. ROSENGREN. 1972. Histamine: Entering physiology. Experientia **28**: 993–1002.
40. ABBOUD, H. E. &T. P. DOUSA. 1983. Renal metabolism and actions of histamine and serotonin. Miner. Electrolyte Metab. **9**: 246–259.
41. LAURENT, B. & R. ARDAILLOU. 1986. Reactive oxygen species: production and role in the kidney. Am. J. Physiol. 251: F765–776.
42. TUNÇEL, N., F. GÜRER, K. UZUNER, Y. AYDIN & C. BAYÇU. 1996. The effect of vasoactive intestinal peptide (VIP) on mast cell invasion/degranulation in testicular interstitium of immobilized + cold stressed and β-endorphin-treated rats. Peptides **17**: 817–824.
43. PAKBAZ, H., H. BERISHA, H. D. FODA, A. ABSOOD & S. I. SAID. 1994. Vasoactive intestinal peptide (VIP) and related peptides: a new class of anti-inflammatory agents? International Symposium on Vasoactive Intestinal Peptide, Pituitary Adenylate Cyclase Activating Polypeptide and Related Regulatory Peptide, G. Rosselin, Ed.: 597–605. Singapore. World Scientific Publishing.
44. SAKAKIBARA, H., J. TAKAMATSU & S. I. SAID. 1990. Vasoactive intestinal polypeptide (VIP) inhibits superoxide anion release from rat alveolar macrophages. Am. Rev. Respir. Dis. **141**: A645.

Enhancement of Systemic and Pulmonary Vasoconstriction by β-Amyloid Peptides and its Suppression by Vasoactive Intestinal Peptide

S.I. SAID,[a] S. RAZA, AND H. I. BERISHA

Department of Veterans Affairs Medical Center, Northport, New York and SUNY Stony Brook, New York 11794-8172, USA

β-amyloid protein (Aβ) has been implicated, both biochemically and genetically, in the pathogenesis of Alzheimer's disease (AD).[1-3] Deposition of Aβ in senile plaques is a pathological hallmark of AD, and focal deposition of Aβ in adult rat cerebral cortex causes neuronal loss.[4] In addition to its own neurotoxicity, Aβ enhances neuronal excitotoxic damage.[3,7-9] Aβ peptides were recently also shown to induce endothelial vascular injury and accentuate vasoconstriction,[10,11] thus potentially contributing to neuronal injury through reduction of cerebral blood flow.

EXPERIMENTAL DESIGN AND METHODS

In this study, we investigated the ability of two peptides, corresponding to amino acid sequences $A\beta_{1-40}$ and $A\beta_{1-42}$, to increase vascular tone and to enhance norepinephrine (NE) -induced contraction of isolated, perifused strips of guinea pig aorta (aortic arch, thoracic or abdominal portions) and pulmonary artery (PA). The tissue strips were prepared and placed in organ baths containing Krebs buffer, equilibrated with 95% O_2–5% CO_2, at 37°C, as described.[12]

In view of the known vasodilator and anti-injury effects of the neuropeptide vasoactive intestinal peptide (VIP),[13-15] we also examined the possible prevention or attenuation of the effects of Aβ peptides by VIP. In two experiments, we also tested the possible attenuation of the response to Aβ by the enzyme superoxide dismutase (SOD), on the basis that the vascular response is mediated by superoxide and other reactive oxygen species.[10] In a typical experiment, the Aβ peptide was added to the tissue bath 10 min before NE. When used, VIP and SOD were given 10 min before the Aβ peptide.

RESULTS AND DISCUSSION

(1) NE, in a concentration range of 5×10^{-8} to 5×10^{-6} M – 10^{-5} M, elicited concentration-dependent contractions of aortic and PA strips (FIG. 1). (2) The contractions were markedly accentuated, in magnitude and duration, at all concentrations of NE

Corresponding author: 17 - 040 HSC, SUNY Stony Brook, NY 11794-8172; Tel.: 516-444-1754; Fax: 516-444-7502; E-mail: SSAID@EPO.SOM.SUNYSB.EDU

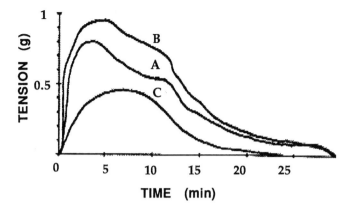

FIGURE 1. β-Amyloid enhancement of contractile response of guinea pig aorta to norepinephrine, and its prevention by VIP. (**A**) Contraction guinea pig aorta strip to 5×10^{-7} M norepinephrine (NE) alone, (**B**) to NE plus 10^{-7} M Aβ$_{1-42}$ (**C**), and to NE plus Aβ plus 10^{-6} M VIP.

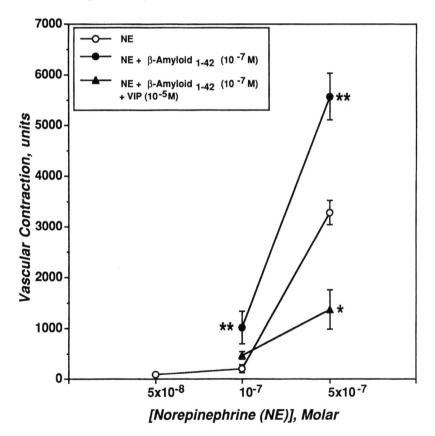

FIGURE 2. Concentration-dependent contractions of guinea pig aortic strips to norepinephrine (NE), their accentuation with β-amyloid, and attenuation by VIP. $*p < 0.05$ vs. NE, $**p < 0.01$ vs. NE. Units plotted on vertical scale are scale units of area under contraction curve.

tested, in the presence of $A\beta_{1-40}$ or $A\beta_{1-42}$, at concentrations of 10^{-7} to 10^{-6} M (FIGS. 1 and 2). (3) Co-treatment with VIP ($10^{-7} - 10^{-5}$ M) reduced or abolished the contraction-enhancing effect of $A\beta$ and, in the absense of $A\beta$, VIP attenuated the contractile response to NE (FIG. 1).

The mechanisms of constrictor potentiation by $A\beta$ are not fully known, but the effect may be mediated by reactive oxygen species, since it can be prevented by SOD.[10] VIP, known to have the ability to scavenge singlet oxygen[16] and hydroxyl radical,[17] may exert its protective action, at least in part, by anti-oxidant mechanisms. In two experiments in which we compared the suppression of the $A\beta$ peptide action by both agents, VIP (10^{-6} M) and SOD (150 U/ml) were approximately equipotent in attenuating the vascular contraction.

SUMMARY AND CONCLUSIONS

(1)$A\beta$ peptides potentiate vasoconstriction, caused by norepinephrine, and possibly other endogenous vasoconstrictors. If this potentiation occurs in the cerebral circulation, close to sites of $A\beta$ deposition in AD brains, the enhanced vasoconstriction could result in neuronal ischemia and death.

(2)By neutralizing this deleterious effect of $A\beta$, and through other neuroprotective mechanisms, VIP may provide an important defense against neuronal loss in AD.

REFERENCES

1. KOWALL, N.W., M. F. BEAL, J. BUSCIGLIO, L. K. DUFFY & B. A. YANKNER. 1991. An *in vivo* model for the neuro-degenerative effects of β amyloid and protection by substance P. Proc. Natl. Acad. Sci. USA **88**: 7247–7251.
2. SCHMECHEL, D. E., A. M. SAUNDERS, W. J. STRITTMATTER, B. J. CRAIN, C. M. HULETTE, S. H. JOO, M.A. PERICAK-VANCE, D. GOLDGABER & A. D. ROSES. 1993. Increased amyloid beta-peptide deposition in cerebral cortex as a consequence of apolipoprotein E genotype in late-onset Alzheimer disease. Proc. Natl. Acad. Sci. USA **15**: 9649–9653.
3. LE, W.-D., L. V. COLOM, G. XIE, G. SMITH, M. ALEXIANU & S. H. APPEL. 1995. Cell death induced by β-amyloid 1-40 in MES 23.5 hybrid clone: the role of nitric oxide and NMDA-gated channel activation leading to apoptosis. Brain Res. **686**: 49–60.
4. SELKOE, D. J. 1993. Physiological production of the β-amyloid protein and the mechanism of Alzheimer's disease. Trends Neurosci. **16**: 403–409.
5. BEHL, C., L. HOVEY, III, S. KRAJEWSKI, D. SCHUBERT & J. C. REED. 1993. Bcl-2 prevents killing of neuronal cells by glutamate but not by amyloid beta protein. Biochem. Biophys. Res. Commun. **197**: 949–956.
6. PREHN, J. H., V. P. BINDOKAS, J. JORDAN, M. F. GALINDO, G. D. GHADGE, R. P. ROOS, L. H. BOISE, C. B. THOMPSON, S. KRAJEWSKI, J. C. REED & R. J. MILLER. 1996. Protective effect of trans-forming growth factor-beta 1 on beta-amyloid neurotoxicity in rat hippocampal neurons. Mol. Pharmacol. **49**: 319–328.
7. KOH, J., L. L. YANG & C. W. COTMAN. 1990. β-amyloid protein increases the vulnerability of cultured cortical neurons to excitotoxic damage. Brain Res. **533**: 315–320.
8. MATTSON, M. P., B. CHENG, D. DAVIS, K. BRYANT, I. LIEBERBURG & R. E. RYDEL. 1992. β-Amyloid peptides destabilize calcium homeostasis and render human cortical neurons vulnerable to excitotoxicity. J. Neurosci. **12**: 376–89.
9. MATTSON, M. P., S. W. BARGER, B. CHENG, I. LIEBERBURG, V. L. SMITH-SWINTOSKY & R. E. RYDEL. 1993. β-Amyloid precursor protein metabolites and loss of neuronal Ca^{2+} homeostasis in Alzheimer's disease.Trends Neurosci. **16**: 409–414.
10. THOMAS, T., G. THOMAS, C. McLENDON, T. SUTTON & M. MULLAN. 1996. β-Amyloid-mediated vasoactivity and vascular endothelial damage. Nature **380**: 168–171.
11. SAID, S. I., H. I. BERISHA & S. RAZA. 1997. Vasoactive intestinal peptide protects against β-amyloid peptide enhancement of systemic vasoconstriction. FASEB J. **11**: A336.
12. HAMASAKI, Y., M. MOJARAD & S. I. SAID. 1983. Relaxant action of VIP on cat pulmonary artery: Comparison with acetylcholine, isoproterenol and PGE_1. J. Appl. Physiol. **54**: 1607–1611.

13. SAID, S. I. 1982. Vasodilator action of VIP: Introduction and general considerations. *In* Vasoactive Intestinal Peptide. S.I. Said, Ed.:145–148. Raven Press. New York.
14. SAID, S. I. 1996. Vasoactive intestinal peptide and nitric oxide: Divergent roles in relation to tissue injury. Ann. N.Y. Acad. Sci. **805**: 379–388.
15. SAID, S. I. 1996. Molecules that protect: The defense of neurons and other cells (Editorial). J. Clin. Invest. **97**: 2163–2164.
16. MISRA, B. R. & H. P. MISRA. 1990. Vasoactive intestinal peptide, a singlet oxygen quencher. J. Biol. Chem. **265**: 15371–15374.
17. KALFIN, R., N. MAULIK, R. M. ENGELMAN, G. A. CORDIS, K. MILENOV, L. KASAKOV & D. K. DAS. 1994. Protective role of intracoronary vasoactive intestinal peptide in ischemic and reperfused myocardium. J. Pharmacol. Exp. Therap. **268**: 952–958.

The Effect of Vasoactive Intestinal Peptide (VIP) and Inhibition of Nitric Oxide Synthase on Survival Rate in Rats Exposed to Endotoxin Shock

N. TUNÇEL AND F. Ç. TÖRE

Department of Physiology, Faculty of Medicine, University of Osmangazi, 26480, Eskişehir, Turkey

Although septic shock cases are increasing, there are no effective cures at present. The treatment of septic shock remains a therapeutic challenge. Mortality rates of 25% to 50% are currently being reported for patients with septic shock.[1]

Administration of lipopolysaccharides (LPS) *in vivo* induces septic shock, characterized by hypotension, vascular injury, and disseminated intravascular coagulation (DIC), which leads to a fatal dysfunction of various organs. Both the early and late phase hypotension after LPS administration are associated with overproduction of nitric oxide (NO). NO seems to be an important mediator of hypotension in septic shock. Increased NO synthesis during experimental endotoxemia has been shown to have both deleterious and beneficial effects.[2] Inhibition of NO synthase has a limited role in the treatment of septic shock.[1] NO synthase inhibitors cause improvement of hemodynamics but the effect of them on survival rate is still obscure.

Elevated plasma levels of vasoactive intestinal peptide (VIP) were also reported in human and animals during septic shock.[3] The release of VIP in septic shock may be attributed to overreacting defense mechanisms. Recently it has been reported that VIP protects many tissues against various injuries such as ischemia-reperfusion, inflammation, and toxins.[4–10] In the case of acute tissue injury, VIP and NO may have opposing roles.[6] In the present study we investigated the effect of VIP and a NO synthase inhibitor, nitro-L-arginine methyl ester (L-NAME), on survival rate of rats exposed to endotoxin shock.

MATERIALS AND METHODS

Forty Sprague-Dawley rats (300–450 g) of either sex were divided into five groups. Group 1: The controls ($N = 8$); Group 2: The septics ($N = 8$), *E. coli* lipopolysaccharide (LPS) was administered (6 mg kg^{-1} intraperitonally); Group 3: VIP was administered (25 ng kg^{-1} intraperitonally) ($N = 8$); Group 4: L-NAME was administered (30 mg kg^{-1} intraperitonally) ($N = 8$); and Group 5: VIP and L-NAME were administered together ($N = 8$).

Experimental Procedures

Rats were anesthetized with urethane (1.5 g kg^{-1} intraperitonally). The left carotid artery was cannulated and connected to a pressure transducer (Narco Biosystem, P1000B) for the measurement of arterial blood pressure, which was displayed a polygraph recorder (Narco Biosystem, MK III S). Body temperature was maintained by a rectal probe at 37±1°C. Upon completion of the surgical procedure, rats were allowed a stabilization period of 20 minutes. After recording baseline arterial pressure, experiments were begun. At the beginning of experiments, LPS was administered. After 30 min of LPS administration, VIP and L-NAME were injected. Control rats received only physiologic saline (0.1 ml intraperitonally) both at the beginning and 30 min later. The rats were monitored for 6 hours.

Statistics

The data presented is mean ± SE, statistical analysis of mean arterial pressure (MAP) levels was performed by using one-way ANOVA and Duncan's multiple range test. Survival rates were performed by using Friedman two-way variance analysis and $p < 0.05$ taken as significant.

RESULTS

The survival results are shown in TABLE 1. The mean survival time of group exposed to septic shock without treatment was 111.2±50 minutes. Death percentage of the animals dramatically increased in groups treated with L-NAME and VIP + L-NAME after 120 min. At the end of 240 min all animals were dead in these two groups. On the other hand, when only VIP was used 75% of the animals were still alive at the 240 min and after 360 min the survival rate dropped to 50%. From these results it was concluded that the VIP group showed the highest survival rate compared to L-NAME and L-NAME + VIP groups. The mean arterial pressure (MAP) data obtained in both postseptic and posttreatment periods are given in FIGURE 1. Average MAP values of the preseptic period for all groups were not significant different from each other. LPS evoked significant reduction in MAP.

At the 30 min, MAP levels for all groups were significantly different from that of control group except for the L-NAME group. MAP levels of the VIP group did not show any statistical difference from that of control group between 60 and 180 min except for 150 min. On the other hand, MAP values of L-NAME and septic groups showed significant differences from that of control group for every time interval. When L-NAME was administered together with VIP, MAP values of the rats significantly differed from that of the control group at only 120 and 150 min.

TABLE 1. The effects of VIP and L-NAME on survival rates of rats exposed to septic shock

Time	Control	Septic	Septic+VIP	Septic + L-NAME	Septic + VIP + L-NAME
0 min	8 / 8	8 / 8	8 / 8	8 / 8	8 / 8
60 min	8 / 8	8 / 7	8 / 8	8 / 8	8 / 8
120 min	8 / 8	8 / 3	8 / 7	8 / 5	8 / 6
180 min	8 / 8	8 / 0	8 / 6	8 / 2	8 / 3
240 min	8 / 8	8 / 0	8 / 6	8 / 0	8 / 0
300 min	8 / 8	8 / 0	8 / 4	8 / 0	8 / 0
360 min	8 / 8	8 / 0	8 / 4	8 / 0	8 / 0

Note: Beginning from 120 min. Control group significantly different from septic group, $p < 0.05$ septic+L-NAME group, $p < 0.05$ septic+L-NAME+VIP group, $p < 0.05$; and Septic+VIP group significantly different from septic group, $p < 0.05$

CONCLUSION

The present study clearly showed that a single bolus injection of LPS led to a decrease in MAP in anesthetized rats. In our study, it was observed that hypotension induced by LPS was blockaded by L-NAME, causing an increase in MAP values to a level higher than controls for every time interval. But this result was at the expense of survival rate. On the other hand, a single bolus injection of VIP kept the MAP values at the level of controls and showed the best survival rate. Thus it was concluded that higher MAP values were not always good enough for the treatment of septic shock. When VIP was used together with an NO synthase inhibitor, MAP values were normal or even in higher than control at the 120 and 150 min. But survival rate was not different from L-NAME–treated group. It indicates that complete inhibition of endogenous NO synthesis does not work to cope with septic shock. For this reason clinical use of NO synthase inhibitors for the treatment of septic shock should be cautiously considered.

We suggest that VIP could be a potential candidate to be used in the treatment of septic shock.

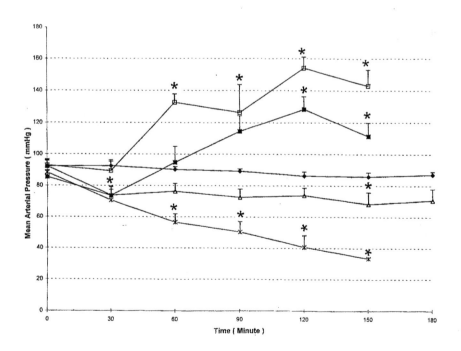

FIGURE 1. The effects of LPS, VIP, and L-NAME on mean arterial pressure levels of rats exposed to septic shock. * $p < 0.05$ vs. control. ◆ control; × septic (LPS); Δ septic + VIP; □ septic + L-NAME; and, ■ septic + VIP + L-NAME. Data are expressed ±SEM, $N = 8$ for each group.

REFERENCES

1. ROBERTSON, F. M., P. J. OFFNER, D. P. CICERI, W. K. BECKER & A. B. PRUITT. 1994. Detrimental hemodynamic effect of nitric oxide synthase inhibition in septic shock. Arch. Surg. **129:** 149–156.

2. KORBUT, R., T. D.WARNER, R. J. GRGYLWESKI & R. J. VANE. 1994. The effect of nitric oxide synthase inhibition on the plasma fibrinolytic system in septic shock in rats. Br. J. Pharmacol. **112:** 289–291.

3. BRANDZAEG, P., O. OKTADALEN, P. KIERULF & P. K. OPSTAD. 1989. Elevated VIP and endotoxin plasma levels in human gram-negative septic shock. Regul. Pept. **24:** 37–44.

4. SAID, S.I. 1996. Molecules That Protect: The defense of neurons. J. Clin. Invest. **97:** 2163–2164.

5. UZUNER, K., N. TUNÇEL, Y. AYDIN, M. TUNÇEL, F. GÜRER, P. BENLI & D. AK. 1995. The effect of vasoactive intestinal peptide (VIP) on superoxide dismutase and catalase activities in renal tissues of rats exposed to hemorrhagic ischemia-reperfusion. Peptides **16:** 911–915.

6. SAID, S. I. 1994. VIP and nitric oxide: Physiological co-transmitters with antagonistic roles in inflammation. Biomed. Res. **15:** 79–84.

7. TUNÇEL, N., H. BASMAK, K. UZUNER, M. TUNÇEL, G. ALTIOKKA, V. ZAIMOGLU, A. ÖZER & F. GURER. 1996. Protection of rat retina from ischemia-reperfusion injury by vasoactive intestinal peptide (VIP): The effect of VIP on lipid peroxidation and antioxidant enzyme activity of retina and choroid. Ann. N.Y. Acad. Sci. **805:** 489–498.

8. TUNCEL, N., S. ERDEN, K. UZUNER, G. ALTIOKKA & M. TUNÇEL. 1996. Ischemic-reperfused rat skeletal muscle: The effect of vasoactive intestinal peptide (VIP) on contractile force, oxygenation and antioxidant enzyme systems. Peptides **18** (2): 269–275.

9. PAKBAZ, H., H. BERISHA, H. D. FODA, A. ABSOOD & S. I. SAID. 1994. Vasoactive intestinal peptide (VIP) and related peptides: A new class of anti-inflammatory agents? International Symposium on Vasoactive Intestinal Peptide, Pituitary Adenylate Cyclase Activating Polypeptide, and Related Regulatory Peptide. Gabriel Rosselin, Ed.: 597–605. World Scientific. Singapore.

10. KALFIN, R., N. MAULIK, M. R. ENGELMAN, A. G. CORDIS, K. MILENOV, L. KASADOV & D. K. DAS. 1994. Protective role of intracoronary vasoactive intestinal peptide in ischemic and reperfused myocardium. J. Pharmacol. Exp. Ther. **268:** 952–958.

Role of PACAP in the Regulation of Gonadotroph Hormone Secretion during Ontogenesis:

A Single Neonatal Injection of PACAP Delays Puberty and Its Intracerebroventricular Administration before the Critical Period of Proestrous Stage Blocks Ovulation in Adulthood[a]

K. KÖVES,[c] J. MOLNÁR, O. KÁNTOR, A. LAKATOS, K. FÓGEL,
M. KAUSZ, M. C. VANDERMEERS-PIRET,[b] A. SOMOGYVÁRI-
VIGH, AND A. ARIMURA[d]

*Department of Human Morphology and Developmental Biology,
Semmelweis University Medical School, Budapest, Hungary H-1094*

[b]*Department of Medicine and Pharmacology, Laboratory of Chemical
Biology and Nutrition, Liberty University of Bruxelles, Belgium*

[d]*US-Japan Biomedical Research Laboratories, Tulane University Hebert
Center, Belle Chasse, Louisiana 70037 USA*

The role of PACAP in the regulation of anterior pituitary functions, including the control of the gonadotroph hormone secretion has been studied in detail. *In vitro* PACAP stimulated adenylate cyclase activity in whole anterior pituitary cell cultures. In a superfusion system this stimulation was associated with moderate, dose-dependent LH release.[1] In another experiment after a 4-h incubation, PACAP stimulated LH and α-subunit release.[2] In a gonadotroph cell line the stimulating effect of PACAP was very slight; however, PACAP potentiated the effect of LHRH.[3] *In vivo* an intravenous infusion of PACAP enhanced the plasma LH levels in male rats[4] but intracerebroventricular administration depressed the amplitude and pulse frequency of the plasma LH levels in ovariectomized ewes.[5] In our previous work we have demonstrated that $PACAP_{1-38}$ administered intracerebroventricularly before the critical period of the proestrous stage inhibited ovulation.[6] In the present experiment we have studied: (1) the effect of $PACAP_{1-38}$, $PACAP_{6-38}$ (a putative PACAP antagonist), and VIP on the maturation of gonadotroph hormone secretion; (2) the effect of $PACAP_{1-38}$, $PACAP_{6-38}$, and VIP on ovulation in adulthood; (3) the presence of PACAP in the retinohypothalamic tract, which mediates the effect of light impulses toward the endocrine hypothalamus; and (4) the effect of removal of the eyes on hypothalamic PACAP immunoreactivity (IR).

[a]This work was partially supported by OTKA grant T020403 (K.K. and M.K.), and National Institutes of Health Grant DK-09094 (A.A.).

[c]Corresponding author: Katalin Köves, M.D., Ph.D., Department of Human Morphology, Semmelweis University Medical School, Tüzoltó u. 58; Budapest, H-1094 Hungary; Tel.: (36)1 - 215 6920; Fax.: (36)1 - 215 3064; E-mail: koves@ana2.sote.hu

METHODS AND RESULTS

Experiment I

Neonatal female rats were injected with 1 μg of $PACAP_{1-38}$ (P_{1-38}), $PACAP_{6-38}$ (P_{6-38}), or VIP in propylene glycol (PG) (50%) subcutaneously. Age-matched controls received only PG. The time of the opening of the vaginal membrane in various groups was recorded. On the following day after the vaginal opening the animals were sacrificed and the ovarian tubes were dissected free. The expelled ova were counted. FIGURE 1 demonstrates the effect of various treatments on the time of the vaginal opening. P_{1-38} more potently delayed the vaginal opening, that is puberty, than P_{6-38} or VIP. The average number of ova was not significantly different in the various groups.

Experiment II

Adult female rats having a regular four-day cycle were used in this experiments. The effects of $PACAP_{1-38}$, $PACAP_{6-38}$, and VIP on ovulation were compared. Before the critical period of proestrous stage (13^{00}–13^{30} h) the animals received the following substances: control, 10 μl saline; P_{1-38}, 10 μg/10 μl; P_{6-38}, 10 μg/10 μl; VIP, 10 μg/10 μl. The peptides were administered intracerebroventricularly in physiological saline. The animals were sacrificed a day later and the tubal ova were counted. P_{1-38} inhibited the expected ovulation in a great

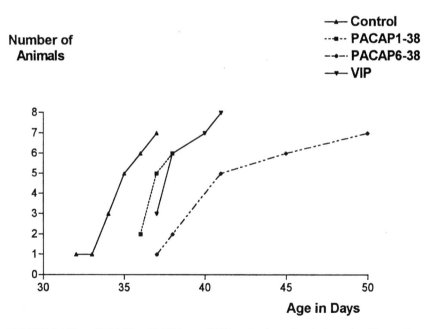

FIGURE 1. Effect of $PACAP_{1-38}$, $PACAP_{6-38}$, and VIP on the time of vaginal opening. The numbers on the abscissa indicate the age of the animals in days, the number on the ordinate indicate the number of animals. The symbols indicate the number of animals in which the vaginal membrane is open in the given day.

percent of the animals (12/15); however, the blocking effects of $PACAP_{6-38}$ and VIP were moderate (1/5 and 2/10, respectively). We may say that $PACAP_{1-38}$ more potently blocked the ovulation than $PACAP_{6-38}$ and VIP.

Experiment III

The presence of PACAP in the retinohypothalamic pathway is supported by the following facts: (1) In the retina PACAP immunoreactivity was observed in part of the ganglion cells (FIG. 2a); (2) PACAP immunoreactive (ir) fibers were seen at the dorsal aspect of the optic nerves and optic chiasm (FIG. 2,c); (3) PACAP ir fiber network was observed in the suprachiasmatic retinorecipient area (FIG. 2,b); and (4) Fast blue (FB), a retrograde fluorescent tracer appeared in the retinal ganglion cells when it was injected through a microcapillary into the suprachiasmatic region (FIG. 2,d and e). Part of these cells also showed PACAP IR.

Experiment IV

The eyes were removed under general anesthesia (enucleation). The effect of this intervention on the hypothalamic PACAP IR was studied by immunohistochemistry and by radioimmunoassay (RIA).

Immunohistochemistry

One group of animals was sacrificed 10 days after the enucleation by perfusion. Cryostat sections of the hypothalamus were stained for PACAP. Enucleation induced the appearance of PACAP ir elements in the magnocellular cell groups (PV and SO) and in the external zone (EZ) of the median eminence where they were not observed in intact rats. Previously we had to use colchicine treatment to demonstrate ir cell bodies.

Radioimmunoassay

Another group of 10 animals was sacrificed by decapitation. 300 μm–thick cryostat sections were prepared. Tissue samples were taken from the PV and ME by a punching technique described by Palkovits.[7] The samples of five animals were pooled. PACAP content was measured by RIA. Ten days after the enucleation the level of PACAP in both PV and ME was at least twice as high as in control rats (242.5 vs. 73.5 pg/mg wet tissue in PV and 269 pg vs. 126 pg in ME).

CONCLUSION

On the basis of our results it was concluded that PACAP is involved in the maturation of the gonadotroph hormone secretion and, in adulthood, in the control of ovulation. This effect is mediated through Type I receptors specific for PACAP, because $PACAP_{1-38}$ was more potent than VIP. PACAP is also involved in the mediation of light impulses to the endocrine hypothalamus and the pituitary gland.

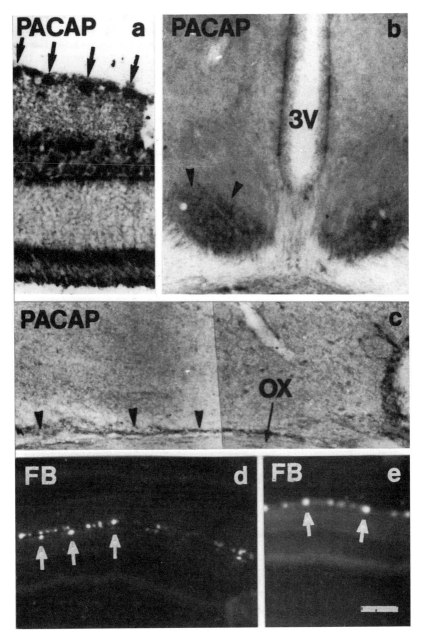

FIGURE 2. Photomicrographs demonstrating PACAP immunoreactive (ir) ganglion cells in the retina (**a**), PACAP ir fibers at the dorsal aspect of the optic nerves and optic chiasm (**c**), and PACAP ir terminal fiber network in the suprachiasmatic nucleus (SCH) (**b**). Fast blue injected in the SCH was retrogradely transported and appeared in retinal ganglion cells (**d** and **e**). *Arrows* indicate cell bodies, *arrowheads* indicate nerve fibers. OX = optic chiasm; 3v = third ventricle. Scale = 50 μm in **a–c,** and 100 μm in **d** and **e.**

REFERENCES

1. MIYATA, A. *et al.* 1989. Isolation of a novel 38 residue-hypothalamic polypeptide which stimulates adenylate cyclase in pituitary cells. Biochem. Biophys. Res. Commun. **164:** 567–574.
2. HART, G. R. *et al.* 1992. Effects of a novel hypothalamic peptide, pituitary adenylate cyclase-activating polypeptide on pituitary hormone release in rats. J. Endocrinol. **134:** 33–41.
3. CULLER, M. D. & C. S. PATSCHALL. 1991. Pituitary adenylate cyclase activating polypeptide (PACAP) potentiates the gonadotropin releasing activity of luteinizing hormone-releasing hormone. Endocrinology **129:** 2260–2262.
4. LEONHARDT, S. *et al.* 1992. Pituitary adenylate cyclase activating polypeptide (PACAP) stimulates pituitary hormone release in male rats. Neuroendocr. Lett. **14:** 313–327.
5. SAWANGJAROEN, K. & J. D. CURLEWIS. 1994. Effects of pituitary adenylate cyclase–activating polypeptide (PACAP) and vasoactive intestinal polypeptide (VIP) on prolactin, luteinizing hormone, and growth hormone secretion in the ewe. J. Neuroendocrinol. **6:** 549–555.
6. KÖVES, K. *et al.* 1996. PACAP participates in the regulation of hormonal events preceding the ovulation. Acta. Biol. Hung. **47:** 239–249.
7. PALKOVITS, M. 1973. Isolated removal of hypothalamic or other brain nuclei of the rat. Brain Res. **59**: 449–450.

Is an Intravenous Bolus Injection Required prior to Initiating Slow Intravenous Infusion of PACAP38 for Prevention of Neuronal Death Induced by Global Ischemia? The Possible Presence of a Binding Protein for PACAP38 in Blood

ANIKÓ SOMOGYVÁRI-VIGH, JAROSLAVA SVOBODA-TEET,
SÁNDOR VIGH, AND AKIRA ARIMURA

Department of Medicine, Tulane University, School of Medicine, New Orleans, Louisiana 70112

US-Japan Biomedical Research Labs, Tulane University Hebert Center, Belle Chasse, Louisiana, 70037 USA

PACAP is a pleiotropic neuropeptide that functions as a hypophysiotrophic hormone, neurotransmitter, neuromodulator, and, most importantly, as a neurotrophic factor. Previous studies showed that natural death or apoptosis of both sensory and motoneurons during development in chick embryos was significantly suppressed by a nanomolar concentration of PACAP.[1]

Furthermore, the cytoprotective action of PACAP was demonstrated in adult male rats subjected to global ischemia by four-vessel occlusion for 15 minutes.[2] An intracerebroventricular infusion of PACAP38, begun immediately after ischemia, significantly attenuated neuronal cell death in the CA1 area of the hippocampus 7 days after ischemia. The dose-response curve was bell-shaped, with maximum cytoprotection at 1 pmol/h over 7 days. The cytoprotective effect of an intravenous infusion of PACAP38 on ischemia-induced neuronal cell death in the hippocampus was also tested. PACAP38 was administered intravenously over one week, starting immediately or 24 h after forebrain ischemia. Two intravenous doses, 16 and 160 pmol/h, were tested. Significant attenuation of neuronal cell death, as a quantitation of viable neurons in the CA1 area of the hippocampus, was observed. In those studies, an intravenous bolus injection of 5 nmol/kg PACAP38 was given immediately prior to the slow intravenous infusion to quickly attain a steady level of this peptide in the blood. It was suspected that the presence of a binding protein for PACAP38 in the blood would prevent PACAP from quickly entering into the brain across the blood-brain barrier, or diminish the activity of PACAP38. In this study, we investigated (1) whether intravenous bolus injection immediately prior to slow infusion was necessary to induce significant neuroprotection in the CA1 region of the hippocampus after four-vessel occlusion and (2) whether a binding protein for PACAP38 exists in the blood.

MATERIALS AND METHODS

Four-Vessel Occlusion

Transient forebrain ischemia was introduced in conscious rats for 15 min by four-vessel occlusion as described by Pulsinelli and colleagues.[3] The criterion for forebrain ischemia is a bilateral loss of righting reflex during insult. Rats that have convulsed during the ischemic or postischemic period were excluded from the study.

Twenty-four hours after ischemia PACAP38 (dissolved in physiological saline containing 0.1% BSA) or vehicle was infused through a jugular cannula at 1 µl/h over one week using an Alzet mini osmotic pump implanted in the subcutaneous space of the nuchal area. One group of animals received a 5 nmol/kg bolus intravenous injection prior to the infusion of 160 pmol/h PACAP38. The second group received only the bolus injection, and the third group received 160 pmol/h of PACAP38 without the bolus. The control group received only the vehicle as a bolus and a long term infusion.

Histological Evaluation

One week after ischemia, the rats were anesthetized with an overdose of Nembutal and perfused intracardially with 4% buffered paraformaldehyde following a brief saline perfusion. The brains were excised and postfixed overnight at 4°C. Then the fixed brains were washed in phosphate-buffered saline, dehydrated, embedded in paraffin, and cut coronally at 7 µm with a microtome. As a reference, the first section was taken immediately anterior to the ventral hippocampus. The reference section and every tenth section anterior to it (four sections altogether) were evaluated for neurological damage. The sections were mounted on gelatin-coated slides, hydrated, and stained with cresyl violet. After dehydration in graded alcohols, the sections were cleared in xylene and covered for light microscopic examination. Four coronal sections from the dorsal hippocampus were examined. Pyramidal cells with a distinct nucleus and nucleolus were regarded as viable neurons. The number of viable neurons per reticle (220 µm diameter) was counted in the center of the CA1 region at × 400, and the mean was used as the viable cell number in each group.

Determination of Adenylate Cyclase Stimulation

Pituitary Cell Culture

Primary pituitary cell cultures were prepared from 225–250 g female CD rats as previously described.[4] The pituitaries were removed after decapitation, the posterior lobes were separated, and the anterior lobes were halved. The anterior lobe halves were enzymatically dispersed with 0.4% collagenase type II and 40 µg/ml deoxyribonuclease II in sterile HEPES buffer for 3 h at 37°C. The dispersed pituitary cells were washed and resuspended in DMEM supplemented with 10% horse serum, 2.5% FBS, and 1% antibiotic-antimycotic. The cells were placed in 24-well plates at a density of 10^5 cells/well, and incubated at 37°C with 95% air and 5% CO_2 for 3 days. 10^{-9} M PACAP38 was incubated in DMEM containing 0.1% BSA with 1:100, 1:500, and 1:1,500 dilutions of normal rabbit serum (NRS) for 30 min at room temperature. The cells were then cultured in these media with PACAP (with or without NRS), NRS alone, or DMEM + 0.1% BSA for control. Following incubation the media was removed and stored at −70°C for cAMP determination.

cAMP Assay

cAMP was measured by radioimmunoassay as described previously.[5] Tyrosyl-cAMP was iodinated by the lactoperoxidase method, and ^{125}I-labeled tyrosyl-cAMP was purified by reverse-phase HPLC using a Vydac C18 column.

Binding of PACAP38 to Serum Proteins

Rabbit, rat, and human sera were examined *in vitro* for the possible presence of binding protein(s) for PACAP38. Five µl serum and 100,000 cpm ^{125}I-PACAP38 in 100 µl PBS (50 mM sodium phosphate, 150 mM NaCl, 0.2% sodium azide, pH 7.4) with 0.1% BSA was incubated at room temperature for 30 minutes. One hundred µl of the mixture was applied on a Sephadex G-50 column (1 × 50 cm) and eluted with PBS at 0.2 ml/min at room temperature. One-ml fractions were collected and counted for radioactivity using an auto-gamma counter.

RESULTS AND DISCUSSION

Neuroprotection by PACAP

Twenty-four hours after forebrain ischemia, an intravenous infusion of PACAP38 or vehicle alone was begun. In rats receiving 5 nmol/kg intravenous PACAP38 bolus injection preceding 160 pmol/h PACAP38 intravenous infusion for one week, statistically significant attenuation of neuronal death was observed one week after the ischemic insult. In those animals receiving only 160 pmol/h PACAP38 intravenous infusion for one week without a bolus injection, the number of viable neurons in the CA1 region of hippocampus was increased as compared to the control group, but did not reach a significant level. A bolus injection alone did not increase the viable cells (FIG. 1).

These results indicate that a bolus injection before the intravenous infusion is important for producing a significant neuroprotective effect of PACAP38. Since the infusion rate was very slow, 1 µl/h, it would have been difficult to quickly attain a stable level of PACAP38 in the blood, unless a bolus dose of the peptide was given before the infusion. In view of the short half-life of PACAP38 in circulation (0.82 min[6]) a priming with a bolus injection of the peptide may be essential. Furthermore, administration of unlabeled PACAP38 with an injection of ^{125}I-PACAP38 resulted in a slowed disappearance of circulating ^{125}I-PACAP38.[6] This could be due to saturation of the binding sites of the peptide in peripheral tissues and in circulation by administration of the unlabeled peptide.

cAMP Accumulation Stimulated by PACAP38

PACAP38 stimulated extracellular cAMP accumulation in the pituitary cell cultures in a dose-dependent manner (FIG. 2). When cultured cells were incubated in the presence of various amounts of NRS with 10^{-9} M PACAP38, the magnitude of cAMP response to PACAP38 decreased in a concentration-dependent manner. The addition of 1:100 diluted NRS decreased the cAMP accumulation to 20% of the level of the response without NRS.

There are two possible explanations for this reduction of bioactivity. First, PACAP38 may be considerably degraded by peptides present in the serum during incubation. Second, the peptide is bound to a binding protein in the serum, which may interfere with the inter-

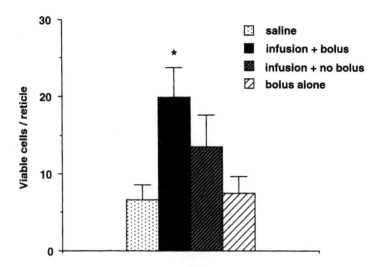

FIGURE 1. Effect of intravenous infusion of PACAP38 on ischemia-induced neuronal cell death. The ordinate shows the mean ± SE ($N = 8$) of the number of viable neurons seven days after ischemia per 220 μm reticle from the center of the CAT. Vehicle-infused rats showed marked neuronal death in the CA1 region of the hippocampus 1 week after ischemia. 160 pmol/h PACAP 38 intravenous infusions following an intravenous bolus (5 nmol/kg) begun 24 h after the ischemic insult. PACAP infusion with bolus significantly prevented neuronal death. *$p < 0.01$ vs. the saline-infused controls. PACAP38 intravenous infusion alone induced neuroprotection, but the effect was not statistically significant.

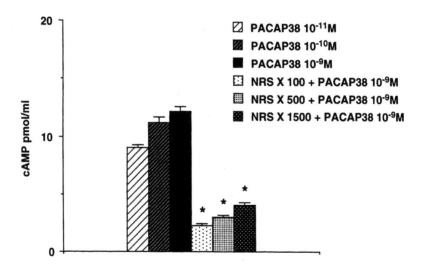

FIGURE 2. Dose-dependent stimulation of extracellular cAMP accumulation by PACAP38 in cultured rat anterior pituitary cells for 3 hours. Various concentrations of NRS significantly suppressed the extracellular cAMP accumulation induced by 10^{-9} M PACAP38. The concentration of extracellular cAMP is expressed as pmol/ml. Each point represents the mean ± SE. *$p < 0.01$ vs. the value of 10^{-9}–10^{-11} M PACAP38 ($N = 4$).

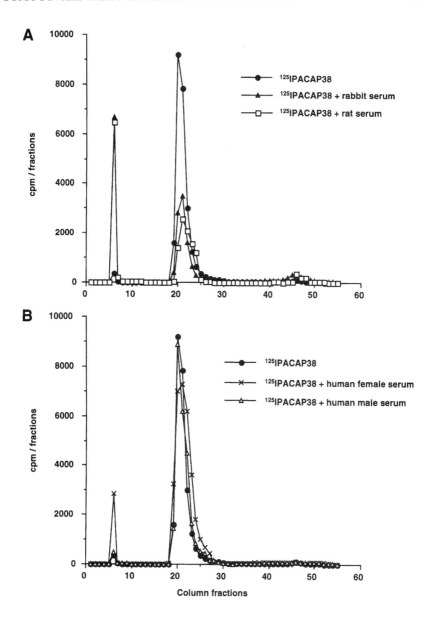

FIGURE 3. Chromatographic profile of rabbit and rat sera (**A**) and human female and male sera (**B**) binding to PACAP38. 10^5 cpm ^{125}I-PACAP38 were incubated for 30 min at room temperature with 5 µl serum or with buffer in a total volume of 100 µl. The samples were subjected to gel chromatography on a Sephadex G-50 column.

the peptide is bound to a binding protein in the serum, which may interfere with the interaction of the ligand and the receptor.

Gel Chromatography

To examine whether PACAP38 is bound to binding proteins in the serum, ^{125}I-PACAP38 was incubated with rabbit serum and then subjected to gel filtration. FIGURE 3A shows chromatographic profiles of ^{125}I-PACAP38 incubated either with buffer or with rabbit serum. Labeled PACAP38 in buffer was eluted in fractions 20–21. When incubated with serum, two radioactive peaks were eluted. The first was in fraction 6, and corresponded to the void volume. The second was in fraction 20–21. The amount of the radioactivity in fraction 6 was 30% of the total radioactivity eluted, indicating that a considerable portion of the radioactivity was bound to a high molecular weight protein, and eluted in the void volume. Rat serum from either sex, when incubated with labeled PACAP38, showed a similar chromatographic pattern. Surprisingly male human serum incubated with ^{125}I-PACAP38 did not show the radioactive peak in fraction 6, but only in fraction 20–21.

These findings suggest that male human serum does not contain the binding protein for PACAP38. Human female serum incubated with ^{125}I-PACAP38 demonstrated radioactive peaks in fractions 6 and 20–21. However, the magnitude of the first peak in fraction 6 was much smaller than those in the corresponding peaks for rabbit and rat sera (FIG. 3B). The findings also indicate that rabbit, rat, and female human sera may contain binding protein(s) for PACAP38 eluted in the void volume of the Sephadex G-50 column. Interestingly, the human male sera tested did not contain the binding protein. This suggests that the amount and/or affinity of the binding protein for PACAP38 varies between species, and possibly between sexes in humans. However, before drawing a conclusion, it is necessary to examine a larger number of serum samples in each group for their ability to bind to PACAP38.

REFERENCES

1. ARIMURA, A., A. SOMOGYVÁRI-VIGH, C. WEILL, R. C. FIORE, I. TATSUNO, V. BAY & D. E. BRENNEMAN. 1994. PACAP functions as a neurotrophic factor. Ann. N. Y. Acad. Sci. **739**: 228–243.
2. UCHIDA, D., A. ARIMURA, A. SOMOGYVÁRI-VIGH, S. SHIODA & W. E. BANKS. 1996. Prevention of ischemia-induced death of hippocampal neurons by pituitary adenylate cyclase activating polypeptide. Brain Res. **736**: 280–286.
3. PULSINELLI, W. A. & J. B. BRIERLY. 1979. A new model of bilateral hemispheric ischemia in the unanesthetized rat. Stroke **10**: 267–272.
4. CULLER, M. D., T. KENJO, N. OBARA & A. ARIMURA. 1984. Stimulation of pituitary cAMP accumulation by human pancreatic GH-releasing factor (1–44). Am. J. Physiol. **247**: E609–E615.
5. TATSUNO, I., A. SOMOGYVÁRI-VIGH, P. E. GOTTSCHALL, K. MIZUNO, H. HIDAKA & A. ARIMURA. 1991. Neuropeptide regulation of interleukin- 6 production from the pituitary: stimulation by pituitary adenylate cyclase activating polypeptide and calcitonin gene-related peptide. Endocrinology **129**: 1797–1804.
6. BANKS, W. A., A. J. KASTIN, G. KOMAKI & A. ARIMURA. 1993. Passage of pituitary adenylate cyclase activating polypeptide 1-27 and pituitary adenyate cyclase activating polypeptide1-38 across the blood-brain barrier. J. Pharmacol. Exp. Ther. **267**: 690–696.

Effect of Rat Glicentin on Intestinal Adaptation in Small Intestine– Resected Rats

Y. HIROTANI,[a,e] M. TAKI,[a] K. KATAOKA,[a] N. KUROKAWA,[a]
T. SATOH,[b] K. SASAKI,[b] C. YANAIHARA,[c] W.Q. LUO,[d] AND
N. YANAIHARA[d]

[a]Laboratory of Pharmaceutical Sciences, Osaka University School of Medicine, Suita Osaka 565, Japan

[b]Pharmaceutical Research Laboratories, Nisshin Flour Milling Co. Ltd., Irumagun Saitama 356, Japan

[c]Department of Pharmacy, Hyogo College of Medicine, Nishinomiya Hyogo 663, Japan

[d]Yanaihara Institute Inc., Fujinomiya Shizuoka 418, Japan

Biological activities of glicentin have not yet been fully established.[1] Our previous study[2] provided synthesis and biochemical characterization of synthetic human glicentin and its related peptides. It has made it possible for us to explore in detail the biological roles of the peptide. The aim of the present study is to investigate the effects of glicentin on intestinal adaptive responses to 70% resection of small intestine in rats, using synthetic rat glicentin and methionyl-rat glicentin[3] (Met-rat glicentin).

MATERIALS AND METHODS

Synthesis of Rat Glicentin

Rat glicentin and its related peptides were synthesized by the Fmoc-strategy using an automated peptide synthesizer (9050 System, PerSeptive Biosystems, Inc., Framingham, MA, USA). The resulting preparations of glicentin and related peptides were purified by reverse-phase HPLC (column: YMC-Pack R-ODS-5, 4.6 × 250 mm, eluent: 0.01 N HCl/CH$_3$CN (95/5 → 40/60, vol/vol linear gradient for 30 min, flow rate: 1 ml/min).

[e]Address for correspondence: Yoshihiko Hirotani, Laboratory of Phamaceutical Sciences, Osaka University School of Medicine, 2-15 Yamadaoka, Suita, Osaka 565, Japan; Tel.: 81-6-879-3371; Fax: 81-6-879-3379; E-mail: yosihiko@drug.med.osaka-u.ac.jp

Recombinant Met-Rat Glicentin

The gene encoding for rat glicentin was constructed from pGL111 according to a modification of the method used for the synthesis of human glicentin.[2]

ELISA for Rat Glicentin

An ELISA system for rat glicentin was developed using rat glicentin as standard, anti-rat glicentin monoclonal antibody 3D5A as the solid-phase antibody, and rabbit anti-rat glicentin (1-32) polyclonal antibody (R984) as the second antibody. The assay was carried out in microtest plates (Nunc, Roskilde, Denmark), essentially according to the procedure described previously for human glicentin ELISA.[2] The minimum detection limit was 1.56 pmol/ml.

Radioimmunoassay for Rat PYY

Radioimmunoassay for rat PYY was carried out using synthetic rat PYY as standard, ^{125}I-synthetic rat PYY as tracer and anti-rat PYY serum (RY073, Yanaihara Institute Inc., Fujinomiya, Japan). The minimum detection limit was 10 fmol/ml.

Animal Experiments

The experiments were performed with male Wistar rats weighing 180–200 g. After a 24-h fast, a part (about 70 cm) of the proximal or distal small intestine was resected. The remaining intestines were joined in an end-to-end anastomosis. Following the operation, the animals received a 5% glucose solution for the first two postoperative days, after which an elemental diet (Elental, Ajinomoto, Tokyo, Japan) and tap water were offered *ad libitum*. From the second day after intestinal resection, Met-rat glicentin (20 µg/day/rat) was continuously given for 28 days using miniosmotic pumps (Alzet model 2001, Palo Alto, CA, USA).

Protein, DNA, and Diamine Oxidase (DAO) Measurements

Mucosal sample obtained from small intestine was homogenized with fivefold weight of ice-cold saline in an ice-bath by using a Bio-Mixter (model BM-1, Nihon Seiki Seisakusho, Tokyo, Japan). The homogenate was used for the following measurements. Protein concentration was determined by the method of Lowry and colleagues[4] and expressed as mg/cm intestine. DNA level was measured by the method of LePecq and Paoletti,[5] using polymerized calf thymus DNA (Sigma Chemical Co., St. Louis, MO, USA) as standard and expressed as µg/cm intestine. DAO activity was assayed according to the method of Suzuki and colleagues.[6] The homogenate above described was centrifuged (10,000 rpm, 30 min) and the supernatant was used for the assay. DAO activity was expressed as mU/mg protein and for plasma sample, as mU/ml plasma.

Student's *t*-test for unpaired data were used for group analysis and the results were given as mean ± SEM. Five to eight rats were used in each experiment.

RESULTS AND DISCUSSION

Plasma enteroglucagon level is known to elevate after small bowel resection[8] and jejuno-ileal bypass operations.[9] The hypersecretion of enteroglucagon in these cases was assumed to be a compensatory mechanism for the deficient absorption in the upper small intestine. Previously reported studies[10,11] with enteroglucagon-producing renal tumor patients suggested that enteroglucagon is a trophic hormone in the small intestine. Glicentin is a major component of enteroglucagon. However, it has not yet been clarified if the peptide could explain such a trophic activity presumably exhibited by enteroglucagon.[1] Limited availability of pure glicentin preparation has hampered detailed study in this respect. To solve the problem, we first prepared glicentin by both chemical and recombinant technologies. Since rats were used in our experiments, the syntheses were carried out according to the rat sequence of glicentin. The recombinant preparation possessed Met residue at the N-terminal of glicentin (Met-rat glicentin). Purity of the final products were assessed by several analytical criteria including HPLC, amino acid analysis, and sequencing. Using synthetic rat glicentin, an ELISA system was developed, which made it possible for the first time to measure directly plasma glicentin. The assay system confirmed that plasma glicentin concentration in small intestine–resected rats under continuous subcutaneous administration of Met-rat glicentin by osmotic pumps for 28 days increased up to levels of $6.10 \pm 3.35 \sim 8.70 \pm 1.50$ pmol/ml during the first 7 days and maintained within the levels from the seventh day on.

TABLE 1 summarizes the effects of Met-rat glicentin on the residual small intestine of distal or proximal small intestine–resected rats. Enteroglucagon including glicentin is localized in the highest concentration in the ileum. To reduce the effects of endogenous enteroglucagon as much as possible, the present study was carried out in 70% distal intestine–resected rats, and 70% proximal intestine–resected rats were used for comparison. Administration of Met-rat glicentin to 70% distal intestine–resected rats resulted in simultaneous significant ($p < 0.01$) increases of the weight of the residual duodenum and its mucosal weight, protein, and DAO activity as well as plasma DAO activity on the 14th day after the initiation of the administration (that is, the 16th day after resection). On the other hand, the residual jejunum showed no significant changes at all as compared with controls during 28 days of the treatment. It is not yet clear whether the stimulation of adap-

TABLE 1. Effects of Met-rat glicentin on residual intestine of small intestine–resected rats

	Distal Resection		Proximal Resection	
	Residual		Residual	
	Duodenum	Jejunum	Duodenum	Ileum
Tissue weight	$114.6 \pm 1.8^{**}$ (14 d)	—	$111.1 \pm 3.6^{*}$ (28 d)	—
Mucosal weight	$123.8 \pm 8.1^{**}$ (14 d)	—	$141.7 \pm 7.7^{**}$ (10 d)	$168.3 \pm 24.6^{**}$ (10 d)
			$129.0 \pm 8.1^{**}$ (28 d)	$187.0 \pm 24.7^{**}$ (10 d)
Mucosal protein	$133.1 \pm 11.0^{**}$ (14 d)	—	$154.1 \pm 10.5^{**}$ (10 d)	
			$145.9 \pm 19.2^{**}$ (28 d)	
Mucosal DNA	—	—	—	$220.1 \pm 39.4^{**}$ (10 d)
Mucosal DAO	$149.9 \pm 14.9^{**}$ (14 d)	—	$179.1 \pm 17.7^{**}$ (2 d)	$297.5 \pm 46.0^{**}$ (21 d)
			$239.3 \pm 32.1^{*}$ (21 d)	
			$136.5 \pm 8.7^{**}$ (28 d)	
Plasma DAO	$144.0 \pm 12.1^{**}$ (14 d)		—	—
	$174.1 \pm 13.3^{**}$ (21 d)			

Note: Figures represent % changes as compared with the values (100%) of control group (mean ± SEM, * $p < 0.05$, ** $p < 0.01$) and only the data significantly different from control values are presented in this table. ( ): days after the initial administration of Met-rat glicentin.

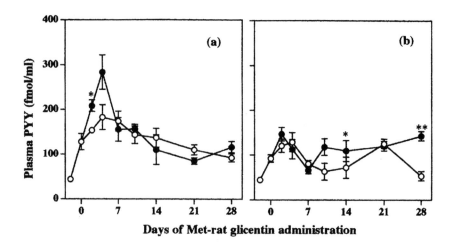

Days of Met-rat glicentin administration

FIGURE 1. Plasma PYY concentration-time courses during Met-rat glicentin (20 μg/day/rat) (●) or saline (○) administration to distal (**a**) or proximal (**b**) small intestine–resected rats. Results are mean ± SEM (*$p < 0.05$, **$p < 0.01$).

tive responses in the duodenum observed is a direct action of Met-rat glicentin or indirect one via some other active element(s). However, the significant stimulating effect of the peptide was clearly localized in the residual duodenum only on day 14 during continuous administration of the peptide for 28 days after operation. This supports not only the trophic activity of glicentin in the system used but also the usefulness of the system for elucidation of the mechanism of the glicentin action. The effects of Met-rat glicentin on the residual duodenum and ileum in 70% proximal intestine–resected rats were more noticeable, but the time profiles of the effects observed were more complicated than those in the case of the distal resection. This may be explained, at least in part, by the postoperative intestinal remnants, especially the ileum that is a main tissue for the production of active peptides such as enteroglucagon, glicentin, PYY,[11] and GLP-2,[12] which are supposed to have trophic actions on the intestine. In fact, significant increases of plasma PYY were observed on day 10 (186.7 ± 29.6%) and day 28 (267.2 ± 19.1%) of Met-rat glicentin treatment (FIG. 1b) concomitantly with significant increases in the several growth parameters of the residual duodenum and ileum examined (TABLE 1). The present results showed clearly that glicentin possesses stimulating activity on the adaptive responses to massive intestinal resection in the rat. Elucidation of the mechanism of the action is now underway in our laboratory. Synthetic rat glicentin and Met-rat glicentin as well as rat glicentin-specific ELISA will be useful tools for elucidation of physiology and pathology of glicentin, which still remains almost unknown.

REFERENCE

1. HOLST, J. J. 1997. Enteroglucagon. Annu. Rev. Physiol. **59:** 257–271.
2. YANAIHARA, N., T. MOCHIZUKI, T. SATOH, S. IMAI, K. SASAKI, H. KAKUYAMA, H. HOSOE, N. TAKATSUKA, K. IGUCHI, Q. W. LUO, N. KUROKAWA & C. YANAIHARA. 1996. Synthesis and biochemical characterization of proglucagon-related peptides. Ann. N. Y. Acad. Sci. **805:** 20–30.
3. HEINRICH, G., P. GROS & G. J. HABENER. 1984. Glucagon gene sequence. Four of six exons encode separate functional domains of rat pre-proglucagon. J. Biol. Chem. **259:** 14082–14087.

4. LOWRY, O. H., N. J. ROSEBROUGH, A. L. FARR & R. T. RANDALL. 1951. Protein measurement with the Folin phenol reagent. J. Biol. Chem. **193:** 265–275.
5. LE PECQ, J.-B. & C. PAOLETTI. 1966. A new fluorometric method for RNA and DNA determination. Anal. Biochem. **17:** 100–107.
6. SUZUKI, T. & T. OKUYAMA. 1967. Estimation of diamine oxidase activity by *o*-dianisidine-peroxidase coupling. **39**(7): 399–404.
7. BLOOM, S. R., H. S. BESTERMAN & T. E. ADRIAN. 1979. Gut hormone profile following resection of large and small bowel[Abstract]. Gastroenterology. **76:** 1101.
8. BESTERMAN, H. S., D. L. SARSON, A. M. BLACKBURN, J. CLEARY, T. R. E. PILKINGTON, J. C. GAZET & S. R. BLOOM. 1978. Gut hormone profile in morbid obesity and after jejunoileal bypass[Abstract]. Gut **19:** A986.
9. GLESSON, M. H., S. R. BLOOM & J. M. POLAK. 1971. Endocrine tumour in kidney affecting small bowel structure motility and absorptive function. Gut **12:** 773–782.
10. BLOOM, S. R. 1972. An enteroglucagon tumour. Gut **13:** 520–523.
11. GOMEZ, G., T. ZHANG, S. RAJARAMAN, K. N. THAKORE, N. YANAIHARA, C. M. TOWNSEND, JR., J. C. THOMPSON & G. H. GREELEY. 1995. Intestinal peptide YY: ontogeny of gene expression in rat bowel and trophic action on rat and mouse bowel. Am. J. Physiol. **268:** G71–G81.
12. DRUCKER, D. J., P. EHRLICH, S. L. ASA & P. L. BRUBAKER. 1996. Induction of intestinal epithelial proliferation by glucagon-like peptide 2. Proc. Natl. Acad. Sci. USA **93:** 7911–7916.

Index of Contributors

Ahrén, B., 441–444, 466–470
Ak, D., 309–322
Amsallem, H., 431–437
Anouar, Y., 92–99
Arakawa, Y., 561–565
Arimura, A., 111–117, 408–411, 427–430,
    438–440, 475–477, 523–528,
    529–532, 590–594, 595–600
Astesano, A., 118–131
Aydin, Y., 570–581

Bali, J. P., 458–462
Bandyopadhyay, A., 226–237
Baranowska, B., 482–485, 556–560
Basille, M., 92–99
Bataille, D., 132–140, 458–462
Baumeister, H., 390–392
Beaudet, M. M., 164–175
Bellancourt, G., 416–419
Berdiev, R. K., 547–550
Berghman, L. R., 471–474
Berisha, H. I., 226–237, 582–585
Bjerre-Knudsen, L., 336–343
Blache, P., 132–140
Bockaert, J., 100–110
Boissard, C., 118–131
Braas, K. M., 164–175, 367–374
Brabet, P., 49–58
Brandenburg, C. A., 367–374
Braslawsky, W., 519–522
Breault, L., 416–419
Brenneman, D. E., 207–212

Cao, Y. J., 82–91
Cardinaux, J-R., 213–225
Carrero, I., 59–63

Chaib, N., 431–437
Chakder, S., 503–511
Chen, D., 197–206
Chepurnov, S. A., 547–550
Chepurnova, N. E., 547–550
Christophe, J., 323–335
Ciani, E., 49–58
Cnudde, J., 247–252
Contesse, V., 416–419
Cottingham, S. L., 420–426
Couvineau, A., 59–63, 378–381, 382–385,
    386–389
Curró, D., 492–494

Dalle, S., 132–140
Darras, V. M., 471–474
Das, D. K., 297–308
Davidson, A., 207–212
De Neef, P., 247–252
De Stefanis, P., 226–237
Deacon, C., 336–343
Dehaye, J. P., 431 –437
Delarue, C., 416–419
Deppert, W., 27–36
Dey, R. D., 226–237
DiCicco-Bloom, E., 274–289
Dickman, K., 226–237
Ding, J. M., 197–206
Dinnis, D. M., 64–72
Dohi, K., 111–117
Du, K., 378–381, 386–389
Dufour, S., 475–477
Duggan, K. A., 515–518
Durda, P. J., 164–175

Eckelman, W. C., 290–296
Eiden, L. E., 10–26, 537–541

Ekblad, E., 393–396
Ekelund, M., 393–396
Emami, S., 118–131
Engelman, R. M., 297–308
Erden, S. H., 570–581
Erkasap, N., 309–322
Ermilov, L. G., 360–366
Eschelbach, A., 181–188

Fahrenholz, F., 82–91
Fahrenkrug, J., 197–206, 259–265,
    375–377, 533–536, 542–546
Feng, J., 1–9
Ferrand, N., 118–131
Filipsson, K., 441–444, 466–470
Fogel, K., 590–594
Fölsch, U. R., 27–36
Forsgren, S., 344–352, 353–359
Forssmann, W. G., xv
Fournier, A., 92–99, 416–419, 475–477,
    478–481
Franzén, L., 344–352
Fridkm, M., 266–273
Frühwald, M. C., 420–426
Fujimiya, M., 495–502

Gallo-Payet, N., 416–419
Ganea, D., 397–407
Garel, J.-M., 118–131
Gaudin, P., 382–385
Gauthier, E. A., 164–175
Gillette, M. U., 197–206
Gilmartin, G. A., 367–374
Glazner, G., 207–212
Gonzalez, B. J., 92–99
Gourlet, P., 247–252, 412–415
Gozes, I., 207–212, 266–273, 290–296
Grzonka, Z., 82–91
Gulbenkian, S., 344–352

Hahm, S. H., 10–26
Hannibal, J., 197–206, 533–536, 542–546

Hansson, M., 344–352
Harakall, S. A., 367–374
Hardwick, J. C., 164–175
Harmar, A. J., 64–72
Hauser, J., 207–212
Hayakawa, T., 463–465
Heinrich, G., 45–48
Henriksson, R., 344–352
Hill, J. M., 207–212
Hino, J., 73–81
Hirai, A., 253–258
Hirotani, Y., 601–605
Höckerfelt, U., 344–352
Hoffmann, A., 49–58
Holmgren, S., 512–514
Holst, J. J., 336–343
Hong, M., 478–481
Houssami, S., 49–58
Hunziker, A., 181–188

Ihle, J. N., 1–9
Ishiguro, H., 463–465
Ito, O., 463–465

Jarrousse, C., 458–462
Jarry, H., 486–491
Jasionowski, M., 82–91
Jiang, X., 397–407
Journot, L., 49–58, 100–110

Kabrè, E., 431–437
Kalfin, R., 297–308
Kang, T., 147–156
Kangawa, K., 73–81
Kántor, O., 590–594
Karganov, M., 519–522
Kataoka, K., 561–565, 601–605
Kausz, M., 529–532, 590–594
Kikuchi, M., 451–457
Kitagawa, M., 463–465
Klimaschewski, L., 181–188
Knudsen, S. M., 259–265, 375–377

Koide, R., 408–411
Kojro, E., 82–91
Kosar, M., 570–581
Köves, K., 529–532, 590–594
Kozicz, T., 523–528
Kruse, M. L., 141–146
Kurokawa, N., 561–565, 601–605
Kuwahara, A., 495–502

Laburthe, M., 59–63, 378–381, 382–385, 386–389
Lakatos, A., 590–594
Lang, L., 290–296
Langouche, L., 471–474
Lankiewicz, L., 82–91
Larsen, P. J., 197–206
Le Brigand, L., 132–140
Le-Nguyen, D., 132–140
Lehoux, J. G., 416–419
Leonhardt, S., 486–491
Leontiv, D., 176–180
Lewin, L. M., 266–273
Leyton, J., 290–296
Lindholm, D., 189–196
Lu, N., 274–289
Luo, W. Q., 601–605
Lyu, R. M., 147–156

Mácsai, M., 566–569
Magistretti, P. J., 213–225
Magous, R., 458–462
Maoret, J. J., 59–63, 382–385
Martin, J.-L., 213–225
Martin, S. C., 45–48
Matsson, L., 353–359
Matsumoto, K., 111–117
Matsuo, H., 73–81
Maulik, N., 297–308
May, V., 164–175, 367–374
McDonald, T. P., 64–72
Meyerhof, W., 390–392
Mikkelsen, J. D., 197–206

Miyata, A., 73–81
Mizushima, H., 111–117
Mochizuki, T., 458–462
Mojsov, S., 45–48
Molnár, J., 590–594
Montero, M., 475–477
Moody, T. W., 290–296
Moran, A., 431–437
Moro, O., 253–258
Morrison, C. F., 64–72
Muroi, M., 438–440

Nakai, Y., 111–117, 408–411, 427–430, 438–440
Nakaja, S., 438–440
Nakajima, M., 463–465
Nakajo, S., 111–117, 427–430
Nakata, M., 451–457
Naruse, S., 37–44, 463–465
Nicole, P., 378–381, 386–389
Nielsen, H. S., 533–536
Nokihara, K., 37–44
Norevall, L. I., 353–359

O'Dorisio, M. S., 420–426
O'Dorisio, T. M., 420–426
Olsson, C., 512–514
Ozawa, H., 111–117

Pakbaz, H., 226–237
Parganas, E., 1–9
Parsons, R. L., 164–175
Paul, S., 238–246
Peeters, K., 471–474
Pelletier, G., 478–481
Perl, O., 266–273
Pessah, M., 118–131
Phan, H., 118–131
Pintar, J. E., 274–289
Pisegna, J. R., 147–156
Preziosi, P., 492–494

Qualman, S. J., 420–426

Radleff-Schlimme, A., 486–491
Radzikowska, M., 482–485, 556–560
Rattan, S., 503–511
Raza, S., 226–237, 582–585
Regnauld, K., 118–131
Robberecht, P., 157–163, 247–252, 412–415
Rodier, G., 458–462
Romanova, G., 519–522
Rosselin, G., 118–131
Rousseau, K., 475–477
Rouyer-Fessard, C., 59–63, 382–385
Rubinraut, S., 266–273

Sachs, G., 147–156
Şahintürk, V., 309–322, 570–581
Said, S. I., xv, 226–237, 582–585
Saito, Y., 253–258
Sakurada, M., 451–457
Sasaki, K., 601–605
Sato, K., 73–81
Satoh, T., 601–605
Schäfer, H., 27–36
Schäfer, M. K. H., 537–541
Schek, E., 566–569
Schiffmann, S. N., 412–415
Schmidt, W. T., 27–36
Schmidt, W. E., 141–146
Schmidt-Choudhury, A., 141–146
Schütz, B., 537–541
Sebens, T., 27–36
Seebeck, J., 141–146
Seki, T., 408–411
Shioda, S., 111–117, 408–411, 427–430, 438–440, 451–457
Shochat, L., 266–273
Skoglösa, Y., 189–196
Somogyvari-Vigh, A., 590–594, 595–600
Spengler, D., 49–58
Stravapodis, D., 1–9
Sundler, F., 393–396

Svoboda-Teet, J., 595–600
Szabó, G., 566–569
Szurszewski, J. H., 360–366

Töre, F. C., 586–589
Tajima, M., 253–258
Takaki, A., 111–117
Takashima, N., 561–565
Takei, N., 189–196
Taki, M., 601–605
Tamakawa, H., 73–81
Tams, J. W., 259–265, 375–377
Tanaka, T., 253–258
Tarshitz, D., 519–522
Tatsuno, I., 253–258
Teglund, S., 1–9
Telegdy, G., 519–522, 566–569
Thierfelder, W., 1–9
Toft-Nielsen, M. B., 336–343
Trauzold, A., 27–36
Tunçel, M., 309–322, 570–581
Tunçel, N., 309–322, 570–581, 586–589

Uchida, D., 253–258

Van Rampelbergh, J., 247–252
Vandermeers, A., 247–252
Vandermeers, M. C., 431–437
Vandermeers-Piret, M. C., 590–594
Vandesande, F., 471–474
Vaudry, D., 92–99
Vaudry, H., 92–99, 416–419, 475–477, 478–481
Vertongen, P., 412–415
Vigh, S., 523–528, 595–600
Vjllalba, M., 100–110

Waelbroeck, M., 157–163, 247–252
Wakade, A. R., 176–180
Walsh, J. H., 147–156
Wang, D., 1–9

Wang, H. Y., 397–407
Wasilewska-Dziubińska, E., 482–485, 556–560
Wei, Y., 45–48
Weihe, E., 537–541
Wen, Y., 147–156
Wollina, U., 551–555
Wong, H., 147–156
Wray, V., 37–44
Wulff, B. S., 259–265
Wuttke, W., 486–491

Yada, T., 427–430, 438–440, 445–450, 451–457

Yaekura, K., 445–450, 451–457
Yamamoto, H., 495–502
Yanagida, K., 445–450
Yanaihara, C., 561–565, 601–605
Yanaihara, N., 561–565, 601–605
Ye, V. Z. C., 515–518
Yon, L., 397–407, 416–419, 475–477, 478–481
Yu, J., 397–407

Zamostiano, R., 266–273
Zeng, N., 147–156
Zhang, J., 274–289
Zhou, C. J., 111–117, 438–440